Explorations in Mathematical Physics

Don Koks

Explorations in Mathematical Physics

The Concepts Behind an Elegant Language

Springer

Cover illustration: The cover shows a spacetime diagram of the famous Twin Conundrum, in the frame of the uniformly accelerated twin. The vertical line is the accelerated twin's worldline, while the solid curve is that of the inertial twin, who ages more in their time apart. Light signals exchanged by the twins on their birthdays follow the dashed curves, demonstrating the position-dependent speed of light in an accelerated frame. Analysing events such as these from an accelerated observer's point of view forms a good and gentle introduction to the physics of gravity.

ISBN-13: 978-1-4419-2168-0
e-ISBN-10: 0-387-32793-2
e-ISBN-13: 978-0-387-32793-8

Printed on acid-free paper.

9 8 7 6 5 4 3 2 1

springer.com

For my parents and family,
both near and far.

Preface

There are many stories of Nasr Eddin, the "wise fool" at the centre of Sufi lore of a thousand years past. In one of these, he spent his life trading across a border post. Each day he'd walk to and fro with his donkeys, carrying bags heavily laden with wool. The border guards were convinced he was smuggling something, but no matter how much they searched, they could never discover anything at all. After many years Nasr Eddin retired, and by this time, the guards, long retired themselves, asked him to tell all. "We knew you were smuggling something Nasr Eddin, but what was it?" they asked. "Oh that's very simple", he replied. "I was smuggling donkeys".

The physics teacher's equivalent of smuggling donkeys has been my aim throughout this book. The donkeys themselves are the language of physics, which carries concepts that allow us to calculate things such as planetary orbits, nuclear scattering statistics, and electromagnetic fields. In the following chapters, you'll find an array of carefully bundled goods, from Fisher information to de Sitter event horizons, decaying nuclei to dashing across the street in the rain. But, underlying all of these vastly different subjects, I wish to show the elegance and unity of the mathematical language of physics and do it in a way that I hope comes across almost unnoticed next to the subjects on centre stage. The mathematical language, like Nasr Eddin's donkeys, is sometimes only thought to carry the goods of real importance; but it can be seen as special in its own right, and it has been constructed and has evolved along paths that give it an allure, as well as presenting a challenge to be seen in ways that suit each individual according to his or her own style of thinking about physics.

Navigating through physical ideas while paying attention to the mathematical language used to describe them can sometimes seem like being lost in the proverbial "forest and trees". The reader should always ask what it is that might be visible ahead. For example, you won't learn much about crystallography and Bragg scattering here, but the discussion of crystallography that you *will* find is there to show how natural the idea of a cobasis is (called a *reciprocal basis* in crystallography), an idea that resurfaces often in tensor analysis. The discussion of Fourier theory using Dirac's brackets is not meant to teach Fourier analysis, which can be found in depth in so many books. Rather, it's designed to teach bracket notation within the familiar surroundings of Fourier analysis, as opposed to the more bizarre realm of quantum mechanics in which Dirac's brackets are usually first encountered. On the other hand, the discussion of Heisenberg's Uncertainty Principle has been taken out of its usual quantum mechanical setting and placed within the perhaps *less* familiar arena of classical wave theory—operators, commutators and all—in order to show that some seemingly quantum mechanical concepts are not really quantum mechanical after all.

Likewise, my discussion of nuclear decay serves not to teach Poisson statistics, but to show how the numbers that crop up in technical discussions of decay and growth have very intuitive meanings that go beyond simply stating an exponential law. In posing the question of how fast we should run across a rainy street to minimise how wet we get, my plan is not to solve the problem for its own sake, but rather to show how the idea of a four-vector arises quite naturally when such an everyday question is put into a relativistic context. In labouring over proving that the frequency–wavenumber is a four-vector, I wish to make the follow-on analysis in terms of covectors that much starker in its simplicity by comparison. And, in asking how the uniformly accelerated space-bound twin in relativity's famous Twin Conundrum observes events, I want to show just how the relevant language turns out to be useful for thinking about gravity.

While following a meandering line through the subject of mathematical physics, I decided to include anything of interest, following the style of what I think is one of the best-ever reads on the history of Ancient Egypt, *Temples, Tombs and Hieroglyphs: A Popular History of Ancient Egypt*, by Barbara Mertz. Hers is a book that tells a rivetting story of civilisation, replete with three thousand years of mankind's striving through triumph and defeat, growth and pain. If a work like Mertz's could be written about the language of physics, it would weave together a host of different subjects far better than I've done here. Even so, it might not tell a story in the way of a history book—unless it was about the history of physics. While physics itself has an interesting history full of all sorts of characters, glorious beginnings, and dead ends, the subject itself is not very linear and becomes difficult to write about without anticipating future results, assuming mathematical know-how, and recalling past theorems.

In any book that tries to weave together different ideas, our need to constantly flick back and forth to check on equations and sections that are pointed to can be tedious on the one hand, but on the other can highlight the coherence of the subject. It's a difficult balance to achieve, and one that I hope I have gotten at least partly right. Questions of which concept is the more fundamental and which should be taught first will always remain, and if nothing else they underline the different ways in which the subject can be viewed or learnt.

So, because of its lack of a simple straight path through many different subjects, this book is best read more than once and not in the selective manner of a textbook, which it is not. If the reader has time to read it more than once, then the first time it should be read like a novel or a history book, from start to finish, in order to assimilate what might be called a large "coherence length" of the chapters and the subject as a whole. Only the second time around should things be followed in more detail.

I have also tried to concentrate on using good, rigorous notation—but not to the extent of adding every last bell and whistle. Properly used, like any language, notation can aid us tremendously. Not only does the symmetry and well-crafted style of good notation suggest ways to proceed in new directions in a calculation, but it can also help to communicate the result to an audience in a way that builds a solid foundation in their own minds. On the other hand, if used incorrectly, notation does not help us solve a task and can become a hindrance; and like any language, sloppy mathematical grammar only creates difficulties in communication that negate the whole point of having a set of rules in the first place.

Mathematical physics is about continually rewriting our approach as we strive for ever-deeper insight, and as it evolves, original ways of viewing the subject are sometimes discarded or lost. A good example of this is one of Maxwell's early theories of electromagnetism, which involved a field built from ghostly meshing gears. Maxwell understood that this idea was provisional, and once it had served its purpose as a scaffold, he discarded it, producing the theory that we learn nowadays. Probably no one now uses his rotating vortices to teach electromagnetism (although they are still worth studying for their mathematical beauty), and instead, the subject is taught from starting points that differ to varying degrees of abstraction. The foundations of physical theories are often obscure enough to allow for this latitude in places from which to begin, and to some extent there is also some latitude in how the *mathematical* language is presented.

But I think this idea of discarding scaffolding can sometimes be taken too far in a mathematical context, where we open a book about the physics of the everyday world, only to find a first page that begins in a business-like way, with a lemma, a theorem, and a corollary. Writing a clear set of postulates for a subject can certainly help to sweep away the wood shavings and put it on a clearer footing, but when taken to an extreme, it can show a kind of snobbery, a refusal to acknowledge that the roots of some subjects are quite indistinct. An edifice is built beautiful and solid, and then the rigging is dismantled; and forever after, it's a mystery to all but a few as to why one corner is topped by a bronze gargoyle, while another side has a multitude of windows that all show blue flags. This suits some, perhaps many, while there are others who are content with a Proof by Repeated Exposure approach to the subject. But there are plenty of aspiring physicists who would like to know the motives for why some subjects are described in the way that they are, and without necessarily having to follow a year-long course in each of the details. This book is for them.

As with all grey areas that separate the easy from the abstruse, it was very difficult to decide what knowledge to assume of the reader in the chapters ahead. There can never be a solid line drawn between what is assumed knowledge and what is not; instead, the boundary is a grey zone within which there are as many simple ideas that are explained as difficult ones that are not.

A note concerning my conventions is appropriate. I have avoided initial capitals on commonly used words that are derived from names, such as Newtonian and Lagrangian, preferring instead to write newtonian and lagrangian. "Newton" is a name, but "Newtonian" is not, and perhaps the lack of a capital does Newton the honour of showing how well a word derived from his name has been assimilated into English. I do, however, use capitals in the mathematics, such as L for the quantity known as the lagrangian. This is akin to the SI system of units, in which all units—even a "newton"—are lowercased, but when derived from a name, their symbols have an initial capital. I also write "Green function" after the manner of Jackson's well-known book of electromagnetism, instead of the more common "Green's function". There are many Green functions, just as there are many Bessel functions, Mozart concertos, and Chaplin movies.

Presumably, every author hopes that their finished book will be, just like Mary Poppins, practically perfect in every way; and contributing to my work on the manuscript were many people who gave feedback or had some input into the chapters ahead to help remove deficiencies and to provide polish. In particular, for listening to my endless talk about the project as well as giving general support, I wish to thank Ine Brummans, Jasmine Day, Suresh Dua, and Juris Grevins. Thanks also go to John Costella (from whose Ph.D. thesis I borrowed the "$\boldsymbol{b}$" notation for canonical momentum) and Eugen Merzbacher for their critical reading of various chapters together with the feedback they provided.

I also thank Armin Ardekani, Eric Bos, Sam Drake, Scott Foster, Alex Kalloniatis, Jim M^c^Carthy, Tim Priest, Rob Purvinskis, Andy Rawlinson, and Alice von Trojan for additional reading and feedback given, both in electronic mail and over many coffee table conversations. I may not have acted on their suggestions every time, but their input was always weighed and valued. I appreciate the fine help given by Unix guru Jonathan Woithe that kept my computer running smoothly throughout the project. I thank Springer for having the confidence that my original raw manuscript could be improved upon greatly, along with Frank Ganz for his LaTeX typesetting help, Hal Henglein for his meticulous proof reading, and Joseph Piliero for designing the cover. Finally, I wish to thank my editors Ron Johnson, Jeanine Jordan, and Francine M^c^Neill for overseeing and producing the finished product.

Adelaide, Australia
July 2006

Don Koks

Contents

1 The Language of Physics

The opening decades of the nineteenth century witnessed a series of experiments that introduced the world to the wonders of electromagnetism. In 1820, Hans Christian Ørsted noticed that every time he switched on the electric current in a wire in order to heat it, a nearby compass needle moved; and this production of magnetism was later confirmed in experiments by André-Marie Ampère. The year 1831 saw the discovery by Michael Faraday of a related, almost complementary principle: that moving magnets produce electric currents.

Electricity and magnetism had been known since ancient times, and the new idea that they were really just two sides of the same coin was one of the truly great advances of modern science. Further work by Faraday and William Thomson elaborated the new theories of electric and magnetic forces, with the idea emerging that they were caused by an ethereal something known as a field.

The idea that electric and magnetic forces might be exerted locally by this field was by no means universally accepted at the time, and researchers such as Ampère and Wilhelm Weber were more comfortable with action-at-a-distance theories. Such theories had been known since Isaac Newton introduced the first theory of gravity a century and a half earlier—although Newton himself had thought it absurd that gravity could be a force able to be communicated through a vacuum without the help of some kind of medium to transfer it.

But action-at-a-distance ideas were doomed to give way to the field concept. As the nineteenth century progressed, there was a creation and a honing of the mathematical description of how fields influence their sources and how those sources in turn modify their fields. The crowning achievement of this work occurred around 1860, when James Clerk Maxwell unified the laws of electric and magnetic fields in a single famous set of equations.

Electromagnetic theory has been a microcosm of the philosophy and principles of mathematical physics over the past two centuries, as well as its difficulties and conundrums. A study of its history yields great insight into the way that physics tends to advance in small steps: first, perhaps, by a lucky experiment such as Ørsted's, and then, one foot after the other, suggested by new ideas such as Faraday's principle of reciprocity between electricity and magnetism. And later, new results are predicted by the mathematical theory that is slowly being built up.

In the coming chapters, we will trace a path that explains some of these advances in the language of mathematical physics. We'll begin by surveying some of the important principles of linear algebra, along with the bracket notation of Dirac that can be used for writing the equations of Fourier analysis and quantum theory. We'll show later how Fourier analysis is used in a more "real-life" statistical context, along with the idea of how all-pervasive the exponential function is within the statistics of the everyday world. Following this, we'll study the theory of three-dimensional rotations—a surprisingly rich subject that's far easier than many approaches to it make out, and unfortunately one beset by latter-day mythology that finds itself propagated widely in the age of the Internet.

From there we'll delve into the geometrical view of special relativity and examine closely life in an accelerated frame, since this forms a good precursor to a study of gravity. The ideas encountered here will give rise to the notion of vectors and tensors, which we will explore further in the chapters that follow, along with a study of curvature that will come in handy when we eventually get to general relativity. We'll then move on to the lagrangian formalism of field theory, and will examine in detail some of the ways and pitfalls of tackling the complex integrals that are used in that subject. Finally, we'll delve into gravity, a subject that draws many of the previous ideas together in an elegant way.

Throughout these travels, our aim will be to show just how the modern notation arises, why it was invented, and why it's generally useful (with a few exceptions along the way!). The standard physics education of today unfortunately finds little time to linger over historical reasons for why things are written in the way that they are. It's often taken for granted that even the most obscure mathematical vocabulary is best taken as given, and will at some later time be seen to justify the initial effort put in to learn it.

Happily, such later appreciation over insights gained does happen, but it may be that a long time passes between our first learning new notation and our seeing just why it is so very useful. And on occasion we are still left wondering just how some sleight of hand was performed. For example, how do matrix determinants find their way into calculus? What do the complicated multiplications of convolution have to do with smoothing data? Why should the four-dimensional objects known as quaternions have anything to do with rotations in three dimensions? Why are real definite integrals sometimes able to be done by turning them into more complicated complex ones? And where did the polar coordinate expression for the laplacian ∇^2 originally come from? Showing that it reduces to the far-simpler cartesian result gives no insight into how it was derived in the first place. By trial and error, perhaps?

This last example is the tip of the iceberg of a whole class of problems that have to do with changing coordinates. These lie at the heart of modern physics, with its emphasis on the requirement that physical laws should be expressible in a way that is coordinate- and frame-independent. So in the chapters to follow, we will not teach mathematical physics from the ground up, but rather will assume a familiarity with the subject along with a desire to explore the reasons that some of its building blocks have been assembled in their well-known and traditional ways.

This, then, is the spirit in which this book should be read. It has many derivations, but not all of them are written for the sake of deriving what is on the surface. Rather, they are aimed at teaching something else: the formalism and ways of thinking about and writing the language of mathematical physics.

Throughout the book, we will use the idea time and again that interesting results often come from an extremisation procedure, or at least are related to the stationary point of some significant quantity. Across the board, from the relationship between the statistical mean and standard deviation to the way in which new physics is predicted by the lagrangian approach to gravity, stationarity is a powerful tool, and one with which we can identify on a human level, since it could be said that everything we ever do as humans is done in order to extremise something such as contentedness. Even when we walk along a twisting path to get from one place to another, we are minimising our effort in what is really a noneuclidean geometry, in the sense that we follow a "shortest" path that itself is not a straight line when drawn on a map.

Finding a good place to start with any subject in physics or mathematics varies with taste. Mathematicians are only too aware that tinkering with the fundamentals of mathematics, such as the famous Axiom of Choice, leads to intriguing paradoxes, such as the Hausdorff–Banach–Tarski theorem.

In its simplest terms, the Axiom of Choice states that if we have any number of bags, each containing some marbles, then it's always possible to choose a marble from each bag. Perfectly obvious, perhaps? But the Hausdorff–Banach–Tarski theorem is concerned with countability, and states that *if* the Axiom of Choice is assumed, then a ball can be dissected into as few as five pieces that can then be reassembled to form two balls identical to the original. The only reason a real ball cannot be dissected in this way is because its atomic makeup ultimately means that it is not a continuum, which is a spanner in the works for the theorem.

Physicists' understanding of many areas of mathematical physics has progressed to the point where there are options for good starting points, and certainly it does not seem reasonable to begin as far back as the Axiom of Choice! As an example of a different point of view, far more emphasis is placed on basis vectors here than is usual in books on mathematical physics. This has been done with the aim of making the invariant nature of vectors, and eventually tensors, much more transparent. The complete vector is not simply an ordered set of numbers in some coordinate system, obeying an unobvious rule concerning coordinate changes, but rather consists of components *and basis vectors* in such a way that by making the roles of the two entities explicit, the invariance of the complete vector under a change of coordinates is more readily seen. This invariance is the key concept. Historically, part of the goal of tensor analysis has been to organise definitions and procedures in such a way that all references to the basis can be dropped. Modern tensor analysis has become slightly less focussed on coordinates, which is a good thing; after all, the main reason that vectors are useful is *because* they include a basis, which allows them to be treated as arrows with an invariant existence of their own. We have taken this idea further in the pages ahead. So the ease with which equations to come, such as (8.225) and (12.41), produce the usual component-wise results shows that tensor expressions that include basis vectors have a useful, and thus very legitimate, role in the calculations.

A Note About Vector and Index Notation

It will be necessary to highlight our vector notation in three and four dimensions. Our vectors in any number of dimensions are normally written in a bold font. Sometimes we'll want to split a four-dimensional vector up into its time and space components (the first and last three components, respectively). This is commonly done in the subject by all authors, because the spatial part is often a familiar vector recognisable from newtonian mechanics—where it would also conventionally be written in bold. To sidestep this potential for a double use of bold fonts, any time that a distinction must be made, the four-dimensional vector will be written with an arrow, such as $\vec{u} = \gamma(1, \boldsymbol{v})$. Alternative approaches have been avoided, such as swapping the functions of arrow and bold to write $\boldsymbol{u} = \gamma(1, \vec{v})$, or even writing $\boldsymbol{u} = \gamma(1, \boldsymbol{v})$. Any possible confusion is one of those things that's probably inevitable when anything is generalised into a higher number of dimensions, but we hope our use of the occasional arrow will be transparent enough.

On the same note, vector indices are written in Greek when they refer to the highest dimensional space, and written in Latin when referring to a subset of that space. This is straightforward enough but needs to be borne in mind when reading about differential geometry in Chap. 9. There the higher space in which surfaces are embedded takes Greek indices as it should, while the surfaces themselves use Latin indices. Again this is fine, but one of the morals of the story in that chapter is that, in the end, the subspace with its Latin indices turns out to be *all* that we need to consider in the physical world, and it becomes recognised as our own spacetime—which of course has used Greek indices elsewhere in the book because of the need to consider subspaces of *that*, confined to space dimensions only. So there is no contradiction in the notation, and any confusion that arises can be cleared up by the reader's keeping this index usage in mind, as shown in Fig. 1.1.

Students of pure mathematics find that as they learn more of their subject, it becomes more and more unified, and the connections between its different branches become increasingly evident. Students of physics, on the other hand, often feel that the reverse is happening: their subject only seems to grow more

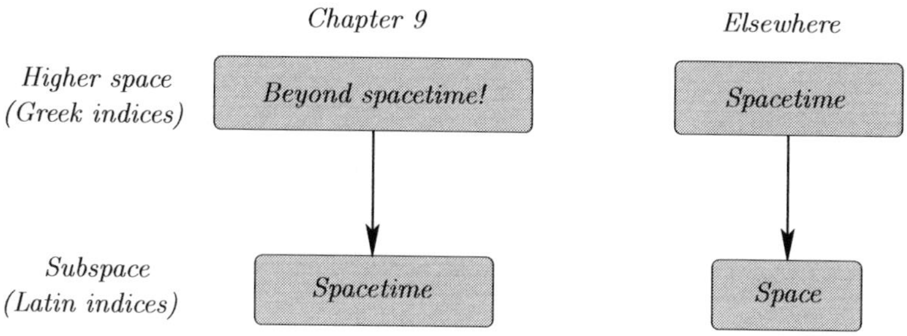

Fig. 1.1. Because there are three levels of dimensions being considered in this book and conventionally just two fonts used to denote their vector indices, one overriding rule is used: Greek indices refer to the higher-dimensional space, while Latin indices refer to the subspace. Spacetime can play either role.

and more tendrils. Perhaps in the chapters to come, we can reverse some of that trend by showing that at the heart of many seemingly diverse areas in mathematical physics, there is a language that is as well thought out as it is elegant.

2 A Trip Down Linear Lane

Mathematical physics was born four centuries ago in the research of Kepler, Galileo, and Newton. For them and their peers, the geometry of the ancients was the royal road to analysis and proof. But the new analyses of motion and gravity created new needs, which were answered by methods such as those of the calculus; and so it was that the calculus soon came to be regarded as the most important subject to be learnt by aspiring analysts of the natural world.

However, recent decades have seen the reappearance of geometry in university physics courses, not so much in its original form but in the more abstract form of linear algebra. The ideas of linearity form the canvas on which we paint many of our current ideas of mathematical physics. At least on a day-to-day level, Nature seems to prefer linearity, such as in her use of electromagnetic and gravitational fields, the principles of circuit theory, and the linear differential equations ubiquitous to mechanics. And in any field of science, linear problems are very often the only ones able to be solved analytically; even when we must resort to numerical techniques, we can still be confident of avoiding the difficulties surrounding the solution of nonlinear equations, such as ill-defined solutions or the onset of chaos.

When analysing these and many other problems, our hard-won techniques of the calculus are often only implemented once a basic concrete foundation has been laid that enables us to form a mental picture of what we are analysing. For despite the abstract nature of much of modern physics, at its heart it still relies on our forming useful mental pictures. Quantities such as vectors and the curvature of higher-dimensional surfaces still rest upon geometrical ideas.

To the theoretician, one thing becomes very obvious from a reading of the history and language of electromagnetism; that is, that the language of mathematics continually changes to keep up with the demands placed upon it by a new science. An example is the expression for the *Lorentz force*, the force exerted on a charge by electric and magnetic fields. In cartesian coordinates, we might write the force in the direction of each axis in the following way:

$$
\begin{aligned}
F^x &= q\left(E^x + v^y B^z - v^z B^y\right), \\
F^y &= q\left(E^y + v^z B^x - v^x B^z\right), \\
F^z &= q\left(E^z + v^x B^y - v^y B^x\right).
\end{aligned}
\tag{2.1}
$$

Of course, the force is seldom written this way. More usual is to employ the geometrical language of vectors:

$$\boldsymbol{F} = q(\boldsymbol{E} + \boldsymbol{v} \times \boldsymbol{B})\,. \tag{2.2}$$

The apparent complication of the three equations in (2.1) belies a symmetry that can more concisely be expressed by the cross product of (2.2). But (2.2) is more than just a tidy way of cramming the intricacies of (2.1) into one line of mathematics. It is built on the idea that the fields are vector quantities: that they add linearly and obey the various theorems that we know the cross product introduces. This rewriting of three equations as one brings with it an increased understanding of electromagnetic field theory, as well as an ability to introduce nonmathematical rules of thumb for how charges move in practice, such as a right-hand rule. Rules such as these are useful mainly because they don't depend on properties of the coordinates, unlike (2.1).

A study of the history of electromagnetic theory shows that it eventually brought forth the ideas of special relativity, and these in turn gave birth to general relativity, which brings gravity into its fold. In the tensor language of relativity, the way of writing the Lorentz force looks completely different:

$$\frac{\mathrm{D}p^\alpha}{\mathrm{d}\tau} = qF^{\alpha\beta}u_\beta\,. \tag{2.3}$$

In fact, this way of writing the force incorporates gravity, although just how that comes about really does need a lot of ink spent in the unwrapping. But it's a definite advance over earlier notation because it combines formerly disparate areas of physics into one language. And at its core, it still uses the language of vectors, adding new insight to the geometrical approach to physics.

Much of this book uses the ideas of linear algebra, and it will be useful in this chapter to survey the subject's general principles. Rather than stopping to prove everything, we will simply point out the basic language, concepts, and notation that are used time and time again in the coming chapters. Any proofs that are omitted can be found in most introductory linear algebra texts.

2.1 Vector Spaces and Matrices

The ideas and philosophy of linearity are contained in the concept of a vector space, a set of abstract entities that obey a small number of useful axioms embodying the notion of linearity. These entities, *vectors*, can be acted upon by some linear operator. We will assume the reader is familiar with the axioms of vector spaces: addition, multiplication by a scalar, the existence of an identity and inverses, and the rules of associativity.

The basic tool for calculations in linear algebra is *gaussian elimination*, which is just the row reduction that converts matrices into other matrices that have similar or identical properties. Again, we assume the reader is familiar with gaussian elimination and the ideas of matrix inversion that are so closely allied with it.

Fundamental to forming a geometric view of linear algebra is the idea of constructing a *basis*, a set of n linearly independent vectors that span an n-dimensional vector space. Every vector in the space can always be written as a unique linear combination of the basis vectors of that space. When this set of coefficients is ordered to match the basis vectors, they form the *coordinate vector* of that element with respect to the chosen basis. This coordinate vector is itself a vector in the vector space of $\mathbb{R}^n$.

The concept of a coordinate vector ensures that all vector spaces can be represented by the "arrows" that we are all familiar with from basic geometry, allowing us to form a very geometrical picture of a vector space. But coordinate vectors are useful for another reason. Known as the *Correspondence Principle* (of linear algebra), it states that even the most abstract linear operator can always be represented by a matrix of numbers together with a set of coordinate vectors in $\mathbb{R}^n$ that forms a basis for the relevant linear space.

The Correspondence Principle here is altogether different from the Correspondence Principle of quantum mechanics that is more well-known to physicists, which states that the expected value of an observable in quantum mechanics behaves in the same way as the observable does in classical mechanics.

Coordinate vectors are important because they are geometrical representations of a vector. But we should not confuse the two, and this is worth stating separately:

> *The coordinate vector is an ordered set of numbers that represents a vector with respect to some chosen basis, allowing us to picture the vector as an arrow.*

Changing the Basis

Just as a vector space will have an infinite number of different basis choices (some more useful than others), each vector in it can be represented by a different coordinate vector. If S and T are two bases of a given vector space, then a vector $\boldsymbol{\alpha}$ is associated with the two coordinate vectors $[\boldsymbol{\alpha}]_S$ and $[\boldsymbol{\alpha}]_T$. Given one of these, we can always produce the other, provided of course that we know the bases. What needs to be done is for one basis to be written as a linear combination of the other. For example, suppose we are given $[\boldsymbol{\alpha}]_S$ and wish to calculate $[\boldsymbol{\alpha}]_T$. If we know how each vector in S can be written as a linear combination of the vectors of T, then the problem is solved. Changing bases is a key idea in linear algebra, so that the ability to find linear relationships between vectors in $\mathbb{R}^n$ is very important and useful. The fundamental and very useful procedure for doing this is known as the *dependency relationship algorithm.* Because we assume a basis set has been found in the various scenarios we will encounter, it makes good sense to examine just how such a set is constructed, given a set that (we hope!) already contains more than we need.

Like many procedures in linear algebra's toolkit, the dependency relationship algorithm depends on the notion of *row equivalence* between two matrices. Two matrices are row equivalent if one can be obtained from the other by a finite series of any of the three elementary row operations:

1. swapping two rows,
2. multiplying a row by a nonzero real number, and
3. adding a nonzero multiple of any row to another row.

The algorithm states that if two matrices are row equivalent, then the linear relationships among the columns of one are identical to the corresponding linear relationships among the columns of the other. An example will illustrate the point. Suppose we are given four coordinate vectors in $\mathbb{R}^3$:

$$\boldsymbol{\alpha}_1 = \begin{bmatrix} 0 \\ 1 \\ 1 \end{bmatrix}, \quad \boldsymbol{\alpha}_2 = \begin{bmatrix} 1 \\ 0 \\ 1 \end{bmatrix}, \quad \boldsymbol{\alpha}_3 = \begin{bmatrix} -1 \\ 2 \\ 1 \end{bmatrix}, \quad \boldsymbol{\alpha}_4 = \begin{bmatrix} 1 \\ 1 \\ 0 \end{bmatrix}. \tag{2.4}$$

Note that if a vector is written bold as $\boldsymbol{\alpha}$, we might better write its coordinate vector as $[\boldsymbol{\alpha}]_{\mathbb{R}^3}$ to emphasise that we are focussing on the three numbers that represent $\boldsymbol{\alpha}$ in $\mathbb{R}^3$. We'll certainly do this in Chap. 8, but there is no need to do so for the following discussion.

First build a matrix with these vectors as its columns and then row reduce it:

$$\begin{bmatrix} 0 & 1 & -1 & 1 \\ 1 & 0 & 2 & 1 \\ 1 & 1 & 1 & 0 \end{bmatrix} \xrightarrow{\text{row reduce}} \begin{bmatrix} 1 & 0 & 2 & 0 \\ 0 & 1 & -1 & 0 \\ 0 & 0 & 0 & 1 \end{bmatrix}. \tag{2.5}$$

Examining the right-hand matrix, we see instantly that $\boldsymbol{\alpha}_1, \boldsymbol{\alpha}_2, \boldsymbol{\alpha}_4$ are linearly independent, while $\boldsymbol{\alpha}_3 = 2\boldsymbol{\alpha}_1 - \boldsymbol{\alpha}_2$ is the dependent member of the set. Of course, $\boldsymbol{\alpha}_3$ is really no more redundant than $\boldsymbol{\alpha}_1$ or $\boldsymbol{\alpha}_2$. If we prefer to choose $\boldsymbol{\alpha}_3$ as a basis vector, then we need only put it farther to the left in the starting matrix of the algorithm, and it will be given priority in the order of linear independence when the set is row reduced.

2.2 Inner Products

To form a completely geometrical picture of a vector space, we need to supply it with the notion of a distance, and this is tied to the concept of a dot product between two vectors. When the space is $\mathbb{R}^n$ and distances within it obey Pythagoras's theorem, the vector space is called *euclidean* and denoted $\mathbb{E}^n$.

Whenever required, we will distinguish between these two spaces to stress that one of them uses this notion of distance.

Although it's not absolutely necessary, dot products are often reserved for operations within a euclidean space, and their generalisation to all vector spaces is the *inner product*. In fact, the two notations are quite interchangeable, and we will certainly sometimes use a dot product for convenience even when the space is not euclidean—as well as using tensor formalism for the same purpose, which will be explained when the need arises.

But the inner product notation has its own uses, especially when generalised to *Dirac bracket notation*, as we'll see in Sect. 2.6. The inner product of two vectors $\boldsymbol{\alpha}, \boldsymbol{\beta}$ is written $\langle\boldsymbol{\alpha}|\boldsymbol{\beta}\rangle$. It has the following basic properties:

1. $\langle\boldsymbol{\alpha}|\boldsymbol{\alpha}\rangle \geqslant 0$, with equality if and only if $\boldsymbol{\alpha} = \mathbf{0}$ (the zero vector).
2. $\langle\boldsymbol{\alpha}|\boldsymbol{\beta}\rangle = \langle\boldsymbol{\beta}|\boldsymbol{\alpha}\rangle^*$, where the asterisk denotes the complex conjugate.
3. Standard mathematics usage is to write, for all complex numbers c,

$$\langle c\,\boldsymbol{\alpha}|\boldsymbol{\beta}\rangle = c\langle\boldsymbol{\alpha}|\boldsymbol{\beta}\rangle\,, \quad \text{in which case } \langle\boldsymbol{\alpha}|c\,\boldsymbol{\beta}\rangle = c^*\langle\boldsymbol{\alpha}|\boldsymbol{\beta}\rangle\,. \tag{2.6}$$

   Physics usage is the other way around:

$$\langle\boldsymbol{\alpha}|c\,\boldsymbol{\beta}\rangle = c\langle\boldsymbol{\alpha}|\boldsymbol{\beta}\rangle\,, \quad \text{in which case } \langle c\,\boldsymbol{\alpha}|\boldsymbol{\beta}\rangle = c^*\langle\boldsymbol{\alpha}|\boldsymbol{\beta}\rangle\,. \tag{2.7}$$

   We will adhere to the physics usage since we intend to use brackets to discuss some quantum mechanics.
4. Finally, a distributive law holds:

$$\langle\boldsymbol{\alpha}+\boldsymbol{\beta}|\boldsymbol{\gamma}\rangle = \langle\boldsymbol{\alpha}|\boldsymbol{\gamma}\rangle + \langle\boldsymbol{\beta}|\boldsymbol{\gamma}\rangle\,. \tag{2.8}$$

Additionally, we note that a vector's squared length $|\boldsymbol{\alpha}|^2$ and the angle θ between two vectors $\boldsymbol{\alpha}$ and $\boldsymbol{\beta}$ are each defined by

$$|\boldsymbol{\alpha}|^2 \equiv \langle\boldsymbol{\alpha}|\boldsymbol{\alpha}\rangle\,, \quad \cos\theta \equiv \frac{\langle\boldsymbol{\alpha}|\boldsymbol{\beta}\rangle}{|\boldsymbol{\alpha}|\,|\boldsymbol{\beta}|}\,. \tag{2.9}$$

Orthogonality and Orthonormality

When the inner product of two vectors is zero, we say they are *orthogonal*. When they are orthogonal and both normalised to unit length, they are called *orthonormal*.

Basis vectors are often combined with each other to produce a new basis composed of orthogonal vectors, enabling simpler inner product calculations to be done with any vector pairs in the space. A useful way to do this is via the *Gram–Schmidt algorithm*. Given a set of vectors $\{\boldsymbol{\alpha}_1, \dots, \boldsymbol{\alpha}_n\}$, we produce an orthogonal set $\{\boldsymbol{\gamma}_1, \dots, \boldsymbol{\gamma}_n\}$ in the following way. First, set $\boldsymbol{\gamma}_1 \equiv \boldsymbol{\alpha}_1$. Now, $\boldsymbol{\alpha}_2$ is a sum of two components, one parallel and one orthogonal to $\boldsymbol{\gamma}_1$. Eliminate the parallel component by subtracting from $\boldsymbol{\alpha}_2$ its projection along the *unit* vector $\widehat{\boldsymbol{\gamma}}_1$ (with the caret denoting unit length):

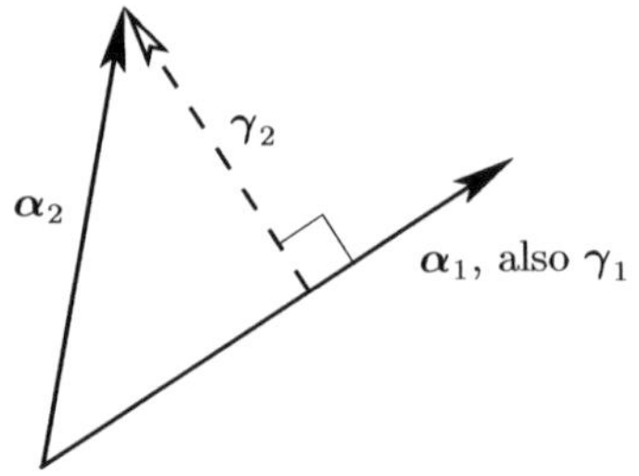

Fig. 2.1. Using the Gram–Schmidt algorithm to orthogonalise the $\{\boldsymbol{\alpha}_1, \boldsymbol{\alpha}_2\}$ basis. In this two-dimensional example, begin by setting the first orthogonalised basis vector to be $\boldsymbol{\gamma}_1 \equiv \boldsymbol{\alpha}_1$. The other orthogonalised basis vector $\boldsymbol{\gamma}_2$ is then formed from $\boldsymbol{\alpha}_2$ by subtracting $\boldsymbol{\alpha}_2$'s projection on $\boldsymbol{\gamma}_1$, producing $\boldsymbol{\gamma}_2 \equiv \boldsymbol{\alpha}_2 - (\boldsymbol{\alpha}_2 \cdot \widehat{\boldsymbol{\gamma}}_1)\,\widehat{\boldsymbol{\gamma}}_1$.

$$\begin{aligned}\boldsymbol{\gamma}_2 &\equiv \boldsymbol{\alpha}_2 - \langle \boldsymbol{\alpha}_2 | \widehat{\boldsymbol{\gamma}}_1 \rangle\, \widehat{\boldsymbol{\gamma}}_1 \\ &= \boldsymbol{\alpha}_2 - \frac{\langle \boldsymbol{\alpha}_2 | \boldsymbol{\gamma}_1 \rangle}{\langle \boldsymbol{\gamma}_1 | \boldsymbol{\gamma}_1 \rangle} \boldsymbol{\gamma}_1 \,. \end{aligned} \tag{2.10}$$

The next vectors are found in the same way by subtracting their projections along the unit vectors already established. For example,

$$\boldsymbol{\gamma}_3 \equiv \boldsymbol{\alpha}_3 - \frac{\langle \boldsymbol{\alpha}_3 | \boldsymbol{\gamma}_1 \rangle}{\langle \boldsymbol{\gamma}_1 | \boldsymbol{\gamma}_1 \rangle} \boldsymbol{\gamma}_1 - \frac{\langle \boldsymbol{\alpha}_3 | \boldsymbol{\gamma}_2 \rangle}{\langle \boldsymbol{\gamma}_2 | \boldsymbol{\gamma}_2 \rangle} \boldsymbol{\gamma}_2 \,, \tag{2.11}$$

and so on. In general,

$$\boldsymbol{\gamma}_n = \boldsymbol{\alpha}_n - \text{orthogonal projection of } \boldsymbol{\alpha}_n \text{ onto space of } \{\boldsymbol{\alpha}_1, \ldots, \boldsymbol{\alpha}_{n-1}\}. \tag{2.12}$$

A geometrical picture using the euclidean inner product is shown in Fig. 2.1. But the algorithm applies to arbitrary inner products, as can be shown via induction. However, the mathematics of higher dimensions in linear algebra is well enough behaved that we can generally use our geometrical intuition without really going awry.

Later in this chapter and in the next, we'll encounter Fourier analysis along with the method of least squares. Both of these are really no more than applying the ideas of orthogonality in practical ways.

2.3 Crystallography and the Cobasis

An important concept of linear algebra and tensor analysis arises in crystallography. One of the great uses of X rays is that their wavelengths tend to be just the right size for use in probing crystal structure through Bragg diffraction. Bragg diffraction can be viewed as a process where incoming rays bounce

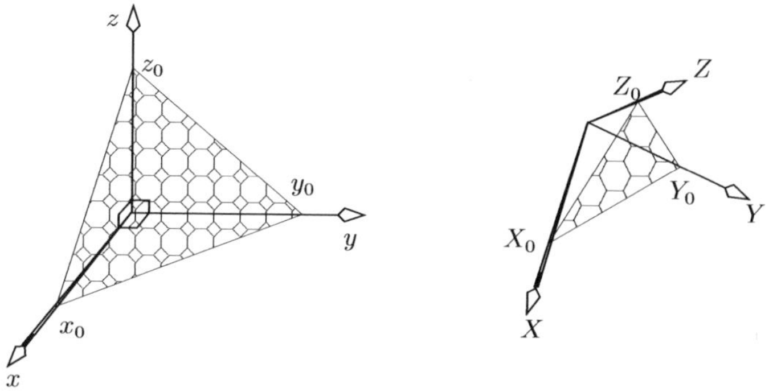

Fig. 2.2. Left: An orthogonal lattice, with cartesian xyz-axes that are cut by a crystal plane at x_0, y_0, z_0. **Right**: A more general lattice, where the natural basis vectors are no longer orthogonal.

off crystal planes to interfere constructively at certain angles that give information about the orientation of these planes, and so the crystallographer's diffraction photographs are really giving information about crystal planes as opposed to the atoms that make up the lattice. Yet it's the atomic lattice that is of real interest, so the crystallographer needs to relate information about planes in the crystal to the positions of its atoms.

To see how this sort of crystal analysis gives rise to something new, consider first the more intuitive case of a lattice whose atoms are arranged in a simple "box" formation, whereby a basis of vectors $\{\boldsymbol{e}_x, \boldsymbol{e}_y, \boldsymbol{e}_z\}$ joining them can be chosen to be orthogonal, as shown on the left-hand side in Fig. 2.2. We'll employ the usual xyz cartesian axes to describe positions in this lattice. A particular plane within the lattice cuts those axes at x_0, y_0, z_0. First, we ask: what is the equation for this plane?

Consider an arbitrary point $\boldsymbol{r} \equiv (x, y, z)$ of the plane. The plane itself has normal vector $\boldsymbol{n} \equiv (n^x, n^y, n^z) = n^x \boldsymbol{e}_x + n^y \boldsymbol{e}_y + n^z \boldsymbol{e}_z$, in which case we observe that

$$(\boldsymbol{r} - x_0\, \boldsymbol{e}_x)\cdot\boldsymbol{n} = (\boldsymbol{r} - y_0\, \boldsymbol{e}_y)\cdot\boldsymbol{n} = (\boldsymbol{r} - z_0\, \boldsymbol{e}_z)\cdot\boldsymbol{n} = 0\,. \tag{2.13}$$

Thus,

$$\boldsymbol{r}\cdot\boldsymbol{n} = x_0\, n^x = y_0\, n^y = z_0\, n^z \equiv N\,. \tag{2.14}$$

Usually "$a \equiv b$" means "a is defined to equal b". In (2.14) it is N that's being defined since it comes at the end of a train of equalities. Throughout this book, when there are several equalities and the last is an "$\equiv$" sign, we will mean that the last term is being defined, so that the "$\equiv$" is then read as "which defines". This will sometimes also be the case for expressions such as $a \equiv b$, where the fact that b is being defined rather than a should be clear from the context.

The normal vector is then

$$\boldsymbol{n} = N\left(\frac{1}{x_0}, \frac{1}{y_0}, \frac{1}{z_0}\right). \tag{2.15}$$

Combining (2.14) with (2.15) gives the equation of the plane as

$$\boldsymbol{r}\cdot\left(\frac{1}{x_0}, \frac{1}{y_0}, \frac{1}{z_0}\right) = 1\,, \quad \text{or} \quad \frac{x}{x_0} + \frac{y}{y_0} + \frac{z}{z_0} = 1\,. \tag{2.16}$$

The normal to the plane is proportional to $(1/x_0, 1/y_0, 1/z_0)$. These three numbers are usefully scaled by their least common multiple and are then called *Miller indices*, integers that characterise the plane, although we need not do that for this discussion. (Integers are useful for the crystallographer because the atomic sites occur at discrete intervals in the plane.)

Now, in general, lattices do not have an orthogonal structure; the natural basis vectors that describe the atomic positions are more like the set on the right-hand side in Fig. 2.2. How can we adapt the calculation of (2.13)–(2.15) to this more general basis? Suppose the axes are now labelled X, Y, Z, with basis vectors $\boldsymbol{e}_X, \boldsymbol{e}_Y, \boldsymbol{e}_Z$. A plane cuts these axes at distances X_0, Y_0, Z_0 from some origin. This plane has normal vector $\boldsymbol{n}$, and an arbitrary point within the plane has position vector $\boldsymbol{r} \equiv X\,\boldsymbol{e}_X + Y\,\boldsymbol{e}_Y + Z\,\boldsymbol{e}_Z$. Analogously to (2.13), we write

$$(\boldsymbol{r} - X_0\,\boldsymbol{e}_X)\cdot\boldsymbol{n} = (\boldsymbol{r} - Y_0\,\boldsymbol{e}_Y)\cdot\boldsymbol{n} = (\boldsymbol{r} - Z_0\,\boldsymbol{e}_Z)\cdot\boldsymbol{n} = 0\,, \tag{2.17}$$

giving

$$\boldsymbol{r}\cdot\boldsymbol{n} = X_0\,\boldsymbol{e}_X\cdot\boldsymbol{n} = Y_0\,\boldsymbol{e}_Y\cdot\boldsymbol{n} = Z_0\,\boldsymbol{e}_Z\cdot\boldsymbol{n}\,. \tag{2.18}$$

But now some care is needed. In the orthogonal case, we were able to write $\boldsymbol{n} \equiv n^x\,\boldsymbol{e}_x + n^y\,\boldsymbol{e}_y + n^z\,\boldsymbol{e}_z$ to conclude that $n^x = \boldsymbol{n}\cdot\boldsymbol{e}_x$ and so on. But such a statement actually makes use of the orthogonality property of $\boldsymbol{e}_x, \boldsymbol{e}_y, \boldsymbol{e}_z$, which the set $\boldsymbol{e}_X, \boldsymbol{e}_Y, \boldsymbol{e}_Z$ does not have.

We can, however, still use orthogonality to calculate the components of $\boldsymbol{r}$ by defining another basis related to the $\boldsymbol{e}_X, \boldsymbol{e}_Y, \boldsymbol{e}_Z$ vectors. Define three new vectors $\boldsymbol{e}^X, \boldsymbol{e}^Y, \boldsymbol{e}^Z$, each of which pairs with one of the subscripted set and is orthogonal to the two others of that set. For example,

$$\boldsymbol{e}^X\cdot\boldsymbol{e}_X \equiv 1\,, \quad \boldsymbol{e}^X\cdot\boldsymbol{e}_Y = \boldsymbol{e}^X\cdot\boldsymbol{e}_Z \equiv 0\,, \tag{2.19}$$

and similarly for Y and Z. These requirements suffice to ensure that

$$\boldsymbol{e}^X = \frac{\boldsymbol{e}_Y \times \boldsymbol{e}_Z}{\boldsymbol{e}_X\cdot(\boldsymbol{e}_Y \times \boldsymbol{e}_Z)}\,, \tag{2.20}$$

with cyclic permutations on this for $\boldsymbol{e}^Y$ and $\boldsymbol{e}^Z$. Note that these three expressions for $\boldsymbol{e}^X, \boldsymbol{e}^Y, \boldsymbol{e}^Z$ all share the same denominator since $\boldsymbol{e}_X\cdot(\boldsymbol{e}_Y \times \boldsymbol{e}_Z)$

is unchanged by cyclic permutations of X, Y, Z. This denominator is actually the *signed volume* of a parallelepiped with sides $\boldsymbol{e}_X, \boldsymbol{e}_Y, \boldsymbol{e}_Z$, which we'll encounter in the next section.

The set of vectors $\{\boldsymbol{e}^X, \boldsymbol{e}^Y, \boldsymbol{e}^Z\}$ is called the *cobasis*; it's also known as the basis *dual* to $\{\boldsymbol{e}_X, \boldsymbol{e}_Y, \boldsymbol{e}_Z\}$. Crystallographers refer to it as the *reciprocal basis* and use a different normalisation, $\boldsymbol{e}^X \cdot \boldsymbol{e}_X \equiv 2\pi$, owing to their emphasis on waves and phases, where factors of 2π are common. Our normalisation (2.19) will be more natural when we meet it later in a tensor context, so we will keep it; in which case the cobasis allows us to write, for any vector $\boldsymbol{v} = v^X \boldsymbol{e}_X + v^Y \boldsymbol{e}_Y + v^Z \boldsymbol{e}_Z$,

$$v^\alpha = \boldsymbol{v} \cdot \boldsymbol{e}^\alpha \quad \text{for } \alpha = X, Y, Z\,, \tag{2.21}$$

so that

$$\boldsymbol{v} = \sum_\alpha \boldsymbol{v} \cdot \boldsymbol{e}^\alpha \, \boldsymbol{e}_\alpha\,. \tag{2.22}$$

Writing the cobasis with "up" indices produces a good symmetry; this can be seen by writing the vector $\boldsymbol{v}$ in terms of the cobasis. For this, define new components v_α:

$$\boldsymbol{v} = \sum_\alpha v^\alpha \, \boldsymbol{e}_\alpha \equiv \sum_\alpha v_\alpha \, \boldsymbol{e}^\alpha\,. \tag{2.23}$$

The general cobasis component v_α is easily calculated by using the orthogonality of the cobasis with the basis:

$$\boldsymbol{v} \cdot \boldsymbol{e}_\alpha = \sum_\beta v_\beta \, \boldsymbol{e}^\beta \cdot \boldsymbol{e}_\alpha = v_\alpha\,, \quad \text{so that } \boldsymbol{v} = \sum_\alpha \boldsymbol{v} \cdot \boldsymbol{e}_\alpha \, \boldsymbol{e}^\alpha\,. \tag{2.24}$$

Expressions such as (2.24) highlight the utility of the cobasis: it brings the benefits of orthogonality to a basis that might not itself be orthogonal. Again, we reiterate that expressions using the basis and cobasis are quite symmetrical. For example,

$$v_\alpha = \boldsymbol{v} \cdot \boldsymbol{e}_\alpha\,, \quad \text{and } v^\alpha = \boldsymbol{v} \cdot \boldsymbol{e}^\alpha\,. \tag{2.25}$$

Using the cobasis means that indices in expressions such as (2.25) always appear either all up or all down, while sums such as (2.23) have their dummy indices arranged diagonally. As we'll see in later chapters, this index bookkeeping is the bread and butter of tensor analysis. The idea of the cobasis will streamline the discussion at the end of Chap. 6, and in Chap. 8 a tensor approach will make the definition of the cobasis vectors more elegant and general.

In the meantime, however, we return to the crystal lattice discussion, using the notion of the cobasis to write $\boldsymbol{n} \cdot \boldsymbol{e}_\alpha = n_\alpha$. Equation (2.18) can now be written as

$$\boldsymbol{r} \cdot \boldsymbol{n} = X_0 \, n_X = Y_0 \, n_Y = Z_0 \, n_Z \equiv N\,, \tag{2.26}$$

giving the normal as

$$n = \sum_\alpha n_\alpha e^\alpha = \frac{N}{X_0} e^X + \frac{N}{Y_0} e^Y + \frac{N}{Z_0} e^Z . \tag{2.27}$$

Finally, (2.26) becomes

$$r \cdot \left(\frac{e^X}{X_0} + \frac{e^Y}{Y_0} + \frac{e^Z}{Z_0} \right) = 1 . \tag{2.28}$$

This looks much like (2.16), of course, except that to continue to make the full identification of the normal with the unscaled Miller indices $(1/X_0, 1/Y_0, 1/Z_0)$, we need to be aware that these indices are actually the *cobasis* components of the plane's normal, not the basis components. Because crystallographers use Miller indices extensively, they constantly use the cobasis to quantify the orientations of planes through the crystal. And because the planes and the cobasis are involved in a Fourier analysis of the crystal lattice, the space spanned by the cobasis (or reciprocal basis) is sometimes called *reciprocal space*. This term might imply a lack of physical reality for the space of the cobasis. But the cobasis vectors are certainly real vectors in the crystal. A difference between the cobasis and the usual basis is that the cobasis vectors are not defined by pointing along actual lines of atoms in the crystal; but nonetheless, they can always be drawn as orthogonal to the set $\{e_X, e_Y, e_Z\}$.

In Chap. 8 we'll remark that the vectors of the cobasis are often replaced by a set of new objects called *one-forms*, which are sometimes visualised as sets of parallel planes. We will certainly neither use one-forms in this book nor have any need of them; it's arguable whether the point of view that constructs them has any real use at all for physics. On the other hand, the cobasis as constructed here is a set of vectors, which can certainly be drawn as arrows. Its elements are used here to describe crystal planes, but they are *not* themselves planes.

As a last observation, it should come as no surprise to find that the euclidean basis is identical to its cobasis. For example,

$$e^x = \frac{e_y \times e_z}{e_x \cdot (e_y \times e_z)} = e_x , \tag{2.29}$$

and similarly for e^y and e^z. No wonder, then, that the cobasis doesn't make itself apparent until we begin to consider nonorthogonal bases such as are naturally found in crystal lattices. We'll meet cobases again in the next section and in later chapters.

2.4 Finding Areas and Volumes: The Use of Determinants

If the ideas of linear algebra are designed to be useful geometrically, then we expect them to be very applicable to concepts such as areas and volumes. So we ask the question: as shown in Fig. 2.3, what is the volume of

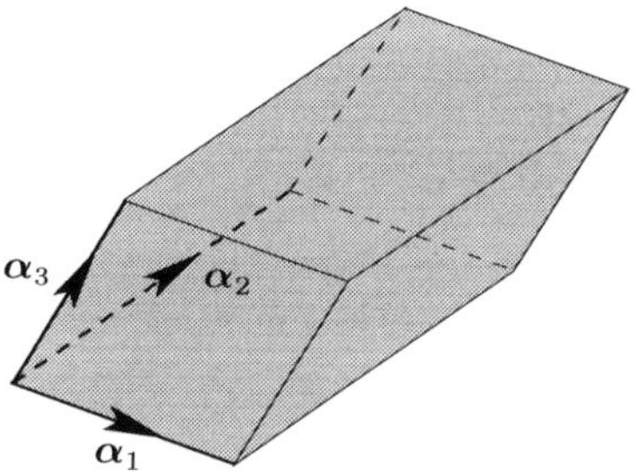

Fig. 2.3. A parallelepiped in three-dimensional euclidean space $\mathbb{E}^3$, delineated by the vectors $\boldsymbol{\alpha}_1, \dots, \boldsymbol{\alpha}_3$. Its volume is a function of the three vectors, and is found in this section. A general n-dimensional volume in $\mathbb{E}^n$ can also be defined, leading to the notion of a determinant.

an n-dimensional parallelepiped in euclidean space $\mathbb{E}^n$, whose sides are the cartesian vectors $\boldsymbol{\alpha}_1, \dots, \boldsymbol{\alpha}_n$? Cartesian coordinates are the easiest to deal with, but we'll encounter more general coordinates in Chap. 8.

The answer will be a function mapping these vectors to the real numbers, so if we build an $n \times n$ matrix that has as its rows each of the vectors $\boldsymbol{\alpha}_1, \dots, \boldsymbol{\alpha}_n$, then this function should act on the matrix to produce the required volume:

$$\begin{bmatrix} \text{—} & \boldsymbol{\alpha}_1 & \text{—} \\ & \vdots & \\ \text{—} & \boldsymbol{\alpha}_n & \text{—} \end{bmatrix} \longrightarrow \text{real numbers.} \tag{2.30}$$

This function also allows for the idea of a negative volume, being simply an indication of which ordering of the vectors $\boldsymbol{\alpha}_1, \dots, \boldsymbol{\alpha}_n$ was used to build the parallelepiped. But more on that later.

A volume in n dimensions can be defined recursively in the same way as the three-dimensional case, where the volume of a parallelepiped is given by the area of its base, determined by two of the vectors, multiplied by its height; this height is the length of a new vector that is the projection of the third vector orthogonally to the base. In n dimensions, we follow the same idea by stipulating that the volume of the n-dimensional parallelepiped be equal to the length of a new vector γ_n that is orthogonal to the space of the other $n-1$ vectors $\boldsymbol{\alpha}_1, \dots, \boldsymbol{\alpha}_{n-1}$, multiplied by the volume of the $n-1$-dimensional parallelepiped that these vectors delineate.

But this new vector γ_n is orthogonal to $\boldsymbol{\alpha}_1, \dots, \boldsymbol{\alpha}_{n-1}$, and can be constructed by the Gram–Schmidt algorithm. Just as the volume is unaltered by changing $\boldsymbol{\alpha}_n$ to γ_n, it is sufficient to find a volume function that is unaffected by linear operations of the matrix rows. Summarising, this function must have several properties:

1. **Linearity in each row**: If a vector's length is doubled, the parallelepiped's volume is doubled.
2. **Indifference to adding a multiple of one row to another row**: This ensures that its value is unchanged by the Gram–Schmidt $\boldsymbol{\alpha}_i \to \boldsymbol{\gamma}_i$ procedure.
3. **The function must be zero if one row is the zero vector**: The parallelepiped's volume must collapse to zero if one of its rows has zero length.
4. **The identity matrix must map to 1**: A box with orthogonal unit sides should have unit volume.

There is actually a unique function with these four properties: the matrix determinant. The relevant proofs of existence and uniqueness are not onerous, but would involve a side excursion over several more pages than we have to spare, being more suited to a pure linear algebra course. Here we'll simply define the determinant and demonstrate some of its properties.

2.4.1 Definition and Properties of the Determinant

Determinants make use of signs that depend on indices being permuted, so their definition is simplified through the use of the *Levi-Civita symbol* $\varepsilon_{ijk\ldots n}$,

$$\varepsilon_{123\ldots n} \equiv 1\,, \tag{2.31}$$

and

$$\varepsilon_{ijk\ldots} \equiv \begin{cases} 0 & \text{if any two subscripts are equal} \\ 1 & \text{if } ijk\ldots \text{ is an even permutation of } 123\ldots \\ -1 & \text{if } ijk\ldots \text{ is an odd permutation of } 123\ldots\,. \end{cases} \tag{2.32}$$

The determinant of a matrix A is then defined as the sum of terms chosen from unique row/column pairs:

$$|A| \equiv \det A \equiv \sum_{\text{all } i,j,k\ldots} \varepsilon_{ijk\ldots} A_{1i}\, A_{2j} \ldots\,. \tag{2.33}$$

In the 1×1 case, the determinant of a matrix $\begin{bmatrix} x \end{bmatrix}$ is just x. More interesting is the 2×2 case, where

$$\begin{vmatrix} A_{11} & A_{12} \\ A_{21} & A_{22} \end{vmatrix} = \sum \varepsilon_{ij} A_{1i}\, A_{2j} = \varepsilon_{12}\, A_{11}\, A_{22} + \varepsilon_{21}\, A_{12}\, A_{21}$$
$$= A_{11}\, A_{22} - A_{12}\, A_{21}\,, \tag{2.34}$$

as is well known.

All of the well-known properties of determinants follow from this definition. We'll illustrate this by showing that $\det A = \det A^t$. Begin with

$$\det A^t = \sum \varepsilon_{ijk\ldots} A_{i1}\, A_{j2} \;\ldots\,. \tag{2.35}$$

Now take each term of this sum and permute the factors so that they have the order $A_{1\,\text{something}}\, A_{2\,\text{something}} \ldots$. For example, consider the term $\varepsilon_{312}\, A_{31}\, A_{12}\, A_{23}$. Rearranging gives $\varepsilon_{312}\, A_{12}\, A_{23}\, A_{31}$. We have really made two rearrangements in parallel here, both with the same number of permutations. We've converted the set of first indices $312 \to 123$, while at the same time also converting the second ones, $123 \to 231$. But this also means that $\varepsilon_{312} = \varepsilon_{231}$, so that

$$\varepsilon_{312}\, A_{31}\, A_{12}\, A_{23} = \varepsilon_{231}\, A_{12}\, A_{23}\, A_{31}\,. \tag{2.36}$$

But the right-hand side of (2.36) is now identical to one of the terms in the expansion of $\det A$, (2.33). All of the terms of (2.35) can be rewritten in this way to correspond with a unique term in (2.33), so we conclude that $\det A = \det A^t$. This equality is very useful because it means that any determinant rule that applies to rows will also apply to columns.

Carrying on two parallel permutations like this is the key to using the Levi-Civita symbol as an aid to proving determinant theorems. The four requirements for the determinant listed on p. 18 can all be shown to be satisfied by such arguments, so we'll assume this has been done and instead discuss the way determinants are calculated in practice. Expansion as a sum using the Levi-Civita symbol is unwieldy for all but the smallest or sparsest of matrices. But the Levi-Civita symbol can be used to derive a simplified way of evaluating determinants, which proves to be highly useful both for numerical work and for proving theorems involving determinants.

This way of reducing the work required to calculate a determinant comes about as follows. Each element of a matrix $A = (A_{ij})$ has associated with it a number called its *cofactor*, or equivalently its *signed minor*. The minor associated with A_{ij} is the determinant of the new matrix formed by crossing out each row and column of A that contain A_{ij}. Here "signed" means that we multiply the minor of A_{ij} by $(-1)^{i+j}$. Thus the cofactor, or signed minor, belonging to A_{11} in a 2×2 matrix is $+A_{22}$, while the cofactor of A_{12} is $-A_{21}$, and so on.

The theory of cofactors has two major theorems of great use to us. The first says that the determinant of a matrix can be evaluated by choosing any one row, or column, multiplying each element in it by its cofactor, and then summing the n terms that result. (We'll defer the second theorem until we really need it, on p. 331.) For example, expand along, say, the second column of the 2×2 matrix in (2.34):

$$\begin{aligned}\det A &= A_{12} \times \text{cofactor of } A_{12} + A_{22} \times \text{cofactor of } A_{22} \\ &= -A_{12}\, A_{21} + A_{22}\, A_{11}\,,\end{aligned} \tag{2.37}$$

again as expected. Now, this expansion by cofactors can be combined with the two most important requirements we discussed earlier that the determinant obeys:

1. If a row is multiplied by a constant k, then the determinant is also multiplied by k.
2. The determinant is unchanged if we add a multiple of one row to another.

These properties of the determinant can be used, first, to simplify the matrix using gaussian elimination, and then to expand along the most convenient row or column by cofactors. An example serves to illustrate the point. By "10 ← row 1" we mean "extract a factor of 10 from row 1":

$$\begin{vmatrix} 10 & 20 & 20 \\ 8 & -3 & 15 \\ 3 & 4 & 6 \end{vmatrix} \overset{10\leftarrow\text{row 1}}{=\!=\!=} 10 \begin{vmatrix} 1 & 2 & 2 \\ 8 & -3 & 15 \\ 3 & 4 & 6 \end{vmatrix} \overset{\text{col. 3}-2\,\text{col. 1}}{=\!=\!=} 10 \begin{vmatrix} 1 & 2 & 0 \\ 8 & -3 & -1 \\ 3 & 4 & 0 \end{vmatrix}$$

$$\underset{\text{along last column}}{\overset{\text{cofactor expand}}{=\!=\!=}} 10 \begin{vmatrix} 1 & 2 \\ 3 & 4 \end{vmatrix} = -20\,. \tag{2.38}$$

Simplifying manipulations such as these are a whole lot easier than dealing with the original matrix from first principles.

The fact that the volume of the parallelepiped is given by a determinant can now be proved. First, we will set the volume of a box in $\mathbb{E}^n$ built from a set of ordered mutually orthogonal vectors $\{\gamma_1, \ldots, \gamma_n\}$ to be $|\gamma_1|\,|\gamma_2|\,\cdots\,|\gamma_n|$. Now suppose that the vectors are written as the rows of a matrix:

$$A \equiv \begin{bmatrix} \text{---} & \gamma_1 & \text{---} \\ & \vdots & \\ \text{---} & \gamma_n & \text{---} \end{bmatrix}. \tag{2.39}$$

Then, using the fact that the determinant of a product equals the product of determinants,

$$\det{}^2 A = \big(\det A\big)\big(\det A^t\big) = \det\big(AA^t\big)\,. \tag{2.40}$$

But AA^t is diagonal:

$$AA^t = \operatorname{diag}\big(\,|\gamma_1|^2, \ldots |\gamma_n|^2\,\big)\,, \tag{2.41}$$

so that

$$|\det A| = \sqrt{\det\big(AA^t\big)} = |\gamma_1|\,|\gamma_2|\,\cdots\,|\gamma_n| = \text{volume of box.} \tag{2.42}$$

So in this basic case of orthogonal vectors, the volume is the absolute value of the determinant. Suppose now that the box has sides given by vectors $\{\boldsymbol{\alpha}_1, \ldots, \boldsymbol{\alpha}_n\}$ that are not necessarily orthogonal, and consider the new matrix

$$\begin{bmatrix} \text{---} & \boldsymbol{\alpha}_1 & \text{---} \\ & \vdots & \\ \text{---} & \boldsymbol{\alpha}_n & \text{---} \end{bmatrix}. \tag{2.43}$$

We know that the volume is not changed if we use the Gram–Schmidt orthogonalisation algorithm to replace $\boldsymbol{\alpha}_1$ by a vector $\boldsymbol{\gamma}_1$ orthogonal to the other vectors and of just the right length:

$$\boldsymbol{\gamma}_1 = \boldsymbol{\alpha}_1 - \text{linear combination of } \boldsymbol{\alpha}_2, \ldots, \boldsymbol{\alpha}_n \,. \tag{2.44}$$

But likewise

$$\begin{vmatrix} \text{---} & \boldsymbol{\alpha}_1 & \text{---} \\ \text{---} & \boldsymbol{\alpha}_2 & \text{---} \\ & \vdots & \\ \text{---} & \boldsymbol{\alpha}_n & \text{---} \end{vmatrix} = \begin{vmatrix} \text{---} & \boldsymbol{\gamma}_1 & \text{---} \\ \text{---} & \boldsymbol{\alpha}_2 & \text{---} \\ & \vdots & \\ \text{---} & \boldsymbol{\alpha}_n & \text{---} \end{vmatrix}, \tag{2.45}$$

since determinants are unchanged by this operation. The argument can be applied over and over to orthogonalise all of the rows, and after each step the determinant is unchanged. But of course the absolute value of the determinant of the final matrix is precisely the volume of the box, which means that the absolute value of the determinant of (2.43) must also be the volume of the box:

$$\boxed{\begin{matrix}\text{Volume of parallelepiped} \\ \text{with sides } \boldsymbol{\alpha}_1, \ldots, \boldsymbol{\alpha}_n\end{matrix} = \text{abs} \begin{vmatrix} \text{---} & \boldsymbol{\alpha}_1 & \text{---} \\ & \vdots & \\ \text{---} & \boldsymbol{\alpha}_n & \text{---} \end{vmatrix} .} \tag{2.46}$$

The absolute value of a number, the norm of a vector, and the determinant of a matrix are all often written $|\cdot|$. If necessary, we'll indicate them explicitly with abs, norm, det.

Permuting the order of the rows (or columns) of a matrix merely changes the sign of its determinant, so some ordering of the vectors $\{\boldsymbol{\alpha}_1, \ldots, \boldsymbol{\alpha}_n\}$ will produce a positive determinant (the required volume), while swapping any of those vectors will only change the determinant's sign. This then gives rise to the idea of a *signed volume* of an ordered set of vectors, being the determinant of the matrix of those vectors laid out in the row order in which they have been specified.

2.4.2 Determinants, Handedness, and the n-Dimensional Cross Product

The sign-changing behaviour of the determinant when its rows or columns are swapped makes it perfect for specifying the *handedness* of a set of ordered, linearly independent vectors in $\mathbb{R}^n$. We define such a set $\{\boldsymbol{\alpha}_1, \ldots, \boldsymbol{\alpha}_n\}$ to be *right handed* if

$$\begin{vmatrix} \text{---} & \boldsymbol{\alpha}_1 & \text{---} \\ & \vdots & \\ \text{---} & \boldsymbol{\alpha}_n & \text{---} \end{vmatrix} > 0 \tag{2.47}$$

and left handed if the determinant is less than zero. (The determinant cannot equal zero if the vectors are linearly independent.) Clearly, swapping any two vectors will change the handedness. A simple example of handedness is that of the xy cartesian axes, which are right handed with that ordering since $\left|\begin{smallmatrix}1&0\\0&1\end{smallmatrix}\right| > 0$.

Cross products are very closely related to determinants. Both are antisymmetric, and both have a geometrical interpretation. While the cross product is usually applied to two vectors in three-dimensional euclidean space $\mathbb{E}^3$ only, the ideas we have been laying out here can be used to generalise it to higher dimensions. If the cartesian basis vectors $\{\boldsymbol{e}_x, \boldsymbol{e}_y, \boldsymbol{e}_z\}$ for $\mathbb{E}^3$ are written $\{\boldsymbol{e}_1, \boldsymbol{e}_2, \boldsymbol{e}_3\}$, or in $\mathbb{E}^n$ are written more generally as $\{\boldsymbol{e}_1, \boldsymbol{e}_2, \ldots, \boldsymbol{e}_n\}$, then we can define the cross product of $n-1$ linearly independent vectors in $\mathbb{E}^n$, $\{\boldsymbol{\alpha}_1, \ldots, \boldsymbol{\alpha}_{n-1}\}$, to be $\boldsymbol{\alpha}_n$ such that

$$\boldsymbol{\alpha}_n = \operatorname{cross}(\boldsymbol{\alpha}_1, \ldots, \boldsymbol{\alpha}_{n-1}) \equiv \begin{vmatrix} \text{---} & \boldsymbol{\alpha}_1 & \text{---} \\ & \vdots & \\ \text{---} & \boldsymbol{\alpha}_{n-1} & \text{---} \\ \boldsymbol{e}_1 & \ldots & \boldsymbol{e}_n \end{vmatrix}. \tag{2.48}$$

In Chap. 8 we'll see that this expression changes in a more general basis by the inclusion of a multiplicative factor, which is 1 for euclidean space. We'll also see that in more general coordinates, the last row of (2.48) should really be the cobasis $\{\boldsymbol{e}^1, \ldots, \boldsymbol{e}^n\}$. Of course, in the euclidean case we are considering here, the basis and cobasis are identical.

The usual cross product in $\mathbb{E}^3$ is often defined with the $\boldsymbol{e}_x\ \boldsymbol{e}_y\ \boldsymbol{e}_z$ as the first row instead of the last. In an odd number of dimensions this makes no difference, but in an even number of dimensions, it will give the wrong sign and so make a left-handed set. Ensuring that the basis vectors form the *last* row will always give the correct sign, in any number of dimensions.

In the common three-dimensional case with vectors $\boldsymbol{a} = (a^x, a^y, a^z)$ and $\boldsymbol{b} = (b^x, b^y, b^z)$, equation (2.48) becomes

$$\begin{aligned}(a^x, a^y, a^z) \times (b^x, b^y, b^z) &= \begin{vmatrix} a^x & a^y & a^z \\ b^x & b^y & b^z \\ \boldsymbol{e}_x & \boldsymbol{e}_y & \boldsymbol{e}_z \end{vmatrix} \\ &= \left(\begin{vmatrix} a^y & a^z \\ b^y & b^z \end{vmatrix}, -\begin{vmatrix} a^x & a^z \\ b^x & b^z \end{vmatrix}, \begin{vmatrix} a^x & a^y \\ b^x & b^y \end{vmatrix} \right).\end{aligned} \tag{2.49}$$

Expanding along the bottom row in cofactors, as done here, is a far easier way of calculating the cross product than the often-used alternative, the slower approach of adding six separate terms calculated from first principles: $a^x b^y \boldsymbol{e}_z + a^y b^z \boldsymbol{e}_x + \cdots$.

For any vectors $\boldsymbol{\alpha}_1, \ldots, \boldsymbol{\alpha}_{n-1}$, along with their cross product $\boldsymbol{\alpha}_n$ defined by (2.48), the set $\{\boldsymbol{\alpha}_1, \ldots, \boldsymbol{\alpha}_n\}$ is always right handed. We can prove this using cofactors by expanding (2.47) along its bottom row. Write the μth element of $\boldsymbol{\alpha}_n$ as α_n^μ:

$$\begin{vmatrix} — & \boldsymbol{\alpha}_1 & — \\ & \vdots & \\ — & \boldsymbol{\alpha}_n & — \end{vmatrix} = \sum_\mu \alpha_n^\mu \times \text{ cofactor belonging to } \alpha_n^\mu \text{ for } \begin{bmatrix} — & \boldsymbol{\alpha}_1 & — \\ & \vdots & \\ — & \boldsymbol{\alpha}_{n-1} & — \\ \alpha_n^1 & \dots & \alpha_n^n \end{bmatrix}$$

$$= \sum_\mu \alpha_n^\mu \times \text{ cofactor belonging to } \boldsymbol{e}_\mu \text{ for } \begin{bmatrix} — & \boldsymbol{\alpha}_1 & — \\ & \vdots & \\ — & \boldsymbol{\alpha}_{n-1} & — \\ \boldsymbol{e}_1 & \dots & \boldsymbol{e}_n \end{bmatrix}$$

$$= \sum_\mu \left(\alpha_n^\mu\right)^2 = |\boldsymbol{\alpha}_n|^2 > 0\,, \quad \text{i.e. right handed; QED.} \tag{2.50}$$

This is very streamlined, yielding the answer far more quickly than if we had started with the basic definition using the Levi-Civita symbol:

$$\operatorname{cross}(\boldsymbol{\alpha}_1, \dots, \boldsymbol{\alpha}_{n-1}) = \sum_{\mu\dots\omega} \varepsilon_{\underbrace{\mu\,\dots\,\omega}_{n \text{ indices}}} \; \alpha_1^\mu \, \alpha_2^\nu \; \dots \; \alpha_{n-1}^\psi \, \boldsymbol{e}_\omega\,. \tag{2.51}$$

We will have more to say about this particular expression in Sect. 8.8.1. Working with the Levi-Civita symbol is really a first-principles approach; using cofactors (if possible) is far more concise. And although we won't stop to give the details, the same sort of cofactor approach used in (2.50) will also easily show that the cross product $\boldsymbol{\alpha}_n$ is orthogonal to $\boldsymbol{\alpha}_1, \dots, \boldsymbol{\alpha}_{n-1}$, as is familiar in three dimensions.

Magnitude of the Cross Product

Equation (2.50) shows that when $\boldsymbol{\alpha}_n = \operatorname{cross}(\boldsymbol{\alpha}_1, \dots, \boldsymbol{\alpha}_{n-1})$, the volume of the right-handed box defined by $\{\boldsymbol{\alpha}_1, \dots, \boldsymbol{\alpha}_n\}$ is $|\boldsymbol{\alpha}_n|^2$. But since $\boldsymbol{\alpha}_n$ is orthogonal to all the other vectors, the volume of the box must also be just $|\boldsymbol{\alpha}_n| \times$ the "area" of the cell composed of $\{\boldsymbol{\alpha}_1, \dots, \boldsymbol{\alpha}_{n-1}\}$. Hence the area of this cell must be $|\boldsymbol{\alpha}_n|$.

$$\boxed{\begin{aligned} &\text{When } \boldsymbol{\alpha}_n \equiv \operatorname{cross}(\boldsymbol{\alpha}_1, \dots, \boldsymbol{\alpha}_{n-1}), \\ &\quad \text{“area” of } \boldsymbol{\alpha}_1, \dots, \boldsymbol{\alpha}_{n-1} = |\boldsymbol{\alpha}_n|\,, \text{ and} \\ &\quad \text{“volume” of } \boldsymbol{\alpha}_1, \dots, \boldsymbol{\alpha}_n = |\boldsymbol{\alpha}_n|^2. \end{aligned}} \tag{2.52}$$

We are familiar with this in two dimensions, where the parallelogram with sides $\boldsymbol{\alpha}_1, \boldsymbol{\alpha}_2$ has area $|\boldsymbol{\alpha}_1 \times \boldsymbol{\alpha}_2|$. This particular expression will be applied to an infinitesimal parallelogram in Chap. 9 when we calculate areas on a curved surface.

2.4.3 Volume of a Parallelepiped in a Higher-Dimensional Space

The area of a parallelogram with side vectors $(1,2)$ and $(5,8)$ is just the absolute value of the determinant $\left|\begin{smallmatrix}1 & 2\\ 5 & 8\end{smallmatrix}\right|$. But what is the area of a parallelogram with sides $(1,2,3)$ and $(5,8,7)$? The determinant of $\left[\begin{smallmatrix}1 & 2 & 3\\ 5 & 8 & 7\end{smallmatrix}\right]$ is not defined; in this case, (2.46) needs *three* vectors. We can certainly calculate the area from the norm of the two vectors' cross product, using (2.48) and (2.52):

$$\begin{aligned}&\text{Area of parallelogram with sides } (1,2,3) \text{ and } (5,8,7)\\ &= \big|(1,2,3)\times(5,8,7)\big| = \text{norm}\begin{vmatrix}1 & 2 & 3\\ 5 & 8 & 7\\ \boldsymbol{e}_x & \boldsymbol{e}_y & \boldsymbol{e}_z\end{vmatrix} = \text{norm}\,(-10,8,-2) = \sqrt{168}\,.\end{aligned} \tag{2.53}$$

But this procedure will not suffice for two vectors in a space of dimension higher than 3, such as our requiring the area of a parallelogram with sides $(1,2,3,4)$ and $(5,8,7,2)$, since we now require *three* vectors to make a cross product, and we only have two. What can be done?

Suppose we are given two vectors $\boldsymbol{\alpha}_1, \boldsymbol{\alpha}_2$ in $\mathbb{E}^4$. The following discussion applies to any number of vectors in any higher-dimensional euclidean space, but we'll stay with this scenario to be concise. The parallelogram spanned by $\boldsymbol{\alpha}_1, \boldsymbol{\alpha}_2$ is confined to a subspace spanned by an orthonormal basis $\Gamma \equiv \{\boldsymbol{\gamma}_1, \boldsymbol{\gamma}_2\}$. (We must use an orthonormal basis since the theory of the last few pages has been built on these, as it assumed the usual cartesian basis of $\mathbb{E}^n$, which is orthonormal.) In that case, the parallelogram's area is the determinant of a 2×2 matrix composed of the coordinate vectors of $\boldsymbol{\alpha}_1, \boldsymbol{\alpha}_2$ with respect to the Γ basis:

$$\text{Area} = \begin{vmatrix} — [\boldsymbol{\alpha}_1]_\Gamma — \\ \\ — [\boldsymbol{\alpha}_2]_\Gamma — \end{vmatrix}. \tag{2.54}$$

We wish to relate this expression to the coordinate vectors of $\boldsymbol{\alpha}_1, \boldsymbol{\alpha}_2$ with respect to the usual cartesian basis of $\mathbb{E}^4$, here called E. These vectors have four components. The clue is to realise that matrix multiplication is all about dot products of rows and columns, and dot products give lengths—which are independent of the orthonormal basis considered.

That the euclidean dot product $\boldsymbol{a}\cdot\boldsymbol{b} \equiv \sum_i a^i\, b^i$ is really independent of the orthonormal basis used can be shown by writing two arbitrary vectors as coordinate vectors with respect to two arbitrary orthonormal bases, and showing that the two sums of the pairwise coordinate products are identical. But it could hardly be otherwise, since we use such an idea all the time when depicting vectors on cartesian axes.

In that case, note that if $P \equiv \begin{bmatrix} — [\boldsymbol{\alpha}_1]_\Gamma — \\ — [\boldsymbol{\alpha}_2]_\Gamma — \end{bmatrix}$, then

$$PP^t = \begin{bmatrix} \boldsymbol{\alpha}_1 \cdot \boldsymbol{\alpha}_1 & \boldsymbol{\alpha}_1 \cdot \boldsymbol{\alpha}_2 \\ \boldsymbol{\alpha}_2 \cdot \boldsymbol{\alpha}_1 & \boldsymbol{\alpha}_2 \cdot \boldsymbol{\alpha}_2 \end{bmatrix}, \tag{2.55}$$

and the area of the parallelogram can be calculated from (2.46):

$$\text{area}^2 = |P|^2 = |P|\,|P^t| = |PP^t|. \tag{2.56}$$

But the dot products in (2.55) could just as well have been formed by dotting the coordinate vectors $[\boldsymbol{\alpha}_1]_E, [\boldsymbol{\alpha}_2]_E$! In other words, define a 2×4 matrix $A \equiv \begin{bmatrix} - [\boldsymbol{\alpha}_1]_E - \\ - [\boldsymbol{\alpha}_2]_E - \end{bmatrix}$ so that $AA^t = PP^t$. This allows us to write (replacing "area" by "volume" to denote arbitrary dimensions being allowed)

$$\boxed{\text{Volume of parallelepiped defined by rows of } A = \sqrt{|AA^t|}\,,} \tag{2.57}$$

where the number of rows of A must be less than or equal to its number of columns. This is the generalisation of (2.46), although we have seen it before for a special case in (2.42). Reassuringly, (2.57) states that when A has just one row, the resulting "volume" of that lone vector is simply its length, given by Pythagoras's theorem. A less trivial example is that of (2.53), which can be redone using this approach:

Area of parallelogram with sides $(1, 2, 3)$ and $(5, 8, 7)$

$$= \sqrt{\left| \begin{bmatrix} 1 & 2 & 3 \\ 5 & 8 & 7 \end{bmatrix} \begin{bmatrix} 1 & 2 & 3 \\ 5 & 8 & 7 \end{bmatrix}^t \right|} = \sqrt{\begin{vmatrix} 14 & 42 \\ 42 & 138 \end{vmatrix}} = \sqrt{168}\,, \tag{2.58}$$

as before. Equation (2.57) will tell us, for example, the four-dimensional volume of the parallelepiped whose sides are given by four vectors in $\mathbb{E}^6$. In this case the matrix A has size 4×6, so that the determinant of the 4×4 matrix AA^t will be required. In general, as long as the sides of the parallelepiped are not overspecified—A should be squat or square, but not tall—then (2.57) gives its volume, and so in a sense extends the concept of the determinant to nonsquare matrices.

The *Gram matrix* of dot products in (2.55) is the real quantifier of volumes in a space, and we'll encounter it in Chap. 8 as the *metric matrix*, when its entries will be basis vectors. We will also encounter (2.57) again in that chapter, in Sect. 8.8, when calculating infinitesimal volume elements. There these will be related to $\sqrt{g}$, where g is the determinant of a metric matrix.

2.4.4 The Cobasis and the Wedge Product

If we dot a vector $\boldsymbol{a}$ with a basis vector that points in the same or opposite direction, what results is the signed length of $\boldsymbol{a}$: positive if the basis vector has the same direction as $\boldsymbol{a}$, and negative otherwise. The same idea applies to higher dimensions, and we'll explore the notation further by remembering from (2.23) that the αth basis component of a vector $\boldsymbol{v}$ is v^α, while the αth cobasis component of the same vector is v_α.

So, for example, the signed area of the parallelogram bounded by the vectors $\boldsymbol{a}$ and $\boldsymbol{b}$ in that order is

$$\begin{vmatrix} — \boldsymbol{a} — \\ — \boldsymbol{b} — \end{vmatrix} = \begin{vmatrix} a^1 & a^2 \\ b^1 & b^2 \end{vmatrix} \stackrel{(2.25)}{=\!=} \begin{vmatrix} \boldsymbol{a}\cdot\mathbf{e}^1 & \boldsymbol{a}\cdot\mathbf{e}^2 \\ \boldsymbol{b}\cdot\mathbf{e}^1 & \boldsymbol{b}\cdot\mathbf{e}^2 \end{vmatrix}, \tag{2.59}$$

where the coordinates are superscripted by 1 and 2. Unlike in the one-dimensional case, this signed area cannot simply be written as a single dot product. But we can mimic the one-dimensional case by using a functional notation, where for example an expression such as $\boldsymbol{a}\cdot\boldsymbol{b}$ could be written as either $\boldsymbol{a}(\boldsymbol{b})$ or $\boldsymbol{b}(\boldsymbol{a})$. Note that we are *not* redefining a vector to be a function of another vector! Functional notation as used here is merely a useful way of combining several entities together, and we'll make more use of it in Chap. 8. In two dimensions, then, write the signed area (2.59) as a *wedge product* of $\boldsymbol{a}$ and $\boldsymbol{b}$ (also known as the *exterior product*):

$$\begin{vmatrix} — \boldsymbol{a} — \\ — \boldsymbol{b} — \end{vmatrix} = \begin{vmatrix} \boldsymbol{a}\cdot\mathbf{e}^1 & \boldsymbol{a}\cdot\mathbf{e}^2 \\ \boldsymbol{b}\cdot\mathbf{e}^1 & \boldsymbol{b}\cdot\mathbf{e}^2 \end{vmatrix} \equiv \boldsymbol{a}\wedge\boldsymbol{b}\,(\mathbf{e}^1, \mathbf{e}^2)\,. \tag{2.60}$$

The wedge product $\boldsymbol{a}\wedge\boldsymbol{b}$ could be viewed as a function of $\mathbf{e}^1$ and $\mathbf{e}^2$, but we take it simply to be a signed area, so that $\boldsymbol{a}\wedge\boldsymbol{b}\,(\mathbf{e}^1, \mathbf{e}^2)$ merely indicates that $\boldsymbol{a}\wedge\boldsymbol{b}$ combines with the ordered pair of $\mathbf{e}^1, \mathbf{e}^2$ to give a real number: the determinant on the left-hand side of (2.60). We can certainly write a new expression $(\boldsymbol{a}\wedge\boldsymbol{b})^{\alpha\beta} \equiv \boldsymbol{a}\wedge\boldsymbol{b}\,(\mathbf{e}^\alpha, \mathbf{e}^\beta)$ and consider this to be the $\alpha\beta^{\text{th}}$ component of $\boldsymbol{a}\wedge\boldsymbol{b}$ over the cobasis. Components with repeated indices such as $(\boldsymbol{a}\wedge\boldsymbol{b})^{11}$ will be zero, since they are given by determinants with repeated columns. Similarly, $(\boldsymbol{a}\wedge\boldsymbol{b})_{\alpha\beta} \equiv \boldsymbol{a}\wedge\boldsymbol{b}\,(\mathbf{e}_\alpha, \mathbf{e}_\beta)$ can be considered as the $\alpha\beta^{\text{th}}$ component of $\boldsymbol{a}\wedge\boldsymbol{b}$ over the basis.

Notice how the sign-changing property of the determinant on a row or column swap implies that

$$\boldsymbol{a}\wedge\boldsymbol{b}\,(\mathbf{e}^1, \mathbf{e}^2) = -\boldsymbol{a}\wedge\boldsymbol{b}\,(\mathbf{e}^2, \mathbf{e}^1) = -\boldsymbol{b}\wedge\boldsymbol{a}\,(\mathbf{e}^1, \mathbf{e}^2) = \boldsymbol{b}\wedge\boldsymbol{a}\,(\mathbf{e}^2, \mathbf{e}^1)\,. \tag{2.61}$$

So just as in one dimension, $\boldsymbol{a}$ is a vector whose signed length is given by dotting it with a unit vector parallel or antiparallel to it, or equivalently supplying it with that unit vector as an argument, so in two dimensions $\boldsymbol{a}\wedge\boldsymbol{b}$ is a *bivector* whose signed area is produced by giving it the two basis vectors as arguments, as shown in Fig. 2.4. Similarly, in three dimensions the signed volume of the parallelepiped determined by vectors $\boldsymbol{a}, \boldsymbol{b}, \boldsymbol{c}$ is

$$\begin{vmatrix} — \boldsymbol{a} — \\ — \boldsymbol{b} — \\ — \boldsymbol{c} — \end{vmatrix} = \begin{vmatrix} \boldsymbol{a}\cdot\mathbf{e}^1 & \boldsymbol{a}\cdot\mathbf{e}^2 & \boldsymbol{a}\cdot\mathbf{e}^3 \\ \boldsymbol{b}\cdot\mathbf{e}^1 & \boldsymbol{b}\cdot\mathbf{e}^2 & \boldsymbol{b}\cdot\mathbf{e}^3 \\ \boldsymbol{c}\cdot\mathbf{e}^1 & \boldsymbol{c}\cdot\mathbf{e}^2 & \boldsymbol{c}\cdot\mathbf{e}^3 \end{vmatrix} \equiv \boldsymbol{a}\wedge\boldsymbol{b}\wedge\boldsymbol{c}\,(\mathbf{e}^1, \mathbf{e}^2, \mathbf{e}^3)\,, \tag{2.62}$$

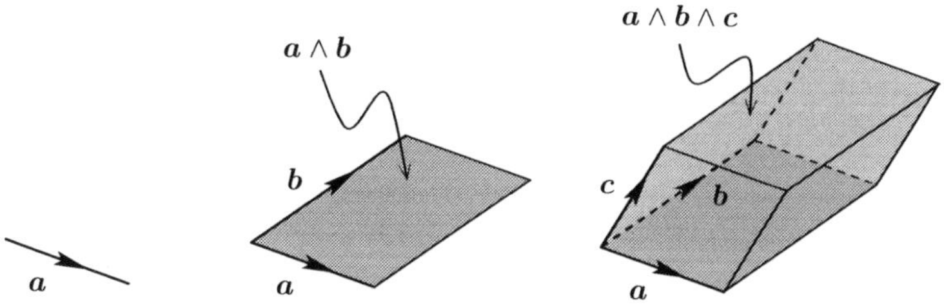

Fig. 2.4. Picturing wedge products. Just as a vector $\boldsymbol{a}$ combines with $\boldsymbol{e}^x$ to give the component $\boldsymbol{a}(\boldsymbol{e}^x) \equiv \boldsymbol{a}\cdot\boldsymbol{e}^x = a^x$ as its signed length in the x-direction, the bivector $\boldsymbol{a} \wedge \boldsymbol{b}$ is a two-dimensional element whose signed area is $\boldsymbol{a} \wedge \boldsymbol{b}\,(\boldsymbol{e}^x, \boldsymbol{e}^y)$, while the trivector $\boldsymbol{a} \wedge \boldsymbol{b} \wedge \boldsymbol{c}$ is a three-dimensional element whose signed volume is $\boldsymbol{a} \wedge \boldsymbol{b} \wedge \boldsymbol{c}\,(\boldsymbol{e}^x, \boldsymbol{e}^y, \boldsymbol{e}^z)$.

so that $\boldsymbol{a} \wedge \boldsymbol{b} \wedge \boldsymbol{c}$ is a *trivector* determining the volume of the parallelepiped. Because the determinant treats all of its rows or columns equally, and hence expressions such as $\boldsymbol{a} \wedge \boldsymbol{b} \wedge \boldsymbol{c}$ look as if they have some assumed associativity, wedge products between vectors and bivectors, or in general between two *multivectors*, are *defined* in order to allow this associativity, in which case $\boldsymbol{a} \wedge (\boldsymbol{b} \wedge \boldsymbol{c}) \equiv (\boldsymbol{a} \wedge \boldsymbol{b}) \wedge \boldsymbol{c} \equiv \boldsymbol{a} \wedge \boldsymbol{b} \wedge \boldsymbol{c}$, and so on.

In two dimensions, the signed area $\boldsymbol{a} \wedge \boldsymbol{b}\,(\boldsymbol{e}^x, \boldsymbol{e}^y)$ of the parallelogram formed from $\boldsymbol{a}$ and $\boldsymbol{b}$ equals the z-component of $\boldsymbol{a} \times \boldsymbol{b}$, or $\boldsymbol{a} \times \boldsymbol{b}\,(\boldsymbol{e}^z)$ using the notation suggested just after (2.59). Thus,

$$\boldsymbol{a} \wedge \boldsymbol{b}\,(\boldsymbol{e}^x, \boldsymbol{e}^y) = \boldsymbol{a} \times \boldsymbol{b}\,(\boldsymbol{e}^z) = \begin{vmatrix} -\,\boldsymbol{a}\,- \\ -\,\boldsymbol{b}\,- \end{vmatrix}. \tag{2.63}$$

Despite the similarity, there is a difference between $\boldsymbol{a} \wedge \boldsymbol{b}$ and $\boldsymbol{a} \times \boldsymbol{b}$. The most obvious difference is that when each is written using functional notation, they take different numbers of arguments. Also, the wedge product $\boldsymbol{a} \wedge \boldsymbol{b}$ is an area element with an associated direction. Since $\boldsymbol{a} \wedge \boldsymbol{b} = -\boldsymbol{b} \wedge \boldsymbol{a}$, swapping the vectors reverses this direction. The cross product $\boldsymbol{a} \times \boldsymbol{b}$ is a vector normal to the area element $\boldsymbol{a} \wedge \boldsymbol{b}$, and it, too, has a direction that reverses when the vectors swap: $\boldsymbol{a} \times \boldsymbol{b} = -\boldsymbol{b} \times \boldsymbol{a}$. So we can consider $\boldsymbol{a} \wedge \boldsymbol{b}$ and $\boldsymbol{a} \times \boldsymbol{b}$ as able to be mapped to each other.

In the next higher dimension, the first line of (2.52) shows that a similar identification of, e.g., $\boldsymbol{a} \wedge \boldsymbol{b} \wedge \boldsymbol{c}$ with cross $(\boldsymbol{a}, \boldsymbol{b}, \boldsymbol{c})$ can be made: $\boldsymbol{a} \wedge \boldsymbol{b} \wedge \boldsymbol{c}$ is a signed volume element in three dimensions, while cross $(\boldsymbol{a}, \boldsymbol{b}, \boldsymbol{c})$ is a vector in four dimensions, orthogonal to the vectors $\boldsymbol{a}, \boldsymbol{b}, \boldsymbol{c}$ and "perpendicular" to their volume element $\boldsymbol{a} \wedge \boldsymbol{b} \wedge \boldsymbol{c}$. The same idea holds in any number of dimensions: cross $(\boldsymbol{a}_1, \boldsymbol{a}_2, \dots, \boldsymbol{a}_n)$ is a vector in $n+1$ dimensions "perpendicular" to the n-dimensional volume element $\boldsymbol{a}_1 \wedge \boldsymbol{a}_2 \wedge \cdots \wedge \boldsymbol{a}_n$. We'll meet these multiple wedge products known as *multivectors* again later, exploring this similarity in Sect. 4.5 and using the distinction when unifying Stokes' and Gauss's theorems in Sect. 8.10. They will also be used in Chap. 12 in an efficient method for calculating the curvature of spacetime.

One new piece of notation will be useful when we do consider Stokes' and Gauss's theorems. That is, because

$$(\boldsymbol{a} \times \boldsymbol{b})\cdot(\boldsymbol{p} \times \boldsymbol{q}) = \begin{vmatrix} \boldsymbol{a}\cdot\boldsymbol{p} & \boldsymbol{a}\cdot\boldsymbol{q} \\ \boldsymbol{b}\cdot\boldsymbol{p} & \boldsymbol{b}\cdot\boldsymbol{q} \end{vmatrix} = \boldsymbol{a} \wedge \boldsymbol{b}\,(\boldsymbol{p}, \boldsymbol{q})\,, \tag{2.64}$$

it makes sense to define a new dot product as a generalisation of that for vectors:

$$(\boldsymbol{a} \wedge \boldsymbol{b})\cdot(\boldsymbol{p} \wedge \boldsymbol{q}) \equiv \boldsymbol{a} \wedge \boldsymbol{b}\,(\boldsymbol{p}, \boldsymbol{q}) = (\boldsymbol{a} \times \boldsymbol{b})\cdot(\boldsymbol{p} \times \boldsymbol{q})\,. \tag{2.65}$$

This extends the analogy between the wedge and cross products. Similarly,

$$(\boldsymbol{a} \wedge \boldsymbol{b} \wedge \boldsymbol{c})\cdot(\boldsymbol{p} \wedge \boldsymbol{q} \wedge \boldsymbol{r}) \equiv \boldsymbol{a} \wedge \boldsymbol{b} \wedge \boldsymbol{c}\,(\boldsymbol{p}, \boldsymbol{q}, \boldsymbol{r}) = \begin{vmatrix} \boldsymbol{a}\cdot\boldsymbol{p} & \boldsymbol{a}\cdot\boldsymbol{q} & \boldsymbol{a}\cdot\boldsymbol{r} \\ \boldsymbol{b}\cdot\boldsymbol{p} & \boldsymbol{b}\cdot\boldsymbol{q} & \boldsymbol{b}\cdot\boldsymbol{r} \\ \boldsymbol{c}\cdot\boldsymbol{p} & \boldsymbol{c}\cdot\boldsymbol{q} & \boldsymbol{c}\cdot\boldsymbol{r} \end{vmatrix}\,. \tag{2.66}$$

The wedge product can seem obscure at first, but the key point to remember is that it relates to volumes in any number of dimensions, and its operations are expressed in the everyday language of numbers via the use of determinants.

2.5 Diagonalisation and Similar Matrices: Changing Spaces

Calculations in mathematical physics are often greatly simplified if we work in some kind of different space. A simple example is a change of frame, such as converting to a centre of mass frame in order to solve a mechanics problem. Another example is the idea of writing a matrix as a product of other matrices that include a diagonal one, since diagonal matrices have useful properties. So for the purpose of our discussion, we'll call such a space "diagonal space". In diagonal space, the equations of physics are simple and elegant. We wish to demonstrate the idea of transforming some complicated expression to that space, solving for whatever needs doing (which by definition is easier in diagonal space), and finally transforming back to the original space.

Linear algebra is useful for showing the general idea. Suppose, as in Fig. 2.5, that in diagonal space a vector will evolve using a very simple operator D (which will be a diagonal matrix). Denote by P the transformation taking diagonal space to laboratory space. Given a vector $\boldsymbol{v}$ in our laboratory that we wish to evolve, we must perform the following steps in order to use the easy evolution of diagonal space.

1. Transform to diagonal space by operating with P^{-1} to get $P^{-1}\boldsymbol{v}$.
2. Evolve this (easy!): $DP^{-1}\boldsymbol{v}$.
3. Now transform back to the laboratory: $PDP^{-1}\boldsymbol{v}$.

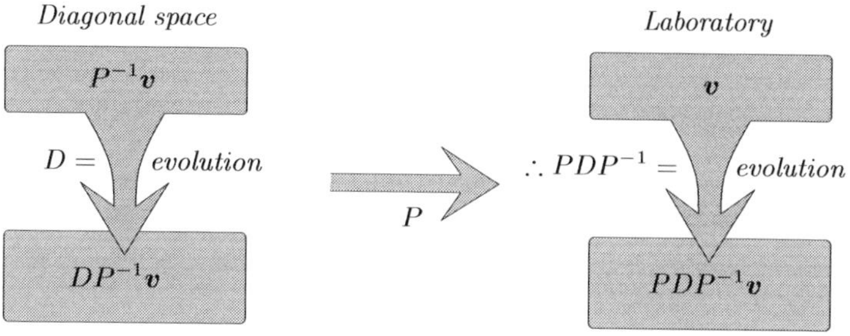

Fig. 2.5. An operator D in "diagonal space" is equivalent to the operator PDP^{-1} in the laboratory. Start with $\boldsymbol{v}$ at the top right, transform to diagonal space (upper left, $P^{-1}\boldsymbol{v}$), evolve this ($DP^{-1}\boldsymbol{v}$), and finally return to the laboratory space, giving $PDP^{-1}\boldsymbol{v}$.

The nett result is that the D of diagonal space is matched with, or in a sense dual to, the PDP^{-1} of the laboratory; matrices representing the two operations are said to be *similar*. Expressions like PDP^{-1} are very common in mathematical physics, where the arenas in which problems are solved are seldom the simplest possible.

Opening a Door and Turning Its Handle

A good example of this procedure appears in three-dimensional rotation theory. We will look more closely at the very rich theory of rotations in Chap. 4, but here we'll briefly examine an aspect of the noncommutivity of rotations around different axes. Rotating an object around two different axes is not commutative, as can be seen by rotating, say, a box around first the spatial x-axis by 90°, then the spatial y-axis by 90°, and then comparing the resulting orientation with what happens when the rotation order is swapped. The two orientations are different. Things are somewhat different when we allow the box to carry the axes along with it. We embed the spatial axes in the box, like sticks, so that when it's turned around the y-axis, the x-axis gets carried along to become a new axis, called the x'-axis. In that case, it turns out that the rotations *can* be thought of as almost commuting in a very restricted way, as long as we're prepared to use several different axes. Let's investigate this using the $A = PDP^{-1}$ procedure above.

Consider what happens to the orientation of a door handle when opening a door and then turning its handle. (This order is not really possible with most doors—that's what handles are for, after all—but we are free to imagine it.) As shown in Fig. 2.6, the door's hinges lie on the spatial z-axis, and when the door is closed, its handle's axle is aligned with the spatial x-axis. Denote the operation of opening the door by a rotation around the z-axis through angle α, written as $R_z(\alpha)$. (Of course, opening the door also rotates the handle by $R_z(\alpha)$, but not around the handle's axis.) Turning the handle through an

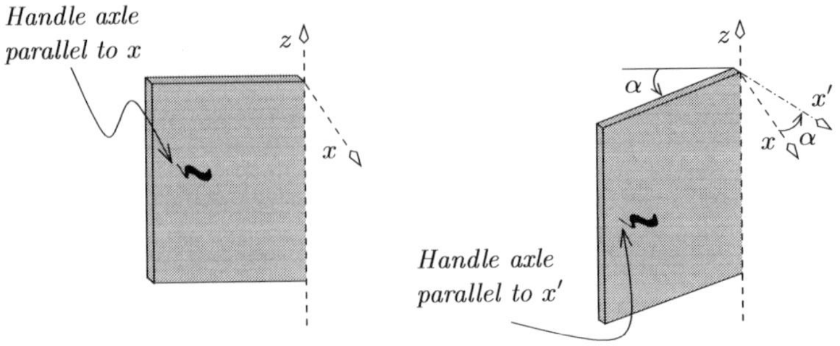

Fig. 2.6. Left: The closed door can open along the z-axis, and its handle turns around the x-axis. **Right**: When opened, the handle's axle now defines a new axis, called the x'-axis.

angle β when the door is closed is equivalent to rotating the handle about the x-axis, $R_x(\beta)$. When the door is opened, the handle axle becomes a new axis, called the x'-axis. The sequence of rotations will here be written from right to left since they can all be treated as operators—and in fact they will be written as matrices in Chap. 4.

Suppose that the door is standing open at an angle α, and we wish to turn the handle through β. That is, we wish to orientate the handle by the rotation operation $R_{x'}(\beta)$. But for some reason we cannot do this; perhaps the handle is in too awkward a position to be turned. Instead, what we *are* able to do is close the door, turn the handle through β, and open it again. This should accomplish the same result. In other words, reading from right to left, close the door (P^{-1}), now easily turn the handle (D), and open it again (P):

$$\underbrace{R_{x'}(\beta)}_{\equiv A} = \underbrace{R_z(\alpha)}_{\equiv P} \underbrace{R_x(\beta)}_{\equiv D} \underbrace{R_z(-\alpha)}_{= P^{-1}} . \tag{2.67}$$

If we postmultiply both sides by P, we have $AP = PD$, or

$$R_{x'}(\beta)\, R_z(\alpha) = R_z(\alpha)\, R_x(\beta) . \tag{2.68}$$

This is all good common sense: the left-hand side of (2.68) is the operation of opening the door and then turning its handle (if that were possible), while the right-hand side is the operation of first turning the handle and then opening the door. Of course, these operations should give the same result. The almost-commutivity of the rotations shows up in the fact that while the order of the angles has been reversed, we are using the x'-axis on one side and the x-axis on the other. When the rotation angle α is small—even if β is not—the x- and x'-axes almost coincide, so that in this limit rotation order is indeed commutative.

There is nothing special about the axes used in this example. They could have been any vectors at all and not necessarily perpendicular. We'll meet up with (2.68) again in Chap. 4 in the context of applied rotation theory, where we'll find that it can help to create either an enlightened view or a confused view of rotations.

2.5.1 Diagonalising a Matrix

Precisely what it is that makes diagonal space preferred is determined by how easily or elegantly a vector can be evolved there or, in other words, how simple D is. The simplest possible D is a diagonal matrix since it does not mix the vector elements, and functions of it such as powers, the exponential, and the logarithm are easily defined and calculated. And just as diagonal evolution matrices D are singled out as preferable, some space-changing matrices P are also special. Particularly useful is a P that preserves length. We'll see more of this in Sect. 2.7.

Given some evolution matrix A in the laboratory, can we convert to diagonal space by writing A as a product PDP^{-1}, where D is diagonal? It turns out that we can, at least for any *square* matrix A that is sufficiently well-behaved. Evolution matrices must be square, and they usually will be well-behaved.

The proof by construction that A can be *diagonalised* runs as follows. We will start with the correct choice of P and then show how it leads to the answer, along with a corresponding D. First, define an *eigenvalue* λ with a corresponding *eigenvector* $\boldsymbol{\alpha}$ of A to be a number and vector such that

$$A\boldsymbol{\alpha} = \lambda\boldsymbol{\alpha}\,. \tag{2.69}$$

Suppose that A is an $n \times n$ matrix with n linearly independent eigenvectors $\boldsymbol{\alpha}_1, \ldots, \boldsymbol{\alpha}_n$, allied in order with eigenvalues $\lambda_1, \ldots, \lambda_n$. They must be linearly independent because we are about to form P by simply writing all of the eigenvectors as columns placed next to each other:

$$P \equiv \begin{bmatrix}\boldsymbol{\alpha}_1 \; \boldsymbol{\alpha}_2 \; \cdots \; \boldsymbol{\alpha}_n\end{bmatrix}. \tag{2.70}$$

If the $\{\boldsymbol{\alpha}_i\}$ are linearly independent, then the dependency relationship algorithm guarantees that P can be row-reduced to the identity matrix; but this means that P is invertible. Certainly, for interesting matrices these eigenvectors will be independent.

Now, given A and its associated P, we can use matrix block multiplication to write the following. (Note that block-multiplying matrices is a powerful tool in linear algebra. It can always be performed, provided that the individual blocks have been marked out in a way that enables *them* to be multiplied.)

$$\begin{aligned}
AP &= A\begin{bmatrix}\boldsymbol{\alpha}_1 \; \boldsymbol{\alpha}_2 \; \cdots \; \boldsymbol{\alpha}_n\end{bmatrix} = \begin{bmatrix}A\boldsymbol{\alpha}_1 \; A\boldsymbol{\alpha}_2 \; \cdots \; A\boldsymbol{\alpha}_n\end{bmatrix} \\
&= \begin{bmatrix}\lambda_1\boldsymbol{\alpha}_1 \; \lambda_2\boldsymbol{\alpha}_2 \; \cdots \; \lambda_n\boldsymbol{\alpha}_n\end{bmatrix}
\end{aligned}$$

$$= \begin{bmatrix}\boldsymbol{\alpha}_1 \; \boldsymbol{\alpha}_2 \cdots \boldsymbol{\alpha}_n\end{bmatrix} \begin{bmatrix}\lambda_1 & \cdots & 0 \\ & \ddots & \\ 0 & \cdots & \lambda_n\end{bmatrix} \equiv PD\,. \tag{2.71}$$

So if D is composed of the eigenvalues of A down its main diagonal, then $AP = PD$, and since P is invertible it follows immediately that $A = PDP^{-1}$. The matrix A has been diagonalised, and in so doing, we have in a sense constructed the diagonal space, accessed via P.

The eigenvectors and eigenvalues of this last proof hold great importance in linear algebra. Eigenvectors form points of unruffled calm amidst the changes brought about by a linear operator. A good example of this is the rotation operator in three dimensions, whose eigenvectors lie along its axis of rotation. A similar idea applies to the derivative operator $\mathrm{d}/\mathrm{d}x$ for functions. A general form for its eigenvector—better called its *eigenfunction* in this case—is the familiar exponential $\mathrm{e}^{\lambda x}$, where λ is the corresponding eigenvalue.

The eigenvalues of any operator are identical to the eigenvalues of its matrix representative in any chosen basis. Likewise, the coordinate vectors of the operator's eigenvectors/eigenfunctions are given by the eigenvectors of the matrix representative. Finding eigenvectors and eigenvalues of a matrix is usually done via determinants. Any matrix A with eigenvector $\boldsymbol{\alpha}$ and associated eigenvalue λ must necessarily be square, and by definition

$$A\boldsymbol{\alpha} = \lambda\boldsymbol{\alpha}\,, \tag{2.72}$$

so that

$$(A - \lambda 1)\,\boldsymbol{\alpha} = \mathbf{0}\,, \tag{2.73}$$

where by 1 we mean the identity matrix. Now if $A - \lambda 1$ were invertible, this last equation would only have the trivial zero vector solution; so we require $A - \lambda 1$ to be noninvertible. But noninvertibility is equivalent to a matrix having zero determinant (for which see the next paragraph). So, for there to be nontrivial solutions to (2.73), we require

$$\det(A - \lambda 1) = 0\,, \tag{2.74}$$

and this enables the eigenvalues to be found, followed by their associated eigenvectors.

Why is noninvertibility equivalent to a matrix having zero determinant? If a matrix is not invertible, then any set of linear equations that it can be used to represent must not be solvable. This implies that its rows cannot be linearly independent, which means that the n-dimensional box whose edge vectors are these rows must have zero volume, since one of its edges has effectively collapsed to zero length. And of course zero volume means zero determinant. This logic works in both directions.

Diagonalisation itself is not just about the choice of a better space in which to perform a calculation. Choosing such a space might imply that we could do the calculation as it stands but perhaps for reasons of efficiency or elegance would prefer not to. But the need to change spaces might be stronger than a search for more elegance. In the next few pages, we'll study four examples of diagonalising matrices: in number theory, differential equations, classical mechanics, and geometry. Examples abound throughout all of physics, but these four are simple enough to examine briefly.

Number Theory: The Fibonacci Sequence

Calculate the nth term of the Fibonacci sequence $1, 1, 2, 3, 5, \ldots$

The Fibonacci sequence is defined by the recurrence relation that relates its nth term u_n to the two terms immediately preceding it. Such a linear problem lends itself very well to a matrix solution. Suppose we write the recurrence relation as

$$\begin{bmatrix} u_{n+2} \\ u_{n+1} \end{bmatrix} = \underbrace{\begin{bmatrix} 1 & 1 \\ 1 & 0 \end{bmatrix}}_{\equiv A} \begin{bmatrix} u_{n+1} \\ u_n \end{bmatrix} . \tag{2.75}$$

We have been a little redundant, inserting the second row which says nothing more than $u_{n+1} = u_{n+1}$; this is purely to make the matrix A square, which is necessary in order to use the diagonalisation technique. Now, the first few terms of the sequence can be found from

$$\begin{bmatrix} u_3 \\ u_2 \end{bmatrix} = A \begin{bmatrix} 1 \\ 1 \end{bmatrix} , \quad \begin{bmatrix} u_4 \\ u_3 \end{bmatrix} = A^2 \begin{bmatrix} 1 \\ 1 \end{bmatrix} , \quad \ldots , \quad \begin{bmatrix} u_n \\ u_{n-1} \end{bmatrix} = A^{n-2} \begin{bmatrix} 1 \\ 1 \end{bmatrix} , \tag{2.76}$$

so that

$$u_n = \begin{bmatrix} 1 & 0 \end{bmatrix} A^{n-2} \begin{bmatrix} 1 \\ 1 \end{bmatrix} \quad (n > 3) . \tag{2.77}$$

(The premultiplication by $\begin{bmatrix} 1 & 0 \end{bmatrix}$ is more than merely a way of specifying that we require "the top element" of the remaining product; rather, since matrix multiplication is associative, this matrix can be multiplied first by A^{n-2} to simplify the calculation straightaway.) Computing an arbitrary power of A might be very difficult, but this is where diagonalisation comes to our aid. Notice that

$$\begin{aligned} A^n &= \left(PDP^{-1}\right)^n = PDP^{-1}\, PDP^{-1} \ldots PDP^{-1}\, PDP^{-1} \\ &= PD^nP^{-1} , \end{aligned} \tag{2.78}$$

and because D is diagonal, its nth power is simply the nth power of each of its terms, so that (2.77) is more usefully written

$$u_n = \begin{bmatrix} 1 & 0 \end{bmatrix} PD^{n-2}P^{-1} \begin{bmatrix} 1 \\ 1 \end{bmatrix} . \tag{2.79}$$

The matrices P and D are built from the eigenvectors and eigenvalues of A, and these are found from our discussion around (2.73). So demand that $A - \lambda 1$ have a zero determinant:

$$\begin{vmatrix} 1-\lambda & 1 \\ 1 & -\lambda \end{vmatrix} = 0 \,. \tag{2.80}$$

This has two solutions, best expressed in terms of the *Golden Ratio* ϕ, a number known to the ancient Greeks from their use of it in mathematics if not architectural aesthetics:[1]

$$\lambda_1 = \frac{1+\sqrt{5}}{2} \equiv \phi \,, \quad \lambda_2 = \frac{1-\sqrt{5}}{2} = -1/\phi = 1-\phi \,. \tag{2.81}$$

What are the associated eigenvectors, $\boldsymbol{\alpha}_1$ and $\boldsymbol{\alpha}_2$? For $\lambda_1 = \phi$, it's easy to solve

$$(A - \phi 1)\,\boldsymbol{\alpha}_1 = \mathbf{0} \tag{2.82}$$

to give

$$\boldsymbol{\alpha}_1 = \begin{bmatrix} \phi \\ 1 \end{bmatrix} . \tag{2.83}$$

(This is really a basis eigenvector; that is, we could take any multiple of it without affecting the final answer.) Similarly, a basis eigenvector corresponding to $\lambda_2 = 1 - \phi$ is found to be

$$\boldsymbol{\alpha}_2 = \begin{bmatrix} 1-\phi \\ 1 \end{bmatrix} . \tag{2.84}$$

As stipulated by (2.70), arrange these eigenvectors as columns to form P, together with the associated eigenvalues making up D:

$$P = \begin{bmatrix} \phi & 1-\phi \\ 1 & 1 \end{bmatrix} , \quad D = \begin{bmatrix} \phi & 0 \\ 0 & 1-\phi \end{bmatrix} , \quad P^{-1} = \frac{1}{\sqrt{5}} \begin{bmatrix} 1 & \phi-1 \\ -1 & \phi \end{bmatrix} . \tag{2.85}$$

The diagonalisation is finished, so we can replace P and D into (2.79) to write

$$\begin{aligned} u_n &= \begin{bmatrix} 1 & 0 \end{bmatrix} P D^{n-2} P^{-1} \begin{bmatrix} 1 \\ 1 \end{bmatrix} \\ &= \begin{bmatrix} \phi & -1/\phi \end{bmatrix} \begin{bmatrix} \phi^{n-2} & 0 \\ 0 & (-1/\phi)^{n-2} \end{bmatrix} \frac{1}{\sqrt{5}} \begin{bmatrix} \phi \\ 1/\phi \end{bmatrix} \\ &= \frac{1}{\sqrt{5}} \left[\phi^n + (-1)^{n-1} \frac{1}{\phi^n} \right] . \end{aligned} \tag{2.86}$$

Although we originally stipulated in (2.77) that $n > 3$, the result (2.86) also gives the correct terms when $n = 1$ and 2, so that it holds for all n.

[1] Intriguingly, since $\pi\sqrt{\phi} \simeq 4$ to an accuracy of one part in a thousand, any architectural method that uses wheels for measurement, along with right triangles, could conceivably introduce what appears to be the Golden Ratio where it was not intended.

Coupled Linear Differential Equations

Our second example of diagonalisation shows a relationship between eigenvalues, eigenvectors, and the exponential number e. Suppose we wish to solve for functions $x(t)$ and $y(t)$ satisfying the coupled linear differential equations (with a prime denoting $\mathrm{d}/\mathrm{d}t$)

$$\begin{bmatrix} x \\ y \end{bmatrix}' = \underbrace{\begin{bmatrix} a & b \\ c & d \end{bmatrix}}_{\equiv A} \underbrace{\begin{bmatrix} x \\ y \end{bmatrix}}_{\equiv \boldsymbol{u}}, \tag{2.87}$$

where a, b, c, d are constants. If A is *diagonable* (meaning able to be diagonalised), then (2.87) becomes

$$\boldsymbol{u}' = A\boldsymbol{u} = PDP^{-1}\boldsymbol{u}, \quad \text{so that } P^{-1}\boldsymbol{u}' = DP^{-1}\boldsymbol{u}. \tag{2.88}$$

Setting $\boldsymbol{v} \equiv P^{-1}\boldsymbol{u}$ allows (2.88) to be written as

$$\boldsymbol{v}' = D\boldsymbol{v}. \tag{2.89}$$

This is now trivial to solve. Since A diagonalises to give P and D,

$$P = \begin{bmatrix} \boldsymbol{\alpha}_1 & \boldsymbol{\alpha}_2 \end{bmatrix}, \quad D = \begin{bmatrix} \lambda_1 & 0 \\ 0 & \lambda_2 \end{bmatrix}, \tag{2.90}$$

it follows that for arbitrary constants c_1, c_2,

$$\boldsymbol{v} = \begin{bmatrix} c_1\,\mathrm{e}^{\lambda_1 t} \\ c_2\,\mathrm{e}^{\lambda_2 t} \end{bmatrix}, \quad \text{with } \boldsymbol{u} = P\boldsymbol{v}. \tag{2.91}$$

The elegance of this solution is made more transparent by keeping the different-eigenvalue parts separate:

$$\begin{aligned} \boldsymbol{u} = P\boldsymbol{v} &= c_1\,\mathrm{e}^{\lambda_1 t}\,P\begin{bmatrix} 1 \\ 0 \end{bmatrix} + c_2\,\mathrm{e}^{\lambda_2 t}\,P\begin{bmatrix} 0 \\ 1 \end{bmatrix} \\ &= c_1\,\boldsymbol{\alpha}_1\,\mathrm{e}^{\lambda_1 t} + c_2\,\boldsymbol{\alpha}_2\,\mathrm{e}^{\lambda_2 t}. \end{aligned} \tag{2.92}$$

It follows that a basis solution of the original set (2.87) is expressed in terms of the eigenvalues and eigenvectors of A:

$$\text{basis solution} = \text{eigenvector} \times \mathrm{e}^{\text{eigenvalue} \times t}, \tag{2.93}$$

a particularly simple result. This study of eigenvalues and eigenvectors can be extended to coupled *non*linear differential equations, by comparing the nonlinear set with an associated linear set. Such a linear set is found by Taylor-expanding the nonlinear equations about a so-called *critical point* of the nonlinear system, at which each derivative equals zero.

Classical Mechanics: Moment of Inertia and Principal Axes

The third application of diagonalisation concerns a solid body's ability to spin smoothly about some chosen axis. Introductory approaches to the mechanics of spin usually begin by analysing only bodies with high symmetry, such as wheels and spheres. Of central importance for such a symmetrical body is its *angular momentum* L, related to its angular velocity ω by a constant of proportionality known as the body's *moment of inertia* I:

$$L = I\omega\,. \tag{2.94}$$

This is entirely equivalent to the $p = mv$ that defines linear momentum in terms of mass and velocity in one dimension.

But even a symmetrical body will not rotate smoothly about just *any* axis; in general it pulls sideways on the axis while spinning, producing wear and tear on the bearings. To deal with such motion, we require a more general expression for angular momentum that can cope with arbitrary axes of rotation. Classical mechanics answers this need by defining the angular momentum of a single particle about a point to be a vector defined as $\boldsymbol{L} \equiv \boldsymbol{r} \times \boldsymbol{p}$, where $\boldsymbol{r}$ is the position vector of the particle relative to the point, and $\boldsymbol{p} = m\boldsymbol{v}$ is its linear momentum. To analyse the spin of a solid body, we imagine breaking it up into particles labelled i, so that

$$\begin{aligned}\boldsymbol{L} &\equiv \sum_i \boldsymbol{L}_i = \sum_i \boldsymbol{r}_i \times \boldsymbol{p}_i \\ &= m_i\,\boldsymbol{r}_i \times \boldsymbol{v}_i = m_i\,\boldsymbol{r}_i \times (\boldsymbol{\omega} \times \boldsymbol{r}_i) \quad \text{(summation assumed)},\end{aligned} \tag{2.95}$$

where the entire body rotates with angular velocity $\boldsymbol{\omega}$, now made into a vector to encode the direction of the axis of rotation.

Double cross products such as in (2.95) are of course expandable as

$$\begin{aligned}(\boldsymbol{a} \times \boldsymbol{b}) \times \boldsymbol{c} &= \boldsymbol{a}\cdot\boldsymbol{c}\,\boldsymbol{b} - \boldsymbol{b}\cdot\boldsymbol{c}\,\boldsymbol{a}\,, \\ \boldsymbol{a} \times (\boldsymbol{b} \times \boldsymbol{c}) &= \boldsymbol{a}\cdot\boldsymbol{c}\,\boldsymbol{b} - \boldsymbol{a}\cdot\boldsymbol{b}\,\boldsymbol{c}\,.\end{aligned} \tag{2.96}$$

Here is a convenient way to remember these identities. The vector $(\boldsymbol{a}\times\boldsymbol{b})\times\boldsymbol{c}$ is perpendicular to both $\boldsymbol{a}\times\boldsymbol{b}$ and $\boldsymbol{c}$, so it must lie in the plane of $\boldsymbol{a}$ and $\boldsymbol{b}$; it must be a linear combination of these. The coefficients of $\boldsymbol{a}$ and $\boldsymbol{b}$ are just the dot products of the two other vectors, where the coefficient of the vector in the middle of the double cross product (i.e. $\boldsymbol{b}$) is given a plus sign and the other a minus sign. This mnemonic also works for $\boldsymbol{a} \times (\boldsymbol{b} \times \boldsymbol{c})$.

Hence, the total angular momentum is

$$\boldsymbol{L} = m_i \left(r_i^2\boldsymbol{\omega} - \boldsymbol{r}_i\cdot\boldsymbol{\omega}\,\boldsymbol{r}_i\right) \quad \text{(with } r_i \equiv |\boldsymbol{r}_i|\text{)}\,, \tag{2.97}$$

whereupon writing $r_i^2 = x_i^2 + y_i^2 + z_i^2$, we can expand all the terms to arrive at

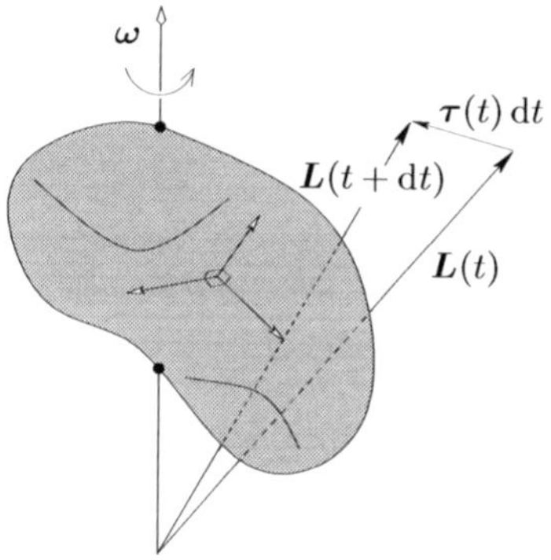

Fig. 2.7. The changing angular momentum of a potato spinning about an axis that does not run through one of its principal axes of inertia. Although the angular velocity $\boldsymbol{\omega}$ is fixed, the angular momentum vector $\boldsymbol{L}$ moves rigidly with the potato and so rotates about $\boldsymbol{\omega}$. The increase in $\boldsymbol{L}$ in an infinitesimal time $\mathrm{d}t$ is $\boldsymbol{\tau}(t)\,\mathrm{d}t$, where $\boldsymbol{\tau}$ is the *torque* required to keep the body from trying to "straighten up" along its axis, such that its moment of inertia would be maximised. The three principal axes of inertia are shown.

$$\boldsymbol{L} = m_i \begin{bmatrix} r_i^2 - x_i^2 & -x_i y_i & -x_i z_i \\ -x_i y_i & r_i^2 - y_i^2 & -y_i z_i \\ -x_i z_i & -y_i z_i & r_i^2 - z_i^2 \end{bmatrix} \boldsymbol{\omega} \equiv I\boldsymbol{\omega}\,. \tag{2.98}$$

This is the rotational equivalent of extending the one-dimensional linear momentum $p = mv$ to the three-dimensional $\boldsymbol{p} = m\boldsymbol{v}$. The big difference, of course, is that while linear momentum $\boldsymbol{p}$ is always parallel to velocity $\boldsymbol{v}$, angular momentum $\boldsymbol{L}$ need not be parallel to angular velocity $\boldsymbol{\omega}$. An example of these vectors for a rotating potato-shaped body is shown in Fig. 2.7.

As an example of using these ideas, the body's total kinetic energy can be written very concisely as

$$E_k \equiv \frac{1}{2} m_i v_i^2 = \frac{1}{2} m_i \boldsymbol{v}_i \cdot (\boldsymbol{\omega} \times \boldsymbol{r}_i) = \frac{1}{2} m_i \boldsymbol{\omega} \cdot (\boldsymbol{r}_i \times \boldsymbol{v}_i) = \frac{1}{2} \boldsymbol{\omega} \cdot \boldsymbol{L} = \frac{1}{2} \boldsymbol{\omega}^t I \boldsymbol{\omega}\,. \tag{2.99}$$

The fact that $\boldsymbol{L}$ is generally not parallel to $\boldsymbol{\omega}$ is what causes the body to pull on its axis as it spins. But all reasonable bodies have a moment of inertia matrix I that is diagonable, which means there can be found a unique set of three orthogonal axes X, Y, Z such that

$$\begin{bmatrix} L_X \\ L_Y \\ L_Z \end{bmatrix} = \begin{bmatrix} I_X & & 0 \\ & I_Y & \\ 0 & & I_Z \end{bmatrix} \begin{bmatrix} \omega_X \\ \omega_Y \\ \omega_Z \end{bmatrix}. \tag{2.100}$$

In other words, $L_X = I_X\,\omega_X$, and similarly for L_Y and L_Z. If the body spins about any of these three *principal axes of inertia*, the angular momentum $\boldsymbol{L}$

will point in the same direction as the angular velocity $\boldsymbol{\omega}$, so that the spin is smooth; the body does not stress its bearings. This is a surprising result; even potato-shaped bodies with no symmetry still have three principal axes—and what is just as remarkable is that these axes are always mutually orthogonal.

Geometry: Plotting Conic Sections

Our last example of diagonalisation involves conic sections. The cartesian-coordinate expressions for conic sections involve squares of the coordinates. This allows matrix theory to be used to analyse these sections. As an example, consider plotting the ellipse $x^2 + xy + y^2 = 6$ in the xy-plane. Because we will eventually introduce a new set of coordinates, rewrite this equation for clarity as $x_1^2 + x_1 x_2 + x_2^2 = 6$, to be plotted in the $x_1 x_2$-plane. An expression such as $x_1^2 + x_1 x_2 + x_2^2$, whose terms are each of second order, is known as a *quadratic form*. The equation to be plotted can be written as

$$\begin{bmatrix} x_1 \ x_2 \end{bmatrix} \underbrace{\begin{bmatrix} 1 & 1/2 \\ 1/2 & 1 \end{bmatrix}}_{\equiv A} \begin{bmatrix} x_1 \\ x_2 \end{bmatrix} = 6 . \tag{2.101}$$

(It will soon become apparent that A is most usefully set to be symmetric, which can always be done.) The fact that the first and last matrices in (2.101) are transposes of one another leads us to suppose that it might be useful to express A in the form PDP^t (as opposed to PDP^{-1}). That way, it will be possible to define new coordinates

$$\begin{bmatrix} y_1 \\ y_2 \end{bmatrix} \equiv P^t \begin{bmatrix} x_1 \\ x_2 \end{bmatrix} , \tag{2.102}$$

enabling (2.101) to be written as

$$\begin{bmatrix} y_1 \ y_2 \end{bmatrix} D \begin{bmatrix} y_1 \\ y_2 \end{bmatrix} = 6 , \tag{2.103}$$

which contains no cross terms and so is trivial to plot. The matrix A can certainly be written as PDP^t using a straightforward technique known as *congruent diagonalisation*, whose result is all that we require here. (The procedure is explained in linear algebra texts and is very similar to inverting a matrix.) One pair of P, D produced by congruent diagonalisation is

$$P = \begin{bmatrix} 1 & 0 \\ 1/2 & 1/2 \end{bmatrix} , \qquad D = \begin{bmatrix} 1 & 0 \\ 0 & 3 \end{bmatrix} . \tag{2.104}$$

We plan to overlay the $y_1 y_2$-axes on the $x_1 x_2$-plane, and plot the ellipse using its simpler form (2.103). Unfortunately, there are two problems with this: [Note that in the next equations we write $P^{-t} \equiv (P^{-1})^t = (P^t)^{-1}$.]

1. We wish to ensure that the ruler defined by the x_1x_2-coordinates is identical to that of the y_1y_2-coordinates, to simplify plotting the ellipse. While not an absolute requirement, this ensures that we do not accidentally deform the ellipse. This ruler is called the *metric*. Unfortunately, here the two coordinate systems do not share the same metric. After all, the distance between two points in the original cartesian x_1x_2-coordinates satisfies Pythagoras's theorem, so that (2.102) gives

$$\Delta x_1^2 + \Delta x_2^2 = \begin{bmatrix}\Delta x_1 & \Delta x_2\end{bmatrix} \begin{bmatrix}\Delta x_1 \\ \Delta x_2\end{bmatrix} = \begin{bmatrix}\Delta y_1 & \Delta y_2\end{bmatrix} P^{-1}P^{-t} \begin{bmatrix}\Delta y_1 \\ \Delta y_2\end{bmatrix}$$
$$= \begin{bmatrix}\Delta y_1 & \Delta y_2\end{bmatrix} \underbrace{\begin{bmatrix}1 & 0 \\ -1 & 2\end{bmatrix}}_{P^{-1}} \underbrace{\begin{bmatrix}1 & -1 \\ 0 & 2\end{bmatrix}}_{P^{-t}} \begin{bmatrix}\Delta y_1 \\ \Delta y_2\end{bmatrix} = \begin{bmatrix}\Delta y_1 & \Delta y_2\end{bmatrix} \begin{bmatrix}1 & -1 \\ -1 & 5\end{bmatrix} \begin{bmatrix}\Delta y_1 \\ \Delta y_2\end{bmatrix}$$
$$\neq \Delta y_1^2 + \Delta y_2^2 \,. \tag{2.105}$$

Thus the y_1y_2-coordinates do not satisfy Pythagoras's theorem.
2. We prefer the y_1-axis to be perpendicular to the y_2-axis. But here they are not, as can be shown by writing them as vectors in x_1x_2-coordinates. The y_1-axis is the set of points for which $y_2 = 0$; these lie along the vector

$$\begin{bmatrix}x_1 \\ x_2\end{bmatrix} = P^{-t} \begin{bmatrix}y_1 \\ 0\end{bmatrix} \propto \text{first column of } P^{-t} = \begin{bmatrix}1 \\ 0\end{bmatrix}. \tag{2.106}$$

Similarly, the y_2-axis is delineated by the second column of P^{-t}, or $\begin{bmatrix}-1 & 2\end{bmatrix}^t$. This is not perpendicular to $\begin{bmatrix}1 & 0\end{bmatrix}^t$.

Fortunately, both of these difficulties can be eliminated at once. The first can be fixed by requiring, if possible, $P^{-1}P^{-t} = 1$, which, by taking its inverse, is equivalent to $P^tP = 1$. Such a P has rows that form an orthonormal set, and also columns that form an orthonormal set, and is known as *orthogonal*. A well-known theorem of linear algebra states that any real symmetric matrix can always be diagonalised with an orthogonal P. In that case, since $P = P^{-t}$, its orthonormal columns also fix the second difficulty above.

For a symmetric (and real) matrix A, the diagonalisation technique described in the last few pages suffices to give a matrix that will serve as P provided that its columns are first orthonormalised, using, say, the Gram–Schmidt algorithm.

In fact, in most cases the columns will already be orthogonal, since it is proved ahead in a slightly different setting [in the discussion around (2.131)–(2.133)] that the eigenvectors belonging to distinct eigenvalues of A are guaranteed to be orthogonal.

Using the A of (2.101), the relevant calculation gives

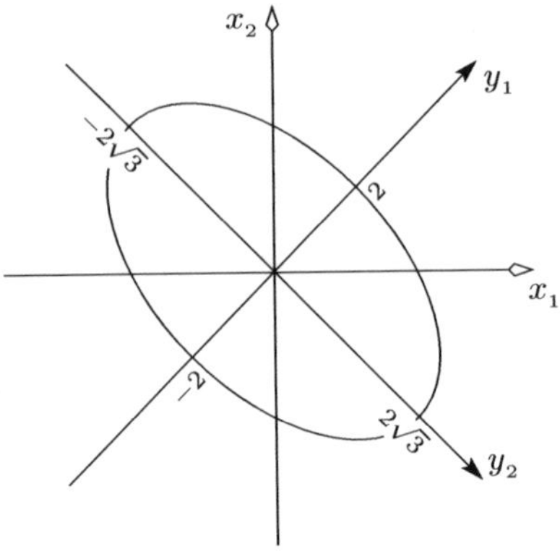

Fig. 2.8. Plotting the ellipse $x_1^2 + x_1x_2 + x_2^2 = 6$ by constructing a more amenable set of coordinate axes, y_1, y_2, in which the plot becomes symmetric and hence more easily drawn.

$$P = \frac{1}{\sqrt{2}} \begin{bmatrix} 1 & 1 \\ 1 & -1 \end{bmatrix}, \quad D = \begin{bmatrix} 3/2 & 0 \\ 0 & 1/2 \end{bmatrix}. \tag{2.107}$$

Now that $P^tP = 1$, the two sets of axes share the same metric, allowing them to be easily superimposed in one plot. The y_1-axis is now the first column of P, or $\begin{bmatrix} 1 & 1 \end{bmatrix}^t$, while the y_2-axis is the second column of P, or $\begin{bmatrix} 1 & -1 \end{bmatrix}^t$, as shown in Fig. 2.8. Which way does each axis point? Although it's not necessary to know this to plot the ellipse, the directions can be found by applying (2.106) more studiously, by calculating the x_1x_2-coordinates of the unit basis vector along each of the y-axes. For the y_1-axis,

$$\begin{bmatrix} x_1 \\ x_2 \end{bmatrix} = P \begin{bmatrix} 1 \\ 0 \end{bmatrix} = \frac{1}{\sqrt{2}} \begin{bmatrix} 1 \\ 1 \end{bmatrix}, \tag{2.108}$$

and similarly for the y_2-axis. These directions are indicated in Fig. 2.8. Finally, the ellipse itself can be plotted. In y_1y_2-coordinates it becomes, from (2.103),

$$\frac{3}{2}\, y_1^2 + \frac{1}{2}\, y_2^2 = 6\,. \tag{2.109}$$

This cuts the y_1-axis at ± 2 and the y_2-axis at $\pm 2\sqrt{3}$, again as shown in the figure.

Drawing the ellipse by constructing a new set of axes has highlighted the notion of *orthogonal diagonalisation.* We began the task with the much simpler idea of congruent diagonalisation, but found that, in fact, this was not sufficient because it introduced a different metric and non-orthogonal axes. Orthogonal diagonalisation solved these problems, and shows why this procedure is very important in other areas of mathematical physics. We'll meet the related idea of an orthogonal operator in Sect. 2.7.

2.6 Dirac's Bracket Notation

The notion of an inner product $\langle \boldsymbol{\alpha} | \boldsymbol{\beta} \rangle$ is central to linear algebra, and following on Dirac's research into quantum mechanics, physicists have extended this use of angle brackets to what has become known as *bracket notation* or *bra-ket notation*. Although this use of brackets is usually reserved purely for writing the equations of quantum mechanics (as we'll see later), it actually is very useful in other areas that involve linearity and summing, such as matrix multiplication, Fourier analysis, and probability theory. To see how it all works, we begin by looking at matrix multiplication using angle brackets.

Consider a general matrix A whose ij^{th} element is A_{ij}, and suppose we write this element in the following way:

$$\langle i|A|j\rangle \equiv A_{ij} \,. \tag{2.110}$$

We have done more here than just write the indices on each side of the matrix. Because matrices go hand in hand with linearity, it becomes possible to consider (2.110) to be a "product" of three things unambiguously: $\langle i| \times A \times |j\rangle$ (that is, the associative law holds, as we shall see). The splitting of the angle brackets in (2.110) leads to $\langle i|$ being called a *bra* while $|j\rangle$ is called a *ket*, so that bracket formalism is sometimes called the language of *bra-kets*.

The operation of *pre*pending with $\langle i|$ singles out the ith row, while the operation of *ap*pending with $|j\rangle$ singles out the jth column, so that $\langle i|A|j\rangle$ becomes the ij^{th} element of A. Thus we can extract the element $\langle i|A|j\rangle$ from A in either of two ways:

1. We can first form $\langle i|A$, the ith row of A, and then extract its jth column (element) by appending $|j\rangle$, or
2. we can first form $A|j\rangle$, the jth column of A, and then extract its ith row (element) by prepending $\langle i|$.

The fact that both ways give the same result is the reason we can split the expression $\langle i|A|j\rangle$, without ambiguity, into a sort of product of three elements: $\langle i|$, A, and $|j\rangle$.

Since premultiplication with the bra $\langle i|$ singles out a row, it must be equivalent to premultiplying by a row vector whose only nonzero entry is a one in the ith position. Similarly, postmultiplying by a ket $|j\rangle$ must be equivalent to postmultiplying by a column vector whose only nonzero entry is a one in the jth position:

$$\langle i| = \begin{bmatrix} 0 & 0 & \cdots & 1 & \cdots & 0 \end{bmatrix} , \qquad |j\rangle = \begin{bmatrix} 0 \\ 0 \\ \vdots \\ 1 \\ \vdots \\ 0 \end{bmatrix} . \tag{2.111}$$

Two new products result immediately from making these identifications. The first is our friend the inner product, now of a bra with a ket, where we simply multiply the two matrices and denote the product as a bracket with one of the vertical bars removed,

$$\text{inner product: } \langle i|j\rangle = \delta_{ij}\,, \tag{2.112}$$

where δ_{ij} is the usual Kronecker delta function. The second, *outer product,* comes from multiplying a ket with a bra:

$$\text{outer product: } |j\rangle\langle i| = \begin{bmatrix} \text{a matrix of zeroes} \\ \text{with a 1 in the } ji^{\text{th}} \text{ position} \end{bmatrix}. \tag{2.113}$$

A first use for this bracket formalism comes by way of matrix multiplication. Remember that the product AB of matrices A and B is defined in the usual way as a matrix of dot products of each row with each column:

$$(AB)_{ij} \equiv \sum_k A_{ik} B_{kj}\,. \tag{2.114}$$

Here k indexes the columns of A and the rows of B. In bracket notation, (2.114) is written as

$$\langle i|AB|j\rangle \equiv \sum_k \langle i|A|k\rangle\,\langle k|B|j\rangle\,. \tag{2.115}$$

Evidently, the core of matrix multiplication is the following identity, known as a *completeness relation*:

$$\sum_k |k\rangle\langle k| = 1\,. \tag{2.116}$$

This identity encapsulates the multiplication because it can be inserted between the A and B on the left-hand side of (2.115) to produce its right-hand side. It also follows from the outer product in (2.113) by setting $i = j = k$ and summing, although we need to remember that the number of columns of A and the number of rows of B must be the same if they are to be multiplied, to ensure that the number of kets in (2.116) really does equal the number of bras so that the sum does indeed make sense. Completeness relations are very common in mathematical physics, and we'll encounter them often in the coming chapters.

Being adventurous now, prepending (2.116) to a ket gives

$$|j\rangle = \sum_k |k\rangle\langle k|j\rangle = \sum_k |k\rangle\,\delta_{kj}\,, \tag{2.117}$$

which is consistent with the Kronecker delta of (2.112). (Appending the summed outer product to a bra produces the same result.) These rules, distilled to the inner product of (2.112) and the summed outer product of (2.116), form the essence of bracket notation. Simple though they are, they form a powerful tool that is especially useful when we realise that the angle brackets can serve as containers to hold labels such as eigenvalues, as we'll see in the next section. This makes them very useful for describing physical states in statistical mechanics and quantum mechanics.

2.7 Brackets and Hermitian Operators

Up until now, we have written $\langle i|A|j\rangle$ for the ij^{th} element of a matrix A, so that it could be read to mean the row vector $\langle i|$ times the matrix A times the column vector $|j\rangle$. Using this associativity, in general the notation can stand for the action of any linear operator sandwiched in between any two vectors:

$$\langle \boldsymbol{\alpha}|A|\boldsymbol{\beta}\rangle \equiv \langle \boldsymbol{\alpha}|A\boldsymbol{\beta}\rangle\,, \tag{2.118}$$

which implies that

$$A|\boldsymbol{\beta}\rangle = |A\boldsymbol{\beta}\rangle\,. \tag{2.119}$$

Bracket notation writes a vector $\boldsymbol{\alpha}$ as $|\boldsymbol{\alpha}\rangle$. This might look like extraneous notation, but the ket can "contain" more than just one symbol. It is especially useful for holding eigenvalues of multiple operators, so that for example $|\ell, m\rangle$ can be understood as the eigenvector corresponding to the eigenvalues ℓ, m of two (commuting) operators.

What can be said about $\langle A\boldsymbol{\alpha}|$? Extracting A from this requires the notion of the *adjoint* $A^\dagger$ of an operator A, defined by

$$\langle \boldsymbol{\alpha}|A^\dagger|\boldsymbol{\beta}\rangle \equiv \langle \boldsymbol{\alpha}|A^\dagger\boldsymbol{\beta}\rangle \equiv \langle A\boldsymbol{\alpha}|\boldsymbol{\beta}\rangle\,, \tag{2.120}$$

so we see immediately that

$$\langle \boldsymbol{\alpha}|A^\dagger = \langle A\boldsymbol{\alpha}|\,. \tag{2.121}$$

Compare this with (2.119). From the basic properties of the inner product, we also see that

$$\langle A^\dagger\boldsymbol{\alpha}|\boldsymbol{\beta}\rangle = \langle \boldsymbol{\beta}|A^\dagger\boldsymbol{\alpha}\rangle^* = \langle A\boldsymbol{\beta}|\boldsymbol{\alpha}\rangle^* = \langle \boldsymbol{\alpha}|A\boldsymbol{\beta}\rangle\,, \tag{2.122}$$

so that the action of taking the adjoint is quite symmetrical: the dagger superscript appears or vanishes when A jumps from one side of the inner product to the other. This symmetry also implies that $A^{\dagger\dagger} = A$, since $\langle \boldsymbol{\alpha}|A^{\dagger\dagger}\boldsymbol{\beta}\rangle = \langle A^\dagger\boldsymbol{\alpha}|\boldsymbol{\beta}\rangle = \langle \boldsymbol{\alpha}|A\boldsymbol{\beta}\rangle$.

It's very helpful to keep the axiomatic associativity of bracket notation in mind when using it. For *any* $\boldsymbol{\alpha}, \boldsymbol{\beta}$ and *any* operator A, the following holds:

$$\boxed{\Big(\langle \boldsymbol{\alpha}|A\Big)\,|\boldsymbol{\beta}\rangle = \langle \boldsymbol{\alpha}|\,\Big(A|\boldsymbol{\beta}\rangle\Big) \equiv \langle \boldsymbol{\alpha}|A|\boldsymbol{\beta}\rangle\,.} \tag{2.123}$$

Of especially simple form are the adjoints of complex numbers under scalar multiplication and those of operator products. For the complex number c,

$$\langle \boldsymbol{\alpha} | c^\dagger \boldsymbol{\beta} \rangle = \langle c\boldsymbol{\alpha} | \boldsymbol{\beta} \rangle = c^* \langle \boldsymbol{\alpha} | \boldsymbol{\beta} \rangle = \langle \boldsymbol{\alpha} | c^* \boldsymbol{\beta} \rangle \,, \tag{2.124}$$

which means that $c^\dagger = c^*$. Thus the adjoint of a complex number under scalar multiplication is just its complex conjugate—which is also true under the alternative inner product usage favoured by mathematicians (see p. 11). For the product AB of two operators, we can write

$$\langle (AB)^\dagger \boldsymbol{\alpha} | \boldsymbol{\beta} \rangle = \langle \boldsymbol{\alpha} | AB\boldsymbol{\beta} \rangle = \langle A^\dagger \boldsymbol{\alpha} | B\boldsymbol{\beta} \rangle = \langle B^\dagger A^\dagger \boldsymbol{\alpha} | \boldsymbol{\beta} \rangle \,, \tag{2.125}$$

giving $(AB)^\dagger = B^\dagger A^\dagger$. Of course, the same proof generalises to a product of any number of operators.

The notion of an adjoint also allows us to view bras and kets as *dual* to each other in the sense of forming a natural pair. Write

$$\langle \boldsymbol{\alpha} | \quad \overset{\text{dual}}{\longleftrightarrow} \quad | \boldsymbol{\alpha} \rangle \,, \tag{2.126}$$

and note that in particular $\langle A\boldsymbol{\alpha} |$ is dual to $| A\boldsymbol{\alpha} \rangle$. Thus we can use (2.119) and (2.121) to rewrite this last duality as

$$\langle \boldsymbol{\alpha} | A^\dagger \quad \overset{\text{dual}}{\longleftrightarrow} \quad A | \boldsymbol{\alpha} \rangle \,. \tag{2.127}$$

These are known as *hermitian conjugates* of each other. The reason is that the hermitian conjugate of a matrix is simply defined to be its conjugate transpose, and for the vectors that usually represent bras and kets, forming the dual means taking the conjugate transpose:

$$\langle \boldsymbol{\alpha} | = \begin{bmatrix} a^* \; b^* \end{bmatrix} \quad \Longleftrightarrow \quad | \boldsymbol{\alpha} \rangle = \begin{bmatrix} a \\ b \end{bmatrix} , \tag{2.128}$$

since this ensures that $|\boldsymbol{\alpha}|^2 \equiv \langle \boldsymbol{\alpha} | \boldsymbol{\alpha} \rangle = |a|^2 + |b|^2$, as required. So the idea of taking the adjoint of an operator is mirrored in its matrix representative by taking the hermitian conjugate of the matrix. (And as a reminder of this, in practice the adjoint is often called the *hermitian* adjoint, although the word hermitian here is superfluous). As an example, the hermitian conjugate of a matrix-vector product $A\boldsymbol{\alpha}$ is $(A\boldsymbol{\alpha})^\dagger = \boldsymbol{\alpha}^\dagger A^\dagger$. In bracket language this is written as $\big(A|\boldsymbol{\alpha}\rangle\big)^\dagger = |A\boldsymbol{\alpha}\rangle^\dagger = \langle \boldsymbol{\alpha} | A^\dagger$. With hindsight, defining the hermitian conjugate of a matrix to be its conjugate transpose is quite reasonable given that the transpose of a matrix product is $(AB)^t = B^t A^t$, reminiscent of the $(AB)^\dagger = B^\dagger A^\dagger$ that we saw just after (2.125).

The duality between bras and kets is entirely equivalent to the duality between the cobasis and basis. Thus (2.22) and (2.24) are equivalent to writing expressions such as

$$|\boldsymbol{v}\rangle = \sum_\alpha |\alpha\rangle\langle\alpha|\boldsymbol{v}\rangle \,, \quad \langle\boldsymbol{v}| = \sum_\alpha \langle\boldsymbol{v}|\alpha\rangle\langle\alpha| \,, \tag{2.129}$$

Table 2.1. Terms describing matrices in the context of orthogonalisation.

A is:	Complex	Real
	A is called:	
$A^\dagger = A^{-1}$	*unitary*	*orthogonal*
$A^\dagger = A$	*hermitian*	*symmetric*

which are dual to each other, and are none other than the completeness relation $\sum_\alpha |\alpha\rangle\langle\alpha| = 1$.

An important class of operator A is one that preserves length. The squared length of a complex vector $\boldsymbol{\alpha}$ is $\boldsymbol{\alpha}^\dagger\boldsymbol{\alpha}$, or $\langle\boldsymbol{\alpha}|\boldsymbol{\alpha}\rangle$. So demanding that length be preserved is equivalent to demanding that $|\boldsymbol{\alpha}|^2 = |A\boldsymbol{\alpha}|^2$ for all vectors $\boldsymbol{\alpha}$, or $\langle\boldsymbol{\alpha}|\boldsymbol{\alpha}\rangle = \langle\boldsymbol{\alpha}|A^\dagger A|\boldsymbol{\alpha}\rangle$ in bracket language. This implies that

$$A^\dagger A = 1\,. \tag{2.130}$$

A matrix or operator A with such a property is called *unitary* (or *orthogonal* if A is real). If A itself, rather than its inverse, is equal to $A^\dagger$, then A is called *hermitian* (or *symmetric* for real A). (These common but perhaps confusing terms are summarised in Table 2.1.) Expressions involving hermitian conjugates arise frequently in quantum mechanics, where such length-preserving matrices are needed to ensure that expressions involving probability are always correctly normalised.

Hermitian Operators in Quantum Mechanics

Hermitian operators—those that are identical to their own adjoints—find major use in quantum mechanics, where the theory seeks out hermitian operators to represent physical observables. To see why this might be, it's useful to derive some properties of hermitian operators that will also demonstrate the economy of using brackets. A good example is a standard theorem of linear algebra that states that the eigenvalues of hermitian operators are real, while the eigenvectors corresponding to distinct eigenvalues are orthogonal. (We saw an example of these orthogonal eigenvectors on p. 39 in the discussion of drawing the ellipse.) To prove this using brackets, consider two eigenvector/eigenvalue pairs of any operator A. The eigenvalues are called m, n, with corresponding eigenvectors $|m\rangle, |n\rangle$, where each ket is labelled by the eigenvalue it holds, and no other vector notation need be introduced (such is the economy of brackets). Since

$$A|m\rangle = m|m\rangle\,, \quad A|n\rangle = n|n\rangle\,, \tag{2.131}$$

it must follow that

$$\begin{aligned} &\langle n|A|m\rangle = m\langle n|m\rangle\,, \\ &\langle m|A|n\rangle = n\langle m|n\rangle \xrightarrow[\text{conjugate}]{\text{complex}} \langle n|A^\dagger|m\rangle = n^*\langle n|m\rangle\,. \end{aligned} \tag{2.132}$$

A subtraction then gives

$$\langle n|A - A^\dagger|m\rangle = (m - n^*)\langle n|m\rangle \,. \tag{2.133}$$

Suppose that A is hermitian: $A = A^\dagger$. Then it follows that $(m - n^*)\langle n|m\rangle = 0$, and the following arguments revolve around this last expression. Setting $m = n$ produces $n = n^*$ since $\langle n|n\rangle \neq 0$, so the eigenvalues are real. Now, for distinct eigenvalues, $m \neq n$, in which case $\langle n|m\rangle = 0$, so that the corresponding eigenvectors are orthogonal and the theorem is proved.

Just as $A^\dagger = A$ defines a hermitian operator, $A^\dagger = -A$ defines an *antihermitian* operator. And by an argument similar to that of the previous paragraphs, it's straightforward to show that antihermitian operators have pure imaginary eigenvalues.

That the eigenvalues of hermitian operators are real, while those of antihermitian operators are pure imaginary, is of central importance in the discussion of Sect. 2.10, where we calculate how the degree of localisation of a wave depends on what variable is used to describe it. But this property of eigenvalues finds especial importance in quantum mechanics, where a measurement of a physical system is postulated to be represented by an operator acting on a *state ket*, a normalised ket that encodes the system's state. This ket can be written as a linear combination of the operator's orthonormal eigenvectors ("eigenkets"), whose eigenvalues are postulated to be the only allowable results of the measurement. Since these results must be real numbers, it is sufficient that operators representing physical measurements be hermitian. Although the logic doesn't actually require it, hermiticity has historically been seen as a requirement for such operators (i.e. hermiticity is not *necessary*, but it is *sufficient*).

The bread and butter of hermitian operators in quantum mechanics are those that act on functions. To see a common example (whose hermiticity is certainly not obvious at first), consider the inner product in function space, defined by

$$\langle f|g\rangle \equiv \int_a^b f^*(x)\, g(x)\, \mathrm{d}x \,, \tag{2.134}$$

where x is real, and f, g are functions with value zero at the integration end points. Suppose that we have a derivative operator $A = c\,\mathrm{d}/\mathrm{d}x$ for some complex number c. This certainly arises in quantum mechanics, and we wish to find any restriction needed if A is to be hermitian (i.e., if A is to represent the action of a measurement). The hermiticity implies that $\langle f|Ag\rangle = \langle Af|g\rangle$, or

$$\int_a^b f^*(x)\, c\, g'(x)\, \mathrm{d}x = \int_a^b c^* f'^*(x)\, g(x)\, \mathrm{d}x \,. \tag{2.135}$$

Integrate the left-hand side of (2.135) by parts to write

$$-\int_a^b f^{*\prime}(x)\, c\, g(x)\, \mathrm{d}x = \int_a^b c^* f'^*(x)\, g(x)\, \mathrm{d}x \,. \tag{2.136}$$

Since $f^{*\prime} = f^{\prime *}$, we infer that $c^* = -c$, so that c must be pure imaginary: $c = \mathrm{i}a$ for some real a. Hence $\mathrm{i}a\,\mathrm{d}/\mathrm{d}x$ is hermitian. Finally, because a whole number power and any real multiple of a hermitian operator are also hermitian (the proofs are straightforward), in particular $\mathrm{d}^2/\mathrm{d}x^2$ must also be hermitian. This last, the laplacian operator, finds frequent use in quantum mechanics.

Quantum mechanics uses the bracket $\langle f|i\rangle$ to denote the *amplitude* for a system to be measured to be in some final state $|f\rangle$ given some initial state $|i\rangle$, where the absolute square of the amplitude gives the probability for that process to occur. So, for example, suppose that we have some sort of quantum mechanical coin that after being tossed is in a superposition of states heads or tails, represented by orthonormal eigenvectors $|H\rangle$ and $|T\rangle$. Its state might be

$$|\psi\rangle = \sqrt{0.4}\,|H\rangle - \mathrm{i}\sqrt{0.6}\,|T\rangle\,, \tag{2.137}$$

for which $\langle\psi|\psi\rangle$ certainly equals one. Possible results of a measurement are represented by prepending the state $|\psi\rangle$ with the appropriate bra, so that the amplitude for the coin to be measured as having landed heads-up is then

$$\langle H|\psi\rangle = \sqrt{0.4}\,, \tag{2.138}$$

which gives a probability of 0.4, while the amplitude for the coin to be measured as having landed tails-up is $\langle T|\psi\rangle = -\sqrt{0.6}\,\mathrm{i}$, giving a probability of 0.6. We'll see more of this notation when we look at the evolution of quantum mechanical states under a measurement in Sect. 10.9.

2.8 Frequency and Wavenumber

The two major attributes of a wave that allow wave motion to be quantified are frequency and the wavenumber vector. Why is it that we never talk about a "wavelength vector" as opposed to the wavenumber vector? It would seem that wavelength is a more basic concept than wavenumber, and yet the issue somehow seems to become complicated by the introduction of the wavenumber. Of course, there *is* a good reason, which will emerge over the next few pages.

Consider a general wave moving along the x-axis without *dispersion* at some velocity v; that is, it consists of an unchanging shape which moves at this velocity. Let $u(t, x)$ measure the amount by which the medium is "waving" (i.e. displaced from its equilibrium position). This quantity might be air pressure, displacement of the air molecules from equilibrium for a sound wave, or transverse displacement for a wave on a string. For light, which has no medium, u might be chosen to be the strength of the electromagnetic field that defines the wave. We can always write

$$u(t, x) = f(x - vt)\,. \tag{2.139}$$

Δ = "increase in", not "change in"

Symbols such as $\partial/\partial x$, d, and Δ are almost always read as "the change in". But what is much more meaningful is to specify whether this change is a gain or a loss; after all, nobody speaks of their bank balance as "changing" by such-and-such an amount. The symbols Δ and d really refer to a *gain*:

$$\begin{aligned} \Delta A &\equiv A_{\text{final}} - A_{\text{initial}} \equiv \text{gain in } A\,, \\ -\Delta A &= \text{loss in } A\,, \end{aligned} \tag{2.140}$$

and similarly for partial derivatives. This is no mere toying with the language. As Richard Feynman once remarked, physics is all about knowing where to put the minus sign; and words such as gain or loss figure in how we translate from a linguistic description of a problem to a useful mathematical statement. Other examples are common. Ohm's rule is written as $V = IR$, but only by remembering that V is a *drop* in potential across a resistance, or $-\Delta\Phi$, are we able to relate this to Maxwell's equations, as well as apply Kirchhoff's laws correctly around an electric circuit. Similarly, in the theory of heat flow the time rate of *loss* of thermal energy, $-\mathrm{d}Q/\mathrm{d}t$, is proportional (with positive constant) to the area A of the surfaces in contact and the space rate of *loss* of temperature $-\nabla T$. Here the minus signs cancel, giving $\mathrm{d}Q/\mathrm{d}t = \kappa A \nabla T$, but our ability to translate the physics into an English statement, and then into mathematics, hinges on an understanding of the signs involved. This can be seen clearly in equations such as (3.91), (10.7), (10.165), and (12.42).

Another example where this understanding is crucial is when we wish to convert a spectrum as a function of frequency, $u(f)$, to a function of wavelength, $\widetilde{u}(\lambda)$. Increasing values of f map to *decreasing* values of λ; so the frequency interval $[f, f + \mathrm{d}f]$ corresponds to the wavelength interval $[\lambda + \mathrm{d}\lambda, \lambda]$, where $\mathrm{d}\lambda$ is *negative*. Given the frequency spectrum $u(f)$, the wavelength spectrum $\widetilde{u}(\lambda)$ is defined by equating infinitesimal areas under the graphs of $u(f)$ and $\widetilde{u}(\lambda)$. Thus the two spectra are related by setting $\widetilde{u}(\lambda) \times -\mathrm{d}\lambda$ equal to $u(f)\,\mathrm{d}f$.

The *phase* of a wave of constant amplitude can be defined using this function. We certainly have an intuitive idea of the phase of a sine wave, meaning the argument of the sine function: the angle that the associated phasor currently makes with a reference axis. At any one moment, the phase ϕ of a sine wave increases with x, and the faster it increases, the more "ripply" the wave is. This suggests that we define a measure of this rippliness called the *wavenumber* k:

$$k \equiv \frac{\partial \phi}{\partial x}\,. \tag{2.141}$$

Choosing now some position x, then as the wave moves, say, to the right, its phase decreases. Define the wave's *circular frequency* (often just called frequency) to be this rate of decrease:

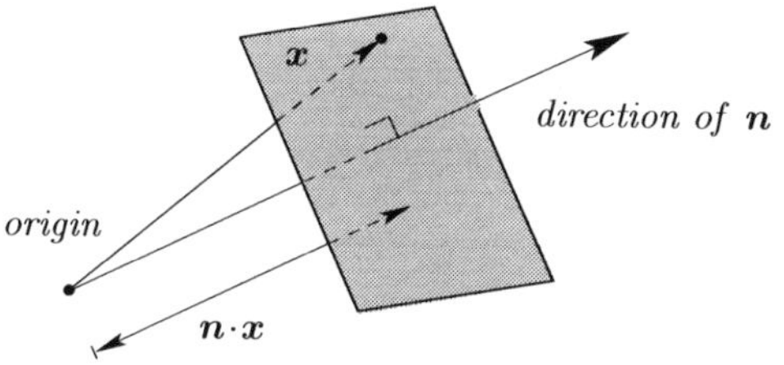

Fig. 2.9. A surface of constant phase for a three-dimensional wave. The wave moves in the direction of its unit normal $\boldsymbol{n}$.

$$\omega \equiv \frac{-\partial \phi}{\partial t}, \tag{2.142}$$

where the minus sign ensures that we are specifying a decrease, not an increase. (See the box on the facing page.)

At first glance, this seems to imply that we have designed ω to be positive for a right-mover and negative for a left-mover, but actually that depends on the sign of the phase. There is a certain amount of ambiguity in our definitions that is really not so important. Also, the circular frequency measures the number of radians per unit time through which the phasor turns, so we must divide this by 2π to get the number of cycles per unit time, which we'll simply call the frequency f. Thus $f = \omega/2\pi$.

From their definitions, a wave with a constant frequency and wavenumber must have a phase of $kx - \omega t$. That is, for some function g,

$$u(t,x) = g(kx - \omega t) = g\big(k[x - vt]\big) \equiv f(x - vt), \tag{2.143}$$

so that $\omega = kv$. This relation between frequency and wavenumber is called the *dispersion relation* of the wave. What are the wavelength and period of the wave? Unlike the wavenumber and frequency, we will really take these two quantities to be always positive, and a simple analysis of how the wave moves produces

$$\lambda = \frac{2\pi}{|k|}, \quad T = \frac{2\pi}{|\omega|}. \tag{2.144}$$

So much for motion in one dimension. A plane wave in three dimensions can be converted to a single dimension by referring to Fig. 2.9, in which we show a side view of a surface of constant phase for the wave, which moves in the direction of its unit normal $\boldsymbol{n}$. Analogously to the one-dimensional case, the function describing the wave has the form

$$u(t,\boldsymbol{x}) = f(\boldsymbol{n}\cdot\boldsymbol{x} - vt). \tag{2.145}$$

With reference to (2.143), we can always define a function g by

$$u = g\big(k(\boldsymbol{n}\cdot\boldsymbol{x} - vt)\big), \tag{2.146}$$

Drawing the Wavenumber $\boldsymbol{k}$ and a Unit Vector $\boldsymbol{n}$

Equation (2.147) defines the wavenumber $\boldsymbol{k}$ and relates it to a unit vector $\boldsymbol{n}$. But an interesting difficulty presents itself here: how can we draw these vectors as arrows? Unlike a position vector $\boldsymbol{x}$, their lengths are not measured in units such as metres. The vector $\boldsymbol{k}$ carries units of, e.g., metres^{-1}, while in a slight linguistic twist, the unit vector $\boldsymbol{n}$ has no units at all!

Having no units, $\boldsymbol{n}$ cannot be drawn as an arrow because its length is not one metre or one light-year or one anything else, but simply one. If the arrow conventionally drawn as representing $\boldsymbol{n}$ is to have a length of one metre, then what we are really drawing is not $\boldsymbol{n}$ but $\boldsymbol{n} \times 1$ metre. Likewise, the arrow drawn as representing $\boldsymbol{k}$ on the same axes is not $\boldsymbol{k}$ at all, but rather $\boldsymbol{k} \times 1$ metre^2.

This apparent clash of what can be drawn and what meaning it carries really presents no problems. After all, we could represent a product such as mass $\times$ acceleration by the area of a rectangle whose sides have lengths "equal" to the mass and acceleration; and yet neither of these has units of metres. It is perfectly well understood that the line segments being drawn are not really mass and acceleration but rather mass $\times$ 1 m/kg and acceleration $\times$ 1 s^2. Units are an integral part of how entities are visualised geometrically, but once a convention of what will be used is established, then we can be confident that units will look after themselves.

In Chap. 8 we'll see that for something like the wavenumber $\boldsymbol{k}$, the actual entity with an existence independent of coordinates is not the set of components (k^x, k^y, k^z) but rather these components *together with a basis.* This is a very important and fruitful idea that will be used frequently in Chap. 8 and later chapters.

where k is defined as in the one-dimensional case. This suggests a three-dimensional version of the wavenumber, a new vector $\boldsymbol{k}$ defined as

$$\boldsymbol{k} \equiv \nabla\phi = k\boldsymbol{n}\,, \quad \text{along with} \quad \omega \equiv \frac{-\partial\phi}{\partial t} = kv \quad \text{as before,} \tag{2.147}$$

so that $u = g(\boldsymbol{k}\cdot\boldsymbol{x} - \omega t)$. We'll make use of this definition of the wavenumber and frequency in Chap. 6 when we join them together to make something new.

Equation (2.147) generalises the wavenumber to a vector. Does it make sense to talk about a wavelength vector $\boldsymbol{\lambda} \equiv \lambda\boldsymbol{n}$? We can always *define* such a thing, but this needs care if the definition is to be useful. To see why, consider a set of plane waves moving along their normal vector $\boldsymbol{n}$, as depicted in Fig. 2.10. If the vector $\boldsymbol{\lambda}$ is to have intuitive meaning, we would surely require that each of its components be the wavelength of the waves as seen on the corresponding axis. But such is not the case! As can be seen in Fig. 2.10, the wavelength on the x-axis, being the distance between points of equal phase, is

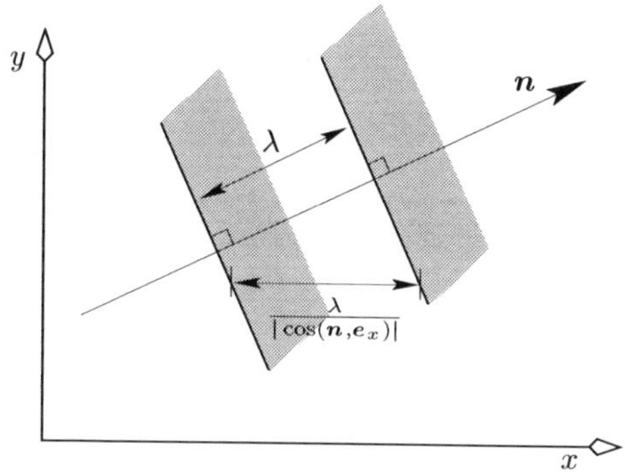

Fig. 2.10. Two phase fronts (surfaces of constant phase) of a three-dimensional plane wave moving along its unit normal $\boldsymbol{n}$. This construction shows why we cannot construct a meaningful wavelength vector as $\boldsymbol{\lambda} \equiv \lambda \boldsymbol{n}$.

$$\frac{\lambda}{|\cos(\boldsymbol{n}, \boldsymbol{e}_x)|}, \tag{2.148}$$

where $\boldsymbol{e}_x$ denotes the unit vector along the x-axis, $(\boldsymbol{n}, \boldsymbol{e}_x)$ denotes the angle between $\boldsymbol{n}$ and $\boldsymbol{e}_x$, and we need to use an absolute value in case this angle is greater than 90°. But contrast (2.148) with the x-component of $\boldsymbol{\lambda}$,

$$\lambda^x = \lambda n^x = \lambda \cos(\boldsymbol{n}, \boldsymbol{e}_x), \tag{2.149}$$

which is *not* the wavelength (2.148) seen on the x-axis! So defining $\boldsymbol{\lambda} \equiv \lambda \boldsymbol{n}$ is problematical and best avoided. On the other hand, the jumping of the cosine from denominator to numerator led us to suspect that it might be better to design a vector based on $1/\lambda$ instead of λ; and this is exactly what the wavenumber $\boldsymbol{k}$ is. For example, the modulus of the wavenumber of the waves along the x-axis is

$$|\boldsymbol{k} \text{ seen on } x\text{-axis}| = \frac{2\pi}{\lambda \text{ on } x\text{-axis}} = \frac{2\pi}{\lambda} |\cos(\boldsymbol{n}, \boldsymbol{e}_x)| = |k \cos(\boldsymbol{n}, \boldsymbol{e}_x)| = |k^x| . \tag{2.150}$$

This is just what we might reasonably expect the x-component of a vector to be: the length of its projection on the x-axis. So here we have a first reason that the vector wavenumber $\boldsymbol{k}$ is so useful: because it accords with our intuition. In Chap. 6 we will encounter a more compelling reason: $\boldsymbol{k}$ pairs very naturally with ω to form a new quantity with new and useful properties.

Similarly, it makes no sense to define a vector $\boldsymbol{\omega} = \omega \boldsymbol{n}$ because the frequency seen on the x-axis is ω, not ωn^x. Again, on the same note, it's also not necessarily a good idea to define a velocity for the wave as $\boldsymbol{v} = v\boldsymbol{n}$ since this has x-component vn^x, while the velocity of the wave as seen on the x-axis is v/n^x. (If this is not clear, remember that the velocity of the wave as seen on the x-axis is the velocity of a wave crest's intersection with that axis. When n^x is very small, the wave is almost parallel to the axis and so this intersection can move arbitrarily quickly.) So while we can always *define* new entities freely, it pays to ensure that new definitions allow us to use our intuition in a useful way. The concept of a wavenumber does just that.

2.9 Deriving the Fourier Transform Using Brackets

Fourier theory is built upon the ideas we have discussed previously: linearity, diagonalisation, and the notions of frequency and wavenumber. It also is naturally expressed using bracket symbolism, which is what we'll show in this section. Fourier theory is concerned with expressing a function as a sum of simple components, each of whose behaviour we are familiar with; so we gain knowledge of the function itself through studying the behaviour of these components.

The starting point for building the Fourier transform is the observation that a function is an entity that exists outside of its functional notation. That is, if $f(x) \equiv \sin x$, and we define a new variable y such that $y = x^2$, then we know it becomes somewhat ambiguous to ask for the value of $f(y)$. Do we

- simply take the sine of the argument, giving $\sin y$, or
- find the value of x that corresponds to y and take the sine of that, giving $\sin \sqrt{y}$?

If by f we simply mean a mechanical procedure that takes the sine of its argument, then we can happily write $f(x) = \sin x$, $f(y) = \sin y$; and this is certainly assumed in mathematics and computer programming. But the meaning given to a function symbol by mathematical physicists is usually a little more subtle. Changes of variables are quite meaningful and common, so we would like the definition of the function to change somehow to be "aware" of this. We could use a new symbol, $\widetilde{f}$, where $\widetilde{f}(y) \equiv f(x) \equiv \sin x = \sin \sqrt{y}$. But now the underlying function has two names: f when its argument is x, and $\widetilde{f}$ when its argument is y. (We did something similar in the box on p. 48, although there the two functions $u(f), \widetilde{u}(\lambda)$ were not actually equal.) More generally applicable names such as $f_y(y) \equiv f_x(x) \equiv \sin x = \sin \sqrt{y}$ are perhaps better and are also used in practice, but they do give the impression of a doubling-up occurring in the notation with x and y.

Often—but not always—physicists will use the single letter f to stand for both forms of the function, and indeed any other instances of it that depend on new variables such as $z = x^2 + x$, etc. There is seldom any confusion. But we have drawn attention to the difficulty because not only can it be fixed by using bracket formalism when occasion demands, but also the use of brackets in this way serves as a natural stepping-off point into the formalism of quantum mechanics.

To apply brackets to the various instances of a function f, associate it with the ket $|f\rangle$. We make a rule that if we prepend this with a bra containing whatever argument we require, then the mechanism of the function will change accordingly. This idea of widening the definition of a function so that it does the right thing depending on what argument it takes is not new. After all, the same symbol "+" is used for addition regardless of whether we are adding numbers or matrices; similarly the exponential of a matrix is completely well-determined without a separate function name being reserved for the exponential with a matrix argument.

With this philosophy in mind, we do away with the multiple names for the various guises of f through defining

$$\langle x|f\rangle \equiv f(x) = \sin x\,, \quad \langle y|f\rangle \equiv \widetilde{f}(y) = \sin\sqrt{y}\,. \tag{2.151}$$

It will also prove useful to define $\langle f|x\rangle$ as the complex conjugate of $\langle x|f\rangle$, so that $\langle f|x\rangle \equiv f^*(x)$.

As an aside, it might be asked why we did not instead define $\langle f|x\rangle$ to equal $f(x)$. One reason is because we are following long-established quantum mechanical formalism, and this formalism, too, must be read from right to left. For example, we have already mentioned the quantum mechanical expression $\langle f|i\rangle$, denoting the amplitude for a system to be found in some final state $|f\rangle$ given that it was originally in some initial state $|i\rangle$. Another reason is that the same convention is followed in probability theory. In Sect. 3.3 we'll see that the probability for some final event f to occur, given some initial event i, is denoted $p(f|i)$ or simply $(f|i)$.

Dirac brackets are thus containers, and examples of what they can hold are coordinates x, y, eigenvalues m, n, or functions f, g. Because all of these arguments belong to different name spaces, there is no problem with widening the definition of the bras and kets to include them all. Another simple example of this "container" use of brackets is the *spherical harmonics* $Y_{\ell m}(\theta, \phi)$, orthonormal eigenfunctions useful in both classical and quantum mechanics when spherical coordinates are used. Writing a spherical harmonic in "pre-coordinate" form as the vector $|\ell\, m\rangle$ allows its usual spherical polar functional form to be denoted

$$\langle\theta\,\phi\,|\,\ell\, m\rangle \equiv Y_{\ell m}(\theta, \phi)\,, \tag{2.152}$$

which would be $\langle x\, y\, z\,|\,\ell\, m\rangle$ in cartesian coordinates. We will see how spherical harmonics' orthonormality relation can be written later in (2.160).

Let's set about defining various expressions involving brackets. First is the fundamental notion of the inner product for functions, as first seen in Sect. 2.7. Fourier analysis needs well-specified limits:

$$\langle f|g\rangle \equiv \int_{-L}^{L} f^*(x)\, g(x)\, \mathrm{d}x = \int_{-L}^{L} \langle f|x\rangle\langle x|g\rangle\, \mathrm{d}x\,. \tag{2.153}$$

Just as a completeness relation was arrived at in (2.116), we can follow the same ideas here by omitting $\langle f|$ and $|g\rangle$ from (2.153) to write

$$\int_{-L}^{L} |x\rangle\langle x| \, \mathrm{d}x = 1 \,. \tag{2.154}$$

This last identity now allows us to write

$$\langle x|f\rangle = \int_{-L}^{L} \langle x|x'\rangle\langle x'|f\rangle \, \mathrm{d}x' \tag{2.155}$$

and so conclude that

$$\langle x|x'\rangle = \delta(x - x') \,. \tag{2.156}$$

Fourier analysis needs the notion of an orthonormal set of basis functions. As usual, a set of functions $\{f_n(x)\}$ is called orthonormal over the domain $[-L, L]$ if $\langle f_m|f_n\rangle = \delta_{mn}$. This set should also be *complete*, meaning that any "well-behaved" function can be written as a linear combination over this set. The restriction to well-behaved functions is needed because Fourier analysis can really only be easily applied to *square-integrable* functions, also called *L^2 functions.* To see what these are, define the L^2 *norm* of a function $f(x)$ over $[-L, L]$ to be $\|f\|$, such that

$$\|f\|^2 \equiv \langle f|f\rangle = \int_{-L}^{L} \langle f|x\rangle\langle x|f\rangle \, \mathrm{d}x = \int_{-L}^{L} |f(x)|^2 \, \mathrm{d}x \,. \tag{2.157}$$

(The L in L^2 has nothing to do with the limits of integration.) In that case, a square-integrable function is any function with a finite L^2 norm; i.e., it decreases sufficiently quickly as $x \to \infty$. There are many choices of orthonormal basis sets used in Fourier theory, but we'll focus on just one:

$$\boxed{\text{Our orthonormal basis on } [-L, L] \text{ is } \left\{\phi_n(x) \equiv \frac{1}{\sqrt{2L}} \, \mathrm{e}^{\frac{\mathrm{i}n\pi x}{L}}\right\}_{n=-\infty}^{\infty}} \tag{2.158}$$

These functions are easily shown to be orthonormal on $[-L, L]$ by carrying out the integration. And although we will not show it here, this set of functions is complete as a basis for square-integrable functions.

For brevity, write these basis functions as $\langle x|n\rangle \equiv \langle x|\phi_n\rangle$, in which case we can express the orthonormality of the basis set using (2.154) as

$$\langle m|n\rangle = \int_{-L}^{L} \langle m|x\rangle\langle x|n\rangle \, \mathrm{d}x = \int_{-L}^{L} \phi_m^*(x) \, \phi_n(x) \, \mathrm{d}x = \delta_{mn} \,, \tag{2.159}$$

a rule that is just the same as (2.112)!

On the same note, the orthonormality relation for the spherical harmonics of (2.152) can be written neatly as

$$\langle \ell \, m \,|\ell' m'\rangle \equiv \int_0^{2\pi} \mathrm{d}\phi \int_0^{\pi} \mathrm{d}\theta \, \sin\theta \, \langle \ell \, m \,|\, \theta \, \phi\rangle\langle \theta \, \phi \,|\, \ell' m'\rangle = \delta_{\ell\ell'} \, \delta_{mm'} \,. \tag{2.160}$$

Why Sinusoids?

Why should we prefer sinusoids to describe waves when any other basis set would be just as valid? Second-order linear differential equations are ubiquitous in models of physical systems, and these often have sinusoidal eigenfunctions. Sinusoids are easy to deal with mathematically, but there is also a physical reason, which can be seen by examining resonance in an electrical circuit. This is fundamentally sinusoidal, so that the excitation of an antenna forms a useful picture. The space around us carries just one vastly complicated electromagnetic field, but this one field contains information from the very lowest frequencies to the very highest; from the one-off movement of a charged piece of amber to the arrival of the highest-energy cosmic ray photons. In between, there are radio and television broadcasts with whose frequency any simple circuit can be set to resonate. So the sinusoidal basis does have a reality; physically, the antenna circuit is resonating with, or picking out, a sine component of the field. In that sense, the capacitance being altered as we tune a radio dial is acting as a very physical projection operator onto the sinusoidal basis.

Now, what meaning can be given to $\langle n|f\rangle$? Again make use of the completeness relation (2.154):

$$\langle n|f\rangle = \int_{-L}^{L} \langle n|x\rangle\langle x|f\rangle \, \mathrm{d}x = \int_{-L}^{L} \phi_n^*(x)\, f(x)\, \mathrm{d}x\,. \tag{2.161}$$

But this last expression is just the nth coefficient of $f(x)$ over the basis (which is trivial to show), or in other words

$$\langle x|f\rangle = \sum_{n=-\infty}^{\infty} \langle x|n\rangle\langle n|f\rangle\,. \tag{2.162}$$

With $f(x)$ being arbitrary, it follows that a new completeness relation must hold:

$$\sum_{n=-\infty}^{\infty} |n\rangle\langle n| = 1\,, \tag{2.163}$$

which by now might not be too surprising. Simple though it might appear, (2.163) encapsulates Fourier analysis and shows how that subject really resembles matrix multiplication. This might come as no surprise, because both subjects are concerned with projecting vectors or functions onto bases. We see why an identity such as (2.163) is called a completeness relation: because it expresses the completeness of the basis $\{|n\rangle\}$ that enables an arbitrary function to be expanded over that basis.

So far, what we have written is all just notation, and it must prove itself useful if we are to give it any real weight. The bracket formalism may well find

Table 2.2. Definitions of bracket entities.

$\lvert f\rangle \equiv$	the unique entity corresponding to a function f.
$\langle x\vert f\rangle \equiv$	$f(x)$ (with the caveat discussed above for another variable y).
$\lvert n\rangle \equiv \lvert\phi_n\rangle =$	a Fourier basis function, where $\langle x\vert n\rangle$ is defined in (2.158).
$\langle f\vert g\rangle \equiv$	$\int_{-L}^{L} f^*(x)\, g(x)\, \mathrm{d}x$.
$\langle a\vert b\rangle \equiv$	$\langle b\vert a\rangle^*$ regardless of what a and b are.

its major use in quantum mechanics, but it has been introduced here in the context of Fourier theory as a way of showing just how brackets work. Later we'll see how it can be used to dovetail quantum mechanics with Fourier theory. But for now, we demonstrate just how useful the formalism is by reproducing the Fourier transform using brackets.

To do this, begin by applying the two completeness relations (2.154) and (2.163) to project a square-integrable function $f(x)$ onto the $\phi_n(x)$ basis:

$$\begin{aligned} f(x) = \langle x|f\rangle &= \sum_n \langle x|n\rangle\, \langle n|f\rangle \\ &= \sum_{n=-\infty}^{\infty} \frac{1}{\sqrt{2L}}\, \mathrm{e}^{\frac{\mathrm{i}n\pi x}{L}}\, \langle n|f\rangle\,. \end{aligned} \tag{2.164}$$

The function $f(x)$ has been transformed to a new one, $\langle n|f\rangle$. (This new function might sometimes be called $\widetilde{f}(n)$ in texts, but now we see the simplicity of using the $|f\rangle$ formalism.) What is the inverse transform?

$$\begin{aligned} \langle n|f\rangle &= \int_{-L}^{L} \langle n|x\rangle\, \langle x|f\rangle\, \mathrm{d}x \\ &= \int_{-L}^{L} \frac{1}{\sqrt{2L}}\, \mathrm{e}^{\frac{-\mathrm{i}n\pi x}{L}}\, f(x)\, \mathrm{d}x\,. \end{aligned} \tag{2.165}$$

The completeness relation of bracket notation expresses the Fourier integral and its inverse very concisely, and the necessary complex conjugation of the basis functions comes about automatically. The relevant expressions used here, plus others we'll use later, have been summarised in Table 2.3.

Normally, we wish to apply the Fourier transform to a function over the whole x-axis, not just the interval $[-L, L]$. This requires considering the limit $L \to \infty$. To do so, first combine (2.164) with (2.165):

$$f(x) = \sum_{n=-\infty}^{\infty} \frac{1}{2L}\, \mathrm{e}^{\frac{\mathrm{i}n\pi x}{L}} \int_{-L}^{L} \mathrm{e}^{\frac{-\mathrm{i}n\pi x'}{L}}\, f(x')\, \mathrm{d}x'\,. \tag{2.166}$$

Table 2.3. Useful Identities of Fourier Theory.

$\langle x|f\rangle \equiv f(x)\,, \quad \langle x|f,t\rangle \equiv f(x,t)\,.$

$\langle x|n\rangle \equiv \langle x|\phi_n\rangle = \dfrac{1}{\sqrt{2L}} \exp \dfrac{\mathrm{i}n\pi x}{L}\,.$

$\langle x|k\rangle \equiv \sqrt{\dfrac{|\alpha|}{2\pi}}\, \mathrm{e}^{\mathrm{i}\alpha kx}$ for any real α.

$\langle f|g\rangle \equiv \displaystyle\int_{-L}^{L} f^*(x)\, g(x)\, \mathrm{d}x\,.$

$\langle a|b\rangle \equiv \langle b|a\rangle^*$ regardless of what a and b are.

$\langle x|x'\rangle = \delta(x-x')\,, \quad \langle m|n\rangle = \delta_{mn}\,.$

$\displaystyle\int_{-L}^{L} |x\rangle\langle x|\, \mathrm{d}x = 1\,, \quad \sum_{n=-\infty}^{\infty} |n\rangle\langle n| = 1\,, \quad \int_{-\infty}^{\infty} |k\rangle\langle k|\, \mathrm{d}k = 1\,.$

Suppose we make a change of variables in the summation:

$$k \equiv \frac{n\pi}{L}\,, \text{ so that } \Delta k = \frac{\pi}{L}\Delta n = \frac{\pi}{L}\,. \tag{2.167}$$

Then (2.166) becomes (leaving n in the summation for now)

$$f(x) = \sum_{n=-\infty}^{\infty} \frac{1}{2\pi}\,\Delta k\; \mathrm{e}^{\mathrm{i}kx} \int_{-L}^{L} \mathrm{e}^{-\mathrm{i}kx'} f(x')\, \mathrm{d}x'\,. \tag{2.168}$$

Now let $L \to \infty$ so that $\Delta k \to 0$ and the sum becomes an integral:

$$f(x) = \int_{-\infty}^{\infty} \frac{\mathrm{d}k}{2\pi}\, \mathrm{e}^{\mathrm{i}kx} \int_{-\infty}^{\infty} \mathrm{e}^{-\mathrm{i}kx'} f(x')\, \mathrm{d}x'\,. \tag{2.169}$$

The equations of Fourier analysis have a certain latitude that allows different conventions to exist. These can all be accommodated by making another change of variables of $k \to \alpha k$ (α real) to write (2.169) as follows:

$$f(x) = \frac{|\alpha|}{2\pi} \int_{-\infty}^{\infty} \mathrm{d}k\; \mathrm{e}^{\mathrm{i}\alpha kx} \int_{-\infty}^{\infty} \mathrm{e}^{-\mathrm{i}\alpha kx'} f(x')\, \mathrm{d}x'\,. \tag{2.170}$$

The main conventions in use are then reproduced by choosing $\alpha = \pm 2\pi$ and $\alpha = \pm 1$. The factor of $|\alpha|/(2\pi)$ can be distributed evenly or otherwise, as two factors, one in front of each integral. This partly corresponds to our defining $\langle x|k\rangle$ as in Table 2.3:

$$\langle x|k\rangle \equiv \sqrt{\frac{|\alpha|}{2\pi}}\, \mathrm{e}^{\mathrm{i}\alpha kx}. \tag{2.171}$$

But (2.170) also allows us to put any factor β in front of the integral in the Fourier transform, provided that $1/\beta$ is also included in front of the integral in the inverse transform.

As a side note, the use of (2.171) gives a meaning to $\int_{-\infty}^{\infty} |k\rangle\langle k|\,\mathrm{d}k$. Begin with

$$\int_{-\infty}^{\infty} \langle f|k\rangle \langle k|g\rangle \,\mathrm{d}k\,, \tag{2.172}$$

insert completeness relations over x and x', and then use (2.171). The result shows that (2.172) can be written as $\langle f|g\rangle$, which means that

$$\int_{-\infty}^{\infty} |k\rangle\langle k|\,\mathrm{d}k = 1\,, \tag{2.173}$$

forming yet another useful completeness relation.

The Fundamental Fourier Identity

If we choose $\alpha = 1$ for simplicity and set $f(x) = \delta(x)$ in (2.170), we obtain the fundamental Fourier identity

$$\int_{-\infty}^{\infty} \mathrm{e}^{ikx}\,\mathrm{d}k = 2\pi\delta(x)\,. \tag{2.174}$$

(Any other choice of α does not give a different identity; the α simply cancels internally.) This is an interesting integral because, strictly speaking, Fourier theory is only valid for square-integrable functions, and a constant is certainly not square integrable. It then must be that the identity (2.174) has invisibly incorporated some sort of limiting procedure. The delta function is a *generalised function* or *functional*, meaning that it is taken as given that another integration will always be carried out with some other square-integrable function, known as a *test function*.

The delta function is not the only functional appearing in Fourier theory. Another comes about by considering the following integral:

$$\int_{0}^{\infty} \mathrm{e}^{ikx}\,\mathrm{d}k = \,? \tag{2.175}$$

We'll have occasion to use this much later when delving into Green function theory in Chap. 11. Its value, whatever that might be (if indeed it exists), should surely incorporate or be related to the delta function in some way, since summing two of these one-sided integrals with an appropriate sign change should yield (2.174) again.

2.10 Commutators and the Indeterminacy Principle

The fundamental Fourier identity (2.174) shows that the delta function and the constant function 1 form a *Fourier pair*, so that the Fourier transform of a spike is a flat, constant function; and vice versa.

Now, just as the Fourier transform establishes a correspondence between position x and wavenumber $k = 2\pi/\lambda$ (being a measure of the spatial rippliness of the wave), it also pairs time t with rotational frequency $\omega = 2\pi f$, a measure of the wave's temporal rippliness. We can appreciate this by asking what frequencies are present in a sound, using (2.174) as a guide to consider two extreme cases of the noisy collision of two objects. When two hard marbles collide, the audible "clack" means that the sound wave $f(t)$ is approximately a delta function. The spread of frequencies $f(\omega)$ produced by the marbles must then be a constant, the Fourier transform of $f(t)$. So the clack of the bouncing marbles is composed of a flat spectrum that has, at least in principle, equal amounts of all frequencies. In practice, the very highest frequencies are unphysical and not present, which means that $f(\omega) \to 0$ for $|\omega| \to \infty$. In hindsight, we might well expect "nearly" all frequencies to be present, because the tiny *dwell time*, or time in contact, of the marbles ensures that only the very highest frequencies of their ringing surfaces will be suppressed during that time.

On the other hand, the collision of two fuzzy objects is anything but a delta function in time, since their long dwell time dampens out all but the very lowest frequencies. In the limit of a very long dwell time, the only surviving frequency is the zero frequency, so that $f(\omega)$ is approximately a delta function. (Thus, the relative amount of the zero frequency present is $\delta(\omega)\, d\omega$, which is 1 for $\omega = 0$, as expected.) In that case, the sound wave $f(t)$ produced is a constant. In that sense it does not really represent a sound at all; it's merely the infinite-wavelength limit of a sound. The limiting-case assumption that only the zero frequency is present is somewhat unphysical, but we see that the resulting sound "wave" is quite flat in time, which is reasonable.

These two cases are actually just the extremes of a more general principle applying to a function and its Fourier transform, in that when one is very localised, the other is spread out. Its application to quantum mechanics is known as the *Heisenberg Uncertainty Principle*, or sometimes the *Heisenberg Indeterminacy Principle*. "Indeterminacy" is perhaps a better title than "Uncertainty" because the principle does *not* refer to some kind of uncertainty that we are obliged to have about an otherwise well-defined property of a system. Rather, the principle refers to properties of a system that are not well defined in the first place. The standard interpretation of quantum mechanics uses the idea that a system might not possess a well-defined value for some quantity of interest until we make a measurement of that quantity. But that discussion can be left until Sect. 10.9. In the meantime, let's see how the same principle also applies to very classical waves. In the process, we will create some new formalism that becomes very useful in quantum mechanics, as well as in more advanced calculations involving linearity in general.

The duality of a function's being localised in one variable while spread out in the Fourier-partner variable is more commonly discussed with wavenumber k and spatial coordinate x, as opposed to the ω and t of the last few paragraphs. We'll follow suit here. If a function has a large *bandwidth*, meaning it is composed of many basis plane waves over a large range of wavenumbers, then the various peaks and troughs can be expected to cancel over large tracts of space, leaving the function nonzero only in very widely spaced regions. Including *all* wavenumbers ensures there is always a trough to balance a peak, so that the function becomes a spike at the origin, the only place where the Fourier basis functions always add constructively. Here we wish to quantify how some useful measure of bandwidth is related to how localised the function is.

To do so, we require a measure of the spatial width of the function. Such an idea is developed to a high degree in statistical theory as the *standard deviation*, examined in great detail in the next chapter. So begin with a normalised function $f(x)$; normalisation makes the notation cleaner. Choosing some probability distribution, define its width to be the standard deviation σ_x of the values of x in its domain. That is, σ_x is the root-mean-squared deviation of x from the mean $\bar{x}$ of x:

$$\sigma_x^2 \equiv \left\langle \Delta x^2 \right\rangle \equiv \left\langle (x - \bar{x})^2 \right\rangle , \tag{2.176}$$

where the deviation of x from its mean is $\Delta x \equiv x - \bar{x}$, and $\langle \cdot \rangle$ denotes the mean (so that $\bar{x} \equiv \langle x \rangle$, but $\bar{x}$ is slightly simpler to write in these expressions). Anticipating a result in quantum mechanics, we'll define the appropriate probability distribution needed to calculate the mean in (2.176) as $|f(x)|^2$, since in general $f(x)$ might be complex.

A mean can be expressed in bracket notation by defining operators with continuous spectra of eigenvalues that are just x and k:

$$\widehat{x}\,|x\rangle \equiv x\,|x\rangle\,, \quad \widehat{k}\,|k\rangle \equiv k\,|k\rangle\,. \tag{2.177}$$

These particular operators are hermitian since they have real eigenvalues. Using them, the mean of x can be written as

$$\begin{aligned}\bar{x} = \langle x \rangle &\equiv \int_{-\infty}^{\infty} x\,|f(x)|^2\,\mathrm{d}x = \int x\,\langle f|x\rangle\,\langle x|f\rangle\,\mathrm{d}x \\ &= \int \langle f|\,\widehat{x}\,|x\rangle\,\langle x|f\rangle\,\mathrm{d}x = \langle f|\,\widehat{x}\,|f\rangle\,. \end{aligned} \tag{2.178}$$

Similarly, $\left\langle x^2 \right\rangle = \left\langle f \middle| \,\widehat{x}^2\, \middle| f \right\rangle$ and so on, while the same expressions also hold for k. The variances in x and k become

$$\sigma_x^2 = \left\langle \Delta x^2 \right\rangle = \left\langle f \middle| \Delta\widehat{x}^2 \middle| f \right\rangle, \quad \sigma_k^2 = \left\langle \Delta k^2 \right\rangle = \left\langle f \middle| \Delta\widehat{k}^2 \middle| f \right\rangle. \tag{2.179}$$

Using bracket notation is more than just a convenient shorthand for the integrations used to calculate the mean. In particular, σ_x^2 is actually an inner product of $\Delta\widehat{x}\,|f\rangle$ with its hermitian conjugate, and similarly for σ_k^2. This fact invokes the *Cauchy–Schwarz inequality* of linear algebra, which finds many applications in mathematical physics. The Cauchy–Schwarz inequality says that for any vectors $\boldsymbol{\alpha}, \boldsymbol{\beta}$,

$$\langle\boldsymbol{\alpha}|\boldsymbol{\alpha}\rangle \langle\boldsymbol{\beta}|\boldsymbol{\beta}\rangle \geqslant |\langle\boldsymbol{\alpha}|\boldsymbol{\beta}\rangle|^2 \tag{2.180}$$

(with strict equality if and only if $\boldsymbol{\alpha}$ and $\boldsymbol{\beta}$ are linearly dependent). The proof of the inequality is actually very simple, but we will not stop to look at it because it's almost identical to the next chapter's proof of a result in the theory of cross correlation: (3.134)–(3.136).

Setting $|\boldsymbol{\alpha}\rangle = \Delta\widehat{x}\,|f\rangle$ and $|\boldsymbol{\beta}\rangle = \Delta\widehat{k}\,|f\rangle$, the Cauchy–Schwarz inequality gives the following relation between the widths of a function and its Fourier transform:

$$\langle f|\Delta\widehat{x}^2|f\rangle\,\langle f|\Delta\widehat{k}^2|f\rangle \geqslant \left|\langle f|\Delta\widehat{x}\,\Delta\widehat{k}|f\rangle\right|^2, \tag{2.181}$$

or in other words

$$\sigma_x^2\,\sigma_k^2 \geqslant \left|\left\langle\Delta x\,\Delta k\right\rangle\right|^2. \tag{2.182}$$

What are we to make of this?

As an aside for the statistics discussion of the next chapter, $\left\langle\Delta x\,\Delta k\right\rangle$ is actually the *covariance* of x and k, also written σ_{xk} (with no exponent 2).

In general, the value of $\left\langle\Delta x\,\Delta k\right\rangle$ depends on f. But it turns out that $\left\langle\Delta x\,\Delta k\right\rangle$ can be split into two parts: one that does not depend on f and another that does—but which can be shown to be zero for one type of function (a gaussian). Hence the part that does not depend on f forms the absolute lower bound for the product $\sigma_x\sigma_k$. This part has to do with *commutator brackets* $[A, B] \equiv AB - BA$, which measure the extent to which the ordering of two operators is important. Together with commutator brackets go *anticommutator braces* $\{A, B\} \equiv AB + BA$, and it's not hard to see that

$$\begin{aligned}\Delta\widehat{x}\,\Delta\widehat{k} &= \frac{1}{2}\left[\Delta\widehat{x},\,\Delta\widehat{k}\right] + \frac{1}{2}\left\{\Delta\widehat{x},\,\Delta\widehat{k}\right\} \\ &= \underbrace{\frac{1}{2}\left[\widehat{x},\,\widehat{k}\right]}_{\text{antihermitian}} + \underbrace{\frac{1}{2}\left\{\Delta\widehat{x},\,\Delta\widehat{k}\right\}}_{\text{hermitian}},\end{aligned} \tag{2.183}$$

where the second line follows from the first by using $\Delta\widehat{x} = \widehat{x} - \bar{x}$ (and similarly for k) and then expanding the brackets. That the operators are antihermitian and hermitian can be shown by, e.g.,

$$\left[\widehat{x},\,\widehat{k}\right]^\dagger = \left(\widehat{x}\,\widehat{k} - \widehat{k}\,\widehat{x}\right)^\dagger = \widehat{k}\,\widehat{x} - \widehat{x}\,\widehat{k} = -\left[\widehat{x},\,\widehat{k}\right], \tag{2.184}$$

with only a marginally different calculation to show that $\left\{\Delta\widehat{x},\,\Delta\widehat{k}\right\}$ is hermitian. The reason that $\left\langle\Delta x\,\Delta k\right\rangle$ has been split in this way, quite apart from the fact that the commutator brackets turn out to give the constant contribution referred to a moment ago, is that because antihermitian operators have pure imaginary eigenvalues, they must give pure imaginary means; and because hermitian operators have real eigenvalues, they must give real means.

Alternatively, these results can be seen by examining a general hermitian operator $\widehat{H}$ belonging to some variable H, through writing

$$\langle H\rangle^* = \langle f|\,\widehat{H}\,|f\rangle^* = \langle f|\,\widehat{H}^\dagger\,|f\rangle = \langle f|\,\widehat{H}\,|f\rangle = \langle H\rangle\,, \tag{2.185}$$

so that $\langle H\rangle$ is real. A similar argument shows that for any antihermitian operator $\widehat{A}$, the mean $\langle A\rangle$ is pure imaginary.

These ideas prove useful when we take the mean of (2.183), which is needed for (2.182):

$$\begin{aligned}\langle\Delta x\,\Delta k\rangle &= \langle f|\Delta\widehat{x}\,\Delta\widehat{k}\,|f\rangle\\ &= \underbrace{\frac{1}{2}\langle f|[\widehat{x},\,\widehat{k}\,]|f\rangle}_{\text{pure imaginary}} \;+\; \underbrace{\frac{1}{2}\langle f|\{\Delta\widehat{x},\,\Delta\widehat{k}\,\}|f\rangle}_{\text{real}}\,.\end{aligned} \tag{2.186}$$

This useful separation lets us use Pythagoras's theorem to calculate the absolute value:

$$|\langle\Delta x\,\Delta k\rangle|^2 = \frac{1}{4}|\langle f|[\widehat{x},\,\widehat{k}\,]|f\rangle|^2 + \frac{1}{4}|\langle f|\{\Delta\widehat{x},\,\Delta\widehat{k}\,\}|f\rangle|^2\,. \tag{2.187}$$

The first term in this sum will be the most important—what is $\langle f|[\widehat{x},\,\widehat{k}\,]|f\rangle$?

$$\langle f|[\widehat{x},\,\widehat{k}\,]|f\rangle = \langle f|\,\widehat{x}\,\widehat{k} - \widehat{k}\,\widehat{x}\,|f\rangle\,. \tag{2.188}$$

Evidently, to proceed we will need to expand the last line using completeness relations for x and/or k. It suffices to choose one basis, or *representation* (the term used in quantum mechanics). Let's choose the x-basis. Expressions such as $\langle x|\widehat{x}|f\rangle$ are no problem, since with $\widehat{x}$ hermitian and thus having real eigenvalues,

$$\langle x|\,\widehat{x}\,|f\rangle = \langle f|\,\widehat{x}\,|x\rangle^* = x\,\langle f|x\rangle^* = x\,\langle x|f\rangle = x\,f(x)\,. \tag{2.189}$$

But what about an expression such as $\langle x|\,\widehat{k}\,|f\rangle$, which will be used to evaluate (2.188)? Remember to read these brackets from right to left using the ideas behind expressions such as (2.151). For example, $\langle x|\,\widehat{x}\,|f\rangle$ tells us to act on a ket $|f\rangle$ with $\widehat{x}$, and then project the resulting ket onto x-space. The result gives the *x-representation* of the operator $\widehat{x}$, which is now an operator on the function $f(x)$—and from (2.189) we see that the x-representation of $\widehat{x}$ is just multiplication by x. Be aware that there are two types of operators here: one acts on a ket, while the other acts on a function. An expression such as (2.189) takes us from one to the other. The distinction between these two operator types becomes more interesting when we ask for the representation of an operator in a different basis. In particular,

$$\langle x|\,\widehat{k}\,|f\rangle = \int \langle x|k\rangle\,\langle k|\,\widehat{k}\,|f\rangle\,\mathrm{d}k \overset{(2.171)}{=\!=\!=} \int\sqrt{\frac{|\alpha|}{2\pi}}\,\mathrm{e}^{\mathrm{i}\alpha kx}\,k\,\langle k|f\rangle\,\mathrm{d}k$$

$$\left\langle x \left| \widehat{x} \left| f \right\rangle = x \left\langle x | f \right\rangle \right. \right. \qquad \left\langle x \left| \widehat{k} \left| f \right\rangle = \frac{-\mathrm{i}}{\alpha} \frac{\mathrm{d}}{\mathrm{d}x} \left\langle x | f \right\rangle \right. \right.$$

$$`` \; \widehat{x} \; \xrightarrow{x\text{-rep.}} \; x \; " \qquad `` \; \widehat{k} \; \xrightarrow{x\text{-rep.}} \; \frac{-\mathrm{i}}{\alpha} \frac{\mathrm{d}}{\mathrm{d}x} \; "$$

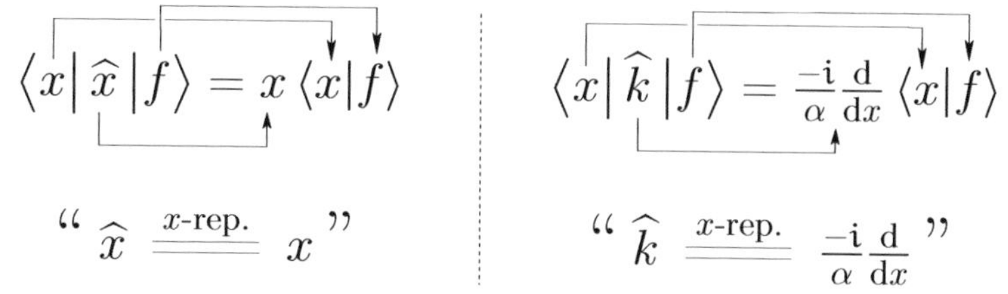

Fig. 2.11. The distinction between an operator in ket space and its representation over some basis in function space. This representation is equivalent to the operator's coordinate vector of p. 9 over that basis. **Left**: The $\widehat{x}$-operator's x-representation is just multiplication by x. **Right**: In contrast, the $\widehat{k}$-operator's x-representation is a differentiation. Alternative language relating each pair of operators is shown in the second line of the figure.

$$= \frac{-\mathrm{i}}{\alpha} \frac{\mathrm{d}}{\mathrm{d}x} \int \sqrt{\frac{|\alpha|}{2\pi}} \, e^{\mathrm{i}\alpha k x} \left\langle k | f \right\rangle \mathrm{d}k \;\; = \frac{-\mathrm{i}}{\alpha} \frac{\mathrm{d}}{\mathrm{d}x} \int \left\langle x | k \right\rangle \left\langle k | f \right\rangle \mathrm{d}k$$

$$= \frac{-\mathrm{i}}{\alpha} \frac{\mathrm{d}}{\mathrm{d}x} f(x) . \tag{2.190}$$

We say "the x-representation of $\widehat{k}$ is $-\mathrm{i}/\alpha \; \mathrm{d}/\mathrm{d}x$". The distinction between the two types of operators is very important and is shown again in Fig. 2.11.

It's worth pointing out a possible source of confusion with these operators. The generic function f had a passive role as a kind of spectator in (2.190), and as a way of shortening the notation, it might be removed from the expression $\left\langle x | \widehat{x} | f \right\rangle = x \left\langle x | f \right\rangle$ and its complex conjugate $\left\langle f | \widehat{x} | x \right\rangle = x \left\langle f | x \right\rangle$ to produce, respectively,

$$\left\langle x | \widehat{x} = x \left\langle x \right| , \qquad \widehat{x} \left| x \right\rangle = x \left| x \right\rangle . \tag{2.191}$$

This is certainly done. The same procedure gives the analogous expression for $\widehat{k}$ in the x-representation, but this requires a little more attention. Removing f from $\left\langle x | \widehat{k} | f \right\rangle = -\mathrm{i}/\alpha \; \mathrm{d}/\mathrm{d}x \left\langle x | f \right\rangle$ and its complex conjugate $\left\langle f | \widehat{k} | x \right\rangle = +\mathrm{i}/\alpha \; \mathrm{d}/\mathrm{d}x \left\langle f | x \right\rangle$ produces

$$\left\langle x | \widehat{k} = \frac{-\mathrm{i}}{\alpha} \frac{\mathrm{d}}{\mathrm{d}x} \left\langle x \right| , \qquad \widehat{k} \left| x \right\rangle = \frac{+\mathrm{i}}{\alpha} \frac{\mathrm{d}}{\mathrm{d}x} \left| x \right\rangle . \tag{2.192}$$

The derivative operator is not really acting on the bra and ket; such an idea makes no sense since differentiation is only defined for functions, not bras or kets. The expressions in (2.192) are really just a way of indicating the x-representation of $\widehat{k}$—they are waiting for a function f to be introduced in the form of $\left| f \right\rangle$ or $\left\langle f \right|$. Note also that the x-representation of $\widehat{k}$ as written in Fig. 2.11 refers to the $\left\langle x | \widehat{k} \right.$ form, *not* the $\widehat{k} \left| x \right\rangle$ form. The difference is a minus sign, and it's important to be aware of it.

Return now to the task of calculating (2.188). The analysis of the last few paragraphs implies that the operators $\widehat{x}, \widehat{k}$ do not commute, because their

x-representations do not commute. This can be seen in detail by letting these representations act on a function $f(x)$:

(The first line in the following equation is suggestive, but not obvious, and is left as an exercise for the reader to prove. It can be shown by using completeness over x, as well as an integration by parts.)

$$\begin{aligned}\langle x|\left[\widehat{x},\widehat{k}\right]|f\rangle &= \left[x, \frac{-\mathrm{i}}{\alpha}\frac{\mathrm{d}}{\mathrm{d}x}\right] f(x) = \frac{-\mathrm{i}}{\alpha}\, x\, f'(x) + \frac{\mathrm{i}}{\alpha}\frac{\mathrm{d}}{\mathrm{d}x}\Big(x f(x)\Big) \\ &= \frac{\mathrm{i}}{\alpha}\, f(x) = \langle x|\, \frac{\mathrm{i}}{\alpha}\, |f\rangle\,, \end{aligned} \tag{2.193}$$

so that the x-representation of $\left[\widehat{x}, \widehat{k}\right]$ is i/α. (Using completeness, it is straightforward to show that this result does not depend on the representation, and so it's more usual to simply say $\left[\widehat{x}, \widehat{k}\right] = \mathrm{i}/\alpha$.) We can now use this to calculate the lower limit in (2.182) by way of (2.187). We can either use it as an x-representation directly to write

$$\begin{aligned}\langle f|\left[\widehat{x}, \widehat{k}\right]|f\rangle &= \int \langle f|x\rangle\langle x|\left[\widehat{x}, \widehat{k}\right]|f\rangle \,\mathrm{d}x \\ &= \int \langle f|x\rangle\, \frac{\mathrm{i}}{\alpha}\, \langle x|f\rangle \,\mathrm{d}x = \frac{\mathrm{i}}{\alpha}\,, \end{aligned} \tag{2.194}$$

or we could write

$$\langle f|\left[\widehat{x}, \widehat{k}\right]|f\rangle = \langle f|\frac{\mathrm{i}}{\alpha}|f\rangle = \frac{\mathrm{i}}{\alpha}\langle f|f\rangle = \frac{\mathrm{i}}{\alpha}\,. \tag{2.195}$$

Either way, the commutator brackets have turned out to give a mean that is independent of the actual function f used, a result that was forecast just after (2.182). Equations (2.182), (2.187), and (2.194) now give

$$\sigma_x^2\, \sigma_k^2 \geqslant \frac{1}{4\alpha^2} + \tfrac{1}{4}\left|\langle f|\{\Delta\widehat{x}, \Delta\widehat{k}\,\}|f\rangle\right|^2. \tag{2.196}$$

We will not concern ourselves with the second term on the right-hand side of (2.196) because it can easily be shown to vanish when $f(x)$ has a gaussian form, since there the standard deviations of x and k can be directly calculated. Hence

$$\sigma_k\, \sigma_x \geqslant \frac{1}{2|\alpha|}\,, \tag{2.197}$$

with equality for a gaussian function—a result related to gaussians' privileged role in the study of wave functions.

As a side note, what we have called $\sigma_x \equiv \sqrt{\langle\Delta x^2\rangle}$ is usually denoted simply Δx in the dialect of quantum mechanics. This differs from our more generally conventional use of Δx in this section. The required quantity *is* a standard deviation, and for the sake of clarity we have used the usual statistical term σ_x for it.

So the product of the widths of a function when expressed over a spatial coordinate x and wavenumber k has a lower limit greater than zero. The constant α is essentially a scale factor, introduced back in (2.170) to account for different Fourier conventions in use. There is also nothing special about the variables x, k used here. Any Fourier pair could have done just as well, so that for example $\sigma_\omega \sigma_t \geqslant 1/(2|\alpha|)$ as well.

What we have shown here is that the more localised a function is in one variable, the more spread out it is in the dual Fourier variable. This agrees with the limiting case of a spike and a constant seen at the start of this section. This duality of spreading and bunching is central to quantum mechanics, where the function f used in this section becomes the wave function $\Psi(\boldsymbol{x}, t)$.

2.11 Evolving Wave Functions in Time

One of the main reasons that Fourier analysis is so useful is because, like diagonalisation, it allows us to transform a difficult problem to a simpler space. Many linear differential equations in all branches of physics and engineering can be solved by beginning with the trial solution $\mathrm{e}^{\mathrm{i}(kx-\omega t)}$. If this is to satisfy the equation, k and ω will need to be related in some way: the *dispersion relation* that we met earlier. A good example is the *beam equation* known to engineers:

$$\frac{\partial^2 \phi}{\partial t^2} + \gamma^2 \frac{\partial^4 \phi}{\partial x^4} = 0 \,. \tag{2.198}$$

Substituting $\phi = \mathrm{e}^{\mathrm{i}(kx-\omega t)}$ into this yields a dispersion relation $\omega^2 = \gamma^2 k^4$. Another example is the more well-known equation of many waves such as light:

$$\frac{1}{c^2}\frac{\partial^2 \phi}{\partial t^2} - \frac{\partial^2 \phi}{\partial x^2} = 0 \,, \tag{2.199}$$

whose dispersion relation is $\omega^2 = c^2 k^2$. The equations' linearity guarantees that any linear combination of $\mathrm{e}^{\mathrm{i}(kx-\omega t)}$ will also be a solution, and this is precisely what a Fourier decomposition is.

We are used to applying Newton's second law $\boldsymbol{F} = m\boldsymbol{a}$ in mechanics problems to produce a differential equation that describes a system. In the same way, $\boldsymbol{F} = m\boldsymbol{a}$ produces a wave equation for a mechanical system with wave properties, such as a taut string. Beginning approaches to the wave equation sometimes reverse this flow of logic, starting from a sine wave and showing that such a wave has the property expressed in (2.199), which is then called a wave equation; but just why we should want to start with the solution to an equation and then work backward to the equation itself might be left unstated. What should be remembered is that the wave equation is always the starting point, since it results from some physical model.

Setting $t = 0$ in such a linear combination of $\mathrm{e}^{\mathrm{i}(kx-\omega t)}$ yields the x-space Fourier transform of the wave function at the initial time, implying that the

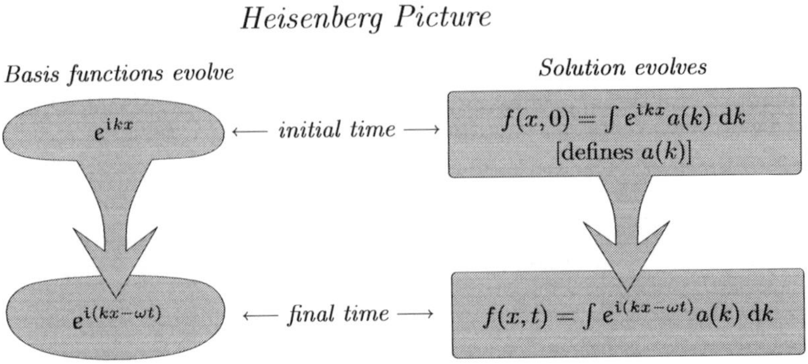

Fig. 2.12. The Heisenberg picture takes the basis functions to evolve in time, while the amount $a(k)$ of each remains constant.

new factor of $e^{-i\omega(k)t}$ for each basis function gives the *time evolution* of that function. This is no different from the diagonalisation of Sect. 2.5. If the matrix A of that section denotes the evolution of a complicated solution to a wave equation, then writing it as PDP^{-1} (and reading this matrix from right to left) corresponds exactly to our Fourier decomposition, evolution, and recomposition:

1. $\boldsymbol{P^{-1}}$ signals going back from the usual laboratory (x-space) to diagonal space (k-space); that is, decomposing the solution at the initial time into basis functions e^{ikx}.
2. **Diagonal matrix** $\boldsymbol{D}$ denotes the "easy" evolution in this space: just multiply each basis function by $e^{-i\omega(k)t}$. (What could be easier?)
3. Finally, $\boldsymbol{P}$ corresponds to Fourier transforming back to the laboratory to rebuild our function, now evolved forward in time as required.

There are actually two ways to view this diagonalisation, known in quantum mechanics as the *Heisenberg* and *Schrödinger* pictures. The view we took in the previous paragraph was the Heisenberg picture shown in Fig. 2.12, where the basis functions e^{ikx} evolve while the amplitude density of each one, $a(k)$, does not. An alternative view is to regard the basis vectors as unchanging, and instead associate the evolution factor $e^{-i\omega t}$ with the amplitude density of each one. This is the Schrödinger picture, shown in Fig. 2.13. Of course, the end result is the same. But small though the difference looks, in quantum mechanics choosing one view over the other gives rise to different, fruitful ways of approaching and extending the physics.

The dispersion relation relating ω to k is the guide to Fourier-analysing the solution to an equation. In general, the following steps are taken:

1. Start with a mechanical analysis for the system of interest and derive an appropriate wave equation.

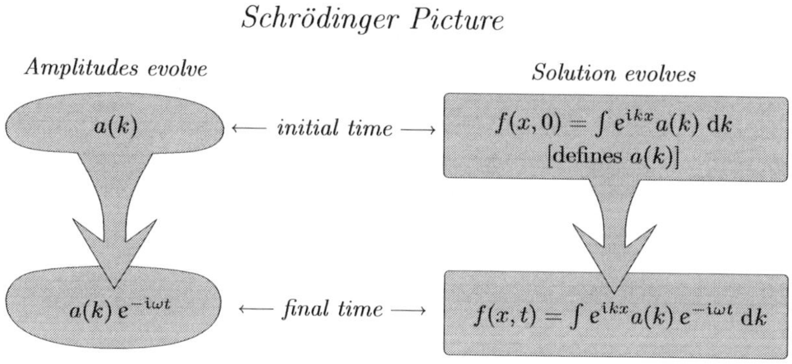

Fig. 2.13. In contrast to the Heisenberg picture, the Schrödinger picture sets the basis functions to be constant, while evolving their amplitudes in time.

2. Substitute a trial solution $e^{i(kx-\omega t)}$ into the wave equation to obtain the dispersion relation. This defines one or more frequencies as functions of wavenumber: say, $\omega_1(k), \ldots, \omega_n(k)$.
3. Superpose these to build a general solution, $\phi(x,t)$, by summing over all ω_j and all allowed values of k. With no physical constraints on the allowed wavenumbers k, all must be used, with infinitesimal weightings $a_j(k)\,dk$:

$$\phi(x,t) = \sum_{j=1}^{n} \int_{-\infty}^{\infty} a_j(k)\, e^{ikx - i\omega_j(k)t}\, dk\,. \tag{2.200}$$

If the physical problem constrains k to discrete values (such as the existence of boundary conditions), the previous expression simplifies to

$$\phi(x,t) = \sum_{j=1}^{n} \sum_{k} a_j(k)\, e^{ikx - i\omega_j(k)t}\,. \tag{2.201}$$

(In this discrete case it's conventional to write a_{jk} for $a_j(k)$ and ω_{jk} for $\omega_j(k)$, but there is nothing deep about this.)

As will be discussed in Sect. 3.7.1, there is no essential difference between the components for positive and negative wavenumbers, but they must both be present. Both signs of k are required as a complete set to build the general wave function solution. Related to this is the fact that complex quantities are often used in physics and engineering to solve what are plainly real-number problems, because complex exponentials can be solutions to real differential equations. They are simple to use, being written as one set (2.158) with just one derivative; whereas if we insisted on using real functions from the outset, we would need to write (2.158) in terms of sines and cosines, as well

Complex Electric Fields?

Complex expressions for the electric field, such as

$$E_y = E_0 \sin k_x x \exp\left[\mathrm{i}\left(\omega t - k_y y\right)\right], \tag{2.202}$$

can look obscure. After all, isn't the field supposed to be real? But (2.202) is really just one simple solution to a *real, linear* wave equation. Such an equation must also have the complex conjugate E_y^* as a solution, since otherwise it would discriminate between i and $-$i, which a real equation cannot do. Thus, we know that any linear combination of E_y and E_y^* must also be a solution. In particular, $\left(E_y + E_y^*\right)/2$ is a solution, and this is none other than the real part of E_y. So the reason that complex-number solutions are routinely used in linear theory is that they are easy to manipulate, and provided everything is linear, two can always be legitimately superposed when a real (physical) quantity is needed—which corresponds to taking the real part.

as writing the derivatives separately (i.e. $\sin \to \cos$, $\cos \to -\sin$). As long as we combine the complex exponentials in such a way as to produce a real solution (which we can certainly do if the wave equation is linear), then it's perfectly acceptable to use a complex approach. Although this is equivalent to "taking the real part of the solution", as it's often described, what might not immediately be apparent is why taking the real part should be a reasonable thing to do at all. What is really happening when we take a real part is that we are superposing two complex solutions to build a real one, and this is a bona fide thing to do if the governing equation is linear.

But some care is required. If the wave function is required to be real, it follows that the sum of the positive and negative frequency components should also be real:

$$a_j(k)\,\mathrm{e}^{\mathrm{i}kx - \mathrm{i}\omega_j(k)t} + a_j(-k)\,\mathrm{e}^{-\mathrm{i}kx - \mathrm{i}\omega_j(-k)t} \quad \text{is real.} \tag{2.203}$$

In this expression, and the resulting one with $t = 0$, examining the real and imaginary parts gives rise to the following two important relations:

$$a_j(-k) = a_j^*(k)\,; \quad \text{this } a_j \text{ is called } \textit{hermitian}, \\ \text{meaning even real and odd imaginary parts;}$$

$$\omega_j(-k) = -\omega_j^*(k)\,; \quad \text{this } \omega_j \text{ is called } \textit{antihermitian}, \\ \text{meaning odd real and even imaginary parts.} \tag{2.204}$$

As an example, for the wave equation (2.199), the dispersion relation $\omega^2 = c^2k^2$ implies that there are two real choices for $\omega(k)$:

$$\omega_1(k) = ck\,, \quad \omega_2(k) = -ck\,. \tag{2.205}$$

These are both odd, consistent with the requirements of (2.204). The full Fourier decomposition is

$$\phi(x,t) = \int_{-\infty}^{\infty} a_1(k)\, \mathrm{e}^{\mathrm{i}k(x-ct)}\, \mathrm{d}k + \int_{-\infty}^{\infty} a_2(k)\, \mathrm{e}^{\mathrm{i}k(x+ct)}\, \mathrm{d}k\,. \tag{2.206}$$

This is consistent with d'Alembert's general solution of the wave equation, which is written as a sum of right- and left-moving waves. If boundary conditions are present whose effect is to quantise the wavenumber, (2.206) becomes

$$\phi(x,t) = \sum_k a_1(k)\, \mathrm{e}^{\mathrm{i}k(x-ct)} + a_2(k)\, \mathrm{e}^{\mathrm{i}k(x+ct)}. \tag{2.207}$$

Is (2.207) reasonable? We can verify that it does indeed give the usual result for a plucked string fixed at both ends. Begin with (2.199), which results from applying Newton's laws to such a string. (The derivation of (2.199) using $\boldsymbol{F} = m\boldsymbol{a}$ is easily found in textbooks, but we also derive it using a lagrangian approach in Chap. 10.) Take the more general expression (2.206) as our solution for the string motion (suspecting nothing as yet about any possible quantisation of k), and produce real functions by converting the integrals to positive wavenumbers only, using the hermiticity of a_1, a_2 in (2.204):

$$\begin{aligned}
\phi(x,t) &= \int_{-\infty}^{\infty} \mathrm{d}k \left[a_1(k)\, \mathrm{e}^{\mathrm{i}k(x-ct)} + a_2(k)\, \mathrm{e}^{\mathrm{i}k(x+ct)}\right] \\
&= \int_0^{\infty} \mathrm{d}k \Big[a_1(k)\, \mathrm{e}^{\mathrm{i}k(x-ct)} + a_1(-k)\, \mathrm{e}^{-\mathrm{i}k(x-ct)} \\
&\qquad + a_2(k)\, \mathrm{e}^{\mathrm{i}k(x+ct)} + a_2(-k)\, \mathrm{e}^{-\mathrm{i}k(x+ct)}\Big] \\
&= \int_0^{\infty} \mathrm{d}k \left[2\,\mathrm{Re}\left\{a_1(k)\, \mathrm{e}^{\mathrm{i}k(x-ct)}\right\} + 2\,\mathrm{Re}\left\{a_2(k)\, \mathrm{e}^{\mathrm{i}k(x+ct)}\right\}\right].
\end{aligned} \tag{2.208}$$

The a_1, a_2 and the complex exponentials can be written in terms of real and imaginary parts. These then multiply to produce all the various sine-cosine combinations, which can then be factored to produce

$$\phi(x,t) = \int_0^{\infty} \mathrm{d}k \left[a_3(k) \sin kx + a_4(k) \cos kx\right] \left[a_5(k) \sin kct + a_6(k) \cos kct\right]. \tag{2.209}$$

Now the string's boundary conditions can be brought in. If one end is fixed at $x = 0$, we can set $a_4 = 0$. Similarly, if the other is fixed at $x = L$, then k can only take the values $k = n\pi/L$ for n in the natural numbers. The integrals, which were only superpositions of solutions, now become sums with noninfinitesimal weightings, and the wave function is

$$\phi(x,t) = \sum_{n=1}^{\infty} \sin \frac{n\pi x}{L} \left[A_n \cos \frac{n\pi ct}{L} + B_n \sin \frac{n\pi ct}{L}\right]. \tag{2.210}$$

For a plucked string, the initial velocity is $\partial\phi(x,0)/\partial t = 0$, which sets all the $B_n = 0$ to give the well-known expression

$$\phi(x,t) = \sum_{n=1}^{\infty} A_n \sin\frac{n\pi x}{L} \cos\frac{n\pi ct}{L} . \tag{2.211}$$

Of course, the wave function above is usually derived using a separation of variables approach, which is certainly simpler for this case of a plucked string. Here we have followed a more roundabout approach that brings together the more general ideas of Fourier analysis via bracket notation. We'll return to the wave function (2.211) for a plucked string in Chap. 10, when we use it as a simple starting point in discussing the quantisation of field theories.

The hermitian/antihermitian nature (2.204) of the weightings and frequencies often seems mysterious when first encountered, especially in the light of dispersion relations that do not seem to fulfill them. An example is the dispersion relation of water waves, $\omega^2 = gk \tanh kh$, where g is the gravitational acceleration and h is the water depth. For deep water, in which the hyperbolic tangent tends to one, this is sometimes rewritten as $\omega = \sqrt{gk}$. But for the exponential basis functions that we are using, this expression does not tell the whole story. Rather, ω has two functional forms: $\omega_\pm = \pm\sqrt{g|k|}\,\operatorname{sgn} k$.

The function sgn is conventionally pronounced "signum". The pronunciation "sign" sometimes encountered only clashes with the sine function.

However, if we rewrite the complex exponentials in terms of sines and cosines, then the integrals over negative k values can be converted to integrals over positive k values. In that case, it *will* be sufficient to write the dispersion relation for water as $\omega = \sqrt{gk}$. But the fact that this reasoning or process is being followed needs to be kept in mind to give meaning to the new, simplified dispersion relation.

On the other hand, in quantum mechanics the wave function solution to the Schrödinger equation need not be real, so we can drop the requirement for a hermitian amplitude and antihermitian frequency.

2.11.1 Brackets and Wave Function Evolution

Given that the theory of evolving wave functions is quite linear, it comes as no surprise to find that brackets describe it very well. First, we need a new definition from Table 2.3:

$$\langle x|f,t\rangle \equiv f(x,t) . \tag{2.212}$$

To describe wave function theory using brackets, begin with the function to be evolved and define the initial amplitudes of the basis functions by writing the initial wave function as

$$f(x,t=0) \equiv \langle x|f,0\rangle = \int \langle x|k\rangle \, \langle k|f,0\rangle \, dk . \tag{2.213}$$

The amplitude of the basis function $\langle x|k\rangle$ is $\langle k|f,0\rangle$. Using a mode of speech that will be useful later when we look at quantum mechanics, we might read this from right to left (as usual) as "the amplitude that the initial wave function $|f,0\rangle$ will have a wavenumber k". (The word "amplitude" here is used grammatically in a similar sense to the word "probability".)

Instead of speaking of the amplitude $\langle k|f,0\rangle$, it is more correct to say amplitude density. The true amplitude is how much of each basis function $\langle x|k\rangle$ there is, which is an infinitesimal: $\langle k|f,0\rangle\,\mathrm{d}k$. The density of this over wavenumber is then $\langle k|f,0\rangle$. But there's no ambiguity in dropping the word density, and this is always done.

This Fourier decomposition of the initial function can now be evolved forward in time using either Fig. 2.12 or Fig. 2.13:

$$\langle x|f,t\rangle = \int \langle x|k\rangle\,\langle k|f,0\rangle\;\mathrm{e}^{-\mathrm{i}\omega(k)t}\,\mathrm{d}k\,. \tag{2.214}$$

The $\langle x|$ common to both sides of this equation really has just a passive role, so it can be omitted to yield an equation relating kets alone. But due to the presence of $\omega(k)$, we cannot go further to remove the integral over k here. However, in the Schrödinger picture we certainly *can* establish the existence of a linear operator $\widehat{L}$ that acts on a ket to produce a new ket whose amplitude to have wavenumber k is

$$\left\langle k\right|\widehat{L}\left|f,0\right\rangle \equiv \langle k|f,0\rangle\;\mathrm{e}^{-\mathrm{i}\omega(k)t}\,, \tag{2.215}$$

so that the k-representation of $\widehat{L}$ is multiplication by $\mathrm{e}^{-\mathrm{i}\omega(k)t}$. Now (2.214) can be rewritten as

$$|f,t\rangle = \int |k\rangle\left\langle k\right|\widehat{L}\left|f,0\right\rangle\mathrm{d}k = \widehat{L}\,|f,0\rangle\,. \tag{2.216}$$

The evolution of a wave function has been encapsulated as $|f,t\rangle = \widehat{L}\,|f,0\rangle$. But is this new operator $\widehat{L}$ anything more than just an idle piece of notation? After all, it depends completely on the wave equation being solved. This question finds a place in the transition from a classical wave theory to quantum mechanics.

2.12 The Transition to Quantum Mechanics

The early history of quantum mechanics was built upon several disparate ideas, all of which were unified by postulating a new way in which the world seems to behave. Central to these new postulates was Louis de Broglie's idea in the early 1920s that particles are associated with, or perhaps in a sense built from, waves, and a plane wave with wavenumber $\boldsymbol{k}$ carries a momentum $\boldsymbol{p} = \hbar\boldsymbol{k}$, where $\hbar$ is a fundamental constant of nature. This plane wave also has an energy of $E = \hbar\omega$. These two equations are related, for in Chap. 6 we'll see that energy and momentum can be joined together naturally to form a new entity, a "four-vector", as can frequency and wavenumber, giving a correspondence between four-vectors: $(E, \boldsymbol{p}) = \hbar(\omega, \boldsymbol{k})$.

Equations (2.215) and (2.216) lend themselves to the language of quantum mechanics, in which the state of a system is denoted by a ket $|\Psi, t\rangle$ that evolves deterministically. Our aim is to find an expression for the operator $\widehat{L}$ in (2.215) and (2.216) that describes this evolution.

Just as Fourier theory builds an arbitrary wave by superposing plane waves, by de Broglie's postulate the quantum mechanical waves associated with particles can be considered as linear superpositions of plane waves, each of which has a well-defined wavenumber $\boldsymbol{k}$ and energy $E = \hbar\omega(\boldsymbol{k})$. Using (2.215), suppose then that we conjecture $\widehat{L}$ acts on the ket $|\boldsymbol{k}, 0\rangle$ describing one of these plane waves at $t = 0$, to give a new ket $\mathrm{e}^{-\mathrm{i}\omega(\boldsymbol{k})t}|\boldsymbol{k}, 0\rangle$. In that case, introduce a hermitian *energy operator* $\widehat{E}$ that acts on that *energy eigenstate* via

$$\widehat{E}\,|\boldsymbol{k}, t\rangle \equiv \hbar\omega(\boldsymbol{k})\,|\boldsymbol{k}, t\rangle \tag{2.217}$$

to return the associated energy eigenvalue (i.e., the wave's energy!). The ket describing a plane wave at time t is then

$$\begin{aligned}
|\boldsymbol{k}, t\rangle &= \widehat{L}\,|\boldsymbol{k}, 0\rangle = \mathrm{e}^{-\mathrm{i}\omega t}\,|\boldsymbol{k}, 0\rangle \\
&= \big[1 - \mathrm{i}\omega t + (-\mathrm{i}\omega t)^2/2! + \cdots\big]|\boldsymbol{k}, 0\rangle \\
&= \big[1 - \mathrm{i}t\widehat{E}/\hbar + (-\mathrm{i}t\widehat{E}/\hbar)^2/2! + \cdots\big]|\boldsymbol{k}, 0\rangle \\
&\equiv \mathrm{e}^{-\mathrm{i}t\widehat{E}/\hbar}|\boldsymbol{k}, 0\rangle\,.
\end{aligned} \tag{2.218}$$

When an arbitrary quantum mechanical state $|\Psi, t\rangle$ is written as a linear superposition of the energy eigenstates $|\boldsymbol{k}, t\rangle$, equation (2.218) applies to each component. The operator $\mathrm{e}^{-\mathrm{i}t\widehat{E}/\hbar}$ doesn't depend on each component's energy so it factors out of the superposition, allowing us to write

$$|\Psi, t\rangle = \mathrm{e}^{-\mathrm{i}t\widehat{E}/\hbar}|\Psi, 0\rangle\,. \tag{2.219}$$

So, time evolution of a quantum state is accomplished by the operator $\mathrm{e}^{-\mathrm{i}t\widehat{E}/\hbar}$. The energy operator $\widehat{E}$ was defined to act on a state of well-defined energy to give that energy as its eigenvalue. This is its *energy* representation; but what is its x-representation—in what sense can $\widehat{E}$ be applied to a wave function $\Psi(\boldsymbol{x}, t) \equiv \langle\boldsymbol{x}|\Psi, t\rangle$? First, write (2.219) as

$$|\Psi, t + \Delta t\rangle = \mathrm{e}^{-\mathrm{i}\widehat{E}\Delta t/\hbar}|\Psi, t\rangle\,. \tag{2.220}$$

Now observe that Taylor's theorem allows us to write the time evolution of an arbitrary function (and specifically the wave function), with $\partial_t \equiv \partial/\partial t$, as

$$\langle\boldsymbol{x}|\Psi, t + \Delta t\rangle \equiv \Psi(\boldsymbol{x}, t + \Delta t)$$

$$
\begin{aligned}
&= \Psi(\boldsymbol{x},t) + \partial_t \Psi(\boldsymbol{x},t)\,\Delta t + \partial_t^2 \Psi(\boldsymbol{x},t)\,\Delta t^2/2! + \cdots \\
&= \left[1 + \Delta t\,\partial_t + \Delta t^2\,\partial_t^2/2! + \cdots\right] \Psi(\boldsymbol{x},t) \\
&\equiv \mathrm{e}^{\Delta t\,\partial_t}\Psi(\boldsymbol{x},t) = \mathrm{e}^{\Delta t\,\partial_t}\langle \boldsymbol{x}|\Psi,t\rangle = \langle \boldsymbol{x}|\mathrm{e}^{\Delta t\,\partial_t}|\Psi,t\rangle\,. \quad (2.221)
\end{aligned}
$$

Omitting the $\langle \boldsymbol{x}|$, we might write the time evolution of a state ket as

$$|\Psi, t + \Delta t\rangle = \mathrm{e}^{\Delta t\,\partial_t}|\Psi, t\rangle\,, \tag{2.222}$$

but this comes with the important proviso that the "invisible" $\langle \boldsymbol{x}|$ is understood to be present on both sides of this equation. After all, ∂_t acts on the wave *function*, while (2.222) is written in terms of state kets only. So (2.222) is a concise mixture of functional and state ket notation, and is not to be taken literally without $\langle \boldsymbol{x}|$ being included.

Now comparing (2.220) with (2.222) gives the x-representation of the energy operator:

$$-\mathrm{i}\widehat{E}\Delta t/\hbar \overset{x\text{-rep.}}{=\!=} \Delta t\,\partial_t\,, \quad \text{so } \widehat{E} \overset{x\text{-rep.}}{=\!=} \mathrm{i}\hbar\,\partial_t\,. \tag{2.223}$$

Normally (2.223) is written simply as an equality, but we should not forget that the energy operator $\widehat{E}$ acts on a state, as opposed to its spatial representation $\mathrm{i}\hbar\,\partial_t$, which acts on a wave function. Equation (2.223) is the evolution operator central to quantum mechanics. In Chap. 10 we'll meet the $\widehat{E}$ operator again in a more general context under the name of the hamiltonian, where it's renamed to the very standard H.

An Expression for the Momentum Operator in Quantum Mechanics

Besides the operator for time evolution, the other important operator in elementary quantum mechanics is that describing a momentum measurement applied to a state. Again, we use de Broglie's idea of momentum being proportional to wavenumber, so that corresponding to the plane waves that make up the Fourier basis functions, we postulate that it is possible to introduce a new set of kets $|p\rangle$ corresponding to plane waves with momentum p (in one dimension for simplicity), forming an orthonormal basis,

$$\langle p|p'\rangle \equiv \delta(p - p')\,, \tag{2.224}$$

in which case a completeness relation holds for momenta, too:

$$\int |p\rangle\langle p|\,\mathrm{d}p = 1\,. \tag{2.225}$$

The quantum mechanical version of (2.171) has a unique conventional form:

$$\langle x|p\rangle \equiv \frac{1}{\sqrt{2\pi\hbar}}\,\mathrm{e}^{\mathrm{i}px/\hbar}\,. \tag{2.226}$$

Now, just as the energy operator $\widehat{E}$ is defined by its return of an energy eigenvalue when applied to an energy eigenket, a hermitian momentum operator $\widehat{p}$ can be defined by its return of a *momentum* eigenvalue when applied to a momentum eigenket, giving the corresponding momentum measurement:

$$\widehat{p}\,|p\rangle \equiv p\,|p\rangle\,. \tag{2.227}$$

As with the energy eigenstates, if an arbitrary wave function is expressed as a linear combination of momentum eigenstates using the momentum completeness relation, then we can write

$$\begin{aligned}\langle p|\,\widehat{p}\,|\Psi,t\rangle &= \langle p|\,\widehat{p}\int |p'\rangle\,\langle p'|\,\Psi,t\rangle\;\mathrm{d}p' = \langle p|\int p'|p'\rangle\,\langle p'|\,\Psi,t\rangle\;\mathrm{d}p' \\ &= p\,\langle p|\,\Psi,t\rangle\,.\end{aligned} \tag{2.228}$$

Alternatively, the $|\Psi,t\rangle$ can be omitted to express this last equation as

$$\langle p|\,\widehat{p} = p\,\langle p|\,, \tag{2.229}$$

which is just the hermitian conjugate of (2.227). Equations (2.228) and (2.229) give a projection onto momentum space by prepending the momentum bra to a momentum operator (which acts on the invisible $|\Psi,t\rangle$). The momentum operator represents a measurement, while prepending everything with $\langle p|$ gives the amplitude for the result to be p.

What is the x-representation of $\widehat{p}$? It's just $\langle x|\,\widehat{p}$, and is derived precisely as was done in (2.190) for $\langle x|\,\widehat{k}$. Write $\partial_x \equiv \partial/\partial x$:

$$\begin{aligned}\langle x|\,\widehat{p}\,|\Psi,t\rangle &= \int \langle x|p\rangle\,\langle p|\,\widehat{p}\,|\Psi,t\rangle\;\mathrm{d}p \overset{(2.226)}{=\!=} \int \frac{\mathrm{e}^{\mathrm{i}px/\hbar}}{\sqrt{2\pi\hbar}}\,p\,\langle p|\,\Psi,t\rangle\;\mathrm{d}p \\ &= -\mathrm{i}\hbar\,\partial_x \int \frac{\mathrm{e}^{\mathrm{i}px/\hbar}}{\sqrt{2\pi\hbar}}\,\langle p|\,\Psi,t\rangle\;\mathrm{d}p \;=\; -\mathrm{i}\hbar\,\partial_x \int \langle x|p\rangle\,\langle p|\,\Psi,t\rangle\;\mathrm{d}p \\ &= -\mathrm{i}\hbar\,\partial_x\,\langle x|\,\Psi,t\rangle\,.\end{aligned} \tag{2.230}$$

So the x-representation of the momentum operator is $-\mathrm{i}\hbar\,\partial_x$. We could write $\widehat{p} \overset{x\text{-rep.}}{=\!=} -\mathrm{i}\hbar\,\partial_x$, although the "$x$-rep." is always taken as understood. Remember that, as shown in Sect. 2.7, $-\mathrm{i}\hbar\,\partial_x$ is hermitian and thus has real eigenvalues, as is required of operators in quantum mechanics. The commutation relation between the position and momentum operators is now easily calculated to be

$$\left[\widehat{x},\widehat{p}\right] = \mathrm{i}\hbar\,. \tag{2.231}$$

In general, the xyz-representation of $\widehat{p}$ for one particle will be $-\mathrm{i}\hbar\,\nabla$.

The discussion around (2.192) is important enough to repeat here for $\widehat{p}$. Equation (2.230) can be abbreviated by omitting $|\Psi,t\rangle$ to write

$$\langle x|\,\widehat{p} = -\mathrm{i}\hbar\,\partial_x\,\langle x| \tag{2.232}$$

(since $|\Psi, t\rangle$ had a passive spectator role), with the implicit understanding that $|\Psi, t\rangle$ must always be re-appended to each side of (2.232) to produce the correct expression (2.230). This is just notation; ∂_x operates on a *function*—it makes no sense to differentiate a bra or a ket. So we must not just equate $\widehat{p}$ and $-\mathrm{i}\hbar\,\partial_x$ blindly. If we were to do so, there might be a temptation to write

$$\text{“}\,\widehat{p}\,|x\rangle = -\mathrm{i}\hbar\,\partial_x\,|x\rangle\,\text{”} \quad \text{(Wrong!)}, \tag{2.233}$$

especially since both $\widehat{p}$ and $-\mathrm{i}\hbar\,\partial_x$ are hermitian. But this equation gives the wrong answer if we prepend both sides with $\langle\Psi, t|$. To see why, write the complex conjugate of (2.230) as

$$\langle\Psi, t|\,\widehat{p}\,|x\rangle = +\mathrm{i}\hbar\,\partial_x\,\langle\Psi, t|x\rangle\,, \tag{2.234}$$

where the sign of i has changed, since these expressions are just numbers. Now omit the $\langle\Psi, t|$ to write

$$\widehat{p}\,|x\rangle = +\mathrm{i}\hbar\,\partial_x\,|x\rangle\,, \tag{2.235}$$

which of course differs from the incorrect (2.233). Equation (2.235) is correct, as long as we realise that ∂_x does not operate on the ket $|x\rangle$! All of this is just formalism, and clearly it pays to be aware of what this abbreviated notation means, and of the difference between an operator and its representative in some basis. Remember that while expressions such as (2.232) and (2.235) combine the derivative of a function with a state ket, the derivative does not simply act on the ket; such an operation has no meaning.

The momentum operator's x-representation appears in a different way that will shed more light on this transition to quantum mechanics in Chap. 8. Suppose that analogously to our use of Taylor's theorem in (2.221) to write the time evolution of $\Psi(\boldsymbol{x}, t)$, we again write the same function's "space translation" as

$$\begin{aligned}\Psi(\boldsymbol{x} + \Delta\boldsymbol{x}, t) &= \Psi(\boldsymbol{x}, t) + (\Delta\boldsymbol{x}\cdot\nabla)\,\Psi + (\Delta\boldsymbol{x}\cdot\nabla)^2/2!\,\Psi + \cdots \\ &= \Big[1 + \Delta\boldsymbol{x}\cdot\nabla + (\Delta\boldsymbol{x}\cdot\nabla)^2/2! + \cdots\Big]\,\Psi(\boldsymbol{x}, t) \\ &\equiv \mathrm{e}^{\Delta\boldsymbol{x}\cdot\nabla}\Psi(\boldsymbol{x}, t)\,.\end{aligned} \tag{2.236}$$

But $\Delta\boldsymbol{x}\cdot\nabla \overset{\boldsymbol{x}\text{-rep.}}{=\!=} \Delta\boldsymbol{x}\cdot\mathrm{i}\widehat{\boldsymbol{p}}/\hbar$, so that $\mathrm{e}^{\Delta\boldsymbol{x}\cdot\nabla} \overset{\boldsymbol{x}\text{-rep.}}{=\!=} \mathrm{e}^{\mathrm{i}\Delta\boldsymbol{x}\cdot\widehat{\boldsymbol{p}}/\hbar}$. Summing up, there is a correspondence between time evolution and space translation in quantum mechanics:

$$\begin{aligned}\mathrm{e}^{-\mathrm{i}\widehat{E}\Delta t/\hbar} &\longleftrightarrow \text{time evolution}, \\ \mathrm{e}^{\mathrm{i}\widehat{\boldsymbol{p}}\cdot\Delta\boldsymbol{x}/\hbar} &\longleftrightarrow \text{space translation}.\end{aligned} \tag{2.237}$$

The associations of energy with evolution, and momentum with translation, are fundamental ones that were known in the lagrangian theory of classical

The Logical Flow of Operators in this Section

The logic of how coordinate representatives of the energy and momentum operators are being related to time evolution and space translation in this section is as follows:

$$\left.\begin{array}{l}\text{Fourier analysis} \to \text{time evolution op.} = \exp(-\mathrm{i}\Delta t\widehat{E}/\hbar)\\ \text{Taylor expansion} \to \text{time evolution op.} = \exp\left(\Delta t\,\partial_t\right)\end{array}\right\} \Rightarrow \widehat{E} = \mathrm{i}\hbar\,\partial_t\,. \tag{2.238}$$

$$\left.\begin{array}{l}\text{Fourier analysis} \to \widehat{\boldsymbol{p}} \overset{\boldsymbol{x}\text{-rep.}}{=\!=} -\mathrm{i}\hbar\,\nabla\\ \text{Taylor expansion} \to \text{space trans. op.} = \exp\left(\Delta\boldsymbol{x}\cdot\nabla\right)\end{array}\right\} \Rightarrow \left\{\begin{array}{l}\text{space trans. op.}\\ \quad = \exp(\mathrm{i}\Delta\boldsymbol{x}\cdot\widehat{\boldsymbol{p}}/\hbar)\end{array}\right. \tag{2.239}$$

mechanics long before they appeared in quantum mechanics, as we'll see in Chap. 10. This knowledge is indeed what drove some of the early research in quantum theory. Energy and momentum are called the "generators" of time evolution and space translation, respectively, and the classical mechanics analogy is the fact that a system displaying time invariance conserves its energy, and a system displaying space invariance conserves its momentum. But, in general, the conserved momentum of classical mechanics is the *canonical momentum*, which we'll meet in Chap. 10. Canonical momentum is a more encompassing version of the $m\boldsymbol{v}$ that itself only applies to the free particles allied with the plane waves central to the last several pages, so it will be the canonical momentum whose spatial representation is $-\mathrm{i}\hbar\nabla$ for a general particle in quantum mechanics.

We will return to these results in Chap. 10 to merge them with others from that chapter to create the Schrödinger equation, the main equation of quantum mechanics, which describes its wave functions in the xyz-representation.

The ideas and techniques of linearity pervade many branches of physics, and the subject is very rich in its applications. As a result, the language that we have encountered in this chapter is very common in mathematical physics, and there will be many opportunities to use it in the coming chapters.

3 The Natural Language of Random Processes

In this chapter, we take a tour of some of the ideas of statistics and signal processing, with a view toward examining how the language and notation revolve around a few well-related ideas that turn up time and again in diverse areas. An example of such a close relationship in statistics is the idea of summing squares, something that is universally used to quantify how well a parameter has been estimated. But why sum squares? Why not sum absolute values, or fourth powers, or any other of an infinite number of functions of the data? In fact, tied to ideas of summing squares is the very intuitive notion of finding an arithmetic mean, and these two things join with the exponential function to form the gaussian distribution that describes so much of the statistics found in Nature. Ideas such as these form the basis of many related entities that we'll examine in the pages to follow.

3.1 From Bar Graphs to Histograms

When first introduced to the ideas of probability and statistics, we learn to plot bar graphs that show probabilities on the y-axis as a function of some discrete variable on the x-axis. Later we advance from discrete to continuous variables, and the graph becomes a *histogram*: still outwardly a bar graph, but now representing the probabilities by the *areas* of the bars. The reason for this switch in the philosophy of how best to plot the data is seldom investigated in texts, which usually discuss the histogram in various ways without giving a unique recipe for how to draw one. And yet the histogram is the important link between the basic idea of a bar graph and the rather sophisticated idea of a probability density curve.

For example, some texts will demand, unnecessarily, that the bars (data bins) of a histogram be of equal width. Others will say, correctly, that different-width bars are certainly allowed but should have a height that is corrected, in some proportional way, for the data that they represent. But even these will probably either give no units to the heights of the histogram bars, or they will plot the bars using the same units as the bar graph from which the histogram was derived. Histograms do *not* in general have the same units as their parent bar graphs, for the simple reason that they hold their information as an area, not a height. This fact is central to the transition in plotting philosophy from bar graphs to densities.

It certainly is a reasonable idea that using areas might be the only way in which probabilities can be represented when the domain is a continuous variable. And yet it's only natural to expect to find a smooth transition in the method of plotting probabilities, for the case where the variable of interest can be made to change smoothly from discrete to continuous. How, then, can the bar graph's height probabilities change in a continuous way to the histogram's area probabilities?

Just how this occurs is best seen through an example. Suppose we are required to find the distribution of lifetimes of the light bulbs produced by a certain factory. The proportion falling in each of five bins is given on the left-hand side of Table 3.1. These are interpreted as probabilities for any particular bulb to burn out in a certain age range; most bulbs are seen to last for around three years. The probabilities can easily be plotted on the bar graph on the left-hand side in Fig. 3.1. Each bar has the same width irrespective of bin size (although this is a pictorial point only; the bin size is the important quantity to consider), and the height of each bar is just the fraction of bulbs burning out within the specified time window. Notice that the 3–5 year window represents a larger number of bulbs compared with the other bins because it spans two years instead of one. Merging bins like this when collecting data exaggerates their weight, and can certainly misrepresent the information in these graphs.

We can fix this problem of a misleading bar height by temporarily doing something quite different: we can plot a cumulative graph of the fraction that burns out after some time, versus that time. These fractions are tabulated on the right-hand side of Table 3.1 and are interpreted as cumulative probabilities. The cumulative graph is a set of points plotted for whole-number years. Nevertheless, we know that it makes complete sense to talk about the fraction of bulbs that burn out within, say, 2.2 years, and that implies that

Table 3.1. The proportions of bulbs burning out in the left-hand table translate to cumulative figures in the right-hand table.

Life expectancy	
Lifetime (years)	Proportion
0–1	5%
1–2	10%
2–3	40%
3–5	35%
> 5	10%
Total	100%

Cumulative proportions	
Lifetime (years)	Proportion
< 0	0%
< 1	5%
< 2	15%
< 3	55%
< 5	90%
$< \infty$	100%

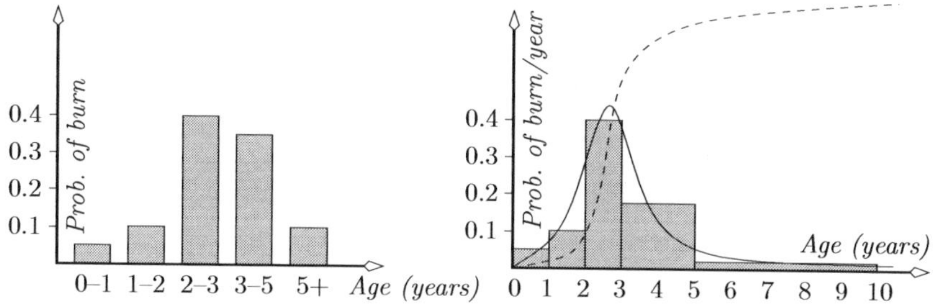

Fig. 3.1. The bar graph on the left shows the probability of a bulb burning out from Table 3.1 for the age ranges in that table. The age 3–5 bin contains two years' worth of data, and so skews the look of the plot somewhat. This is cured by plotting the histogram at the right: probabilities per year on a continuous time axis (and setting the last bin to have a five year width). Now the data are spread across the bins more democratically. An estimate of the actual probability density curve (solid) is superimposed on this graph and uses the same units as the bars. The cumulative curve (dotted) is also superimposed, but it carries different units so its scale is arbitrary here.

the cumulative graph should really be continuous. Of course, we could always interpolate a smooth curve through the half dozen data points, but to produce a more accurate graph, the width of the time intervals that make up the data bins really needs to be decreased. That means the corresponding bars on the probability bar graph will get narrower, and the height of each will also shrink. After all, if we halve the bin width, then we are also roughly halving the fraction of bulbs in each bin.

If we are really to construct a continuous cumulative curve, the bin size must shrink to zero. But in that case the height of each bar shrinks to zero—which makes it impossible to draw. A simple way to remedy this is to magnify these ever-shrinking heights. We do this by plotting not the fraction of bulbs in each bin, but rather that fraction divided by the bin width. That way the two quantities (fraction of bulbs and bin width) are decreasing in tandem, so that their ratio is largely unaffected. When we halve the bin widths, the fraction in each bin in also roughly halved, so that the quantity

$$\frac{\text{fraction in bin}}{\text{width of bin}}$$

is roughly unchanged—and it changes less and less as the bin width decreases to zero. This process of magnification is really the essence of differential calculus: we are able to plot something whose size drops to zero by continuously magnifying it.

For example, consider a bin that covers the period from $t = 2.2$ years to $t = 2.2 + 10^{-6}$. This bin contains roughly one millionth of the fraction of bulbs burning out in the period $t = 2$ years to $t = 3$ years. Its height is fairly constant over its width, and its height is just the fraction of bulbs burning out in this interval *divided* by the bin width 10^{-6}. So dividing by 10^{-6} has thus magnified its height a million times, which has been necessary to keep the scale of the graph more or less unchanging.

So instead of plotting fractions (probabilities), we are doing something new: we're plotting *probabilities per unit bin width*, or in this example probabilities per year. And that is the unique and simple recipe for plotting a histogram. We take the parent bar graph's heights, divide each height by the associated bin width, and plot the result. Although the name given to the graph changes from bar graph to histogram, it's more usually called a *probability density*, because a probability per unit time is a temporal density of probability. (The analogous way of plotting, say, the mass of a string versus its length would be to plot the linear mass *density* versus its length parameter x. The mass between two points x_1, x_2 on the string would then be the area under the curve.)

The histogram of the data in Table 3.1 is shown on the right-hand side in Fig. 3.1. Note that while the numbers on the two vertical axes look the same, their units are quite different. The most prominent new feature is that the height of the bar in the range 3–5 years has been halved to compensate for splitting that bin into two halves. Also, the last bin in Table 3.1 has, in principle, infinite width, so to be realistic we have treated it as representing 5–10 years (and then divided the corresponding height in the bar graph by 5 years); perhaps the manufacturer has not been taking data for longer than that. The solid curve on the right-hand axes shows the limit of the density as the bin width goes to zero (if we have more data to approximate this), while the dashed curve approximates the cumulative plot. This plot has different units from the histogram and density curve, and is merely shown for comparison.

Now, because by the definition of the histogram

$$\text{height of bar} = \frac{\text{fraction of bulbs}}{\text{width of bar}}\,, \tag{3.1}$$

it follows that

$$\begin{aligned}\text{fraction of bulbs} &= \text{height of bar} \times \text{width of bar} \\ &= \text{area of bar}\,,\end{aligned} \tag{3.2}$$

so that the probability of a bulb's surviving for any interval, or equivalently the fraction of bulbs expiring in that interval, will be given by the total area of the histogram bars covering that interval. So the way of expressing the information has changed in a histogram. As the bin width drops to zero, we cease bothering to draw vertical lines separating the strips, and the graph becomes truly continuous.

The transition between bar graph and histogram now becomes clear: they are visually identical—but with different y-axis units—when the bin width is one unit. This changeover, accompanied by an important change of y-axis units and the introduction of the idea of a density, is perhaps seldom given the emphasis that it deserves in textbooks that present histograms from first principles. These books tend to concentrate on plotting probabilities alone, with a dramatic switch in philosophy for continuous variables, justified in hindsight because areas seem to be necessary for representing continuous data. But for any continuous quantity, if we begin by plotting a probability per unit bin width, then the modified bar graph, now called a histogram, will embody probability as an area from the outset; it will already be a density graph, and will tend continuously to a density *curve* as the bin width shrinks to zero.

The relationship between the probability density and its associated cumulative distribution embodies the Fundamental Theorem of Calculus, which says that finding the sum of an infinite number of infinitesimal quantities (*integration*) is the reverse process of finding a rate of increase (*differentiation*). To see this, suppose we ask what fraction of bulbs expires in the interval t to $t + \Delta t$ years. Divide this interval into an infinite number of strips and add all the strip areas. With a strip height of $p(t)$ and an associated infinitesimal width of $\mathrm{d}t$, the summed area is written using the Old English symbol $\int$ for S ("sum"):

$$\text{fraction of bulbs burning out in } t \to t + \Delta t = \int_t^{t+\Delta t} p(t')\,\mathrm{d}t' . \tag{3.3}$$

The next step asks: what is the cumulative plot, and how is it related to $p(t)$? Call this plot $C(t)$. Then, by definition, $C(t)$ must equal the fraction of bulbs burning out at any age less than t. In other words,

$$C(t) \equiv \int_0^t p(t')\,\mathrm{d}t' . \tag{3.4}$$

Being cumulative, $C(t)$ cannot decrease: it starts from 0 and climbs to 1. The slope of the cumulative distribution is

$$\frac{\mathrm{d}C}{\mathrm{d}t} \equiv \lim_{\Delta t \to 0} \frac{C(t + \Delta t) - C(t)}{\Delta t} . \tag{3.5}$$

But

$$\begin{aligned} C(t + \Delta t) &= \text{fraction of bulbs lasting} < t + \Delta t\,, \\ C(t) &= \text{fraction of bulbs lasting} < t\,, \end{aligned} \tag{3.6}$$

in which case it follows that

$$C(t + \Delta t) - C(t) = \text{fraction of bulbs lasting from } t \text{ to } t + \Delta t\,. \tag{3.7}$$

So $\mathrm{d}C/\mathrm{d}t$ is the small-bin limit of the fraction of bulbs in the $t \to t + \Delta t$ bin divided by that bin's width. But this was just how we constructed $p(t)$! Consequently,

$$\frac{dC}{dt} = p(t)\,. \tag{3.8}$$

Equation (3.8), taken together with (3.4), shows how a density is related to a cumulative plot:

$$\begin{aligned} p &= \text{slope of } C, \\ C &= \text{area under } p. \end{aligned} \tag{3.9}$$

And, of course, this is nothing more than the Fundamental Theorem of Calculus: slopes and areas (or sums) form a natural pair.

3.2 The Privileged Sum of Squares

One of the most basic concepts taught in statistics courses is the idea of gaining insight into a set of data by defining measures of its centre and spread. Most useful for this are the data's *mean* and *standard deviation.* The mean seldom calls for any motivation or explanation, being identical to the average, an everyday intuitive idea. Nevertheless, there are an infinite number of possible choices for the mean m, such as any expression that treats all n data points homogeneously. So, given any function $f(x)$, we could define the mean m in such a way that

$$n\, f(m) \equiv \sum_{i=1}^{n} f(x_i)\,. \tag{3.10}$$

The simplest choice of the function is also the one universally used: a constant $f(x) \equiv 1$.

In contrast with the mean, the standard deviation can appear quite mysterious. At first, the manipulations involved in calculating it seem a little arbitrary and laborious. Eventually the student comes to accept the great utility of this *root-mean-squared*, or *rms*, deviation of the numbers from their mean, and finds it used extensively throughout physics, along with sums of squares in general.

However, as a measure of spread, why not just use an average distance of the data points from the mean? On the face of it, this is much simpler than finding the square root of the mean of a squared distance. We would only need to ensure that a *distance* is used (a positive number by definition) instead of a *deviation* (which might be positive or negative: i.e., distance $\equiv$ |deviation|). This is because the average deviation of the data points from their mean will always be zero. (Why? Because the mean of the deviations is the mean of $x_i - m$, which is the mean of all the x_i minus the mean of m, which equals $m - m$, or zero.)

It is certainly possible to work with the mean of the distances of all the data from their mean; the result is called the *absolute* deviation:

$$\text{absolute deviation} \equiv \frac{\sum |x_i - m|}{n} . \tag{3.11}$$

But the absolute deviation tends not to be used in practice. The reason is sometimes given that the absolute values in (3.11) are difficult to treat mathematically, since the function $y = |x|$ has a kink at $x = 0$. Certainly this lack of differentiability at the origin makes the absolute value function difficult to treat analytically, but this does not explain why a root-mean-square approach to quantifying spread should be more useful or meaningful than an absolute value approach.

Let's turn the question on its head. Conventionally, the mean m of a sample is first defined and is then used to define the sample's standard deviation s:

$$m \equiv \frac{\sum x_i}{n} , \quad \text{then } s \equiv \sqrt{\frac{\sum (x_i - m)^2}{n}} . \tag{3.12}$$

(The square of the standard deviation is called the *variance*.) Suppose instead that we define the standard deviation first without any reference to a mean, and then use that standard deviation to define an appropriate mean. That also makes sense: we already have an intuitive idea of the "spread" of a set of numbers as being, for example, the distance from the minimum to the maximum of the set, which does not rest on any concept of a mean. But such a trivial quantity is not very useful. For a more sophisticated approach, define the spread with reference to this as yet undefined mean in such a way that the spread is *minimised* for some appropriate choice of the mean. At this stage we have no idea what the mean should be. It is not necessarily the sum of the data divided by their total number as in (3.12); in our new approach, (3.12) doesn't exist yet. The mean will be a number to be determined purely through requiring that it minimise some choice of the spread. Relating the result of such a minimisation to meaningful processes in Nature is an idea we'll encounter often in the chapters ahead.

So our programme is to choose several different definitions of the spread and see what mean results from each. One simple definition is the absolute deviation (3.11). Consider this now to be a function of some variable λ:

$$s(\lambda) \equiv \frac{\sum |x_i - \lambda|}{n} . \tag{3.13}$$

The choice of mean corresponding to this definition of spread is the value of λ that minimises $s(\lambda)$ for some set of data points. At first thought, we might try a calculus approach by setting $s'(\lambda) = 0$. But that will fail because the absolute value function is not everywhere differentiable. The failure is the same problem as that of minimising $y = |x|$ by calculating $\mathrm{d}y/\mathrm{d}x$ and solving for where it equals zero:

$$\frac{\mathrm{d}y}{\mathrm{d}x} = \frac{|x|}{x} = 0 . \tag{3.14}$$

Of course the minimum occurs at $x = 0$, but this is just the value of x for which (3.14) fails! So we cannot use the calculus to minimise $s(\lambda)$ with respect to λ in (3.13). A different approach is needed. For simplicity, assume the x_i are all different and plot them on a number line:

Focus now on the variable λ which lies anywhere on the line. We'll move it back and forth until the absolute deviation $s(\lambda)$ is minimised.

For example, suppose λ lies far to the right of all the data, so that we decrease its value by a small amount ε. This decreases its distance equally from every data point. But what happens once it moves left of the rightmost data point x_n?

Once this happens, as we continue to move λ farther to the left, its distance from the points $x_2 \ldots x_{n-1}$ still decreases, but now it recedes from x_n by just the same amount that it approaches x_1. So the values of the outer two data points no longer have any effect on the value of the spread $s(\lambda)$.

Similarly, as the choice of λ moves ever farther to the left, the deviation keeps decreasing as, pair by pair, the outside data pairs that form a kind of zero-dimensional shell outside of λ no longer contribute to the spread $s(\lambda)$, while all pairs on the inside exert a kind of pull that lessens this spread.[1] Once the position of λ is within the innermost pair (if the number of data points is even), moving it still farther to the left inside the innermost pair will no longer change $s(\lambda)$. Or, if the number of data points is odd, then the minimum of $s(\lambda)$ is reached when λ equals the middle value, the so-called *median* of the data. So for an odd number of different points, we have found a unique value to be called the mean m: the median or midpoint of the data. If the number of points is even, then we can choose m to be any value within the innermost pair.

What this demonstrates is that medians and absolute deviations form a natural pair. The absolute deviation will be a minimum provided that we understand the notion of mean to be the median of the data (or some midpoint found in the way of the previous paragraph if some data points are the same or there are an even number). This apparently back-to-front approach of defining the mean *after* the spread brings together these two previously disparate quantities. We have found that a choice of median for the mean implies that the absolute deviation has real meaning for the spread.

[1] There is a good analogy to the gravity produced by a spherical shell here. Outside of it, its gravity is the same as if all of its mass were concentrated at its centre. However, everywhere inside the shell, its gravity is zero.

But as we have just seen, if the number of data points is even, then the mean (i.e., the midpoint) is not particularly well defined. So more sophistication in defining the spread is called for.

Our next measure of the spread is slightly more complicated than (3.13): we will square the distances before adding them, and retain the correct units for the spread by including a square root in its definition. So define the new spread $s(\lambda)$ as follows:

$$s(\lambda) \equiv \sqrt{\frac{\sum(x_i - \lambda)^2}{n}} \,. \tag{3.15}$$

This function will always be differentiable, so we can safely use a calculus approach to minimise it. Minimising $s^2(\lambda)$ is equivalent but easier since it eliminates the square root. The resulting λ will be the new choice of mean m:

$$\frac{\mathrm{d}}{\mathrm{d}\lambda} s^2(\lambda) = \frac{-2\sum(x_i - \lambda)}{n} = 0 \,, \quad \text{so that } \lambda = \frac{\sum x_i}{n} \equiv m \,. \tag{3.16}$$

What appears is the familiar average, or *arithmetic mean* (cf. the *geometric mean*, defined for n numbers as the nth root of their product; the geometric mean is always less than or equal to the arithmetic mean). This definition of the mean, then, pairs naturally with the root-mean-square definition of spread in (3.15), now called the *standard* deviation because it's so closely linked with our intuitive notion of the mean, (3.16).

Finally, it's apparent that the reason that the standard deviation involves *squared* deviations is because by *starting* with the reasonable, tractable, and useful requirement that the expression for the mean be linear, the minimisation procedure we have just discussed, followed in reverse, will then require a measure of spread involving powers that are one more than linear: i.e., squares.

We now have two natural pairings of measures of mean and spread:

$$\begin{aligned} \text{median} &\longleftrightarrow \text{absolute deviation,} \\ \text{arithmetic mean} &\longleftrightarrow \text{standard deviation.} \end{aligned} \tag{3.17}$$

Choosing more complicated definitions of the spread can of course be done, but this only serves to produce ever more complicated expressions for the mean. There are really an infinite number of different means and deviations able to be defined for a data set, but the most useful are the usual (arithmetic) mean and standard deviation. And as we'll show next, these two partners completely characterise the *normal distribution*, the statistical distribution so frequently observed in Nature.

The Mean, Standard Deviation, and Normal Distribution

The normal, or gaussian, distribution is the "top dog" of statistics, both as a distribution in its own right and as an approximation of other distributions. We have already encountered one of its important characteristics in

n or $n-1$ in the Denominator?

Should the denominator of the standard deviation in (3.12) be n or $n-1$? The standard deviation is an averaged deviation, and an average is defined to use n. So the denominator of (3.12) must be n. But the basic aim of statistical theory is to estimate parameters describing the population knowing only something about a sample; any number produced from a sample that can help us to estimate a parameter of the population is called a *statistic*. We wish to estimate the standard deviation σ of the whole population, given the standard deviation s of a sample (i.e. s is the relevant statistic). In practice, it tends to be impossible or impractical to determine the population spread σ exactly—that is, after all, why statistical theory was invented in the first place. It so happens that when the population is much larger than the sample size, the *best estimate* of the population variance, here denoted $(\sigma^2)_{\text{be}}$, that can be inferred from a sample of size n is related to that sample's variance s^2 through

$$(\sigma^2)_{\text{be}} = s^2 \frac{n}{n-1}\,. \tag{3.18}$$

The best estimate of the population variance, $(\sigma^2)_{\text{be}}$, almost but not quite equals the square of the best estimate of the population standard deviation, $(\sigma_{\text{be}})^2$; thus we can only say

$$\sigma_{\text{be}} \simeq s\sqrt{\frac{n}{n-1}}\,, \quad \text{which means } \sigma_{\text{be}} \simeq \sqrt{\frac{\sum (x_i - m)^2}{n-1}}\,. \tag{3.19}$$

σ_{be} differs from s because the expression for s uses m, which itself is only an estimate of the population mean μ—albeit the best estimate of μ (i.e. $\mu_{\text{be}} = m$). (With hindsight, it's reasonable that s might be slightly smaller than σ since the value of s is such that the sum of squares of deviations from m is a minimum.) So the best estimate of the population standard deviation has an $n-1$ in its denominator. This is a derived expression for σ_{be}; there is nothing fundamental about it. It's not a standard deviation but rather only the best estimate of one based on a limited sample. The expression for a standard deviation, whether s for a sample or σ for a population, *always* has an n in its denominator.

the previous chapter, where it was noted that a measure of the width of a gaussian is as economical as it can be, in the sense that (2.197) is an equality for gaussians. We will derive the gaussian distribution here using the same technique that was used to derive the sum-of-squares expression in (3.16). That is, we'll consider the mean of a data set to be a quantity whose value is to be determined via a minimisation.

Consider, then, a situation in which the true value μ of a quantity x is the mean of a large number of observed values x_i. (We are now considering an entire population. The m and s relate to sample mean and standard deviation,

Gaussian Integrals

Gaussian integrals are frequently needed in physics, and it is very useful to have a general expression capable of handling any one that presents itself. The following is a rare and useful identity that uses (and essentially defines) the *error function* erf x:

$$\int e^{-ax^2+bx}\,dx = \frac{1}{2}\sqrt{\frac{\pi}{a}}\, e^{\frac{b^2}{4a}}\,\operatorname{erf}\left(\sqrt{a}\,x - \frac{b}{2\sqrt{a}}\right). \tag{3.20}$$

This expression is true for all values of a and b (even complex ones). In particular, erf is an odd, strictly increasing function over the reals, that when plotted resembles an inverse tangent function, but with $\operatorname{erf}\infty = 1$. Hence

$$\int_{-\infty}^{\infty} e^{-ax^2+bx}\,dx = \sqrt{\frac{\pi}{a}}\, e^{\frac{b^2}{4a}}. \tag{3.21}$$

When dealing with a complex integration, as on p. 437, it's useful to remember that $\operatorname{erf}(-z) = -\operatorname{erf} z$ for all complex z, and $\operatorname{erf} z \to 1$ as $|z| \to \infty$, as long as $|\arg z| < \pi/4$.

An analogous definite integral in higher dimensions is also useful. If the n variables are held in a column vector $\boldsymbol{x} \equiv [x_1 \ldots x_n]^t$, while A is a real symmetric $n \times n$ matrix and $\boldsymbol{b}$ is a column vector, then

$$\int_{-\infty}^{\infty} \exp\left(-\boldsymbol{x}^t A\boldsymbol{x} + \boldsymbol{b}^t\boldsymbol{x}\right) d^n x = \frac{\pi^{n/2}\exp\left(\boldsymbol{b}^t A^{-1}\boldsymbol{b}/4\right)}{\sqrt{\det A}}. \tag{3.22}$$

while the corresponding values for a population, usually unknowable, are μ and σ.) This is entirely consistent with what our intuition demands, of course; after all, if the mean of a large number of values did not tend toward the true value of the quantity, we would suspect some bias in our measurements, or a badly defined mean. When there is no bias, we can write[2]

$$\lim_{n\to\infty} \sum_{i=1}^{n} (x_i - \mu) \overset{\text{req.}}{=\!=} 0 \quad \text{for all choices of the data } x_i\,. \tag{3.23}$$

Suppose that we are required to find the probability density $p(x)$ associated with the random variable x. This can always be written as a function $f(x-\mu)$, which we do here since it will prove useful in a comparison with (3.23):

$$p(x) = f(x - \mu)\,. \tag{3.24}$$

The probability of obtaining the set of measurements x_i must be the product of individual probabilities:

[2] In (3.23) and throughout this book, we use $a \overset{\text{req.}}{=\!=} b$ to mean "a is required to equal b". This is a *demand*, as opposed to a simple statement of equality.

$$p(x_1, \ldots, x_n, \mu) = \prod_i p(x_i) = \prod_i f(x_i - \mu)\,. \tag{3.25}$$

Now, as before, at this point we take the mean to be the value we must fix for a variable λ. Consider (3.25) as giving a sort of generalised probability as a function of λ:

$$p(x_1, \ldots, x_n, \lambda) = \prod_i f(x_i - \lambda)\,. \tag{3.26}$$

We demand that x be distributed in such a way that its most probable value is μ itself, so that $p(x)$ is a maximum at $\lambda = \mu$. In that case, $\partial p/\partial\lambda \overset{\text{req.}}{=} 0$ when $\lambda = \mu$. Take the natural logarithm of both sides of (3.26) to facilitate the differentiation, writing

$$\frac{1}{p}\frac{\partial p}{\partial \lambda}\bigg|_{\lambda=\mu} = \sum_i \frac{-f'(x_i - \mu)}{f(x_i - \mu)} \overset{\text{req.}}{=} 0 \quad \text{for all choices of } x_i\,. \tag{3.27}$$

The only way that (3.23) and (3.27) can both hold for any arbitrary set of x_i as $n \to \infty$ is if

$$\frac{-f'(x - \mu)}{f(x - \mu)} \propto x - \mu\,. \tag{3.28}$$

Writing $\xi \equiv x - \mu$, this means that for some proportionality constant k,

$$\frac{f'(\xi)}{f(\xi)} = k\xi\,, \tag{3.29}$$

which has a normalised solution of

$$f(\xi) = \sqrt{\frac{-k}{2\pi}}\; \mathrm{e}^{k\xi^2/2} \tag{3.30}$$

using (3.21). We can express k in terms of the population variance σ^2 of x (which equals the population variance of ξ since shifting the data does not alter its spread), by using an alternative form for the variance: the variance of a data set equals the mean of the squares of the data values minus the square of the mean of the data values (which is easily proved from first principles). Denoting the mean by $\langle\cdot\rangle$,

$$\sigma^2 = \langle\xi^2\rangle - \underbrace{\langle\xi\rangle^2}_{=0} = \int_{-\infty}^{\infty} \xi^2 f(\xi)\,\mathrm{d}\xi = \frac{-1}{k}\,. \tag{3.31}$$

This allows us to write

$$f(\xi) = \frac{1}{\sigma\sqrt{2\pi}} \exp\frac{-\xi^2}{2\sigma^2}\,. \tag{3.32}$$

Finally, since the function f is just the required probability density p shifted to centre on μ, the required probability density must be

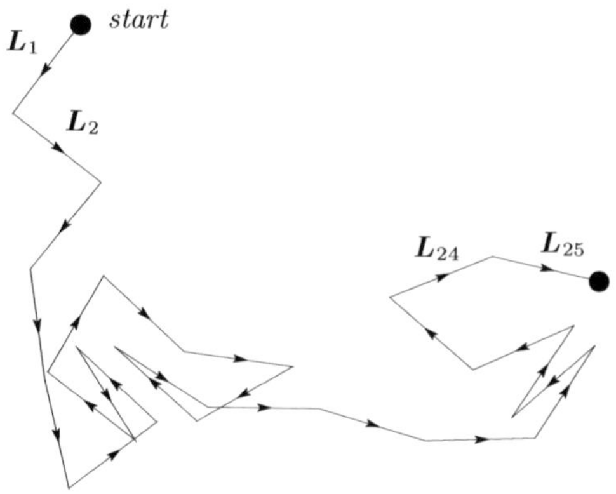

Fig. 3.2. A random walk of 25 steps, each of length L. Over an ensemble of these walks, the root-mean-square distance from start to end positions will be $5L$. The actual distance in the single trial pictured is about $5.2L$.

$$p(x) = \frac{1}{\sigma\sqrt{2\pi}} \exp \frac{-(x-\mu)^2}{2\sigma^2} . \tag{3.33}$$

As can be seen, a few basic assumptions about a probability density lead naturally to a function that incorporates the population mean and standard deviation, and this elevates the sum of squares to a leading status in statistical theory. Again, we can see that squares arise quite naturally because, as in (3.29), we frequently find ourselves writing a differential equation for a quantity that appears linearly in that equation. So it's only natural that the solutions appearing should contain the integral of this first power, which is of course a square.

3.2.1 Sums of Squares and the Random Walk

Sums of squares feature in the classic random walk scenario. If a drunken man stumbles away from a start point, taking steps of uniform length L in arbitrary directions, then how far from the start will he be after n steps? The problem, being statistical, can only produce some sort of average answer. So we need to take the view that there is a whole collection, an *ensemble*, of drunken men all walking away from their own start points. We'll calculate an average over this ensemble.

With the start point taken as the origin, let the ith step be a vector $\boldsymbol{L}_i$. After n steps, the man's position relative to the start point is given by the vector sum

$$\text{position} = \boldsymbol{L}_1 + \cdots + \boldsymbol{L}_n \,, \tag{3.34}$$

as shown in Fig. 3.2. The square of the final distance from the start point is

Calculating the rms and Mean Speeds of Ideal Gas Particles

A good example of the differences in difficulty and approach to calculating the *rms* value of a quantity, versus calculating its *mean* value, lies in analysing the spread of speeds of gas particles. Their rms speed can be found by the procedure of Sect. 12.4, in which we consider how many gas particles bounce off the walls of the gas's container in some time interval, transferring a certain momentum that relates to the pressure exerted, which can then be related to the gas's temperature via the Ideal Gas Law, $PV = nRT$. The expression for the momentum transferred contains a velocity term, as does the expression for the time taken for the transferral. Thus velocity appears in the equations as a square. So when appropriate averages are taken, the mean of the square of the particle speeds appears, giving an rms speed of $\sqrt{3kT/m}$, where k is *Boltzmann's constant*.

In contrast, calculating the mean speed of the particles requires an application of statistical mechanics, along with the *Boltzmann factor* $e^{-E/(kT)}$, which gives the spread of particle energies. We'll discuss this in Sect. 3.5.2. The velocity vectors of the gas particles are gathered together, and by considering the density at which they fan out from the origin of *velocity space*, we can calculate the number of gas particles per unit speed interval. This is the famous *Maxwell distribution* of gas particle speeds. Being a simple function of speed (in fact, proportional to the square of the speed and a gaussian function of the speed), it is easy to deal with analytically, allowing the mean speed to be calculated as $\sqrt{8kT/(\pi m)}$.

Yet another related quantity is the *most probable* speed of the particles, where the Maxwell distribution peaks. This turns out to be $\sqrt{2kT/m}$.

$$\begin{aligned}\text{distance}^2 \equiv |\text{position}|^2 &= (\boldsymbol{L}_1 + \cdots + \boldsymbol{L}_n)\cdot(\boldsymbol{L}_1 + \cdots + \boldsymbol{L}_n) \\ &= nL^2 + \boldsymbol{L}_1\cdot\boldsymbol{L}_2 + \boldsymbol{L}_1\cdot\boldsymbol{L}_3 + \cdots . \end{aligned} \tag{3.35}$$

In averaging over the ensemble, the cross terms are just as likely to be positive as negative since there is no correlation from one step to the next. They therefore make no contribution to the average, and the mean of the squared distances becomes

$$\left\langle \text{distance}^2 \right\rangle = nL^2 . \tag{3.36}$$

Thus the root-mean-squared distance from start to end is

$$\text{rms distance} \equiv \sqrt{\left\langle \text{distance}^2 \right\rangle} = \sqrt{n}\, L\, . \tag{3.37}$$

Here we have the famous result that the rms distance from the start point only increases as the *square root* of the number of steps. Calculating the true mean distance is a good deal more complicated, and that's why the rms value is usually used.

The random walk is featured in the constructive interference of light, and so it figures in the explanation of why laser light is so bright. Suppose we have n *incoherent* light sources, such as incandescent bulbs. By "incoherent" we mean that, like the steps of the drunken man, phase information is not preserved over time or space, and this is the case for incandescent bulbs that use a hot filament to generate their light. At any particular place and time, the strength of the electric field due to light from the ith bulb is represented by a phasor (a vector $\boldsymbol{E}_i$ of magnitude E). Light intensity is proportional to the square of the amplitude of the total electric field. The total electric field is found by adding the phasors as vectors, so the total intensity at the point of interest must be

$$I_{\text{tot}} \propto \left|\sum_i \boldsymbol{E}_i\right|^2 . \tag{3.38}$$

We are back to a sum of squares and the random walk (where $\boldsymbol{E}_i$ has the role of the step $\boldsymbol{L}_i$), since the incoherent light bulbs produce electric vectors $\boldsymbol{E}_i$ that have no relation to each other. Over the integration time of the human eye, the total intensity smoothens out just like the steps of the random walk, and its average I_{tot} becomes proportional to nE^2, or n times the average intensity due to one light bulb alone. This is expected, of course, from our everyday experience, and it shows why the concepts of rms value and the random walk are so closely allied with our physical perceptions.

If the light sources are coherent—which is to say that the "steps" are correlated from one to the next, such as in a laser (the drunk sobers up!)—then the random walk picture no longer applies. If n lasers are carefully tuned to be in phase with one another, their electric field vectors at the screen will add constructively, so that

$$I_{\text{tot}} \propto \left|\sum_i \boldsymbol{E}_i\right|^2 = |n\boldsymbol{E}_1|^2 = n^2E^2 . \tag{3.39}$$

The average I_{tot} is now n^2 times the average intensity due to one laser alone, as contrasted with the n-fold multiplication for incandescent bulbs. In a lone laser, the sources are really lasing atoms, so that here again the sum of squares shows why the coherency of a single laser makes it so very bright.

Fluctuation, Dissipation, and $\sqrt{\mathrm{dt}}$

One of the odd features of random walks is that they lead us to write differential equations that involve square roots of infinitesimals. To see why this might be, consider the simple case of a random walk in which each step has length ΔL and takes a time Δt. In a unit time, the number of steps taken is $1/\Delta t$, so that the rms distance from the origin is, from (3.37),

$$\text{rms distance} = \sqrt{\text{number of steps}} \times \Delta L = \frac{\Delta L}{\sqrt{\Delta t}} . \tag{3.40}$$

Suppose we require that this rms distance tend to a constant k as $\Delta t \to 0$. In that case, the speed of the particle is given by

$$\frac{dL}{dt} \to \frac{k\sqrt{dt}}{dt} = \frac{k}{\sqrt{dt}} . \qquad (3.41)$$

The limit idea of this *Wiener process* really describes a series of ever-finer models of *brownian motion.* This is the motion of a particle that is much larger or more massive than the molecules that surround it, while still small or light by laboratory standards. In (3.41), the particle has been required to move at whatever speed is necessary to match real-world observations of the slow, steady drift characteristic of brownian motion.

A more detailed analysis of such motion produces a model in which the particle's velocity can certainly change. This, a *stochastic* model, represents the environment's effect on the particle by a statistical distribution. The force on the particle is composed of two terms: a zero-mean *fluctuation* term that's uncorrelated over time, and a *dissipation* term that describes a steady drag. The fluctuation term is stochastic, meaning that its value is drawn from a probability distribution. Thus Newton's force-acceleration law itself becomes stochastic when approximated in this brownian motion realm. The effect is that this law might appear to suffer from some amount of indeterminacy, but that is purely a result of the model; there is no implication of anything inherently unpredictable about the physical laws that the meandering particle obeys.

The $\sqrt{dt}$ that appeared in (3.41) finds its way into the force-acceleration law obeyed by a brownian particle. In one dimension, the infinitesimal increase $m\,dv$ in the particle's momentum after a time dt can be written as

$$m\,dv = \underbrace{\sqrt{2\gamma kT}\,N(0,1)\,\sqrt{dt}}_{\text{fluctuation term}} - \underbrace{\gamma v\,dt}_{\text{dissipation term}} , \qquad (3.42)$$

where the constant γ is the *drag coefficient,* k is Boltzmann's constant, T is the temperature of the bath through which the mass m is moving, and the fluctuations' stochastic nature is represented by including a term drawn from a normal distribution $N(0,1)$ with zero mean and unit standard deviation. Should the dt term be omitted as negligible compared with the $\sqrt{dt}$ term? No, because the fluctuation has zero mean and so over a long time produces about as much effect as the dt term, which does not fluctuate and so is more steady in its effect. Finally, the fact that the drag coefficient γ is present in both terms shows that fluctuation and dissipation must both be present; we cannot have one without the other. The theory of stochastic processes—which is really just an abbreviated name for the theory of stochastic *models* of completely deterministic processes—builds on expressions such as (3.42), using them to make predictions for situations that are too complicated in their details to be studied by more classical methods. We'll encounter fluctuation–dissipation again in Sect. 11.5.

3.3 Least Squares Analysis, Bayes' Theorem, and the Matrix Pseudo Inverse

In Sect. 3.2, we saw how a sum-of-squares definition for the spread of data points ties naturally with the mean, leading to the normal distribution that governs so many statistical aspects of Nature. In this section, we will examine how best to fit a curve to a set of data points along with what the idea of a best fit really means. The idea of fitting a curve to a set of data is fundamental in statistical theory, and it turns out that a sum of squares is the best tool for the job. In order to study this a little more closely and see why it should be so, we pause to look back at the completeness relation that we first met in the previous chapter.

Completeness Relations Again

As we have seen, a completeness relation concisely expresses a great many results of linearity, and so appears in matrix theory, Fourier analysis, and quantum mechanics. A very similar-looking identity can also be written down for probability theory. Suppose an event a can occur with probability $p(a)$. Since we are only dealing with probabilities here, the ubiquitous p can be omitted, and we write (a) for the probability of the outcome a. Besides saving some effort, this will also allude to the bracket notation of the previous chapter.

Consider a set of events $\{b_i\}$ that are separate from a, and ask the question: what is the probability (a, b_k) of both outcomes a and b_k? This is the product of two probabilities: the probability $(a|b_k)$ of event a occurring given that b_k occurred, times the probability (b_k) that b_k itself occurs:

$$(a, b_k) = (a|b_k)(b_k)\,. \tag{3.43}$$

As a further aid with the notation, introduce a statement that is true by definition, or equivalently an event that occurs with absolute certainty. Call this event "1". Then it must certainly be true that $(a, 1) = (a)$ and also that $(a|1) = (a)$. Equation (3.43) writes this as

$$(a, 1) = (a|1)(1)\,, \tag{3.44}$$

which is certainly true since (1) must be equal to 1. Our notation is self-consistent so far! But now notice something. If the set of b_i's is exhaustive, so that $\sum(b_i) = 1$, then the probability that event a occurs is just the sum over all the probabilities that a happens together with one of the set of b_i's:

$$(a) = \sum_i (a, b_i)\,, \tag{3.45}$$

which using (3.43) can be written as

$$(a) = \sum_i (a|b_i)(b_i)\,; \tag{3.46}$$

and this in turn can be rewritten by introducing event 1:

$$(a|1) = \sum_i (a|b_i)(b_i|1)\,. \tag{3.47}$$

At this point we allow ourselves some licence with the notation, to match the brackets of the previous chapter. The quantity $(p|q)$ will be considered to be a "product" of two new entities, $(p|$ and $|q)$, and also each of these can be "multiplied" by the number 1 without being changed. In that case, the essence of (3.47) becomes

$$\sum_i |b_i)(b_i| = 1\,. \tag{3.48}$$

So this is really just a completeness relation, and links together conditional probabilities such as in (3.46). Thus, it enables us very quickly to write down (3.46) with only a moment's consideration about whether the set of b_i's is complete; but its real significance is in the fact that it encapsulates the linearity of probability theory, so that the mathematics of conditional probability is well represented by matrix formalism.

As an aside, if $(a|1) = (a)$, then what meaning does $(1|a)$ have? The probability that event 1 occurs given some a must be 1, but we can also derive it as follows. Notice that $(b_k, a) = (a, b_k)$, so (3.43) gives

$$(b_k|a)(a) = (a|b_k)(b_k)\,, \tag{3.49}$$

or

$$(b_k|a) = \frac{(a|b_k)(b_k)}{(a)} = \frac{(a|b_k)(b_k)}{\sum_i (a|b_i)(b_i)}\,. \tag{3.50}$$

This is *Bayes' theorem*, which has many applications in probability theory and *signal processing*, the modern field that is concerned with studying how noisy signals can be processed to extract a maximum of information from them. Setting $b_k = 1$ in Bayes' theorem indeed yields the expected $(1|a) = 1$.

To see Bayes' theorem in action, consider the following scenario. A certain disease is known to strike 0.01% of the population. A test has been developed to screen for it. Apparently it is quite a successful test because it correctly identifies 99 out of 100 people who have the disease, while returning a false alarm result for just one person in a thousand. We ask the question: if the test returns a positive result (i.e. disease detected), what is the chance that the person tested really has the disease?

The above scenario describes two errors that the test can make. The *probability of a false alarm* is 1/1000 ("one person in a thousand"). Contrast this with the *probability of a missed detection*, which is $1 - 99/100 = 1/100$. Statisticians traditionally call a false alarm a *Type 1 error*, and call a missed detection a *Type 2* error. These uninsightful labels obscure the fact that we are constantly balancing the chances of making these two errors in every aspect of life. In general, any process that lowers the chance of our making one of these errors will raise the chance of our making the other error. In the courtroom, these two errors are the acts of *finding an innocent man guilty*, and *finding a guilty man innocent*, respectively. (This is not to imply that statistical theory is necessarily understood by the court system; in fact, a lack of understanding of it can easily put innocent people in jail.)

An everyday example of this balance lies in the simple act of setting the volume of a transistor radio. Set it to loud, and we hear all the music, but we also hear noise crackles (high probability of false alarms). Set it to soft, and we hear no crackles—but now, listening to the music is more difficult (high probability of missed detections).

Let h stand for a healthy state, d stand for the disease being present, and + stand for a positive test result. We require the probability that the disease is present given a positive test result, or $(d\,|+)$. We know that

$$(d) = 0.01\%\,, \quad (+|d) = 99\%\,, \quad (+|h) = 0.1\%\,. \tag{3.51}$$

Bayes' theorem gives

$$\begin{aligned}(d\,|+) &= \frac{(+|d)(d)}{(+)} = \frac{(+|d)(d)}{(+|d)(d) + (+|h)(h)} \\ &= \frac{0.99 \times 10^{-4}}{0.99 \times 10^{-4} + 0.001 \times (1 - 10^{-4})} \\ &\simeq 9\%\,. \end{aligned} \tag{3.52}$$

Perhaps surprisingly, when the test returns a positive result, there is only a 9% chance that the patient really has the disease. What has happened here is that although the test gives a false positive for only 0.1% of patients, the overwhelming number of patients will be healthy, providing many opportunities for the test to register a false alarm. So the difference between $(d\,|+)$ and $(+|d)$ is crucial to an understanding of the power of the test. Here we see a good example of how the old saying "There are lies, damned lies, and statistics" can apply when a proper analysis of statistical statements is not made.

3.3.1 Least Squares Analysis for Curve Fitting

Another example of this difference between $(d\,|+)$ and $(+|d)$ occurs when we require the best estimate of an unknown quantity x given a set of data. There are divided opinions among statisticians as to what exactly constitutes a best estimate. Two important definitions of the best value of x are:

***Maximum likelihood* estimate of x**: the value of x that maximises $(\text{data}\,|\,x)$;

***Maximum a posteriori* estimate of x**: the value of x that maximises $(x\,|\,\text{data})$.

Bayes' theorem states that there can be a difference between these two estimates. As an example, consider fitting a gaussian to one peak of an X ray spectrum produced by bombarding a rock sample with energetic protons. The protons knock out inner electrons from elements in the sample, exciting the atoms which quickly decay by emitting characteristic X and γ rays. These rays are then counted by detectors subject to statistical fluctuations, producing plots of photon numbers detected versus their energies. Different elements in the sample produce peaks of detected photons at different energies, allowing the rock's elemental makeup to be inferred. By fitting an appropriate curve to each peak, we are able to isolate the peaks and so subtract whatever remains as *noise*. How are these curves to be fitted?

Suppose each curve specifies some sort of mean energy E_0 and a characteristic width σ. Consider each of the following fitting methods in turn.

Maximum Likelihood Fit. Consider first the maximum likelihood estimates of E_0, σ. Given these, we wish to maximise the probability $(\text{data} \,|\, E_0, \sigma)$ that the measured data could result, when the expected number of counts in each energy bin E is given by the fitted curve $f_{E_0,\sigma}(E)$. There are two different statistical processes to keep track of here. The fit predicts the counts in each bin based on the fitting function $f_{E_0,\sigma}(E)$, but we are concerned with the *departure* of the measured number of counts in each bin from this predicted number. Generally each bin itself introduces gaussian statistics, and the value of $(\text{data} \,|\, E_0, \sigma)$ is given by a product of gaussian expressions, one for each of the energy bins. As shown in Fig. 3.3, we must therefore multiply exponentials of negative squares of deviations from the predicted fit (one for each bin), and then maximise that product over all values of E_0, σ. This is equivalent to summing the squares themselves, and then *minimising* this sum.

But this is just the *method of least squares*, and so we have come full circle. Earlier in this chapter we showed that a measure of spread that incorporates a sum of squares (the standard deviation) forms a natural pair with the usual definition of a mean; then we found that the ubiquitous normal distribution is expressed in terms of these two quantities; and finally we used this normal distribution to show again that a natural way to best-fit a curve is by minimising a sum of squares.

Maximum a Posteriori Fit. The maximum a posteriori estimate of E_0, σ chooses the best estimate of the fitting parameters E_0, σ to be those that maximise $(E_0, \sigma \,|\, \text{data})$. Bayes' theorem (3.50) helps us along here:

$$(E_0, \sigma \,|\, \text{data}) = \frac{(\text{data} \,|\, E_0, \sigma)(E_0, \sigma)}{\sum_{ij} (\text{data} \,|\, E_{0i}, \sigma_j)(E_{0i}, \sigma_j)} \,. \tag{3.53}$$

Now, at this point we ask: what is the value of (E_{0i}, σ_j), the probability of occurrence of any two parameters E_{0i}, σ_j? This is called the *prior* probability

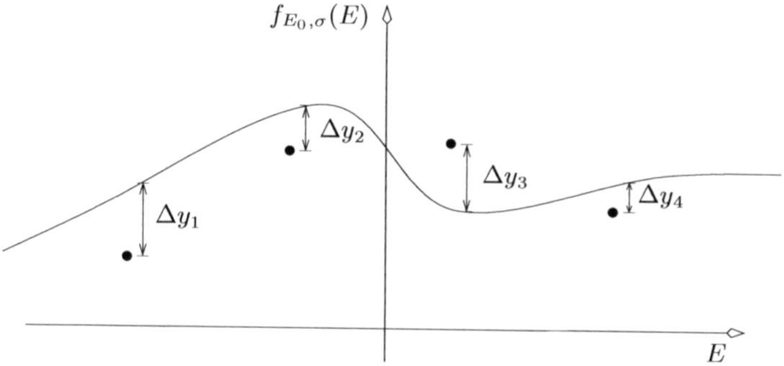

Fig. 3.3. Fitting a curve to a set of points using the maximum likelihood philosophy. Suppose we have a model that predicts that due to statistical variation, the points will depart from the curve with a gaussian spread (which is typically the case). Given the arbitrary hypothesised curve shown, the probability that the points really could have resulted is the product of the probabilities that each could have resulted separately. Assuming that the points are independent, this probability is $(\text{data} \mid \text{curve}) \propto \mathrm{e}^{-k\Delta y_1^2}\, \mathrm{e}^{-k\Delta y_2^2} \ldots \mathrm{e}^{-k\Delta y_4^2}$ for some k. Adjusting the curve in order to maximise this product is equivalent to minimising $\Delta y_1^2 + \Delta y_2^2 + \cdots + \Delta y_4^2$, which is none other than the method of least squares.

(or just prior for short): the chance that the rock sample really does have an element that could produce a peak of emitted photons at energy E_{0_i} with width σ_j. This is information we probably do not have in any real situation, and it's even debatable to what extent these priors can be considered as well-defined probabilities at all.

Without such prior information, the most even-handed approach is to set all the priors equal, in which case they will cancel in (3.53), producing a maximum a posteriori estimate exactly the same as the maximum likelihood estimate. On the other hand, if we do have knowledge of the priors, perhaps because we know in advance some of the elements that the rock sample contains, then the curve fitted by the maximum a posteriori method can be different from that found via maximum likelihood.

Linear Least Squares Fitting: More than Just Straight Lines

Least squares analysis is one area that tends to be seen as thoroughly within the domain of calculus, being all about minimisation. In fact, it forms a good example of the interface between linear algebra and calculus, because minimising a sum of squares is equivalent to minimising a distance in euclidean space. So the data to be fitted can be arranged in a matrix, which makes for very concise and efficient computation. This use of matrices to shuffle data about in highly efficient ways is the bread and butter of modern signal processing theory. Here we use the example of least squares fitting to give a flavour of some of the ideas.

A common misconception on first encountering least squares theory is to think that it concerns only fitting a line to a set of points. This is not so; the linearity concerns not the data but the coefficients of the fit: *these* must occur linearly in the equation of the curve to be fitted. So a parabola $y = ax^2 + bx + c$ can be fitted just as well as the straight line $y = mx + c$, since in both cases the coefficients a, b, c or m, c occur linearly. This linear requirement is a second reason why matrices have a natural place in least squares theory.

Imagine that we are given a set of data points $(x_1, y_1), \dots, (x_n, y_n)$ to which a curve is to be fitted, say $y = ax^2 + bx + c$. If the curve did fit the points perfectly, then it would certainly be true that

$$\underbrace{\begin{bmatrix} y_1 \\ \vdots \\ y_n \end{bmatrix}}_{\equiv \boldsymbol{y}} = \underbrace{\begin{bmatrix} x_1^2 & x_1 & 1 \\ \vdots & \vdots & \vdots \\ x_n^2 & x_n & 1 \end{bmatrix}}_{\equiv A} \underbrace{\begin{bmatrix} a \\ b \\ c \end{bmatrix}}_{\equiv \boldsymbol{z}} . \tag{3.54}$$

More generally, if the curve does not fit the points exactly, then the two sides of (3.54) won't be identical. There will be some small noise present represented by the addition of a vector $\boldsymbol{\nu}$, so that (3.54) becomes $\boldsymbol{y} = A\boldsymbol{z} + \boldsymbol{\nu}$. This, then, is the essence of a least squares fit: we must choose a, b, c so that the noise $\boldsymbol{\nu} = \boldsymbol{y} - A\boldsymbol{z}$ is minimised in a least squares sense. That means that the sum of the squares of the elements of $\boldsymbol{\nu}$ is minimised.

But such a sum of squares is really a distance in euclidean geometry, the length of $\boldsymbol{\nu}$, so that given $\boldsymbol{y}$ and A, we are required to find the value of $\boldsymbol{z}$, comprised of "three least squares estimates of a, b, c", that will minimise the distance between the n-dimensional vectors $\boldsymbol{y}$ and $A\boldsymbol{z}$.

As usual, rather than deal directly with the length $|\boldsymbol{y} - A\boldsymbol{z}|$, it's easier to minimise its square—an entirely equivalent problem. This is easily written in matrix form:

$$\begin{aligned} |\boldsymbol{y} - A\boldsymbol{z}|^2 &= (\boldsymbol{y} - A\boldsymbol{z})^t (\boldsymbol{y} - A\boldsymbol{z}) = \left(\boldsymbol{y}^t - \boldsymbol{z}^t A^t\right)(\boldsymbol{y} - A\boldsymbol{z}) \\ &= \boldsymbol{y}^t\boldsymbol{y} - \boldsymbol{z}^t A^t \boldsymbol{y} - \boldsymbol{y}^t A\boldsymbol{z} + \boldsymbol{z}^t A^t A\boldsymbol{z} \\ &= \boldsymbol{y}^t\boldsymbol{y} - 2\boldsymbol{y}^t A\boldsymbol{z} + \boldsymbol{z}^t A^t A\boldsymbol{z} \,. \end{aligned} \tag{3.55}$$

(The last line follows because all of the terms in the line before are scalars and so are equal to their transposes.) Minimising the right-hand side of the last line with respect to each of the elements of $\boldsymbol{z}$ requires that its gradient be calculated with respect to those elements and then set equal to zero. For this, two theorems are useful. The first says:

$$\textit{For all column vectors } \boldsymbol{a}, \quad \nabla_{\boldsymbol{z}} \left(\boldsymbol{a}^t \boldsymbol{z}\right) = \boldsymbol{a}^t , \tag{3.56}$$

where $\nabla_{\boldsymbol{z}}$ is the row vector of derivatives with respect to the elements of $\boldsymbol{z}$. The proof is very straightforward. For tidiness, use the convention that repeated indices are summed over.

$$\nabla_{\boldsymbol{z}}\left(\boldsymbol{a}^t\boldsymbol{z}\right) = \nabla_{\boldsymbol{z}}(a_i z_i) = [a_1, \ldots, a_n] = \boldsymbol{a}^t. \quad \text{QED.} \tag{3.57}$$

The second theorem is very similar but involves matrices instead of vectors:

$$\textit{For all square matrices } B, \quad \nabla_{\boldsymbol{z}}\left(\boldsymbol{z}^t B\boldsymbol{z}\right) = \boldsymbol{z}^t\left(B + B^t\right). \tag{3.58}$$

The proof is a little more involved but is done in exactly the same way: we calculate the kth element of the gradient on the left-hand side of (3.58), and use it to build the right-hand side of (3.58):

$$\begin{aligned}\left[\nabla_{\boldsymbol{z}}\left(\boldsymbol{z}^t B\boldsymbol{z}\right)\right]_k &= \frac{\partial}{\partial z_k} B_{ij}\, z_i\, z_j = B_{ij}\left(\delta_{ik}\, z_j + z_i\delta_{jk}\right) = B_{kj}\, z_j + B_{ik}\, z_i \\ &= (B\boldsymbol{z})_k + (\boldsymbol{z}^t B)_k\,.\end{aligned} \tag{3.59}$$

Now, $B\boldsymbol{z}$ is a column vector and $\boldsymbol{z}^t B$ is a row vector. We want the gradient to be a row vector, so we should write $(B\boldsymbol{z})_k = (\boldsymbol{z}^t B^t)_k$, which then gives

$$\left[\nabla_{\boldsymbol{z}}\left(\boldsymbol{z}^t B\boldsymbol{z}\right)\right]_k = (\boldsymbol{z}^t B^t)_k + (\boldsymbol{z}^t B)_k\,, \tag{3.60}$$

so that

$$\nabla_{\boldsymbol{z}}\left(\boldsymbol{z}^t B\boldsymbol{z}\right) = \boldsymbol{z}^t B^t + \boldsymbol{z}^t B = \boldsymbol{z}^t\left(B^t + B\right). \quad \text{QED.} \tag{3.61}$$

Returning to the original expression in (3.55), we use the two theorems (3.56) and (3.58) to write

$$\begin{aligned}\nabla_{\boldsymbol{z}}\left|\boldsymbol{y} - A\boldsymbol{z}\right|^2 &= \nabla_{\boldsymbol{z}}\left(\boldsymbol{y}^t\boldsymbol{y} - 2\boldsymbol{y}^t A\boldsymbol{z} + \boldsymbol{z}^t A^t A\boldsymbol{z}\right) \\ &= -2\boldsymbol{y}^t A + 2\boldsymbol{z}^t A^t A\,,\end{aligned} \tag{3.62}$$

and this must be equated with the zero matrix to yield the required minimum, defining the best fit $\boldsymbol{z}$ in the process as $\bar{\boldsymbol{z}}$. Thus

$$\bar{\boldsymbol{z}}^t A^t A = \boldsymbol{y}^t A\,. \tag{3.63}$$

Transposing both sides renders this equation somewhat cleaner, in that the result is exactly the same as what would have been written had we (mistakenly!) written the equations to be solved as $A\boldsymbol{z} = \boldsymbol{y}$ (an overspecified set) and then premultiplied both sides by A^t:

$$A^t A\bar{\boldsymbol{z}} = A^t\boldsymbol{y}\,. \tag{3.64}$$

(That's an elegant point that makes it easy to recreate (3.64) from memory.) Finally, $A^t A$ will be invertible for any reasonable set of data, so

$$\bar{\boldsymbol{z}} = (A^t A)^{-1} A^t\boldsymbol{y} \equiv A^{\#}\boldsymbol{y}\,. \tag{3.65}$$

The expression $(A^t A)^{-1} A^t$ is known as the *pseudo inverse* of A (written $A^{\#}$), since it left-multiplies A to give the identity matrix; the same cannot be guaranteed to happen for right-multiplication. The pseudo inverse is an apt name because, as stated above, in a sense we have started with the over-specified equation $A\boldsymbol{z} = \boldsymbol{y}$ and then "inverted" it as best we could, by using not A^{-1} but $A^{\#}$ (cf. if A were square, then $A^{-1} = A^{\#}$). The resulting expression (3.65) is not only an elegant use of matrices, but also is used very often in the field of signal processing when data are being fitted, such as in tracking moving objects, where we are presented with a lot of data and need to find the best solution to a set of equations that might result from a trigonometric analysis of the scenario.

Actually, the pseudo inverse is defined a little more generically than this, but certainly $(A^t A)^{-1} A^t$ satisfies the more general definition. The definition used here can be badly behaved numerically for certain *ill-conditioned* data sets, and building better alternatives is an area of linear algebra that is especially important to numerical computation software.

Covariance Matrices and Correlation

When doing calculations involving lots of sums, it is very helpful to realise that the operation of summation is linear. This means that for constants a, b and variables x, y, we have $\sum (ax_i + by_i) = a \sum x_i + b \sum y_i$, as is easily proved by writing out the sums term by term. Of course, this also implies that the calculation of a mean $m(\cdot)$ is linear: $m(ax + by) = a\,m(x) + b\,m(y)$, since these means are calculated using summations.

In contrast, the standard deviation is *not* linear, since it involves a square root, which has no linear properties. In that case, might the process of calculating a variance $s^2(\cdot)$ be linear instead? Let's calculate $s^2(ax + by)$ and see how it relates to $s^2(x)$ and $s^2(y)$.

Note that $s^2(x)$ and $s^2(y)$ are more usually called s_x^2, s_y^2. We have used parentheses here and in the mean above merely to stress the fact that the mean and variance are being considered as *operators* on a set of data.

We will suppress all summation indices on x, y, and denote the means of x, y by m_x, m_y, respectively:

$$\begin{aligned} s^2(ax + by) &\equiv \frac{1}{n} \sum (ax + by - am_x - bm_y)^2 \\ &= \frac{1}{n} \sum \left[a\,(x - m_x) + b\,(y - m_y) \right]^2 \\ &= a^2 s^2(x) + b^2 s^2(y) + \frac{2ab}{n} \sum (x - m_x)(y - m_y) \,. \end{aligned} \tag{3.66}$$

The variance certainly is not linear. Along with the coefficients a, b being squared, an extra term combining the x and y data has appeared. This extra term (without the $2ab$) is called the *covariance* s_{xy} of x and y:

$$s_{xy} \equiv \frac{1}{n} \sum (x - m_x)(y - m_y) \,. \tag{3.67}$$

Nonlinear Least Squares Fitting: The Gauss–Newton Algorithm

Suppose that the set of data points $(x_1, y_1), \ldots, (x_n, y_n)$ is to be fitted not by some linear-coefficient curve such as $y = a_1 x^2 + a_2 x + a_3$, but rather by one in which the coefficients occur nonlinearly: $y = f(x; a_1, \ldots, a_n) \equiv f(x; \boldsymbol{z})$. There are various approaches to finding the function $y = f(x; \boldsymbol{z})$, and in general the problem can be quite difficult, especially if the fit is destined to be poor. One approach utilising the work of the preceding pages, known as the *Gauss–Newton* algorithm, linearises $f(x; \boldsymbol{z})$ about some starting estimate $\boldsymbol{z}_0$ of the coefficients $a_1, \ldots, a_n$ (where $\boldsymbol{z}, \boldsymbol{z}_0$ are column vectors). Begin by relating the data to the fit, with a noise term $\boldsymbol{\nu}$ added, and then Taylor-expand to first order:

$$\begin{bmatrix} y_1 \\ \vdots \\ y_n \end{bmatrix} = \begin{bmatrix} f(x_1; \boldsymbol{z}) \\ \vdots \\ f(x_n; \boldsymbol{z}) \end{bmatrix} + \boldsymbol{\nu} \simeq \begin{bmatrix} f(x_1; \boldsymbol{z}_0) + \nabla_{\boldsymbol{z}} f(x_1; \boldsymbol{z}_0)(\boldsymbol{z} - \boldsymbol{z}_0) \\ \vdots \\ f(x_n; \boldsymbol{z}_0) + \nabla_{\boldsymbol{z}} f(x_n; \boldsymbol{z}_0)(\boldsymbol{z} - \boldsymbol{z}_0) \end{bmatrix} + \boldsymbol{\nu}\,. \tag{3.68}$$

In other words,

$$\begin{bmatrix} y_1 - f(x_1; \boldsymbol{z}_0) \\ \vdots \\ y_n - f(x_n; \boldsymbol{z}_0) \end{bmatrix} \simeq \begin{bmatrix} \nabla_{\boldsymbol{z}} f(x_1; \boldsymbol{z}_0) \\ \vdots \\ \nabla_{\boldsymbol{z}} f(x_n; \boldsymbol{z}_0) \end{bmatrix} (\boldsymbol{z} - \boldsymbol{z}_0) + \boldsymbol{\nu}\,, \tag{3.69}$$

which when solved for $\boldsymbol{z}$ will give a new estimate of $\boldsymbol{z}$, since truncating the Taylor series means the results are only approximate. Equation (3.69) now resembles (3.54), so that an updated estimate of $\boldsymbol{z}$ can immediately be written down via a least squares solution using the pseudo inverse defined on p. 99:

$$\boldsymbol{z} \simeq \boldsymbol{z}_0 + \begin{bmatrix} \nabla_{\boldsymbol{z}} f(x_1; \boldsymbol{z}_0) \\ \vdots \\ \nabla_{\boldsymbol{z}} f(x_n; \boldsymbol{z}_0) \end{bmatrix}^{\#} \begin{bmatrix} y_1 - f(x_1; \boldsymbol{z}_0) \\ \vdots \\ y_n - f(x_n; \boldsymbol{z}_0) \end{bmatrix}. \tag{3.70}$$

Equation (3.70) can be iterated by replacing $\boldsymbol{z}_0$ with the new estimate, and will often converge to a solution. For a similar approach to solving nonlinear simultaneous equations, see the box on p. 367. In practice, nonlinear least squares fits often involve weighting the contributions from some points more than others, which might better reflect the way the data have been gathered.

(s_{xy} conventionally is not written as a square; it suffices to have two indices.) We can see immediately that the covariance of x with itself is the variance of x: $s_{xx} = s_x^2$, so the covariance is a generalisation of the variance to multiple variables. If the covariance of x and y is zero, meaning the two variables are *uncorrelated*, then the expression for the variance $s^2(ax + by)$ comes as close to being linear as it can:

$$s^2(ax+by) \underset{\text{uncorr.}}{\overset{x,y}{=\!=}} a^2 s^2(x) + b^2 s^2(y)\,. \tag{3.71}$$

When more than two variables are involved, the set of covariances among all the pairs can be written as a matrix P:

$$P = \begin{bmatrix} s_x^2 & s_{xy} & s_{xz} \cdots \\ & s_y^2 & s_{yz} \cdots \\ \text{(symmetric)} & & s_z^2 \cdots \end{bmatrix}, \tag{3.72}$$

which finds a major use in the field of signal processing when estimating the values of variables based on noisy data.

A simple expression for the covariance can be found by expanding the parentheses in (3.67) to obtain

$$s_{xy} = m_{xy} - m_x\, m_y\,, \tag{3.73}$$

which is similar to the simplified expression for the variance $s_x^2 = m_{x^2} - m_x^2$ that we noted in the paragraph before (3.31). Two variables are thus uncorrelated when the mean of their product equals the product of their means.

Note that this is not quite the same as the variables being *independent*, by which is meant that the probability of obtaining given values of x and y can be factored: $p(x,y) = p_1(x)\,p_2(y)$. Independence certainly does imply noncorrelation, but the converse is not true.

The idea of the covariance appears under the different name of *regression analysis*, which is concerned with how well a least squares line fits a set of data. Let's look briefly at how this comes about.

Quantifying Least Squares Fits of Lines Using the Covariance

Two variables are defined to be uncorrelated if their covariance is zero. Not surprisingly, this idea of correlation also relates to the idea of plotting a set of data points and fitting a least squares line through it. Suppose we plot y versus x for the set $(x_1, y_1), \ldots, (x_n, y_n)$, and then fit a least squares line $y = mx + c$ by writing, similar to (3.54),

$$\underbrace{\begin{bmatrix} y_1 \\ \vdots \\ y_n \end{bmatrix}}_{\equiv\, \boldsymbol{y}} \simeq \underbrace{\begin{bmatrix} x_1 & 1 \\ \vdots & \vdots \\ x_n & 1 \end{bmatrix}}_{\equiv\, A} \begin{bmatrix} m \\ c \end{bmatrix}, \tag{3.74}$$

with the "$\simeq$" sign showing that we are aware that (3.74) is not exact but can be "pseudo-inverted" to give an exact solution for m and c using the ideas of the previous few pages. In particular, (3.65) gives

$$\begin{bmatrix} m \\ c \end{bmatrix} = A^{\#}\boldsymbol{y} = (A^t A)^{-1} A^t \boldsymbol{y} = \begin{bmatrix} s_{xy}/s_x^2 \\ m_y - m_x\, s_{xy}/s_x^2 \end{bmatrix}. \tag{3.75}$$

This equation is not symmetric in x and y because the least squares fit seeks to minimise the sum of the squared *vertical* distances of the data points from the line, as opposed to, say, horizontal distances or closest distances. How well does the line fit? We need only calculate this sum I^2 of squared vertical deviations; for a perfect fit it will be zero:

$$I^2 = \frac{1}{n} \sum (y_i - mx_i - c)^2 \overset{(3.75)}{=\!=} s_y^2 \left(1 - \frac{s_{xy}^2}{s_x^2\, s_y^2}\right)$$
$$\equiv s_y^2 \left(1 - r^2\right) , \quad \text{where } r \equiv \frac{s_{xy}}{s_x\, s_y} . \tag{3.76}$$

The parameter r is called the *regression coefficient* for the data, and is really just a normalised covariance. We can see this by noting that

$$s_{xy}^2 \leqslant s_x^2\, s_y^2 , \tag{3.77}$$

which is certainly true, because expanding each side of (3.77) in terms of x and y deviations from their means shows that the right-hand side of (3.77) is just equal to its left-hand side with the addition of positive cross terms. As a consequence, $r^2 \leqslant 1$, so that the regression coefficient r is nothing more than a normalised covariance.

When x and y are uncorrelated, $r = 0$, and the rms deviation of the points from the line is then $I = s_y$. But this is exactly to be expected; after all, $r = 0$ implies that $m = 0$ and $c = m_y$, so that the best-fit line is horizontal and divides the y data exactly at their mean. Also, when $r = \pm 1$ the rms deviation is $I = 0$. Thus the points all lie on the line: they are perfectly correlated. Finally, since m and r share the same sign, the sign of r determines whether the points are *positively* or *negatively* correlated (i.e., positive for $r > 0$).

As a last note, if we plot the points the other way around (x versus y), the least squares fit is still done by minimising the sum of squared *vertical* deviations, so that the roles of x and y simply interchange in (3.74)–(3.77). It follows that the product of the slopes of the best-fit lines (x versus y and y versus x) will be r^2. So if the points are perfectly positively correlated ($r = 1$), then interchanging the roles of x and y will flip the best-fit line about the $y = x$ line, as we might expect. (Likewise, if the points are perfectly negatively correlated, $r = -1$ and the best-fit line flips about $y = -x$). But if the points are perfectly uncorrelated, then $r = 0$, and the best-fit line is horizontal in *both* plots.

Fisher Information and a Special Covariance Matrix

A special covariance matrix that finds a use in signal processing is the inverse of the *Fisher information matrix*, and here we give a brief discussion about what it is and why it should be useful, omitting the relevant proofs. Suppose the results of some measurements form a set of numbers $\boldsymbol{x} \equiv x_0, \ldots x_n$ from which we require to extract a parameter θ of interest. For example, θ might be the average age of a population, the position of a spacecraft, or the predicted cost of a commodity. To make use of the data, we wish to find an *estimator* $\widehat{\theta} \equiv \widehat{\theta}(\boldsymbol{x})$. An *unbiassed* estimator $\widehat{\theta}$ is one that takes on the correct value θ on the average, or, in statistical language, the *expected value over the data set* is $\mathbb{E}_{\boldsymbol{x}}\{\widehat{\theta}\} = \theta$. (Note that in the box on p. 86, μ_{be} and σ_{be} are unbiassed.)

The probability that, given some particular θ, the measurements $\boldsymbol{x}$ will be observed is the *likelihood* $(\boldsymbol{x}|\theta)$ referred to on p. 95, although of more use here is the negative logarithm of the likelihood: $L \equiv -\ln(\boldsymbol{x}|\theta)$. It can be shown in a few lines of algebra that a well-behaved likelihood will be *regular*, meaning that

$$\mathbb{E}_{\boldsymbol{x}}\{\partial L/\partial\theta\} = 0 \quad \text{for all } \theta, \tag{3.78}$$

which is really a statement that differentiation with respect to θ and summing over the values $x_0, \ldots, x_n$ commute. The Fisher information F is then defined as

$$F \equiv \mathbb{E}_{\boldsymbol{x}}\left\{\frac{\partial^2 L}{\partial\theta^2}\right\} \underset{\text{regular}}{\overset{p_\theta(\boldsymbol{x})}{=\!=\!=}} \mathbb{E}_{\boldsymbol{x}}\left\{\left(\frac{\partial L}{\partial\theta}\right)^2\right\}. \tag{3.79}$$

The Fisher information acts to set a lower limit on how well the system giving the data set can be characterised. If F is near zero, then the variance of any unbiassed estimator $\widehat{\theta}$ must necessarily be large, meaning that it's impossible to estimate the parameter θ very well. This is an indeterminacy principle, just like the more well-known Heisenberg Indeterminacy Principle in quantum mechanics. The actual principle is given by the *Cramér–Rao theorem*, which states that the variance $\operatorname{var}\widehat{\theta}$ of any unbiassed estimator $\widehat{\theta}$ must satisfy

$$\operatorname{var}\widehat{\theta} \geqslant 1/F \equiv \text{“\textit{Cramér–Rao lower bound}”}, \tag{3.80}$$

where the derivative in the Fisher information is evaluated at the true value of θ. An unbiassed estimator can even be found that gives exact equality in (3.80), as long as a function $\widehat{\theta}(\boldsymbol{x})$ exists that satisfies

$$\frac{\partial L}{\partial\theta} = F\left[\theta - \widehat{\theta}(\boldsymbol{x})\right], \tag{3.81}$$

in which case the estimator will then be $\widehat{\theta}(\boldsymbol{x})$.

As an example, suppose we are modelling the incoming data as gaussian with unknown mean θ but known variance σ^2. How well can we estimate the mean θ? The likelihood is

$$(\boldsymbol{x}|\theta) = \frac{1}{\left(\sigma\sqrt{2\pi}\right)^n} \exp\left[\frac{-1}{2\sigma^2}\sum_{i=1}^{n}(x_i-\theta)^2\right], \tag{3.82}$$

so that

$$L = \frac{1}{2\sigma^2}\sum_{i=1}^{n}(x_i-\theta)^2 + \text{constant}. \tag{3.83}$$

The Fisher information is

$$F = \mathbb{E}_{\boldsymbol{x}}\left\{\frac{\partial^2 L}{\partial\theta^2}\right\} = \mathbb{E}_{\boldsymbol{x}}\left\{\frac{n}{\sigma^2}\right\} = \frac{n}{\sigma^2}\,, \tag{3.84}$$

from which it follows that more data implies more Fisher information. Equation (3.80) becomes $\operatorname{var}\widehat{\theta} \geqslant \sigma^2/n$. So no matter what weird and wonderful expression could possibly be concocted to model the parameter θ, when we use it to create many estimates of θ—one from each data set that we have sampled—the variance of those estimates *must* be at least as large as σ^2/n. And we can certainly find an unbiassed estimator $\widehat{\theta}$ that actually attains the Cramér–Rao lower bound. Write (3.81) as

$$\underbrace{\frac{n}{\sigma^2}\left(\theta - \frac{\sum x_i}{n}\right)}_{\partial L/\partial\theta} = \underbrace{\frac{n}{\sigma^2}}_{F}\left(\theta - \widehat{\theta}\right), \tag{3.85}$$

which sets the best estimator of the mean to be $\widehat{\theta} \equiv \sum x_i/n$—the usual arithmetic mean—which is probably what we expected all along. Again, this underlines the dominance of the arithmetic mean in statistics.

If both the mean *and* the variance of the data in the previous example are required to be modelled by parameters θ_1 and θ_2, then the Fisher information becomes a 2×2 matrix with ij^{th} component

$$F_{ij} \equiv \mathbb{E}_{\boldsymbol{x}}\left\{\frac{\partial^2 L}{\partial\theta_i\,\partial\theta_j}\right\} \overset{\text{regularity}}{=\!=\!=} \mathbb{E}_{\boldsymbol{x}}\left\{\frac{\partial L}{\partial\theta_i}\frac{\partial L}{\partial\theta_j}\right\}, \tag{3.86}$$

where the derivatives are evaluated at the true values of θ_1, θ_2. Now the Cramér–Rao theorem becomes

$$\operatorname{var}\widehat{\theta}_i \geqslant \left(F^{-1}\right)_{ii}. \tag{3.87}$$

The Cramér–Rao bound F^{-1} is a covariance matrix, and is the "best" one obtainable from an unbiassed estimator of θ_1, θ_2. This means that no matter what expressions we can produce to estimate θ_1, θ_2, the covariance matrix P of those estimates will be "greater than or equal to" F^{-1}. In matrix language this means that $P - F^{-1}$ will be *positive semidefinite*: any nonzero vector $\boldsymbol{v}$ will satisfy $\boldsymbol{v}^t\left(P - F^{-1}\right)\boldsymbol{v} \geqslant 0$.

So the Fisher information sets a fundamental limit on how much information can be gleaned from a set of data, and in certain situations (usually related to gaussian densities!) that limit can be attained in practice.

3.4 Time Constants to Describe Growth and Decay

In the previous sections, we saw how the exponential number e appears quite naturally in the normal distribution so applicable to the statistics of the everyday world, and how it's tied closely in that distribution with the mean and standard deviation, which are themselves tied to ideas of linearity. Linearity and the intuitive notion of a mean lifetime also relate e to ideas of growth and decay. We will explore this relationship further in this section, where we concentrate on radioactive decay and growth.

3.4.1 The Poisson Statistics of Radioactive Decay

The decay of a radioactive element forms a good example of an exponential law, and the various terms involved with the process, such as *time constant*, *mean life*, and *probability per second*, shed light on how intuitive ideas are attached to mathematical language.

The *half life* of a radioactive element is, of course, the time required for half of it to decay. A half life of one hour means that one half of the element will decay in one hour. But the remaining half does not decay after another hour, which raises the question: is the element somehow slowing the internal processes that govern its decay rate, over time?

On a macroscopic level, many processes in Nature tend to go to their completion only asymptotically. The most predictable and well-behaved are those that are built from a large number of discrete microscopic processes. Perhaps the most well known of these is the characteristic time for radioactive decay; another example is the characteristic distance of penetration of uncharged particles through matter. Both of these processes are concerned with exceptionally large numbers of particles with a sort of binary nature that either do or don't behave in a certain way. In the radioactive element, any particular atom may or may not decay in a set time interval, while in a beam of neutral particles (such as light passing through matter), any particular particle may or may not interact in a set distance interval, being completely removed from the beam when it does interact.

This binary behaviour is not guaranteed for all large-number processes. A beam of *charged* particles passing through matter travels a well-defined distance at which more or less all of the particles come to rest (although this point is not as well defined for less massive particles such as electrons). This is because charged particles do not interact with the medium in a binary way; they lose tiny fractions of their energy in multiple collisions, so that all lose energy at a similar rate, which brings them all to rest at about the same distance from their point of entry. An exponential type of behaviour is only applicable to particles acting in an all-or-nothing way.

Given the lack of a well-defined end to such a binary process (at least in principle), the ideas of a half life, and the mean life that we'll encounter soon, are a useful way of quantifying how fast the process occurs. While we focus on radioactive decay and growth here, examples from other fields can be found, such as the recovery of a muscle after some exertion, the cooling of a hot body, mixing problems that use constant flow rates, the filling of a cistern, and the behaviour of currents and voltages in resistor-capacitor and resistor-inductor circuits in electrical theory. A system need not even behave strictly

To Explain or to Predict?

Science is often thought to proceed by our logically deducing the laws that govern the world. But there are limits to what we can deduce, especially about things in which we cannot directly participate, such as radioactive decay. We cannot use a microscope to watch the events that make an element decay; the process is quite mysterious. But what we can do is make a theory of how decay might work, and then use that theory to predict what measurements we can expect. If sometimes these predictions turn out to be wrong, then the theory needs improvement, perhaps to be relegated to one particular regime of application or perhaps discarded outright.

So the hallmark of a good scientific theory is not what it seems to *explain*, but rather what it *predicts*. A theory can always be built that explains, in some sense, whatever we want. But if it is unable to make any predictions, then from a scientific point of view it has no use, because it contains nothing that allows its truth to be tested. On the other hand, while it's arguable that the theory of quantum mechanics explains anything at all, it certainly does *predict* a huge number of different phenomena that have been observed; and that is just what makes it a very useful theory.

exponentially in order to have a characteristic time or distance associated with it.

Deriving the exponential expression for radioactive decay is a straightforward exercise in introductory calculus, and is usually quickly passed over in favour of looking at some real-world examples. And yet it shows the classic paradigm of how research in science is carried out. Regardless of how we might *expect* an element to behave—where perhaps the second half might be expected to decay in the same amount of time as the first half—this simply does not happen. We must search for a theory that predicts this.

For radioactive decay, our theory is that the nuclei decay quite spontaneously, for reasons unknown. We postulate that the nuclei decay independently of whether their neighbours are decaying, and also that their tendency to decay is independent of how old they are. A given nucleus might decay after one microsecond or one million years; however long it has survived makes no difference to its ability to decay right now. If the mechanism behind its decay is strong, in the sense that the nucleus has a large chance of decaying, then chances are that it will soon decay. After all, the chance that it does not decay in some time interval is small, so the chance that it survives for any appreciable amount of time is then even smaller. This is an exercise in probability; we simply multiply together, for a string of time intervals, the probabilities that the nucleus does not decay in each interval.

The statistics of decay, such as the mean and standard deviation of the number of nuclei decaying in a given time interval, were measured soon after radioactivity was discovered. They were found to match those predicted by this idea of random decay, called *Poisson statistics.* Whenever a binary event has a small chance of happening but has lots of opportunities to occur, Poisson statistics result.

A mnemonic for the Poisson decay law is the word "mnemonic" itself. If the probability of a decay in a given time interval is very small, but there are a large number of nuclei so that the mean number of decays expected is m, then the probability of n decays in that interval is $\boldsymbol{m}$ raised to $\boldsymbol{n}$ (times) e raised to $-\boldsymbol{m}$ **o**ver $\boldsymbol{n}$ factorial (an upside-down "i"):

$$P(n \text{ decays}) = \frac{m^n e^{-m}}{n!}\,. \tag{3.88}$$

(The last mnemonic letter, "c", wasn't used—but then again perhaps we could rescale everything by the speed of light.)

Although processes such as cooling and muscle recovery follow exponential laws, there is a fundamental difference between them and radioactivity. Cooling involves an interaction with an environment: the more a hot body cools to approach the temperature of its environment, the slower is the rate of energy transfer to that environment. Similarly, with muscle recovery, the rate of chemical transfer to recharge a muscle is determined by the total system of muscle plus biochemical environment. Again, the filling of a cistern is governed by an interaction with a water bath that slowly closes a valve. But experimentally, radioactive decay does not depend on an element's environment. It proceeds at a constant rate regardless of temperature, gravitational force, or electrical gradients, and so seems to be governed by something internal to the nucleus that has no "contact" with the outside world, so to speak.

Certainly no experiment or theoretical result has proven that this stand-alone-nucleus theory of nuclear decay is valid. The logical statement is that *if* nuclei decay randomly, *then* Poisson statistics result. Experiments show that Poisson statistics do indeed result, but logically this does not imply that nuclei decay randomly. Nevertheless, the way of science is that we do postulate that nuclei decay randomly, until a further experiment calls this into question—which has yet to happen.

This reverse use of logic has a good pedigree in the field of mechanics. Ideas of gravity, mass, and acceleration were originally produced by Newton through the same process: because they predicted planetary orbital periods that could be verified experimentally. Because of this great success, expressions such as $F = ma$ and $F = GMm/r^2$ came to be canonical in physics. The logic was indeed being used in reverse; but no one was surprised when, three centuries later, one of the Apollo astronauts dropped a feather and a hammer together in the Moon's vacuum and found that they both fell at the same rate (although it was still beautiful and dramatic to watch!). That reverse logic had, after all, allowed him to get to the Moon in the first place. So this way of conducting science works very well.

The Factorial and the Gamma Function

The factorial function $n!$ finds wide application, not just in combinatorics and statistical theory, but in the whole of mathematics and physics; and it has been studied by many mathematicians over the centuries. When the theory of interpolation was coming to the fore at the start of the eighteenth century, one of the areas that attracted interest was how to generalise the factorial to real numbers. The exclamation mark in $x!$ is impractical for writing expressions such as derivatives, and as every student of mathematical physics knows, eventually the *gamma function* $\Gamma(x)$ became established. For positive x,

$$\Gamma(x) \equiv \int_0^\infty e^{-t} t^{x-1}\, dt = (x-1)! \,. \tag{3.89}$$

The most obvious thing here is that apart from equalling $(x-1)!$, the function $\Gamma(x)$ is *defined* in terms of $x-1$. This apparently anomalous unit shift seems to be due to Euler, but its reason appears to be unknown. Certainly the great majority of the common expressions involving the gamma function would be cleaner if this shift by 1 had never occurred in its definition. As an alternative to $x!$, the functional notation $\Pi(x)$ for the factorial is sometimes used, eliminating the unit shift in (3.89):

$$\Pi(x) \equiv \int_0^\infty e^{-t} t^{x}\, dt = x! \,. \tag{3.90}$$

Despite the neatness of $\Pi(x)$, still it is $\Gamma(x+1)$ that has mostly continued to be written today as the generalisation of the factorial to real and complex numbers. But in most work involving factorials, converting all appearances of $\Gamma(x+1)$ to $\Pi(x)$ simplifies the expressions, aiding readability. This unit shift appears to be nothing more than a historical quirk.

A "People" Experiment to Simulate Radioactive Decay

A simple test of this theory of radioactive decay uses a group of people numerous enough to assure good statistics. Put 1000 people into a hall and give each a coin. Each person represents a radioactive nucleus, and the coin represents their ability to decay. They must each toss their coin once per minute. If the result is heads, then the person should immediately leave the hall, corresponding to the nucleus decaying. If tails, they take no action, except to wait for another minute to elapse, and then flip their coin again.

What happens? After one minute, roughly half of the people get up and walk out because their coins landed heads up. After another minute, everyone again flips their coin, and about half the remainder will walk out. Of course, we don't expect *everyone* to leave after the second minute; as each minute goes by, roughly half of the group walks out. So this simple model of random behaviour has produced a half life; this particular "element" has a half life of one minute. For a smoother simulation, we might ask everyone to flip their coin continually while stipulating a different probability, such as leaving the hall only if four heads in a row result.

Any particular person will flip their coin for a length of time unaffected by how many of their neighbours have now left the hall. They might even sit for years, endlessly flipping their coin only to find that it always lands tails up. The chance that they leave the hall at any time is completely independent of how long they have been there.

3.4.2 The Mean Life of the Decaying Nuclei

Suppose we are given a radioactive element and wish to analyse the statistics of its decay. If we hypothesise that, in a time interval $\mathrm{d}t$, the fractional loss in the number of nuclei N present is $-\mathrm{d}N/N$ with positive proportionality constant λ (see the note on p. 48 for a discussion of the signs of these quantities), then the resulting differential equation and its solution are straightforward:

$$-\mathrm{d}N/N = \lambda\,\mathrm{d}t\,; \quad \text{solution: } N = N_0\,\mathrm{e}^{-\lambda t}. \tag{3.91}$$

The decaying nuclei have a spread of lifetimes, but their mean life can certainly be calculated. In a time interval $\mathrm{d}t$, the number of nuclei that decay is $-\mathrm{d}N$, and each of these had a lifetime equal to t. We need only sum these lifetimes and divide by the original number to get the mean life:

$$\text{mean life } \equiv \tau = \frac{1}{N_0}\int_{t=0}^{\infty} -t\;\mathrm{d}N = \frac{1}{\lambda}\,. \tag{3.92}$$

This simple result allows the concept of the mean to be tied to the number e in a particularly simple way:

$$N = N_0\,\mathrm{e}^{-t/\tau}. \tag{3.93}$$

After one mean life τ, the number N of nuclei remaining falls by a factor of e. So e has a very close relationship to the concept of the arithmetic mean, since any other choice of base would complicate the analogous expression to (3.93) by introducing a factor of the natural logarithm of that base.

Any general process that follows an $\mathrm{e}^{-t/\tau}$ law might have nothing to do with a large number of individual entities, in which case τ will not be a mean life. Nevertheless, τ is still a characteristic time for the process, called the *time constant*. But even in such a process, the time constant still has an intuitive meaning, and once again we use the language of nuclear decay to see what this is.

Plotting the amount of a radioactive element remaining as a function of time shows that it drops in the manner of Fig. 3.4. Suppose that the laws of Nature were such that the element decayed not exponentially but linearly, so that its rate of decay was always equal to its initial rate. In such a world, the

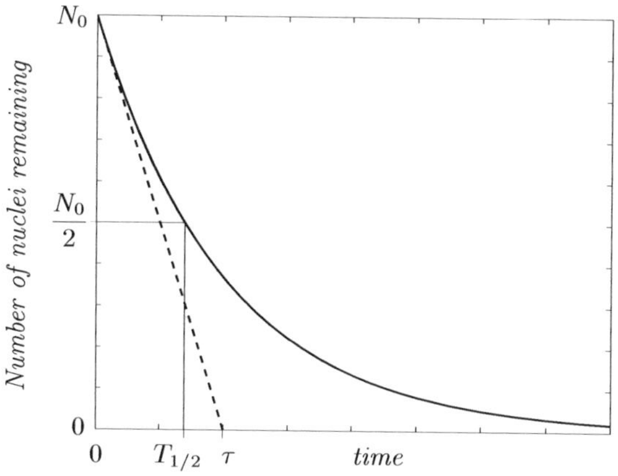

Fig. 3.4. The mean life τ is not just an average lifetime; it also quantifies an initial rate of decay, which is very different from the meaning of the half life $T_{1/2}$. *If* the element kept decaying at its initial rate, it would vanish completely after a time τ.

second half really would decay in the same length of time as the first half. For this idealised case, how long would the whole take to vanish completely? That is, suppose we have 1000 radioactive nuclei, and initially there are 10 decays per second. *If* they were to keep decaying at this rate, then how long would they take to vanish completely? The answer is, of course, 100 seconds. But notice that the initial rate of decay is

$$\text{initial decay rate} = -\mathrm{d}N/\mathrm{d}t\,\Big|_{t=0} = N_0/\tau\,, \tag{3.94}$$

which implies that, in such a linear world, all N_0 nuclei would decay after a time τ in Fig. 3.4. So the time constant of an arbitrary process—which happens to equal the mean life in the case of decaying nuclei—has a very intuitive meaning: it is the time the process would take to terminate *if* it evolved at a constant rate equal to its initial rate. So for the 1000 nuclei decaying at 10 decays per second, we see immediately that $\tau = 100$ seconds. As an afterthought, we can also surmise straightaway that the real element's half life is 69.3 seconds, since $T_{1/2} = \tau \ln 2 \simeq 100 \times 0.693$.

The naïvely reasonable idea that the "whole life" of a radioactive element might be twice its half life suggests that there can be a certain linearity in our ideas of how the world works. And so it turns out that to some extent we *can* appeal to this sort of linear intuition. If the nuclei were to behave in this simple linear way, then the time needed for all of them to decay would be exactly equal to the mean life of the actual real-world element.

3.4.3 The Notion of a "Probability per Second"

The concept of an initial rate of change also applies to the growth of the daughter element. Suppose for simplicity that the daughter is not radioactive. Initially there are no daughter nuclei, but gradually their numbers grow as the parent nuclei decay. The daughter element's growth must eventually flatten out over the same time that it takes for the parent element to fully decay. The sum of the parent and daughter nuclei is a constant, so that the daughter's growth curve is essentially just the parent's decay curve flipped upside down. In that case, *if* the growth of the daughter were to continue linearly at its initial rate, then the time it would take for the sample to be completely composed of daughter atoms would be just the mean life of the parent.

This way of using the time constant for the growth of the daughter produces a turn of phrase that can at first seem quite mysterious. Given a single radioactive nucleus, what is the probability that it decays before a time t? This probability will grow, tending toward one as $t \to \infty$. The actual expression turns out to be $1 - \mathrm{e}^{-t/\tau}$, as can be shown by examining the probability for decay in an infinitesimal time interval and applying the rules of probability as follows. The meaning of "the probability per second of decay is p" is: "in the next ε seconds, the probability of decay is $p\varepsilon$". In that case, what is the chance of a given nucleus decaying in the time interval $0 \to t$? Writing the probability of x occurring as (x), divide t into N equal intervals of length ε so that $t = N\varepsilon$, in which case the chance of decay for the given nucleus is

$$\begin{aligned}
&\lim_{\varepsilon\to 0} \,(\text{decay in } 0 \to \varepsilon) + (\text{no decay in } 0 \to \varepsilon)(\text{decay in } \varepsilon \to 2\varepsilon) \\
&\qquad + (\text{no decay in } 0 \to 2\varepsilon)(\text{decay in } 2\varepsilon \to 3\varepsilon) + \cdots \\
&\qquad + \big(\text{no decay in } 0 \to (N-1)\varepsilon\big)\big(\text{decay in } (N-1)\varepsilon \to N\varepsilon\big) \\
&= \lim_{\varepsilon\to 0} \, p\varepsilon + (1-p\varepsilon)\,p\varepsilon + (1-p\varepsilon)^2 p\varepsilon + \cdots + (1-p\varepsilon)^{N-1}\, p\varepsilon \\
&= \lim_{\varepsilon\to 0} \, 1 - (1-p\varepsilon)^N = \lim_{N\to\infty} \, 1 - \left(1 - \frac{pt}{N}\right)^N = 1 - \mathrm{e}^{-pt}.
\end{aligned} \tag{3.95}$$

This means that if the fraction of nuclei decaying in the interval $0 \to t$ is $1-\mathrm{e}^{-pt}$, so that the fraction left over must be e^{-pt}, which we already know is $\mathrm{e}^{-t/\tau}$; thus the decay probability per second is $p = 1/\tau$. The plot of $1 - \mathrm{e}^{-t/\tau}$ is shown in Fig. 3.5.

Now, although the chance that the nucleus has decayed after one second is $1 - \mathrm{e}^{-1/\tau}$, this is not entirely illuminating. So, as before, we ask a different (but related) question. *If* the world was such that the probability of decaying within time t continued to increase linearly at its initial rate, then how long would it take for the nucleus to have definitely decayed—after what time would this probability reach one? Again, simple differentiation determines

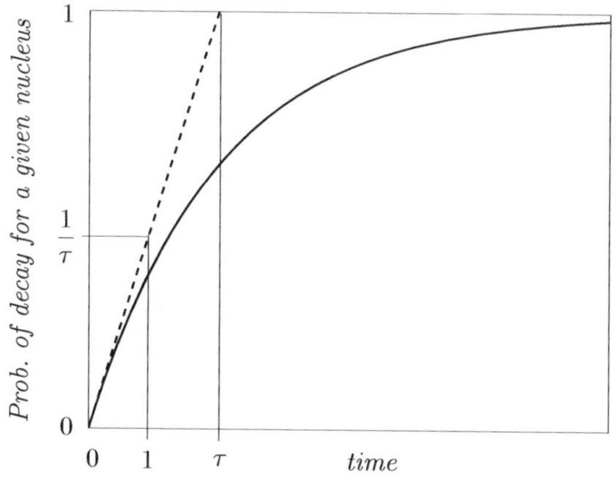

Fig. 3.5. Analogously to the decay case, the time constant or mean life here carries the meaning of an initial rate of growth of the daughter element. *If* the (nondecaying) daughter kept growing at its initial rate, then the sample would be entirely composed of daughter atoms after a time τ.

this time interval to be just the mean life τ. So if our simplified "linear nucleus" attains a decay probability of one after a time τ, then it must reach probability $1/\tau$ after a unit time.

In the example above with 1000 nuclei and initially 10 decays per second, we concluded that the mean life was $\tau = 100$ seconds. So, *if* the chance that any particular nucleus had decayed was to keep increasing uniformly at its initial rate, then after one second, the chance that it had decayed would be $1/\tau$, or 1/100. This makes sense: 1/100 of the initial 1000 nuclei is the 10 nuclei measured to have decayed after each second. So we say "the nucleus has a probability of decay of 1/100 per second". Of course this does not mean that after 100 seconds our nucleus will definitely have decayed! That would only be true if the nucleus were to behave in a linear way. It's analogous to stating the speed of a trolley as 1/100 metres per second. After 100 seconds the trolley will have covered one metre only if it continues to move linearly, and this is understood in the definition of speed. Unlike the trolley, however, the "speed" of the nucleus is immutable, set by the laws of Nature.

The fact that the nucleus's probability of having decayed is in a sense "slowing down", is just like this trolley starting out with a speed of 1/100 metres per second, but decelerating due to friction. If there were no friction, so that it moved at a constant velocity, then it would take 100 seconds to travel one metre. It doesn't do this, of course—it might never cover a metre because it is decelerating—but $1/100\,\text{ms}^{-1}$ refers to its *initial* rate of distance increase. And likewise for our generic nucleus that represents the whole population of nuclei, "1/100 per second" refers to its *initial* rate of "decay-probability increase". In an average sense for the whole population, that decay probability will never quite reach one, although it will eventually reach one for any particular nucleus.

Finally, then, a time constant can be defined for a system that grows or decays in a nonexponential way, in accord with our intuition, by being set equal to the time it would take the system to grow or decay fully *if* it kept evolving at its initial rate. Nevertheless, Nature's simplest and most basic systems often do follow an exponential law.

3.5 Logarithms and Exponentials in Statistical Mechanics

The other side of the exponential coin, the logarithm, is usually seen as being useful in describing quantities that range over several orders of magnitude, or allowing products to become sums in order to deal with them more easily. But sometimes this notion of additivity can give insight into quantities that Nature herself seems to regard as fundamentally additive. Here we look at the concept of entropy as an example, and examine how the famous Boltzmann factor of statistical mechanics arises.

3.5.1 Entropy and Heat Flow Define Temperature

Statistical mechanics regards as fundamental the concept of *entropy.* Entropy is defined as a logarithm, but this is not because we want to reduce large numbers in size. The definition is quite natural and paves the way for the idea that entropy is something tangible that can be identified with the notion of heat.[3]

The motivation for defining entropy comes from the worldview of statistical mechanics, which considers a system at any one time as occupying one of a possibly large number of *microstates.* A microstate is simply a quantifiable configuration of the system's basic constituents. To fully describe the microstate of a gas at any one moment, for example, would entail specifying the positions and velocities of each of its molecules. Needless to say, this might be for all intents and purposes impossible, so that it becomes the job of statistical mechanics (and on a more macroscopic level, thermodynamics) to describe such systems. Generally we are not concerned with quantifying each microstate. Instead, we are usually content to consider only the set of all microstates that have equal amounts of some easily quantified properties, such as pressure and volume for a gas. This set of all "similar" microstates is called a *macrostate* of the system. But it is microstates that play the key

[3] There has long been discussion over the nature of heat in physics, and in particular whether it is a verb or a noun. The idea that it be considered as a noun synonymous with entropy is put forward in *The Dynamics of Heat* by H.U. Fuchs (1996, Springer).

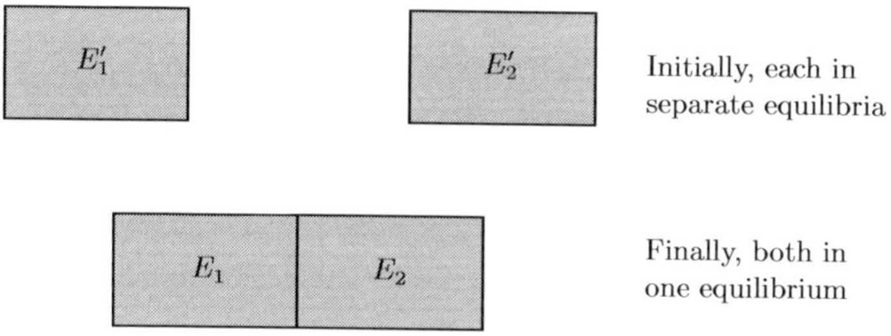

Fig. 3.6. Two systems with initial energies E'_1, E'_2 are placed in contact via a *diathermic wall*, so that only energy can be exchanged between them. Their final energies E_1, E_2 can be different, though of course the total is conserved.

role in developing the subject's basic ideas. The fundamental assumption of statistical mechanics is that:

> *A closed system is equally likely to be in any of the microstates accessible to it.*

Just which microstates are accessible is governed by the system's total energy. For any given energy E, a closed system might have a large number of microstates it can occupy. This important number is called the *multiplicity* of its microstates, $g(E)$.

Suppose that two systems with energies E'_1, E'_2 are placed in contact in such a way that energy, but no particles, can flow between them, as in Fig. 3.6. Their final energies are E_1, E_2, with the total unchanged: $E \equiv E'_1 + E'_2 = E_1 + E_2$. Each system is in some microstate. When they are placed in contact, another of the fundamental principles of statistical mechanics is called upon. This one is an extremisation principle that defines thermodynamic equilibrium: we postulate that the total energy will redistribute itself in such a way as to maximise the multiplicity of microstates in the combined configuration:

$$\text{At equilibrium, } g(E) = g_1(E_1)\, g_2(E_2) \text{ is maximised.} \tag{3.96}$$

Since the total energy E is constant, express (3.96) using just one variable, which allows us to maximise $g(E) = g_1(E_1)\, g_2(E - E_1)$ with respect to E_1:

$$\frac{dg}{dE_1} = g'_1(E_1)\, g_2 - g_1\, g'_2(E_2)\,. \tag{3.97}$$

This is required to equal zero when the two systems are in equilibrium. Thus, when equilibrium has been reached,

$$g'_1(E_1)\, g_2 = g_1\, g'_2(E_2)\,, \tag{3.98}$$

so that

$$\frac{g_1'(E_1)}{g_1} = \frac{g_2'(E_2)}{g_2}. \tag{3.99}$$

This implies that

$$\frac{\mathrm{d}}{\mathrm{d}E_1} \ln g_1 = \frac{\mathrm{d}}{\mathrm{d}E_2} \ln g_2 . \tag{3.100}$$

If we define the *statistical entropy* of each system to be a measure of its multiplicity of states,

$$\text{statistical entropy } \sigma \equiv \ln g , \tag{3.101}$$

then (3.96) implies that this entropy will be maximised in thermal equilibrium, and the quantity $\mathrm{d}\sigma/\mathrm{d}E$ will be identical for each system.

A notion of statistical entropy allows the *statistical temperature* τ of a system to be defined:

$$\frac{1}{\tau} \equiv \frac{\partial \sigma}{\partial E} \tag{3.102}$$

(where partial derivatives are needed because σ might depend on other state variables), so that in equilibrium the temperatures of the two systems are necessarily equal. In terms of everyday physical units, the quantity τ/k is more useful, where k is Boltzmann's constant; so we define the thermodynamic temperature T and the thermodynamic entropy S by

$$T \equiv \tau/k , \quad S \equiv k\sigma . \tag{3.103}$$

Experiments and thermodynamical considerations show that T is identical to the everyday temperature measured by thermometers.

For our purposes, the main point to note is that since $g = g_1 \, g_2$, it must follow that $\sigma = \sigma_1 + \sigma_2$. Thermodynamically, an additive quantity is interesting since the sum $\sigma_1 + \sigma_2$ suggests that entropy is something quite tangible: the total entropy of two systems before they come into contact is just the sum of their individual entropies. However, when they do come into contact and their temperatures begin to equalise, their total entropy increases. Entropy comes into existence from nowhere.

A simple example of this entropy increase occurs when the two systems are boxes of coins to be shaken together, shown in Fig. 3.7. Since the fundamental assumption of statistical mechanics is that a system is equally likely to be found in any of its accessible microstates, if the number of ways we can order, say, half of the coins to show heads is large (i.e. if the multiplicity of that configuration is large), then a box of coins will very likely have about half of its heads up. If there are many coins in each box, then a calculation using the binomial distribution shows that the number of combinations of having a 50/50 mix of heads and tails in each box is much larger than any other fraction, such as a 51/49 configuration—and stupendously larger when the number of coins approaches real-world numbers such as the number of molecules in a gas. So physical systems with around Avogadro's number of particles (say, 10^{24} particles) fulfill this "half-heads" condition very well. But to keep the calculation simple, imagine that each box holds just ten coins, with five heads up, and that when all twenty coins are mixed randomly together, ten heads turn up as a result. This mimics a large physical system very well.

The total entropy for both boxes, each with five heads up before coming into contact, is $2 \ln {}^{10}C_5$, or about 11. When the boxes are put together and shaken up, and we assume that ten heads face up, the total entropy is now $\ln {}^{20}C_{10} \simeq 12$. So there has been a slight increase in entropy, in line with thermodynamical ideas that the universe's total entropy always increases over time. This is the statistical basis for the famous *Arrow of Time*. At their heart, the equations of physics are, for the most part, time-reversible (an exception involves the weak interaction), but the growth of entropy in any process makes it clear which way time's arrow is pointing—whether the movie is being run forward or backward, so to speak.

3.5.2 The Boltzmann Factor: Chief Star of Statistical Mechanics

Having spoken of the ubiquity of exponentials and logarithms in statistics, it is an easy step to derive the famous *Boltzmann factor*. This factor appears in the central result of statistical mechanics. Suppose we have a system S that's able to be placed in any of several macrostates of different energies. These macrostates define the various *energy levels* of S. Suppose we place S in contact with a huge heat reservoir R that is so large that its fundamental temperature $\tau = kT$ is essentially fixed, regardless of what energy it imparts to S. We ask the question: what is the chance that S will be forced into any given *micro*state of energy E?

The entire closed system of $S+R$, with total (fixed) energy U, is postulated to be equally likely to occupy any of its available microstates. That means that regardless of how many microstates of S have energy E, the chance that S occupies a particular one of them is determined by how easily the reservoir R allows itself to lose that energy E, which is given by the number of ways in which this loss can be accomplished. That is, the probability that S occupies a particular microstate is proportional to the multiplicity of R at energy $U - E$, or $g_R(U - E)$. From (3.101), this equals $\exp \sigma_R(U - E)$, where $\sigma_R(U - E)$ is the entropy of the reservoir at energy $U - E$:

$$\sigma_R(U - E) = \sigma_R(U) - E\,\frac{\partial \sigma_R}{\partial U} + \cdots . \tag{3.104}$$

Since $E \ll U$, and the huge reservoir's temperature is approximately constant, so that $1/\tau = \partial\sigma_R/\partial U$ is approximately constant, (3.104) becomes

$$\sigma_R(U - E) \simeq \sigma_R(U) - \frac{E}{kT} . \tag{3.105}$$

Thus, the multiplicity of R at energy $U - E$ is

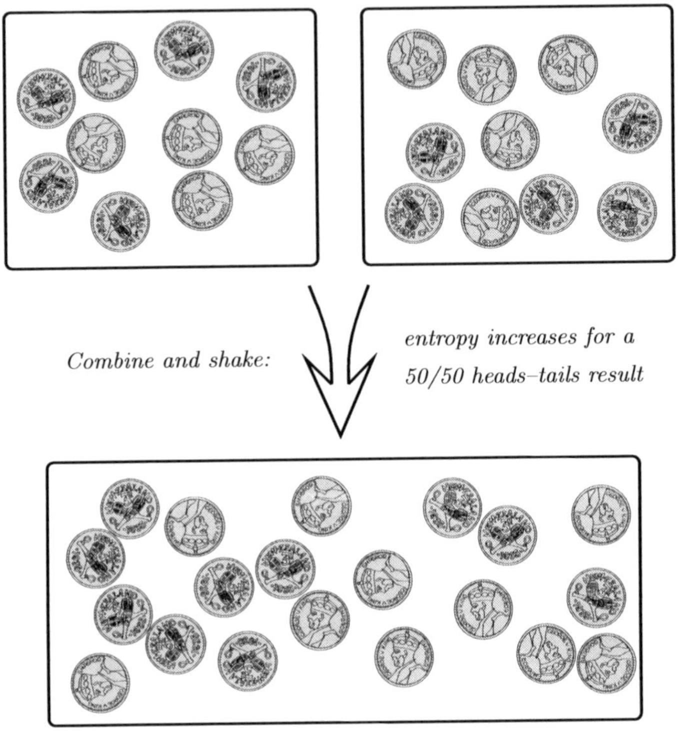
Combine and shake:
entropy increases for a
50/50 heads–tails result

Fig. 3.7. Top: Two boxes, each with ten coins, form a simplified representation of a system that involves a stupendously large number of objects. Half of the coins in each box show heads, and so in each box there are ${}^{10}C_5$ ways that five heads can appear. This gives each box an entropy of $\ln {}^{10}C_5$, making a total entropy of $2 \ln {}^{10}C_5 \simeq 11$. **Bottom**: A good representation of random mixing happening in a very large system is if all the coins were to be mixed together randomly into one box and again half were to show heads, which could happen in ${}^{20}C_{10}$ different ways. The total entropy would now be $\ln {}^{20}C_{10} \simeq 12$. Mixing the coins has created a small amount of entropy, suggesting a forward direction in time.

$$g_R(U-E) = \mathrm{e}^{\sigma_R(U-E)} \simeq \exp\left[\sigma_R(U) - \frac{E}{kT}\right] \propto \exp\frac{-E}{kT}\,. \tag{3.106}$$

Since the probability that S occupies the given microstate is proportional to $g_R(U-E)$, we have shown the following:

> *The probability that a system in contact with a large heat reservoir is found in a given microstate of energy E is proportional to the famous Boltzmann factor $\mathrm{e}^{-E/(kT)}$.*

This is, in fact, the central result of statistical mechanics, and as in so many of the major equations of physics, the exponential function plays an important role in shaping its character.

3.5.3 Logarithms and Decibels

Logarithms are useful not only for giving insight into a physical quantity by emphasising its additivity (such as with entropy), but they also are able to reduce numbers that span many orders of magnitude, in a way that might imitate the response of some measuring apparatus. Engineers make good use of a logarithmic scale when keeping track of a signal's power as it works its way through a system that introduces gains and losses, a procedure known as a power budget in telecommunications.

More everyday examples of logarithmic scales are those used to measure hearing and visual responses. Our hearing relates roughly logarithmically to sound intensity, it will be useful to reduce the huge spectrum of sound intensity levels down to manageable numbers by taking logarithms. So, conventionally, a unit loudness is assigned to a sound that causes an intensity of $I_0 \equiv 10^{-12}\,\mathrm{Wm^{-2}}$ at our ears. Another sound, of intensity I, will then have a loudness of

$$m \equiv \log_{10} \frac{I}{I_0}\,, \quad \text{so that } I = 10^m\, I_0\,. \tag{3.107}$$

Historically, this unitless "loudness magnitude" m has been assigned a unit of the *bel*, an alternative name for an old unit of logarithms called the *dex* (so that for example 6 dex equals the number 10^6). While the dex is simply the logarithm of any number, the use of the bel acknowledges the fact that in acoustics, the number whose logarithm is being taken is really a ratio: the absolute intensity divided by $10^{-12}\,\mathrm{Wm^{-2}}$. So we speak of a sound whose intensity is, say, $10^{-8}\,\mathrm{Wm^{-2}}$ as having a loudness of 4 bels, or more commonly 40 decibels (40 dB) since the bel is a large unit and more conveniently divided into tenths. Thus, if a sound is described as being -20 dB, we know immediately that this is just -2 bels, or -2 dex, or 10^{-2} times the base intensity, or finally $10^{-14}\,\mathrm{Wm^{-2}}$.

Just as our ears relate logarithmically to sound intensity, our eyes relate at least approximately logarithmically to brightness, and a logarithmic approach has been used from ancient times to quantify the brightness of stars. In modern times, this old concept of *apparent magnitude* has been quantified in the following way. As usual, the arbitrary-looking constants are chosen to make the mathematical definition of magnitude conform closely to its historical use. Astronomers assign a base intensity I_0 to a star delivering $2.48 \times 10^{-8}\,\mathrm{Wm^{-2}}$ to the top of Earth's atmosphere. A star of intensity I then has an apparent magnitude defined as

$$m \equiv -2.5 \log_{10} \frac{I}{I_0}\,, \quad \text{so that } I = 10^{-0.4m}\, I_0 \simeq 2.512^{-m}\, I_0\,. \tag{3.108}$$

"Bar" Notation in Logarithms

Notation used for computing logarithms to base 10 is no longer known to a generation of scientists brought up on electronic calculators. When computing decibels mentally, *bar notation* is a useful tool worth reviewing.

The notation rests on our separating the logarithm into its integer and decimal parts. Specifically, if the logarithm is negative, then we wish to be able to write it quickly with a positive decimal part. The integer part will still be negative, and is written with an overbar to remind us of that. Thus, for example,

$$\bar{2}.3 \equiv -2 + 0.3 = -1.7\,. \tag{3.109}$$

This is a useful way to handle logarithms, because it keeps the fractional part positive and so economises on what logarithms need memorising to allow us to perform common calculations quickly. If we need to visualise just how large, say, $-17\,\text{dB}$ is, we need only realise that

$$-17\,\text{dB} = -1.7\,\text{B} = 10^{-1.7} = 10^{\bar{2}.3} = 10^{-2} \times 10^{0.3} \simeq 2/100\,, \tag{3.110}$$

which in practice is a very quick mental process. The reverse is also useful when converting to decibels:

$$0.005 = 5 \times 10^{-3} = \bar{3}.7\,\text{B} = -2.3\,\text{B} = -23\,\text{dB}. \tag{3.111}$$

(A coincidental closeness often causes the two numbers 2.5 and 2.512 to be confused.) This scale is useful and intuitive after a little use, though it might seem archaic and obscure when first encountered. Unlike acoustic magnitudes, which decrease as the loudness drops, stellar magnitudes *increase* as the brightness drops—a historical oddity that is perfectly justified, since over time astronomers have had to catalogue ever-fainter objects, and there is after all no sense in having to carry around minus signs all the time. While the Sun has an apparent magnitude of about −27 and the full Moon about −12, the brightest stars have magnitudes of around 0 or 1. Conventionally, the faintest ones visible to the naked eye were once around magnitude 6, but the current levels of city light pollution and backyard motion-sensor lights that unfortunately do so much to extinguish backyard astronomy have reduced this to 3 or 4. The faintest objects yet photographed are around magnitude 30.

Engineers use logarithms to enable easy bookkeeping in systems that amplify signals. Suppose an electrical signal loses 90% of its strength when sent down a one metre line, so that it has dropped from one unit to 1/10 of a unit at the end: a drop from, say, 0 B to −1 B. Equivalently, it loses 10 dB along that metre length. But suppose it's now sent down a line of length five metres. If it drops to 1/10 of its power along each metre, then it must be reduced to $1/10^5$ of its original strength at the very end, which corresponds to a drop of 50 dB. That is, the multiplicative drop in power is more easily

Logarithms and Prime Numbers

One of the fascinations of number theory is the way in which all manner of seemingly unrelated functions appear in unexpected places. Logarithms are very closely related to prime numbers, and here we show two intriguing examples, both of which use a double logarithm. The first is the curious result that the nth prime is *asymptotically* given by

$$n^{\text{th}} \text{ prime} \sim n(\ln n + \ln\ln n - 1)\,, \tag{3.112}$$

where by the asymptotic convergence $f(n) \sim g(n)$ of two functions $f(n)$ and $g(n)$ is meant that the ratio $f(n)/g(n) \to 1$ as $n \to \infty$; however, it does *not* imply that $f(n) - g(n) \to 0$.

The second example of a double logarithm concerns the question: what is the number of prime factors $\Omega(n)$ of any given natural number n, where these factors need not be distinct? So, for example, $\Omega(20) = 3$ since $20 = 2^2 \times 5$.

Plotting $\Omega(n)$ versus the natural numbers n shows that, not unexpectedly, the number of factors bounces up and down extremely erratically. It can be high: $\Omega(1024) = 10$, but it certainly comes back down to one infinitely often (whenever n is prime), so that over a large enough domain the plot looks very noisy. The plot can be smoothened by changing the question above slightly: what is the *average* number of prime factors for all natural numbers less than or equal to a given natural number n? This turns out to be an example of the fact that many problems involving natural numbers are rendered more tractable by altering them to involve the real numbers.

It can be shown that as n grows, this average number of prime factors very quickly asymptotically approaches a double logarithm:

$$\frac{1}{n}\sum_{i=1}^{n} \Omega(i) \sim \ln\ln n + 1.03465\ldots \tag{3.113}$$

As a consequence, the average number of prime factors increases stupendously slowly over the natural numbers. For example, over all the numbers up to one thousand million, this average number of prime factors is still only about four.

expressed as an additive loss in decibels; so the line loss is 10 dB per metre, making it especially easy for a back-of-the-envelope calculation to develop a *link power budget* over a communications line, for example.

Again, suppose a signal loses 10 dB in one medium and 35 dB in another. In that case, the combined loss when the signal traverses both media in succession is 45 dB. So by quantifying power drops in decibels, we can easily add simple numbers to quantify a whole chain of power absorbers, which in practice is far easier than multiplying several factors together. And, of course, the same idea holds when discussing the exponential absorption of light in a medium.

In this context, another way the decibel is used is when the reference level is built into the new unit created. Thus we have "dBW", meaning decibels referred to a power level of one watt, or "dBm", referring them to one milliwatt. Thus 30 dBm equals one watt, equals 0 dBW, and so on.

3.6 Signal Processing and the z-Transform

Transform methods form an important part of signal processing theory, but apart from the well-known use of Fourier analysis, the uses of other transforms are not always so transparent at the outset. Transforms are useful fundamentally because their linearity allows us to invert their equations easily. This promotes them in other areas that might be more purely mathematical. In this section, we will discuss an example of one such transform, known as the *z-transform*, used both in signal processing and, as we'll see, in pure mathematics.

Signal processing concerns itself with extracting useful data from a noisy signal. Signals are, in their rawest form, just sequences of numbers. The goal of a signal processor is to reveal any hidden trends that might be lurking in this sequence: to check for the presence of real information that might be hiding amongst or perhaps even below the ambient noise level.

Whether we are dealing with a sequence of numbers that forms a data set, or perhaps with a purely mathematical sequence, it's of very great use to be able to characterise that sequence in a compact way. A very simple way of doing this represents the sequence of numbers by a polynomial having those numbers as coefficients, and this is just what the z-transform does. There is a one-to-one relationship between sequences and polynomials: given one, we can always reconstruct the other. This relationship is linear; for example, adding two sequences term by term is equivalent to adding their polynomials. Mapping a sequence to a function allows clever operations to be done, such as differentiating the function, which are equivalent to operating on the sequence in some way. While transforming the set of numbers $\{3, -2, 5\}$ to the polynomial $3x^2 - 2x + 5$ might seem like a pointless exercise, the utility of the z-transform appears when the sequence has infinite extent in one or both directions:

$$\{x_n\}_{n=-\infty}^{\infty} \equiv \ldots,\ x_{-2},\ x_{-1},\ x_0,\ x_1,\ x_2,\ \ldots\,. \tag{3.114}$$

The z-transform of this sequence is defined to be the sum

$$\mathcal{Z}\big(\{x_n\}\big) \equiv X(z) \equiv \cdots + x_{-2}z^2 + x_{-1}z + x_0 + \frac{x_1}{z} + \frac{x_2}{z^2} + \cdots\,. \tag{3.115}$$

(An alternative convention sets $s \equiv 1/z$ in (3.115).) The variable z is a placeholder only and doesn't take on any values, but its real use is that now the polynomial can be treated analytically, and we hope even summed to produce a compact expression.

3.6.1 Deriving the Fibonacci Sequence from the z-Transform

An idea of the z-transform's usefulness can be obtained by using it to again find an expression for the nth term of the Fibonacci sequence, as an alternative to the approach of Sect. 2.5. The sequence is described by

$$x_{n+2} = x_{n+1} + x_n\,, \quad \text{with } x_0 = x_1 = 1\,. \tag{3.116}$$

This relation suggests that adding the Fibonacci sequence term by term to a copy of itself shifted one to the left will result in another copy of itself shifted two to the left:

$$\begin{array}{cccccccc} & 1 & 1 & 2 & 3 & 5 & \dots \\ + & 1 & 2 & 3 & 5 & 8 & \dots \\ \hline = & 2 & 3 & 5 & 8 & 13 & \dots \end{array} \tag{3.117}$$

So if the z-transforms of the shifted sequences can be related to the transform $X(z)$ of the original sequence, then (3.116) will yield an equation that can be solved for $X(z)$; inverting $X(z)$ will then produce an expression for the nth term of the original sequence, and the task will be finished.

How, then, do we relate the shifted sequence to the original one? Begin by transforming the Fibonacci sequence:

$$x_0, x_1, x_2, \dots \xrightarrow{\text{transform}} x_0 + \frac{x_1}{z} + \frac{x_2}{z^2} + \dots \equiv X(z)\,. \tag{3.118}$$

Equations (3.114) and (3.115) show that shifting the sequence by one and two places changes its z-transform in a simple way:

$$\begin{aligned} x_1, x_2, x_3, \dots &\xrightarrow{\text{transform}} x_1 + \frac{x_2}{z} + \dots = z[X(z) - x_0]\,, \\ x_2, x_3, x_4, \dots &\xrightarrow{\text{transform}} x_2 + \frac{x_3}{z} + \dots = z^2\left[X(z) - x_0 - \frac{x_1}{z}\right]. \end{aligned} \tag{3.119}$$

Equation (3.117) sums two sequences, and the z-transform of this sum is found by transforming each sequence and adding the results:

$$X(z) + z[X(z) - 1] = z^2[X(z) - 1 - 1/z]\,, \tag{3.120}$$

which gives

$$X(z) = \frac{z^2}{z^2 - z - 1}\,. \tag{3.121}$$

It now remains to recover the actual Fibonacci sequence from $X(z)$, which requires finding the inverse transform of $X(z)$. Inspecting (3.115), we see the need to convert $X(z)$ to a function of $1/z$ since we know that there are no "negative index" terms. Follow this with a separation by partial fractions to write (3.121) in terms of the Golden Ratio ϕ as

$$X(z) = \frac{1}{1 - 1/z - 1/z^2} = \frac{1}{\sqrt{5}}\left[\frac{-1}{-1/\phi + 1/z} + \frac{1}{\phi + 1/z}\right]. \tag{3.122}$$

The reason for this separation is that now each of the two bracketed terms in (3.122) is a geometric series and so is easily inverted. And because the inverse transform is also linear, each term can be inverted separately and the results added. Invert by using the usual formula for summing a geometric series:

$$\frac{a}{b+1/z} = \frac{a/b}{1+\frac{1}{bz}} = \frac{a}{b} + \frac{a}{b}\left(\frac{-1}{bz}\right) + \frac{a}{b}\left(\frac{-1}{bz}\right)^2 + \cdots . \tag{3.123}$$

The coefficient of $1/z^n$ in this last series is $-a(-1/b)^{n+1}$. Denoting the inverse transform by $\mathcal{Z}^{-1}$, the relation between this particular z-transform and its series becomes

$$\mathcal{Z}^{-1}\left(\frac{a}{b+1/z}\right) = \left\{-a(-1/b)^{n+1}\right\}_{n=0}^{\infty} . \tag{3.124}$$

This allows (3.122) to be inverted to give

$$\begin{aligned} \mathcal{Z}^{-1}(X) &= \frac{1}{\sqrt{5}}\left[\mathcal{Z}^{-1}\left(\frac{-1}{-1/\phi+1/z}\right) + \mathcal{Z}^{-1}\left(\frac{1}{\phi+1/z}\right)\right] \\ &= \frac{1}{\sqrt{5}}\left\{\phi^{n+1} - (-1/\phi)^{n+1}\right\}_{n=0}^{\infty} \\ &= \frac{1}{\sqrt{5}}\left\{\phi^{n} - (-1/\phi)^{n}\right\}_{n=1}^{\infty} , \end{aligned} \tag{3.125}$$

where we have re-indexed the last line to start from $n = 1$ to give the sequence a more intuitive feel, and to allow a comparison with the same result arrived at previously in Sect. 2.5. The last line of (3.125) shows the nth term of the Fibonacci sequence, obtained with only a small amount of effort and investment. It shows the power of the transform idea.

3.6.2 Convolving to Smoothen a Signal

A common signal processing task is the smoothening of a signal. Suppose that some property of a system is measured by the noisy reading on a gauge as a sequence of numbers, produced once per second. Our task is to look for trends in this incoming data from which we hope to make predictions about the future behaviour of the system.

Fluctuations in the data can be smoothened by the use of a moving mean. Such an idea is simple but powerful. For example, make a pass through the data, replacing each number with the average of a set of, say, three numbers: itself and the number on each side. (At the start and end, this procedure fails, so we are limited to averaging two numbers. This sort of edge effect is common in signal processing but won't concern us here.) While trends in the data do not tend to change their signs quickly, the noise certainly will,

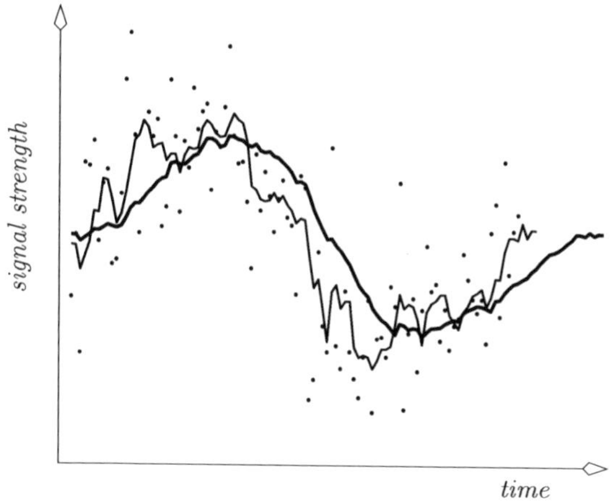

Fig. 3.8. A noisy signal, smoothened by moving means of two widths. The signal is composed of 100 points of a sinusoid, with added gaussian random noise. The lighter, jagged curve is the result of applying a moving mean of width 5 to these data, meaning that every five neighbouring data points are averaged to produce one point of the curve. (These new points are not shown but have been joined, which is why the curve is apparently continuous.) The heavier, smoother curve is the result of applying a moving mean of width 20 to the original data (again, it connects points which themselves are not shown). The sinusoid buried in the noisy data is quite evident in this heavily smoothened curve. The two curves extend beyond the data simply because the moving mean is beginning to "fall off the end", so to speak, so that the last smoothened points are not really relevant to the data themselves.

so this averaging will generally serve to add fluctuations of opposite signs, thereby mostly removing them from the signal. An example of noisy data with moving means of two different widths applied is shown in Fig. 3.8.

We could average more than three numbers at a time, of course. Choosing such a number appropriately is an art, but it should not be too large, since that will not only cancel noisy fluctuations from one number to the next, but will also begin to destroy any real trends that should be left intact. (This art of choosing the right averaging window is a good example of balancing the two types of error described on p. 95.) For the purpose of an example, we'll use a three-term moving mean, which can be represented by a *smoothening sequence* such as $\{1/3, 1/3, 1/3\}$, each number being the factor used to multiply one of the numbers of the data sequence. Additionally we will allow a different bias to be given to older data; perhaps there is an inbuilt correlation that such a bias can offset or accentuate.

Begin with a sequence x of observations infinitely long at both ends to reduce any irrelevant edge effects. We are at liberty to take element zero of each sequence to be any of its members, since the only effect of such a shift is to introduce an unimportant final shift in all the elements of the smoothened sequence:

$$x \equiv \ldots,\ x_{-2},\ x_{-1},\ x_0,\ x_1,\ x_2,\ \cdots, \tag{3.126}$$

which we'll smoothen in a weighted way with another sequence such as

$$\ldots,\ 0,\ 0,\ 0,\ \frac{3}{6},\ \frac{2}{6},\ \frac{1}{6},\ 0,\ 0,\ 0,\ \ldots. \tag{3.127}$$

In general, this three-term smoothening sequence can be written as

$$s \equiv \ldots,\ 0,\ 0,\ 0,\ s_0,\ s_1,\ s_2,\ 0,\ 0,\ 0,\ \ldots, \tag{3.128}$$

which is required to produce the following moving mean:

$$\ldots,\ s_0x_{-2}+s_1x_{-1}+s_2x_0,\quad s_0x_{-1}+s_1x_0+s_2x_1,\quad s_0x_0+s_1x_1+s_2x_2,\ \ldots. \tag{3.129}$$

This new sequence is denoted in signal processing as $s \star x$ (or sometimes by $x \star s$). The multiplication of terms from each sequence suggests that a related task might be to multiply the corresponding z-transforms. But it quickly becomes apparent that for this to work, we need to reverse the order of the s-sequence:

$$\tilde{s} \equiv \ldots,\ 0,\ 0,\ 0,\ s_2,\ s_1,\ s_0,\ 0,\ 0,\ 0,\ \ldots. \tag{3.130}$$

The reason for this becomes clear when we actually multiply the transform of x with that of $\tilde{s}$. To see why, first transform each sequence to produce $\widetilde{S}(z)$ and $X(z)$, respectively:

$$\begin{aligned} \widetilde{S} &= s_2 + \frac{s_1}{z} + \frac{s_0}{z^2}, \\ X &= \cdots + x_{-2}z^2 + x_{-1}z + x_0 + \frac{x_1}{z} + \frac{x_2}{z^2} + \cdots. \end{aligned} \tag{3.131}$$

On multiplying these two series, the coefficient of each power of z is found by pairing terms *crosswise*, which is why we needed to reverse the s-sequence:

$$\begin{aligned} \widetilde{S}X &= \cdots + \left(s_0x_{-3} + s_1x_{-2} + s_2x_{-1}\right) z + s_0x_{-2} + s_1x_{-1} + s_2x_0 \\ &\quad + \left(s_0x_{-1} + s_1x_0 + s_2x_1\right) \frac{1}{z} + \cdots. \end{aligned} \tag{3.132}$$

It is evident that the sequence corresponding to this last expression is exactly what we set out to calculate: the moving mean shown in (3.129). This smoothened sequence is known as the *convolution* of the two sequences $\tilde{s}$ and x, and is also written $\tilde{s} * x$ or $x * \tilde{s}$; convolution is commutative because the corresponding multiplication is commutative ($\widetilde{S}X = X\widetilde{S}$).

The two notations just described are related by $s \star x = \tilde{s} * x$. Insight into the various permutations of what can be smoothened with what is obtained by listing these smoothenings with their associated notation:

- the result of smoothening x with s is written $s \star x = \tilde{s} * x$,
- the result of smoothening s with x is written $x \star s = \tilde{x} * s$,
- the result of smoothening x with $\tilde{s}$ is written $\tilde{s} \star x = s * x$, and
- the result of smoothening s with $\tilde{x}$ is written $\tilde{x} \star s = x * s$.

In particular, since $s * x = x * s$, it follows from the last two statements that smoothening x with the reverse of s gives the same result as smoothening s with the reverse of x.

In summary, to apply a moving mean s to a sequence x, we convolve $\tilde{s}$ (the reversed version of s) with x. And to convolve two sequences, we simply multiply their z-transforms and take the inverse transform. The convolution of two sequences is really nothing more mysterious than using the reverse of either sequence to smoothen the other via a moving mean.

In fact, the z-transform is really just an intermediate step in the whole process; the convolution of two sequences x and y can simply be written in terms of its nth element as

$$(x * y)_n \equiv \sum_{k=-\infty}^{\infty} x_k \, y_{n-k} \,, \quad \text{while} \;\; (x \star y)_n = \sum_{k=-\infty}^{\infty} x_k \, y_{n+k} \,. \tag{3.133}$$

Although it might not be apparent at first glance why combining two sequences via (3.133) should be a meaningful thing to do, we've seen that there is nothing more mysterious underlying this procedure than an everyday intuitive smoothening of noisy data, along with the powerful fact that convolution and multiplication of transforms go hand in hand.

Cross Correlation of Sequences

The identities in (3.133) can be used to show various interesting theorems that involve the use of convolution, such as the fact that $x \star y$ is just the reverse of $y \star x$ (which accounts for the two ways in which the $\star$ is commonly defined). But let's ask an interesting question: what results when we smoothen a sequence with itself to give $x \star x$? It's not difficult to show that term zero of $x \star x$ is its largest. We can see this by beginning with the expression

$$\sum_k \left(x_k + \lambda x_{n+k}\right)^2 > 0 \,, \tag{3.134}$$

which is trivially true for all λ and n, except when $\lambda = -1$ and $n = 0$ simultaneously. Now expand (3.134) and refer to (3.133) to write it as

$$(x \star x)_0 + 2\lambda \, (x \star x)_n + \lambda^2 (x \star x)_0 > 0 \,. \tag{3.135}$$

If the left-hand side of this inequality is to have no roots over λ for $n \neq 0$, its discriminant must be negative, and that implies

$$\left|(x \star x)_{n \neq 0}\right| < (x \star x)_0 \,, \tag{3.136}$$

in which case we conclude that $(x \star x)_0$ must be positive. Equation (3.136) is what we set out to prove (and is actually an example of the Cauchy–Schwarz inequality used in the previous chapter). It says that when the sequence "matches up with itself", the moving mean is globally maximised, and this indicates a best fit in the sense of a correlation. This motivates us to call $x \star x$ the *autocorrelation* of x, while $x \star y$ is the *cross correlation* of x and y.

Autocorrelation and cross correlation are very important ideas in signal processing. Comparing a signal x with a template y by calculating $x \star y$ allows us to quantify the extent to which the template is present in the signal, so that details of signals can often be extracted even when they are well below the noise level. This sort of processing makes heavy demands on computing power, which is why signal processing has only really come of age in the last few decades; the very elegant but complex analogue electronic circuitry of former years has nowadays given way to digital processing of data, which in turn has opened up new uses for signal processing.

3.7 The Discrete Fourier Transform

The correspondence between convolution of sequences and multiplication of their transforms appears more generally in other transform contexts, such as the Fourier transform. And just as the z-transform acts on real data (discrete numbers as opposed to a continuous function), the Fourier transform of Chap. 2 can be remodelled to act on a sequence of data, where it goes by the name of the *discrete Fourier transform.* This discrete transform is very useful for analysing data; after all, it's one thing to transform a continuous, well-behaved function on paper, but quite another to do something meaningful with a set of numbers that has been generated in the laboratory! And, in recent years, the use of computers together with a very fast and efficient way of calculating the discrete Fourier transform, known as the *fast Fourier transform*, has revolutionised signal processing, a field that makes heavy use of transform methods.

Because the Fourier transform projects an arbitrary function onto its basis functions $\phi_n(x)$—each of which has a well-defined frequency—what it's doing is separating a signal into a spectrum of frequencies, just as a prism does for light. In practice, the only signal we might have at our disposal is one that has been sampled (i.e., a function whose values we know only at discrete values of x). What we wish to do in order to extract the spectrum from such a collection of data is in fact the reverse of what we did in going from (2.166) to (2.169). That is, rather than change the sum in (2.166) to an integral, we will instead convert the integral to a sum.

But it must be realised that a finite number of samples can never reconstitute a function exactly. It turns out, however, that if we sample a continuous signal at some given rate, then we *can* know all of the frequencies present with values up to half of that sampling rate. This is *Nyquist's theorem.* Let's see how it comes about.

3.7.1 Sampling Using Nyquist's Theorem

Frequency, like many words in the physicist's vocabulary, has a slightly more specific meaning than that of its everyday use, which takes frequencies to be only positive. In signal processing they can also be negative, being set by the direction of rotation of the phasor describing the relevant oscillation. In order to distinguish between frequencies and their moduli, we'll refer to the positive value as the *repetition rate*:

$$\text{Repetition rate} \equiv |\text{frequency}| \,. \tag{3.137}$$

A given repetition rate corresponds to both positive and negative frequencies, in the same way that we cannot distinguish the rotation direction of a spinning phasor by only viewing its projection on one axis in the phasor plane. For example, if between measurements a phasor appears to advance by $^3\!/_4$ of a clockwise revolution, then it might be said that $^3\!/_4$ of a cycle had passed. However, we could add any integer to this value since the extra one or more cycles occurs between measurements, rendering those cycles effectively invisible. So the phasor might well have turned through $1^3\!/_4$ cycles, or $-^1\!/_4$ cycle, or $-1^1\!/_4$ cycles, and so on. We cannot know whether the phasor has really turned clockwise or counterclockwise.

Speak, then, of repetition rate instead of frequency, and ask: for a given sampling rate, what is the highest repetition rate that we can detect? (The answer to this question will constitute Nyquist's theorem.) As an example, consider the cycle of the number of sunspots, which happens to have a period of 11 years. How often must we count the number of sunspots in order to be able to detect this period (i.e., a repetition rate of $1/11\ \text{year}^{-1}$)?

$$\text{Repetition rate} = |\text{freq.}| = \frac{|\text{cycles}|}{\text{unit time}} = \frac{|\text{cycles}|}{\text{no. samples}} \times \frac{\text{no. samples}}{\text{unit time}} \,. \tag{3.138}$$

Call the number of samples taken per unit time the sampling rate f_s. Equation (3.138) then becomes

$$\text{Repetition rate} = f_s \,|\text{cycles/sample}| \,. \tag{3.139}$$

Nyquist's question then becomes: what is the largest repetition rate we can know about, given a sampling rate f_s?

$$\text{Largest repetition rate we can detect} = f_s \,|\text{cycles/sample}|_{\text{max}} \,. \tag{3.140}$$

In that case, what is the maximum number of cycles per sampling period that is able to be detected? Suppose that we make one sunspot count per year: $f_s = 1\ \text{year}^{-1}$. Our question becomes: what is the finest detail that we can see in these data? If we can detect a large number of cycles per year, then we are doing very well—we are observing fine detail in the Sun's behaviour. So let's make a wild guess:

- Is this repetition rate, this number of cycles per year, perhaps, say, 10? No, it cannot be 10 because the frequencies corresponding to this number, 10 and −10, "look" like zero, being integers. If the phasor turns through 10 revolutions in either direction while our backs are turned, then we have no way of inferring that it has done so when we look at it again.

- So the repetition rate, the number of cycles per year, must be smaller than any integer. If it were not—if it were really, say, 1.2—then in between samplings we would effectively look away and then look again to find that it had advanced 0.2 of a revolution. But we would have no way of knowing whether this advance was any of 0.2, 1.2, 2.2, etc. So the number of cycles per year that we can unequivocally say occurred cannot even be 1.2, let alone 10. It must be less than 1.

- Could it be 0.9? If $|\text{cycles/sample}| = 0.9$, then $\text{cycles/sample} =$ either 0.9 or −0.9. But what looks like 0.9 might really be −0.1; if the phasor has advanced 0.9 of a turn, then we might think it really has gone backward 0.1 of a turn. So we cannot detect 0.9 of a turn. And similarly −0.9 looks like +0.1. So a repetition rate of 0.9 is indistinguishable from a rate of 0.1 (i.e., 0.9 gets *aliased* to 0.1).

- Similarly, 0.8 is indistinguishable from 0.2, and 0.7 is indistinguishable from 0.3. It follows that the largest number of cycles per sample that we can measure is 1/2; anything higher is ambiguous.

In that case we have, from (3.140),

$$\text{Largest repetition rate we can detect} = f_s/2\,. \tag{3.141}$$

So the highest repetition rate that can be extracted from sampled data is just half of the frequency that was used to obtain that data. (Or, by taking reciprocals, the smallest period we can know about is twice the sampling period.) In fact, this statement is actually just half of Nyquist's theorem. Not only is $f_s/2$ the largest repetition rate that can be detected, but it's also the largest repetition rate that *will* be detected. That is, while we have shown that sampling at frequency f_s is necessary to detect a repetition rate of $f_s/2$, it also turns out that sampling at this frequency is *sufficient* to detect that repetition rate, but we will not stop to prove this; it can be found in books on signal processing.

Thus, if we wish to verify that the Sun indeed has an 11 year periodicity in its sunspot numbers, then we need to sample solar activity at least once every 5.5 years. Of course, if we sample at a higher rate, say every year, then

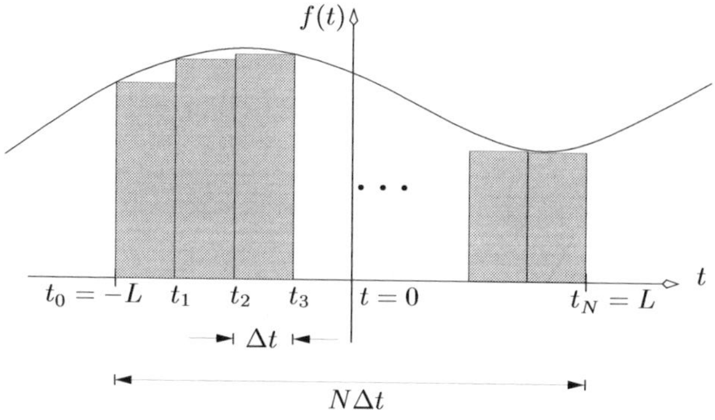

Fig. 3.9. Approximating the integral in (2.166) by a Riemann sum, as used in (3.142). The interval $t = -L \to L$ is divided into N strips, each of width Δt.

we'll do even better. Not only will we see the 11 year cycle, but we'll also detect shorter cycles down to a 2 year period. Any cycles with a period less than this cannot be detected unless we sample more frequently.

3.7.2 Discretising the Fourier Transform

Once an appropriate sampling rate has been determined to analyse the frequency spectrum of something like the solar cycle, we can begin to analyse the measured data by returning to the Fourier theory of Chap. 2. But because a real signal generally occurs over time rather than space, the discrete Fourier transform is normally couched in terms of t instead of x.

So begin by writing (2.166) with the integral approximated by a sum and with $x \to t$. Do this in the usual Riemann way shown in Fig. 3.9. Divide the domain into N steps of length Δt, and define the sequence $t_0, \ldots, t_N$, where

$$\begin{aligned} t_0 &\equiv -L\,, & t_N &= L\,, \\ N\Delta t &= 2L\,, & t_k &= t_0 + k\Delta t\,. \end{aligned} \tag{3.142}$$

In that case, (2.166) approximates to

$$\begin{aligned} f(t) &\simeq \sum_{n=-\infty}^{\infty} \frac{1}{2L} \exp\frac{\mathrm{i}n\pi t}{L} \sum_{k=0}^{N-1} f(t_k)\frac{2L}{N} \exp\frac{-\mathrm{i}n\pi(t_0 + k\Delta t)}{L} \\ &= \sum_{n=-\infty}^{\infty} \exp\frac{\mathrm{i}n\pi t}{L}\, e^{\mathrm{i}n\pi} \times \underbrace{\frac{1}{N}\sum_{k=0}^{N-1} f(t_k) \exp\frac{-\mathrm{i}2\pi nk}{N}}_{\substack{\equiv \text{ element } n \text{ of discrete Fourier} \\ \text{transform of } \{f(t_k)\}_{k=0}^{N-1}, \text{ up} \\ \text{to some conventional factor } \alpha \\ \text{in (3.144)}}}\,. \end{aligned} \tag{3.143}$$

Writing $f_k \equiv f(t_k)$, the discrete Fourier transform, or DFT, maps a sequence of N numbers $\{f_0, \dots, f_{N-1}\}$ into another sequence (whose size is yet to be determined),

$$F_n \equiv \frac{\alpha}{N} \sum_{k=0}^{N-1} f_k \exp \frac{-\mathrm{i}2\pi nk}{N}, \tag{3.144}$$

where we have inserted an arbitrary *real* factor called α to accommodate other common definitions of the DFT in the discussion that follows, where for example α might equal N or $\sqrt{N}$. Thus (3.143) becomes

$$f(t) \simeq \frac{1}{\alpha} \sum_{n=-\infty}^{\infty} (-1)^n F_n \exp \frac{\mathrm{i}n\pi t}{L}. \tag{3.145}$$

It might appear from (3.143) that the sequence $\{F_n\}$ must be infinitely long, but in fact it can also be reduced to length N, as we'll see shortly.

The DFT in (3.143) is the coefficient of a particular spectral component indexed by n, so that it tells us how much of that component is present in the signal. Nyquist's theorem stipulates how often we need to sample in order to extract information about the frequencies present. So, in principle, we could vary the value of the number of samples N depending upon the index n of the spectral component we wish to isolate. But, in practice, that requires a different set of samples for each spectral component, and that is just not practical in a real situation. We are seldom at liberty to sample a signal every which way we choose. Even with such freedom, producing set after set of evenly spaced samples, with sizes determined by which frequency we wish to study, would necessitate a huge amount of sampling. In practice, there is almost always just one set of sampled values of the signal available, so that we must let N be the size of this set, using it in each sum over k in (3.143). Nyquist's theorem then tells us that there is a maximum value of n of which we are able to have any knowledge.

Given N, what is this value of n, called $n_{\max}$? The sampling frequency is f_s, where (3.142) gives

$$f_s = \frac{1}{\Delta t} = \frac{N}{2L}. \tag{3.146}$$

If n in (3.145) were *continuous* with a maximum value $\widehat{n}$ set by Nyquist's theorem, then the $\widehat{n}\pi/L$ in that equation would be the angular frequency corresponding to half of f_s:

$$\frac{\widehat{n}\pi}{L} = 2\pi \frac{f_s}{2}, \tag{3.147}$$

giving

$$\widehat{n} = N/2. \tag{3.148}$$

Since n takes on integer values only, and its maximum is positive, $n_{\max}$ must be the largest natural number less than or equal to $\widehat{n}$, written as

$$n_{\max} = [\,\widehat{n}\,] = [N/2]\,. \tag{3.149}$$

So, for any given N, the highest frequency component that can be detected corresponds to $n = [N/2]$. When N is odd, this presents no problems, but when N is even, $[N/2]$ lies just on the boundary between frequencies that are detectable and those that are not. So we can expect to have to tread carefully in that case, as will be evident in the next few pages.

Incorporating $n_{\max} = [N/2]$ into (3.145) might indicate that the best approximation to the signal $f(t)$ that can be made is

$$f(t) \simeq \frac{1}{\alpha} \sum_{n=-n_{\max}}^{n_{\max}} (-1)^n F_n \exp \frac{in\pi t}{L}\,. \tag{3.150}$$

But the two approximations that have been made (truncating the sum and discretising the integral) turn out not to give the best approximation to the signal, and in fact we can do better in encapsulating a data set by a more useful set of numbers.

To see how, let's go back to the drawing board. Equation (3.150) rewrites the set of N data points in terms of another set of approximately N numbers. (We say "approximately" since we have not stopped to count properly by checking what redundancy might be present in the new set of numbers.) Of course, it's trivially possible to encapsulate the information contained in the N data numbers by a set of N numbers: just use the data! More analytically, however, it is always possible to find the N coefficients of the unique $N-1^{\text{th}}$-order polynomial that passes through N points; we need only write this polynomial down with its N unknown coefficients to be determined. Fitting it to the N points gives a set of N linear equations, and these can be solved.[4] Something has been gained: although we have simply swapped N data numbers for N polynomial coefficients, the polynomial is smooth—and that allows it to interpolate the data more meaningfully.

In fact, a similar fit can be made to the sequence $f_0, \ldots, f_{N-1}$, but using sinusoids (complex exponentials) instead of a polynomial. Remember that Nyquist's theorem only allows knowledge of frequencies corresponding to $n = -[N/2] \to [N/2]$, so restrict the DFT for this sequence to just those values. This might look like the DFT sequence has, for example, five numbers for both $N = 4$ and $N = 5$, but in fact this is not so. When N is even, the DFT sum (3.144) is calculated for n running from $-N/2$ to $N/2$, or $N+1$ numbers; but a moment's inspection makes it clear that the sum for the last value $n = N/2$ is the same as that for the first, $n = -N/2$:

$$N \text{ even} \Longrightarrow F_{-N/2} = F_{N/2}\,. \tag{3.151}$$

[4] As a side note, this polynomial can always be written down immediately—with no equations to be solved—in a particularly intuitive and straightforward way, in a form known as a *Lagrange polynomial.*

Thus the last value $F_{N/2}$ is extraneous, and the DFT need really only be calculated N times. So it transforms N numbers to another set of N numbers. On the other hand, when N is odd, the DFT sum uses n running from $n = -(N-1)/2$ to $(N-1)/2$, or N numbers; but the sum (3.144) for $n = (N-1)/2$ is *not* the same as that sum for $n = -(N-1)/2$ and so is not extraneous. Thus the DFT can always be used to transform N numbers to N numbers. For that reason, write the discrete Fourier transform as

$$\boxed{F_n = \frac{\alpha}{N} \sum_{k=0}^{N-1} f_k \exp \frac{-\mathrm{i}2\pi nk}{N}, \quad \text{for } N \text{ values of } n = -[N/2], \ldots} \tag{3.152}$$

This set of equations can be inverted to express f_k as a function of the set of F_n in the usual way, through making use of the orthogonality of the basis functions. Write

$$F_n \exp \frac{\mathrm{i}2\pi n\ell}{N} = \frac{\alpha}{N} \sum_{k=0}^{N-1} f_k \exp \frac{\mathrm{i}2\pi n(\ell-k)}{N}, \quad 0 \leqslant \ell \leqslant N-1\,. \tag{3.153}$$

When N is even, sum both sides over the N numbers of the DFT:

$$\sum_{n=-N/2}^{N/2-1} F_n \exp \frac{\mathrm{i}2\pi n\ell}{N} = \frac{\alpha}{N} \sum_{k=0}^{N-1} f_k \underbrace{\sum_{n=-N/2}^{N/2-1} \exp \frac{\mathrm{i}2\pi n(\ell-k)}{N}}_{=N\delta_{k\ell}} = \alpha f_\ell\,. \tag{3.154}$$

The underbraced sum in (3.154) equates to the Kronecker delta by virtue of being a geometric series. Finally, we obtain

$$f_\ell = \frac{1}{\alpha} \sum_{n=-N/2}^{N/2-1} F_n \exp \frac{\mathrm{i}2\pi n\ell}{N}, \quad 0 \leqslant \ell \leqslant N-1\,. \tag{3.155}$$

Likewise, (3.152) can be inverted when N is odd: repeat the procedure, again summing over the N numbers of the DFT, $n = -(N-1)/2$ to $(N-1)/2$. The proof follows the same steps and is omitted. Both even and odd N can be written in one formula, known as the *inverse discrete Fourier transform* (IDFT):

$$\boxed{f_\ell = \frac{1}{\alpha} \sum_{n=-[N/2]}^{N \text{ numbers}} F_n \exp \frac{\mathrm{i}2\pi n\ell}{N}, \quad 0 \leqslant \ell \leqslant N-1\,.} \tag{3.156}$$

Compare this with (3.152). The IDFT maps the N numbers of the DFT back to the N data points. What has been gained by transforming N numbers to N numbers? The basis exponentials are smooth functions of ℓ if ℓ is extended from integer values to real values; so besides showing frequencies present in the data, the DFT and IDFT allow the data to be interpolated, which we focus on next.

3.7.3 Interpolating Real Data with the DFT

Although the main use of the DFT lies in analysing a signal for its frequency spectrum, studying how it can be used to interpolate real data is a useful exercise in familiarity with Fourier theory in general. How can (3.152) and (3.156) be used to interpolate such a set? It turns out that we need to consider odd and even N separately here, so let's look at each in turn.

Simpler Case: When N is Odd

For odd N, the IDFT is

$$\alpha f_\ell = \sum_{n=-(N-1)/2}^{(N-1)/2} F_n \exp \frac{i2\pi n\ell}{N}, \quad 0 \leqslant \ell \leqslant N-1 . \tag{3.157}$$

Since (3.152) gives $F_{-n} = F_n^*$ when both the data and α are real (which we'll always assume), (3.157) incorporates positive and negative frequencies symmetrically, and so will always be real (remember the discussion of Sect. 2.11). To interpolate to any value of t, write

$$t_\ell - t_0 = \ell\,\Delta t\,, \tag{3.158}$$

in which case

$$\alpha f(t_\ell) = \sum_{n=-(N-1)/2}^{(N-1)/2} F_n \exp \frac{i2\pi n(t_\ell - t_0)}{N\Delta t} . \tag{3.159}$$

Because this is real and returns the data points exactly, it can be used for an interpolating function f_{int} for any t in the interval $[t_0, t_N]$:

$$\alpha f_{\text{int}}(t) \equiv \sum_{n=-(N-1)/2}^{(N-1)/2} F_n \exp \frac{i2\pi n(t-t_0)}{N\Delta t}$$

$$= F_0 + \sum_{n=1}^{(N-1)/2} 2\,\text{Re}\left(F_n \exp \frac{i2\pi n(t-t_0)}{N\Delta t}\right) . \tag{3.160}$$

Harder Case: When N is Even

When N is even, a minor complication occurs that we predicted might happen in the discussion following (3.149). Now the IDFT is (3.155), but this is *not* quite symmetric in its positive and negative frequencies—the first term is an odd one out, and so the sum will not always be real. It will certainly be real for integral ℓ since such values just return the (real) data points. But for general interpolation of the data, ℓ need not be an integer, and so the IDFT need not be real. Even so, the remedy is straightforward. We are free to substitute any other term in the sum (3.155) as long as, ultimately, the data points are still returned. So we need only ensure that this new term is identical to the one being replaced, when ℓ is an integer.

The first term of (3.155), which is to be replaced, is

$$\frac{1}{\alpha} F_{-N/2} \exp \frac{\mathrm{i}2\pi(-N/2)\ell}{N} = \frac{1}{\alpha} F_{-N/2} (-1)^{\ell} . \tag{3.161}$$

Equation (3.152) ensures that $F_{-N/2}$ is real. In that case, replace (3.161) with a term that is real for all ℓ and agrees with (3.161) for integral ℓ. Such a term is

$$\frac{1}{\alpha} F_{-N/2} \cos \pi\ell . \tag{3.162}$$

Now replace the first term of (3.155) with (3.162), which gives a real function that also returns the data points exactly. Thus, in the even-N case the interpolating function for any value of t in $[t_0, t_N]$ is

$$\begin{aligned} \alpha f_{\text{int}}(t) &\equiv F_{-N/2} \cos \frac{\pi(t-t_0)}{\Delta t} + \sum_{n=-N/2+1}^{N/2-1} F_n \exp \frac{\mathrm{i}2\pi n(t-t_0)}{N\Delta t} \\ &= F_{-N/2} \cos \frac{\pi(t-t_0)}{\Delta t} + F_0 + \sum_{n=1}^{N/2-1} 2 \operatorname{Re} \left(F_n \exp \frac{\mathrm{i}2\pi n(t-t_0)}{N\Delta t} \right) . \end{aligned} \tag{3.163}$$

This expression is not as complicated as it looks, as can be seen in a simple example.

Example: Transforming a Small Data Set. Suppose we use the DFT in (3.152) to transform a simple signal, such as the four numbers $\{2, 3, 5, 7\}$. This is a useful exercise in exploring what the resulting four numbers mean. Here $N = 4$ with the Fourier index n running from -2 to 1. We will set $\alpha = 1$.

$$\begin{aligned} &\{2, 3, 5, 7\} = \{f_0, f_1, f_2, f_3\} \\ &\xrightarrow{\text{DFT (3.150)}} \left\{ {-3}/4,\ {-3}/4 - \mathrm{i},\ 17/4,\ {-3}/4 + \mathrm{i} \right\} = \{F_{-2}, F_{-1}, F_0, F_1\} . \end{aligned} \tag{3.164}$$

This new sequence then inserts into (3.163) to give

$$f_{\text{int}}(t) = \frac{-3}{4} \cos \frac{\pi(t-t_0)}{\Delta t} + \frac{17}{4} - \frac{3}{2} \cos \frac{\pi(t-t_0)}{2\Delta t} - 2 \sin \frac{\pi(t-t_0)}{2\Delta t} . \tag{3.165}$$

The value of Δt is unspecified; it can be anything set by the original scenario. So, for example,

$$f_{\text{int}}(t_3) = f_{\text{int}}(t_0 + 3\Delta t) \overset{(3.165)}{=\!=} 7 , \tag{3.166}$$

as expected. Finally, (3.165) is a function that reproduces the sampled data $\{2, 3, 5, 7\}$ as well as interpolating and expressing it in terms of Fourier basis functions that show something of the frequencies present. Comparing each of the sinusoid arguments in (3.165) with a template of $\omega t \equiv 2\pi f t$, we see that the frequencies that Nyquist's theorem allows us to know about are

$$\frac{1}{2\Delta t}\,, \quad 0\,, \quad \text{and} \quad \frac{1}{4\Delta t}\,, \tag{3.167}$$

where the zero frequency belongs to the constant $^{17}/_4$, called the *DC component* in analogy with the time dependence of direct and alternating currents. Four data points might not really constitute a meaningful signal, but they serve to illustrate what the numbers comprising the DFT really mean.

One final point worth noting is that several conventions for calculating the DFT and its inverse are widely used. For example, a change of variables in the sum over n in (3.155) might be done so that the sum is over the values $0, \ldots, N-1$. These changes must be made with care when working with a computer programme that implements the DFT using a different convention. Here we have followed a minimalist approach, by working from the first principles of Chap. 2 and changing the notation as little as possible.

A More Realistic Signal

A more realistic scenario might generate data that really do have inbuilt periodicities. Consider sampling a 5 Hz signal. (We could use a signal with multiple frequencies, but that would produce more peaks in the plotted spectra of Figs 3.10 and 3.11, which would only obscure the main point of this discussion.) By Nyquist's theorem, we need to sample the signal at higher than 10 Hz to actually detect this frequency to be present in the Fourier transform. Sampling for two seconds amounts to ten periods, which might be sufficient for an accurate analysis; plotting the DFT will be the real test of whether this is so. Results are shown in Fig. 3.10 for two choices of sampling frequency. The horizontal axes in these plots measure frequency. How were their units chosen? The basis functions of (3.143) are

$$\exp\frac{\mathrm{i}n\pi t}{L} \equiv \mathrm{e}^{\mathrm{i}\omega t} = \mathrm{e}^{\mathrm{i}2\pi f t}\,, \tag{3.168}$$

so that frequency f and Fourier index n relate via $f = n/(2L)$. Thus the data points increase in frequency steps of

$$\Delta f = \frac{\Delta n}{2L} = \frac{1}{2L} = \frac{1}{\text{width of sampled interval}}\,. \tag{3.169}$$

The plots are symmetrical about zero frequency, so we plot the points likewise and set the frequency interval from one point to the next to be as in (3.169). The plots show the 5 Hz peak very well, together with its Fourier partner of −5 Hz. It's clear that sampling at a higher rate gives less ambiguity in the measured frequencies present: we do expect two sharp peaks.

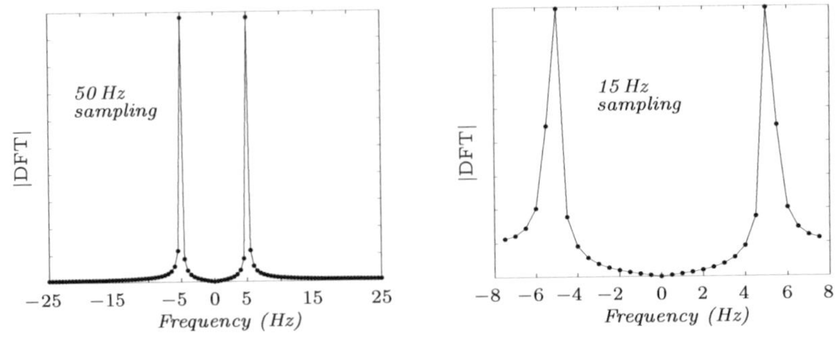

Fig. 3.10. Sampling a 5 Hz signal for two seconds at rates that are above the Nyquist lower limit for correct detection of the 5 Hz frequency. The elements of the DFT are complex, so their absolute values have been plotted (but are not important here). **Left**: Sampling at 50 Hz produces sharp peaks at the expected values of ±5 Hz. The total spectral domain encompasses the Nyquist frequency of 25 Hz, and so is −25 Hz to 25 Hz. **Right**: Sampling at a lower rate of 15 Hz produces coarser peaks in frequency. The domain is now −7.5 Hz to 7.5 Hz.

Aliasing: Peaks Under Assumed Names (or Frequencies)

What frequency spectra result if the signal is sampled less frequently than the 10 Hz demanded by Nyquist's theorem? We can expect problems when sampling at such lower frequencies. Figure 3.11 shows the results for sampling at 8 Hz (left) and 3 Hz (right). These rates of sampling are so low that we have sampled for 20 seconds to produce this figure. These plots show that below the Nyquist sampling rate, the ±5 Hz peaks masquerade as other frequencies. When sampling falls below the Nyquist 10 Hz rate, the actual frequencies present in the signal are aliased to lower ones: they shift by integer multiples of the "allowed" spectrum width until they fall into it. For example, the left plot of Fig. 3.11 shows the result of sampling at 8 Hz. This sampling rate can only give us information about frequencies from −4 to 4 Hz, by Nyquist's theorem. The 5 Hz signal thus shifts down by 8 Hz to become −3 Hz, while the −5 Hz signal shifts up by 8 Hz to become +3 Hz. The two peaks at ±3 Hz can be seen in the plot.

Similarly, sampling at a still lower frequency of 3 Hz can only yield information about frequencies from −1.5 to 1.5 Hz. The 5 Hz signal now shifts down by 6 Hz to −1 Hz, and this peak along with its +1 Hz companion is visible in the right-hand plot of Fig. 3.11.

Aliasing certainly can make it seem that an unexpected frequency is present in the signal, and we have no way of knowing otherwise. The only way around the problem is to prevent frequencies higher than half the sampling rate from being present in the processed signal from the start. In practice, such high-frequency filtering is included in the design of the hardware that collects the signals. Of course, this is fine when we are the ones who collect

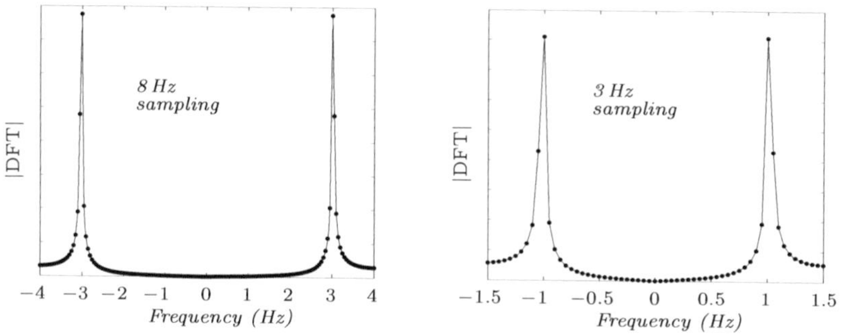

Fig. 3.11. Sampling a 5 Hz signal for 20 seconds, now at rates that are *below* the Nyquist lower limit for correct detection of the 5 Hz frequency. Again the absolute values of the DFT elements are plotted versus frequency. We expect, and get, spurious results here. **Left**: Sampling at 8 Hz shows peaks at spurious frequencies of ± 3 Hz. **Right**: Sampling at 3 Hz shows peaks at other spurious frequencies, this time at ± 1 Hz.

the data, but if they have been bequeathed to us by history, then we don't have such freedom.

Eliminating the Negative Frequencies

The negative frequencies in the spectra of Figs 3.10 and 3.11 are useful for helping us understand the mathematics behind the discrete Fourier transform. For example, including negative frequencies ensures that the DFT of a sampled gaussian function gives a result that again looks like a gaussian, since it peaks at zero frequency. This transformation of a gaussian to a gaussian is just what we expect of the Fourier transform; we have come across it already in (3.21).

Even so, negative frequencies do not really give any more information than is contained in the positive frequencies. In fact, the sequence of DFT elements is often calculated in a way that swaps its two halves (i.e., swaps its negative- and positive-frequency domains). This brings the zero frequency to the left-hand end of the DFT sequence, but it then shifts the negative frequencies to higher positive ones. The frequency spectrum is still symmetrical about its middle, but the entire set of negative frequencies is now shifted by the width of the sampled interval to high positive ones—all higher than the Nyquist frequency. This Nyquist frequency is now in the middle of the plot, if indeed the new right-hand half of the spectrum is plotted at all; there is no need for it to be. For example, the -5 Hz peak in the left-hand plot of Fig. 3.10 gets shifted to $-5 + 15 = 10$ Hz, while the same peak in the right-hand plot of Fig. 3.10 gets shifted to $-5 + 50 = 45$ Hz. These are higher than the Nyquist frequency and so can be discounted. They have not really been *aliased* to higher frequencies, since we may well have sampled above the Nyquist frequency. The shift is purely a choice in presenting the DFT, and indeed there is no real necessity to show this right-hand part of the frequency spectrum.

The intricacies of swapping the two halves of the DFT sequence are controlled by the fact that the exponentials in the DFT (3.144) have a modulo N equivalence:

$$\exp \frac{-\mathrm{i}2\pi(n+N)k}{N} = \exp \frac{-\mathrm{i}2\pi nk}{N} \quad \text{when } k \text{ is an integer.} \tag{3.170}$$

So, instead of choosing $-[N/2] \leqslant n < [N/2]$, we can choose it as follows. The negative values of n are increased by N, while the zero and positive values remain unaltered. This way the original interval of N numbers is divided into two halves, which are then swapped. For example, if $N = 4$, the set of $n = -2, -1, 0, 1$ becomes $n = 2, 3, 0, 1$, whose two halves are swapped to give a plot with a domain of $n = 0, 1, 2, 3$. Although the DFT sequence is often written in this fashion, we need to remember that its right-hand half is then really the negative frequencies shifted upward, beyond the Nyquist frequency.

The discrete Fourier transform is the bread and butter of much sampling theory, although it has only become really useful in recent years with the growth of computer processing power. But its ability to give information on how the frequency spectrum of a signal changes over time is limited. This is because we must "window" the data, in the sense of analysing short pieces of it for their frequency content. But how short is short? For this type of analysis, other techniques such as the use of *wavelets* have recently grown in popularity. Wavelets are essentially a different set of basis functions that fall to exactly zero for different widths. Thus they have their own inbuilt ability to window the data, and so can be used to inspect the data on various scales, unlike the infinite-extent basis functions of Fourier theory. But no matter what techniques are used to sample and analyse data, the idea of sampling lies at the interface between the smoothly continuous functions of physical theory and the noisy data of the real world.

3.8 Correct and Convincing: Presenting Solutions to Problems

We wish to end this chapter on a lighter and completely different note (though still in an experimental vein), having to do with the pedagogy of working a numerical exercise. Two examples have been chosen here. The first has to do with presenting a version of a formula that incorporates some set of desired units. The second example is concerned with a nuclear scattering experiment in which protons are fired at an iron foil. Given appropriate information, we must predict the number of scattered protons striking a detector.

Solving problems can be an underrated area in a physics degree. While it's possible to learn lots of formalism, how to actually solve a problem is so much more subjective, and different people have very different approaches.

In the everyday world, when we're given such a well-defined task as this, we will probably assemble all manner of concepts that relate to the task and the information we have been given, and then endeavour somehow to knit everything together into an overall structure that just might get to the answer. Unless we are thoroughly familiar with the task at hand, we might experiment with different concepts, try various approaches, and hopefully converge on the solution.

Even so, getting there is only half the job. Regardless of how the solution was arrived at, we must still convince our audience that the problem has indeed been solved correctly. Presenting the analysis in a logical way can do just that, whether or not it was the approach we actually used. The two problems worked through in this section are given as examples of a structured way of presenting an analysis.

3.8.1 Tailoring a Formula to a Given Set of Units

The first example of presenting a solution in a structured way is the following. Given the period of a pendulum

$$T = 2\pi\sqrt{L/g}\,, \tag{3.171}$$

we wish to present a version of this equation that specifies T in years, L in feet, and g in kilometres per hour squared.

Calculating and presenting a structured solution to this sort of "units" problem revolves around using a clear method of indicating that each of the quantities is indeed specified in its required units. Writing "T (in years)" is appropriate for the final formula, but it's of no help in deriving that formula. Luckily, there is a simple piece of notation that does everything for us. Just as "6/2" means "the number of twos in six", "$T/(1\,\text{yr})$" means "the number of years in T". This is a pure number—it carries no units. In such an expression, T does not even have to be specified in years because the expression itself is a mini-formula that converts T to years.

In that case, to convert (3.171) to the required units, we need only rewrite it with T replaced by $T/(1\,\text{yr})$, L replaced by $L/(1\,\text{ft})$, and g replaced by $g/(1\,\text{km}\,\text{hr}^{-2})$. The following shows the calculation with part of the scaffolding still in place to show some of the intermediate steps. All we are doing is dividing and multiplying by whatever is necessary to preserve the form of the original expression.

$$T = 2\pi\sqrt{\frac{L}{g}}\,, \quad \text{so}$$

$$\frac{T}{1\,\text{yr}}\cdot 1\,\text{yr} = 2\pi\sqrt{\frac{L}{1\,\text{ft}}\cdot 1\,\text{ft}\cdot\frac{1}{\dfrac{g}{1\,\text{km}\,\text{hr}^{-2}}\cdot 1\,\text{km}\,\text{hr}^{-2}}}$$

$$= 2\pi\sqrt{\frac{L/(1\,\mathrm{ft})}{g/(1\,\mathrm{km\,hr^{-2}})}\cdot\frac{1\,\mathrm{ft}}{1\,\mathrm{km}}\cdot 1\,\mathrm{hr}^2}$$

$$= 2\pi\sqrt{\frac{L/(1\,\mathrm{ft})}{g/(1\,\mathrm{km\,hr^{-2}})}\cdot\frac{1\,\mathrm{ft}}{1\,\mathrm{km}}}\cdot 1\,\mathrm{hr}\,. \tag{3.172}$$

Thus,

$$\frac{T}{1\,\mathrm{yr}}\cdot\frac{1\,\mathrm{yr}}{1\,\mathrm{hr}} = 2\pi\sqrt{\frac{L/(1\,\mathrm{ft})}{g/(1\,\mathrm{km\,hr^{-2}})}\cdot\frac{1\,\mathrm{ft}}{1\,\mathrm{km}}}\,. \tag{3.173}$$

Now, "$1\,\mathrm{yr}/(1\,\mathrm{hr})$" is a pure number, the number of hours in a year, or about 8742 (depending on how we define a year!). Likewise, "$1\,\mathrm{ft}/(1\,\mathrm{km})$" is also a pure number, the number of kilometres in a foot:

$$1\,\mathrm{km} = 1000\,\mathrm{m} \simeq 3281\,\mathrm{ft}\,, \quad \text{so} \quad \frac{1}{3281} \simeq \frac{1\,\mathrm{ft}}{1\,\mathrm{km}}\,. \tag{3.174}$$

In that case, (3.173) becomes

$$\frac{T}{1\,\mathrm{yr}}\cdot 8742 \simeq 2\pi\sqrt{\frac{L/(1\,\mathrm{ft})}{g/(1\,\mathrm{km\,hr^{-2}})}\cdot\frac{1}{3281}}\,, \tag{3.175}$$

or finally

$$\frac{T}{1\,\mathrm{yr}} \simeq 1.25\times 10^{-5}\sqrt{\frac{L/(1\,\mathrm{ft})}{g/(1\,\mathrm{km\,hr^{-2}})}}\,. \tag{3.176}$$

This allows us to write

$$\text{“}T \simeq 1.25\times 10^{-5}\sqrt{L/g}\,, \quad \text{with } T \text{ in years, } L \text{ in feet, } g \text{ in km hr}^{-2}\text{”}. \tag{3.177}$$

Of course, presenting such a line-by-line solution might seem tedious, and it has been elaborated here for clarity. But as far as convincing a skeptical audience is concerned, it presents all the information and gives ample structure for any subsequent changes or investigation into difficulties encountered.

3.8.2 Calculating a Nuclear Scattering Rate

Scattering experiments have a long history in nuclear physics, going back to the work of Rutherford, Marsden, and Geiger in the early twentieth century. At that time, the *Thomson model* of the atom was in vogue, in which the atom was believed to be a sphere of positive charge with embedded electrons. To test this theory, Rutherford and his colleagues directed alpha particles onto gold foil, finding to their complete amazement that some of the alpha particles bounced completely backward. This could not have occurred with any reasonable likelihood if the Thomson model were correct. It seemed that there must be an almost point-like nucleus with which alpha particles could collide head-on, and so be repelled from very strongly.

The statistics of scattering particles into some given solid angle is handled using the concept of a *cross section*. The cross section of a target is defined to be an area, in such a way that this area divided by the total uniformly illuminated area of the target is the probability that an incident particle will be scattered into a specified solid angle. Let's see how it works, and in the process write down the solution to the following question in a very structured way.[5]

> *A piece of ^{54}Fe foil is bombarded by a 100 nA current of 60 MeV protons. The foil has a mass per unit area of 10 mg cm^{-2}, and the incident protons scatter inelastically from its iron nuclei. Some of them are scattered through a 40° angle into a detector placed 10 cm from the foil. The detector has an area of 0.1 cm^2.*
>
> *Given that the differential cross section per target nucleus (i.e. cross section per target nucleus per unit solid angle) for scattering at 40° is* $\mathrm{d}\sigma/\mathrm{d}\Omega = 1.3$ *mbarn sr^{-1}, find the number of protons per second that strike the detector. (Note that 1 barn* $\equiv 10^{-28}$ *m^2.)*

How might we go about tackling this? There is a lot of information with which we might not be immediately familiar, and assembling it into a coherent structure to get at the answer may well be a difficult task. The job can be approached in various ways, but whatever approach is used, here is a very logical way of at least writing down the analysis and perhaps approaching it in the first place. We ask what it is we wish to calculate, and begin by making this clear first of all. We then examine its constituents and burrow down through layers like those of an onion, to see what lies beneath in terms of more basic concepts. Hopefully the scenario will contain enough information to feed the requirements of each of those layers and, if all goes well, home in on the answer very directly.

The scenario is shown in Fig. 3.12. We wish to find the rate at which protons leave the foil to strike the detector: the scattering rate. Working backward, we know that not every proton incident on the foil will end up being scattered at around 40°, so we need to know both the incident rate and the probability that a scattering event will occur:

$$\text{scattering rate} = \text{incident rate} \times \begin{bmatrix} \text{probability of a} \\ \text{scatter occurring} \end{bmatrix}. \tag{3.178}$$

First, the incident proton rate is indirectly given in the specifications, being equal to the incident current divided by the charge of a single proton:

[5] The example problem to be solved in this section has been paraphrased from *Quantum Physics of Atoms, Molecules, Solids, Nuclei, and Particles* by R. Eisberg and R. Resnick, 2nd edition (1985, John Wiley and Sons). Reprinted with permission of John Wiley and Sons, Inc.

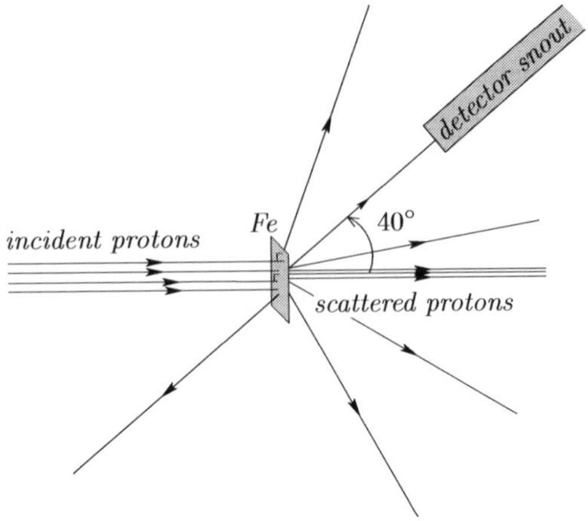

Fig. 3.12. Setup for the scattering experiment. Protons normally incident on the iron foil are scattered in all directions. We wish to predict how many will arrive at the detector.

$$\text{scattering rate} = \frac{\text{inc. current}}{\text{proton charge}} \times \begin{bmatrix} \text{probability of a} \\ \text{scatter occurring} \end{bmatrix} . \tag{3.179}$$

The probability of a scatter is expressed in terms of the target's cross section, where the required probability is given by the cross section of the target as a fraction of its real area illuminated by the beam. So we write

$$\text{scattering rate} = \frac{\text{inc. current}}{\text{proton charge}} \times \frac{\text{cross section of target}}{\text{target area (i.e. illuminated area)}} . \tag{3.180}$$

The illuminated target area is unknown—so we hope it will cancel somewhere! Let's look more closely at the target's cross section. The scattering is something that happens on an individual nucleus basis, so we express this cross section as the cross section per target nucleus times the number of nuclei in the target. (Remember, too, that the target is just the illuminated area of the foil, which we take to be uniform; anything outside that region does not count as part of the target.)

$$\text{scattering rate} = \frac{\text{inc. current}}{\text{proton charge}} \times \frac{\begin{bmatrix} \text{cross section per} \\ \text{target nucleus} \end{bmatrix} \times \begin{bmatrix} \text{number of nuclei} \\ \text{in target} \end{bmatrix}}{\text{target area}} . \tag{3.181}$$

The cross section per target nucleus $\sigma(\theta)$ is a function of the scattering angle θ, which is specified as 40°. Since the process of scattering is axially symmetric, we need only ask how many protons are scattered into an infinitesimally narrow cone with solid angle $d\Omega$, whose axis lies at 40° to the main beam. So we write the cross section per target nucleus $\sigma(\theta)$ as the integral across the detector area of an infinitesimal quantity: the cross section per nucleus per unit solid angle $(\mathrm{d}\sigma/\mathrm{d}\Omega)$ multiplied by an infinitesimal solid angle $\mathrm{d}\Omega$ subtended in the direction of the detector:

$$\text{scattering rate} = \frac{\text{inc. current}}{\text{proton charge}} \times \int \frac{\mathrm{d}\sigma}{\mathrm{d}\Omega}\,\mathrm{d}\Omega \times \begin{bmatrix}\text{number of nuclei}\\ \text{per unit area}\end{bmatrix}. \tag{3.182}$$

At this point we inject some knowledge of the physics of the experiment to realise that the differential cross section $\mathrm{d}\sigma/\mathrm{d}\Omega$ does not change dramatically with angle over the area of a typical detector. So, since we can always write an integral in terms of an average $\langle\cdot\rangle$ over the region of integration,

$$\left\langle \frac{\mathrm{d}\sigma}{\mathrm{d}\Omega} \right\rangle \Delta\Omega \;\equiv\; \int \frac{\mathrm{d}\sigma}{\mathrm{d}\Omega}\,\mathrm{d}\Omega\,, \tag{3.183}$$

we need only recognise that this average of $\mathrm{d}\sigma/\mathrm{d}\Omega$ over the small area subtended by the detector is quite accurately the value at 40° that was supplied in the original statement of the problem. So now we can drop the reference to the averaging and use this supplied value, writing

$$\text{scattering rate} \simeq \frac{\text{inc. current}}{\text{proton charge}} \times \frac{\mathrm{d}\sigma}{\mathrm{d}\Omega}\,\Delta\Omega \times \begin{bmatrix}\text{number of nuclei}\\ \text{per unit area}\end{bmatrix}. \tag{3.184}$$

The number of nuclei per unit target area is just the mass per unit area of the target (given) divided by the mass per nucleus (easily worked out):

$$\text{scattering rate} \simeq \frac{\text{inc. current}}{\text{proton charge}} \times \frac{\mathrm{d}\sigma}{\mathrm{d}\Omega}\,\Delta\Omega \times \frac{\text{mass per unit area}}{\text{mass per nucleus}}. \tag{3.185}$$

Finally, the solid angle subtended by the small area of the detector at the scattering centre is just approximately that area divided by the square of its distance from the scattering centre:

$$\begin{aligned}\text{scattering rate} \simeq\ & \frac{\text{inc. current}}{\text{proton charge}} \times \frac{\mathrm{d}\sigma}{\mathrm{d}\Omega} \times \frac{\text{detector area}}{(\text{det. distance from foil})^2}\\ & \times \frac{\text{mass per unit area}}{\text{mass per nucleus}}.\end{aligned} \tag{3.186}$$

It only remains to insert the supplied parameters. We are given the incident current, the proton charge is a standard value, the differential cross section at 40° is supplied, the detector's illuminated area is given, and the detector distance from the foil is given. So is the mass per unit area of the foil, and the mass of a nucleus can be worked out from the mass number of iron (where to the accuracy required here, the proton and neutron rest masses can be considered to be the same). Using the following values:

$$\begin{aligned} \text{proton charge} &\simeq 1.6 \times 10^{-19}\,\text{C}, \\ \text{nucleon mass} &\simeq 1.7 \times 10^{-27}\,\text{kg}, \end{aligned} \tag{3.187}$$

the scattering rate is found by inserting the relevant numbers into (3.186), using SI units for simplicity. The one place where we need not convert to SI units is when calculating the solid angle $\Delta\Omega$, since this is unitless. The scattering rate is approximately

$$\frac{10^{-7}}{1.6 \times 10^{-19}} \times 1.3 \times 10^{-31} \times \frac{0.1\,\text{cm}^2}{(10\,\text{cm})^2} \times \frac{10 \times 10^{-6}/(10^{-2})^2}{54 \times 1.7 \times 10^{-27}} \simeq 90\,. \tag{3.188}$$

So the detector sees about 90 protons per second. Working through this exercise has, on paper at least, been a smooth process of peeling away layers. But it doesn't necessarily follow what tends to be done in the everyday world, which might be more haphazard. Still, regardless of how the answer was originally obtained, presenting the analysis in a very structured way is much more likely to convince an audience of its correctness.

Generally, of course, physics research is not about finding a single number. It is more about following one's nose to see where it will lead, and perhaps the end product is the reverse of any of the neat flows of logic above. The real process of research is often quite erratic, and the same can be said for the history of physics in general, with its many blind alleyways and not always useful notations, which might not be discarded as the subject evolves. Like a chess game, the path to be followed is seldom clear or unique beyond the next few steps. But the beauty of mathematical physics can be seen by asking questions that we suspect might have interesting answers.

4 A Roundabout Route to Geometric Algebra

When first encountering matrix theory, one of the most useful things we learn is how to rotate a vector in the xy-plane, by multiplying it by a 2×2 matrix derived from the angle through which it is turned. We quickly see by rotating a vector through two angles α and β in succession, and representing each rotation by a matrix, the very useful result that quantities such as $\sin(\alpha + \beta)$ can be written in terms of the sine and cosine of α and β individually.

This is all easy to understand, mostly because these rotations are easy to draw. In contrast, rotations in three dimensions often seem more difficult, perhaps because it can be hard to make a faithful drawing of the rotation. Another reason that three-dimensional rotations carry an aura of obscurity is because, unlike the two-dimensional case, rotations in three dimensions generally do not commute, and there are also more ways to describe and think about three-dimensional rotations. But they are really not difficult at heart—and perhaps the best way to come to understand them is to refrain from trying to make too many drawings! We will leave that to the artists, because pictures of all manner of arrows and axes in three dimensions are not overly necessary for understanding three-dimensional rotations. Some imagination is needed, and a set of axes fashioned from wire is very handy for helping to visualise how a body is orientated by multiple rotations.

There are two results from the theory of three-dimensional rotations that we will study in this chapter. The first is the purely mathematical exploration that comes from considering the entities involved, how they interact, how they can be simplified, and what they might tell us about the world when applied to physics in a novel way. Following this, we'll take a moment to look at how rotations are applied in some situations, because in an age of computers, being able to rotate a vector around a given axis is an old art that is being used in new ways, but not always correctly, as we'll see.

Three-dimensional rotation theory is centred on one result that will be proved shortly. Suppose, as in Fig. 4.1, that an axis is defined by a unit vector $\boldsymbol{n}$, and originating on this axis is a vector $\boldsymbol{r}$. To rotate $\boldsymbol{r}$ about $\boldsymbol{n}$ in a right-handed sense to produce a new vector $\boldsymbol{r}'$, a matrix multiplication can be used, where $\boldsymbol{n}, \boldsymbol{r}, \boldsymbol{r}'$ are columns:

$$\boldsymbol{r}' = R_{\boldsymbol{n}}(\theta)\, \boldsymbol{r}\,, \quad \text{where}$$
$$R_{\boldsymbol{n}}(\theta) = (1 - \cos\theta)\, \boldsymbol{n}\boldsymbol{n}^t + \cos\theta\, 1 + \sin\theta\, n^{\times}\,, \tag{4.1}$$

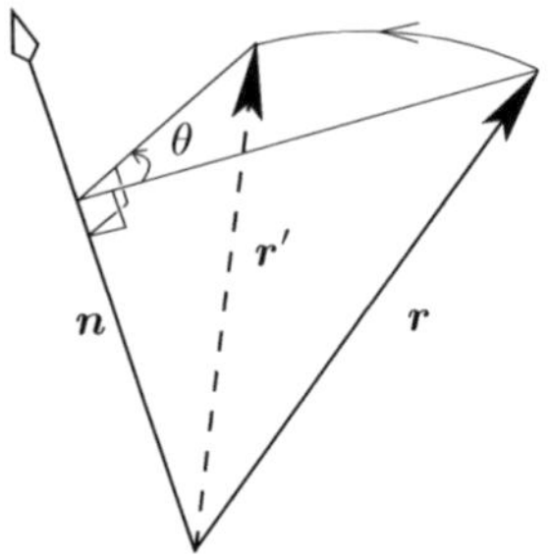

Fig. 4.1. The central result of three-dimensional rotation theory is a matrix multiplication that rotates a vector $\boldsymbol{r}$ in a right-handed sense around the vector $\boldsymbol{n}$ to produce $\boldsymbol{r}'$.

in which "1" is the 3×3 identity matrix, and $n^{\times}$ is a very simple matrix derived from $\boldsymbol{n}$, as we'll see later. All of rotation theory is aimed at either producing, using, or analysing this result. Let's see how it all comes about.

4.1 Matrix Representation of an Orientation

In the two dimensions of the xy-plane, rotating a body while keeping one point fixed is entirely equivalent to changing its orientation. But, in three dimensions, the relationship between rotation and orientation is not so obvious. How are these two ideas related, if at all?

Begin by leaving rotations aside for a moment, and instead ask how to represent any arbitrary orientation of a body in three dimensions. Just as for orientations in the plane, we can only do it in a relative way, by specifying how the body has been moved from some initial, or base, position. Suppose that the body is not moved sideways; some point in it stays in the same position, but otherwise the body can be moved in some arbitrary way from its initial position.

In the initial position, we specify three linearly independent vectors $\boldsymbol{e}_1, \boldsymbol{e}_2, \boldsymbol{e}_3$ to arbitrary points in the body, and ask how they change as the body's orientation is changed. That is, under some change in orientation, the three column vectors $\boldsymbol{e}_1, \boldsymbol{e}_2, \boldsymbol{e}_3$ become $\boldsymbol{e}'_1, \boldsymbol{e}'_2, \boldsymbol{e}'_3$, respectively. Now, suppose we construct a matrix A from the six column vectors:

$$A \equiv \left[\boldsymbol{e}'_1\ \boldsymbol{e}'_2\ \boldsymbol{e}'_3\right] \left[\boldsymbol{e}_1\ \boldsymbol{e}_2\ \boldsymbol{e}_3\right]^{-1}. \tag{4.2}$$

In that case, A correctly changes the orientation of $\boldsymbol{e}_1, \boldsymbol{e}_2, \boldsymbol{e}_3$:

$$A \left[\boldsymbol{e}_1\ \boldsymbol{e}_2\ \boldsymbol{e}_3\right] = \left[\boldsymbol{e}'_1\ \boldsymbol{e}'_2\ \boldsymbol{e}'_3\right]. \tag{4.3}$$

The reason is that matrix multiplication proceeds column by column, so that the last equation is equivalent to $A\boldsymbol{e}_1 = \boldsymbol{e}'_1$, and similarly for $\boldsymbol{e}_2$ and $\boldsymbol{e}_3$.

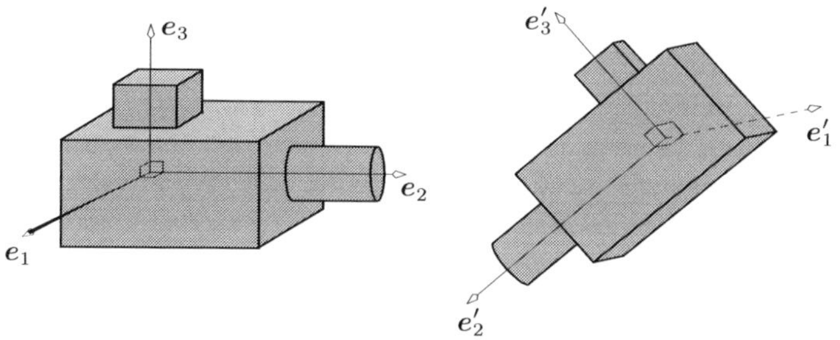

Fig. 4.2. Specifying a change in orientation of a body. **Left**: The base orientation is described by any three linearly independent vectors e_1, e_2, e_3 in some frame that is *fixed* for all changes in orientation. The vectors here have been chosen to lie along axes of symmetry for clarity, but need not be. **Right**: In the new orientation, the three vectors have been mapped to e'_1, e'_2, e'_3, respectively, in the same frame as before.

If the body is rigid, then linearity ensures that A will suffice to describe the orientation of the whole body. By this we mean that any other point in the base-positioned body that's described by a certain linear combination of the basis vectors $\boldsymbol{e}_1, \boldsymbol{e}_2, \boldsymbol{e}_3$, will be described by the *same* linear combination of the new basis vectors $\boldsymbol{e}'_1, \boldsymbol{e}'_2, \boldsymbol{e}'_3$ in the new orientation because

$$A\left(\alpha \boldsymbol{e}_1 + \beta \boldsymbol{e}_2 + \gamma \boldsymbol{e}_3\right) = \alpha \boldsymbol{e}'_1 + \beta \boldsymbol{e}'_2 + \gamma \boldsymbol{e}'_3 \,. \tag{4.4}$$

So A multiplies any vector to produce its orientated form. In particular, if $\boldsymbol{e}_1, \boldsymbol{e}_2, \boldsymbol{e}_3$ are chosen to be the three orthonormal cartesian basis vectors $\boldsymbol{e}_x, \boldsymbol{e}_y, \boldsymbol{e}_z$, which are orientated to produce $\boldsymbol{e}'_x, \boldsymbol{e}'_y, \boldsymbol{e}'_z$ respectively, then the matrix describing the orientation will be

$$A = \left[\boldsymbol{e}'_x \; \boldsymbol{e}'_y \; \boldsymbol{e}'_z\right]. \tag{4.5}$$

Since any element of the A in (4.5) is a dot product of an initial and a final basis vector of the two orthonormal sets, it will be the cosine of the angle between the corresponding axes. Hence this A is often called a *direction cosine matrix.*

As mentioned at the start of this chapter, a familiar example of using (4.5) to construct A in two dimensions is when a planar body is orientated by rotating it through an angle θ in the xy-plane, keeping one point fixed. The orthonormal basis vectors then map as

$$\boldsymbol{e}_x = \begin{bmatrix}1\\0\end{bmatrix} \longrightarrow \begin{bmatrix}\cos\theta\\ \sin\theta\end{bmatrix} \equiv \boldsymbol{e}'_x\,, \qquad \boldsymbol{e}_y = \begin{bmatrix}0\\1\end{bmatrix} \longrightarrow \begin{bmatrix}-\sin\theta\\ \cos\theta\end{bmatrix} \equiv \boldsymbol{e}'_y\,, \tag{4.6}$$

giving the familiar rotation matrix

$$A = \left[\boldsymbol{e}'_x \; \boldsymbol{e}'_y\right] = \begin{bmatrix}\cos\theta & -\sin\theta\\ \sin\theta & \cos\theta\end{bmatrix}. \tag{4.7}$$

4.1.1 Describing an Orientation by a Rotation

In two dimensions, orientating a body while keeping one point fixed can only be a rotation, but whether a change of orientation in three dimensions can always be represented by one or more rotations is not so obvious; the transformation (4.2) simply produces a final orientation given some initial orientation. Nevertheless, it is not hard to show that any final orientation can be produced from some initial orientation by just *one* rotation about some axis, where that axis need not be any of the x-, y-, or z-axes (and in general won't be). This is known as *Euler's theorem of rotation*. The matrix A in (4.2) describes this rotation, although what is not obvious is what the axis and angle are. But we'll come to that in due course.

The proof of Euler's rotation theorem involves some elegant linear algebra. Set the origin to be the body's fixed point as it changes orientation, and since the body is rigid, all distances from the origin are preserved as the body changes orientation. That is, for any position vector $\boldsymbol{r}$ in the body (again written as a column), $|A\boldsymbol{r}| = |\boldsymbol{r}|$. Squaring both sides gives

$$\boldsymbol{r}^t A^t A \boldsymbol{r} = \boldsymbol{r}^t \boldsymbol{r}\,, \tag{4.8}$$

or $A^t A = 1$. We encountered this in Chap. 2; the matrix A is orthogonal. (Since its rows and columns clearly form orthonormal sets, calling it orthonormal would be more apt, but orthogonal is the conventional term.) Taking determinants then gives

$$\det\left(A^t A\right) = \det{}^2 A = 1\,, \tag{4.9}$$

so that $\det A = \pm 1$. But A should vary smoothly from the identity matrix as the body is orientated out of its base position, and since the identity matrix has determinant $+1$, then A must have this determinant, because there is no reason why the sign of $\det A$ should change abruptly at any orientation. Thus we infer that $\det A = +1$ always. Now notice that since $A^t A = 1$, we can write

$$A^t A - A = 1 - A\,, \quad \text{or} \quad \left(A^t - 1\right) A = 1 - A\,. \tag{4.10}$$

Taking the determinant of each side of the last equation, and noting that

$$\det\left(A^t - 1\right) = \det\left[(A-1)^t\right] = \det(A-1)\,, \tag{4.11}$$

produces

$$\begin{aligned}\det(A-1) &= \det(1-A)\\ &= (-1)^n \det(A-1) \quad \text{for } n \text{ dimensions.}\end{aligned} \tag{4.12}$$

In three dimensions, we can then infer that $\det(A-1) = 0$. But standard linear algebra tells us that

$$\det(A-1) = 0 \iff (A-1)\,\boldsymbol{n} = 0 \quad \text{for some nonzero } \boldsymbol{n}. \tag{4.13}$$

For a discussion of why this is so, see the small text on p. 32. A zero determinant of $A - 1$ implies linear dependence of the rows of $A - 1$, in which case it's trivial to solve $(A - 1)\,\boldsymbol{n} = 0$ for some nonzero $\boldsymbol{n}$.

Thus $A\boldsymbol{n} = \boldsymbol{n}$ (so A has an eigenvalue of 1 with corresponding eigenvector $\boldsymbol{n}$). Since the body is rigid, the only way that $\boldsymbol{n}$ can be unchanged by A is if A causes a rotation around $\boldsymbol{n}$. This proves Euler's theorem.

Euler's theorem allows us to "compose" (i.e. add) rotations. Since any two rotations simply change a body's orientation, thus producing a new A matrix, they must be equivalent to a single rotation. This being the case, we can say that the set of matrices describing rotations about arbitrary axes forms a group, known as the *special* (+1 determinant) *orthogonal group in three dimensions*: SO_3, or O_3^+. (Showing that a product of rotations is another rotation was the main requirement for showing that rotations form a group; the other requirements, of associativity and the existence of an identity and inverses, are straightforward.)

4.2 Calculating the Matrix for an Arbitrary Rotation

The matrix A in (4.2) describes the rotation that changes a body's orientation, provided we know how the basis vectors change their orientation. But usually we do not know the final vector positions. Since an arbitrary orientation can be described by just one rotation, it is useful to determine how to rotate an arbitrary vector about an arbitrary axis. It will turn out that this rotation can be accomplished by a matrix multiplication, and is the central result of rotation theory.

So let's calculate this matrix that describes a rotation through some angle θ about an arbitrary axis along a unit vector $\boldsymbol{n}$ using the right-hand rule, as shown on the left-hand side of Fig. 4.3. (Demanding that $\boldsymbol{n}$ have unit length will simplify the following expressions.) Suppose that the rotation acts on a vector $\boldsymbol{r}$ to give $\boldsymbol{r}'$. Writing $\boldsymbol{r} = \boldsymbol{r}_{/\!/} + \boldsymbol{r}_\perp$, a sum of components parallel and perpendicular to $\boldsymbol{n}$, as on the right-hand side of Fig. 4.3, we see that only the perpendicular component is rotated. Hence

$$\boldsymbol{r}' = \boldsymbol{r}_{/\!/} + \boldsymbol{r}_\perp \text{ rotated through } \theta\,. \tag{4.14}$$

Rotating $\boldsymbol{r}_\perp$ through θ is equivalent to reducing its length by a factor of $\cos\theta$ while introducing a new component in the direction of $\boldsymbol{n} \times \boldsymbol{r}$ with length $|\boldsymbol{r}_\perp| \sin\theta$. Thus

$$\boldsymbol{r}' = \boldsymbol{r}_{/\!/} + \boldsymbol{r}_\perp \cos\theta + |\boldsymbol{r}_\perp| \sin\theta \, \frac{\boldsymbol{n} \times \boldsymbol{r}}{|\boldsymbol{n} \times \boldsymbol{r}|}\,. \tag{4.15}$$

The two components of $\boldsymbol{r}$ are easily expressed in terms of $\boldsymbol{r}$ and $\boldsymbol{n}$, where the expressions are simplest if we stipulate that $\boldsymbol{n}$ is a unit vector.

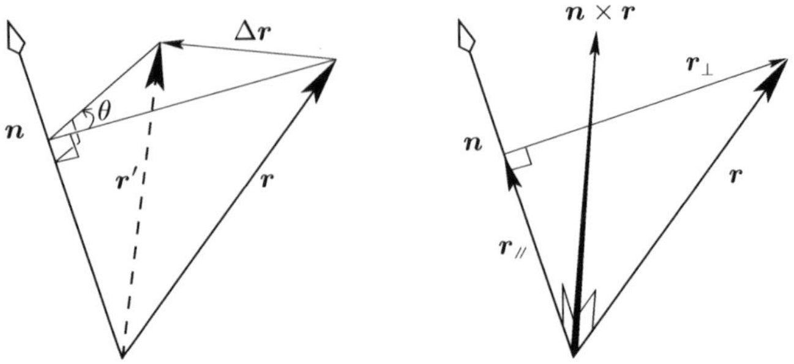

Fig. 4.3. Left: Rotating $\boldsymbol{r}$ around $\boldsymbol{n}$ by angle θ to produce $\boldsymbol{r}'$. **Right**: Resolving $\boldsymbol{r}$ into components parallel and perpendicular to $\boldsymbol{n}$.

$$\boldsymbol{r}_{/\!/} = \boldsymbol{r}\cdot\boldsymbol{n}\,\boldsymbol{n} \quad \text{and} \quad \boldsymbol{r}_{\perp} = \boldsymbol{r} - \boldsymbol{r}\cdot\boldsymbol{n}\,\boldsymbol{n}\,; \tag{4.16}$$

$$|\boldsymbol{n}\times\boldsymbol{r}| = |\boldsymbol{r}|\sin(\boldsymbol{n},\boldsymbol{r}) = |\boldsymbol{r}_{\perp}|\,, \tag{4.17}$$

where $(\boldsymbol{n},\boldsymbol{r})$ denotes the angle between $\boldsymbol{n}$ and $\boldsymbol{r}$. Using these last identities, (4.15) can be written as

$$\boldsymbol{r}' = \boldsymbol{r}\cdot\boldsymbol{n}\,\boldsymbol{n} + (\boldsymbol{r} - \boldsymbol{r}\cdot\boldsymbol{n}\,\boldsymbol{n})\cos\theta + \boldsymbol{n}\times\boldsymbol{r}\ \sin\theta\,. \tag{4.18}$$

So $\boldsymbol{r}'$ is a linear combination of the components of $\boldsymbol{r}$, and by writing (4.18) in terms of its vectors' components, we arrive at a matrix form for the rotation (where the vectors are columns):

$$\boldsymbol{r}' = R_{\boldsymbol{n}}(\theta)\,\boldsymbol{r}\,. \tag{4.19}$$

Writing

$$\boldsymbol{n} = \begin{bmatrix} n^1 \\ n^2 \\ n^3 \end{bmatrix} \tag{4.20}$$

allows the rotation matrix to be written as

$$R_{\boldsymbol{n}}(\theta) = (1-\cos\theta)\begin{bmatrix} (n^1)^2 & n^1n^2 & n^1n^3 \\ n^2n^1 & (n^2)^2 & n^2n^3 \\ n^3n^1 & n^3n^2 & (n^3)^2 \end{bmatrix} + \cos\theta\ 1 + \sin\theta \begin{bmatrix} 0 & -n^3 & n^2 \\ n^3 & 0 & -n^1 \\ -n^2 & n^1 & 0 \end{bmatrix}, \tag{4.21}$$

which can be expressed more concisely as

$$\boxed{R_{\boldsymbol{n}}(\theta) = (1-\cos\theta)\,\boldsymbol{n}\boldsymbol{n}^t + \cos\theta\ 1 + \sin\theta\ n^{\times},} \tag{4.22}$$

where "1" is unambiguously the 3×3 identity matrix and

$$n^{\times} \equiv \begin{bmatrix} 0 & -n^3 & n^2 \\ n^3 & 0 & -n^1 \\ -n^2 & n^1 & 0 \end{bmatrix}. \tag{4.23}$$

The last matrix is so named because

$$n^\times \boldsymbol{r} = \boldsymbol{n} \times \boldsymbol{r}\,, \tag{4.24}$$

with the n written nonbold on the left to emphasise that $n^\times$ is a matrix, not a vector. It appears generally in rotational calculations and satisfies useful identities, such as

$$\boldsymbol{n}\boldsymbol{n}^t n^\times = 0 \quad \text{and} \quad (n^\times)^2 = \boldsymbol{n}\boldsymbol{n}^t - 1\,. \tag{4.25}$$

These identities make it easy to check that the rotation matrix as written in (4.22) is indeed orthogonal. The matrix $\boldsymbol{n}\boldsymbol{n}^t$ is sometimes called a *dyadic* in older literature and written as $\boldsymbol{n}\boldsymbol{n}$. This of course doesn't fit with modern matrix terminology, and the term and notation have mostly disappeared now.

Writing the fundamental rotation equation (4.18) in matrix form is very useful because it brings with it the full power of matrix algebra when doing rotations. As an example of the use of (4.22), rotate the vector $(0, 1, 1)$ by $90°$ about the y-axis. What vector results? We need $R_y(90°)$. Equation (4.22) gives it as

$$R_y(90°) = \boldsymbol{n}\boldsymbol{n}^t + n^\times = \begin{bmatrix} 0 \\ 1 \\ 0 \end{bmatrix} \begin{bmatrix} 0 & 1 & 0 \end{bmatrix} + \begin{bmatrix} 0 & 0 & 1 \\ 0 & 0 & 0 \\ -1 & 0 & 0 \end{bmatrix} = \begin{bmatrix} 0 & 0 & 1 \\ 0 & 1 & 0 \\ -1 & 0 & 0 \end{bmatrix}. \tag{4.26}$$

(In this very simple example, we can calculate $R_y(90°)$ alternatively using (4.5). Simply note that the basis vectors rotate as

$$\begin{bmatrix} 1 \\ 0 \\ 0 \end{bmatrix} \longrightarrow \begin{bmatrix} 0 \\ 0 \\ -1 \end{bmatrix}, \quad \begin{bmatrix} 0 \\ 1 \\ 0 \end{bmatrix} \longrightarrow \begin{bmatrix} 0 \\ 1 \\ 0 \end{bmatrix}, \quad \begin{bmatrix} 0 \\ 0 \\ 1 \end{bmatrix} \longrightarrow \begin{bmatrix} 1 \\ 0 \\ 0 \end{bmatrix}, \tag{4.27}$$

so that (4.5) yields $R_y(90°)$ trivially.) The required rotated vector is then

$$R_y(90°) \begin{bmatrix} 0 \\ 1 \\ 1 \end{bmatrix} = \begin{bmatrix} 0 & 0 & 1 \\ 0 & 1 & 0 \\ -1 & 0 & 0 \end{bmatrix} \begin{bmatrix} 0 \\ 1 \\ 1 \end{bmatrix} = \begin{bmatrix} 1 \\ 1 \\ 0 \end{bmatrix}, \tag{4.28}$$

as expected.

4.2.1 Deriving the Rotation Matrix $R_{\boldsymbol{n}}(\theta)$ via Diagonalisation

An alternative way of deriving (4.22) brings out the elegance of the diagonalisation procedure that we covered in Sect. 2.5. Figure 4.3 shows that in the limit of an infinitesimal rotation $\mathrm{d}\theta$, the increase $\mathrm{d}\boldsymbol{r}$ from the original vector $\boldsymbol{r}$ to the rotated vector $\boldsymbol{r}'$ is perpendicular to both $\boldsymbol{n}$ and $\boldsymbol{r}$, as well as having length $|\boldsymbol{r}_\perp|\,\mathrm{d}\theta$. So it must be true that

$$\mathrm{d}\boldsymbol{r} = \mathrm{d}\theta\, \boldsymbol{n} \times \boldsymbol{r}\,. \tag{4.29}$$

This means we can write the infinitesimally rotated vector in the following way, with "1" meaning the identity matrix:

$$\boldsymbol{r}' = \left(1 + \mathrm{d}\theta\, n^{\times}\right) \boldsymbol{r} \,. \tag{4.30}$$

(In hindsight, this expression agrees with the limits of either (4.18) or (4.22) to first order in $\mathrm{d}\theta$, but of course we're assuming that neither of those equations are at our disposal.)

So an infinitesimal rotation is obtained by multiplying by $1 + \mathrm{d}\theta\, n^{\times}$. If a *non*infinitesimal rotation through θ is interpreted as the result of composing a large number N of rotations through θ/N (with $N \to \infty$), then we can multiply N infinitesimal rotation matrices to write the now *non*infinitesimally rotated vector $\boldsymbol{r}'$ as

$$\boldsymbol{r}' = \lim_{N\to\infty} \left(1 + \frac{\theta}{N}\, n^{\times}\right)^{N} \boldsymbol{r} \,. \tag{4.31}$$

This resembles the well-known useful identity for real numbers,

$$\lim_{N\to\infty} \left(1 + \frac{x}{N}\right)^{N} = \mathrm{e}^{x} \,, \tag{4.32}$$

and provided that we assume a similar expression also holds for matrices, equation (4.31) produces

$$R_{\boldsymbol{n}}(\theta) = \mathrm{e}^{\theta\, n^{\times}} , \tag{4.33}$$

which is certainly a very compact form for the rotation matrix! A matrix exponential is defined in the expected way:

$$\mathrm{e}^{A} \equiv 1 + A + A^2/2! + A^3/3! + \cdots \,. \tag{4.34}$$

This looks to be a difficult sum for matrices, but we remember from (2.78) that a diagonalisation enables us to write $A^n = PD^nP^{-1}$, so that an arbitrary matrix exponential can be expressed as

$$\begin{aligned} \mathrm{e}^{A} &= PP^{-1} + PDP^{-1} + PD^2P^{-1}/2! + \cdots \\ &= P\mathrm{e}^{D}P^{-1}. \end{aligned} \tag{4.35}$$

This is very useful because we know from (4.34) that, for any diagonal matrix,

$$\exp \begin{bmatrix} d_1 & \cdots & 0 \\ & \ddots & \\ 0 & \cdots & d_n \end{bmatrix} = \begin{bmatrix} \mathrm{e}^{d_1} & \cdots & 0 \\ & \ddots & \\ 0 & \cdots & \mathrm{e}^{d_n} \end{bmatrix} . \tag{4.36}$$

So we need only diagonalise the matrix $\theta\, n^{\times}$. The process is lengthy, but after finding its eigenvalues and eigenvectors, we can write $\theta\, n^{\times} = PDP^{-1}$, where

$$P = \begin{bmatrix} n^1 & -n^2 + \mathrm{i}n^1n^3 & -n^2 - \mathrm{i}n^1n^3 \\ n^2 & n^1 + \mathrm{i}n^2n^3 & n^1 - \mathrm{i}n^2n^3 \\ n^3 & -\mathrm{i}\left[1 - (n^3)^2\right] & \mathrm{i}\left[1 - (n^3)^2\right] \end{bmatrix} , \quad D = \mathrm{diag}(0,\, \mathrm{i}\theta,\, -\mathrm{i}\theta) \,, \tag{4.37}$$

and $R_{\boldsymbol{n}}(\theta) = P\mathrm{e}^{D}P^{-1}$. Inverting P and multiplying the factors yields (4.22), as expected. While this was not the most straightforward way of deriving (4.22), it's reassuring to know that it works!

4.2.2 Are Rotations Vectors?

A rotation is sometimes loosely represented by a vector whose direction defines the rotation axis (via a right-hand rule), and whose length is the angle turned through. This is not quite valid, for the simple reason that rotations through noninfinitesimal angles do not commute, whereas vector addition certainly does commute. The noncommutivity of rotations was discussed previously in Sect. 2.5. There we saw that rotations through infinitesimal angles *do* commute in some sense, and this makes them candidates for being represented by vectors. We can show this more quantitatively as follows. Define $\mathrm{d}\boldsymbol{\theta} \equiv \mathrm{d}\theta\, \boldsymbol{n}$, enabling (4.29) and (4.30) to be written more compactly as

$$\mathrm{d}\boldsymbol{r} = \mathrm{d}\boldsymbol{\theta} \times \boldsymbol{r}\,, \quad \boldsymbol{r}' = (1 + \mathrm{d}\theta^{\times})\,\boldsymbol{r}\,. \tag{4.38}$$

In that case, two infinitesimal rotations of $\mathrm{d}\theta_1, \mathrm{d}\theta_2$ around different axes $\boldsymbol{n}_1, \boldsymbol{n}_2$ combine to give

$$\begin{aligned} R_1 R_2 &= \left(1 + \mathrm{d}\theta_1^{\times}\right)\left(1 + \mathrm{d}\theta_2^{\times}\right) \\ &= 1 + \mathrm{d}\theta_1^{\times} + \mathrm{d}\theta_2^{\times} \text{ to first order,} \\ &= R_2 R_1\,. \end{aligned} \tag{4.39}$$

So infinitesimal rotations do commute, which means that $\mathrm{d}\boldsymbol{\theta}$ can be considered as a candidate for a vector. But (4.39) also implies that, to first order,

$$R_1 R_2 = 1 + (\mathrm{d}\theta_1 + \mathrm{d}\theta_2)^{\times}\,, \tag{4.40}$$

so that after the two infinitesimal rotations, the increase in $\boldsymbol{r}$ is

$$\mathrm{d}\boldsymbol{r} = R_1 R_2\, \boldsymbol{r} - \boldsymbol{r} = (\mathrm{d}\boldsymbol{\theta}_1 + \mathrm{d}\boldsymbol{\theta}_2) \times \boldsymbol{r}\,. \tag{4.41}$$

Comparing this with (4.38) shows that the resultant rotation is given by simply adding the vectors $\mathrm{d}\boldsymbol{\theta}_1, \mathrm{d}\boldsymbol{\theta}_2$ of the individual rotations! This is why an infinitesimal rotation can be treated as a vector.

In the everyday world, where infinitesimals are not used, situations involving rotational motion can usefully be described by rescaling the infinitesimal rotation to construct the *angular velocity vector* $\boldsymbol{\omega} \equiv \mathrm{d}\boldsymbol{\theta}/\mathrm{d}t$, which finds frequent application in classical mechanics. One such use occurs in the study of vectors in rotating frames, such as vectors on Earth that represent global phenomena such as wind flow. Because the frame of the rotating Earth is not inertial, applying Newton's laws to such phenomena can be difficult. What is much simpler is to calculate within the frame of the solar system, which to a good approximation *is* inertial. In that case, consider the motion of a small air cell. Suppose this cell is at a position $\boldsymbol{r}$ in Earth's frame (relative to some origin that can be any point on Earth's axis). In a small time interval $\mathrm{d}t$, any step $\mathrm{D}\boldsymbol{r}$ taken by the cell in the solar system's inertial frame is the sum of the step $\mathrm{d}\boldsymbol{r}$ it took on Earth, and the step $\mathrm{d}\boldsymbol{\theta} \times \boldsymbol{r}$ that Earth's rotation provided in the solar system frame:

$$\underbrace{\mathrm{D}\boldsymbol{r}}_{\substack{\text{cell's step in}\\ \text{solar system}}} = \underbrace{\mathrm{d}\boldsymbol{r}}_{\substack{\text{cell's step}\\ \text{on Earth}}} + \underbrace{\mathrm{d}\boldsymbol{\theta} \times \boldsymbol{r}}_{\substack{\text{step due to Earth's}\\ \text{rotation in solar system}}} . \tag{4.42}$$

Thus

$$\frac{\mathrm{D}\boldsymbol{r}}{\mathrm{d}t} = \frac{\mathrm{d}\boldsymbol{r}}{\mathrm{d}t} + \boldsymbol{\omega} \times \boldsymbol{r} . \tag{4.43}$$

So the wind velocity $\mathrm{D}\boldsymbol{r}/\mathrm{d}t$ in the inertial frame of the solar system differs from the velocity $\mathrm{d}\boldsymbol{r}/\mathrm{d}t$ on the noninertial Earth by the addition of a cross product. We see here a kind of "inertial differentiation":

$$\frac{\mathrm{D}}{\mathrm{d}t} \equiv \frac{\mathrm{d}}{\mathrm{d}t} + \omega^{\times} . \tag{4.44}$$

The $\omega^{\times}$ is, in a sense, a correction to restore inertiality, and in Chap. 8 we'll encounter a similar idea under the name *covariant differentiation*. There we will find a kind of analogy to the $\omega^{\times}$ matrix: the "Christoffel symbols", which play a major role in tensor calculus and differential geometry, as well as encoding the gravitational field in general relativity.

In order to write down Newton's laws for wind motion on Earth, we might start out by differentiating (4.43) again to produce accelerations, which can then be related to forces. But we must be careful to again use the inertial differentiation, since Newton's force–acceleration relation only applies to inertial frames. Writing $\boldsymbol{v}$ and $\boldsymbol{a}$ for velocity and acceleration, (4.43) becomes

$$\boldsymbol{v}_{\text{inertial}} \equiv \frac{\mathrm{D}\boldsymbol{r}}{\mathrm{d}t} = \frac{\mathrm{d}\boldsymbol{r}}{\mathrm{d}t} + \boldsymbol{\omega} \times \boldsymbol{r} \equiv \boldsymbol{v}_{\text{Earth}} + \boldsymbol{\omega} \times \boldsymbol{r} , \tag{4.45}$$

so that the next inertial differentiation gives, with an overdot $\dot{}$ meaning $\mathrm{d}/\mathrm{d}t$,

$$\begin{aligned} \boldsymbol{a}_{\text{inertial}} &\equiv \frac{\mathrm{D}\boldsymbol{v}_{\text{inertial}}}{\mathrm{d}t} = \left(\mathrm{d}/\mathrm{d}t + \omega^{\times}\right) \left(\boldsymbol{v}_{\text{Earth}} + \boldsymbol{\omega} \times \boldsymbol{r}\right) \\ &= \frac{\mathrm{d}\boldsymbol{v}_{\text{Earth}}}{\mathrm{d}t} + \dot{\boldsymbol{\omega}} \times \boldsymbol{r} + 2\boldsymbol{\omega} \times \boldsymbol{v}_{\text{Earth}} + \boldsymbol{\omega} \times (\boldsymbol{\omega} \times \boldsymbol{r}) \\ &\equiv \boldsymbol{a}_{\text{Earth}} + \dot{\boldsymbol{\omega}} \times \boldsymbol{r} + 2\boldsymbol{\omega} \times \boldsymbol{v}_{\text{Earth}} + \boldsymbol{\omega} \times (\boldsymbol{\omega} \times \boldsymbol{r}) . \end{aligned} \tag{4.46}$$

If we wish to enforce Newton's law that force = mass × acceleration in both frames—even the noninertial Earth frame—then we must introduce a *fictitious force* to "fix up" this law on the noninertial Earth frame. The role of this new force is purely one of keeping the books balanced, so to speak. Rearranging (4.46) and multiplying by the mass m of the air cell produces

$$m\,\boldsymbol{a}_{\text{Earth}} = m\,\boldsymbol{a}_{\text{inertial}} \underbrace{- m\,\dot{\boldsymbol{\omega}} \times \boldsymbol{r} - 2m\,\boldsymbol{\omega} \times \boldsymbol{v}_{\text{Earth}} - m\,\boldsymbol{\omega} \times (\boldsymbol{\omega} \times \boldsymbol{r})}_{\equiv\ \text{required fictitious force enabling Newton's laws to hold}} . \tag{4.47}$$

Earth's rotation is constant, so $\dot{\boldsymbol{\omega}} = 0$. Although there is really only one fictitious force in (4.47), its two remaining nonzero terms are conventionally split into a "Coriolis force" (which depends on the air cell's velocity over Earth) and a "centrifugal force" (which does not):

$$\begin{aligned} \text{Coriolis force} &\equiv -2m\,\boldsymbol{\omega}\times\boldsymbol{v}_{\text{Earth}}\,, \\ \text{Centrifugal force} &\equiv -m\,\boldsymbol{\omega}\times(\boldsymbol{\omega}\times\boldsymbol{r})\,. \end{aligned} \tag{4.48}$$

This is a direct analogy with the conventional splitting of the electromagnetic force on a charge into one that depends on its velocity (the magnetic force) and one that does not (the electric force). There, as here, only one force is really present. The Coriolis force only arises when the air cell moves, whereas the centrifugal force is always present. They are both simply forces that we invent to enable us to continue to apply Newton's laws in the noninertial frame of a laboratory tied to a rotating body, such as Earth's surface.

4.3 Combining Two Rotations

Earlier, we invoked Euler's theorem of rotation to say that two rotations will always combine to give a new rotation. This is, of course, equivalent to multiplying the two rotation matrices. But what are the axis and angle of the new rotation? For example, suppose we rotate a vector twice:

1. First rotate it by 90° about the x-axis (call the matrix R_x).
2. Then rotate the result by 90° about the y-axis (call the matrix R_y).

What is the corresponding axis and angle of the combined rotation $R_y R_x$? The two rotation matrices and their product are easily found:

$$\text{First, } R_x\text{: } \theta = 90°,\ \boldsymbol{n} = \begin{bmatrix}1\\0\\0\end{bmatrix},\quad \text{so } R_x = \begin{bmatrix}1&0&0\\0&0&-1\\0&1&0\end{bmatrix}.$$

$$\text{Second, } R_y\text{: } \theta = 90°,\ \boldsymbol{n} = \begin{bmatrix}0\\1\\0\end{bmatrix},\quad \text{so } R_y = \begin{bmatrix}0&0&1\\0&1&0\\-1&0&0\end{bmatrix}.$$

$$\text{Result: } R_y R_x = \begin{bmatrix}0&1&0\\0&0&-1\\-1&0&0\end{bmatrix}. \tag{4.49}$$

To find the angle and axis of the single rotation that is equivalent to $R_y R_x$, we can of course compare each element of $R_y R_x$ with the general expression (4.21), but this is arduous and inelegant. It is much easier to employ (4.21) to write down two properties of the general rotation matrix $R_{\boldsymbol{n}}(\theta)$ that involve its trace and transpose:

All Rotations Have One Basic Procedure

Occasionally, users of rotation matrices can trip up on what is meant by the second rotation in (4.49): whether it's supposed to mean a rotation about the spatially fixed y-axis or about the latest incarnation of the y-axis—the result of the original y-axis being carried along with the body during the first rotation.

There need never be any confusion, since each rotation is about a basis vector and we have used a *single* frame to write the coordinates of all of the vectors, which is good practice in general; using multiple coordinate systems here is a recipe for disaster. So there is only one y-axis, the spatially fixed one, which is what the second rotation in (4.49) is about. If we really wished to rotate about the new axis (call it the y'-axis), then we would need to calculate the vector $\boldsymbol{n}$ representing the y'-axis. But that is not being done in (4.49).

Always remember that the fundamental procedure of rotation is that *one vector is rotated about another*, and that the matrices being multiplied to combine several rotations must all be calculated in the same frame. We will discuss this in more detail in Sect. 4.6.1.

$$\begin{aligned} \operatorname{tr} R_{\boldsymbol{n}}(\theta) &= 1 + 2\cos\theta\,, \\ R_{\boldsymbol{n}}(\theta) - R^t_{\boldsymbol{n}}(\theta) &= 2\sin\theta\, n^\times\,. \end{aligned} \tag{4.50}$$

In that case, the combined rotation is about some unit vector $\boldsymbol{n}$ through angle θ, where

$$1 + 2\cos\theta = 0\,, \quad 2\sin\theta\, n^\times = \begin{bmatrix} 0 & 1 & 1 \\ -1 & 0 & -1 \\ -1 & 1 & 0 \end{bmatrix}. \tag{4.51}$$

There appear to be two solutions for $\boldsymbol{n}$ and θ here, but they both represent the same rotation so that we can always choose the unique positive value of θ together with its corresponding $\boldsymbol{n}$. In this case,

$$\theta = 120°\,, \quad \boldsymbol{n} = (1, 1, -1)/\sqrt{3} \tag{4.52}$$

(where $\boldsymbol{n}$ should really be written as a column vector in actual use, but we are saving space here). This simple example makes it evident that the single angle and axis resulting from combining two rotations generally do not bear any obvious relation to the two original angle–axis pairs.

4.4 Rotations Lead to Complex Numbers and Quaternions

Historically, rotations have been closely related to complex numbers. To see why this should be so, suppose that we rotate a vector in the xy-plane about the origin through an angle θ:

$$\begin{bmatrix} x' \\ y' \end{bmatrix} = \begin{bmatrix} \cos\theta & -\sin\theta \\ \sin\theta & \cos\theta \end{bmatrix} \begin{bmatrix} x \\ y \end{bmatrix} = \begin{bmatrix} x\cos\theta - y\sin\theta \\ x\sin\theta + y\cos\theta \end{bmatrix}. \tag{4.53}$$

There is a redundancy here with the repeated sine and cosine. Perhaps we could just isolate the essential parts of (4.53) to write a new kind of vector multiplication,

$$\begin{aligned}(x', y') &= (\cos\theta, \sin\theta)\,(x, y) \\ &\equiv (x\cos\theta - y\sin\theta,\; x\sin\theta + y\cos\theta)\,, \end{aligned} \tag{4.54}$$

where this multiplication of the vectors (a^1, a^2) and (b^1, b^2) is defined by

$$(a^1, a^2)\,(b^1, b^2) \equiv (a^1b^1 - a^2b^2,\; a^1b^2 + a^2b^1)\,. \tag{4.55}$$

Alternatively, if we dislike the idea of introducing a new type of multiplication for vectors, we might choose to employ a placeholder, a new symbol "i" together with *conventional* multiplication, so that (4.54) would become

$$x' + \mathrm{i}y' = (\cos\theta + \mathrm{i}\sin\theta)(x + \mathrm{i}y)\,, \quad \text{provided that } \mathrm{i}^2 \equiv -1. \tag{4.56}$$

Of course, the rotation $\cos\theta + \mathrm{i}\sin\theta$ is usually written $\mathrm{e}^{\mathrm{i}\theta}$, and in this form a connection can be made between (4.56) and the general rotation matrix in (4.33).

The new, complicated, multiplication has now been completely invested in the much simpler relation $\mathrm{i}^2 = -1$, while the presence of the i also acts as a way of keeping the vector components separate. The requirement that $\mathrm{i}^2 = -1$ is the *algebra* of our planar rotation formalism, in the sense of being a rule to be applied mechanically that allows (4.56) to give the right answer when rewritten as a vector. Figure 4.4 shows the rotation in the plane, now christened the *complex plane.*

Another alternative to eliminating the redundancy of the original matrix multiplication (4.53) is to rewrite it to resemble (4.56):

$$\begin{bmatrix} x' \\ y' \end{bmatrix} = \left(\cos\theta\, 1 + \sin\theta \begin{bmatrix} 0 & -1 \\ 1 & 0 \end{bmatrix}\right) \begin{bmatrix} x \\ y \end{bmatrix}, \tag{4.57}$$

which makes it clear that $\left[\begin{smallmatrix} 0 & -1 \\ 1 & 0 \end{smallmatrix}\right]$ is a matrix representation of i. (Try squaring this matrix, which corresponds to two successive rotations through 90° in the plane that combine to give one rotation of 180°.)

In what is one of the most important equalities in physics, the rotation factor $\cos\theta + \mathrm{i}\sin\theta$ is identical to $\mathrm{e}^{\mathrm{i}\theta}$, if $\mathrm{e}^{\mathrm{i}\theta}$ is defined by a series with the same form as (4.34). Leaving aside any subtleties of the finer points of this identification, equating the two allows us to bring any amount of complex analysis to bear on a rotation scenario. But is there any similar idea that works for an arbitrary rotation in three dimensions? We can reasonably expect something like the set $\{1, \mathrm{i}\}$, where the elements are placeholders to

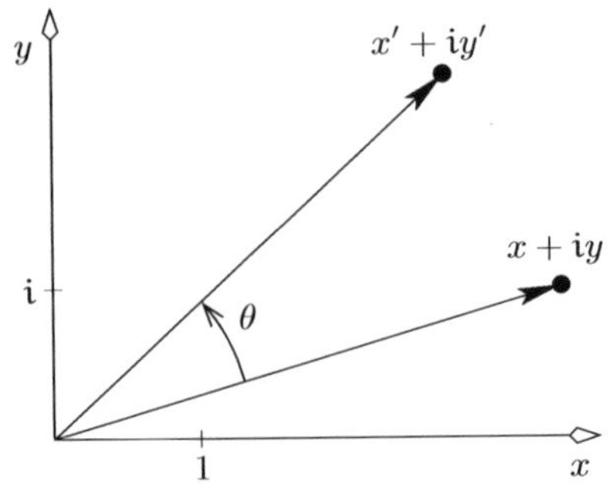

Fig. 4.4. The rotation of (4.56) can be viewed as assigning a number $x + iy$ to the vector (or point) (x, y). The rotation of the vector by θ is then effected by multiplying the number $x + iy$ by $\cos\theta + i\sin\theta$ to give the number $x' + iy'$ corresponding to the vector (x', y'). In order for this to work using the usual rules of multiplication, the symbol i requires the property $i^2 = -1$.

keep the two vector components apart, and their algebra provides the correct combinations of those components. Different algebras are possible, the major criterion being how useful and elegant they are. Without trying to be exhaustive, here we will develop one such system created a century and a half ago by William Hamilton.

In two dimensions, there are only the two quantities $\cos\theta, \sin\theta$ to consider. But three dimensions brings fully nine different and complicated coefficients of x, y, z to work with: the entries of $R_{\boldsymbol{n}}(\theta)$. All of these must be derived from θ, n^1, n^2, n^3. Of course, we can always define a multiplication analogous to (4.54), but there is a question of whether such a procedure is useful, and this is a path we won't take. Instead, we notice from (4.21) that a major part of $R_{\boldsymbol{n}}(\theta)$ consists of terms quadratic in the set of n^i. This suggests that we might be able to write down some new sort of rotation factor more simply than that of (4.21), as long as we are prepared to multiply twice by an expression that only has the n^i linearly. Two multiplications suggest using $\theta/2$, since products of half-angles will give the required $\sin\theta, \cos\theta$ terms. (For example, $\sin\theta = 2\sin{}^{\theta}/_{2}\cos{}^{\theta}/_{2}$.)

The presence of both dot and cross products in (4.18) suggests that a way forward is to make use of some sort of identity involving both of these products. To this end, consider three abstract placeholders $\sigma_1, \sigma_2, \sigma_3$, and two vectors $\boldsymbol{a}, \boldsymbol{b}$. For brevity, also write $\boldsymbol{\sigma} \equiv (\sigma_1, \sigma_2, \sigma_3)$. This is not a vector, but merely a convenient way to write the three placeholders in one ordered set. We will analyse the expression $\boldsymbol{a}\cdot\boldsymbol{\sigma}\ \boldsymbol{b}\cdot\boldsymbol{\sigma}$, granting the $\sigma_1, \sigma_2, \sigma_3$ whatever properties they need in order to create both a dot and a cross product.

$$\boldsymbol{a}\cdot\boldsymbol{\sigma}\ \boldsymbol{b}\cdot\boldsymbol{\sigma} = \sum_{ij} a^i\sigma_i\, b^j\sigma_j$$

$$= \underbrace{\sum_i a^i\, b^i\, \sigma_i^2}_{\Uparrow} \qquad + \qquad \underbrace{\sum_{ij,\ i\neq j} a^i\, b^j\, \sigma_i\, \sigma_j}_{\Uparrow}\ . \tag{4.58}$$

A dot product, as long as the σ_i^2 are all equal. So set $\sigma_i^2 \equiv \alpha$, a constant.

Will be a cross product, as long as swapping i, j swaps the sign of $\sigma_i\,\sigma_j$. So require $\sigma_i\,\sigma_j = -\sigma_j\,\sigma_i$.

An identity linking dots and crosses now begins to take shape:

$$\begin{aligned}\boldsymbol{a}\cdot\boldsymbol{\sigma}\ \boldsymbol{b}\cdot\boldsymbol{\sigma} &= \alpha\,\boldsymbol{a}\cdot\boldsymbol{b}\ +\ (a^2\,b^3 - a^3\,b^2)\,\sigma_2\,\sigma_3\\ &\qquad\qquad\ +\ (a^3\,b^1 - a^1\,b^3)\,\sigma_3\,\sigma_1\\ &\qquad\qquad\ +\ (a^1\,b^2 - a^2\,b^1)\,\sigma_1\,\sigma_2\\ &= \alpha\,\boldsymbol{a}\cdot\boldsymbol{b}\ +\ (\boldsymbol{a}\times\boldsymbol{b})\cdot\underbrace{(\sigma_2\,\sigma_3,\ \sigma_3\,\sigma_1,\ \sigma_1\,\sigma_2)}_{\Uparrow}\end{aligned}$$

Define $\sigma_1\,\sigma_2 \equiv \beta\sigma_3$ and cyclic permutations, where β is a constant.

$$= \alpha\,\boldsymbol{a}\cdot\boldsymbol{b}\ +\ \beta(\boldsymbol{a}\times\boldsymbol{b})\cdot\boldsymbol{\sigma}\,. \tag{4.59}$$

We will try to use this identity to rewrite (4.18) in a simpler way. Because we plan to make use of half-angles, define for brevity

$$s \equiv \sin\frac{\theta}{2}\,, \quad c \equiv \cos\frac{\theta}{2}\,. \tag{4.60}$$

Equation (4.18) becomes

$$\begin{aligned}\boldsymbol{r}' &= \boldsymbol{r}\cdot\boldsymbol{n}\ \boldsymbol{n}(1-\cos\theta) + \boldsymbol{r}\cos\theta + \boldsymbol{n}\times\boldsymbol{r}\sin\theta\\ &= \boldsymbol{r}\cdot\boldsymbol{n}\ \boldsymbol{n}\ 2s^2 + \boldsymbol{r}(c^2 - s^2) + \boldsymbol{n}\times\boldsymbol{r}\ 2sc\,.\end{aligned} \tag{4.61}$$

Dotting both sides of this with $\boldsymbol{\sigma}$ produces

$$\boldsymbol{r}'\cdot\boldsymbol{\sigma} = c^2\boldsymbol{r}\cdot\boldsymbol{\sigma} + 2sc(\boldsymbol{n}\times\boldsymbol{r})\cdot\boldsymbol{\sigma} + s^2\boldsymbol{n}\cdot\boldsymbol{r}\ \boldsymbol{n}\cdot\boldsymbol{\sigma} - s^2(\boldsymbol{r} - \boldsymbol{n}\cdot\boldsymbol{r}\ \boldsymbol{n})\cdot\boldsymbol{\sigma}\,. \tag{4.62}$$

The last term, $\boldsymbol{r} - \boldsymbol{n}\cdot\boldsymbol{r}\ \boldsymbol{n}$, can be written as a double cross product (see the discussion on p. 36); we will also add an extra term of $\alpha(\boldsymbol{n}\times\boldsymbol{r})\cdot\boldsymbol{n}$, which is zero but is used to facilitate the following lines by making use of (4.59):

$$\begin{aligned}\boldsymbol{r}'\cdot\boldsymbol{\sigma} &= c^2\boldsymbol{r}\cdot\boldsymbol{\sigma} + \frac{sc}{\beta}\left[\alpha\boldsymbol{n}\cdot\boldsymbol{r} + \beta(\boldsymbol{n}\times\boldsymbol{r})\cdot\boldsymbol{\sigma}\right] - \frac{sc}{\beta}\left[\alpha\boldsymbol{r}\cdot\boldsymbol{n} + \beta(\boldsymbol{r}\times\boldsymbol{n})\cdot\boldsymbol{\sigma}\right]\\ &\quad + s^2\boldsymbol{n}\cdot\boldsymbol{r}\ \boldsymbol{n}\cdot\boldsymbol{\sigma} - \frac{s^2}{\beta}\left\{\alpha(\boldsymbol{n}\times\boldsymbol{r})\cdot\boldsymbol{n} + \beta\left[(\boldsymbol{n}\times\boldsymbol{r})\times\boldsymbol{n}\right]\cdot\boldsymbol{\sigma}\right\}.\end{aligned} \tag{4.63}$$

This can be rewritten in the form

$$\begin{aligned} \boldsymbol{r}'\!\cdot\!\boldsymbol{\sigma} &= c^2\boldsymbol{r}\!\cdot\!\boldsymbol{\sigma} + \frac{sc}{\beta}\boldsymbol{n}\!\cdot\!\boldsymbol{\sigma}\,\boldsymbol{r}\!\cdot\!\boldsymbol{\sigma} - \frac{sc}{\beta}\boldsymbol{r}\!\cdot\!\boldsymbol{\sigma}\,\boldsymbol{n}\!\cdot\!\boldsymbol{\sigma} \\ &\quad - \frac{s^2}{\beta^2}\left[-\beta^2\boldsymbol{n}\!\cdot\!\boldsymbol{r} + \beta(\boldsymbol{n}\times\boldsymbol{r})\!\cdot\!\boldsymbol{\sigma}\right]\boldsymbol{n}\!\cdot\!\boldsymbol{\sigma}\,. \end{aligned} \tag{4.64}$$

We still have the freedom to relate α and β; so set $\alpha = -\beta^2$, which allows the brackets in (4.64) to be written as $\boldsymbol{n}\!\cdot\!\boldsymbol{\sigma}\,\boldsymbol{r}\!\cdot\!\boldsymbol{\sigma}$. In that case

$$\begin{aligned} \boldsymbol{r}'\!\cdot\!\boldsymbol{\sigma} &= c^2\boldsymbol{r}\!\cdot\!\boldsymbol{\sigma} + \frac{sc}{\beta}\boldsymbol{n}\!\cdot\!\boldsymbol{\sigma}\,\boldsymbol{r}\!\cdot\!\boldsymbol{\sigma} - \frac{sc}{\beta}\boldsymbol{r}\!\cdot\!\boldsymbol{\sigma}\,\boldsymbol{n}\!\cdot\!\boldsymbol{\sigma} - \frac{s^2}{\beta^2}\boldsymbol{n}\!\cdot\!\boldsymbol{\sigma}\,\boldsymbol{r}\!\cdot\!\boldsymbol{\sigma}\,\boldsymbol{n}\!\cdot\!\boldsymbol{\sigma} \\ &= \underbrace{\left(c + \frac{s}{\beta}\boldsymbol{n}\!\cdot\!\boldsymbol{\sigma}\right)}_{\equiv Q_{\boldsymbol{n}}(\theta)}\,\boldsymbol{r}\!\cdot\!\boldsymbol{\sigma}\,\underbrace{\left(c - \frac{s}{\beta}\boldsymbol{n}\!\cdot\!\boldsymbol{\sigma}\right)}_{= Q_{\boldsymbol{n}}(-\theta) = Q_{\boldsymbol{n}}^{-1}(\theta)}\,. \end{aligned} \tag{4.65}$$

Hence

$$\boldsymbol{r}'\!\cdot\!\boldsymbol{\sigma} = Q_{\boldsymbol{n}}(\theta)\,\boldsymbol{r}\!\cdot\!\boldsymbol{\sigma}\,Q_{\boldsymbol{n}}^{-1}(\theta)\,, \quad \text{where } Q_{\boldsymbol{n}}(\theta) = \cos\frac{\theta}{2} + \frac{1}{\beta}\boldsymbol{n}\!\cdot\!\boldsymbol{\sigma}\sin\frac{\theta}{2}\,. \tag{4.66}$$

Equation (4.66) is an alternative version of the basic rotation equation (4.22), and uses half angles along with the placeholders $\sigma_1, \sigma_2, \sigma_3$. The only demands we made on these placeholders are that they anticommute and that

$$\sigma_1^2 = \sigma_2^2 = \sigma_3^2 = -\beta^2 \quad \text{and } \sigma_1\sigma_2 = \beta\sigma_3 \text{ (and cyclic permutations)}, \tag{4.67}$$

which is what forms their algebra. What choices can be made for β? Two that are generally used are $\beta = 1$ and $\beta = \mathrm{i}$, as in Table 4.1. Historically, the $\beta = 1$ case of *quaternions* was the first example of this algebra, invented by Hamilton in 1843. Hamilton used the notation $\mathrm{i}^2 = \mathrm{j}^2 = \mathrm{k}^2 = -1$ to emphasise that he was trying to extend (4.56) to form a superset of the complex numbers.

Table 4.1. Choices of the algebra in (4.59), where $\alpha = -\beta^2$.

	Quaternions	*Pauli matrices*
$\beta = ?$	1	i
$\boldsymbol{a}\!\cdot\!\boldsymbol{\sigma}\;\boldsymbol{b}\!\cdot\!\boldsymbol{\sigma} = ?$	$-\boldsymbol{a}\!\cdot\!\boldsymbol{b} + (\boldsymbol{a}\times\boldsymbol{b})\!\cdot\!\boldsymbol{\sigma}$	$\boldsymbol{a}\!\cdot\!\boldsymbol{b} + \mathrm{i}(\boldsymbol{a}\times\boldsymbol{b})\!\cdot\!\boldsymbol{\sigma}$
Algebra (both are cyclic and anticommutative):	$\sigma_1^2 = -1$ $\sigma_1\sigma_2 = \sigma_3$	$\sigma_1^2 = +1$ $\sigma_1\sigma_2 = \mathrm{i}\sigma_3$
$R = ?$	$\cos\frac{\theta}{2} + \boldsymbol{n}\!\cdot\!\boldsymbol{\sigma}\,\sin\frac{\theta}{2}$	$\cos\frac{\theta}{2} - \mathrm{i}\boldsymbol{n}\!\cdot\!\boldsymbol{\sigma}\,\sin\frac{\theta}{2}$
Alternative representation:	$\mathrm{i} \equiv \sigma_1$ $\mathrm{j} \equiv \sigma_2$ $\mathrm{k} \equiv \sigma_3$	$\sigma_1 = \begin{bmatrix} 0 & 1 \\ 1 & 0 \end{bmatrix},\ \sigma_2 = \begin{bmatrix} 0 & -\mathrm{i} \\ \mathrm{i} & 0 \end{bmatrix},\ \sigma_3 = \begin{bmatrix} 1 & 0 \\ 0 & -1 \end{bmatrix}$

If we choose to use quaternions in an effort to reproduce (4.56), we find that (4.66) becomes

$$x'\mathrm{i} + y'\mathrm{j} = \left(\cos\frac{\theta}{2} + \mathrm{k}\sin\frac{\theta}{2}\right)(x\,\mathrm{i} + y\,\mathrm{j})\left(\cos\frac{\theta}{2} - \mathrm{k}\sin\frac{\theta}{2}\right). \tag{4.68}$$

This is more complicated than we might have hoped. But the placeholders $\{\mathrm{i}, \mathrm{j}, \mathrm{k}\}$ are, in a sense, more evenly balanced than the choice of $\{1, \mathrm{i}\}$ because i, j, and k all have similar properties; for example, they *all* square to give -1. So they reinforce the fact that the vector components (x, y, z) all share an equal footing. Equation (4.68) also shows that the quaternion i is *not* the extension to three dimensions of the complex number i. Rather, the quaternion k has this role because a rotation in the xy-plane, written as $x+\mathrm{i}y$ in (4.56), is really a rotation about the z-axis. There is nothing mysterious about this; it is just the way the naming in the subject evolved. If the imaginary number i had been called k from the start, there would be no confusion, but, without the wisdom of hindsight, the naming conventions bestowed by history are not always straightforward.

Perhaps, since $(\boldsymbol{a}\cdot\boldsymbol{\sigma})^2 = -\beta^2|\boldsymbol{a}|^2$, it might be more natural to set $\beta = \mathrm{i}$. As we'll see in Sect. 4.5, modern usage certainly favours this choice, having been heavily influenced by the matrix representation of spin in quantum mechanics. That formalism employs a matrix representation of $\sigma_1, \sigma_2, \sigma_3$ called *Pauli matrices*, which allows everyday multiplication to be used, as discussed previously in the context of (4.57). The Pauli matrices are not the only matrix choice that can represent the algebra of (4.59), but they are commonly used. We will meet other choices in the next section.

Although (4.66) splits a rotation into two steps, each using half the rotation angle, we are *not* free to interpret this to mean that $Q_{\boldsymbol{n}}(\theta)$ rotates a vector by $\theta/2$. This sort of idea surfaces in the spin formalism of quantum mechanics. The quantum mechanical concept of spin is usually described using Pauli matrices that multiply vector representations of spin kets, known as *spinors*. Equation (4.66) shows that to rotate a vector, these matrices must be used in pairs. But in spin formalism they appear singly. Thus, acting on a spin ket with, say, $Q_z(360°)$ is equivalent to implementing only one step of the two needed to actually rotate the vector through 360°. However, this is *not* the same as rotating the vector through 180°; equation (4.66) cannot be interpreted in that way.

4.4.1 Tidying Up the Placeholders

Obviously, we never write a numerical multiplication by including all of the factors of 10, such as

$$12 \times 34 = (1 \times 10 + 2)(3 \times 10 + 4) = \text{etc.} \tag{4.69}$$

So, also, the tedious algebra of the quaternions σ_i is much more usefully hidden inside a new multiplication rule akin to the two-dimensional case (4.55), by defining a four-component entity that is also always called a quaternion,

$$(a^0, \boldsymbol{a}) \equiv a^0 + \boldsymbol{a}\cdot\boldsymbol{\sigma} \,. \tag{4.70}$$

(Perhaps the set of $\{1, \sigma_1, \sigma_2, \sigma_3\}$ might better be called *basis* quaternions, but there is no real confusion if they are not.) Multiply these quaternions using standard rules of algebra:

$$\begin{aligned}(a^0, \boldsymbol{a})\,(b^0, \boldsymbol{b}) &\equiv (a^0 + \boldsymbol{a}\cdot\boldsymbol{\sigma})\,(b^0 + \boldsymbol{b}\cdot\boldsymbol{\sigma}) \\ &= a^0 b^0 - \beta^2 \boldsymbol{a}\cdot\boldsymbol{b} + \left(a^0 \boldsymbol{b} + b^0 \boldsymbol{a} + \beta\, \boldsymbol{a}\times\boldsymbol{b}\right)\cdot\boldsymbol{\sigma} \\ &= \left(a^0 b^0 - \beta^2 \boldsymbol{a}\cdot\boldsymbol{b},\;\; a^0 \boldsymbol{b} + b^0 \boldsymbol{a} + \beta\, \boldsymbol{a}\times\boldsymbol{b}\right). \end{aligned} \tag{4.71}$$

Although not commutative, this multiplication certainly is associative. It is a natural extension of the complex number case (4.55). Choosing $\beta = 1$ defines what is conventionally called *quaternion multiplication*. To rotate a vector $\boldsymbol{r}$ through θ around unit vector $\boldsymbol{n}$ using quaternions, (4.66) becomes

$$\boxed{(0, \boldsymbol{r}') = \left(\cos\frac{\theta}{2}, \boldsymbol{n}\sin\frac{\theta}{2}\right)\;(0, \boldsymbol{r})\;\left(\cos\frac{\theta}{2}, -\boldsymbol{n}\sin\frac{\theta}{2}\right).} \tag{4.72}$$

As an example, on p. 153 we used a matrix to rotate $(0, 1, 1)$ by 90° about the y-axis, giving $(1, 1, 0)$. In contrast, the quaternion approach gives

$$\begin{aligned}(0, \boldsymbol{r}') &= \left(\frac{1}{\sqrt{2}}, \frac{1}{\sqrt{2}}(0,1,0)\right)\;(0,0,1,1)\;\left(\frac{1}{\sqrt{2}}, \frac{-1}{\sqrt{2}}(0,1,0)\right) \\ &= \frac{1}{2}\,(1,0,1,0)\;(0,0,1,1)\;(1,0,-1,0) \\ &= (0,1,1,0)\,, \end{aligned} \tag{4.73}$$

so that the result is $(1, 1, 0)$ as expected.

Combining two rotations using quaternions is straightforward. A rotation of Q_1 followed by Q_2 gives

$$(0, \boldsymbol{r}') = Q_2\, Q_1\, (0, \boldsymbol{r})\, Q_1^{-1}\, Q_2^{-1} = Q_2\, Q_1\, (0, \boldsymbol{r})\, (Q_2\, Q_1)^{-1}, \tag{4.74}$$

so that the corresponding rotation quaternion is $Q_2 Q_1$ (i.e., simply a product, just as for matrices). This allows us to use quaternions to redo the example of Sect. 4.3 in which two 90° rotations around the x- and y-axes were combined into a single resultant:

$$\text{resultant quaternion} = \left(\frac{1}{\sqrt{2}}, \frac{1}{\sqrt{2}}(0,1,0)\right)\;\left(\frac{1}{\sqrt{2}}, \frac{1}{\sqrt{2}}(1,0,0)\right)$$

$$= \frac{1}{2}\,(1,0,1,0)\;(1,1,0,0) = \frac{1}{2}\,(1,1,1,-1)$$

$$= \left(\frac{1}{2}, \frac{\sqrt{3}}{2}\frac{(1,1,-1)}{\sqrt{3}}\right) = \left(\cos 60°, \frac{(1,1,-1)}{\sqrt{3}} \sin 60°\right), \qquad (4.75)$$

which implies $\theta = 120°$ and $\boldsymbol{n} = (1,1,-1)/\sqrt{3}$, just as we found in (4.52).

Quaternion Identity and Length. It can easily be seen that the identity for the multiplication of (4.71) is $(1, \boldsymbol{0})$ regardless of β, in which case $(a^0, \boldsymbol{a})$ has multiplicative inverse

$$(a^0, \boldsymbol{a})^{-1} = \frac{(a^0, -\boldsymbol{a})}{(a^0)^2 + \beta^2 |\boldsymbol{a}|^2}. \qquad (4.76)$$

If we define the *conjugate* of $(a^0, \boldsymbol{a})$ to be $(a^0, \boldsymbol{a})^* \equiv (a^0, -\boldsymbol{a})$, then by analogy to complex numbers, the product $(a^0, \boldsymbol{a})(a^0, \boldsymbol{a})^*$ should be the squared length of $(a^0, \boldsymbol{a})$ times the identity, and this serves to define the length $|(a^0, \boldsymbol{a})|$:

$$(a^0, \boldsymbol{a})(a^0, \boldsymbol{a})^* = (a^0, \boldsymbol{a})(a^0, -\boldsymbol{a}) = \left[(a^0)^2 + \beta^2 |\boldsymbol{a}|^2\right](1, \boldsymbol{0}) \equiv \left|(a^0, \boldsymbol{a})\right|^2 (1, \boldsymbol{0}). \qquad (4.77)$$

In that case, (4.76) can be written as

$$(a^0, \boldsymbol{a})^{-1} = \frac{(a^0, \boldsymbol{a})^*}{|(a^0, \boldsymbol{a})|^2}. \qquad (4.78)$$

(It's written in this way with the conjugate to highlight the similarity with the reciprocal of a complex number z, being $z^{-1} = z^*/|z|^2$.) With the choice $\beta = 1$ for quaternion multiplication, $|(a^0, \boldsymbol{a})|^2 = (a^0)^2 + |\boldsymbol{a}|^2$, which is just the squared length of a vector in four-dimensional euclidean space $\mathbb{E}^4$. In particular, rotation quaternions $(\cos {}^{\theta}\!/_2, \boldsymbol{n} \sin {}^{\theta}\!/_2)$ have unit length, forming the quaternion analogy to the unit determinant and orthogonality of a rotation matrix. This will have an important use in Sect. 4.6.3, where we treat quaternions as vectors in $\mathbb{E}^4$.

Matrix Representatives of Quaternions

The choice of $\beta = 1$ for quaternions in Table 4.1 can be combined with a matrix representation of the placeholders σ_1 to σ_3 to produce a 2×2 matrix representation of quaternions, simply by writing the quaternion $(a^0, \boldsymbol{a})$ as the sum in (4.70). We could use the Pauli choice of matrices, but there are others. A more symmetric choice in terms of a natural grouping of the quaternion elements a^0 to a^3 results if we choose

$$\sigma_1 = \mathrm{i}\begin{bmatrix} 1 & 0 \\ 0 & -1 \end{bmatrix}, \quad \sigma_2 = \mathrm{i}\begin{bmatrix} 0 & -\mathrm{i} \\ \mathrm{i} & 0 \end{bmatrix}, \quad \sigma_3 = \mathrm{i}\begin{bmatrix} 0 & 1 \\ 1 & 0 \end{bmatrix}, \qquad (4.79)$$

which again obey $\sigma_1\sigma_2 = \sigma_3$ and cyclic permutations, along with $\sigma_i^2 = -1$, the correct algebra for quaternions. In that case, (4.70) becomes

$$(a^0, \boldsymbol{a}) = a^0 + \boldsymbol{a}\cdot\boldsymbol{\sigma} = \begin{bmatrix} a^0 + \mathrm{i}a^1 & a^2 + \mathrm{i}a^3 \\ -a^2 + \mathrm{i}a^3 & a^0 - \mathrm{i}a^1 \end{bmatrix} = \begin{bmatrix} z & w \\ -w^* & z^* \end{bmatrix},$$

$$\text{where } z \equiv a^0 + \mathrm{i}a^1\,, \quad w \equiv a^2 + \mathrm{i}a^3\,. \tag{4.80}$$

Rotation quaternions having unit length implies that $|z|^2 + |w|^2 = 1$. These numbers z and w are known as *Cayley–Klein parameters*, used historically in rotational motion calculations.

With this 2×2 matrix representation, the rotation example in (4.73) is written as

$$\begin{aligned}(0, \boldsymbol{r}') &= \left(\frac{1}{\sqrt{2}}, \frac{1}{\sqrt{2}}(0,1,0)\right)(0,0,1,1)\left(\frac{1}{\sqrt{2}}, \frac{-1}{\sqrt{2}}(0,1,0)\right) \\ &= \frac{1}{2}(1,0,1,0)\,(0,0,1,1)\,(1,0,-1,0) \\ &= \frac{1}{2}\begin{bmatrix} 1 & 1 \\ -1 & 1 \end{bmatrix}\begin{bmatrix} 0 & 1+\mathrm{i} \\ -1+\mathrm{i} & 0 \end{bmatrix}\begin{bmatrix} 1 & -1 \\ 1 & 1 \end{bmatrix} = \begin{bmatrix} \mathrm{i} & 1 \\ -1 & -\mathrm{i} \end{bmatrix} \\ &= (0,1,1,0)\,, \end{aligned} \tag{4.81}$$

so that the rotated vector is $(1, 1, 0)$ as before.

Finally, quaternions can be represented by 4×4 matrices (along with the usual matrix multiplication), such as

$$(a^0, \boldsymbol{a}) = \begin{bmatrix} a^0 & a^1 & a^2 & a^3 \\ -a^1 & a^0 & -a^3 & a^2 \\ -a^2 & a^3 & a^0 & -a^1 \\ -a^3 & -a^2 & a^1 & a^0 \end{bmatrix}, \tag{4.82}$$

although this is a very redundant way to encode the information necessary for a rotation. Even so, this last matrix is seen to be composed of four blocks: $a^0, \boldsymbol{a}, -\boldsymbol{a}^t$, and $a^0 + a^\times$, the last of which shows its rotation pedigree.

4.5 Producing a "Geometric" Algebra

Although we have used the language of vectors to develop quaternion ideas, historically it happened the other way around. Quaternion algebra was invented first, by Hamilton, and enjoyed a great popularity as well as giving birth to much important work in physics. Vectors came later, and with the invention of the dot and cross products for vectors, users of quaternions and vectors split into two camps. This is partly an accident of history, since it's only natural that the beginnings of this type of analysis should have seen different versions of what were essentially the same structures. Confusion and arguments over how everything was related saw quaternion algebra begin to fade into obscurity.

Adding to the emerging view that quaternions were "just" a mathematical curiosity was the invention of an eight-dimensional set of objects called *octonions*, which happened immediately after quaternions first appeared and was directly due to them. But while quaternions are not commutative, octonions are neither commutative nor associative. This lack of associativity makes them very difficult to use (try inverting a simple octonion equation!), and what role octonions might play in physics, if any, remains unclear.

But the idea of reuniting quaternion and vector ideas has been worked on heavily in recent years by David Hestenes, and in this section we wish to outline some introductory ideas of the *geometric algebra* that he has popularised.[1]

We saw earlier in this chapter that the Pauli matrices can be used to represent the set $\{\sigma_1, \sigma_2, \sigma_3\}$, and together with the identity matrix they form a basis for 2×2 matrices. But $\sigma_1, \sigma_2, \sigma_3$ are abstract entities that obey a certain algebra and are not matrices per se; we can represent them by anything useful. For example, representing them by matrices allows matrix multiplication to be used in their algebra, making the various theorems of matrix analysis available to us.

But it's not hard to show that if we are prepared to invent a new sort of multiplication, we can represent $\sigma_1, \sigma_2, \sigma_3$ by the familiar basis vectors of $\mathbb{R}^3$. The many expressions in Sect. 4.4 with the form $\boldsymbol{a}\cdot\boldsymbol{\sigma}$, as well as the approach of writing quaternions as four-component entities, suggest that $\sigma_1, \sigma_2, \sigma_3$ might usefully be considered as basis vectors in some new space, so that $\boldsymbol{a}\cdot\boldsymbol{\sigma} = a^1\sigma_1 + a^2\sigma_2 + a^3\sigma_3$ implies that the components of the vector $\boldsymbol{a}$ in this new space are (a^1, a^2, a^3). In that case, (4.59) could be written much more elegantly as

$$\boldsymbol{ab} \equiv -\beta^2 \boldsymbol{a}\cdot\boldsymbol{b} + \beta\, \boldsymbol{a} \times \boldsymbol{b}\,. \tag{4.83}$$

This amounts to a new form of vector multiplication. The Pauli choice of $\beta = \mathrm{i}$ gives one form of the *geometric vector product* (which we'll refine on p. 169):

$$\boldsymbol{ab} \equiv \boldsymbol{a}\cdot\boldsymbol{b} + \mathrm{i}\, \boldsymbol{a} \times \boldsymbol{b} \quad \text{("early" definition).} \tag{4.84}$$

The fact that we are adding a number to a (complex) vector presents no difficulty, since this equation just encapsulates everything we have done up until now but in a tidier format. Adding dissimilar entities is no different from adding real and imaginary parts when forming complex numbers; such an addition is simply a device to process two entities in parallel using techniques with which we are comfortable, such as real-number multiplication. Equation (4.84) forms the entry point into the modern field of geometric algebra, which makes one of its aims the simplification and unification of diverse fields

[1] See, for example, *New Foundations for Classical Mechanics* by David Hestenes (1986, D. Reidel Publishing), and also Hestenes' paper "Oersted Medal Lecture 2002: Reforming the mathematical language of physics" in the American Journal of Physics, **71**, 104–121 (February 2003).

in mathematical physics. As a simple example, the rotation of (4.66) can now be written

$$\begin{aligned} \boldsymbol{r}' &= \left(\cos\frac{\theta}{2} - \mathrm{i}\boldsymbol{n}\sin\frac{\theta}{2}\right) \boldsymbol{r} \left(\cos\frac{\theta}{2} + \mathrm{i}\boldsymbol{n}\sin\frac{\theta}{2}\right) \\ &= \mathrm{e}^{-\mathrm{i}\boldsymbol{n}\theta/2}\, \boldsymbol{r}\, \mathrm{e}^{\mathrm{i}\boldsymbol{n}\theta/2}\,, \end{aligned} \tag{4.85}$$

where the exponential form follows by considering the relevant exponential series, making use of the geometric vector product.

In the geometric vector product (4.84), $\sigma_1, \sigma_2, \sigma_3$ have disappeared entirely and need not be considered again; their algebra has been taken over by the basis vectors $\boldsymbol{e}_1, \boldsymbol{e}_2, \boldsymbol{e}_3$, since, for example, (4.84) gives

$$\begin{aligned} \boldsymbol{e}_k^2 &= \boldsymbol{e}_k \cdot \boldsymbol{e}_k + \mathrm{i}\,\boldsymbol{e}_k \times \boldsymbol{e}_k = 1 \quad \text{for all } k, \text{ and} \\ \boldsymbol{e}_1\boldsymbol{e}_2 &= \boldsymbol{e}_1 \cdot \boldsymbol{e}_2 + \mathrm{i}\,\boldsymbol{e}_1 \times \boldsymbol{e}_2 = \mathrm{i}\,\boldsymbol{e}_3 \ \text{(and cyclic permutations)}, \end{aligned} \tag{4.86}$$

exactly mimicking the algebra of $\sigma_1, \sigma_2, \sigma_3$. Thus, we can even discard the idea of a new space mentioned just before (4.83). The vectors exist in the usual space with basis vectors $\boldsymbol{e}_1, \boldsymbol{e}_2, \boldsymbol{e}_3$, and the new algebra (4.86) of the basis is what supports the geometric product.

The Geometric Product and Maxwell's Equations

The electromagnetic field forms a good instance of the unifying power of the geometric product. In Chap. 8 we'll see that the nabla operator ∇ (also called del) actually behaves as a vector, so in that case write

$$\nabla \boldsymbol{E} = \nabla \cdot \boldsymbol{E} + \mathrm{i} \nabla \times \boldsymbol{E}\,. \tag{4.87}$$

Nabla acting on a cartesian vector is usually understood to act on each of its components individually; that is, usually $\nabla \boldsymbol{E}$ is understood to mean $(\nabla E^x, \nabla E^y, \nabla E^z)$. But here we are using it in the novel way of geometric algebra. If we now write Maxwell's equations,

$$\begin{aligned} \nabla \cdot \boldsymbol{E} &= \frac{\varrho}{\varepsilon_0}\,, & \nabla \times \boldsymbol{E} &= \frac{-\partial \boldsymbol{B}}{\partial t}\,, \\ \nabla \cdot \boldsymbol{B} &= 0\,, & \nabla \times \boldsymbol{B} &= \frac{\boldsymbol{j}}{\varepsilon_0 c^2} + \frac{1}{c^2}\frac{\partial \boldsymbol{E}}{\partial t}\,, \end{aligned} \tag{4.88}$$

we see that they combine to give

$$\nabla \boldsymbol{E} = \frac{\varrho}{\varepsilon_0} - \mathrm{i}\frac{\partial \boldsymbol{B}}{\partial t}\,, \quad \nabla \boldsymbol{B} = \mathrm{i}\frac{\boldsymbol{j}}{\varepsilon_0 c^2} + \frac{\mathrm{i}}{c^2}\frac{\partial \boldsymbol{E}}{\partial t}\,. \tag{4.89}$$

Suppose we form a new vector $\boldsymbol{F} \equiv \boldsymbol{E} + k\boldsymbol{B}$, where k is some constant, so that $\nabla \boldsymbol{F}$ is determined by (4.89). It's useful to mimic the wave equation for the electromagnetic field, albeit with first derivatives instead of second ones, and to do this we wish to relate $\nabla \boldsymbol{F}$ to $\partial \boldsymbol{F}/\partial t$. Calculating $\nabla \boldsymbol{F}$ from (4.89) shows that this can be arranged if $k^2 = -c^2$, so set $k = \mathrm{i}c$ and write

$$\boldsymbol{F} \equiv \boldsymbol{E} + \mathrm{i}c\boldsymbol{B}\,, \tag{4.90}$$

in which case Maxwell's equations can be written as one expression:

$$\left(\nabla + \frac{1}{c}\frac{\partial}{\partial t}\right)\boldsymbol{F} = \frac{1}{\varepsilon_0}\left(\varrho - \frac{\boldsymbol{j}}{c}\right). \tag{4.91}$$

The goal of the modern field of geometric algebra is to gain insight into physical laws based on such simple unifying expressions. As an example of the usefulness of $\boldsymbol{F}$, consider mimicking the expression for the electric field density $\varepsilon_0|\boldsymbol{E}|^2/2$, but now for $\boldsymbol{F}$, where we make use of the complex conjugate $\boldsymbol{F}^* \equiv \boldsymbol{E} - \mathrm{i}c\boldsymbol{B}$:

$$\begin{aligned}\frac{\varepsilon_0}{2}\boldsymbol{F}\boldsymbol{F}^* &= \frac{\varepsilon_0}{2}\left(\boldsymbol{F}\cdot\boldsymbol{F}^* + \mathrm{i}\,\boldsymbol{F}\times\boldsymbol{F}^*\right)\\ &= \underbrace{\frac{\varepsilon_0}{2}\left(|\boldsymbol{E}|^2 + c^2|\boldsymbol{B}|^2\right)}_{=\text{ energy density}} \;+\; \underbrace{\varepsilon_0 c\,\boldsymbol{E}\times\boldsymbol{B}}_{=\text{ Poynting vector}/c}\,.\end{aligned} \tag{4.92}$$

Besides the energy density for the *entire* field now appearing, what has also emerged for free is the *Poynting vector*, the flux density of field energy! (Flux density will be defined in Chap. 6, where again it will be seen to pair naturally with a spatial density.) Of course, we cannot know without experimental input that this new vector term should very well be a flux density, but the fact that it has appeared hand in hand with the spatial energy density is very suggestive that some sort of unification has taken place. We'll meet a similar idea with the *continuity equation* in (10.165). And, in the coming chapters, we will investigate the tensor approach to unifying spatial and flux densities, such as those of rainfall, along with the charge and current densities ϱ and $\boldsymbol{j}$. We'll also see the tensor approach to unifying the fields $\boldsymbol{E}$ and $\boldsymbol{B}$.

A More Refined Geometric Product

In fact, the basic definition of the product used in geometric algebra is a little different from (4.84). Using the wedge product that we first met in Sect. 2.4.4, the geometric product can be defined more generally as

$$\boxed{\boldsymbol{a}\boldsymbol{b} \equiv \boldsymbol{a}\cdot\boldsymbol{b} + \boldsymbol{a}\wedge\boldsymbol{b}\,,} \tag{4.93}$$

and like the wedge product, is defined to be associative. How does this definition relate to (4.84)? Write $\boldsymbol{a}$ and $\boldsymbol{b}$ as linear combinations over their bases, so that their product is

$$\boldsymbol{a}\boldsymbol{b} = \sum_{ij} a^i\boldsymbol{e}_i\, b^j\boldsymbol{e}_j = \sum_{ij} a^i b^j\boldsymbol{e}_i\boldsymbol{e}_j\,. \tag{4.94}$$

However,

$$\boldsymbol{e}_1\boldsymbol{e}_2 = \boldsymbol{e}_1 \wedge \boldsymbol{e}_2 = -\boldsymbol{e}_2 \wedge \boldsymbol{e}_1 = -\boldsymbol{e}_2\boldsymbol{e}_1 \tag{4.95}$$

along with cyclic permutations, so that swapping basis vectors changes the sign of the product. Also, $\boldsymbol{e}_i^2 = 1$ for all i, so that we can write

$$\boldsymbol{e}_1\boldsymbol{e}_2 = \boldsymbol{e}_1\boldsymbol{e}_2\boldsymbol{e}_3\boldsymbol{e}_3\,. \tag{4.96}$$

But the product $\boldsymbol{e}_1\boldsymbol{e}_2\boldsymbol{e}_3$ can be squared by permuting its factors to bring like ones together, and then using the fact that $\boldsymbol{e}_i^2 = 1$ for all i:

$$\boldsymbol{e}_1\boldsymbol{e}_2\boldsymbol{e}_3\,\boldsymbol{e}_1\boldsymbol{e}_2\boldsymbol{e}_3 = -\boldsymbol{e}_1\boldsymbol{e}_2\boldsymbol{e}_3\boldsymbol{e}_3\boldsymbol{e}_2\boldsymbol{e}_1 = -\boldsymbol{e}_1\boldsymbol{e}_2\boldsymbol{e}_2\boldsymbol{e}_1 = -\boldsymbol{e}_1\boldsymbol{e}_1 = -1\,. \tag{4.97}$$

So because $(\boldsymbol{e}_1\boldsymbol{e}_2\boldsymbol{e}_3)^2 = -1$, we make the identification

$$\boldsymbol{e}_1\boldsymbol{e}_2\boldsymbol{e}_3 = \mathrm{i}\,. \tag{4.98}$$

This identity allow (4.96) and its cyclic permutations to be written as

$$\boldsymbol{e}_1\boldsymbol{e}_2 = \mathrm{i}\boldsymbol{e}_3\,,\quad \boldsymbol{e}_2\boldsymbol{e}_3 = \mathrm{i}\boldsymbol{e}_1\,,\quad \boldsymbol{e}_3\boldsymbol{e}_1 = \mathrm{i}\boldsymbol{e}_2\,, \tag{4.99}$$

which we have already seen in (4.86). The product (4.94) becomes, for example with $(\boldsymbol{a} \times \boldsymbol{b})^1$ denoting the x-component of $\boldsymbol{a} \times \boldsymbol{b}$,

$$\begin{aligned}\boldsymbol{ab} &= \boldsymbol{a}\cdot\boldsymbol{b} + \left(a^1b^2 - a^2b^1\right)\boldsymbol{e}_1\boldsymbol{e}_2 + \left(a^3b^1 - a^1b^3\right)\boldsymbol{e}_3\boldsymbol{e}_1 + \left(a^2b^3 - a^3b^2\right)\boldsymbol{e}_2\boldsymbol{e}_3\\ &= \boldsymbol{a}\cdot\boldsymbol{b} + (\boldsymbol{a}\times\boldsymbol{b})^3\,\mathrm{i}\boldsymbol{e}_3 + (\boldsymbol{a}\times\boldsymbol{b})^2\,\mathrm{i}\boldsymbol{e}_2 + (\boldsymbol{a}\times\boldsymbol{b})^1\,\mathrm{i}\boldsymbol{e}_1\\ &= \boldsymbol{a}\cdot\boldsymbol{b} + \mathrm{i}\,\boldsymbol{a}\times\boldsymbol{b}\,,\end{aligned} \tag{4.100}$$

which now matches (4.84)! So the identification of $\boldsymbol{e}_1\boldsymbol{e}_2\boldsymbol{e}_3$ with i forms the bridge that relates the two definitions (4.84) and (4.93) of the geometric product. Such correspondences between different concepts form the bread and butter of geometric algebra.

The language and ideas of geometric algebra have gained in acceptance in the few decades since it was first popularised, and a large amount of mathematical physics has been converted to its language. Hopefully the insights it provides will find themselves incorporated into the "common" physics tongue, helping it to evolve in a useful direction.

4.6 Rotations in Popular Usage

(The rest of the material in this chapter is very applied in its nature, and apart from briefly mentioning the idea of parallel transport, does not introduce any new concepts that are central to mathematical physics. However, it has been included to address some points and misconceptions that are common in rotation literature. For some, these misconceptions even inspire strange ideas of a breakdown of the relevant mathematics, but it is nothing more than a result of rotation theory being applied incorrectly.)

Users of rotation theory in applied work, such as the aerospace industry and computer graphics programming, rarely have the luxury of delving into the subject's intricacies. They usually need to perform rotations for quite complicated scenarios and often have no more than recipes to guide them. What they find in the literature is a large amount of information about rotations around fixed axes, nonfixed axes, active rotations, passive rotations, and apparent problems with rotations, together with intricacies of the use of matrices and quaternions, conjugate operations, and any number of other ideas all of which act to obfuscate the subject and make it seem harder than it really is. In this section we will briefly consider some of these ideas in the light of the fundamental building block of rotation that was given at the start of this chapter and stressed in the box on p. 158: that every rotation consists simply of turning one vector about another.

4.6.1 Describing an Orientation by Using *Three* Rotations

An important concept in aerospace theory is the fact that an arbitrary orientation of a body can be produced by at most three successive rotations around the space-fixed x, y, z-axes, through angles known as *Euler angles.* To prove this, it's perhaps easier to visualise the process in reverse. Start by visualising the orientated body, where arrows aligned with the basis vectors $\boldsymbol{e}_x, \boldsymbol{e}_y, \boldsymbol{e}_z$ were frozen into the body in its base position, and they had eventually been transformed into $\boldsymbol{e}_X, \boldsymbol{e}_Y, \boldsymbol{e}_Z$ as the body's orientation was changed. Return the body to its base position, and ask how these $\boldsymbol{e}_X, \boldsymbol{e}_Y, \boldsymbol{e}_Z$ vectors can be mapped back onto the $\boldsymbol{e}_x, \boldsymbol{e}_y, \boldsymbol{e}_z$ vectors using three rotations. We have not drawn a picture, since Euler angle pictures of three-dimensional rotations are notoriously difficult to comprehend; what *is* very useful is for the reader to make a set of wire axes with labels to help follow the sequence of rotations.

1. First rotate the body around the z-axis such that $\boldsymbol{e}_X$ comes to rest in the xz-plane.
2. Now rotate the result around the y-axis such that $\boldsymbol{e}_X$ lines up with the x-axis.
3. Finally, rotate the result around the x-axis such that the newly rotated $\boldsymbol{e}_Y, \boldsymbol{e}_Z$ line up with the y- and z-axes, respectively.

This procedure will return the body to its base orientation in three rotations around the space-fixed axes. The reverse operation yields the correct order that we have chosen to rotate through the Euler angles. In practice, there are many different conventions for specifying Euler angles, but three angles are always sufficient. Some authors leave one axis out and repeat another, so that the orientations are performed around, for example, the x-axis, then the y-axis, and then the x-axis again. All similar choices of three distinct rotations will also work, such as z, y, z, etc.—but not, e.g., z, y, y, since this is really only two distinct rotations.

A set of three Euler angles around fixed axes is fine for specifying the orientation of a body. But turning through these angles is certainly not what the pilot of a manoeuvring aircraft does. An aircraft carries three orthogonal basis vectors with it as it flies. Conventionally,

- $\boldsymbol{e}_x$ points through its nose in the forward direction,
- $\boldsymbol{e}_y$ points out along its right wing, and
- their cross product $\boldsymbol{e}_z$ points below the aircraft.

These three vectors are specified in some fixed, global frame within which all calculations can be performed, and they can change continuously in that frame. The pilot can *roll* around $\boldsymbol{e}_x$, *pitch* around $\boldsymbol{e}_y$, and *yaw* around $\boldsymbol{e}_z$. Any combination of yaw–pitch–roll angles will reorientate the aircraft, and these three angles are also referred to as Euler angles, even though they describe rotations around the *latest* positions of $\boldsymbol{e}_x, \boldsymbol{e}_y, \boldsymbol{e}_z$, not the space-fixed axes. A more everyday example of this type of Euler set is provided by orientating a telescope. We might sight a star by first rotating the telescope to the correct azimuth (the astronomical term for bearing), and then rotating about the *latest* position of the elevation axis to set the correct elevation.

The various manoeuvres of an aircraft are easy to follow in a global, fixed frame, such as the *Earth-centred, Earth-fixed* frame. For example, when the aircraft rolls, we simply rotate each of its basis vectors around the latest $\boldsymbol{e}_x$, giving three new vectors (one of which, $\boldsymbol{e}_x$, is unchanged, but is still part of the new trio). Similarly, a pitch entails rotating each basis vector around the latest $\boldsymbol{e}_y$ to give three new ones, and a yaw does the same for the latest $\boldsymbol{e}_z$. Finally and importantly, the coordinates of, say, the Moon *in the frame of the aircraft* are then simply found by forming three dot products of the vector from the aircraft to the Moon (in the *global* frame), with the basis vectors that specify the latest orientation of the aircraft in the global frame. In practice, these three dot products can be calculated using a 3×3 matrix multiplication.

Confusing Which Frame is Being Used for the Rotation

This is a convenient place to point out some rotation terminology. Rotating a vector in, say, the xy-plane is sometimes called an *active* rotation. Equivalently, we might leave the vector fixed and rotate the x- and y-axes in the opposite direction; this is sometimes called a *passive* rotation. In the author's opinion the distinction is neither natural nor useful, and in more complicated scenarios it is not even particularly well defined. We have dealt solely with active rotations in this chapter (and will continue doing so). Ultimately, *every* rotation is implemented by rotating one vector around another, and this includes rotating the basis vectors that determine the axes of a frame. Thus any passive rotation must be implemented by an active rotation, so we will not use these terms any further.

Now, consider rotating the set of axes that are tied to the body of a manoeuvring aircraft. In the example above of finding the position of the

Moon in the manoeuvring aircraft's frame, we rotated the aircraft's basis vectors in a global frame, and found the Moon's coordinates with respect to the reorientated basis vectors by using dot products. This is necessary if we are to use Newton's laws to follow the motion of the aircraft in an inertial, or approximately inertial, frame attached to Earth. On the other hand, if we know how the aircraft manoeuvres, we might mentally sit inside it, treating it as stationary, and rotate the Moon around it.

So, when we remain in the aircraft's frame, which is by definition fixed, the vector from this frame's origin to the Moon turns around the aircraft. After each yaw, pitch, and roll, the updated version of this vector is found by rotating it around one of three *fixed* axes—because the aircraft does not move in its own frame. (The Moon will also translate as the aircraft flies, but this is not important to the current discussion.) Thus, the Earth frame's view of rotating the plane's basis vectors around latest versions of each other, and then finding dot products of these with the vector pointing from the aircraft to the Moon, is identical to rotating this "aircraft-to-Moon" vector around *space-fixed axes in the same order but in the opposite direction.* All that we have done is switch our frame from, e.g., an Earth-centred Earth-fixed frame to that of the aircraft.

Many users of rotation theory rotate vectors in the aircraft's frame, so that their rotation matrices tend to be the well-known simple ones obtained from (4.22) by setting $\boldsymbol{n}$ to be each of the aircraft's axes in its own frame. These are, by definition, the *unchanging* basis vectors $\begin{bmatrix}1 & 0 & 0\end{bmatrix}^t$, $\begin{bmatrix}0 & 1 & 0\end{bmatrix}^t$ and $\begin{bmatrix}0 & 0 & 1\end{bmatrix}^t$. But despite the fact that the three resulting rotation matrices deal with a manoeuvring aircraft *from the frame of the aircraft*, they are commonly described as rotating around the *latest* versions of moving axes. This is misleading. The three matrices rotate around an unchanging set of three body axes; it is just that the frame being used is that of the aircraft, and this frame defines these body axes to be fixed in (the body's) space. The world is now considered to rotate around the aircraft. This is an important point to be aware of when reading aerospace rotation literature. Such literature might describe an aircraft being rotated in an Earth-centred, Earth-fixed frame, and then, in an apparently contradictory fashion, will derive results using matrices for the three simple space-fixed axes a few lines up (which imply computations in the aircraft's frame). Mixing the use of the two frames in such a way requires a careful description of the procedure being followed. This is seldom, if ever, given. Relevant here is the discussion in the box on p. 158.

Making Rotation Almost Commutative

Let's revisit a rotation result first calculated in Sect. 2.5. There, it was shown that two noninfinitesimal rotations can be swapped in a restricted sense, with the proviso that we modify one of the axes. The result was expressed in (2.68):

$$\boxed{R_{x'}(\beta)\, R_z(\alpha) = R_z(\alpha)\, R_x(\beta)\,.} \tag{4.101}$$

In the case of rotating any vector in an aircraft, and working in, say, an Earth-centred, Earth-fixed frame, (4.101) describes

- via its left-hand side, a yaw around the below-aircraft vector $\boldsymbol{e}_z$ (i.e. a rotation $R_z(\alpha)$ of any vector fixed to the aircraft), followed by a roll $R_{x'}(\beta)$ around the aircraft's latest nose direction, $\boldsymbol{e}_{x'}$, the result of which is identical to
- the rotations of its right-hand side, which are both about *initial* basis vectors: first a roll around the nose direction $\boldsymbol{e}_x$, and then a rotation around the *initial* below-aircraft direction, $\boldsymbol{e}_z$, which, if the roll was nonzero, is *not* a yaw.

Although the rotations have been changed from being around latest axes to being around initial axes, the procedure of (4.101) is not the same idea as the use of the above Earth and aircraft frames involving the Moon, since the rotations of (4.101) rotate the *same* vector in the *same* frame, and have *reversed* their order from one side of that equation to the other. In contrast, the scenario involving the Moon was different. It dealt with rotating *different* vectors in *different* frames, but keeping the order of rotations unchanged:

- rotating the aircraft's basis vectors around latest versions of one another in an Earth-centred, Earth-fixed frame, versus
- rotating the aircraft-to-Moon vector, in the same order but in opposite directions, around body axes in the aircraft's frame. These axes are, by definition, space-fixed in that frame.

The rotation reordering of (4.101) can be extended to any number of Euler rotations. For example, suppose the pilot yaws through angle α about the below-aircraft direction, $\boldsymbol{e}_z$, then pitches by β about the *new* right wing direction, $\boldsymbol{e}_{y'}$, and then rolls by γ about the latest forward direction, $\boldsymbol{e}_{x''}$:

$$R_{x''}(\gamma)\, R_{y'}(\beta)\, R_z(\alpha)\,. \tag{4.102}$$

We can swap rotations in pairs by way of (4.101) to reverse this sequence:

$$\begin{aligned} R_{x''}(\gamma)\, R_{y'}(\beta)\, R_z(\alpha) &= R_{y'}(\beta)\, R_{x'}(\gamma)\, R_z(\alpha) \\ &= R_{y'}(\beta)\, R_z(\alpha)\, R_x(\gamma) \\ &= R_z(\alpha)\, R_y(\beta)\, R_x(\gamma)\,. \end{aligned} \tag{4.103}$$

So, the final orientation of the aircraft could also have been obtained by rotating about the *initial* vectors $\boldsymbol{e}_x, \boldsymbol{e}_y, \boldsymbol{e}_z$ in that order, with the angle order reversed. No pilot flies this way, but (4.103) proves the important point that the order of specifying Euler angles is somewhat reversible, as long as all of the initial axes are used in place of always rotating about the latest ones. This theorem is useful, but it can cause confusion if invoked without warning. Certainly rotation literature is not above speaking of rotation about the latest basis vectors, just as a pilot would fly—and then using rotation formalism that only uses the *initial* vectors, along with a rotation order that has been reversed for no apparent reason. Or, it might keep the original rotation order but now rotate in the opposite direction around space-fixed axes (as in the above Moon discussion)—with the switched direction perhaps hidden behind a left-handed rotation convention. We see now why it all works, although probably there are times in the literature where rotations have been ordered in whatever ad hoc way will give the "right" answer!

Additionally, a convention is sometimes used that reads expressions such as $R_x(\alpha)\, R_y(\beta)$ as acting from *left to right*, which can make it appear as if the rotation order of the previous paragraph had not been reversed after all. The matrix expressions of the whole rotation formalism might also be transposed from what we have written in this chapter, which not only transposes each matrix, but reverses their order, because $(AB)^t = B^t A^t$. The resulting matrix then *post*multiplies a *row* vector to rotate it. Not surprisingly, confusion is common in this topic due to different conventions not always being defined. The bottom line is that we should always make clear precisely what is being done when a three-dimensional rotation is carried out: what vectors are being rotated about what axis vectors, in what frame, and in what order.

An Alternative View of Rotation Noncommutivity

Picturing the various rotations involved with Euler angles can be difficult, whether rotating around original axes or new axes. Easier to visualise is the method shown in Fig. 4.5. In this figure, we show the idea of this restricted commutivity for y and z rotations through angles α (like a latitude on Earth) and ω (like a longitude).[2] Path ① consists of first rotating each of $\boldsymbol{e}_x, \boldsymbol{e}_y, \boldsymbol{e}_z$ around $\boldsymbol{e}_z$ by ω (equivalent to rotating them all around the z-axis in the figure), and then rotating the resulting vectors $\boldsymbol{e}_{x_1}, \boldsymbol{e}_{y_1}, \boldsymbol{e}_{z_1}$ around $\boldsymbol{e}_{y_1}$ by $-\alpha$; the minus sign is immaterial, being an artifact of the sign convention for latitude.

In contrast, path ② consists of first rotating the initial vectors $\boldsymbol{e}_x, \boldsymbol{e}_y, \boldsymbol{e}_z$ around $\boldsymbol{e}_y$ by $-\alpha$ to give $\boldsymbol{e}_{x_2}, \boldsymbol{e}_{y_2}, \boldsymbol{e}_{z_2}$, and then rotating each of these around $\boldsymbol{e}_z$ (*not* $\boldsymbol{e}_{z_2}$!) by ω. The results are identical:

$$\underbrace{R_{y_1}(-\alpha)\, R_z(\omega)}_{\text{path ①}} = \underbrace{R_z(\omega)\, R_y(-\alpha)}_{\text{path ②}} . \tag{4.104}$$

This just matches (4.101), as we might expect. This way of visualising rotations can be applied to any sort of convoluted rotation path connecting the initial and final basis sets. There can be all manner of zig-zags and switch-backs to the path. We can also picture an additional rotation: the initial and final bases being rotated by a bearing angle β around their x- and X-axes. To do the three rotations via path ①, some study of the figure shows that

[2] Since latitude and longitude both start with the same letter, we have chosen the Greek variables to match the second letter of each word.

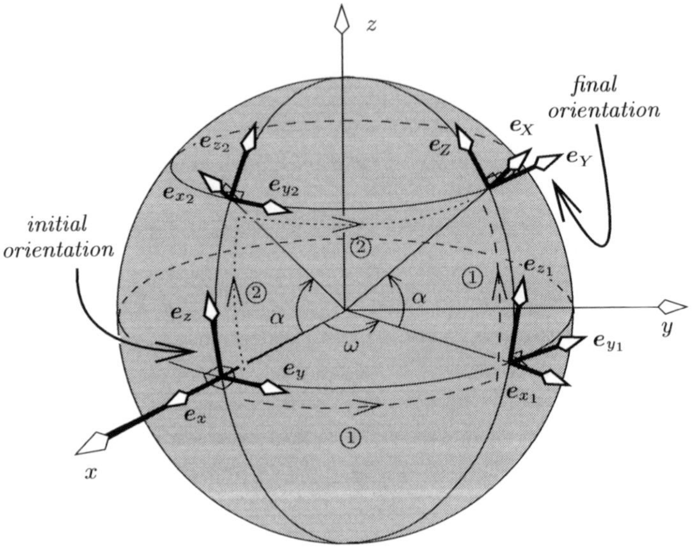

Fig. 4.5. The "almost-commutivity" of rotations can be seen by picturing a rotation as sliding the basis-vector set over a sphere. Compare dashed path ① with dotted path ② as representing alternative rotation methods converting the set $\boldsymbol{e}_x, \boldsymbol{e}_y, \boldsymbol{e}_z$ to $\boldsymbol{e}_X, \boldsymbol{e}_Y, \boldsymbol{e}_Z$. Both paths give the same result but by way of different rotations, producing the intermediate basis sets $\boldsymbol{e}_{x_1}, \boldsymbol{e}_{y_1}, \boldsymbol{e}_{z_1}$ and $\boldsymbol{e}_{x_2}, \boldsymbol{e}_{y_2}, \boldsymbol{e}_{z_2}$. Path ① is actually an example of *parallel transport*, which we'll meet again in Chap. 9.

this bearing turn must be left until last, while in path ② it must be done prior to any other rotation. Hence,

$$\underbrace{R_X(\beta)\, R_{y_1}(-\alpha)\, R_z(\omega)}_{\text{path ①}} = \underbrace{R_z(\omega)\, R_y(-\alpha)\, R_x(\beta)}_{\text{path ②}} . \tag{4.105}$$

As expected, the rotation orders of (4.105) match the reversal seen in (4.103). And as a side note, the movement of vectors along path ① is actually an example of the notion of *parallel transport*, an idea that we'll investigate properly in Chap. 9.

4.6.2 Confusing Euler Angle Orientation with Incremental Rotation

As we have seen, an object's orientation can be specified by three Euler angles through which the object can be turned in some preset order from a base position, which will result in the required orientation. This is useful and unambiguous, but it leads to much confusion when implemented in a particularly rigid way.

The situation can be shown as follows. Suppose we have written a computer programme that draws some object that we require to rotate at will. Attached to the computer are three levers each of which can be moved along a scale that reads 0° to 360°. These levers allow us to rotate the object about each of the fixed x, y, z-axes. Moving each lever by itself should have the effect of rotating the object about the appropriate axis. We can move any of these levers at any time, and the software, using very fast processing, will calculate the rotations so quickly that the response is effectively immediate. Modern computers do this very easily. To simplify the discussion, we will assume that the computer only redraws the object after the lever being moved has come to rest. (In practice, the lever positions are sensed almost continuously by the computer, so that the object seems to move continuously. But our simplification suffices for the following discussion.)

The scenario is shown in Fig. 4.6. When we start the programme running, the object is drawn in its base position, and each of the three levers is automatically parked at its home position, corresponding to 0°. If we now move the x-lever to read 10° and leave the two other levers at zero, then the programme should read the 10° and apply a rotation of $R_x(10°)$ to the object. If we wish to rotate the object through a further 5°, we move the x-lever to 15°. The programme should sense the *increment* of 5° and apply a rotation of $R_x(5°)$ to the already rotated object. If it does this, the object will rotate to match our intuition. So moving the x-lever backward and forward along its whole range will have the required effect: the object will rotate backward and forward about the x-axis. Similarly, if we shift only the y-lever back and forth, or only the z-lever, the same sort of intuitive results occur.

Suppose we now shift two of the levers; we'll ignore the z-lever, as it's not needed in the following discussion. Start with all levers at zero and with the object drawn in its base orientation. Then move the x-lever to 10°. The rotation is $R_x(10°)$, so the object rotates around x by 10°. Now set the y-lever to 10°. This rotates the already rotated object about the y-axis by 10°. So far, the behaviour matches what we expect. The history of the rotations can be summed up in the factor $R_y(10°)\,R_x(10°)$, rotating the object from its *base* orientation. Of course, there is no need to rerotate the object through the whole history from its base position each time we move a lever; the latest lever increment is all that is needed. But keeping a close eye on the history is useful for understanding what happens next.

Now suppose we wish to rotate the object a further 5° about the x-axis. So, we move the x-lever a little further over to the 15° mark. The rotation history now should be the factor $R_x(5°)\,R_y(10°)\,R_x(10°)$, again rotating the object from its base orientation.

Unfortunately, and perhaps surprisingly, computer programmes that purport to do the above in fact often do something quite different. What they do is interpret the three lever positions as Euler angles that are to be used to specify an orientation from an unchanging base position. Each time any one

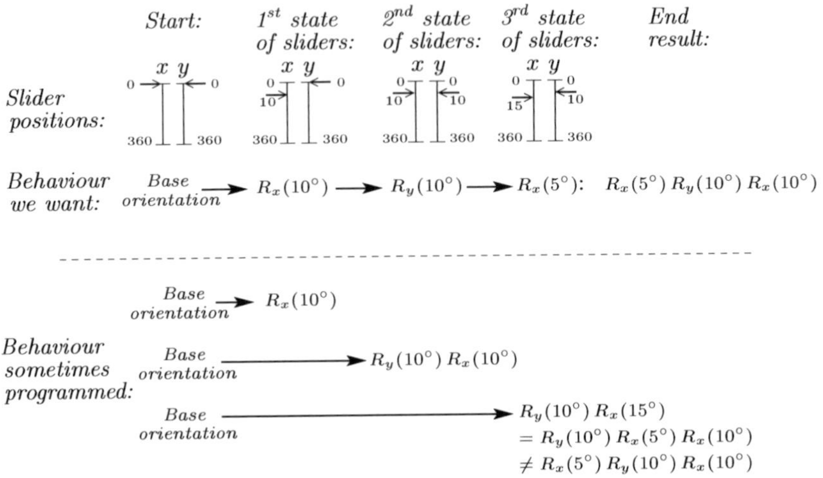

Fig. 4.6. Widespread confusion over the use of Euler angles to display and change an object's orientation within a graphical computer programming environment is shown by the two scenarios depicted, above and below the dashed line. **Above**: What the programmer requires, which corresponds to good common sense, is the ability to alter sliders that change the orientation of an object around the x- and y-axes. Changing one slider at a time performs one *incremental* rotation of the *current* orientation. The results always match our intuition, and no problems are encountered. **Below**: What is typically programmed into software is that every time one of the sliders is moved, *all* slider settings are read and treated as Euler angles, and the resulting three rotations (or two in the example shown) are then applied in an unchanging order (x, then y above) to the *base* orientation. This is an unnatural way to allow a computer user to change an object's orientation, and because it does not apply the rotations in the order that the user applied them to the sliders, wrong orientations will eventually result. This simple mistake in applying rotation order is conventionally blamed on Euler angles instead of where the fault really lies: with an incorrect use of rotations.

lever is moved, the programme rereads *all* of the current lever positions. It then applies all of the rotations in a set order, arbitrarily chosen but rigidly adhered to.

Suppose that this order is first the x-rotation, then the y-rotation (and finally z, which we're ignoring here). In the case above, the first two lever movements work well. The rotations performed are first $R_y(0^\circ)\,R_x(10^\circ)$ on the base orientation and then $R_y(10^\circ)\,R_x(10^\circ)$ on the base orientation. Now, what happens with the third slider movement? It produces a rotation of $R_y(10^\circ)\,R_x(15^\circ)$ on the base orientation. But this is not what we wanted! This rotation is equivalent to $R_y(10^\circ)\,R_x(5^\circ)\,R_x(10^\circ)$ on the base orientation, which differs from what we wanted $[R_x(5^\circ)\,R_y(10^\circ)\,R_x(10^\circ)]$. A careful comparison shows that two of the rotations have become swapped. And because rotations are not generally commutative, such a swap will only result in behaviour that differs from what we wanted. Also, the high processing speed of modern computers means that this wrong behaviour is so responsive to the sliders that it's all too easy to think that the object really is being rotated according to the *incremental* slider adjustments, as opposed to their latest positions.

It should be clear that the more we push the levers to and fro, the more the resulting rotations of the object will depart from what we require the programme to be doing. Although the mathematics and the computer software have done exactly what was asked of them, the programme does not match what we normally expect levers to do, especially after we have experimented with each lever individually. Using this approach to changing an object's orientation interactively is like adding up a restaurant bill using the modulo-12 system of clock arithmetic: while the mathematics is valid and correct, and is doing what it's asked to do with no singularities and no problems, the answer that we will get is completely unrelated to what we set out to compute. But the blame certainly does not lie with clock arithmetic.

If the programme has really been written not to rotate incrementally, but rather to read all three lever positions and apply a rotation about the base orientation (i.e., giving unintuitive results), then any confusion over the effects of the levers will probably reach its worst if now all three levers are used. The x-lever is set to an angle α, the y-lever is set to $-90°$, and the z-lever is set to γ; these can be set in any order because after each lever is moved, the programme rotates the *base* orientation by the latest values it reads from all of the lever settings. So when the third lever is moved, irrespective of which one it was, the programme rotates the *base* orientation by $R_z(\gamma)\, R_y(-90°)\, R_x(\alpha)$. However, because the following identity holds for all angles α, γ,

$$R_z(\gamma)\, R_y(-90°)\, R_x(\alpha) = \begin{bmatrix} 0 & -\sin(\alpha+\gamma) & -\cos(\alpha+\gamma) \\ 0 & \cos(\alpha+\gamma) & -\sin(\alpha+\gamma) \\ 1 & 0 & 0 \end{bmatrix}, \tag{4.106}$$

the total rotation—being a function only of the *sum* of the x- and z-lever angles—will not distinguish between the positions of the x- and z-levers. Thus, moving either of the x- and z-levers has the same effect, making it appear that we have lost one rotation axis somewhere, as if, to use an oft-quoted phrase, "the x-axis has been rotated onto the z-axis" (!) But the x-, y-, and z-axes are always perpendicular of course; they are not collapsible, and the idea that one has somehow been rotated onto another is nonsense. The software is doing exactly what it was designed to do, but unfortunately that design often matches neither our intuition nor what the programmer really intended in the first place.

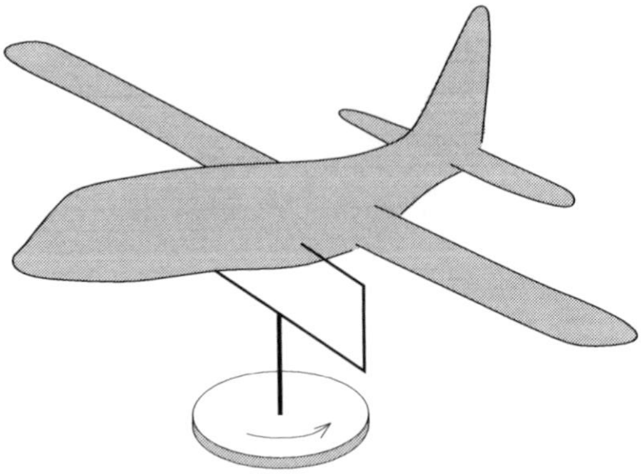

Fig. 4.7. A spinning flywheel forms a simple inertial navigation system, where the flywheel's axis defines the vertical direction used by the autopilot. From the attitude pictured, the aircraft can pitch (nose up/down) and yaw (nose left/right), but any attempt to roll (wing tips up/down) will force the flywheel out of its spin plane, causing the inertial navigation system to react strongly. This is a very schematic and simplified example of *gimbal lock*, which is sometimes wrongly thought to be a fault inherent in describing rotation about fixed axes. Rather, gimbal lock is purely a mechanical problem, albeit one that lends its name to numerical instabilities in some algorithms used to perform complex rotational calculations.

Additionally, because Euler angles were used incorrectly in performing what we thought would be *incremental* rotations, they routinely get blamed for what is purely a wrong use of rotation theory.

The apparent loss of one axial degree of freedom resulting from this procedure is often erroneously labelled *gimbal lock*, being confused with a real mechanical problem suffered by some gyroscopes that give directional stability to aircraft. In a real gyroscope, axes *are* set on hinges and *can* line up; this can sometimes act to remove a degree of freedom from the system that uses the gyroscope.

To picture why this might be, imagine a very simple (and not very useful) example of a system that helps an aircraft be stable: a horizontal spinning flywheel carried on a yoke attached to hinges at each side of the fuselage as in Fig. 4.7, and held underneath the aircraft as it flies straight and level. The flywheel's spin axis determines the up/down direction used by the autopilot to help keep the aircraft on course. The aircraft can pitch up and down and change its heading from straight and level without affecting the flywheel's motion. But if the human pilot tries to tilt one wingtip up and the other down, the flywheel will be forced out of its spin plane, with perhaps disastrous consequences as the autopilot tries to correct what it thinks is a serious instability. This is an oversimplification of a real system, because we have not provided a whole set of flywheels on movable axes ("gimbals") of which two *can* line up. But it certainly does encapsulate the problem that can happen with some gyroscopes when the aircraft's attitude reaches some extreme position.

> An everyday example of gimbal lock occurs when we tighten a nut with a universal joint ratchet spanner. These spanners incorporate two small hinges at right angles to each other, and are very useful for tightening nuts around corners, where the axis of the ratchet motion does not line up with the nut's rotation axis. But as the relative position of these two axes approaches a right angle, it becomes more difficult to turn the spanner; and when the axes are at a right angle, the mechanism locks up entirely.

Two Euler angles also describe how we sight a telescope with a standard "altazimuth" tripod. When the telescope is pointed nearly overhead, it becomes difficult to move easily because the closer we come to pointing it at the zenith, the more its azimuthal motion (motion about the vertical axis) constrains its tube to making smaller movements. This means that small movements around the zenith translate to large changes in the azimuthal angle, which can be difficult to achieve evenly. *Mechanically,* this constriction of the movement is also an example of gimbal lock.

If the telescope is tracking a fast-moving aircraft, rapid changes in the azimuth angle can be difficult to cope with in any numerical processing that goes with the tracking, so the term gimbal lock has also come to be applied to instabilities or difficulties in any numerical algorithms used. The problem is purely one of numerical stability when one of the angles changes quickly. As shown in Fig. 4.5, Euler angles are no different to using latitude and longitude for specifying locations on Earth. When flying near a pole, the longitude of an aircraft changes quickly. That requires attention to the numerics of the coordinates, but it does not render latitude–longitude a bad coordinate set for specifying locations on Earth.

So, gimbal lock is a mechanical problem inherent in some gyroscopes, and the term is also used to describe the numerical instabilities of specifying an orientation using Euler angles. But there is no such problem with Euler angles in software rotations such as described above if implemented properly, and it's wrong to blame gimbal lock and Euler angles for their misapplication in that way, or for natural numerical difficulties that form the bread and butter of numerical analysis. Gimbal lock's supposed appearance in the software rotations just described has become one of the great enduring myths of three-dimensional rotation theory.

Unfortunately, historically these misinterpretations went hand in hand with using matrices to perform rotations, with the result that matrices and Euler angles are sometimes seen as being intrinsically badly behaved in rotation theory. This is quite incorrect. Euler angles need care in numerical analysis when they change rapidly, but attention to rapidly changing numbers in numerical analysis across the whole of physics is normal and accepted.

Part of the ongoing confusion lies with the fact that comparisons are often made between quaternions and Euler angles in rotation literature. This is like comparing apples and oranges. Quaternions should be compared with *matrices*, since both can be used to rotate one vector about another. In contrast, an orientation is specified by how the body has been rotated from some base orientation. The rotation, or set of rotations, is a recipe for reconstructing the body's orientation from the base orientation. Some ways in which this can be done are by specifying the following quantities.

- The angle and axis of a single rotation from the base orientation (by Euler's theorem), carried out using either a quaternion or a matrix, and called the *angle–axis* representation. The angle and axis are trivially used to construct a quaternion $(\cos{}^{\theta}/_2,\ \boldsymbol{n}\sin{}^{\theta}/_2)$ for this rotation, so that it is sometimes called the "quaternion representation".
- Three rotations from the base position (Euler angles) around either space-fixed or carried-along axes. Each of these three rotations can be specified and performed using either a quaternion or a matrix.
- The three basis vectors in the final orientation, being the final images of the three initial basis vectors. Equations (4.2)–(4.5) tell us that a matrix comprised of these is the rotation matrix taking initial to final bases. It is unique, so encodes both the angle–axis representation and is the product of the Euler matrices.

So quaternions and matrices are tools for rotating a vector about another vector, whereas angle–axis numbers, Euler angles, and final basis vectors specify an orientation, which is a different, though related, concept.

This mixing of different concepts is one reason why rotation literature is notoriously difficult to follow. Books on rotation theory are full of recipes that work in practice, and each has its disciples. But the reasons why these recipes work do not always follow logically from the explanations those books provide. Being very clear in our own minds about what is being rotated is the essential ingredient to understanding rotation theory.

4.6.3 Quaternions Used in Computer Graphics

Computer graphics routines utilise rotations very heavily and commonly employ both matrices and quaternions for the job. One problem that all computer algorithms suffer from is the inevitable creeping in of numerical errors due to rounding. The effect this has on a varying rotation matrix $R_{\boldsymbol{n}}(\theta)$ is to make it slowly depart from being orthogonal, so that in practice a periodic correction for these rounding errors needs to be applied, which calls for extra computer processing. Rotation quaternions, which are normally more complicated to implement, have an advantage here because as their length slowly departs from one due to rounding errors, it can always easily be reset to one simply by dividing the four numbers of the quaternion by it, as calculated in (4.77).

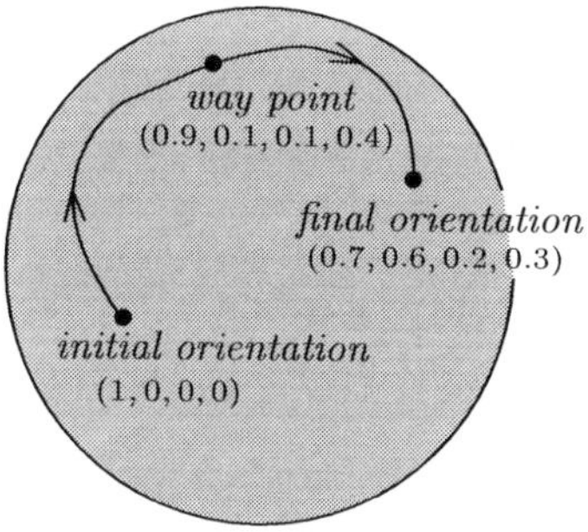

Fig. 4.8. Schematic representation of a sphere in four dimensions, each surface point of which represents a quaternion. Orientations of some system are represented by quaternions that rotate, say, the initial orientation. Given way-point orientations, the task is to find a path connecting them that is described by a set of intermediate quaternions, that can then be used to generate smoothly changing intermediate orientations of the system by always multiplying the *initial* orientation by each quaternion on the path. That is, incremental changes to the orientation of the system are *not* required here.

> Of course, in principle, matrices and quaternions can always be converted one to the other for the best of all worlds by calculating the rotation angle and axis vector, but in practice numerical inaccuracies can muddy the process. For example, in (4.50) we calculated the rotation angle by taking the trace of the rotation matrix, but the trace only makes use of the three diagonal entries. This is fine in a perfect world but is not guaranteed to give the best result when numerical inaccuracies are present.

An important use of rotation theory lies in generating a smooth fly-through of a scene by interpolation of camera orientations between specified key locations (known as way points). If a camera's orientation is specified at each way point by a rotation matrix or quaternion that has generated that orientation by rotating some base orientation, then the question arises how best to interpolate the matrices or quaternions, in order to generate rotations that act on the base orientation to give acceptable intermediate orientations of the camera. Matrices are not so easy to interpolate. In contrast, the unit length of quaternions allows them to be treated as points on the surface of a unit-radius "sphere" in four dimensions, as shown in Fig. 4.8. In this figure, the initial orientation is taken as the base orientation, in which case it must be represented by the quaternion $(1, 0, 0, 0)$ because this is the identity quaternion that does not rotate anything, as shown by applying (4.72):

$$(0, \boldsymbol{r}') = (1, 0, 0, 0)\ (0, \boldsymbol{r})\ (1, 0, 0, 0) = (0, \boldsymbol{r})\,. \tag{4.107}$$

We wish to visit the way-point orientation $(0.9, 0.1, 0.1, 0.4)$ and arrive at the final orientation $(0.7, 0.6, 0.2, 0.3)$. The very simplest way is to make a linear

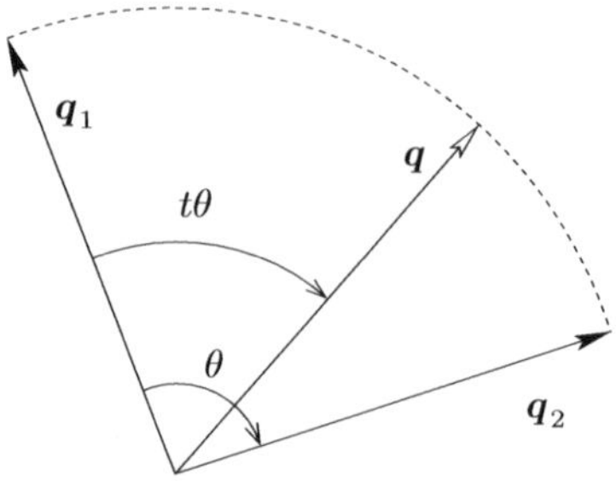

Fig. 4.9. The quaternion $\boldsymbol{q}$ is interpolated between quaternions $\boldsymbol{q}_1$ and $\boldsymbol{q}_2$ using (4.109).

interpolation in angle between way points on the sphere's surface, known as *spherical linear interpolation*, or SLERP. To see how SLERP is used to interpolate between two quaternions, draw them as vectors $\boldsymbol{q}_1$ and $\boldsymbol{q}_2$ lying in a plane in Fig. 4.9. The angle θ between them is given by the usual dot product in $\mathbb{E}^4$,

$$\cos\theta = \frac{\boldsymbol{q}_1 \cdot \boldsymbol{q}_2}{|\boldsymbol{q}_1|\,|\boldsymbol{q}_2|} = \boldsymbol{q}_1 \cdot \boldsymbol{q}_2\,. \tag{4.108}$$

Use a parameter t that runs from zero to one, values of which generate intermediate quaternions such as the $\boldsymbol{q}$ shown in Fig. 4.9. Straightforward two-dimensional vector analysis and a use of the sine rule results in

$$\boldsymbol{q} = \frac{\sin(1-t)\theta}{\sin\theta}\,\boldsymbol{q}_1 + \frac{\sin t\theta}{\sin\theta}\,\boldsymbol{q}_2\,. \tag{4.109}$$

This is the standard formula for generating intermediate orientations using SLERP. In Fig. 4.8 we would first interpolate for various values of t between the initial orientation and the way point, and then do the same for the way point and the final orientation. Note that the process is *not* incremental. That is, each quaternion generated does *not* multiply the orientation just produced; rather, it multiplies the initial orientation.

SLERP is popular in computer graphics programming, but there are certainly other ways of interpolating the quaternions, each of which gives a different flavour to the "fly-through". But that is a specialised subject, the domain of computer graphics programmers.

The subject of three-dimensional rotations is a very fruitful one. Its main entity is the rotation matrix (4.22), and with this we can rotate anything and describe any orientation. This is a good ability for a mathematical physicist to have because the calculations of physics sometimes demand rotations in space. At its heart, the subject is simpler than is often thought, but it's certainly very rich in its consequences.

5 Special Relativity and the Lorentz Transform

5.1 Deriving the Doppler Shift from an Invariance

Deriving the nonrelativistic Doppler shift for sound is a common elementary physics problem, where we follow waves from a possibly moving source on their way to a possibly moving observer, carefully accounting for the different path lengths followed by successive waves. The calculation changes depending on who is moving relative to the air that carries the sound—and perhaps the air is moving, too. The resulting equation has terms that depend on these relative velocities. What we learn from working through examples of Doppler shifts shows that observers and emitters measure different frequencies for the sound, and this manifests as the different pitches that characterise the Doppler shift.

There is another way of analysing the Doppler shift that uses a very simple fact. Suppose we could make the air visible so that all the ripples of the sound waves' compressions and rarefactions could be seen. In that case, we would notice that regardless of who is moving in what direction, *both emitter and observer must always agree on the wavelength of the sound*, because the wavelength is a physical property of the air. The distance between neighbouring ripples is a well-defined number, independent of who is moving where and with what velocity. Such a frame-independent quantity is called a *scalar*; it's more than just a number such as was used in previous chapters. This, then, is our first contact with the utility of knowing how a quantity transforms from one frame to another. The scalar is the simplest case of this—it does not change at all.

Knowing that the wavelength of the sound emitted is a scalar enables us to derive the Doppler shift in a simple way by realising that wavelength is just equal to wave speed divided by frequency.

> *Question: A car horn is built to sound a frequency of 400 Hz. What is the frequency received (heard) by a stationary observer whom the car moves directly toward at 30 m/s? Take the speed of sound to be 340 m/s with no wind blowing.*

Since both observer and emitter measure the same wavelength for the sound, we have

$$\frac{v_e \text{ (sound vel. in emitter frame)}}{f_e \text{ (freq. in emitter frame)}} = \frac{v_r \text{ (sound vel. in receiver frame)}}{f_r \text{ (freq. in receiver frame)}}, \quad (5.1)$$

so that

$$f_r = \frac{f_e v_r}{v_e}. \quad (5.2)$$

We know that $f_e = 400\,\text{Hz}$ and $v_r = 340\,\text{m/s}$ in the Earth frame, where the air is at rest. Since the emitter (the car horn) then feels a wind moving over it at 30 m/s, retarding the progress of its sound waves, we must have $v_e = 340 - 30\,\text{m/s}$. Finally,

$$f_r = \frac{400 \text{ Hz} \times 340 \text{ m/s}}{310 \text{ m/s}} = 439 \text{ Hz}. \quad (5.3)$$

This is a much simpler way of doing the standard calculation, and is based on the useful notion that the wavelength of sound waves is a scalar.

5.2 The Postulates of Special Relativity

Up until now, implicit in our discussion of the Doppler shift for sound has been the fact that in the Earth frame, it really does make a difference whether the source moves and the observer is at rest, whether the observer moves while the source is at rest, or whether they both move. The reason is due to the air that carries the sound waves. If the observer is at rest on Earth with no wind, then the speed of sound he measures will be independent of the source's velocity, purely because the sound is carried by the air, and the mechanism for that is independent of the motion of the source. But if the observer *moves* in the Earth frame, then he feels that air flowing past him and so measures a different speed for the sound waves that it carries.

Why is it that since its earliest days relativity theory has always been constructed using a discussion of light? Surely, couldn't it be constructed from the same sort of textbook arguments but applied to sound instead? In fact, it turns out that we would only be successful in doing so if we could be confident that the calculation above with the 400 Hz car horn was correct to the highest accuracy. But experiment tells us that it is *not* quite correct. It is certainly very accurate, but it turns out—with hindsight, and based purely on experiment—that, to be fully accurate, we are not allowed to just subtract the 30 m/s from the 340 m/s to obtain the sound velocity in the emitter frame. Intuition suggests that surely we can do this, but experiment proves that we cannot. If we knew just what calculation was needed to replace the subtraction of 30 m/s from 340 m/s, then we certainly *could* obtain the equations of relativity by thinking about sound waves. But it turns out that the necessary calculation for sound is completely unobvious, and it only becomes transparent if we consider light instead of sound because the relevant laws are simpler for light.

A textbook discussion of special relativity therefore generally begins by considering light because it has the wisdom of hindsight. And as every textbook shows, special relativity is built on a set of postulates. In the absence of gravity, what we find are experimental results that have given rise to these postulates. Now what exactly is a postulate? It's not necessarily an experimental observation. It is also not an axiom, which is a rule that a mathematical system is built upon and is correct by definition; an axiom *defines* the mathematical system. In a sense, a postulate is a kind of "physical axiom" in that the postulate is taken to be a foundation upon which to build a theory, while being something that may or may not later be proven correct. It is a statement that, given its truth, allows us to build a theory that can then be tested. Perhaps the postulates will need to be modified, but that doesn't matter, because it can certainly be done if necessary. The kinematics of special relativity are built upon three postulates. It is quite customary to give two at the outset, leaving the third for a more advanced discussion. In fact, the third postulate is very necessary in that it forms a stepping stone into general relativity, but we'll leave that for later. Additionally, the dynamics of relativity—the notions of energy and momentum—require additional postulated forms for momentum and the calculation of energy.

The first two postulates of special relativity deal with *inertial frames*. An inertial frame is, by definition, one in which Newton's laws hold without the need to introduce a gravity force. That is, the tell-tale sign that the frame *is* inertial is that a thrown ball moves away from us with a constant velocity. That means the everyday frame that each of us inhabits on Earth's surface is not inertial, although we can treat it as inertial by introducing a gravity force to account for "why" the ball curves as it flies through the air. But that complicates the issue, since we must then also ask how gravity affects all the other laws of physics—which is not an easy thing to do, and a subject we'll postpone until later. Begin, then, with the first two postulates of special relativity:

1. The laws of physics are the same in all inertial frames.
2. The speed of light is the same in all inertial frames.

On these is built the *Lorentz transform* that relates times and positions across inertial frames.

5.3 The Lorentz Transform

Most relativity texts derive the Lorentz transform by using the two postulates above to produce a set of equations that converts times and positions in one inertial frame to those of another with the same orientation, moving with constant velocity relative to the first. The procedure is straightforward, and we will not rederive it here. It begins by writing a generic set of equations that are linear in both sets of coordinates, since linearity ensures their

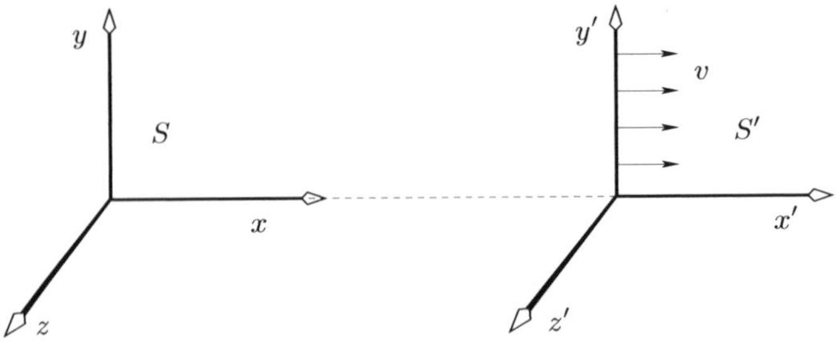

Fig. 5.1. The two-frame setup used in describing the basic Lorentz transform (5.5). The S'-frame moves with velocity v relative to S. All events are given spatial coordinates in either frame together with time coordinates t, t'.

invertibility—and by the first postulate, we know they should be invertible simply by changing the sign of the velocity wherever it appears. The unknown coefficients can then be fixed by considering just a few scenarios in each frame: the motion of a particle at rest in the other frame, a set of simultaneous events in one frame, and the motion of a light ray in one frame. These allow both postulates to be used, and the transform that results is the following. Suppose we have two inertial frames, S and S', as drawn in Fig. 5.1. An event is assigned coordinates (t, x, y, z) in S, and (t', x', y', z') in S'. The frames have their axes aligned so that, in the S-frame, S' moves with velocity v along the x-axis. With the definition

$$\gamma \equiv 1/\sqrt{1 - v^2/c^2}\,, \tag{5.4}$$

the Lorentz transform relates the two frames' coordinates by

$$\begin{aligned} t' &= \gamma\left(t - vx/c^2\right) + \text{constant}, \\ x' &= \gamma\left(x - vt\right) + \text{constant}, \\ y' &= y + \text{constant}, \\ z' &= z + \text{constant}. \end{aligned} \tag{5.5}$$

One part of the Lorentz transform that texts sometimes leave unexplained is the derivation of the y and z equations in (5.5). Why are lengths perpendicular to the direction of motion unchanged?

We can see why they should be unchanged by considering two hollow cylinders of slightly different diameters, each with its rotation axis along the coincident x-axis ($\equiv x'$-axis), with one tube sliding within the other and maintaining contact throughout. If relativity required a dilation in the radial direction (i.e., the y- and z-coordinates), then there would be a contradiction in the measurements made by observers at rest on one tube with those of observers at rest on the other. One set would predict a tightening and the other a loosening of the tubes. This cannot be, so we infer that there can be no dilation in the radial direction.

Usually, the frames are calibrated to coincide at $t = t' = 0$, so that the constants can all be set to zero. But, after all, when solving real-life problems in relativity using the Lorentz transform, there is no reason that clocks at the coincident origins should both read zero. It's an arbitrary choice that we might make of when time "starts", or when the clocks read zero. The same can be said for our choice of calibrating the space axes. To be more general, we should only consider *changes* in coordinates. This is completely normal; in everyday life we are only concerned with lengths of objects (changes in space coordinates) or the time it takes to do something (changes in the time coordinate). Specifying some absolute origin in space or some absolute start point in time is of no concern to us. Consequently, even if nonzero constants do appear in (5.5), we can always eliminate them by writing the Lorentz transform using increments in the coordinates:

$$\boxed{\begin{aligned} \Delta t' &= \gamma\left(\Delta t - v\,\Delta x/c^2\right), \\ \Delta x' &= \gamma(\Delta x - v\,\Delta t), \\ \Delta y' &= \Delta y, \quad \Delta z' = \Delta z. \end{aligned}} \tag{5.6}$$

This transformation can always be inverted by simply swapping the sign of v, which we'll do freely in the pages to follow.

5.3.1 Paradoxes or Conundrums?

Special relativity is famous for its so-called paradoxes, but the word paradox is misleading since it implies unresolved problems. Although there *are* everyday phenomena that depend for their understanding on the principles of relativity, such as the way a magnet works, life is lived at speeds that reduce the Lorentz transform to its nonrelativistic limit of $c \to \infty$ in (5.5), known as the *Galilei transform*. Our lack of experience with high speeds renders relativistic effects very unintuitive. But so as not to perpetuate the idea that there is something wrong with the theory, we will instead refer to these apparently illogical results as conundrums. They rest on three well-known consequences of the Lorentz transform. We will review the first two only briefly, concentrating instead on the third—probably the most important, yet the least discussed. Lorentz transform calculations, such as the following, always hinge on our specifying very carefully two events, 1 and 2, and we will always define the increase Δ to mean "2 minus 1"; e.g., $\Delta t \equiv t_2 - t_1$.

First: Time Dilation

An observer measures the period of a clock passing him at velocity v. If the clock has period T_0 in its rest frame (i.e. T_0 is the "factory setting", called the *proper period*), what period does the observer measure? We place the observer in S and the clock at rest in S'. Define two events: event 1 is a particular tick of the clock, and event 2 is the next tick. By definition, $T_0 \equiv t'_2 - t'_1 \equiv \Delta t'$. We require Δt:

$$\Delta t = \gamma \left(\Delta t' + v\Delta x'/c^2\right) = \gamma T_0 \,, \tag{5.7}$$

since $\Delta x' = 0$ (the clock is at rest in its own frame). So the S-observer measures the moving clock to be ageing slowly by a factor of γ.

Second: Length Contraction

An observer measures the length of a rod moving past him along its length at velocity v. If the rod has length L_0 in its rest frame (also called its *proper length*), what length does the observer measure? As before, the observer is in S, with the rod at rest in S'. Define two events: event 1 is a particular observation of the x-coordinate of the trailing end of the rod, and event 2 is the simultaneous observation of the x-coordinate of the leading end. So we require Δx, where $\Delta x' = L_0$ and $\Delta t = 0$:

$$\Delta x = \gamma \left(\Delta x' + v\Delta t'\right) . \tag{5.8}$$

Unfortunately, we don't know $\Delta t'$. But now try the analogous equation from the inverse transform:

$$\begin{aligned} \Delta x' &= \gamma \left(\Delta x - v\Delta t\right) , \\ \text{or } L_0 &= \gamma \Delta x \,, \end{aligned} \tag{5.9}$$

which yields $\Delta x = L_0/\gamma$. The rod's length is contracted.

These two effects are the first two surprising results of the Lorentz transform usually encountered in textbooks. But the effect considered next is by far the most dominant when it comes to resolving the famous conundrums of special relativity.

Last but Most Important: The Lack of Synchronicity

Clocks that are fixed along a rod and synchronised in the rod's rest frame will be measured to be out of synchronisation by an observer past whom the rod moves along its length.

To better illustrate this, describe the rod as a train that moves past us as we stand on the station platform (the S-frame). The train is at rest in the S'-frame and passes us by, moving along our x-axis in the direction of increasing x with velocity v. If a clock in the front of the train reads time zero, what does a clock in the rear of the train read? The clocks are synchronised as far as the train's passengers are concerned, but they turn out to be unsynchronised for us in S.

To show this, again define two events. Call the trailing clock on the last carriage "clock 1" and the driver's clock "clock 2". We require the difference in coordinates of two simultaneous events:

Event 1 = our observation of the coordinates of clock 1.
Event 2 = our observation of the coordinates of clock 2.

Note carefully that by "observation", we are *not* talking about sighting the clocks, and having to account for the noninfinite speed of the light signal that we receive from each clock. The observations that we make need not depend on such complications, and if they do, we must certainly account for things such as signal transit times in the calculations. Each observation could simply be the recording of a clock's coordinates by a device placed arbitrarily close to that clock, and we can read these coordinates later at our leisure.

Given $t'_2 \equiv 0$, we are required to find t'_1, or just $-\Delta t'$ (since $\Delta t' \equiv t'_2 - t'_1$). We dive right into the Lorentz transform to write

$$t'_1 = -\Delta t' = -\gamma\left(\Delta t - v\,\Delta x/c^2\right) = \gamma v\,\Delta x/c^2\,, \quad \text{since } \Delta t \equiv 0\,. \tag{5.10}$$

Now we need Δx. Remember that if the *proper distance* between the clocks (i.e. the distance as measured in their rest frame S') is L_0, then we must measure their distance to be contracted, so that $\Delta x = L_0/\gamma$, and finally

$$t'_1 = vL_0/c^2. \tag{5.11}$$

So the trailing clock *leads* in time by vL_0/c^2. In a manner of speaking, the trailing clock is always older than the driver's clock.

This calculation is correct but a little inelegant, since we had to remember that the train is contracted in the S-frame. An alternative approach begins with the *other* frame rather than that which our intuition might expect. Since we wish to find $\Delta t'$, begin with the expression for Δt:

$$\Delta t = \gamma\left(\Delta t' + v\,\Delta x'/c^2\right), \tag{5.12}$$

in which case

$$0 = \gamma\left(\Delta t' + vL_0/c^2\right), \tag{5.13}$$

so that $t'_1 = -\Delta t' = vL_0/c^2$ again. Not only is this approach shorter, but it eliminates the potential source of error if we forget about taking the length contraction into account. "Start with the *other* frame" is a good rule of thumb for applying the Lorentz transform.

The fact that trailing clocks lead in time by vL_0/c^2 is well worth remembering when dealing with exercises involving simultaneity, such as:

> *Two clocks are synchronised in their rest frame and are held 40 light-minutes apart. If they are viewed from a frame moving along their line of separation at 0.8c and the spatially leading clock reads 12:00 pm, what does the spatially trailing clock read?*

The trailing clock leads in time by an amount vL_0/c^2. A light-minute is the distance that light travels in one minute, or $c \times$ one minute. So we know that

$$v = 0.8\,c\,, \quad \text{and } L_0 = 40\ c \text{ minutes.} \tag{5.14}$$

Thus, the time by which the trailing clock leads is

$$\frac{0.8\ c \times 40\ c \text{ minutes}}{c^2} = 32 \text{ minutes.} \tag{5.15}$$

The trailing clock reads 12:32 pm. We'll use this example in the following discussion.

5.3.2 How Does Each Frame Measure the Other as Ageing Slowly?

Probably the earliest and most famous conundrum of special relativity encountered by the student is the question of how it is possible that two frames can each measure the other as ageing slowly. Along with the Twin Conundrum, which we'll meet in Chap. 7, the notion that both frames surely cannot measure each other as ageing more slowly has led to many attempts over the last century to show that relativity just cannot be correct. Actually, while time dilation and length contraction both play a role in explaining the conundrum, by far the most important cause is the loss of synchronicity of clocks in motion despite these clocks being held synchronised in their own shared rest frame.

The following discussion uses Fig. 5.2. Consider two pairs of clocks mounted on the ends of two *identical* rods. The rods ensure that each clock pair is held rigidly separated by 40 light-minutes in its rest frame. Each pair of clocks is synchronised in its own rest frame. The rods are passing each other at a speed of 0.8 c (so $\gamma = 5/3$), and one of them has a jagged edge to distinguish it in our discussion. We will analyse the situation from the jagged rod's frame, and this will suffice to explain the conundrum.

First the straight rod begins to pass the jagged rod at a "jagged time" of 0. Jagged observers measure the straight clocks to be out of synchronisation, with the trailing straight clock reading a time of 32 minutes, as was calculated in (5.14) and (5.15). Since the straight rod is Lorentz-contracted in the jagged frame, the left-hand clocks line up *before* the right-hand clocks do. The straight rod's length is $40/\gamma = 24$ light-minutes, so it takes a jagged time of $24/0.8 = 30$ jagged minutes for the left ends to align, during which time each clock on the straight rod has aged by $30/\gamma = 18$ minutes.

Finally, the right-hand clocks line up. The jagged time for the top rod to traverse the bottom rod is 40 c minutes/(0.8 c) = 50 minutes, so that the jagged clocks have aged 50 minutes in total. In contrast, the straight clocks aged just $50/\gamma = 30$ minutes.

So far, there is no problem with this scenario, since we have used time dilation to ensure only that the jagged clocks measure the straight ones as ageing slowly. But how are we to conclude that the *straight* clocks measure the *jagged* ones as ageing slowly? The analysis of the straight observer must

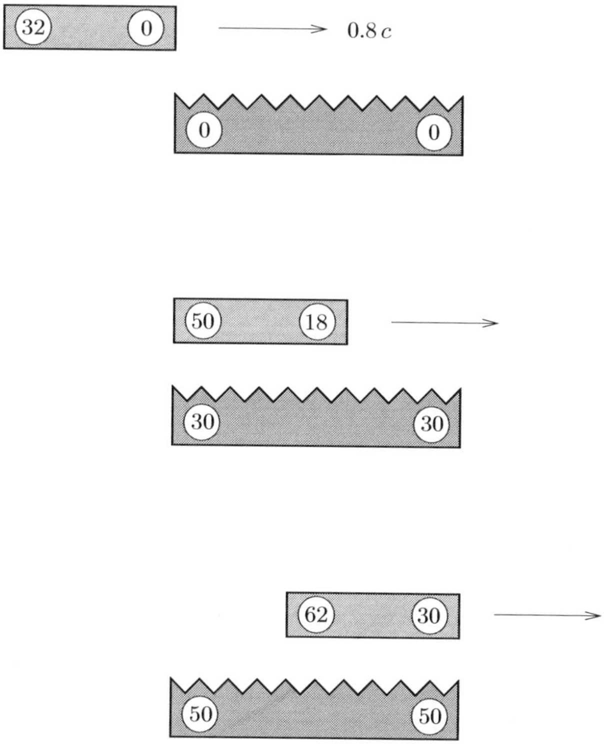

Fig. 5.2. Clocks used in Sect. 5.3.2. The three time steps (top to bottom) are drawn in the "jagged frame", in which the jagged rod is at rest. The straight rod moves from left to right past the jagged rod. This figure ties together the three phenomena of length contraction, time dilation, and loss of simultaneity and shows how they work together to give a consistent picture across the two frames.

be identical to that of the jagged observer, of course, so let's assume a straight observer's role and follow the progress of the left-hand jagged clock. As the top picture of Fig. 5.2 shows, the straight observer measures the left-hand jagged clock to begin to pass by when both the coincident clocks read zero.

We are using here the fact that two events that are virtually coincident in space and time—such as the close passage of the straight and jagged clocks both reading zero—are agreed by *everyone* to be simultaneous. After all, we could arrange for each event to leave a mark on a passing (third) clock. In such a case, all observers *would have to agree* that the two events left their marks simultaneously, because the evidence is right there on the third clock.

Now remember that the straight ruler's clocks, while not synchronised in the jagged frame of Fig. 5.2, *are* synchronised in the straight frame. Then, as shown in the middle picture of Fig. 5.2, the left-hand jagged clock finishes its transit of the straight ruler when it reads 30 (jagged) versus the close-by straight clock's 50. Since these clocks are adjacent, there is no complication introduced by light-travel times; so because the straight clocks are synchronised *in the straight frame*, the straight observer concludes that the left-hand jagged clock aged 30 minutes during 50 straight minutes. The straight observer does indeed conclude that the jagged clocks are ageing slowly—by the same factor, $\gamma = 5/3$.

The way we resolved the conundrum was to enforce two points. The first was that in order to determine how moving clocks are ageing, we must ensure that the time reading on *the same clock* is measured, no matter which clock it is. The jagged observer chose to concentrate on the right-hand straight clock for this, while the straight observer concentrated on the left-hand jagged clock. There is nothing special about these clocks, of course. The second point we enforced was that for an observer to make pronouncements easily about a moving clock, he may wish to avoid the complication of having to account for light-travel times. For this very reason, we only considered adjacent clocks in the analysis above.

This analysis highlights the need to make a distinction between *seeing*—the stuff of optical illusions, which does not take into account light-travel times—and *measuring* or *observing*, which certainly does account for light-travel times. In special relativity, the standard depiction of a frame is a lattice of clocks held together by rigid rods, where the clocks are used to note the time of events happening in their vicinity. In fact there is really a continuum of clocks, so that an alternative way of picturing a frame populates it with a continuum of observers, each of whom is at rest relative to all the others. Each of them makes notes carefully indexed by the time on their clock. This time is global for the frame, but the observers do not look at anyone else's clock since that involves the needless hard work of correcting for light-travel times. So to eliminate the need for these time corrections, each observer only records what happens right next to him and sends his set of notes to a "headquarters" of the frame, where everything is collated and the frame's global view of events is reconstructed. This is a very fruitful idea, which we'll meet again in Chap. 7. There it will be very important in helping us to make sense of life in an accelerated frame, forming a bridge to the ideas of general relativity and gauge theory.

An easy way to understand the distinction between seeing and measuring is through the familiar red- and blueshifts. We *see* a receding clock ageing slowly by the redshift factor of $\gamma(1 - v/c)$ (where $v > 0$) when it recedes, and likewise we *see* it ageing quickly by this same factor ($v < 0$) when it approaches. But if we account carefully for light-travel times, in both cases we *measure* the clock to be ageing slowly by a factor of γ.

Rescuing Our Intuition

Our intuitive idea that the frames should be asymmetric with regard to how they measure each other as ageing does have some correspondence with what each *sees*. Consider the left-hand jagged observer again in Fig. 5.2 (i.e. the one who stands next to the left-hand jagged clock), who notes that as the jagged clocks show 0, the straight clock opposite him also shows 0. The *right-hand* jagged observer then notes that 50 jagged minutes later, this same straight clock shows 30 minutes, corresponding to the adage of "moving clocks run slow". This is a good measurement, since the two jagged observers are synchronised, and they are seeing only what a specific straight clock reads and only when it's adjacent to each of them.

Contrast this with an altogether different scenario. Again, the left-hand jagged observer notes the jagged time to be 0 when the straight clock opposite him shows 0. He then closes his eyes for 30 jagged minutes. Now he opens his eyes and looks across to the straight clock adjacent to him, noting that it now reads 50 minutes. He concludes that the moving clocks are actually ageing *quickly* by a factor of γ! But his mistake lies in comparing apples with oranges: he is comparing the time on the straight right-hand clock (0) with the time on the straight left-hand clock (50) and concluding that 50 straight minutes have elapsed. He is comparing two different clocks that, in his jagged frame, were never synchronised! What he should have realised was that the trailing straight clock always led by 32 minutes, so that it only aged 18 minutes when his clock aged 30 minutes. The leading straight clock also aged just 18 minutes—and this ratio 30/18 is just the expected γ slowing factor. So, in a sense, the moving frame appears to age in fast motion by a factor of γ, and it's only by ensuring that we keep track of the same clock that we discover that the real ageing is slow, not fast, by the same factor.

5.4 The Symmetry of the Lorentz Transform

Conventionally in relativity, factors of c are absorbed into time and velocity, because time always appears multiplied by c while velocity always appears divided by c. We will do the same here. The t and v that we have been using up until now have had conventional units and so will be renamed $t_{\text{conv}}, v_{\text{conv}}$. Their new definitions are

$$\begin{aligned} t &\equiv c\, t_{\text{conv}} && \text{(units of distance),} \\ v &\equiv v_{\text{conv}}/c && \text{(no units).} \end{aligned} \tag{5.16}$$

It is as if c has been set to equal 1, but that is not the real reason for writing (5.16). The change is not made just to avoid writing repetitive factors of c, and neither is it made because c is in some sense a "big" number. Rather, absorbing c reflects a fundamental symmetry in the Lorentz transform. With the redefinitions above, the Lorentz transform (5.5) and (5.6) becomes

$$\begin{aligned} t' &= \gamma(t - vx) + \text{constant}, \\ x' &= \gamma(x - vt) + \text{constant}, \\ y' &= y + \text{constant}, \quad z' = z + \text{constant}. \end{aligned} \tag{5.17}$$

$$\begin{aligned} \Delta t' &= \gamma(\Delta t - v\,\Delta x)\,, \\ \Delta x' &= \gamma(\Delta x - v\,\Delta t)\,, \\ \Delta y' &= \Delta y\,, \quad \Delta z' = \Delta z\,. \end{aligned} \tag{5.18}$$

There is now a complete symmetry between the space coordinate x whose axis S' moves along, and the time coordinate t.

This symmetry between time and space also appears in the ideas of simultaneity that we have been studying. For example, consider the following statement familiar from everyday life:[1]

> *Two events that occur at the same place at different times for one observer occur at different places for another observer moving relative to the first.*

This accords with our everyday experience, but it can also be seen by applying the Lorentz transform: if S observes the events with $\Delta x = 0$, $\Delta t \neq 0$, then (5.18) shows that S' observes $\Delta x' \neq 0$ (as well as $\Delta t' \neq 0$).

But now swap "time" and "place" in the above statement, as suggested by the symmetry in the Lorentz transform:

> *Two events that occur at the same time at different places for one observer occur at different times for another observer moving relative to the first.*

This is proved by again applying (5.18). S observes the events with $\Delta t = 0$, $\Delta x \neq 0$, while S' observes $\Delta t' \neq 0$ (as well as $\Delta x' \neq 0$). Though quite unobvious, this statement is also true and embodies the very important idea of the lack of simultaneity between two frames.

The Lorentz Transform Using a Complex Rotation

An alternative view of the Lorentz transform is based around the notion of quaternions. Defining $\theta \equiv \tanh^{-1} v$ allows (5.18) to be written as

$$\begin{bmatrix} \mathrm{i}\,\Delta t' \\ \Delta x' \\ \Delta y' \\ \Delta z' \end{bmatrix} = \begin{bmatrix} \cos \mathrm{i}\theta & -\sin \mathrm{i}\theta & 0 & 0 \\ \sin \mathrm{i}\theta & \cos \mathrm{i}\theta & 0 & 0 \\ 0 & 0 & 1 & 0 \\ 0 & 0 & 0 & 1 \end{bmatrix} \begin{bmatrix} \mathrm{i}\,\Delta t \\ \Delta x \\ \Delta y \\ \Delta z \end{bmatrix}. \tag{5.19}$$

The 4×4 matrix applies a rotation through a complex angle—whatever that might actually mean. The rotation can be written in terms of quaternions by

[1] This example of space–time symmetry is paraphrased from *Differential Geometry and Relativity Theory* by R.L. Faber (1983, Marcel Dekker). Reproduced by permission of Routledge/Taylor & Francis Group, LLC.

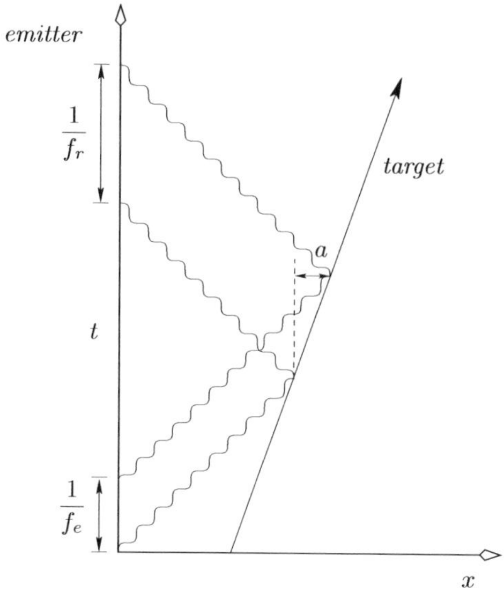

Fig. 5.3. Using radar in the derivation of the Lorentz γ-factor. Two pulses of light are shown, each of which bounces off a possibly moving target to return to the emitter.

the theory of Chap. 4. This idea of using quaternions is commonly known as the *spinor approach* to the Lorentz transform. It is a topic addressed in the study of geometric algebra, although we will not pursue it further here.

5.5 Using Radar to Derive Time Dilation

Absorbing factors of c into the Lorentz transform allows for a geometrisation of space and time. To see this, consider the following scenario. We are aboard a possibly moving radar carrier and wish to locate an object by bouncing radio waves from it and measuring the time for their return trip. But perhaps our target is moving, and if so, then Doppler information can be obtained and we can calculate its velocity. So let's disregard range information and instead concentrate on the target's velocity, asking for the Doppler shift in frequency sent out versus that received after the radio waves have bounced off the target.

To derive the relationship between target velocity and emitted radar frequency, we draw a time versus distance diagram. For no particular reason, in relativity these diagrams are usually drawn with the time axis pointing up and the space axis to the right, as in Fig. 5.3. This diagram shows our frame with the emitter moving radially away from us—only one space dimension need be drawn.

These time–distance diagrams are conventionally calibrated so that a ray of light makes a track, a *worldline*, at 45° to each axis. So if one unit of (conventional) time is a year, then one unit of time as redefined in (5.16) will be a light-year, and one unit of the space axis can be conveniently chosen to be a light-year also. Plotting this redefined time ensures that both axes have units of distance, and thus the figure becomes purely geometrical. This is a very powerful way to visualise relativity calculations.

Suppose that the target recedes radially with velocity $v > 0$. (In the approach case of $v < 0$, the diagram changes a little, but otherwise the calculation is almost unchanged and we will not consider it.) We send out pulses of radio waves at a *pulse repetition frequency* f_e, and receive them Doppler-shifted at a new pulse repetition frequency f_r. Consider two of these pulses. Since the time between pulses is the reciprocal of the pulse repetition frequency, we can immediately write down the times between emission and reception of the pulses in Fig. 5.3.

Now work closely with the geometry of the figure, in particular noting that constant velocities mean straight lines, and all light rays make 45° angles with the axes. In that case, the received interpulse period $1/f_r$ relates to the emitted interpulse period $1/f_e$ by the amount a that the target moves away in the interval of time between its reception of the two pulses:

$$\frac{1}{f_r} = \frac{1}{f_e} + 2a\,. \tag{5.20}$$

But that amount a is also just the target speed multiplied by the time between successive receptions:

$$a = v\left(\frac{1}{f_e} + a\right). \tag{5.21}$$

Eliminating a from these two equations yields

$$f_r = f_e\,\frac{1-v}{1+v}\,. \tag{5.22}$$

This is the required Doppler shift, so knowing the emitted and received frequencies tells us the velocity of the target.

Now something important emerges. This scenario has been a two-way affair: we emitted signals, they bounced off the target, and then they returned to us. We should be able to solve the problem in two stages by considering first the emission and then the reception of the pulses. By the postulates of relativity, the two stages should be entirely symmetrical. If the emitted frequency has been multiplied by a factor of $(1-v)/(1+v)$ by the time we get the pulses back again, then it must have been multiplied by $\sqrt{(1-v)/(1+v)}$ on each leg, outbound and inbound. In other words, the target must *see* the frequency of our emitted radio waves multiplied by a factor of $\sqrt{(1-v)/(1+v)}$.

But this implies something remarkable: that we measure a moving target's clock to be timing (ageing) slowly. To see why this should be so, refer to Fig. 5.4. This figure shows a one-way transmission of two light pulses from us to the target, which moves at velocity v as before.

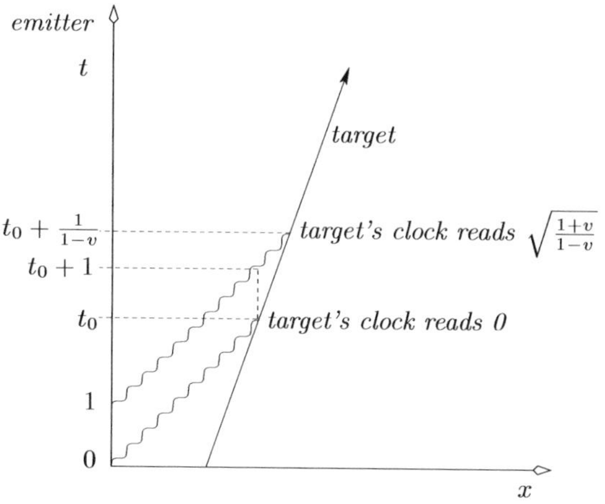

Fig. 5.4. Deriving the γ-factor by following the passage of two pulses of light, keeping careful track of the readings on the emitter and target clocks.

Suppose we send out the pulses at one-second intervals. The target sees these at time intervals of the reciprocal of the frequency $\sqrt{(1-v)/(1+v)}$ above, or $\sqrt{(1+v)/(1-v)}$. Now, given that the target's clock shows this time interval from one pulse to the next as marked in Fig. 5.4—and making no assumption that the target's clock runs at the same rate as our own—we ask what time interval we measure between the target's times zero and $\sqrt{(1+v)/(1-v)}$. From the geometry of the figure, it must be $1/(1-v)$ since this gives the correct slope of $1/v$ for the target's worldline. So when the target's clock ages $\sqrt{(1+v)/(1-v)}$ seconds, our clock ages $1/(1-v)$ seconds. The ratio of these is just γ:

$$\frac{\text{our clock rate}}{\text{target clock rate}} = \frac{1}{1-v} \bigg/ \sqrt{\frac{1+v}{1-v}} = \frac{1}{\sqrt{1-v^2}} \,. \tag{5.23}$$

That is,

$$\text{our clock rate} = \gamma \times \text{target clock rate}. \tag{5.24}$$

The radar approach used here to derive the γ-factor was first introduced by Hermann Bondi under the name of *k-calculus*. It can be extended to derive the entire Lorentz transform, but that is not our aim here. We have only wished to show the idea of using the geometry of a space–time diagram to facilitate calculations involving light signals.

It should be noted that although this analysis must also apply to the rate at which the target measures its clock to be running relative to ours, the target cannot glibly say that for it this ratio is $\sqrt{(1+v)/(1-v)}$: 1 as opposed to γ : 1, because this time interval $\sqrt{(1+v)/(1-v)}$ marks what the target *sees*, which is complicated by the fact that the light signals it receives are Doppler-shifted. We need to remember the distinction between the target's *seeing* our clock age one second while it ages $\sqrt{(1+v)/(1-v)}$ seconds, as shown in Fig. 5.4, and the target's *measuring* or *observing* our clock to age one second while it ages $\gamma \neq \sqrt{(1+v)/(1-v)}$ seconds. This distinction is not evident from the figure but must result from the analysis above from our point of view, coupled with the principle of relativity that says the target's viewpoint is just as good as our own. This point was alluded to previously on p. 194.

The analysis above makes it evident why the mathematics of relativity, such as the γ-factor and the Lorentz transform, is always derived using light signals. As discussed earlier, if we did try to use sound signals, we would hit some snags along the way. First, since sound is carried by air, we would be unable to make use of the symmetry above between the target's view of the world and our own to derive the square root in the discussion following (5.22), since there would be no such symmetry. We and the target would measure different speeds for sound, since we cannot both be at rest in the air.

The second problem would arise if we decided to be very careful and take into account the fact that the air that carries the sound waves must be moving in at least one of the frames, ours or the target's. At some point, we would need to add the velocity of the sound waves in the air to the velocity of the air itself (with respect to, say, the target). If we just added these two velocities as vectors in the usual (pre-relativity) way, then the result would disagree with the analysis above that uses light waves; we then could only surmise that something somewhere was wrong. But since experiment verifies the result derived through using light signals, the conclusion would have to be that velocities cannot be added with the usual vector addition in special relativity. And indeed the set of correct equations for the relativistic addition of velocities is one of the triumphs of the Lorentz transform. Equation (5.18) gives

$$\frac{\mathrm{d}x'}{\mathrm{d}t'} = \frac{\gamma\,(\mathrm{d}x - v\,\mathrm{d}t)}{\gamma\,(\mathrm{d}t - v\,\mathrm{d}x)} = \frac{\mathrm{d}x/\mathrm{d}t - v}{1 - v\,\mathrm{d}x/\mathrm{d}t}\,. \tag{5.25}$$

Contrast this with the nonrelativistic result of the Galilei transform, which is $\mathrm{d}x'/\mathrm{d}t' = \mathrm{d}x/\mathrm{d}t - v$.

5.6 Space–Time Becomes Spacetime

In normal euclidean geometry, the distance between two points is independent of the coordinate system used to calculate it. Two cartesian systems that apply Pythagoras's theorem to the coordinate differences will agree on the points' separation $\Delta\ell$, even though they might disagree on the coordinates ascribed to the points. Thus,

$$\Delta\ell^2 = \Delta x^2 + \Delta y^2 + \Delta z^2 = \Delta x'^2 + \Delta y'^2 + \Delta z'^2\,. \tag{5.26}$$

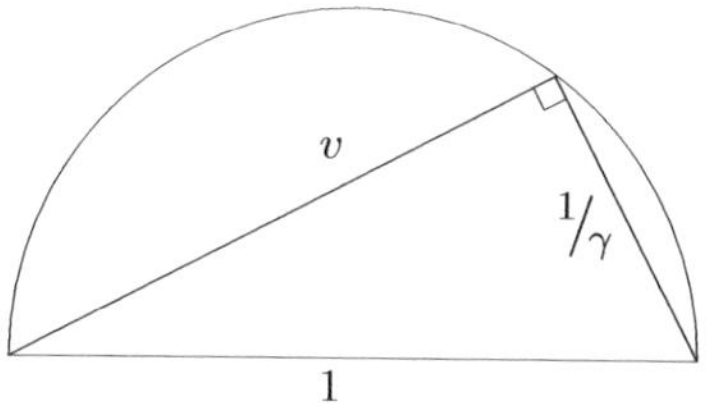

Fig. 5.5. Here is a geometrical way to view the γ-factor. Because $v^2 + 1/\gamma^2 = 1$, it follows that if we draw a right-angled triangle of unit hypotenuse length, then if one side has length v, the other will have length $1/\gamma$. Since any triangle drawn inside a semicircle must always be right angled, we have here a good way of visualising how γ changes as the speed of an object approaches that of light (i.e. $v \to 1$). As v approaches one, even though it changes less and less (the "v-arm" in the figure is mostly just rotating and only slightly increasing its length toward one), the length of the other arm, $1/\gamma$, is changing dramatically. So even a small change in v translates to a very big change in γ, and finally γ increases without limit as $v \to 1$.

In other "curvilinear" systems where the basis vectors depend upon position, such as polar coordinates, no such simple relation for any two points holds. However, we *can* always find such a relation if we consider the distance between two infinitesimally separated points, in which case, for example,

$$d\ell^2 = dx^2 + dy^2 + dz^2 = dr^2 + r^2 d\theta^2 + r^2 \sin^2\theta \, d\phi^2 . \tag{5.27}$$

This invariant quantity is called the *line element* for the coordinate system used, while the coefficients of the infinitesimals are called the *metric*. These two terms tend to be used interchangeably. We first met the idea of a metric when setting out to plot the quadratic form on p. 38. The line element has an analogue in special relativity, where, owing to length contraction, the distance between two events is definitely not frame-independent. What *is* frame-independent is the following expression, again called the metric or *interval*, and easily verified by inspection of (5.18) (or also more generally the Lorentz transform (5.45) in an arbitrary direction, as will be derived in Sect. 5.8):

$$\Delta t^2 - \Delta x^2 - \Delta y^2 - \Delta z^2 = \Delta t'^2 - \Delta x'^2 - \Delta y'^2 - \Delta z'^2 . \tag{5.28}$$

In preparation for dealing with different coordinate systems, as in (5.27), we will write all expressions for the metric using infinitesimals:

$$dt^2 - dx^2 - dy^2 - dz^2 = dt'^2 - dx'^2 - dy'^2 - dz'^2 . \tag{5.29}$$

In particular, the interval is related to the *proper time* $d\tau$, the time that elapses on the unique inertial clock that is present at both events. Because the inertial clock measures the events' spatial separation to be zero, that separation contributes nothing to the interval, in which case this clock measures the interval to be just $d\tau^2$. Since all frames agree on the interval, we have

$$\boxed{\mathrm{d}\tau^2 = \mathrm{d}t^2 - \mathrm{d}x^2 - \mathrm{d}y^2 - \mathrm{d}z^2\,.} \tag{5.30}$$

So the interval is just the square of the proper time, and the amount of ageing of an inertial clock visiting each of the two events becomes the analogue, for those events, of the notion of distance for two points in standard geometry. Also, in a frame in which two events are simultaneous, the interval between them is

$$\mathrm{d}\tau^2 = -\mathrm{d}x^2 - \mathrm{d}y^2 - \mathrm{d}z^2 \equiv -\mathrm{d}\ell^2, \tag{5.31}$$

where $\mathrm{d}\ell$ is the *proper distance* between the two events. The sequence of signs in the metric $(+---)$ is loosely referred to as its *signature* (being very closely related to an analogous use of the term in linear algebra). We are of course free to multiply the interval by any constant; -1 is frequently chosen, which flips the signs of its signature but unfortunately has the very confusing effect of reversing some—but not all!—signs in all manner of equations. No one standard sign convention is preferred by the mathematical physics community.

Because all inertial observers agree on the value of the interval between any two events, we can think of the three space axes together with the time axis as the axes of a fundamental new entity called *spacetime*. The metric quantifies the geometry of this spacetime, and for the type of spacetime we are considering, (5.30) is called the *Minkowski metric*. So it is that spacetime is *not* just the depiction of time and space axes that has always been such a fundamental tool for describing kinematics—for how projectiles move, and so on. Ideologically, the interval is what separates the mere "space–time" plot of Fig. 5.3 from the unified entity spacetime, which of course is graphed using the same axes.

In the next chapter we'll see how this invariance of the interval allows the spatial vectors of three-dimensional euclidean geometry to be generalised to spacetime vectors in four dimensions.

5.7 Spacetime Diagrams and Hyperbolic Geometry

Given a spacetime diagram for the S-frame with its t- and x-axes, how do we go about plotting the S'-frame's axes on the same diagram?

Consider just the space coordinate x' for ease of drawing the diagrams—the other space coordinates are unaffected by the Lorentz transform. For simplicity, set the frames to share a common origin:

$$(t, x) = (0, 0) \Longleftrightarrow (t', x') = (0, 0)\,. \tag{5.32}$$

In that case the Lorentz transform (5.5) becomes

$$\begin{aligned} t' &= \gamma(t - vx)\,, \\ x' &= \gamma(x - vt)\,. \end{aligned} \tag{5.33}$$

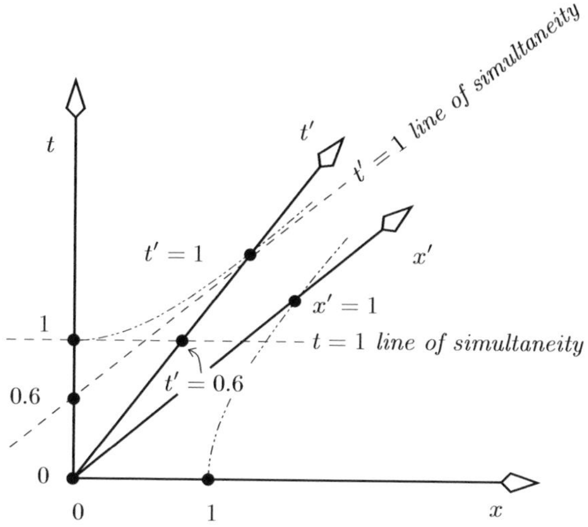

Fig. 5.6. Calibrating the t'- and x'-axes. The length of a one-second "tick" on the x-axis cannot simply be transferred to the x'-axis. The key to calibrating the primed axes correctly is to make use of the invariance of the interval, as in (5.34).

The primed axes are constructed in the following way:

The t'-axis is the set of all events that have $x' = 0$. Thus S draws it as the line $t = x/v$, which is a line through the origin of the spacetime diagram with slope $1/v$. This is just the worldline of the S'-origin. *Any time axis on a spacetime diagram is the worldline of an observer with that time coordinate.*

The x'-axis is the set of all events with $t' = 0$, so S draws it as the line $t = vx$, which is a line through the origin of the spacetime diagram with slope v.

These primed axes, of slopes $1/v$ and v, are "orthogonal", by which is meant that they both make equal angles to the line at 45° to the t- and x-axes. The unprimed axes are orthogonal, too; all pairs of spacetime axes must be. This orthogonality ensures that, in the S-frame, the worldline of a photon will bisect the angle between the t'- and x'-axes, just as it does for the t- and x-axes. This demonstrates and ensures that, like S, the observers of S' will also measure light to have unit speed (i.e. c).

At any one time, each observer connects all events that are simultaneous by a line parallel to their space axis; for example, the set of events for which $t' = 0$ is the x'-axis. This is an important device for making sense of spacetime diagrams and for unravelling the typical conundrums that arise in relativity. An important device is the technique that different observers use to calibrate their axes, shown in Fig. 5.6. There, the S'-frame is moving to the right in S with velocity $v = 0.8$. Because the S'-clocks run slowly in S, the "tick" interval on the x'-axis must be longer than that of the x-axis. We can calibrate the x'-axis graphically as follows. The invariance of the interval between the origin and any point of interest gives

$$t'^2 - x'^2 = t^2 - x^2 . \tag{5.34}$$

Short pieces of the hyperbolae $t^2 - x^2 = \pm 1$ are drawn in Fig. 5.6. The $t^2 - x^2 = +1$ hyperbola must cut the t-axis at $t = 1$ since at that point $x = 0$. But, for the same reason, it must also cut the t'-axis at $t' = 1$ since that hyperbola is also the set of all events for which $t'^2 - x'^2 = 1$. This calibrates the t'-axis. The same idea using $t^2 - x^2 = -1$ serves to calibrate the x'-axis, although it's easier to borrow the t'-axis tick length and simply transfer it to the x'-axis.

The immediate consequence of these differing calibrations is that each frame measures the other to be ageing slowly. To see why this should be, note that the S line of simultaneity for $t = 1$ intersects the t'-axis at $t' = 0.6$, while the S' line of simultaneity for $t' = 1$ intersects the t-axis at $t = 0.6$. This geometric view has a simplicity and symmetry that are not so evident when the same scenario is described using trains passing each other, as was drawn in Fig. 5.2.

If a rule of the form (note the plus sign) $\Delta t^2 + \Delta x^2 = constant$ were obeyed by all of the events on spacetime diagrams, spacetime geometry would simply be euclidean. But with the invariant interval $\Delta t^2 - \Delta x^2 = constant$, the geometry is instead called hyperbolic or *lorentzian*. In the next chapter, we'll meet hyperbolic geometry again in the guise of the *energy–momentum* diagram. Spacetime diagrams with lines of simultaneity will be used in much more detail in Chap. 7, where the ideas of inertial frames, the interval, and simultaneity will be extended to serve as the stepping stone to the more advanced ideas of general relativity.

5.8 The Lorentz Transform in an Arbitrary Direction

In the next chapter, we use the Lorentz transform to analyse a scenario involving three inertial frames. To make life easy there, we have chosen all motions at right angles to each other, so that by suitable cyclic permutations of the axis labelling, we can still apply (5.18). But frames' relative velocities are seldom so contrived. How can we extend the Lorentz transform to arbitrary directions without having to rederive it each time?

Do it as follows. Suppose that at $t = t' = 0$ the axes line up on top of each other. The S'-frame is moving with velocity $\boldsymbol{v}$ in S,

$$\boldsymbol{v} \equiv (v^1, v^2, v^3) \equiv v\, \boldsymbol{n} , \tag{5.35}$$

where $\boldsymbol{n}$ is a unit vector with components (n^1, n^2, n^3). We need not pay attention to the sign of v; the following discussion and equations are unchanged if $v, \boldsymbol{n} \to -v, -\boldsymbol{n}$.

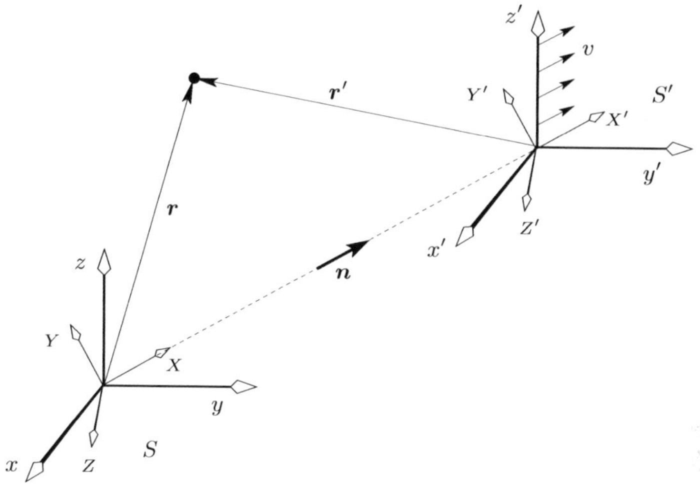

Fig. 5.7. Deriving the general Lorentz transform for the case where S' moves in an arbitrary direction $\boldsymbol{n}$ in S. Two sets of intermediate axes are constructed whose X- and X'-axes coincide, to which (5.17) will certainly apply. These are then related to the xyz- and $x'y'z'$-axes by simple geometry.

Define a new S-frame axis parallel to $\boldsymbol{v}$ along which the S'-frame moves, which allows us to make use of (5.17). We'll call this axis X, with the corresponding axis X' in S', so that X and X' are always collinear. By supplementing these with other axes T, Y, Z to form a right-handed spatial coordinate system, together with the analogous set T', Y', Z', we know that (5.17) will relate T, X, Y, Z to T', X', Y', Z'. So it is sufficient to specify how T, X, Y, Z relate to t, x, y, z.

First, $T \equiv t$ since we're only rotating the space axes. Now put $\boldsymbol{r} \equiv (x, y, z)$, and use the notation $(\boldsymbol{a}, \boldsymbol{b}) \equiv$ the angle between vectors $\boldsymbol{a}$ and $\boldsymbol{b}$.

$$
\begin{aligned}
&\text{Then } X = \boldsymbol{r}\cdot\boldsymbol{n}\,, \quad \text{while } \sqrt{Y^2+Z^2} = |\boldsymbol{r}|\sin(\boldsymbol{r},\boldsymbol{n}) = |\boldsymbol{r}\times\boldsymbol{n}|\,. \\
&\text{Also } X' = \boldsymbol{r}'\cdot\boldsymbol{n}\,, \quad \text{and } \sqrt{Y'^2+Z'^2} = |\boldsymbol{r}'\times\boldsymbol{n}|\,.
\end{aligned}
\tag{5.36}
$$

The new axes obey the usual Lorentz transform:

$$X' = \gamma(X - vT) = \gamma(X - vt)\,, \tag{5.37}$$

$$\sqrt{Y'^2+Z'^2} = \sqrt{Y^2+Z^2}\,, \tag{5.38}$$

$$t' = \gamma(t - vX)\,. \tag{5.39}$$

We need only convert these three equations back to $\boldsymbol{r}, \boldsymbol{r}'$ language:

$$\boldsymbol{r}'\cdot\boldsymbol{n} \overset{(5.36,5.37)}{=\!=\!=} \gamma(\boldsymbol{r}\cdot\boldsymbol{n} - vt)\,, \tag{5.40}$$

$$|\boldsymbol{r}'\times\boldsymbol{n}| \overset{(5.36,5.38)}{=\!=\!=} |\boldsymbol{r}\times\boldsymbol{n}|\,,\ \text{so that}\ \boldsymbol{r}'\times\boldsymbol{n}=\boldsymbol{r}\times\boldsymbol{n}$$
$$\text{(since these vectors are parallel)}, \tag{5.41}$$

$$t' \overset{(5.36,5.39)}{=\!=\!=} \gamma(t - v\,\boldsymbol{r}\cdot\boldsymbol{n})\,. \tag{5.42}$$

These are linear equations for the S'-coordinates in terms of the S-coordinates. They can easily be solved to give

$$\boxed{\begin{aligned} t' &= \gamma(t - \boldsymbol{v}\cdot\boldsymbol{r})\,,\\ \boldsymbol{r}' &= \left[I + (\gamma-1)\boldsymbol{n}\boldsymbol{n}^t\right]\boldsymbol{r} - \gamma\boldsymbol{v}t\,, \end{aligned}} \tag{5.43}$$

where in the last equation $\boldsymbol{r}$, $\boldsymbol{r}'$, $\boldsymbol{v}$, and $\boldsymbol{n}$ are column vectors, I is the 3×3 unit matrix, and $\boldsymbol{n}\boldsymbol{n}^t$ is a 3×3 matrix with ij^{th} element $n^i n^j$. Two alternative representations of (5.43) each have their own advantages, depending on the problem to be solved:

$$\begin{aligned} t' &= \gamma(t - \boldsymbol{v}\cdot\boldsymbol{r})\,,\\ x' &= x - n^1\boldsymbol{n}\cdot\boldsymbol{r} + \gamma n^1(\boldsymbol{n}\cdot\boldsymbol{r} - vt)\,,\\ y' &= y - n^2\boldsymbol{n}\cdot\boldsymbol{r} + \gamma n^2(\boldsymbol{n}\cdot\boldsymbol{r} - vt)\,,\\ z' &= z - n^3\boldsymbol{n}\cdot\boldsymbol{r} + \gamma n^3(\boldsymbol{n}\cdot\boldsymbol{r} - vt)\,; \end{aligned} \tag{5.44}$$

$$\begin{bmatrix} t'\\ x'\\ y'\\ z' \end{bmatrix} = \begin{bmatrix} \gamma & -\gamma v^1 & -\gamma v^2 & -\gamma v^3\\ & 1+\left(n^1\right)^2(\gamma-1) & n^1n^2(\gamma-1) & n^1n^3(\gamma-1)\\ & & 1+\left(n^2\right)^2(\gamma-1) & n^2n^3(\gamma-1)\\ & \text{(symmetric)} & & 1+\left(n^3\right)^2(\gamma-1) \end{bmatrix} \begin{bmatrix} t\\ x\\ y\\ z \end{bmatrix}. \tag{5.45}$$

We will use (5.45) in the next chapter, which employs matrix formalism for the Lorentz transform.

5.9 Energy and Momentum in Special Relativity

Using the Lorentz transform, it's easy to concoct a scenario where our usual idea of momentum conservation fails. Because of this, momentum must be redefined in special relativity, and such a modification has a flow-on effect on our notion of energy.

The scenario we'll analyse is the very standard one shown in Fig. 5.8. Two identical particles collide in S in a symmetrical way that allows us to immediately write down their final velocities. The x-component of each is unchanged, while the y-component of each changes sign. Each particle has a mass m.

In Fig. 5.8 we have been careful in drawing the arrows that indicate motion. The *initial* direction of each arrow is the direction of increase of the coordinate, while the *final* direction of each arrow is more heuristic: it's drawn to show the direction in which each mass moves, component-wise. If the arrows were not folded, it would be easy to confuse the calculation. Suppose, for example, that we drew mass number 1 with a downward-pointing arrow. What would we write next to the arrow? It's natural to point the arrow down to indicate that the mass's y-component is decreasing as it moves downward. But if u^y is drawn next to this arrow and subsequently we find that $u^y = 1\,\mathrm{m/s}$, does that mean the mass has a velocity of $1\,\mathrm{m/s}$ downward? It is easy to think so—and is potentially even more misleading if the $1\,\mathrm{m/s}$ is explicitly written next to the downward arrow. But $u^y = 1\,\mathrm{m/s}$ is a *velocity*, and being positive really means the mass is moving upward. Writing $-1\,\mathrm{m/s}$ next to the downward arrow would then be more reasonable but is never done. The directions of arrows can easily cause such confusion here and in any similar diagram of mechanics, and we have used an approach of combining the axis direction and the actual motion in the one arrow.

Construct the S'-frame in the usual way by moving with velocity v along the x-axis of S. The interaction looks different now. The velocities are calculated from (5.18):

$$\frac{\mathrm{d}x'}{\mathrm{d}t'} = \frac{\gamma\,(\mathrm{d}x - v\,\mathrm{d}t)}{\gamma\,(\mathrm{d}t - v\,\mathrm{d}x)} = \frac{\mathrm{d}x/\mathrm{d}t - v}{1 - v\;\mathrm{d}x/\mathrm{d}t}\,,$$

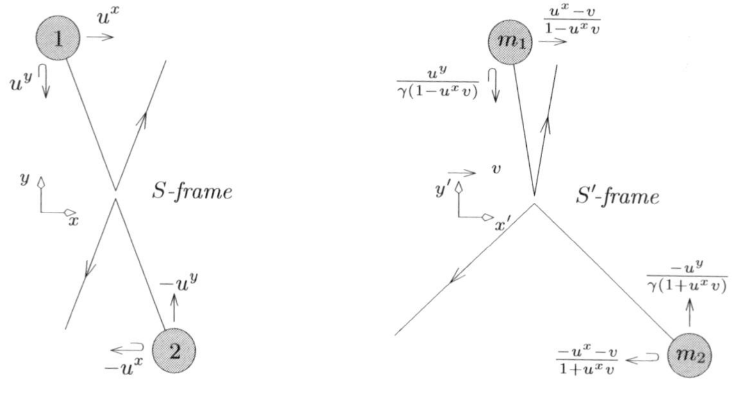

Fig. 5.8. Left: A symmetrical collision in S of two identical particles. The collision's symmetry tells us how the particles bounce from each other. **Right**: The same collision seen from S', with velocities calculated from (5.46). If we insist on defining momentum as mass × velocity, then the idea of mass must be altered in order for momentum to be conserved. The meaning of the curved arrows is described in the text.

$$\frac{\mathrm{d}y'}{\mathrm{d}t'} = \frac{\mathrm{d}y}{\gamma\,(\mathrm{d}t - v\,\mathrm{d}x)} = \frac{\mathrm{d}y/\mathrm{d}t}{\gamma\,(1 - v\;\mathrm{d}x/\mathrm{d}t)}\,. \tag{5.46}$$

In S, the masses are identical, so the total y-momentum is zero since the particles' y-velocities differ only by a sign. Thus it's conserved in the collision. But, in S', if we take the momentum of each particle to be m times its velocity, then the total y-momentum before colliding will not be zero since the velocities are now quite different from each other, which means that when this momentum flips sign in the collision, it cannot be conserved—unless we redefine momentum in some way.

To do this, suppose we call the magnitude of momentum divided by that of velocity the *relativistic mass* of a particle, in which case the relativistic masses of the particles 1 and 2 are denoted m_1, m_2 in S'. We will demand a definition of momentum that ensures it is conserved in the collision. Since symmetry dictates that the y-velocities in S' switch sign in the collision, the total y-momentum can only be zero. In that case,

$$\frac{m_1 u^y}{\gamma(1 - u^x v)} + \frac{m_2(-u^y)}{\gamma(1 + u^x v)} = 0\,, \quad \text{or} \quad \frac{m_1}{m_2} = \frac{1 - u^x v}{1 + u^x v}\,. \tag{5.47}$$

If we "freeze" mass number 1 by setting $v = u^x$ and taking the limit $u^y \to 0$, then we expect that $m_1 \to m$, so that

$$m_2 = \frac{1 + v^2}{1 - v^2}\,m = \frac{m}{\sqrt{1 - \alpha^2}}\,, \quad \text{where } \alpha \equiv \frac{2v}{1 + v^2}\,, \tag{5.48}$$

and α is also the *speed* of mass number 2, which is now only moving horizontally to the left in S'. So if the relativistic mass m_{rel} of a body moving with any arbitrary velocity (vector) $\boldsymbol{V}$ is defined to be the γ-factor for that velocity multiplied by the body's *rest mass* m, then we might have a good candidate for a body's relativistic momentum by defining it to be

$$\boldsymbol{p} \equiv m_{\text{rel}}\boldsymbol{V} = \frac{m\boldsymbol{V}}{\sqrt{1 - V^2}} = \gamma(V)m\boldsymbol{V}\,. \tag{5.49}$$

Certainly, the ratio of the relativistic masses defined in this way for the velocities of Fig. 5.8 will be as in (5.47) for any frame velocity v. In other words, this definition implies that momentum will be conserved in *every* frame S', no matter what v is. The definition of momentum (5.49) has only been derived from a very symmetrical interaction, but experiments suggest that, when so defined, momentum is conserved quite generally.

Force and Newton's Second Law

Does $\boldsymbol{F} = m\boldsymbol{a}$ still hold relativistically? Clearly it cannot since a constant force applied to a body would give it an arbitrarily large speed. Speeds higher than c are ruled out by relativity because they produce causality problems, as can be shown by drawing lines of simultaneity on a spacetime diagram. But Newton's second law in the form $\boldsymbol{F} = \mathrm{d}\boldsymbol{p}/\mathrm{d}t$ is not prone to such a problem, since the relativistic momentum of a particle *can* become arbitrarily large. So the relativistic force on a body is defined to be the rate of increase of its relativistic momentum; but, of course, whether this means that the usual expressions for force still hold is another thing entirely!

What, then, is the relativistic version of $\boldsymbol{F} = m\boldsymbol{a}$? In Newton's theory, the force on a mass is always parallel to the resulting acceleration. The corresponding equation in special relativity is a little more complicated. It turns out that the force $\boldsymbol{F}$ is not always parallel to the acceleration $\boldsymbol{a}$! As usual, set m to be the rest mass and $\boldsymbol{v}$ as its velocity (a column vector), and let I be the 3×3 identity matrix. Then if $\boldsymbol{F} \equiv \mathrm{d}\boldsymbol{p}/\mathrm{d}t \equiv m\,\mathrm{d}(\gamma\boldsymbol{v})/\mathrm{d}t$, some algebra produces

$$\boldsymbol{F} = \gamma m \left(I + \gamma^2 \boldsymbol{v}\boldsymbol{v}^t\right) \boldsymbol{a}\,, \tag{5.50}$$

which inverts to give

$$\boldsymbol{a} = \frac{I - \boldsymbol{v}\boldsymbol{v}^t}{\gamma m} \boldsymbol{F}\,. \tag{5.51}$$

(There is actually a good correspondence with the one-dimensional case here. Just as γ^2 can be written as $1 + \gamma^2 v^2$, and its reciprocal is $1 - v^2$, the matrix $I + \gamma^2 \boldsymbol{v}\boldsymbol{v}^t$ has determinant γ^2 as well as inverse $I - \boldsymbol{v}\boldsymbol{v}^t$.)

In (5.50) and (5.51), the rest mass m is always accompanied by a factor of γ. This suggests that the relativistic mass γm, which is normally seen as just something derived from the rest mass m, really does have a fundamental existence or identity of its own. On the other hand, (5.50) and (5.51) show that it's easier to accelerate a mass perpendicular to its motion than to accelerate it in the direction of its motion. To see this, choose the velocity $\boldsymbol{v}$ to be along the x-axis for convenience, in which case (5.50) becomes

$$\boldsymbol{F} = m \begin{bmatrix} \gamma^3 a^x \\ \gamma a^y \\ \gamma a^z \end{bmatrix}. \tag{5.52}$$

This leads to a *longitudinal mass* being defined as $\gamma^3 m$, with a *transverse mass* defined as γm. Thus, as the particle's speed approaches that of light, its resistance to acceleration in its direction of motion $\gamma^3 m$ will be much higher than its resistance γm to acceleration transverse to this direction.

These sorts of ideas of relativistic mass are useful in that they help us develop an intuition of how a mass responds to a force. Conventionally, when doing calculations, the mass is taken to be the invariant quantity m, and any directional information is put into the separate matrix factors of (5.50) and (5.51). (We will always use m to mean the rest mass.) A matrix such as $\gamma m \left(I + \gamma^2 \boldsymbol{v}\boldsymbol{v}^t\right)$ that relates force to acceleration accounts for why mass is sometimes said to have a *tensor* character in special relativity. We'll meet tensors later in situations where they have more meaning, but for now it suffices to say that such a statement about mass is not especially useful, and merely serves to indicate that force and acceleration vectors relate via a matrix multiplication.

5.9.1 Einstein's Relation of Mass and Energy

Prior to any notion of relativity theory, we know that if a force accelerates a body from velocity $\boldsymbol{v}_1$ to $\boldsymbol{v}_2$, during which it moves from position $\boldsymbol{r}_1$ to $\boldsymbol{r}_2$, then the force does work on the body by an amount equal to

$$\begin{aligned}\int_{\boldsymbol{r}_1}^{\boldsymbol{r}_2} \boldsymbol{F}\cdot \mathrm{d}\boldsymbol{r} &= m\int_{\boldsymbol{r}_1}^{\boldsymbol{r}_2} \frac{\mathrm{d}\boldsymbol{v}}{\mathrm{d}t}\cdot \mathrm{d}\boldsymbol{r} = m\int_{\boldsymbol{v}_1}^{\boldsymbol{v}_2} \boldsymbol{v}\cdot \mathrm{d}\boldsymbol{v} \\ &= m\int_{\boldsymbol{v}_1}^{\boldsymbol{v}_2} \left(v^x \mathrm{d}v^x + v^y \mathrm{d}v^y + v^z \mathrm{d}v^z\right) = \frac{1}{2}m\Big[\left(v^x\right)^2 + \left(v^y\right)^2 + \left(v^z\right)^2\Big]_{\boldsymbol{v}_1}^{\boldsymbol{v}_2} \\ &= \frac{1}{2}m\left[v^2\right]_{\boldsymbol{v}_1}^{\boldsymbol{v}_2} = \frac{1}{2}mv_2^2 - \frac{1}{2}mv_1^2 . \end{aligned} \tag{5.53}$$

This line of thought associates an energy $mv^2/2$ with a body that has speed v. (For simplicity we are ignoring any internal energy the body might have.) Since this energy vanishes when the body is at rest, it must be an energy of motion: kinetic energy.

In relativity, the same sort of idea is applied. This time, *define* the work done as $\boldsymbol{F}\cdot\mathrm{d}\boldsymbol{r}$, where now we must be careful to use the relativistic form of the force:

$$\text{work done} = \int_{\boldsymbol{r}_1}^{\boldsymbol{r}_2} \boldsymbol{F}\cdot \mathrm{d}\boldsymbol{r} = \int_{\boldsymbol{r}_1}^{\boldsymbol{r}_2} \frac{\mathrm{d}(\gamma m\boldsymbol{v})}{\mathrm{d}t}\cdot \mathrm{d}\boldsymbol{r} = m\int_{\boldsymbol{v}=\boldsymbol{v}_1}^{\boldsymbol{v}_2} \boldsymbol{v}\cdot \mathrm{d}(\gamma\boldsymbol{v}) . \tag{5.54}$$

The last integral can be evaluated by noting that

$$\boldsymbol{v}\cdot \mathrm{d}(\gamma\boldsymbol{v}) = \boldsymbol{v}\cdot(\mathrm{d}\gamma\,\boldsymbol{v} + \gamma\,\mathrm{d}\boldsymbol{v}) , \tag{5.55}$$

so that using

$$\mathrm{d}\gamma = \mathrm{d}\left[(1-\boldsymbol{v}\cdot\boldsymbol{v})^{-1/2}\right] = \gamma^3 \boldsymbol{v}\cdot \mathrm{d}\boldsymbol{v} , \tag{5.56}$$

we obtain

$$\boldsymbol{v}\cdot \mathrm{d}(\gamma\boldsymbol{v}) = \gamma^3 \boldsymbol{v}\cdot \mathrm{d}\boldsymbol{v} = \mathrm{d}\gamma . \tag{5.57}$$

Looking carefully, we see a suggestion of the longitudinal mass of (5.52) here, care of the γ^3 term. This should come as no surprise; the fact that a dot product was used in the expression for work (5.54) has the consequence that the transverse mass will not enter into the calculation.

The total work done is therefore

$$m\int_{\boldsymbol{v}=\boldsymbol{v}_1}^{\boldsymbol{v}_2} \boldsymbol{v}\cdot \mathrm{d}(\gamma\boldsymbol{v}) = m\left[\gamma\right]_{\boldsymbol{v}_1}^{\boldsymbol{v}_2} = \frac{m}{\sqrt{1-v_2^2}} - \frac{m}{\sqrt{1-v_1^2}} . \tag{5.58}$$

Comparing this result with (5.53) shows that we might associate an energy $\gamma(v)\, m$ with a body that has speed v. But this is just the relativistic mass, which reduces to the rest mass when $v = 0$. Einstein interpreted this relativistic mass (times c^2 for the correct units) as the body's total energy E, so that even when at rest, the body has energy equal to its rest mass. Notice that the square of the total energy minus the square of the momentum is

$$E^2 - p^2 = (\gamma m)^2 - (\gamma m v)^2 = m^2, \tag{5.59}$$

and of course m^2 is a constant, being the body's rest mass. So, like the interval, all observers agree on the value of $E^2 - p^2$. This is an idea we will investigate more fully in the next chapter when we use the Lorentz transform in a very different context. There we'll find that an important language called *covariant notation* arises quite naturally when we use relativity to describe the everyday idea of crossing the street in the rain.

6 Four-Vectors and the Road to Tensors

When crossing a rainy street, should we walk or run? Walking takes time, during which we get very wet, but while running reduces our time spent in the rain, it does force us to sweep through the raindrops more quickly, since the rain will be falling at a different angle and thus pounding into the front of our body in a way that it wasn't when we walked slowly. Probably most people would choose to run as fast as possible every time, and that's what we'll find, too, for the case of vertical rain and a simple "boxy" model of a human in the coming pages. (For other angles of rainfall, such as when it ploughs into our back, and for other bodily angles of running, we might adopt a different strategy. But the case of vertical rain will suffice for our needs here.)

The problem of just how fast we should walk in the rain is interesting because it introduces two quantities that become fused into one when we take a relativistic viewpoint. These quantities are *number density*, which is how many raindrops there are in a unit volume (the more there are, the wetter we are going to get), and *flux density*, which is how many of those drops are crossing a unit area in a unit time (pouring rain delivers more water than light rain).

That these two types of density become fused into one might only be of passing interest if it were not for the fact that this new quantity, a *four-vector*—as well as the bigger family of which it is a part, called tensors—also appears in a great many other diverse areas of physics. Understanding how four-vectors arise makes it possible to see how useful they are in those other areas, too.

So we'll look more closely at these number and flux densities. This question of how fast to run in the rain is a great example of the utility of taking a whimsical problem to an extreme—in this case, the relativistic limit. Despite the apparent irrelevance of taking such a limit, we find that when we do, this fusion of two quantities into one happens, and we learn something interesting about Nature. Let's do just that.

6.1 Number Density and Flux Density

First, we need to define these two densities. If we have a collection of particles such as raindrops, their number density n is the number that inhabit a unit

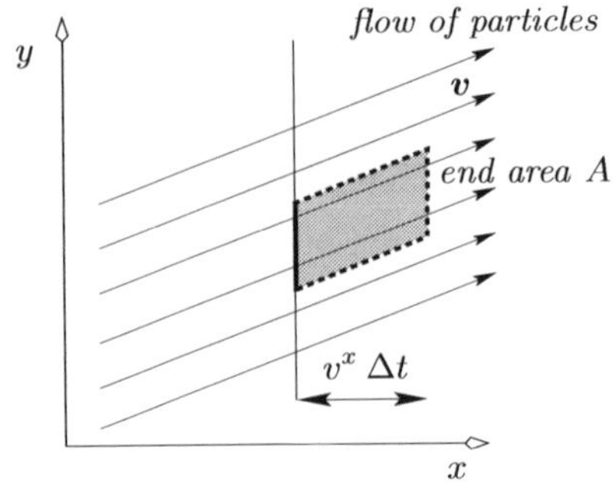

Fig. 6.1. The thin vertical line is the edge of a plane perpendicular to the x-axis, and through this plane we wish to count the number of particles flowing. A tube has been drawn that follows the passage of drops in some area A in a time interval Δt.

volume. Their flux density is a little more complicated, being a vector. Its x-component is the number of particles flowing across a unit plane area at constant x in unit time, and likewise for its y- and z-components. For instance, if all of the particles are moving along the x-axis in a direction of increasing x, then their flux density will equal some positive number times $(1, 0, 0)$.

The flux density vector is useful because its three numbers determine the flux through any surface at all, no matter what its orientation. And such a surface need not be planar, since we can always use calculus to break such a curved surface into infinitesimal elements that can be treated as planar. To calculate the three components, imagine as in Fig. 6.1 a plane drawn at right angles to the x-axis with some area A, through which the rain moves with velocity $\boldsymbol{v}$. The x-component of the flux density will be the number of particles passing through this surface in a time Δt divided by its area A and the time interval Δt. This is

$$x\text{-component of flux density} =$$

$$\frac{\begin{bmatrix}\text{static number}\\ \text{density}\end{bmatrix} \times \begin{bmatrix}\text{volume of swept tube}\\ \text{of end area } A\end{bmatrix}}{A\,\Delta t} = \frac{n\,v^x\,\Delta t\,A}{A\,\Delta t} = nv^x\,. \tag{6.1}$$

Similarly, the two other components are nv^y, nv^z, so the total flux density vector is $n\boldsymbol{v}$.

These expressions for the number and flux densities, n and $n\boldsymbol{v}$, will only be valid nonrelativistically since at high speeds of rainfall, the length contraction of special relativity changes the relevant geometry; but we'll get to the relativistic versions later. In the meantime:

$$\begin{aligned}\text{(nonrelativistic) number density} &= n\,,\\ \text{(nonrelativistic) flux density is } \boldsymbol{\Phi} &= n\boldsymbol{v}\,.\end{aligned} \tag{6.2}$$

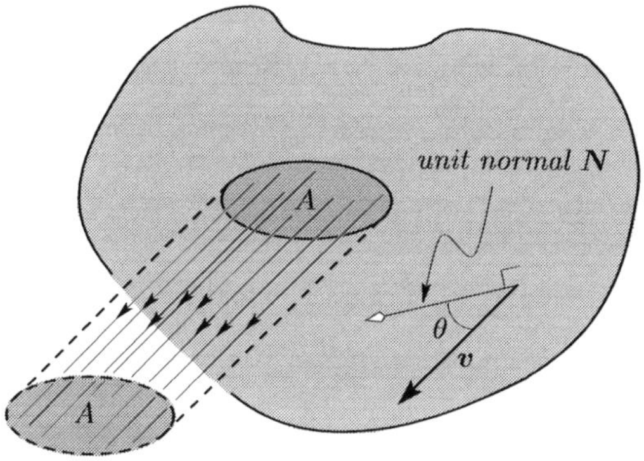

Fig. 6.2. A general plane across which a tube of particles flows. The tube has end area A, and the particles flow with velocity $\boldsymbol{v}$ for a time Δt.

Now imagine that the surface through which the particles flow is a plane orientated such that its unit normal vector is the $\boldsymbol{N}$ of Fig. 6.2. There are, of course, two choices for $\boldsymbol{N}$, but we will show in a moment that each is as good as the other. The flux density through this plane is given by the number of particles passing through its area A in a time Δt divided by $A\,\Delta t$. Again, this is the number density times the volume of a swept tube of end area A divided by this area and the time interval Δt. With θ the angle between $\boldsymbol{v}$ and $\boldsymbol{N}$, we have

$$\begin{aligned}&\text{flux density through plane} = \\ &\qquad \frac{n \times \begin{bmatrix}\text{perpendicular height} \\ \text{(not length!) of tube}\end{bmatrix} \times A}{A\,\Delta t} = \frac{n\,v\,\Delta t\,\cos\theta\,A}{A\,\Delta t} \\ &\quad = nv\cos\theta = n\boldsymbol{v}\cdot\boldsymbol{N}\,. \end{aligned} \tag{6.3}$$

So we have shown that the flux density $n\boldsymbol{v}$ suffices to give the flux in *every* direction $\boldsymbol{N}$, which is what makes it useful.

It makes no difference whether we choose the unit normal vector $\boldsymbol{N}$ to point "into" or "out of" the surface. Changing the sign of $\boldsymbol{N}$ will change the sign of the dot product $n\boldsymbol{v}\cdot\boldsymbol{N}$, which just embodies the reversal of the choice of preferred direction. That is, a flux of, say, 5 particles in one direction through a surface is exactly the same as a flux of -5 particles in the opposite direction.

Once we know the flux density of the rain, we can calculate how many drops will flow through any surface at all. The number density n and flux density $\boldsymbol{\Phi} = n\boldsymbol{v}$ completely determine how much rain there is and how fast it is falling. Summing up,

$$n = \text{number of drops per unit volume,}$$
$$\boldsymbol{\Phi}\cdot\boldsymbol{N} = \text{number of drops passing through unit area in unit time.} \quad (6.4)$$

Now that we have written the flux density as a vector, the calculation becomes easy to do for the case where the rain falls at an arbitrary angle. So such an angle is no more complicated, but it also gives no more insight, and so we'll stay with vertical rain.

6.2 Running Nonrelativistically

The surface in Fig. 6.2 through which we drew the rain falling was not moving. Because of this restriction, we need to analyse the rainfall scenario in our own frame as we cross the road, because we never move in our own frame.

When changing frames, we ask: what quantities will change? If the speeds of our walk and the rainfall are slow enough, then the whole exercise will be nonrelativistic, and that of course means that we'll perceive there to be no change in the properties of space and time. The number density of the raindrops will not change because the volume we measure them to inhabit has not changed. On the other hand, their flux density *will* change because, when we run, the rain will fall at an angle in *our* frame instead of falling vertically as it does in the street frame.

So let's calculate how many drops we encounter as a function of our speed across the street. Once this has been done, we'll repeat the calculation relativistically to find the promised insight by fusing the number and flux densities into one.

To analyse the rainfall scenario, use the three frames shown in Fig. 6.3:

- the **street frame**, S_{street}, since the whole scenario takes place there;
- the **rain frame**, S_{rain}, in which the number density and flux density have their simplest forms; and
- the **body frame**, S_{body}, since the expression $\boldsymbol{\Phi}\cdot\boldsymbol{N}$ requires the surface of interest to be at rest. ($\boldsymbol{\Phi}\cdot\boldsymbol{N}$ was developed in the frame of the surface through which the rain travels.)

We'll also use the following notation:

$$\begin{aligned} \boldsymbol{v}_{A\leftarrow B} &\equiv \text{velocity of } A \text{ relative to } B \\ &= \boldsymbol{v}_{A\leftarrow C} - \boldsymbol{v}_{B\leftarrow C} \quad \text{(where } C \text{ is any third point)} \\ &= \boldsymbol{v}_{A\leftarrow C} + \boldsymbol{v}_{C\leftarrow B}\,. \end{aligned} \quad (6.5)$$

We require the flux density of the rain in the body frame, or $n\boldsymbol{v}_{\text{rain}\leftarrow\text{body}}$, where

$$\begin{aligned} \boldsymbol{v}_{\text{rain}\leftarrow\text{body}} &= \boldsymbol{v}_{\text{rain}\leftarrow\text{street}} - \boldsymbol{v}_{\text{body}\leftarrow\text{street}} \\ &= (0, v_r) - (v_b, 0) \quad \text{(refer to Fig. 6.3)} \\ &= (-v_b, v_r)\,. \end{aligned} \quad (6.6)$$

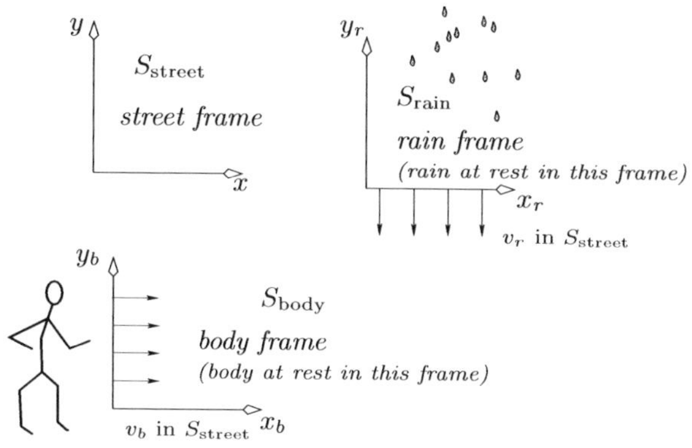

Fig. 6.3. The street and body frames are used for the nonrelativistic case, while the rain frame is really only needed later for the relativistic calculation. The street frame is at rest on the printed page, so to speak.

The flux density of the rain in the body frame is then

$$\boldsymbol{\Phi} = n(-v_b, v_r)\,. \tag{6.7}$$

Assume our head to be flat with area A_h, and our front to be flat with area A_f. How many raindrops will hit our head? This is just the absolute value of the flux through that flat surface. Refer to (6.4):

$$\begin{aligned} &\text{number of drops hitting our head} = |\text{flux across head}| \\ &= |\text{flux density across head} \times A_h \times \text{time to cross street}| \\ &= |n(-v_b, v_r)\cdot(0,1) \times A_h \times \text{time to cross street}| \\ &= |v_r/v_b|\; n\, A_h\, s \end{aligned} \tag{6.8}$$

since the time to cross the street is $s/|v_b|$. Similarly,

$$\begin{matrix}\text{number of drops} \\ \text{stopped by our front}\end{matrix} = \left| n(-v_b, v_r)\cdot(1,0)\, A_f \,\frac{s}{|v_b|}\right| = n\, A_f\, s\,. \tag{6.9}$$

The total number of drops we intercept is, of course, the sum of (6.8) and (6.9):

$$\text{total number intercepted} = ns\Big(|v_r/v_b|\, A_h + A_f\Big)\,. \tag{6.10}$$

The total number of drops that strike us is a decreasing function of our running speed $|v_b|$, reducing asymptotically to nsA_f as we run ever faster. So we should run as fast as possible!—although there is a minimum amount of rain that must hit us no matter how fast we run (nonrelativistically). This limiting amount is just the constant number of drops that hit our front, which is reasonable since the faster we run, the more the falling drops are "frozen" in flight and thus the fewer of them will hit our head. The fact that the number of drops hitting our front is constant shows that while we do sweep through more of them as we run faster, we also spend less time in the rain, and these two factors cancel.

6.3 Running Relativistically

Having solved the rainfall problem nonrelativistically, it's time to ask what insight might be gained by considering the relativistic version. The basic calculation is much the same, provided we amend the definitions of number density and flux density to incorporate the effect of relativity. It is understood that the rain, or our speed, need not actually be relativistic for the following calculation to apply.

In a relativistic treatment, lengths will be contracted, implying that particle densities must increase. The number density is easily corrected: if the n particles inhabit a box of unit volume in their own rest frame, then we who measure them as moving with some speed v will measure that box's volume to have decreased by a factor of $\gamma = 1/\sqrt{1-v^2}$ because its side in the direction of its motion is Lorentz-contracted by γ. So we'll observe an increased number density of γn.

The γ-factor also finds its way into (6.1) for the same reason, since the number density n there must be replaced by the relativistic value γn. The volume of the swept tube in Fig. 6.1 is unchanged; it just contains more raindrops. Including the extra factor of γ in (6.1) then produces the fully relativistic definitions of the densities:

$$\boxed{\begin{aligned} \text{fully correct number density} &= \gamma n\,, \\ \text{fully correct flux density } \boldsymbol{\Phi} &= \gamma n \boldsymbol{v}\,. \end{aligned}} \tag{6.11}$$

These quantities allow a relativistic analysis with the approach of (6.6), where the velocities are now to be transformed relativistically.

However, while the corrected densities in (6.11) suffice, we can gain further insight by taking a slightly different path. Just as the Lorentz transform *mixes* space and time from one frame to another, it must also mix the number and flux densities, since the first is defined using space (the volume of a box) while the second is defined using time (the flow of the raindrops). So let's examine just how the Lorentz transform mixes number and flux densities.

6.3.1 Combining Number and Flux Densities into Something New

In Chap. 5, we wrote down the Lorentz transform that relates time and space between inertial frames. Its form used in this chapter can be written from (5.45), but we will restrict all motion to the x- and y-directions, making the much simpler (5.17), together with its corresponding version for motion along the y-axis, quite adequate for our needs.

However, as was pointed out in Chap. 5, by treating *increases* in the coordinates, we can bypass having to keep track of arbitrary zero calibrations of the axes. And, in fact, to be fully general for the noninertial case of acceleration (where v is changing, as described in the next chapter), we really wish to consider the infinitesimal form of the Lorentz transform. For example, if the primed frame moves with velocity v in the positive-x direction of the unprimed frame, (5.18) becomes

$$\begin{aligned} dt' &= \gamma(dt - v\,dx)\,, \\ dx' &= \gamma(dx - v\,dt)\,, \\ dy' &= dy\,, \quad dz' = dz\,. \end{aligned} \tag{6.12}$$

Just how dt, dx, dy, dz transform between frames is the central result of special relativity, and we mark this by considering the four infinitesimals to be a single unit called a *four-vector*. If this name was reserved for these four quantities alone, there would hardly be any point in inventing it. But the surprise is that we can find other quantities that also transform between frames in just the same way as dt, dx, dy, dz. We also bundle those together and call the new ordered set a four-vector. Obeying the Lorentz transform is precisely what defines a four-vector. Once we have investigated the properties of the fundamental four-vector (dt, dx, dy, dz), we will have begun to explore *all* of the four-vectors of physics.

It must be said here that we'll reserve the name four-vector for an ordered set of numbers such as (dt, dx, dy, dz), with nothing yet being said about why such a set should be called a vector, apart from the fact that its spatial part has the components of a vector in euclidean 3-space. In Chap. 8, we will look more closely at this important point and introduce basis vectors that, when combined with the four-vector components, produce a more general object called a *proper vector*, which itself will be a special case of the larger set called tensors. But, for now, it suffices to be aware that the four-vector is really just an ordered set of components that obey the Lorentz transform.

Some further points should be noted. First, a euclidean vector with four components is not a four-vector. A four-vector is a physical quantity; it transforms in a certain well-defined way when we change frames. Euclidean vectors are not obliged to obey the Lorentz transform! Second, the term four-vector doesn't naturally imply that there are also five-vectors, six-vectors, and so on. A five-vector certainly could be defined by joining dt to a *euclidean* four-component vector (dx, dy, dz, dw) together with an appropriate transform. This is done in some descriptions of cosmological models, but it's not an important idea for our purposes, and we need not consider it.

Last, the set of coordinates of an event (t, x, y, z) is *not* a four-vector, since it does not obey the Lorentz transform in general. Usually, in special relativity, the origins of the two inertial frames are chosen to coincide at $t = t' = 0$, in which case (t, x, y, z) certainly will obey the Lorentz transform. But while pedagogically useful, this is a very restrictive condition that cannot be expected to hold in general. In contrast, the components of true four-vectors obey *all* Lorentz transforms.

The fundamental four-vector $(\mathrm{d}t, \mathrm{d}x, \mathrm{d}y, \mathrm{d}z)$ involves infinitesimals. A useful exercise is to factor one of these out, leaving the rest as ratios and hence noninfinitesimal, so that we are dealing with as many physical quantities as possible. The natural choice is to factor out $\mathrm{d}t$ since the remaining ratios will then be well-understood velocities.

So picture two events separated by time and space intervals $\mathrm{d}t, \mathrm{d}x, \mathrm{d}y, \mathrm{d}z$. We have arranged for a clock to travel at constant velocity from the first event to the second. By definition, the clock will tick out the proper time $\mathrm{d}\tau$ between the events. This is the time interval that the clock shows, and all observers must agree on that regardless of their state of motion (because there is only one clock that can move in such a way, and it reads $\mathrm{d}\tau$!). So $\mathrm{d}\tau$ must be a frame-independent quantity, another example of a scalar. We measure the time between the two events to be $\mathrm{d}t = \gamma\,\mathrm{d}\tau$, where $\gamma = 1/\sqrt{1 - v^2}$ relates to the speed v of the clock connecting the events since "moving clocks run slow".

Alternatively, we can see that $\mathrm{d}t = \gamma\,\mathrm{d}\tau$ without applying the "moving clocks run slow" rule of thumb by applying the inverse of (6.12) to the moving clock, which inhabits the primed frame and moves with its x'-axis joining the two events. The inverse of (6.12) for the time coordinate is $\mathrm{d}t = \gamma(\mathrm{d}t' + v\,\mathrm{d}x')$. Since the events take place at the same spot for that clock, $\mathrm{d}x' = 0$ and $\mathrm{d}t' \equiv \mathrm{d}\tau$. Hence $\mathrm{d}t = \gamma\,\mathrm{d}\tau$. We might have been inclined to apply (6.12) as it is, since the required $\mathrm{d}t' \equiv \mathrm{d}\tau$ is already isolated on one side of the first equation. This actually would not be such a good idea, leading to more effort than we spent by using the *other* frame above [i.e. first inverting (6.12)]. As discussed in Sect. 5.3.1, the moral here is that in special relativity analyses, we should be prepared to apply the Lorentz transform not from the frame that might seem intuitively more direct or simple, but from the *other* frame.

In the unprimed frame, the clock travelling between the two events has velocity $\boldsymbol{v} \equiv (\mathrm{d}x/\mathrm{d}t, \mathrm{d}y/\mathrm{d}t, \mathrm{d}z/\mathrm{d}t)$. Factoring out $\mathrm{d}t$ gives

$$\begin{aligned}(\mathrm{d}t, \mathrm{d}x, \mathrm{d}y, \mathrm{d}z) &= \mathrm{d}t\,(1, v^x, v^y, v^z) \\ &= \text{the scalar } \mathrm{d}\tau \times \gamma\,(1, v^x, v^y, v^z)\,. \end{aligned} \tag{6.13}$$

Now realise that if four numbers transform in the way of (6.12), then a scalar multiple of those numbers will also transform in that way because the equations are linear. So because $\mathrm{d}\tau$ is a scalar, we have established an important fact:

$$\boxed{\vec{u} \equiv \gamma\,(1, \boldsymbol{v}) \text{ is a four-vector.}} \tag{6.14}$$

Not surprisingly, this four-vector is an important one, and it is given the name *four-velocity* or *proper velocity* $\vec{u}$. The arrow is a reminder that we are dealing with four dimensions, as opposed to a bold font denoting three dimensions. (In later chapters, where the concept of a vector is generalised, a bold font will be sufficient to denote a vector in *any* number of dimensions, but here we wish to be more explicit while the concepts are introduced.) The four components of $\vec{u}$, such as $u^t = \gamma$, $u^x = \gamma v^x$, etc., are conventionally indexed by a Greek superscript, so that although the four-vector is really $\vec{u}$, we tend to call it u^α by referring to all of its components at once. The indices can equally well be referred to by the numbers 0 to 3, which is equivalent to renaming the coordinates

$$t, x, y, z \longrightarrow x^0, x^1, x^2, x^3. \tag{6.15}$$

However, the generic names of x^0, x^1, x^2, x^3 can also be used to denote any other system of spacetime coordinates, such as t, r, θ, ϕ.

The appearance of γ and $\gamma \boldsymbol{v}$ in (6.14) is beginning to look suspiciously like the number and flux densities of (6.11). With an eye on this, we try something new and bold: we put the number and flux densities together into a set of four numbers:

$$\begin{aligned}(\text{number density}, \text{flux density}) &= \gamma n\,(1, \boldsymbol{v}) \\ &= n \times \text{four-velocity}.\end{aligned} \tag{6.16}$$

But n is a scalar; the density of raindrops in those drops' rest frame is a number that all observers must agree upon. That means that this *number–flux density*, $n\vec{u}$, is a four-vector! So the Lorentz transform gives the correct recipe for how number and flux densities transform in special relativity.

This new four-dimensional density allows us to calculate how many raindrops hit our head and front in the relativistic case, by using the number–flux density in our body frame. Its value in the rain frame, easily calculated, is converted to our body frame via the street frame, because only in the street frame do we know the relevant velocities. So we need two Lorentz transforms:

$$S_{\text{rain}} \longrightarrow S_{\text{street}} \longrightarrow S_{\text{body}}$$

$$\boldsymbol{v}_{\text{rain}\leftarrow\text{street}} = (0, v_r, 0)\,, \qquad \therefore\ \boldsymbol{v}_{\text{street}\leftarrow\text{rain}} = (0, -v_r, 0)\,.$$

Using (5.45), the transform is

$$\begin{bmatrix} \gamma_r & 0 & \gamma_r v_r & 0 \\ 0 & 1 & 0 & 0 \\ \gamma_r v_r & 0 & \gamma_r & 0 \\ 0 & 0 & 0 & 1 \end{bmatrix}.$$

$$\boldsymbol{v}_{\text{body}\leftarrow\text{street}} = (v_b, 0, 0)\,.$$

The transform is

$$\begin{bmatrix} \gamma_b & -\gamma_b v_b & 0 & 0 \\ -\gamma_b v_b & \gamma_b & 0 & 0 \\ 0 & 0 & 1 & 0 \\ 0 & 0 & 0 & 1 \end{bmatrix}.$$

(6.17)

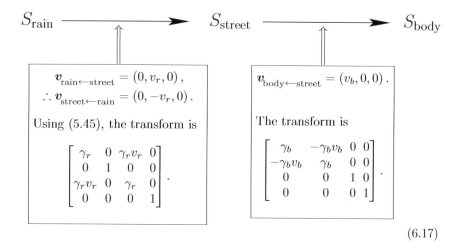

The number density in the rain frame is simply n, while the flux density in that frame is the zero vector. (The rain does not move in its own frame.) Thus, the number–flux density in the body frame is built from two Lorentz transforms:

$$\begin{bmatrix} n_b \\ \boldsymbol{\Phi}_b \end{bmatrix} = \begin{bmatrix} \gamma_b & -\gamma_b v_b & 0 & 0 \\ -\gamma_b v_b & \gamma_b & 0 & 0 \\ 0 & 0 & 1 & 0 \\ 0 & 0 & 0 & 1 \end{bmatrix} \begin{bmatrix} \gamma_r & 0 & \gamma_r v_r & 0 \\ 0 & 1 & 0 & 0 \\ \gamma_r v_r & 0 & \gamma_r & 0 \\ 0 & 0 & 0 & 1 \end{bmatrix} \begin{bmatrix} n \\ 0 \\ 0 \\ 0 \end{bmatrix} = n\gamma_r \begin{bmatrix} \gamma_b \\ -\gamma_b v_b \\ v_r \\ 0 \end{bmatrix}. \tag{6.18}$$

Finally, we take the flux density and use it to calculate how many raindrops the body intercepts, in the way of (6.8) to (6.10):

$$\begin{aligned} &\text{number of drops hitting our head} = |\text{flux across head}| \\ &= |\text{flux density across head} \times A_h \times \text{time to cross street}| \\ &= \left| n\gamma_r \begin{bmatrix} -\gamma_b v_b \\ v_r \\ 0 \end{bmatrix} \cdot \begin{bmatrix} 0 \\ 1 \\ 0 \end{bmatrix} \times A_h \times \text{time to cross street} \right|, \end{aligned} \tag{6.19}$$

where the time to cross the street is

$$\frac{\text{contracted road width}}{\text{running speed}} = \frac{s}{\gamma_b |v_b|}. \tag{6.20}$$

> Care is needed in remembering to contract the road width, but even in a more complex problem we would hardly be distracted from doing this because we would need to apply a Lorentz transform to calculate the time interval. Here, we don't need the full machinery because the problem is simple enough, but we do need to remember the end result of that machinery: length contraction.

Combining (6.19) with (6.20) gives

$$\text{number of drops hitting our head} = \frac{\gamma_r}{\gamma_b} \left| \frac{v_r}{v_b} \right| n\, A_h\, s\,. \tag{6.21}$$

In a similar way,

$$\text{number of drops hitting our front} = \gamma_r\, n\, A_f\, s\,. \tag{6.22}$$

The total number of drops we encounter is the sum of (6.21) and (6.22):

$$\text{Total number} = ns\gamma_r \left(\left| \frac{v_r}{v_b} \right| \frac{A_h}{\gamma_b} + A_f \right). \tag{6.23}$$

Compare this with the nonrelativistic (6.10). The factor $\gamma_b |v_b|$ increases as we run faster, so the conclusion is identical to that of the nonrelativistic case: cross the street as fast as possible. This is all well and good, but the real jewel here is not about keeping dry, but rather the new idea of the number–flux density four-vector and the knowledge of how it can be calculated in any frame when known in one. This knowledge of how numbers change between frames is the fundamental idea behind four-vectors and their larger tensor family.

6.3.2 The "Length" of the Four-Velocity

Four-vectors have different components in different frames, and although they are four-dimensional entities, their euclidean length is not invariant from one frame to the next. Rather, what is invariant is the new measure of length as determined by the idea of the metric that we encountered in Sect. 5.6. In general, the squared length of any four-vector $\vec{a}$ is defined to be

$$|\vec{a}|^2 \equiv {a^t}^2 - {a^x}^2 - {a^y}^2 - {a^z}^2 . \tag{6.24}$$

Applying this to the four-velocity (6.14) shows that its length is always one, regardless of frame:

$$|\vec{u}|^2 \equiv \gamma^2 \left(1 - {v^x}^2 - {v^y}^2 - {v^z}^2\right) = \gamma^2 \left(1 - v^2\right) = 1 . \tag{6.25}$$

We'll use the invariance of the vector length in the next section.

As discussed in the previous chapter, we have not joined space and time together artificially; they really were always two parts of a single structure with a metric $\mathrm{d}\tau^2 = \mathrm{d}t^2 - \mathrm{d}x^2 - \mathrm{d}y^2 - \mathrm{d}z^2$. Written with infinitesimals, the metric applies to *all* observers, even noninertial ones—as we'll see when we study an accelerated observer in the next chapter.

6.4 Examples of Other Four-Vectors

Besides the four-velocity and number–flux density, other four-vectors arise in physics with important uses, and in this section we will study three notable examples. The first, energy–momentum, will be familiar from the end of the previous chapter. The second example is the charge–current density, of importance in studying electromagnetism. The last four-vector, frequency–wavenumber for light, will not quite fit the mould and will force us to extend our ideas of four-vectors.

Energy–Momentum

In Sect. 5.9 we saw how a particle of rest mass m is postulated to have total energy and momentum of

$$E = \gamma m , \quad \boldsymbol{p} = \gamma m \boldsymbol{v} . \tag{6.26}$$

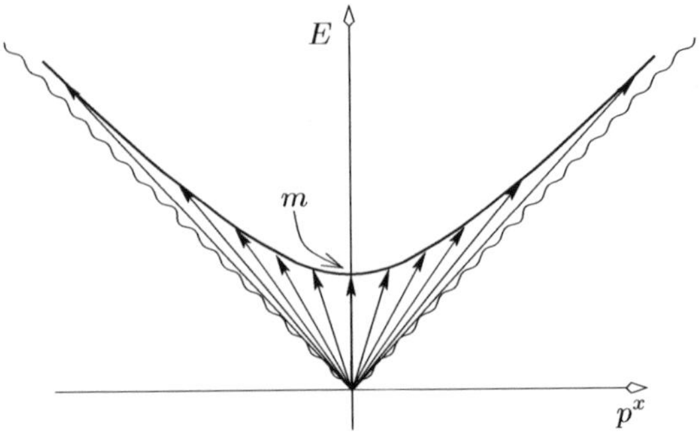

Fig. 6.4. In one space dimension, all particles of rest mass m are represented by four-vectors whose heads lie on the positive-energy branch of the hyperbola $E^2 - p^{x\,2} = m^2$.

It follows immediately that $(E, \boldsymbol{p})$ is a four-vector since it equals a scalar m times the four-velocity:

$$\vec{p} \equiv (E, \boldsymbol{p}) = \gamma\, m(1, \boldsymbol{v}) = m\, \vec{u}\,. \tag{6.27}$$

This, the *energy–momentum four-vector*, is written as $\vec{p}$ with components p^α. Thus we can straightaway write down how energy and momentum transform in special relativity since they mimic the four-velocity. For example, when the S'-frame moves along the x-axis with velocity v, refer to (6.12) to write

$$\begin{aligned} E' &= \gamma\,(E - v\,p^x)\,, \\ p'^x &= \gamma\,(p^x - v\,E)\,, \\ p'^y &= p^y\,, \quad p'^z = p^z\,. \end{aligned} \tag{6.28}$$

Knowledge of the energy–momentum $\vec{p}$ is used frequently in relativity and particle physics, because it allows an analysis of dynamic processes in any frame we choose. To see why, we introduce a somewhat under-used device in the theory of interactions, called the *energy–momentum diagram*, shown in Fig. 6.4. Because the summed energies and momenta of all the particles in an interaction are conserved, any interaction on such a diagram can be depicted as a vector addition of the energy–momentum four-vectors for the interacting particles, in a way akin to a spacetime plot. We just substitute energy and momentum for time and space, respectively, now with the speed of light c absorbed into the definition of energy. Because a photon satisfies $E = pc$, or just $E = p^x$ in one space dimension (with c absorbed), it is represented on such a plot as a 45° wavy line, just as it is on the usual t–x spacetime diagram.

The axes of the energy–momentum diagram need calibration. After all, the lengths of its vectors are meant in the scalar sense of (6.24). Recall from (5.59)

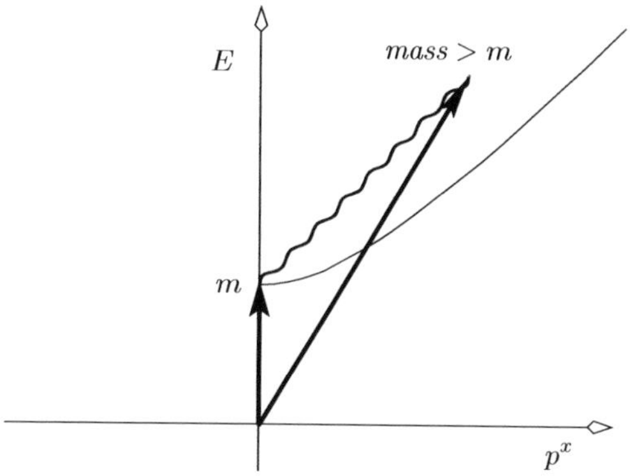

Fig. 6.5. A particle's energy–momentum plus a photon's energy–momentum gives a resulting vector that lies off the $E^2 - p^{x\,2} = m^2$ hyperbola, or *off shell*. Thus, if the photon is to disappear in the interaction, the particle's rest mass must change, regardless of its resulting motion.

that $E^2 - p^{x\,2} = m^2$, a scalar, so that the rest mass m is an energy–momentum vector's length in the hyperbolic geometry of the energy–momentum diagram. Further, since all known particles have real mass, they must all obey the relation $E > p$, so that their vectors on the energy–momentum diagram must be sloped more steeply than that of light, which itself has zero rest mass ($E = p$). This forms a good analogy to the fact that all worldlines on a spacetime diagram slope more steeply than a light ray's worldline. And it implies that all particles with rest mass m—regardless of their motion—can be represented as vectors, with tail at the origin and head on the hyperbola $E^2 - p^2 = m^2$, as shown in Fig. 6.4. We see on that diagram that the energy of a particle at rest ($p^x = 0$) is indeed its rest mass m.

A good example of the power of energy–momentum diagrams in analysing collisions is shown in the following scenario. Our task is to prove that a particle cannot simply absorb a photon that strikes it, without some change occurring to the particle's internal structure. All we need do is work in the rest frame of the particle before the collision, adding its four-vector to that of the photon as shown in Fig. 6.5. The particle at rest has zero momentum, and energy equal to its rest mass m. When it absorbs the photon, the resulting excited particle must have a different rest mass because the head of the total vector simply can no longer lie on the hyperbola $E^2 - p^{x\,2} = m^2$. So the particle resulting from the collision must have a different rest mass, implying there has been a restructuring of its internal constituents. QED! An energy–momentum diagram lets us balance energy and momentum in one fell swoop.

Charge–Current Density

The second four-vector that we'll consider has great use in electromagnetism, and can be built out of the number–flux density. Suppose that we are examining not raindrops but charge carriers, each with charge q, that flow in a current. In their rest frame, they have a number density of n charges per unit volume and hence a proper charge density of $\varrho_0 \equiv nq$. In that case, the charge density ϱ in our frame is $\varrho = \gamma nq$. It must be larger than ϱ_0 by a factor of γ since the charges now occupy a box with side length contracted by γ. (Why is q a scalar? Because everyone agrees on its value. *Experimentally*, charge is not observed to change from one reference frame to the next.)

The *current density* $\boldsymbol{j}$ is the flux density of the charges times their charge, or $\gamma n\boldsymbol{v} \times q$. So now we have a new four-vector, the *charge–current density* $\vec{\jmath}$:

$$\vec{\jmath} \equiv (\varrho, \boldsymbol{j}) = (\gamma nq, \gamma n\boldsymbol{v}q) = q \times \gamma n(1, \boldsymbol{v}) = \varrho_0 \vec{u}\,, \tag{6.29}$$

with its four components $j^\alpha = \varrho_0 u^\alpha$. The charge–current density $\vec{\jmath}$ is just a scalar q times the number–flux density $n\vec{u}$, or equivalently a scalar ϱ_0 times the four-velocity $\vec{u}$.

We now know how charge and current densities transform between frames. Because charges produce the electric field and currents the magnetic field, much of the machinery is now in place to begin to analyse how electric and magnetic fields mix under a Lorentz transform. We'll see later in Sects 10.4.1 and 11.3 when deriving and solving Maxwell's equations, that if the electrostatic potential Φ is combined with the vector potential $\boldsymbol{A}$, then the result is closely related to $\vec{\jmath}$ in a way that makes a new four-vector,

$$\vec{A} \equiv (\Phi, \boldsymbol{A})\,, \tag{6.30}$$

called simply the *electromagnetic potential.* This combination of charge and current densities, as well as the two potentials of electromagnetism, is a great unification, and these ideas will be used throughout the coming chapters.

Frequency–Wavenumber for Light

Our last four-vector will be something quite different: one that does not lean on the four-velocity for its definition. With the tools we've been using until now, the calculations of this section will be a little onerous; but this is deliberate because a way to streamline them through the use of *covariant* notation will emerge. This new notation will then serve to introduce a new set of numbers closely related to the idea of the cobasis that was introduced in Sect. 2.3.

So consider the frequency and wavenumber of a wave as defined in terms of the phase ϕ in (2.147):

$$\omega \equiv \frac{-\partial\phi}{\partial t}\,, \quad \boldsymbol{k} \equiv \nabla\phi\,. \tag{6.31}$$

The frequency depends on time, while the wavenumber depends on space. It might not come as a surprise, then, to find that frequency and wavenumber can be joined to make a new four-vector $\vec{k} \equiv (\omega, \boldsymbol{k})$. To prove that $\vec{k}$ is indeed a four-vector, we need to show that its components k^α transform correctly through the Lorentz transform. A little more work is required here than for the charge–current density $\vec{\jmath}$, since $\vec{k}$ is not proportional to any four-velocity in the way that $\vec{\jmath}$ was in (6.29).

Here is what we do. The Lorentz transform for a general velocity $\boldsymbol{v} = v\boldsymbol{n}$ was written in (5.45). Remember that to eliminate the need to include any calibration constants in our axes, the matrix in (5.45) can be applied to differentials of the coordinates. So write (5.45) in the following way:

$$\begin{bmatrix} \mathrm{d}t' \\ \mathrm{d}x' \\ \mathrm{d}y' \\ \mathrm{d}z' \end{bmatrix} = \begin{bmatrix} \gamma & -\gamma v_1 & -\gamma v_2 & -\gamma v_3 \\ & 1 + n_1^2(\gamma - 1) & n_1 n_2(\gamma - 1) & n_1 n_3(\gamma - 1) \\ & & 1 + n_2^2(\gamma - 1) & n_2 n_3(\gamma - 1) \\ & (\text{symmetric}) & & 1 + n_3^2(\gamma - 1) \end{bmatrix} \begin{bmatrix} \mathrm{d}t \\ \mathrm{d}x \\ \mathrm{d}y \\ \mathrm{d}z \end{bmatrix}. \tag{6.32}$$

This matrix equation can be expressed more conveniently by writing it in terms of its components, while introducing a new and very useful notation where the indices indicate a frame change:

$$\mathrm{d}x^{\alpha'} = \Lambda^{\alpha'}_{\beta}\, \mathrm{d}x^\beta. \tag{6.33}$$

Such compact notation needs some explanation. We used the usual vector notation of a raised index, but the prime was placed not on the $\mathrm{d}x$ but instead on its index. The 4×4 Lorentz transform matrix in (6.32) has been denoted by its $(\alpha, \beta)^{\text{th}}$ component, again with the prime placed over the upper index, indicating that it transforms unprimed indices to primed ones (i.e. frames $S \to S'$).

Also making its debut in (6.33) is the *Einstein summation convention*, which requires any index that appears both raised and lowered to be automatically summed over. This is of immediate use here because matrix multiplication uses a sum over adjoining repeated indices, such as in (2.114). To annotate the entries of a general matrix A, we denote A^i_k as the element of row i, column k, so that Einstein's summation convention would then allow the generic matrix multiplication (2.114) to be written as

$$(AB)^i_j = A^i_k\, B^k_j\,. \tag{6.34}$$

The practice of putting the prime on the index eliminates the need to write a plethora of different symbols. For example, the four-vector of components uses the symbols $\mathrm{d}x$ in each frame, but the prime tells us that $\mathrm{d}x^{\alpha'}$ refers to the S'-frame while $\mathrm{d}x^\alpha$ refers to the S-frame. If we want to write the four-vectors as complete entities, without explicit mention of components, we then *do* need to write the prime explicitly. In that case, we might write $\overrightarrow{\mathrm{d}x} \equiv (\mathrm{d}t, \mathrm{d}\boldsymbol{x}) \equiv (\mathrm{d}t, \mathrm{d}x, \mathrm{d}y, \mathrm{d}z)$ and $\overrightarrow{\mathrm{d}x'} \equiv (\mathrm{d}t', \mathrm{d}\boldsymbol{x}') \equiv (\mathrm{d}t', \mathrm{d}x', \mathrm{d}y', \mathrm{d}z')$.

So $\mathrm{d}x^{\alpha'}$ is really the αth component of $\overrightarrow{\mathrm{d}x'}$. Calling it the α'^{th} component is usually harmless, but can sometimes be unwise if we want to keep track of the indices when working with matrix expressions. [For an example of where care is needed, see the discussion around (8.160).] When all is said and done, the four-vector $\overrightarrow{\mathrm{d}x}$ is usually just referred to by its general component $\mathrm{d}x^{\alpha}$. But much more will be said about this in Chap. 8.

The term $\Lambda^{\alpha'}_{\beta}$ is also understood to refer to the whole Lorentz transform matrix, despite really referring to just its $(\alpha,\beta)^{\text{th}}$ component. What is the inverse transform? We could work it out the hard way by inverting (6.32), but we can make do with much less effort simply by switching the signs of all velocity terms.

We denote the fact that the Lorentz transform matrix has been inverted, by writing the inverse with the prime now only on the subscript, since the inverse matrix transforms the primed frame to the unprimed frame:

$$\mathrm{d}x^{\alpha} = \Lambda^{\alpha}_{\beta'}\,\mathrm{d}x^{\beta'}\,, \tag{6.35}$$

with

$$\Lambda^{\alpha}_{\beta'} = \begin{bmatrix} \gamma & \gamma v_1 & \gamma v_2 & \gamma v_3 \\ & 1+n_1^2(\gamma-1) & n_1 n_2(\gamma-1) & n_1 n_3(\gamma-1) \\ & & 1+n_2^2(\gamma-1) & n_2 n_3(\gamma-1) \\ & \text{(symmetric)} & & 1+n_3^2(\gamma-1) \end{bmatrix} = \left(\Lambda^{\alpha'}_{\beta}\right)^{-1}. \tag{6.36}$$

A close look at (6.33) and (6.35) reveals something else. It follows from these equations that the partial derivative of any coordinate of one frame with respect to any coordinate of the other must be the appropriate Lorentz transform matrix element:

$$\frac{\partial x^{\alpha'}}{\partial x^{\beta}} = \Lambda^{\alpha'}_{\beta}\,, \quad \text{and similarly} \quad \frac{\partial x^{\alpha}}{\partial x^{\beta'}} = \Lambda^{\alpha}_{\beta'}\,. \tag{6.37}$$

The product of these matrices that represent opposite Lorentz transforms must be the unit matrix, since one Lorentz transform undoes the other:

$$\Lambda^{\alpha'}_{\beta}\,\Lambda^{\beta}_{\gamma'} = \delta^{\alpha'}_{\gamma'}\,. \tag{6.38}$$

Here, the frame-independent Kronecker delta could equally well have been written as δ^{α}_{γ}, but by writing the indices as α',γ', with each holding its up-down position on both sides of the equation, a sort of positional invariance of the components allows for an efficient visual bookkeeping to check that no mistakes have been made.

The product (6.38) is true for the physical reason of representing opposite Lorentz transforms, but it also holds due to the chain rule for the partial derivatives of (6.37). For the same reason, with general changes of coordinates (not just Lorentz transforms), the product of these matrices of partial derivatives, known as *jacobian matrices*, will always be one (i.e., the multiplicative identity matrix). We will see more of jacobian matrices in Chap. 8.

Returning to our candidate four-vector k^α, we wish to show that it transforms under a Lorentz transform in just the way that a four-vector should, as

$$k^{\alpha'} = \Lambda^{\alpha'}_{\beta}\, k^\beta . \tag{6.39}$$

We'll do so by starting with the left-hand side of (6.39). The following proof is inelegant, but deliberately so. What it shows is that in the case of the frequency–wavenumber, something is missing in the four-vector formalism, which will be remedied at the end of the proof in a very simple and elegant way.

The phase of the wave must be a scalar since all observers must agree on how far through its cycle a sine wave is; so the phase ϕ needs no prime, and the resulting primed frequency–wavenumber is

$$k^{\alpha'} = \left(\frac{-\partial\phi}{\partial t'}, \frac{\partial\phi}{\partial x'}, \frac{\partial\phi}{\partial y'}, \frac{\partial\phi}{\partial z'}\right). \tag{6.40}$$

The minus sign in the first term will make the following calculation quite awkward. It has upset the symmetry, requiring us to treat the temporal part separately, and that can only be a bad sign as far as our goal of unifying space and time goes! Using the chain rule for partial derivatives along with (6.37), we write

$$k^{t'} = \frac{-\partial\phi}{\partial t'} = \frac{-\partial\phi}{\partial x^\beta}\, \Lambda^\beta_{t'} . \tag{6.41}$$

Comparing this with (6.39), it seems that the t' index is in the wrong place here. The Einstein summation convention, along with the notation of having some indices raised and some lowered, really demands that all equal indices should be either raised or lowered. The t' "should" be raised on the right-hand side of (6.41) just as it is on the left. We seem to have been forced into using the *inverse* Lorentz transform matrix.

The only way out of our predicament is to employ some symmetries in the matrices (6.32) and (6.36) to try to raise that t' index. To write them, here and throughout this book we'll use Greek indices to denote all of the coordinates, in this case 0 to 3, while Latin indices will denote a subset, which here will be the space coordinates 1 to 3. (This is a fairly standard but not universal convention, and in particular some authors will swap the roles of Greek and Latin.) The necessary symmetries are (where t is of course time, and not an index)

$$\begin{aligned} \Lambda^t_{t'} &= \Lambda^{t'}_t , & \Lambda^a_{t'} &= -\Lambda^{t'}_a , \\ \Lambda^t_{a'} &= -\Lambda^{a'}_t , & \Lambda^b_{a'} &= \Lambda^{a'}_b . \end{aligned} \tag{6.42}$$

Returning to (6.41), we treat its time and space components separately to try to make sense of those minus signs:

$$k^{t'} = \frac{-\partial\phi}{\partial t}\, \Lambda^t_{t'} - \frac{\partial\phi}{\partial x^a}\, \Lambda^a_{t'} = k^t\, \Lambda^t_{t'} - \sum_a k^a\, \Lambda^a_{t'}$$

$$= k^t \Lambda_t^{t'} + k^a \Lambda_a^{t'} \quad \text{[using the symmetries (6.42)]}$$
$$= k^\beta \Lambda_\beta^{t'} . \tag{6.43}$$

So far, so good: we see that (6.39) holds for $\alpha = t$. The space components are treated in just the same way, so omitting the details, the result for them is

$$k^{a'} = k^\beta \Lambda_\beta^{a'} . \tag{6.44}$$

Finally, combine (6.43) with (6.44) to give Greek indices throughout:

$$k^{\alpha'} = k^\beta \Lambda_\beta^{\alpha'} . \tag{6.45}$$

The proof is finished. $\vec{k}$ changes between frames via the Lorentz transform $\Lambda_\beta^{\alpha'}$ and so is indeed a four-vector.

This four-vector character relates two of the fundamental postulates of quantum mechanics: $E = \hbar\omega$ and $\boldsymbol{p} = \hbar\boldsymbol{k}$. We have shown that both of these involve four-vectors:

$$(E, \boldsymbol{p}) = \hbar(\omega, \boldsymbol{k}) , \quad \text{or } \vec{p} = \hbar\vec{k} , \quad \text{or } p^\alpha = \hbar k^\alpha . \tag{6.46}$$

But this last expression cannot be used to show that $(\omega, \boldsymbol{k})$ is a four-vector based on our earlier result that $(E, \boldsymbol{p})$ is a four-vector, because (6.46) is a *postulate*: a statement of physics. Each of the entities $E, \boldsymbol{p}, \omega, \boldsymbol{k}$ is well defined independently of quantum mechanics, and it makes no sense to use any postulate to establish a relationship between them.

A useful corollary to the result that $\vec{k}$ is a four-vector is the elegance with which the well-known expression for the relativistic Doppler shift can be derived:

Calculate the relativistic radial redshift. That is, a spacecraft is receding from us at velocity v and it sends us a signal. How is its frequency changed?

The frequency change is usually derived by considering how the light waves change in frequency as they "wash over" us, being careful to remember that we also measure the spacecraft's clock to be timing slowly! But the power of four-vectors can be used to produce the result far more elegantly and economically now that we have done the groundwork.

We on Earth occupy the frame S, while the spacecraft occupies S', moving along our x-axis (which is radial) with velocity v (which can be negative—the craft can be approaching us). Since the spacecraft has been designed to emit a frequency f' in its own frame, we can immediately write down the components of the frequency–wavenumber for S' (remembering that $\boldsymbol{k}$ is pointing back to us):

$$\omega' = 2\pi f' , \quad \boldsymbol{k}' = -(2\pi/\lambda', 0, 0) = (-2\pi f', 0, 0) , \tag{6.47}$$

using $c = 1$ as usual. Thus $k^{\alpha'} = (2\pi f', -2\pi f', 0, 0)$, and it's a simple exercise to transform to the S-frame using the inverse Lorentz transform:

$$\omega = k^t = \gamma(k^{t'} + vk^{x'}) = \gamma 2\pi f'(1 - v) \,. \tag{6.48}$$

This very straightforward derivation shows that the Doppler-shifted frequency received on Earth is $f = f'\gamma(1 - v)$.

6.5 Introducing Covectors and Fully Covariant Notation

There were two awkward spots in the proof of (6.39)–(6.45) that k^α is a four-vector. The first was that the appearance of the minus sign in the definition of ω in (6.31) seemed to throw a spanner in the works, forcing us to treat time and space components separately. The second was that we needed to take advantage of symmetries in the Lorentz transform. Although we succeeded, our aim ultimately is to discuss how physical laws are subject to general changes of coordinates, not just the Lorentz transform. In those more general cases, we cannot count on there being symmetries to take advantage of.

We will try to combine ω and $\boldsymbol{k}$ again, but this time changing the troublesome sign of ω. So define something new, called $K^\alpha \equiv (-\omega, \boldsymbol{k})$, which together with (6.31) means that $K^\alpha = \partial\phi/\partial x^\alpha$. This cannot be a four-vector since we can already relate ω and $\boldsymbol{k}$ by a Lorentz transform, and we would need to switch *all* of the signs to keep that relation correct, not just the sign of ω. How then does K^α transform? Use the chain rule for partial derivatives in (6.37):

$$K^{\alpha'} = \frac{\partial\phi}{\partial x^{\alpha'}} = \frac{\partial\phi}{\partial x^\beta}\,\Lambda^\beta_{\alpha'} = \sum_\beta K^\beta\,\Lambda^\beta_{\alpha'} \,. \tag{6.49}$$

The transformation is very straightforward, with time and space components needing no separation—except that K^α transforms by the *inverse of the transpose* of the Lorentz transform! In a sense it's a kind of inverse to the four-vector k^α, so we write it as k_α with a lowered index to emphasise this point, discarding the capital K to save on notation. Now (6.49) is written much more elegantly:

$$k_{\alpha'} = k_\beta\,\Lambda^\beta_{\alpha'} \,. \tag{6.50}$$

These lowered-index quantities might remind us of the cobasis components in Chap. 2. Just as the ordered set of numbers k^α is called a four-vector, the ordered set of numbers k_α is called a *covector*. And just as the name four-vector was reserved on p. 219 for the numbers k^α only, with no mention of any basis vectors, we will reserve the name covector for the ordered set of k_α, again with no mention of any basis. In Chap. 8 it will be shown that the four-vector components k^α and the covector set k_α are in fact the components of the *same* object in two different bases that are related in just the same orthonormal way as were the basis and cobasis discussed in Sect. 2.3. But for now and the remainder of this book, four-vectors and covectors are understood to mean ordered sets of numbers only; when we eventually introduce the appropriate bases, the new objects combining components with bases will be called proper vectors (or just simply vectors).

Note that in many books the term "covector" is synonymous with an object called a "one-form", which we'll meet briefly in Chap. 8 but will not be useful to us. The two words covector and one-form are *not* synonymous in this book. We take a covector to denote an ordered set of numbers with the properties investigated here.

Just as four-vectors are produced naturally from quantities related to infinitesimals of spacetime coordinates (such as the four-velocity), covectors are produced naturally from quantities that result from *differentiation* with respect to those coordinates. If we want to write (6.50) as a matrix equation, we need to bear in mind that it indicates matrix multiplication only as long as the repeated index β is written on the "inside" as in (2.114). Because of this, a covector set should be written as a row vector to enable k_β to be interpreted as its $(1,\beta)^{\text{th}}$ element, which is the only way to keep the β's adjacent to each other as per (2.114). So four-vectors are naturally written as columns, and covectors as rows. Analogously, if A is a one-row and B a one-column matrix, then their product has a single element, so (6.34) can be written

$$AB = A_k B^k. \tag{6.51}$$

This reinforces the idea that sets of numbers described by lowered indices (covectors) are represented by single-row matrices, while sets of numbers described by raised indices (four-vectors) are represented by single-column matrices.

By denoting four-vectors with raised indices and covectors with lowered indices, expressions such as (6.39) and (6.50) preserve the up or down positions of like indices, which makes them easy to check quickly for correct index bookkeeping. They also take advantage of the simplicity of the Einstein summation convention. This highly compact notation is termed *covariant*, and forms the main language of mathematical physics when we are dealing with changes in frames and coordinates. In the next chapter, we'll reinforce covariant notation by showing how it helps us apply the Lorentz transform to analyse frames with changing velocities.

7 Accelerated Frames: Onward to the Principle of Covariance

Two chapters back, in (5.11), we saw that when two clocks synchronised in their shared rest frame move past us with constant velocity v, the spatially trailing clock leads its partner by a time vL_0/c^2, where their proper separation is L_0.

But on some reflection this leads to a difficulty. Suppose that one of the synchronised clocks sits by our side, while the other lies in a distant galaxy, one thousand million light-years away. Leaving aside questions as to the meaning of distance on a cosmological scale, set their proper separation to be $L_0 = 10^9$ light-years. Now suppose that we on Earth pace slowly to and fro, moving with speed $v = 10^{-9}c$ in each direction.

In that case, the clock next to us alternately leads and trails its partner by one year. But because it's right next to us, we can see that its time is not changing at all by any more than the few seconds we spend pacing. That implies that the clock in the distant galaxy is alternately jumping ahead of us by one year and then suddenly dropping behind by the same amount; and this see-sawing continues for as long as we pace to and fro. Can this apparently nonsensical state of affairs really be happening?

Yes, the time we measure on the distant clock really is swinging back and forth wildly, as we accelerate periodically to switch our walking direction. This is a symptom of the fact that as we pace to and fro, we no longer inhabit a single inertial frame; in fact, we no longer inhabit even a single *accelerated* frame.

Inertial frames were the subject of the previous two chapters and form the backbone of special relativity. They are the only type of frame in which physical processes take on a certain simplicity. For instance, a thrown object will maintain a constant velocity, as will a beam of light.

For very small distances and times, the frame of a laboratory is approximately inertial. In a small laboratory, there is almost no deviation between the paths of two neighbouring falling objects, and those objects maintain a fairly constant velocity over a short time period. The classic undergraduate physics task of analysing the explosion of a projectile in terms of the momenta of the resulting fragments is able to be done because in the very small time interval of the explosion, gravity has very little effect on the projectile, both before and after it breaks apart. The calculation can then be done assuming an inertial frame, and gravity need not be taken into account—unless we wish to track the fragments over longer times.

But our day-to-day experience of the world is that we are not inertial. Here on Earth, an object thrown in the air soon falls down, and the speed of light turns out to be influenced by gravity. It's as if we are living in a rocket permanently accelerated at one Earth gravity. This is the heart of Einstein's *Equivalence Principle*, which states that when we confine our attention to a small volume of space for a short period of time, the gravitational field there will be identical in its effects to a uniformly accelerated frame.

So apparently we can imagine that on Earth we inhabit a uniformly accelerated frame. As such, it is the most natural frame we have. Apart from the few occasions on which we might jump into the air (during which we are weightless), our whole life is lived within the confines of an accelerated frame, not an inertial one. So it is that we would like to investigate accelerated frames by deriving the transformation equations that relate their description of events to an inertial frame's description of the same events.

Our aim here is to use special relativity to build the basic framework that allows us to write the laws of physics in an accelerated frame, and to use this to get an introductory feel for some of the relativistic effects of gravity via the Equivalence Principle.

Accelerated frames are seldom explored in any depth in textbooks. This is mostly because the laws of physics quickly become complicated even in this first departure from the comfortable inertial frame. But discussing the form taken by those laws in an accelerated frame is very useful for furthering an understanding of relativity.

What exactly constitutes a frame? A frame is a collection of observers, each taking notes about what happens in their immediate vicinity. Why only their immediate vicinity? When we are analysing the events of the world around us, we should not worry about distant events ourselves—information from them takes time to reach us, and if we don't know how far away they are, then we cannot correct for this time interval. Our frame consists of an infinite number of synchronised observers, all at rest with respect to us, and each only concerned with writing down the time that an event occurs provided only that it happens right under their nose. None of them care about distant events. Each only looks after their own vicinity and reports back to us periodically, allowing us to collate their observations and so construct a coherent picture of the events happening throughout the frame.

It is sometimes said that to describe physics properly in an accelerated frame, special relativity is insufficient, and the full machinery of general relativity is necessary for the job. This is quite wrong. Special relativity is entirely sufficient to derive the physics of an accelerated frame.

In the frame local to Earth's surface, in which we live our everyday lives, we notice a constant push on our bodies from the ground. Anything unsupported or throw in the air will fall down, and inside a closed room we can well imagine that we are not really on Earth at all. Rather, we could well be in a rocket that constantly accelerates at one Earth gravity in deep space. The rocket does not form an inertial frame, so we would like to go beyond the Lorentz transform to discover the equations that relate measurements in our rocket frame to those of inertial observers.

The first step in constructing this new transform is to locate the events of our rocket frame in the "outside" inertial frame. The simplest scenario is to take our acceleration as truly constant, which means we have been accelerating forever—if this is allowed by special relativity. Fuel considerations aside, such an acceleration *is* possible. It's made possible by the third postulate of special relativity, called the *Clock Postulate.*

7.1 The Clock Postulate

We saw the first two postulates of special relativity on p. 187. First, all inertial frames are equally valid, and second, light travels at the same speed in all inertial frames. But what happens if we wish to use special relativity to describe an object that doesn't have a constant velocity?

As we accelerate in our rocket, our worldline in the inertial frame of an outside observer is not straight, not even in any limit. Its curvature does not go away as we focus on smaller and smaller segments of the worldline, just as the dizzying acceleration that we feel on a roller-coaster ride does not go away just because we might want to concentrate on shorter and shorter segments of the ride.

Nevertheless, the Clock Postulate *does* state that our curved worldline can be considered to be, in a sense, a kind of joining together of a series of inertial frames, each one different, but with no acceleration terms appearing at all. It says that the physics we observe to be happening around us can always be analysed by asking only what is observed at some instant by an inertial observer who at that moment shares our position and velocity: a *momentarily comoving inertial observer*, who is at rest in a *momentarily comoving inertial frame*, or MCIF. (It turns out that this postulate also applies when gravity is present, as we'll see in Chap. 12.) An accelerated observer, together with one of his MCIFs, is shown in Fig. 7.1.

So, for instance, our accelerated clock's rate is identical to the clock rate in the MCIF. We can imagine this frame as holding an inertial clock that for a brief moment is moving alongside us, so that our relative velocity is momentarily zero. At that moment, both accelerated and inertial clocks are ticking at the same rate—and both must agree on this fact, since they can always be placed together so that their mechanisms are working in unison, where we could arrange an experiment where they physically interact with each other to confirm this. A moment later, our accelerated clock has a new MCIF, again one that matches our velocity just at that moment, and there is a new inertial clock that briefly slows to a stop alongside us and our accelerated

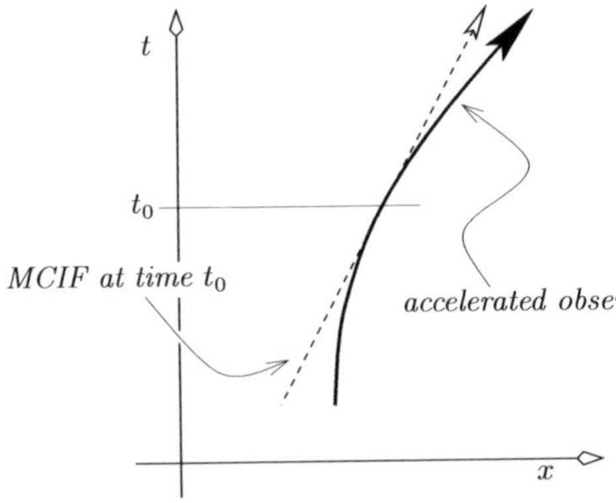

Fig. 7.1. The Clock Postulate states that the physics described by an accelerated observer at time t_0 is identical to that described by a momentarily comoving inertial observer at this time. That observer, as shown, is only *momentarily* comoving at t_0 and so forms a valid MCIF only at that moment. For earlier and later times, a different observer and MCIF must be used.

clock. And again, the rates of our accelerated clock and the new inertial one will be momentarily the same.

An alternative way of stating the Clock Postulate is to say that the timing rate of the accelerated clock slows compared with an inertial clock according to the usual γ-factor of special relativity, $\gamma = 1/\sqrt{1-v^2}$, where v is the speed that the inertial observer measures our accelerated clock to have. In other words, our acceleration does not enter into the calculation at all. This is a postulate because we cannot be sure that the "correct" expression for γ is really lacking an acceleration term that goes to zero in the usual derivations of special relativity, since these derivations only deal with nonaccelerating observers. We just postulate that γ is not a function of acceleration.

Not only does the postulate say that a clock's timing is not affected by its acceleration, but it also states that neither do *higher* derivatives of velocity have any effect. The γ-factor of the accelerating clock is now a function of time, but at any instant it equals precisely the γ-factor of its MCIF. And as well as time intervals, the Clock Postulate also applies to the shortening of rods and the relativistic mass of a moving object, since these are both measurements set by the MCIF's γ-factor.

Of course, the ultimate test of any postulate is whether or not it stands up to experiment. The Clock Postulate does phenomenally well. It has stood the test of experiment with accelerations as high as 10^{18} times Earth's gravity.

We should understand the Clock Postulate well. While it says that the timing rate of an accelerated clock doesn't depend on its acceleration, it does *not* say that this rate is *unaffected* by the clock's acceleration. The timing rate will certainly be affected if the acceleration changes the clock's speed of motion, because its speed determines how fast it counts out its time—i.e., by the γ-factor. But the timing rate won't be affected by the accelerations felt along non-straight-line motion at constant speed.

If this choice of words sounds paradoxical, consider an everyday analogy. When riding your bicycle on an icy morning, you get very cold due to the wind chill factor: the faster you go, the colder your hands get. This wind chill is a function of your speed but not your acceleration. Nevertheless, it *is* affected by your acceleration when your acceleration changes your speed. But irrespective of whether you have a low or a high acceleration, the only thing that matters as far as your cold hands are concerned is your current speed. So, for example, in circular motion, two cyclists who follow different-diameter circles at the same speed will feel the same wind chill, even though they have different accelerations.

Another example of such a dependence comes from electromagnetism. The force on a charge q when moving with velocity $\boldsymbol{v}$ in electric and magnetic fields $\boldsymbol{E}$ and $\boldsymbol{B}$ is $\boldsymbol{F} = q(\boldsymbol{E} + \boldsymbol{v} \times \boldsymbol{B})$. This force is independent of the charge's acceleration. But any acceleration that the charge might have will certainly change its velocity and so change the force. So although the force is not an explicit function of the acceleration, it certainly is affected by the acceleration.

A last simpler example concerns the value of the local civilian times that we encounter while flying around the world. This is a function of our position but not our flight speed. But again, it certainly is affected by our flight speed because that speed changes our position.

The Clock Postulate is not meant to be obvious, and it cannot be proved. It is just a postulate. For instance, we cannot magically verify it by noting that the Lorentz transform is only a function of speed, because the Lorentz transform is something that's built before the Clock Postulate enters the picture. Also, we cannot simply wave our arms and maintain that an acceleration can be treated as a sequence of constant velocities, each of which exists only for an infinitesimal time interval. After all, an accelerating body (away from gravity) feels a force, while a constant-velocity body does not. Although the Clock Postulate does speak in terms of constant velocities and infinitesimal time intervals, there is no a priori reason why that should be meaningful or correct. This is just like the fact that even though a 1000-sided polygon looks a lot like a circle, a small piece of a circle cannot always be treated as an infinitesimal straight line. For, no matter how small the circular arc is that we take from the circle, it will always have the same radius of curvature as the circle, whereas a straight line has an infinite radius of curvature. Also, it won't do to simply *define* a clock to be a device whose timing is unaffected by its acceleration, because it's not clear what such a device has to do with the real world; that is, it's not clear how well it approximates the mechanism that chimes on the wall, or that we wear on our wrist.

The Clock Postulate is important and deserves more attention than it usually gets because it allows us to geometrise relativity, and this is the first step toward creating the theory of general relativity. And as we shall see at the end of this chapter, it also allows us to develop further the notion of covariance introduced at the end of the previous chapter, which enables the equations of physics to be written in a frame-independent way.

7.1.1 The Interval for Noninertial Observers

In Sect. 5.6, we saw the idea of the interval or metric as applied to inertial observers. But what is the time elapsing on a *noninertial* (i.e. accelerated) clock connecting two widely separated events? We don't expect an expression as simple as $\Delta t^2 - \Delta x^2 - \Delta y^2 - \Delta z^2$ because that expression is only a function of the endpoints, and encodes no information about the path that the clock took and its speed at various points. The assumption we make, that the Clock Postulate holds, is something that really leans on experiment for its ultimate justification. It was first postulated by Einstein in the following way. He considered a curved worldline to be the limiting case of part of a polygon, in essence considering two neighbouring observers: one traversing the curved worldline and the other traversing the polygonal worldline. These observers are always very close to each other, with a relative speed that tends to zero in the limit as the number of polygonal segments tends to infinity. It's unreasonable to expect that the measurements of one should be different from those of the other. Additionally, the polygonal observer's clock is inertial while it traverses each straight segment of that polygon, and the only effect the acceleration could possibly have would have to occur at the vertices where it changes direction. But such an effect can only last for the zero time that the clock takes to change direction at the vertex. So, no matter how many vertices we give the polygonal worldline, any effect of acceleration would happen for a total time of zero, and so a clock travelling on such a worldline cannot be affected by those accelerations. The real step into the unknown that Einstein took was to postulate that the curved worldline really can be considered to be the limiting case of a polygonal worldline in such a physical way.

Postulating this, we can infer that when an accelerating clock moves from one event to another that is infinitesimally close, the infinitesimal time $\mathrm{d}\tau$ elapsed is given by

$$\mathrm{d}\tau^2 = \mathrm{d}t^2 - \mathrm{d}x^2 - \mathrm{d}y^2 - \mathrm{d}z^2 \tag{7.1}$$

since this is just the time that elapses on a clock in the MCIF. And we can now integrate this $\mathrm{d}\tau$ along the accelerated clock's worldline to get the actual elapsed time $\Delta\tau$ that it reads. The fact that we *can* calculate the time elapsed on a moving clock purely by reference to its path through spacetime allows spacetime to be imbued with a structure of its own, and this is why the postulate opens up the possibility of the geometrisation of spacetime that leads to general relativity.

This sort of idea is exactly analogous to the idea of calculating the length $\Delta\ell$ of a space curve by dividing it into a large number of short segments, each of which is "almost straight", and then summing the lengths $d\ell$ of each of those using Pythagoras's theorem:

$$\Delta\ell = \int d\ell\,, \quad \text{where } d\ell^2 = dx^2 + dy^2 + dz^2. \tag{7.2}$$

That, of course, is a basic tenet of integral calculus. Despite the fact that we are treating each short segment as straight, we know that it definitely is not, not even in the limit. It still has some curvature that does not go away as we divide it into ever-smaller segments. But the usual theorems of calculus show that the total error introduced, as we approximate the segments as straight, gets ever smaller when the number of segments becomes larger and larger, each one tending to zero length.

So, the idea of calculating the total time elapsed on a clock that accelerates from one event to another is just the same as calculating the total length of a curve with an integration. We are dividing the clock's worldline into small segments that the Clock Postulate lets us work with. And just as the curvatures of small curve segments in the euclidean geometry never become zero (and yet we can ignore them), the acceleration of the clock along its worldline never becomes zero either—but analogously, we ignore it, too.

This, then, is why we can give some structure to spacetime, because it's possible to talk about the "length" of a curved worldline as being just the time shown on a clock moving along it, even though the clock itself is accelerating. If the "length" of such a curve depended upon its curvature, then it would not be an intrinsically geometric thing. The Clock Postulate geometrises spacetime.

Ultimately, this is why the interval is usually written using infinitesimals. It has nothing to do with notions of curved spacetime; we use infinitesimals even if the spacetime is flat. We use them because in this infinitesimal form the interval embodies the Clock Postulate. But now the grand thing is that this idea of spacetime structure enables us to speak of spacetime as a separate entity with its own metric—an idea that allows the eventual transition to general relativity to occur when gravity is considered in more detail.

The great success of general relativity lends plausibility to the parts that make it up—and one of these is the Clock Postulate. Even if the postulate had never been tested experimentally, we would still have confidence in it because the theory towards which it paves the way—general relativity—has been tested experimentally and verified to an extraordinarily high accuracy. (But see the discussion of reverse logic on p. 108.) Although once thought to be a theory that only found application in rarefied areas such as light bending around the Sun and eccentric planetary orbits, general relativity is now used and verified daily all over the world, since its application forms an integral part of the global satellite systems that determine the positions of all manner of vehicles, from aircraft to taxis.

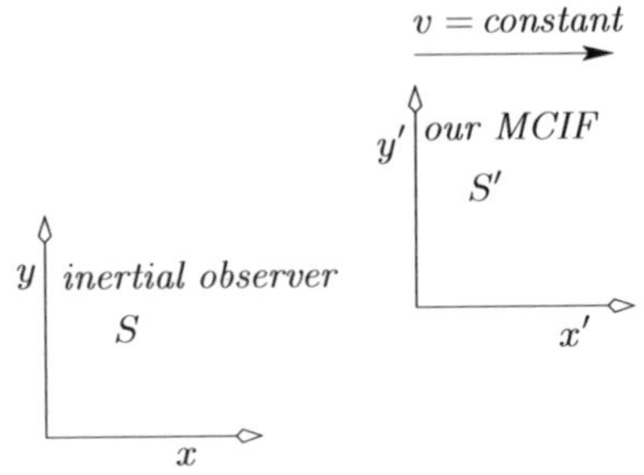

Fig. 7.2. The fixed inertial frame together with our MCIF. Note that both of these frames are inertial: v is constant.

7.2 Coordinates for the Accelerated Frame

Finding a set of coordinates that relate our accelerated frame to an inertial one is a very useful exercise, partly as a gentle introduction to the ideas of general relativity in a special relativity setting, and partly because the resulting coordinates have a lot in common with the description of a *Schwarzschild black hole*, as we'll see later in Fig. 12.6 and also when comparing Figs 7.6 and 12.5.

So the plan is to find these coordinates that relate an accelerated frame to an inertial one. The very first thing to do is to establish our worldline in that inertial frame.

What does it really mean to have a constant acceleration forever? Is there no problem with exceeding the speed of light? There would certainly be a problem if the inertial observer measured us to be accelerating constantly. But this actually is not what we are demanding. We wish for our rocket's engine to apply a constant force, felt by us as a constant weight that presses us into our seats with a force of, say, one Earth gravity. What this means is that *our acceleration as measured in our MCIF is constant*, since in the MCIF our tendency is to coast at constant velocity as dictated by Newton's first law of motion. Anything on top of this we will feel as an applied force. So our acceleration must be referred to our MCIF, and there is certainly no problem at all with firing our rocket motors in such a way that we always accelerate in our MCIF.

To analyse such a motion, refer to Fig. 7.2, which shows the "fixed" inertial frame S of the outside observer along with our MCIF S' at some moment. We wish to draw our worldline in the inertial frame S, which we'll do by finding the S coordinates t, x of our motion parametrised as functions of the time τ shown on our clock. For us to feel a constant force, our acceleration must be some constant g in the MCIF:

$$\frac{\mathrm{d}^2x'}{\mathrm{d}t'^2} = g\,. \tag{7.3}$$

If we can express t', x' in terms of t, x, then the resulting differential equation will describe our motion in S; we can do that through the Lorentz transform that relates S to S'. Note carefully that γ and v must be treated as constants here because we are only considering inertial frames. The MCIF we have drawn in Fig. 7.2 might only be momentarily comoving, but it is certainly an inertial frame forever, by definition. Our motion is described by $x(t)$, so write, from (5.17) or (6.12) with an overdot meaning $\mathrm{d}/\mathrm{d}t$,

$$\frac{\mathrm{d}x'}{\mathrm{d}t'} = \frac{\mathrm{d}x'/\mathrm{d}t}{\mathrm{d}t'/\mathrm{d}t} = \frac{\dot{x} - v}{1 - v\dot{x}} \,. \tag{7.4}$$

Similarly, the next derivative is

$$\frac{\mathrm{d}^2x'}{\mathrm{d}t'^2} = \frac{\mathrm{d}\left(\frac{\dot{x}-v}{1-v\dot{x}}\right)/\mathrm{d}t}{\mathrm{d}t'/\mathrm{d}t} = \frac{\ddot{x}}{\gamma^3(1 - v\dot{x})^3} \,. \tag{7.5}$$

The next step is to realise that the MCIF is just that: *momentarily* comoving, so to incorporate the succession of our MCIFs we must replace v by $\dot{x}$. If it seems odd that we are replacing a constant by something that is not constant, realise that what we are doing is first applying the Lorentz transform, which applies to the MCIF (which by definition has a constant v), and only then are we incorporating the bigger picture of a succession of MCIFs by replacing v with the nonconstant $\dot{x}$.

Equations (7.3) and (7.5) together give the final differential equation to be solved:

$$g = \frac{\ddot{x}}{\left(1 - \dot{x}^2\right)^{3/2}} \,. \tag{7.6}$$

This has the parametrised solution in terms of hyperbolic functions

$$\begin{aligned} t &= g^{-1}\,\mathrm{sh}\, a\lambda + b\,, \\ x &= g^{-1}\,\mathrm{ch}\, a\lambda + c\,, \end{aligned} \tag{7.7}$$

for some constants a, b, c and some parameter λ.

The hyperbolic functions are commonly written sinh, cosh, tanh, and sometimes sh, ch, th, as here. These abbreviations are not aimed at saving ink; rather, such notation is all about developing a useful mathematical style when we need to write calculations by hand over many pages. After all, writing cosh as ch loses no information and is not confusing. On the other hand, severe shortening of notation can certainly be confusing when it takes over notation that already exists, as we'll see with the case of the exterior derivative at the end of Chap. 8.

Ideally, the parameter λ is best related to the time shown on our clock, since that is the one obvious parameter that increases along our worldline. From moment to moment, the time elapsing on our clock is the proper time $\mathrm{d}\tau$, where (7.7) gives

$$\mathrm{d}\tau^2 = \mathrm{d}t^2 - \mathrm{d}x^2 = \frac{a^2}{g^2}\,\mathrm{d}\lambda^2 . \tag{7.8}$$

There is freedom to set $a \equiv g$ and $\lambda \equiv \tau + k$ for some constant k. Note that

$$\mathrm{d}x/\mathrm{d}t = \mathrm{th}[g(\tau + k)] , \tag{7.9}$$

and also, calibrate our clock by ensuring that

$$\tau = \tau_0 \Longleftrightarrow (t, x) = (t_0, x_0) . \tag{7.10}$$

In that case, (7.7) can be written in terms of these initial conditions as

$$\begin{aligned} t &= g^{-1}\,\mathrm{sh}\,[g(\tau + k)] + t_0 - g^{-1}\,\mathrm{sh}\,[g(\tau_0 + k)] , \\ x &= g^{-1}\,\mathrm{ch}\,[g(\tau + k)] + x_0 - g^{-1}\,\mathrm{ch}\,[g(\tau_0 + k)] , \\ &\text{where } \mathrm{d}x/\mathrm{d}t|_{\tau=\tau_0} = \mathrm{th}[g(\tau_0 + k)] \text{ determines } k. \end{aligned} \tag{7.11}$$

Choose some simplifying conditions. At $t = t_0 \equiv 0$, our velocity is zero and our clocks read $\tau = 0$. Further, set $x_0 \equiv 1/g$ to give the simplest possible expression for our worldline [but see the aside around (7.30)]:

$$\begin{aligned} t &= g^{-1}\,\mathrm{sh}\,g\tau , \\ x &= g^{-1}\,\mathrm{ch}\,g\tau . \end{aligned} \tag{7.12}$$

These equations describe the hyperbolic worldline of a uniformly accelerated observer, ourselves, who are firing our rocket motors to accelerate forever. That is, we start out far in the past and move toward the origin $x = 0$ of the inertial frame S, always firing our rocket to produce a fixed acceleration away from this origin along the positive-x direction. Eventually we slow to a stop at $t = \tau = 0, x = 1/g$, reverse direction, and pick up speed again, now moving away from the origin. Our worldline is shown in Fig. 7.3.

Figure 7.3 also shows the MCIFs at various times. These have been drawn by reference to the discussion in Sect. 5.7. They are the coordinate axes in S of a moving inertial observer, whose S-frame worldline must be straight. Any inertial frame such as the single-primed MCIF in Fig. 7.3 will be described by the Lorentz transform, and in that case the primed axes will be such that S will draw the t'-axis with slope $1/v$, where v is the velocity of S' (as referred to on p. 203). Thus the t'-axis is just the worldline of the MCIF at the event where $t' = x' = 0$. Likewise, S draws the x'-axis with slope v. These primed axes are orthogonal in the sense of Sect. 5.7. So at any point on our hyperbolic worldline, the time axis of the MCIF is just the tangent to our worldline since we are, by definition, always momentarily at rest in our MCIF. And orthogonal to this time axis is the MCIF's space axis.

Now that we have derived (7.12), which describes our accelerated worldline in S in terms of the time τ on our clock, we need to find coordinates $(\bar{t}, \bar{x})$ relating our frame $\bar{S}$ to the inertial frame S, if indeed such a set can be defined. These barred coordinates are *not* those of any MCIF such as S', S'', S'''.

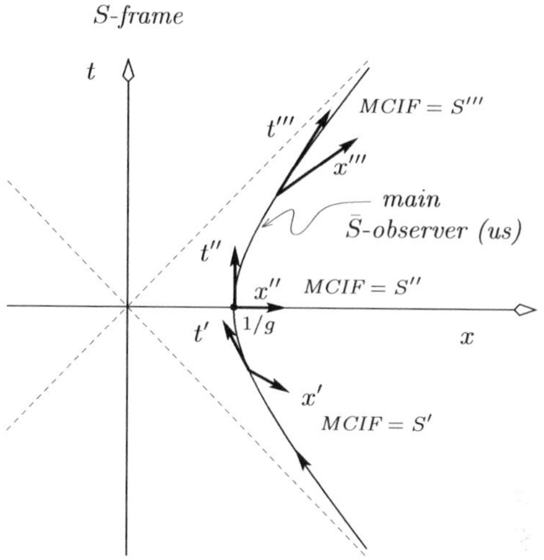

Fig. 7.3. Our hyperbolic worldline in the outside inertial frame S together with axes of some of the MCIFs, S', S'', S''', as described in the text. The positive-x direction is "up" for us, in the sense that we feel a force pressing us into our seats "downward" in the negative-x direction.

Rather, they form a single, global set of coordinates describing our frame $\bar{S}$, and as yet there is no guarantee that they even exist at all.

How do we go about finding such a coordinate set? As with constructing any frame, we need to find a set of observers all of whom agree with us on the simultaneity of events. If nobody can be found who agrees with our timing of events, then a global coordinate system $(\bar{t}, \bar{x})$ cannot be constructed. But happily it turns out that there *is* such a set of observers, and we can find their worldlines in the following way.

First, remember that the line of simultaneity of an inertial observer at any event P, which connects and *defines* the set of events that are simultaneous with P, is just the inertial space axis drawn at P, because it is precisely for all events along this space axis that the inertial time coordinate has a constant value. All lines of simultaneity in S are thus horizontal in Fig. 7.3, while those of our MCIFs (and therefore ourselves) are orthogonal to our worldline's tangent at each point.

Now, let's draw this line of simultaneity through an arbitrary point of our hyperbolic worldline. The tangent has slope $\mathrm{d}t/\mathrm{d}x$, so the corresponding orthogonal line has slope $\mathrm{d}x/\mathrm{d}t$. From (7.12), $\mathrm{d}x/\mathrm{d}t = t/x$, in which case the line of simultaneity for any event on our worldline must pass through the origin of S, $(t, x) = (0, 0)$.

This common intersection point of every line of simultaneity gives rise to a wildly odd state of affairs, since it means that we find the event $(t, x) = (0, 0)$

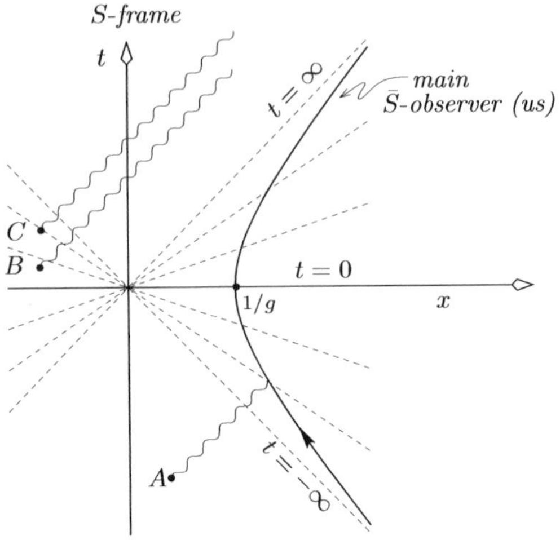

Fig. 7.4. Our worldline in the outside inertial frame S, together with some events of interest. We can *see* event A but cannot ascribe a time to it! On the other hand, we can never see events B and C, but in our frame C happens before B, even if B caused C.

to be simultaneous with *every* event on our worldline, past and future. But the situation gets worse! Consider event A in Fig. 7.4. We can see this event; light from it reaches us just before we come to a halt and reverse direction. But we cannot ascribe a time to it because it is not simultaneous with *any* time on our clock! And more is to come: although we can never see or be affected by events B and C—since light from them can never reach us—our line of simultaneity in the region of those two events is actually tilting more and more downward (increasing its slope) as we age, implying that we must insist that C happened *before* B.

The record of events from the inertial S-frame shows no such bizarre behaviour, and such gentle manners give inertial frames their preferred status among frames in general. But we who accelerate will say that C precedes B, even if B plainly caused C. The fact that we cannot be influenced by those two events seems to be scant consolation for the fact that their causal relationship is wrong.

This worrying state of affairs is partly rescued by virtue of the extremely unphysical nature of our worldline. To say that we are accelerating forever is a very strong statement about the entire universe and time in general. By changing our worldline sometime in our far past, we can certainly arrange for our line of simultaneity to pass through A at *some* time, making it possible to ascribe a time to event A. Similarly, changing our motion in our distant future can undo the wrong causality of B and C. Our line of simultaneity

The Four-Acceleration a^α

Suppose a particle is moving on some arbitrary worldline. At any event, an MCIF can be drawn; its time axis will be tangent to the worldline, and its space axis or axes will be orthogonal to this. We can show that these axes point along four-vectors: the time axis along the particle's four-velocity, and (in one dimension) the space axis along the particle's *four-acceleration*, in the following way. In the absence of gravity and using t, x, y, z-coordinates, define the particle's four-acceleration a^α by

$$a^\alpha = \frac{\mathrm{d}u^\alpha}{\mathrm{d}\tau}, \quad \text{where } u^\alpha \text{ is the particle's four-velocity.} \tag{7.13}$$

(This will be modified in Sect. 12.3.2 when we consider arbitrary coordinates and gravity, but the new expression there reduces to (7.13) in the absence of gravity, such as we have here.) The four-acceleration is a four-vector because since $u^{\alpha'} = \Lambda^{\alpha'}_{\beta} u^\beta$, it follows that

$$a^{\alpha'} = \frac{\mathrm{d}}{\mathrm{d}\tau}\left(\Lambda^{\alpha'}_{\beta} u^\beta\right) = \Lambda^{\alpha'}_{\beta}\frac{\mathrm{d}u^\beta}{\mathrm{d}\tau} = \Lambda^{\alpha'}_{\beta} a^\beta . \tag{7.14}$$

(Remember that the Lorentz transform $\Lambda^{\alpha'}_{\beta}$ is not being differentiated because it uses the (constant) MCIF velocity, not the changing particle velocity.) Now, recall from (6.25) that the length of the four-velocity is always one, irrespective of which inertial frame is used:

$$u^{0\,2} - u^{1\,2} - u^{2\,2} - u^{3\,2} = 1\,, \tag{7.15}$$

so that differentiating both sides with respect to τ yields

$$u^0 a^0 - u^1 a^1 - u^2 a^2 - u^3 a^3 = 0\,. \tag{7.16}$$

This relation holds for any four-velocity. In every one of the particle's MCIFs, its three-velocity is $\boldsymbol{v} = \boldsymbol{0}$ so that $\gamma = 1$, implying that any MCIF measures the particle's four-velocity to be $u^\alpha = \gamma(1, \boldsymbol{v}) = (1, 0, 0, 0)$. Thus u^α is a unit vector pointing along the time axis of any MCIF. But since (7.16) holds in any inertial frame, in particular it means that, in the MCIF, the time component of the four-acceleration is $a^0 = 0$, where superscript 0 means the time component in that MCIF's coordinates. That implies u^α and a^α are orthogonal; and since u^α is tangent to the worldline, a^α must always point along the MCIF's space axis, which is also the MCIF's line of simultaneity at every event. (In two and three spatial dimensions, this line becomes a plane and a volume, respectively.) We'll apply this same orthogonality argument again in a different context in (9.6).

can even see-saw across B and C several times, depending on how we change our acceleration. (See the further discussion of this on p. 262.)

There is, then, a fundamental difference between accelerating from the infinite past to the infinite future, and having accelerated only for some finite time. Relevant to this is an old problem in classical electrodynamics concerning the question of whether an electric charge that is constantly accelerated in a straight line will radiate. Charges with a nonuniform acceleration—such as oscillating electrons—do radiate, producing light. Also, charges moving with constant speed in a circle radiate, producing *synchrotron radiation*. But there is as yet no clear agreement as to whether a *constantly* linearly accelerated charge also radiates. Besides the calculation being very difficult and the definition of just what constitutes radiation being complex, part of the problem is that we must make this distinction between really accelerating forever and accelerating only for the last million years or so. As has been stated, in the former case we can never allocate a time to event A, while in the latter case we certainly can.

Questions of this nature aside, the fact that events like B and C of Fig. 7.4 are recorded by us to be happening in reverse causal order is a problem inherent in ascribing coordinates to accelerated frames. But at least those events cannot influence us! With this caveat in mind, let's search for a set of observers who will help us to make observations of the events around us, because such a set does exist.

The members of this set of observers must agree with us and each other about the synchronicity of events: what happened when. Remember that our $\bar{S}$-frame is a collection of observers all of whom report back to us about what is happening in their vicinity. That way we need not concern ourselves with the complexities of allowing for light-travel times from events we see visually (and as we shall find, light has a position-dependent speed in an accelerated frame). If everyone were to disagree on the timing of events, then they would really be useless as far as being "observing agents" for us is concerned, since we would have no chance of making sense of their reports.

While such a set of observers in general won't exist for an arbitrary rocket motion, constant acceleration is actually quite a special case. Three of these observers' worldlines are drawn in Fig. 7.5. Just as our worldline is defined by the fact that we come to rest at $x = 1/g$ (where g is our acceleration), we can populate space with observers labelled i, each of whom has a constant acceleration g_i in S, such that their turnaround point is $x = 1/g_i$ (or c^2/g_i in conventional units). If this can be arranged, then they will all share lines of simultaneity (the crucial requirement), since these lines are just the ones that pass through the origin of S. At any one event that they and we all agree is happening at the same moment for us all, their four-velocities are all parallel to ours (since all of these four-velocities are orthogonal to the common line of simultaneity). So in each observer's MCIF, the worldlines of all the other observers maintain a constant position, and so a constant separation. Thus

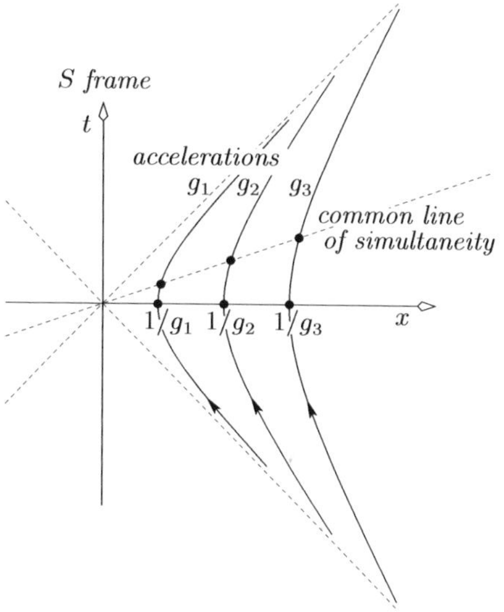

Fig. 7.5. Three of the continuum of observers who help us make measurements in our accelerated frame. If we give them accelerations such that they cross the x-axis at positions given by those accelerations' reciprocals (really c^2/g_1, etc., in conventional units), then the geometry of hyperbolae guarantees that these observers will always share a common line of simultaneity, which is precisely what a single frame's observers require.

they all always measure each other to be at rest with respect to them and ourselves. In other words, as far as they are concerned, they form a rigid lattice of observers; a perfect way for a frame to be!

The clocks of the set of observers who can constitute a frame for us now need calibration. We command the ship: our clock dictates what everyone else's should read. When our clock reads time zero [$\tau = 0$, which happens at $t = 0$ from (7.12)], we and all of the observers of our frame are crossing the x-axis, and all agree on this. So set all of their clocks to read zero when $t = 0$, and set $\bar{x}_i$, the unchanging position of each observer in our frame $\bar{S}$, to be their value of x at this time. This value of x is just $1/g_i$, so that observers stationed far "above" us (large x or $\bar{x}$) need to accelerate much less strongly than ourselves.

When our clock reads one second ($\tau = 1$), again all of our barred observers agree on this, so we'll calibrate their clocks so each reads one second ($\bar{t} \equiv 1$). Thus, shared simultaneity implies that we can provide all observers with one *coordinate* time $\bar{t}$—not one proper time. Each has a different proper time, but they all share the same coordinate time. As for position, there is no need to change what we have decided to be the observers' positions because none of them is moving with respect to any of the others or ourselves.

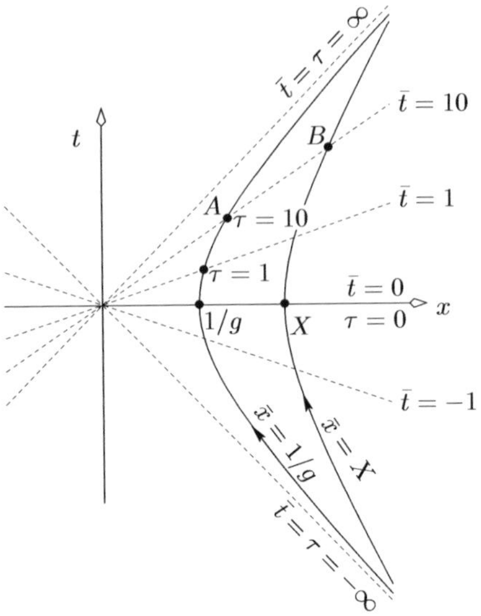

Fig. 7.6. The $\bar{t}$-coordinate that our team of observers ascribes to a set of simultaneous events is defined to be the proper time τ shown on *our* clock when these events occur for them and us. The $\bar{x}$-coordinate they ascribe to any event is given by the x-coordinate, at $t = 0$, of the observer who is present at that event.

Now, when our clock reads two seconds, what will the clocks of our barred observers read? (We hope two seconds also, if they are to form a meaningful frame!) Refer to Fig. 7.6 and consider the events A and B, which are simultaneous in $\bar{S}$. Of course, in S they are not simultaneous and so are labelled (t_A, x_A) and (t_B, x_B). Since their common line of simultaneity passes through the S-origin, it must be true that

$$\frac{t_A}{x_A} = \frac{t_B}{x_B} . \tag{7.17}$$

But if the accelerations of the observers present at A and B are g_A, g_B, then (7.12) gives

$$\begin{Bmatrix} t_A \\ x_A \end{Bmatrix} = \frac{1}{g_A} \begin{Bmatrix} \text{sh} \\ \text{ch} \end{Bmatrix} g_A \tau_A \quad \text{and} \quad \begin{Bmatrix} t_B \\ x_B \end{Bmatrix} = \frac{1}{g_B} \begin{Bmatrix} \text{sh} \\ \text{ch} \end{Bmatrix} g_B \tau_B , \tag{7.18}$$

where τ_A is the time shown on the clock present at event A and similarly for B. Thus (7.17) and (7.18) together imply that

$$\text{th}\, g_A \tau_A = \text{th}\, g_B \tau_B , \quad \text{or } g_A \tau_A = g_B \tau_B , \quad \text{and so } \frac{\mathrm{d}\tau_B}{\mathrm{d}\tau_A} = \frac{g_A}{g_B} = \frac{\bar{x}_B}{\bar{x}_A} . \tag{7.19}$$

But $\mathrm{d}\tau_B/\mathrm{d}\tau_A$ is the rate at which the observer at event B ages compared with the observer at event A, so the observer at B ages g_A/g_B times as fast as the observer at event A. Of course, here $d\tau_B/d\tau_A = \tau_B/\tau_A$, but this is only true for our choice of origin. In general, the ageing rate is $d\tau_B/d\tau_A$ and not simply τ_B/τ_A.

Let's take the observer at A to be ourselves. In principle, B's clock is counting out its seconds faster than ours by this factor of g_A/g_B; but we are still free to make its reading always agree with our own by gearing it down by the same factor. Then it will read two seconds when ours does, as required. For example, if we and observer B cross the x-axis at $x = 1/g$ and $x = 3/g$, respectively, then B's clock is accelerating at one third of our acceleration, and it counts out its seconds three times as quickly, ageing three times as fast as us. So we gear it down by a factor of three. How about a clock at $x = 1/(3g)$? This accelerates three times as fast as us, so it counts out its time one third as fast as our clock—i.e., it ages one third as quickly as us. In that case we need to gear it *up* by a factor of three.

As we consider altering the gearing of clocks closer and closer to the S-origin ($1/g$ "below" us), we need to make them tick faster and faster if they are to agree with our own as required. Nothing can be done to alter their rate of *ageing*, which will be very slow indeed below us and very fast for clocks above us; but that does not prevent us from gearing them all to match what our clocks read. By so doing, we have ensured that $\bar{t}$ is a global time coordinate.

How much the clocks really age is their proper time τ, which is different for each clock. What is shown on their dials—the result of internal gearing up or down relative to ours—is the coordinate time $\bar{t}$ (identical for each clock), which we have defined to be *our proper* time. We could do this because all of our observers share a common standard of simultaneity with us. This idea of a clock's ageing, as opposed to the simple timing procedure of counting out seconds, is an important one. There is nothing to stop us from gearing the clocks in the accelerated frame by some possibly weird position-dependent factor. The time they show is just a label that we are free to make evolve in whatever way we choose, although some choices will be more useful and meaningful than others. But by the clock's *ageing* we also mean the physical, biological ageing of the observer who holds the clock:

$$\begin{aligned} \tau &= \text{ageing: physical, biological time,} \\ \bar{t} &= \text{clock display produced by gearing.} \end{aligned} \tag{7.20}$$

According to the Clock Postulate, the real, biological age of an observer is the sum of the age increments of his series of MCIFs:

$$\Delta\text{Age} \equiv \Delta\tau = \int_{\substack{\text{observer's} \\ \text{worldline}}} d\tau\,. \tag{7.21}$$

To reiterate, (7.19) shows that an observer at $\bar{x} = 3/g$ really ages three times as fast as ourselves at $\bar{x} = 1/g$:

$$\frac{d\tau_{\text{obs. at } 3/g}}{d\tau_{\text{obs. at } 1/g}} \overset{(7.19)}{=\!=} \frac{g_{\text{obs. at } 1/g}}{g_{\text{obs. at } 3/g}} = \frac{g}{g/3} = 3\,, \tag{7.22}$$

so that

$$\Delta\tau_{\text{obs. at } 3/g} = 3\Delta\tau_{\text{obs. at } 1/g}\,. \tag{7.23}$$

So the rates of ageing of two observers A and B are set by $d\tau_A/d\tau_B$. The biological, proper time τ can be expressed in terms of the coordinate times t or $\bar{t}$, but the coordinate times themselves have no bearing on biological ageing. We are free to define our coordinates in other ways, but the relationship of time coordinates to space coordinates—the metric—will adjust itself accordingly to absorb the new definition and keep the elapsed proper time $d\tau$ unaltered, as it should be. After all, biological ageing has nothing to do with coordinate choices. Later we will express this central fact in tensor notation as $d\tau^2 = g_{\alpha\beta}\, dx^\alpha dx^\beta$.

Our final task is to write down a set of coordinates for our $\bar{S}$-frame relating $(\bar{t}, \bar{x})$ of Fig. 7.6 to (t, x). Note that (7.18) and (7.19) imply that if we are the observer present at event A, then

$$\begin{Bmatrix} t_B \\ x_B \end{Bmatrix} = \frac{1}{g_B}\begin{Bmatrix} \text{sh} \\ \text{ch} \end{Bmatrix} g_B\tau_B = \frac{1}{g_B}\begin{Bmatrix} \text{sh} \\ \text{ch} \end{Bmatrix} g\tau_A = \bar{x}_B \begin{Bmatrix} \text{sh} \\ \text{ch} \end{Bmatrix} g\bar{t}_B\,, \tag{7.24}$$

since $\bar{x}_B$ is defined to be the value of x where the observer at B crossed the x-axis (i.e. $1/g_B$), and $\bar{t}_B$ is defined to be the value of τ shown on *our* clock at the event A (i.e. τ_A), which we and our observers agree is simultaneous with event B. And the observer at B has geared his clock to match ours, to give a single time coordinate $\bar{t}$ for the whole frame. Finally, since event B is quite arbitrary, the sought-after transform relating inertial and accelerated frames is

$$\boxed{\begin{aligned} t &= \bar{x}\,\text{sh}\, g\bar{t}\,, \\ x &= \bar{x}\,\text{ch}\, g\bar{t}\,, \\ y &= \bar{y}\,, \quad z = \bar{z}\,, \end{aligned}} \tag{7.25}$$

where the y- and z-coordinates are unaffected by our motion perpendicular to their axes in the Lorentz transform, hence in the MCIF, and hence in (7.25). The inverse transform to (7.25) is

$$\begin{aligned} \bar{t} &= \frac{1}{g}\,\text{th}^{-1}\frac{t}{x} = \frac{1}{2g}\ln\frac{x+t}{x-t}\,, \\ \bar{x} &= \sqrt{x^2 - t^2}\,, \\ \bar{y} &= y\,, \quad \bar{z} = z\,. \end{aligned} \tag{7.26}$$

The metrics for the two coordinate systems are

$$\boxed{\begin{aligned} d\tau^2 &= dt^2 - dx^2 - dy^2 - dz^2 \\ &= g^2\bar{x}^2 d\bar{t}^2 - d\bar{x}^2 - d\bar{y}^2 - d\bar{z}^2. \end{aligned}} \tag{7.27}$$

Close to us (i.e. when $\bar{x} \approx 1/g$), the accelerated frame's metric is approximately Minkowski, for then $g\bar{x} \approx 1$ and

$$d\tau^2 \simeq d\bar{t}^2 - d\bar{x}^2 - d\bar{y}^2 - d\bar{z}^2. \tag{7.28}$$

(The y- and z-coordinates are really extraneous to this discussion, so we'll drop further mention of them—equivalent to setting them to have particular constant values.) We can use this familiar form of the metric as long as we do not try to quantify events too far away from our position. The meaning of "too far" is the length scale $1/g$, or c^2/g in more conventional length units. For a comfortable one Earth-gravity acceleration, c^2/g turns out to be just under one light-year:

$$\frac{c^2}{g} \simeq \frac{9 \times 10^{16}\ \mathrm{m^2 s^{-2}}}{10\ \mathrm{ms^{-2}}} = 9 \times 10^{15}\ \mathrm{m} \simeq 0.97\ \text{light-years}. \tag{7.29}$$

The form (7.28) of the accelerated frame's metric also clearly shows that locally we measure the speed of light to have the usual value of 1: since light connects events whose interval $d\tau^2$ is zero, (7.28) implies that $c = d\bar{x}/d\bar{t} \simeq 1$. In general, setting the interval to zero in (7.27) means that we will measure the speed of light at some arbitrary $\bar{x}$ to be $g\bar{x}$.

How fast do we and all of the observers who make up our $\bar{S}$-frame age compared with what our clocks show? This ageing rate can be found by reference to clocks that are stationary in our frame and is just $\partial\tau/\partial\bar{t}$, given by (7.27) as $g\bar{x}$, coincidentally the same as the light speed.

The coordinate system we have set up for the accelerated frame takes us, based at $\bar{x} = 1/g$, to be "in command" in the sense that we have dictated what all the clocks of the frame read. Our clock is the only one requiring no gearing to change its display. It can be useful to make a coordinate change to place ourselves at the origin of the frame, a simple shift in space coordinates by $1/g$ that defines double-barred coordinates:

$$\begin{aligned} \bar{\bar{x}} &\equiv \bar{x} - 1/g\,, \\ \bar{\bar{t}} &\equiv \bar{t}\,, \\ d\tau^2 &\overset{(7.27)}{=\!=} (1 + g\bar{\bar{x}})^2\, d\bar{\bar{t}}^2 - d\bar{\bar{x}}^2. \end{aligned} \tag{7.30}$$

We'll give these new coordinates to the traveller Eve in Sect. 7.3.2.

Imagine that we are in deep space far from any gravity, in a spaceship with a one Earth-gravity acceleration, where the natural notion of "up" defines the positive-$\bar{x}$ direction. Our spaceship is the heart of the $\bar{S}$-frame, so our position in this frame is $\bar{x} = 1/g \simeq 1$ light-year, as shown in Fig. 7.4. One light-year below us lies the mysterious origin of our $\bar{S}$-frame, really a plane containing events that are simultaneous with everything we do. Closer and closer to that plane (i.e. as $\bar{x} \to 0$) time is passing ever more slowly for all physical processes, and the clocks of our frame in that region must be geared to run ever faster to keep up with our own. One light-year below us, time has stopped completely; the events on that plane are eternally frozen.

Farther than one light-year below us, where the events B and C lie on Fig. 7.4, time runs in reverse. We cannot know anything about events there, however; signals from them will never reach us. This idea of a kind of "edge" to what we can see or know about gives rise to the term *event horizon* for the plane one light-year below us. It is the set of events that divide all events we can eventually see from those we can *never* see. If our acceleration were, say, n Earth gravities, then the event horizon would lie only about $1/n$ light-years below us. The stronger our acceleration, the closer below us lies the event horizon, where time slows to a stop.

But while below us time slows, above us it quickens. One light-year above us, $\bar{x} \simeq 2$ light-years, so (7.27) says that $\partial\tau/\partial\bar{t} \simeq 2$. Clocks there are ageing—*all* physical processes are occurring—at about twice the rate of our own. Of course, we have geared those clocks down to keep them ticking at the same pace ($\bar{t}$) as our own, as mentioned several paragraphs back. But *time* (τ) still "runs" faster for them; they age more quickly than we do. Two light-years above us, $\bar{x} \simeq 3$ light-years and $\partial\tau/\partial\bar{t} \simeq 3$, so clocks are ageing at three times the rate of our own but again are geared down to always agree with our displayed time, and so on.

As we saw in Fig. 7.4, when uniformly accelerated, we cannot see all of spacetime. The region covered by the barred coordinates is called *Rindler spacetime* (or Rindler space) and is characterised by its event horizon and position-dependent ageings. But as we have seen, Rindler spacetime is really just a different coordinatisation of a part of the *Minkowski spacetime* of Chap. 5. In particular, we have been focussing on the *right-hand Rindler wedge*, the section between $t = +x$ and $t = -x$ shown in most of the figures in this chapter. A more graphic picture of Rindler spacetime will be drawn in Fig. 7.12, and it's the construction and use of such a figure that we turn to next.

7.3 The Twin Conundrum

Although physical laws are generally complicated in any noninertial frame, the transformation equations of an accelerated frame can be used to develop a better feel for the famous Twin Conundrum.

The Twin Conundrum reads like this. Twins Adam and Eve wish to test what their special relativity textbook tells them about how moving systems age slowly. They arrange for Adam to stay within an inertial frame, which is usually taken to be Earth. The fact that frames on Earth are not quite inertial is irrelevant to the story, so we can and will take Adam's frame to be inertial. (We can always replace Earth by a far-flung space station where gravity is essentially zero.) Eve, the intrepid space traveller, then blasts off from Earth and travels to a distant star before turning around and heading

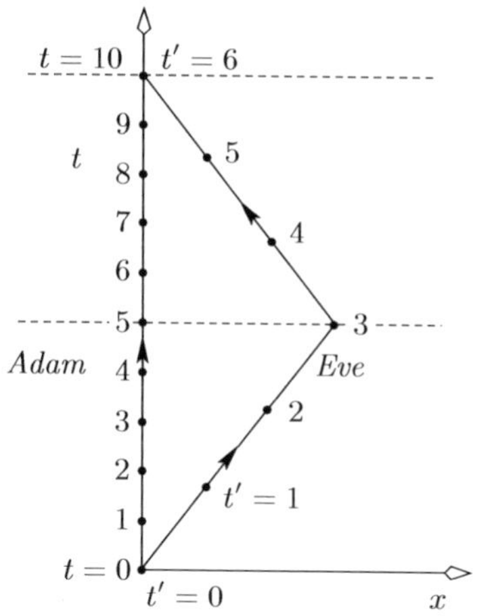

Fig. 7.7. The worldlines of Adam and Eve. Eve has speed $0.8\,c$ on both outbound and inbound legs. Adam's horizontal lines of simultaneity when he has aged 5 and 10 years are shown as dashed.

back to Earth to be reunited with Adam some years later. Apart from small periods of acceleration and deceleration, her speed is a constant $0.8\,c$. The worldlines of the twins are drawn in Fig. 7.7.

Adam reasons that on both outbound and inbound legs, Eve will age more slowly than he does by a factor of $1/\gamma$, or $3/5$ in this case. If Eve is gone for ten of Adam's years, then on her return she should have aged only six years. This is correct, and is a consequence of the fact that the twins have travelled along two different paths in spacetime.

That their ageings differ is reasonable when we remember that special relativity places space and time on an equal footing, as was discussed in Sect. 5.4. Eve *travels* a different number of light-years than Adam, so it is not surprising that she also *ages* a different number of years than he does. Our intuition has no problem with the idea that the length of a journey between two cities will change if a side trip to a third city is included along the way. We need to realise that Eve's "side trip" involved movement through space and time, while Adam's trip only involved movement through time.

The famous conundrum states that surely Eve might want to consider Adam as the one who moves away and back at $0.8\,c$, so that in her six years away *he* will be the one to have aged less: $3/5$ of 6 years, or 3.6 years. They cannot both be right!

Textbooks that explore this well-known story of the twins who decide to test special relativity for themselves usually, if not always, make the worldlines of the travelling twin straight, with a single kink at the point of return. In a

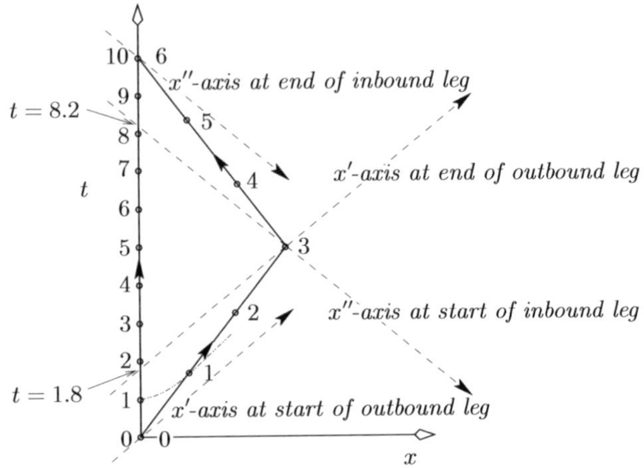

Fig. 7.8. Eve has t', x'-axes at the start and end of her outbound leg, and t'', x''-axes at the start and end of her inbound leg. Note that her x'-axis at the end of her outbound leg is *not* the same as the x''-axis she switches to at the start of her inbound leg. Her time axes are just her worldlines, while her space axes are her lines of simultaneity. A piece of the axis-calibrating hyperbola is shown, as discussed in the text.

moment we will see why this can actually serve to obscure the discussion rather than clarify it.

First, however, we'll follow that standard easy approach and analyse the problem while neglecting the periods during which Eve accelerates. We know she does have periods of acceleration, but we will consider her trip to consist of two legs, outbound from and inbound to Adam, on both of which her speed is absolutely constant at $0.8c$. Thus she accelerates only for infinitesimal periods at the start, midpoint, and end of her trip. Hence all worldlines drawn in Adam's inertial frame will be straight.

In Fig. 7.7, Adam's lines of simultaneity are always horizontal, so that after 5 of his years, he will know that Eve has aged 3, and after 10 years Eve will rejoin him 6 years older. To see Eve's point of view, we need to draw her lines of simultaneity. Just how to do this was explained on p. 203, and Eve's lines of simultaneity (her space axes) have been plotted in Fig. 7.8. These always have a slope equal to her current velocity.

Care must be used when calibrating Eve's axes, which is done by the procedure of Sect. 5.7. A short piece of the calibrating hyperbola $t^2 - x^2 = 1$ has been drawn in Fig. 7.8. This cuts both Adam's and Eve's time axes at $t = 1$ and $t' = 1$, respectively.

It is immediately apparent from Fig. 7.9 that Adam and Eve will each measure the other as ageing slowly. We have encountered this line of rea-

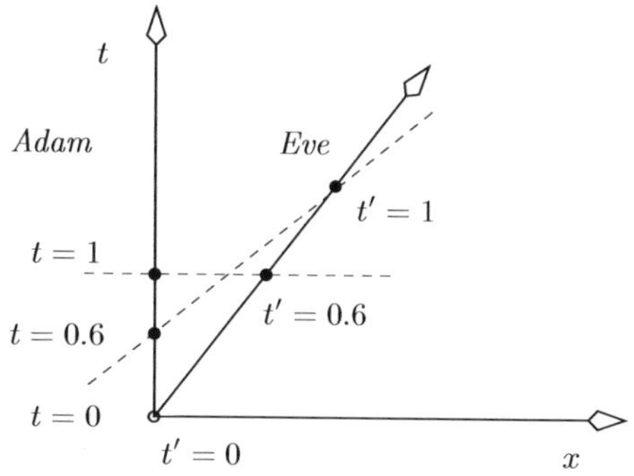

Fig. 7.9. A detail of the beginning of Eve's trip. When Adam has aged one year, his line of simultaneity indicates that Eve has aged only 0.6 years. But Eve has a different standard of simultaneity, and after one of her years she notes that *Adam* is the one to have aged only 0.6 years.

soning already in Chap. 5. Adam's line of simultaneity for $t = 1$ intersects Eve's worldline (the t'-axis) at $t' = 0.6$, expressing the Lorentz transform's insistence that Eve ages 3/5 as fast as Adam. Likewise, when Eve has aged one year, *her* line of simultaneity with $t' = 1$ intersects Adam's worldline (the t-axis) at $t = 0.6$. Each measures the other as ageing slowly simply because each has their own personal line (i.e. standard) of simultaneity at every event.

This idea of differing lines of simultaneity is the key to resolving the Twin Conundrum. Figure 7.8 shows that on Eve's outbound leg, Adam ages by 5 years and measures Eve to age by 3 years. Likewise, after Eve has travelled for 3 years and is about to turn around, she measures Adam to have aged by $3/\gamma = 1.8$ years. She then quickly decelerates, turns, and accelerates to $0.8\,c$ again, now on the inbound leg. Again Adam will measure her to age by 3 years for his 5. On the other hand, when Eve swaps from outbound to inbound, she changes inertial frames. In her inbound frame, Adam is now no longer 1.8 years older than he was on their parting, but 8.2 years older!

Eve ascribes this odd change in affairs to having been handed a new set of bookkeeping figures by the observers who make up her new $t''x''$-frame on the inbound leg. Unlike her, these observers always have been, and always will be, inertial. Eve inherits a new set of observers whenever she changes frames. The set of these observers that Eve inherits at the start of her inbound leg is entirely different from the $t'x'$-set that she left behind at the end of her outbound leg, and it is these two sets of observers who record different values for Adam's age, which is not surprising because they comprise different frames.

Still, Adam's age has jumped by 6.4 years during Eve's turnaround, a fact that unsettles her. Nevertheless, she accepts it and on her 3 year return leg again measures Adam to age by 1.8 years. So on her arrival he is indeed

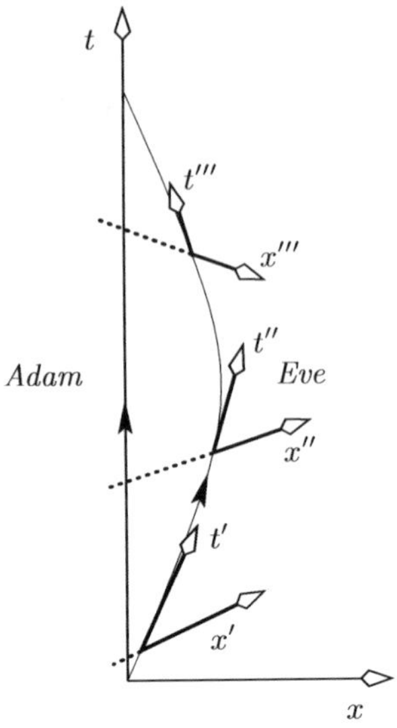

Fig. 7.10. Eve now travels with a constant acceleration. At each event on her worldline, we can draw the space and time axes of her MCIF analogously to Fig. 7.3.

$1.8 + 6.4 + 1.8 = 10$ years older than when she last saw him, while she is just 6 years older—and both she and Adam agree that their bookkeeping tallies. But while Adam noted nothing unusual during Eve's trip, Eve can only scratch her head in puzzlement at why Adam's age jumped by 6.4 years at the moment when she turned around to begin her trip home.

7.3.1 Making Eve Accelerate Uniformly

Eve can gain insight into Adam's "missing years" by redoing her trip in a way that goes against our intuition of what might serve to simplify the problem: by accelerating more realistically than has just been described. Figure 7.10 shows a redrawing of Fig. 7.7, this time allowing Eve's worldline to curve gently all the way throughout her trip. (We could also allow a gentle acceleration on her departure and return to Earth, but one piece of acceleration will be enough to shed light on the conundrum.)

The Clock Postulate says that at any moment, Eve can analyse events by drawing the space and time axes of her MCIF. Her line of simultaneity is, as usual, her MCIF's space axis, which is orthogonal to her MCIF's time axis, which itself is always tangent to her worldline. *This* is the crucial idea that allows Eve to better understand her bookkeeping as she accelerates throughout her trip. Her calculations are more complicated now because she depends on a continuous succession of different MCIFs to supply her bookkeeping entries.

On the other hand, the insight gained is that Adam now ages quite smoothly, never mysteriously jumping ahead by 6.4 years. Eve records his age at any moment to be the time of the particular event on Adam's worldline that lies on her line of simultaneity.

Although Eve is only consulting with her MCIFs—each of which always measures Adam to be ageing slowly—the nett result is that she measures him to be

1. first ageing slowly as her line of simultaneity slowly slides up the t-axis in a mostly translational way,
2. then ageing quickly as she slows, the distance between them increases, and her line of simultaneity begins to rotate somewhat and sweep very quickly up the t-axis, and
3. finally, ageing slowly again as their reunion approaches, and the sweep of the simultaneity line is again mostly translational.

So there are two competing effects that determine the rate at which Eve notes Adam to be ageing: the normal slowing that occurs because of his motion relative to her, and a quickening because of the fact that Eve's line of simultaneity is now rotating through spacetime.

This is all very satisfying for Eve because she can keep continuous track of Adam's age all the way throughout her trip. But what she cannot do is demand that in fact she never accelerated and that rather it was *Adam* who accelerated. This becomes evident if she and Adam are each given a bucket of water on their farewell. Adam holds onto the bucket for ten years, and the water sits placidly still. Eve, on the other hand, finds that the water in her bucket mysteriously climbs up the bucket's side and might slosh over her rocket's floor. One thing is for sure: she and Adam have very different experiences with their buckets.

It's important to realise that the oft-quoted but wrong aphorism "all motion is relative" was never a part of Einstein's relativity. Einstein was well-aware of the simple fact that if I spin around while you remain stationary, then I am the one who becomes dizzy. I cannot maintain that because you circled around me for the duration, you should be dizzy instead. I might postulate that when I decided to spin around, a complex force field suddenly permeated the universe that pulled dizzyingly on the fluids within my inner ear responsible for passing body-orientation signals to my brain, and this field was arranged in such a way that it had no effect on anything turning around me. That would explain why the rest of the universe that circled about me did not become dizzy. But it is certainly a very strange thing for such a field to appear instantaneously throughout all of space just on my whim. It also turns out to be impossible to establish a set of coordinates that correctly describes the physics of such a rotating frame, because there exists no set of observers who all agree on the simultaneity of events.

It is not true, then, that all motion is relative. Eve knows this. She is aware that while accelerating, she is no longer inertial and so cannot use simple inertial frame machinery, such as the Lorentz transform, to describe the physics around her (unless, of course, she uses our analysis of the last few pages—but our analysis was really performed in *Adam's* frame, not Eve's). But she does find that everything works out right if she continuously consults with her MCIF, since her textbook's Lorentz theory *does* apply to that frame.

This is all very well, but Eve does not really want to consult with a continuously changing set of inertial observers. That's just too hard to do in practice. As far as she is concerned, she would like to be able to quantify her trip in terms of just one set of coordinates: the $\bar{S}$-coordinates of her accelerated frame, in which she is always at rest. And so she can, if she is *uniformly* accelerated—as we have established in this chapter. So we now ask how a uniformly accelerated Eve views the events of her trip.

7.3.2 How the Twins Record Each Other's Trips

A good way to explore the asymmetric ageing of the twins is to record the trip from both of their points of view. We plotted Eve's voyage in Adam's frame in Fig. 7.7, but only for the case where her travel consists solely of two constant-velocity legs. Now we'll make her accelerate uniformly, but in such a way that she still ages six years while Adam ages ten.

The scenario of the trip is modelled upon the simpler one already used in Fig. 7.7. We wish to plot the two worldlines in both frames with yearly intervals marked out, as well as light signals that Adam and Eve send to each other at the start of every new year. To prevent confusion with what we have already done, Adam's inertial coordinate system is now relabelled t_a, x_a, while Eve's is t_e, x_e. Both distance and time will be measured in years, so that a distance of, say, two years can be directly interpreted to mean two light-years. The twins part when $t_a = t_e = 0$ and reunite when $t_a = 10, t_e = 6$.

The details of setting up the two sets of coordinates are based upon what we have already covered in (7.25) and (7.26), and though not difficult, they detract from the main discussion here. We will simply quote them and leave the work of translations and scaling as an appendix in Sect. 7.6.

Eve's Trip as Observed in Adam's Frame

As shown in Sect. 7.6, Eve's worldline in Adam's frame turns out to be given by

$$\begin{aligned} t_a &= 5 + \frac{1}{g}\,\mathrm{sh}\,[g(t_e - 3)]\,, \\ x_a &= b - \frac{1}{g}\,\mathrm{ch}\,[g(t_e - 3)]\,, \end{aligned} \tag{7.31}$$

where t_a, x_a now label each point of Eve's worldline (in *Adam's* frame—hence the "a" subscript), and t_e parametrises Eve's motion, going from 0 to 6 years. The constant b and Eve's acceleration g are approximately

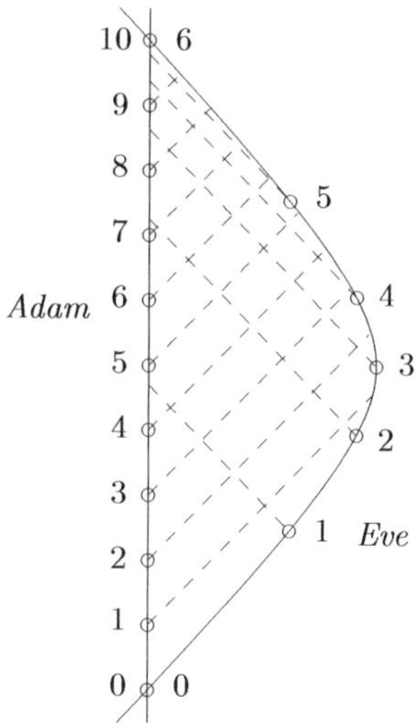

Fig. 7.11. Eve's trip from the viewpoint of Adam's inertial frame. Also shown are light signals that Adam and Eve send to each other once per year. Standards of simultaneity are not shown: Adam's is determined by all the horizontal lines, while Eve's is determined by (7.53) with t_e held constant, which also produces lines that are everywhere orthogonal to Eve's worldline.

$$b \simeq 5.26 \text{ years}\,, \quad g \simeq 0.613 \text{ years}^{-1}. \tag{7.32}$$

To get a feel for just how strongly Eve must accelerate, express her acceleration g in conventional units by multiplying by c:

$$g_{\text{conventional}} \equiv gc = \frac{0.613 \text{ yr}^{-1}}{31.5 \times 10^6 \text{ s/yr}} \times 3 \times 10^8 \text{ ms}^{-1} \simeq 5.8 \text{ ms}^{-2}, \tag{7.33}$$

or a little less than the familiar force of gravity on Earth. Her worldline in Adam's inertial frame is shown in Fig. 7.11. Each year the twins have arranged to send a light signal to each other, and these are shown as 45° dashed lines. Adam does not see Eve's first signal until almost five years have passed for him; that is, he sees her signals as redshifted. Eve's second signal only reaches Adam just after his year 7, with the remaining four of Eve's signals coming in more frequently in the last three of Adam's years, forming the familiar blueshift.

In contrast, Eve does not see Adam's first signal until 2.6 years have elapsed for her, and thereafter she sees him ageing with increasing rapidity. We should point out here a commonly held misconception that relativity theory is all somehow a trick of the light, something to do with light's finite speed and the apparent reordering of spatially separated events solely due to the differing transit times of the light signals they send out. Surprisingly, this naïve and quite wrong idea finds its way into physics books and encyclopaedias. To stress this point we again distinguish what Adam "sees" from what he "observes" or "measures", as was done originally on p. 194. From Fig. 7.11 we know that Adam will *see* Eve to be 1 year older when he is 4.7 years older. This is just a trick of the light caused by Eve's distance from him. What he *observes* or *measures*, however, is that she had aged 1 year when he was about 2.5 years older. (Remember that Adam's line of simultaneity is horizontal in his frame.) When he *sees* Eve to be 1 year older, he knows she is really about 2.8 years older and that her large distance from him has merely delayed the light from her current age in reaching him.

Just how Adam makes his observations in practice might be difficult technically, but he does not base them simply on what he sees. Rather, he must collate what each of his band of synchronised observers sees, as explained on p. 234. *They* can certainly rely on their sight alone since they only concern themselves with events that happen right under their noses, so that there are no complications introduced by nonzero light-travel times.

Adam's "Trip" as Observed in Eve's Frame

Adam might not feel that he went anywhere (a reflection of the fact that he never departed from his original inertial frame), but Eve considers that he certainly is the one who went away and came back. Her record of his worldline as recorded in her frame is given by the set of events (t_e, x_e). From Sect. 7.6 these turn out to be

$$\begin{aligned} t_e &= \frac{1}{g}\,\mathrm{th}^{-1}\,\frac{t_a - 5}{b} + 3\,, \\ x_e &= \sqrt{b^2 - (t_a - 5)^2} - 1/g\,, \end{aligned} \tag{7.34}$$

where t_a parametrises Adam's motion. Note carefully that these equations are *not* the inverse of (7.31)! This is fine because it reflects the fact that Adam and Eve have different standards of simultaneity. Adam's worldline is plotted in Fig. 7.12. (We have drawn him in Fig. 7.12 to be moving to the right just as we did for Eve in Fig. 7.11, to emphasise the similarity between the two frames' points of view.)

Eve remains at rest in her frame for 6 years while Adam departs, returning to her when he is 10 years older. Again they swap signals, and a comparison of Figs 7.11 and 7.12 shows the consistency of those signals' arrival times, as must be the case. After all, Adam's signal saying "I am one year older" is received by Eve after 2.6 of her years regardless of whose frame we consult; her age on reception of this message must be frame-independent. (Just how the curved light signals have been drawn in Eve's frame is explained shortly.)

Adam's "trip" in Eve's frame looks similar to her trip in his frame, except that he begins to age faster the farther he travels "above" Eve. What is this ageing rate that Eve measures for all physical processes? As discussed on p. 249, this is given by the rate of ageing of an observer stationary in

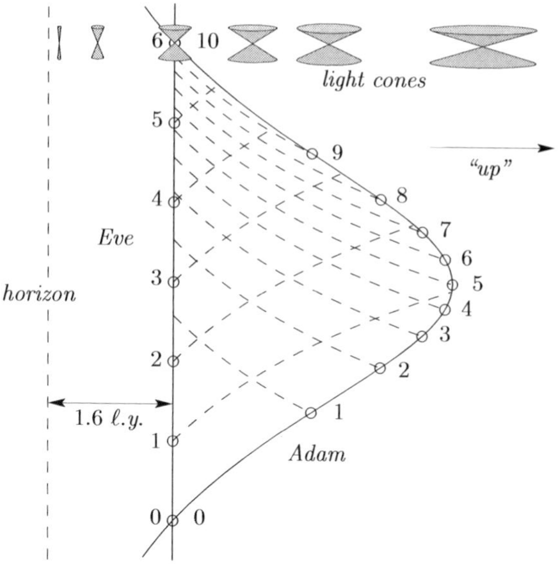

Fig. 7.12. Adam's "trip" from Eve's noninertial frame. (This diagram has been orientated so that, as in Fig. 7.11, the travelling twin is on the right, making for a more suggestive comparison between the two figures.) The light signals that they send to one another now have a position-dependent velocity! Everything is drawn to scale, including the light cones at the top of the picture. There is a horizon at Eve's space origin, 1.6 light-years "below" her, where time and light slow to a stop, closing up the cones. Simultaneity standards are not shown. Eve's standard consists of all horizontal lines, while Adam's is determined by (7.54) with t_a held constant, and is now a set of curves (cf. the inertial case where all simultaneity is mapped out by lines). Note that Eve's signals to Adam speed up as they climb "upward" in what seems to her to be something like a gravitational field, while the signals that Adam sends to her slow down as they "fall" toward Eve (the opposite of what might at first be expected). All of the light signals (dashed curves) in the picture, if continued "below" Eve, would approach the horizon asymptotically.

Eve's frame compared with Eve's own rate of ageing, or $\mathrm{d}\tau/\mathrm{d}\tau_{\mathrm{Eve}}$. But Eve's ageing rate *is* just her time coordinate by construction, so the required ratio is just $\partial\tau/\partial t_e$. Equation (7.30) gives the metric as

$$\mathrm{d}\tau^2 = \mathrm{d}t_a^2 - \mathrm{d}x_a^2 = (1 + gx_e)^2\mathrm{d}t_e^2 - \mathrm{d}x_e^2\,, \tag{7.35}$$

so that

$$\partial\tau/\partial t_e = 1 + gx_e\,. \tag{7.36}$$

The rate of flow of time in Eve's frame is then $1 + gx_e$. (The negative choice of the square root giving $\partial\tau/\partial t_e = -(1 + gx_e)$ applies to the backward flow of time of the inaccessible events that lie below Eve's event horizon.) That is, since Eve remains at $x_e = 0$, her clock is ageing (by $d\tau$) just exactly as fast as it is timing (by dt_e). Clocks "higher up" in her frame ($x_e > 0$) are ageing ($d\tau$) faster than they are timing (dt_e), since their timekeeping mechanisms have been slowed to keep pace with Eve's clock. We see the reason that Eve gives for Adam's accelerated ageing as he moves away from her: he is moving through a part of her frame where time flows more quickly. And below Eve ($x_e < 0$), clocks are ageing ($d\tau$) slower than they are timing (dt_e) since their timing mechanisms have been mechanically sped up to keep pace with Eve's clock.

In Fig. 7.12 are also shown representative light cones. The worldlines of light rays in Eve's frame are the *null curves* of her metric; that is, curves along which the interval is zero. In Adam's frame, we can write

$$d\tau^2 = dt_a^2 - dx_a^2 = 0\,, \text{ so that speed of light} = |dx_a/dt_a| = 1 \qquad (7.37)$$

as expected, since his frame is inertial. In Eve's frame, we write

$$d\tau^2 = (1+gx_e)^2 dt_e^2 - dx_e^2 = 0\,, \text{ so that speed of light} = |dx_e/dt_e| = 1+gx_e\,. \qquad (7.38)$$

In Eve's vicinity ($x_e \simeq 0$), light has the familiar speed of 1, but far above her (in Adam's direction) this speed is greater than 1, indicated by the light cones opening up in Fig. 7.12. Equation (7.38) is a differential equation that can be solved analytically for the worldlines of light signals. They turn out to be simple exponentials, and that's just how they were drawn in Fig. 7.12.

Below Eve is an event horizon where the light cones close up as the light speed tends toward zero and time slows to a halt. This horizon at $x_e = -1/g$, or 1.6 light-years below Eve, is intriguing. Eve observes not just Adam but the whole visible universe to be falling toward but never quite reaching it, since every worldline must lie within the light cones, and these are closing up the nearer they are to the horizon. On this mysterious plane, she observes that time itself has stopped, and although she cannot see what lies beyond, she knows from our previous discussion that behind the horizon time is flowing in reverse!

What happens if Eve has a short-lived change of heart about returning to Earth? A short while after commencing her return trip, she fires her rockets and accelerates for a time away from Earth again, before reversing her decision and again heading back for a reunion with Adam, so that her worldline in his frame now has a $\lesseqgtr$ shape. If she accelerates strongly enough away from Adam, then we might think that an event horizon should form between them for the duration. In fact, no horizon forms because that would imply that Adam had become invisible to her. But he never does because Eve does not accelerate forever, which is the only way that she would be assured of always outrunning his light signals. The event horizon is a *global* phenomenon; it cannot be something that forms locally for a short time.

Even so, on such a vacillating trip Eve's line of simultaneity will seesaw backward and forward, and she will conclude that for a while Adam will be growing younger—but that he'll grow older again on her return at a compensating rate, so that on their reunion he is still the older of the two.

This wild see-sawing of Eve's line of simultaneity brings us full circle to where we began this chapter. It's a problem inherent in constructing a frame for an observer with a changing acceleration (as measured in some inertial frame). Even a uniformly accelerated Eve is aware of the backward flow of time below her horizon, but in practice this does not present much of a problem. She cannot be influenced by any of those acausal events, since signals from them can never reach her. And if she or Adam *were* to move with arbitrarily complicated accelerations, we can see by drawing arbitrary worldlines that they both would always *see* each other ageing in the normal forward fashion, because neither can ever travel faster than light. This visible ageing of each might have some variable rate depending on how the other moves, but it still only happens in the forward direction. Eve might conclude that Adam is ageing in reverse for part of her trip, but she never sees such weird events unfold.

Psychological Versus Physical Ageing

Throughout this chapter, we have implicitly assumed that psychological ageing is tied to physical ageing. This is conventional in relativity, reflecting an assumption that consciousness is anchored to brain chemistry, a physical process that must slow down in a moving frame to obey the postulates of relativity, that prefer no one inertial frame to another.

But perhaps consciousness actually favours one particular frame—be it that of Earth, or some frame set by the distant stars, or indeed something more esoteric that is not part of this physical universe at all. Current time dilation experiments using decaying subatomic particles and clocks on board aircraft and satellites do not shed any light on this, and probably will not for the foreseeable future.

7.4 A Glance Ahead to Gauge Theory

The asymmetry of the twins' frames suggests a new principle that we will explore in Chap. 10. When we change from one inertial frame to another via the Lorentz transform, the change is linear and global, and there is not a great deal of difference in the way the world looks in one frame or the other. Quantities that are conserved in one frame are conserved in both, and indeed this existence of conservation principles is related to such a *global* transformation.

But the general transform (7.25) between an inertial and uniformly accelerated frame is not linear, with the result that a noninertial Eve experiences events quite differently from an inertial Adam, depending on where in her frame they occur. Clocks "below" her run slow, while those "above" her run fast. The speed of light depends on its height above her, and so on. The most obvious manifestation of this more local aspect to the transformation between inertial and accelerated frames is that Eve feels a force that determines a down direction for her. Adam would refer to it as a pseudo force, but for Eve it is real enough. She can only keep cherished notions about such things as Newton's laws if she holds that this force is responsible for driving free particles off their constant-velocity courses. So this local character to the Adam–Eve frame transformation has introduced for Eve the idea of a force that prevents a free particle's momentum from being conserved in her frame.

We might reasonably ask whether it is possible to write down other transformations that might have nothing to do with changing frames but that do give rise to other forces of Nature. It turns out that in quantum mechanics we can. There, a constant phase ϕ of a wave function can be chosen arbitrarily, and transforming the wave function by multiplying it by $e^{i\phi}$ does not alter the physics. All constant values of ϕ are equally valid, in the same way that all inertial frames are equally valid for describing a physical system. But, as we shall see in Chap. 10, making the phase of a wave function *depend* on position introduces a local character to the transformation that results in an interaction such as electromagnetism, which is reminiscent of the switch to an accelerated frame that introduces a pseudo-gravity force. The study of such transformations is known as *gauge theory*. So accelerated frames are not just a stopping point on the way to general relativity, but they also serve to point the way to gauge theory.

7.5 Covariant Notation and Generalising the Clock Postulate

We now bid farewell to Adam and Eve and, aware of the Clock Postulate and with more insight into frame changes, return once again to the subject of covariant analysis. In Chap. 6, covariant notation was introduced as a neat and concise way of showing how four-vectors and covectors transform from one inertial frame to another. We wish now to show how this notation can be extended to cover the relationships between quantities when one or both of the frames is *not* inertial, as a first step toward writing the language of physics in a way that applies to all frames, noninertial as well as inertial.

The Clock Postulate can be generalised to say something about measurements made in a noninertial frame. First, it tells us that any measurements made in a noninertial frame that use rods and clocks will be identical to measurements made in the MCIF. (It should be remembered that although different regions of the noninertial frame might have different MCIFs, a measurement is necessarily something that happens locally.) But we now choose to extend the postulate to include *all* measurements—though perhaps it can be argued that, at their very heart, all measurements only ever use rods and clocks anyway.

Consider, then, a general four-vector field whose unprimed components A^α have been determined by measurements that were made in an inertial frame S. We know from the previous chapter that because A^α is a four-vector, it transforms from S to another inertial frame S' through the Lorentz transform $\Lambda^{\alpha'}_{\beta}$:

$$A^{\alpha'} = \Lambda^{\alpha'}_{\beta} A^\beta . \tag{7.39}$$

The question is, using this language, what are the four-vector field components in a *non*inertial frame?

Earlier we obtained the coordinate transform (7.25) for an accelerated frame by using the MCIF as an intermediate step, after which it was discarded. We would like to do the same here. Suppose that we are using a noninertial frame $\bar{S}$ and wish to calculate the components $A^{\bar{\alpha}}$, given that the components are A^α in S. The extended Clock Postulate says that the field in our frame, $A^{\bar{\alpha}}$, will be identical to the field $A^{\alpha'}$ measured by our MCIF (call this S'). Because S' is inertial, we know how to calculate $A^{\alpha'}$: it's just given by (7.39). We wish to transform coordinates in the order

$$S \longrightarrow S'(\text{MCIF}) \longrightarrow \bar{S} . \tag{7.40}$$

The Clock Postulate says that measurements of distance and time in $\bar{S}$ are identical to those of our MCIF S':

$$\mathrm{d}x^{\bar{\alpha}} = \mathrm{d}x^{\alpha'} . \tag{7.41}$$

But this is just another way of saying that for any α, β,

$$\frac{\partial x^{\bar{\alpha}}}{\partial x^\beta} = \frac{\partial x^{\alpha'}}{\partial x^\beta} \overset{(6.37)}{=\!=} \Lambda^{\alpha'}_{\beta} . \tag{7.42}$$

The field components in $\bar{S}$ can now be written in the following way:

$$\begin{aligned} A^{\bar{\alpha}} &= A^{\alpha'} && \longleftarrow \text{(field we measure = MCIF field} \\ & && \qquad \text{by the extended Clock Postulate)} \\ &= \Lambda^{\alpha'}_{\beta} A^\beta && \longleftarrow \text{(Lorentz transform from } S \text{ to } S'\text{)} \\ &= \frac{\partial x^{\bar{\alpha}}}{\partial x^\beta} A^\beta && \longleftarrow \text{(by (7.42), the Clock Postulate again).} \end{aligned} \tag{7.43}$$

In the last line of (7.43), the MCIF S' has disappeared entirely! In (6.37) we related Lorentz transform coefficients Λ to partial derivatives, but now we choose to generalise the Λ notation to mean a partial derivative between *any* two sets of coordinates, with no need for either set to be inertial:

$$\Lambda^{\bar{\alpha}}_{\beta} \equiv \frac{\partial x^{\bar{\alpha}}}{\partial x^\beta} . \tag{7.44}$$

This more general definition of Λ allows the last line of (7.43) to be written in a familiar way:

$$A^{\bar{\alpha}} = \Lambda^{\bar{\alpha}}_{\beta} A^{\beta}. \tag{7.45}$$

Equation (7.45) is just like the Lorentz transform between two inertial frames of (7.39), except that now it applies to an *arbitrary* frame $\bar{S}$, which need not be inertial. There is still the inertial frame S present in (7.45), but we can even remove all reference to that. Consider another noninertial frame $\bar{\bar{S}}$, in which the field must have components

$$A^{\bar{\bar{\alpha}}} = \Lambda^{\bar{\bar{\alpha}}}_{\beta} A^{\beta}. \tag{7.46}$$

This equation can be inverted easily by multiplying both sides by $\Lambda^{\gamma}_{\bar{\bar{\alpha}}}$ and summing over the repeated index (called *contracting* both sides with $\Lambda^{\gamma}_{\bar{\bar{\alpha}}}$), and using the chain rule for partial derivatives:

$$\Lambda^{\gamma}_{\bar{\bar{\alpha}}} A^{\bar{\bar{\alpha}}} = \Lambda^{\gamma}_{\bar{\bar{\alpha}}} \Lambda^{\bar{\bar{\alpha}}}_{\beta} A^{\beta} = \delta^{\gamma}_{\beta} A^{\beta} = A^{\gamma}. \tag{7.47}$$

Finally, combine (7.45) and (7.47) to give

$$A^{\bar{\alpha}} \overset{(7.45)}{=\!=} \Lambda^{\bar{\alpha}}_{\beta} A^{\beta} \overset{(7.47)}{=\!=} \Lambda^{\bar{\alpha}}_{\beta} \Lambda^{\beta}_{\bar{\bar{\gamma}}} A^{\bar{\bar{\gamma}}} = \Lambda^{\bar{\alpha}}_{\bar{\bar{\gamma}}} A^{\bar{\bar{\gamma}}}, \tag{7.48}$$

which now relates the field components between two completely arbitrary frames. Neither need be inertial!

Since (7.48) looks so much like the Lorentz transform, it might be taken for granted as being trivially correct. But it is not trivial at all. Yes, four-vectors transform like infinitesimals between inertial frames, but we cannot expect a priori that four-vectors will also transform in the same way between *arbitrary* frames. However, we have shown that they do, provided only that the Clock Postulate holds.

The covariant language of up and down indices pays no heed to the nature of the frames, and this makes it so useful for writing physical laws in a frame-independent way. That's why covariance is a natural language of physics.

7.6 Appendix: Details of Setting Up Adam's and Eve's Coordinates

Our last task in this chapter is to describe the details of drawing Eve's worldline in Adam's frame and vice versa, as required in Sect. 7.3.2.

Begin by plotting Eve's worldline in Adam's frame, in which the events take on their simplest form. Refer to Fig. 7.13 for the details. As in Sect. 7.3.2, rename Adam's coordinates t, x to be t_a, x_a.

The basic coordinate transform in (7.25) and (7.26) relates to the accelerated observer of Fig. 7.3. Eve's worldline is found by reflecting the hyperbola in Fig. 7.3 left to right and shifting it somewhat. So introduce an intermediate set of axes T, X in Fig. 7.13 since these match the unbarred (inertial) coordinates of (7.25) and (7.26):

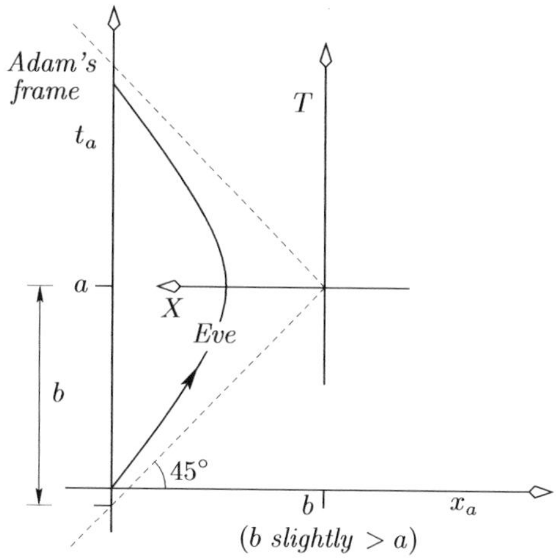

Fig. 7.13. Using an intermediate T-X frame to help define a useful set of coordinates for Eve.

$$T \equiv t_a - a\,, \quad X \equiv b - x_a\,. \tag{7.49}$$

T and X must be used in place of t and x in (7.25) and (7.26). In the interests of symmetry, it's also preferable if, like Adam, Eve has a space coordinate of zero in her own frame—not the $1/g$ that we used in Figs 7.3, 7.4, and 7.6. In that case, form a new space coordinate x_e for her by subtracting $1/g$ from the $\bar{x}$-coordinate of (7.25): $x_e \equiv \bar{x} - 1/g$. Equations (7.25) and (7.26) then become

$$\begin{aligned} t_a &= a + (x_e + 1/g)\,\mathrm{sh}\,g\bar{t}\,, \\ x_a &= b - (x_e + 1/g)\,\mathrm{ch}\,g\bar{t} \end{aligned} \tag{7.50}$$

and

$$\begin{aligned} \bar{t} &= \frac{1}{g}\,\mathrm{th}^{-1}\,\frac{t_a - a}{b - x_a}\,, \\ x_e &= \sqrt{(b - x_a)^2 - (t_a - a)^2} - 1/g\,. \end{aligned} \tag{7.51}$$

The time coordinate $\bar{t}$ still does not quite serve for Eve, since again, in the interests of symmetry, we wish Adam's worldline in Eve's frame to resemble her worldline in his frame—so that he leaves Eve not only at his time zero but also at her time zero. But the $\bar{t}$ of (7.50) and (7.51) won't lead to this; it was defined somewhat differently. This difference is just a shift by a constant, since if we draw Adam's worldline in Eve's $\bar{t}x_e$-frame by using (7.51) to plot $\bar{t}, x_e$ as functions of t_a for $x_a = 0$, we find that Adam's worldline is symmetrical about the x_e-axis and cuts the $\bar{t}$-axis at $\bar{t} = -1/g\ \mathrm{ch}^{-1}\,gb$. So shift his worldline up to put the departure at Eve's time of zero, which means we'll need to add $1/g \text{ ch}^{-1} gb$ onto the $\bar{t}$-coordinate to make a new coordinate t_e. Remember, this is purely a shift in time origin for Eve to make her picture of events look more analogous to Fig. 7.7:

$$t_e \equiv \bar{t} + \frac{1}{g} \text{ ch}^{-1} gb\,. \tag{7.52}$$

Now we have coordinates t_a, x_a for Adam and t_e, x_e for Eve. The final transformations relating the two frames are

$$\begin{aligned} t_a &= a + (x_e + 1/g) \text{ sh} \left(gt_e - \text{ch}^{-1} gb\right), \\ x_a &= b - (x_e + 1/g) \text{ ch} \left(gt_e - \text{ch}^{-1} gb\right) \end{aligned} \tag{7.53}$$

and

$$\begin{aligned} t_e &= \frac{1}{g} \left[\text{th}^{-1} \frac{t_a - a}{b - x_a} + \text{ch}^{-1} gb\right], \\ x_e &= \sqrt{(b - x_a)^2 - (t_a - a)^2} - 1/g\,, \end{aligned} \tag{7.54}$$

which are inverse to one another. For the following calculations, it's helpful to use $\text{sh}^{-1} ag = \text{ch}^{-1} bg = \text{th}^{-1} a/b$, and in the given scenario these equal $3g$. Adam and Eve are together at the start of the journey, so that the departure event has coordinates in the two frames:

$$(t_a, x_a) = (0, 0)\,, \quad (t_e, x_e) = (0, 0)\,. \tag{7.55}$$

Similarly, the reunion event has coordinates

$$(t_e, x_e) = (6, 0)\,, \quad (t_a, x_a) = (10, 0)\,. \tag{7.56}$$

Then, either (7.53) or (7.54) leads to

$$\sqrt{b^2 - 1/g^2} = a = 5 \quad \text{(from Fig. 7.13)}\,, \tag{7.57}$$

and incorporating this into (7.54) for the reunion event (7.56) gives

$$\sqrt{b^2 - 25} \text{ th}^{-1} \frac{5}{b} = 3\,. \tag{7.58}$$

Finally, (7.57) and (7.58) yield

$$b \simeq 5.26 \text{ years}, \quad g \simeq 0.613 \text{ years}^{-1}. \tag{7.59}$$

Eve's worldline in Adam's frame is given by (7.53) with $x_e = 0$:

$$t_a = 5 + \frac{1}{g} \text{ sh} \left[g(t_e - 3)\right], \quad x_a = b - \frac{1}{g} \text{ ch} \left[g(t_e - 3)\right], \tag{7.60}$$

and is shown in Fig. 7.11.

The equations describing Adam's worldline in Eve's frame are given by (7.54) with $x_a = 0$:

$$t_e = \frac{1}{g}\,\mathrm{th}^{-1}\frac{t_a - 5}{b} + 3\,, \quad x_e = \sqrt{b^2 - (t_a - 5)^2} - 1/g\,, \tag{7.61}$$

as shown in Fig. 7.12. Equations (7.60) and (7.61) are *not* simply inverses of each other, since in each case only one variable of the relevant coordinate pair is being used as a parameter for the worldline.

The calculations of this appendix underline the careful bookkeeping that must be used to show how an inertial-frame scenario looks in an accelerated frame. While an attention to detail is required, the results, Figs 7.11 and 7.12, are highly illuminating.

8 The Elegance and Power of Tensor Notation

8.1 Back to Vectors, in a More Generic Way

Using the Lorentz transform given in the last two chapters, changing inertial frames in special relativity turned out to be identical to simply changing coordinates in a particular way in four dimensions. This suggests that it might be useful to study general coordinate transformations in the hope that they'll prove useful in expanding our view of physical laws. The main demand to make is that the physics we are describing must be independent of our choice of coordinates. For example, although cartesian and polar coordinates are two different tools for describing the motion of a particle, that motion in itself cannot depend upon which of these coordinate systems we choose to work within.

Previously, we evolved the concepts of four-vectors and covectors, sets of quantities that transform under a change of coordinates in such a way that, in a sense, they can be considered to describe a "something" that has a reality of its own. This property makes them independent of our coordinate choice, and so makes them ideal contenders with which to describe physical laws.

The independence of coordinate choice is illustrated in Fig. 8.1. A function $\boldsymbol{X}(u, v)$ takes two parameters and specifies a position relative to the origin of the uv-coordinate system. Any increment in u and v results in a new position, $\boldsymbol{X}(u+\Delta u, v+\Delta v)$. Likewise, the same can be done for another function $\widetilde{\boldsymbol{X}}$ for a different coordinate system, x-y. The start and end points drawn as dots in Fig. 8.1 are now specified by $\widetilde{\boldsymbol{X}}(x, y)$ and $\widetilde{\boldsymbol{X}}(x+\Delta x, y+\Delta y)$. The two functions $\boldsymbol{X}, \widetilde{\boldsymbol{X}}$ are, in general, quite different. But the *displacement* between the two points is a well-defined entity, irrespective of the different coordinate systems used. Its physical relevance is due to the fact that it embodies our very intuitive notion that the dimensions and age of an object are independent of where we choose to place our ruler's origin or our clock's zero time. The physically meaningful quantities are *differences* in the coordinates, not the coordinates themselves.

However, what is more useful is to consider only infinitesimal displacements, since these have a far greater range of applicability to physical systems, and are also more easily described when the two coordinate systems are related nonlinearly. Some of the geometrical structure that can be given to

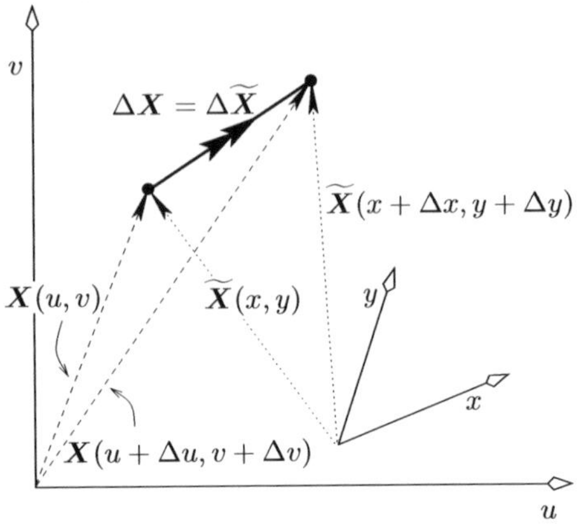

Fig. 8.1. Coordinate values of the start and end positions of the solid-line "displacement arrow" depend on the choice of coordinates. But the arrow itself has an independent existence, unlike the dashed arrows that denote its start and end points, which depend fully on the origin of their coordinate systems. In the limit of infinitesimal length, the solid displacement arrow becomes $\mathrm{d}\boldsymbol{X} = \mathrm{d}\widetilde{\boldsymbol{X}}$ and is called a *proper vector*, while the dashed arrows are *position vectors*, or *relative vectors.*

physical laws relates to our considering what happens along paths that connect two points in spacetime. Discarding information about events between the points—which is what the Δ is effectively doing—is not necessarily very useful. Rather, by focussing our attention on infinitesimal displacements, we are better able to describe the geometry of a space and to quantify laws that are a function of paths that join those widely separated points. These infinitesimal displacements are called *proper vectors.*

Contrast these proper vectors with the dotted arrows of Fig. 8.1, all of which must anchor their tails at the origins of either the uv-axes or the xy-axes. They are vectors in the uv and xy vector spaces. These vector spaces are not so important physically because the origins of their coordinate systems have generally been chosen arbitrarily. The dotted arrows (corresponding to the functions $\boldsymbol{X}$ and $\widetilde{\boldsymbol{X}}$) are called *relative vectors* or *position vectors.* They are of secondary importance to proper vectors.

Proper vectors, the subject of this chapter, are always specified with respect to some coordinate transform. In Fig. 8.1, we can see that the arrow $\Delta\boldsymbol{X}$ moves as an arrow should if the scene is rotated about a single point or translated by some amount. Compare this behaviour with that of the position vector $\boldsymbol{X}$ or $\widetilde{\boldsymbol{X}}$ to any one of the points, which always extends outward from its origin regardless of how the point in positioned relative to a new set of axes. (E.g., $\boldsymbol{X}$ will change to $\widetilde{\boldsymbol{X}}$ under the coordinate change in the figure.) That is, $\Delta\boldsymbol{X} = \Delta\widetilde{\boldsymbol{X}}$, but $\boldsymbol{X} \neq \widetilde{\boldsymbol{X}}$. This "less absolute" behaviour of position vectors makes them less useful for describing physical laws.

Proper vectors are something like arrows pointing from each star to its neighbour. Although observers standing on planets orbiting a multitude of stars each describe their sky differently, their descriptions are *consistent*, and indeed differ in exactly the way we would predict if we were to assume that the starry heavens exist independently of where each observer stands. All of the observers certainly recognise the arrows connecting the stars. Contrast this with the stars' position vectors, which are simply arrows pointing to the stars from each observer's planet. Each observer has a unique set of position vectors, and although that set is useful, it applies to that observer alone.

We'll find that four-vectors can be considered as examples of proper vectors under the Lorentz transform (a transform must always be specified, as we'll see), provided we can find a suitable basis for them. That will be done in this chapter (amongst other things).

In the familiar three dimensions, the coordinate-independence of physical quantities is typified by the velocity field that describes ocean currents or air flows, commonly seen in meteorology. We recognise that each vector of such a field describes the motion of an element of the fluid and so has an existence independent of the coordinate system. Nevertheless, any calculation concerning this field requires us to quantify the vectors within a particular system of coordinates x^α. How might this be done?

Suppose the velocity causes an element to change its position by a vector $\mathrm{d}\boldsymbol{X}$ in a time $\mathrm{d}t$. This might happen, for example, in an ocean current, where the fluid element is changing its position over time. The element's velocity $\boldsymbol{v} \equiv \mathrm{d}\boldsymbol{X}/\mathrm{d}t$ is a vector proportional to its infinitesimal displacement $\mathrm{d}\boldsymbol{X}$; that is, this displacement has been rescaled to a noninfinitesimal size. Velocity, of course, concerns itself with *infinitesimal* changes in the element's position; this is just the same reasoning as was applied back in Sect. 3.1 when we spoke of how an ever-decreasing quantity (the fraction of light bulbs in each bin in that section) can be rescaled by dividing it by another quantity (the bin width) that is also decreasing as part of the same process. This is also the same reasoning that led us to use an infinitesimal version of the metric in our study of the Clock Postulate.

At any point in space, the velocity vector points in the direction of the flow, with a magnitude that is just the flow speed at that point. (The only reason velocity vectors are drawn with small lengths in, for example, a picture of fluid flow is because longer ones would snag each other and clutter the diagram. But they certainly don't have an infinitesimal length.) It will prove very useful to sift out and pair up the components of the velocity with their associated directions that the velocity field codifies at each point, and this can be done using the chain rule of partial derivatives. Letting α sum over the three space coordinates (the Einstein summation convention), the velocity is

$$\boldsymbol{v} \equiv \frac{\mathrm{d}\boldsymbol{X}}{\mathrm{d}t} = \underbrace{\frac{\partial \boldsymbol{X}}{\partial x^\alpha}}_{\text{direction}} \times \underbrace{\frac{\mathrm{d}x^\alpha}{\mathrm{d}t}}_{\substack{\text{velocity} \\ \text{components } v^\alpha}} . \tag{8.1}$$

The velocity components are just the usual components v^α of the velocity vector, while the directions with which they are paired are given by the *basis vectors* $\boldsymbol{e}_\alpha$:

$$\boxed{\boldsymbol{e}_\alpha \equiv \frac{\partial \boldsymbol{X}}{\partial x^\alpha} .} \tag{8.2}$$

The index α, being a subscript on the left-hand side of (8.2), is also considered to be a subscript on the right-hand side. The $\mathrm{d}t$ in (8.1) is really just a scaling factor. As such, we'll stop writing it, saying simply that the small increase $\mathrm{d}\boldsymbol{X}$ in position equals $\mathrm{d}x^\alpha \boldsymbol{e}_\alpha$, a linear combination of basis vectors $\boldsymbol{e}_\alpha$. Of course, we are always free to put the $\mathrm{d}t$ back in to convert the quantities back to macroscopic ones able to be used in numerical calculations—such as the velocity $\boldsymbol{v}$ in this case—just as we did when introducing the four-velocity back in (6.13). The fact that our rescaling has produced a noninfinitesimal arrow that points along an element of flow, even right out of the fluid, gives rise to the idea that $\boldsymbol{v}$ is a *tangent vector* to the flow.

Although $\boldsymbol{v}$ has an existence of its own independent of any coordinate system, in order to specify it in component form we need to specify the set of axes referred to by these components. So, the v^α go together with the set of basis vectors $\boldsymbol{e}_\alpha$. In the flow example above, the basis vectors might change depending on where the fluid element is (i.e. where $\boldsymbol{X}$ is), but at that point they determine the local set of axes.

8.1.1 Honing the Vector Idea

Up until now, we've taken $\boldsymbol{X}$ to be the position of a point on some surface. Quantify this by making $\boldsymbol{X}$ a function that maps one set of coordinates to another. An example is shown in Fig. 8.2, where the function $\boldsymbol{X}\colon \mathbb{R}^2 \to \mathbb{E}^3$ maps the two variables u, v into a euclidean 3-space, producing a *2-surface* (since it needs just two variables to describe it). If $\boldsymbol{X}$ is known, then the basis vectors $\boldsymbol{e}_u$ and $\boldsymbol{e}_v$ are calculated using (8.2).

An instructive way to study the basis vectors is by way of the more concrete example of a sphere of radius R, parametrised by the usual spherical polar coordinates that resemble latitude and longitude, in Fig. 8.3. The mapping function is $\boldsymbol{X}(\theta, \phi) = (x, y, z)$:

$$\begin{aligned} x &= R \sin\theta \cos\phi , \\ y &= R \sin\theta \sin\phi , \\ z &= R \cos\theta . \end{aligned} \tag{8.3}$$

Equation (8.2) tells us that in xyz-space, the basis vectors are therefore

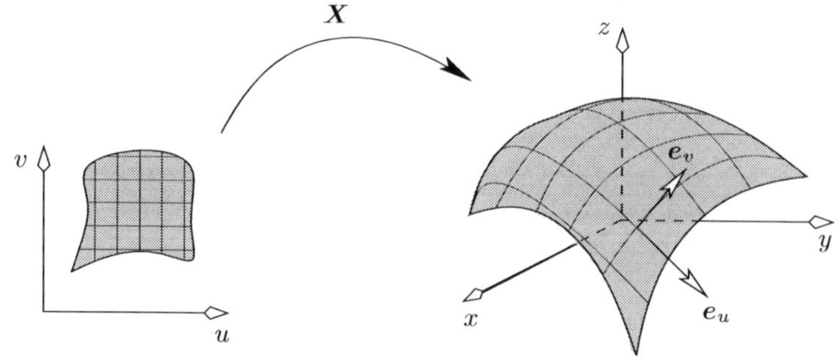

Fig. 8.2. The function $\boldsymbol{X}$ maps the uv-plane into a euclidean 3-space, and the basis vectors $\boldsymbol{e}_u, \boldsymbol{e}_v$ track the motion of the point $\boldsymbol{X}(u,v)$ under an increase in the coordinates u, v. So, for example, $\boldsymbol{e}_u$ points in the direction in which $\boldsymbol{X}(u,v)$ increases for an infinitesimal increase in u.

$$\boldsymbol{e}_\theta = \frac{\partial \boldsymbol{X}}{\partial \theta} = \frac{\partial}{\partial \theta}\begin{bmatrix} x(\theta,\phi) \\ y(\theta,\phi) \\ z(\theta,\phi) \end{bmatrix} = \begin{bmatrix} R\cos\theta\cos\phi \\ R\cos\theta\sin\phi \\ -R\sin\theta \end{bmatrix},$$

$$\boldsymbol{e}_\phi = \frac{\partial \boldsymbol{X}}{\partial \phi} = \frac{\partial}{\partial \phi}\begin{bmatrix} x \\ y \\ z \end{bmatrix} = \begin{bmatrix} -R\sin\theta\sin\phi \\ R\sin\theta\cos\phi \\ 0 \end{bmatrix}. \tag{8.4}$$

These calculations aside, a very important point in visualising $\boldsymbol{e}_\theta, \boldsymbol{e}_\phi$ as shown in Fig. 8.3 is that we do *not* need to do any calculation like (8.4) to find their direction. For example, to visualise $\boldsymbol{e}_\theta$ we need only follow what happens to

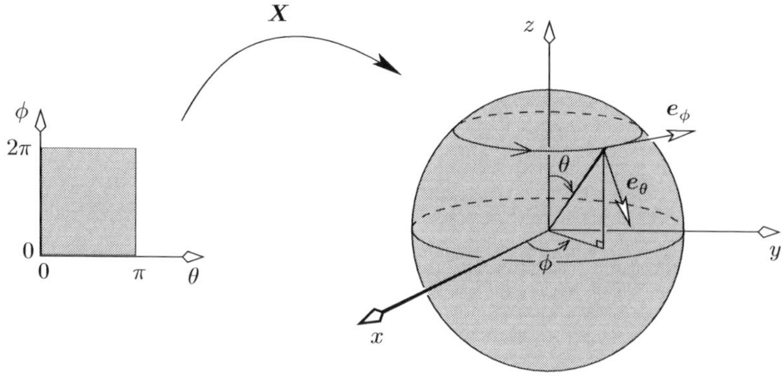

Fig. 8.3. Calculating the basis vectors $\boldsymbol{e}_\theta, \boldsymbol{e}_\phi$ for a sphere. Here, the relevant map is $\boldsymbol{X}(\theta,\phi) = (x,y,z)$. If the sphere is imagined to be Earth, with the z-axis parallel to Earth's rotation axis, then at any point on the sphere $\boldsymbol{e}_\theta$ points south and $\boldsymbol{e}_\phi$ points east.

the position vector $\boldsymbol{X}$ as θ increases by a small amount:

$$\boldsymbol{e}_\theta \equiv \frac{\partial \boldsymbol{X}}{\partial \theta} = \frac{\mathrm{d}\boldsymbol{X}\big|_{\phi \text{ const.}}}{\mathrm{d}\theta} \propto \mathrm{d}\boldsymbol{X}\big|_{\phi \text{ const.}}, \tag{8.5}$$

with a positive constant of proportionality if $\mathrm{d}\theta$ is positive. Thus $\boldsymbol{e}_\theta$ points in the direction in which the head of the position vector $\boldsymbol{X}$ moves when θ increases by a positive infinitesimal amount. If the sphere were Earth, with the z-axis parallel to Earth's rotation axis, then θ would increase along a meridian heading south, so that $\boldsymbol{e}_\theta$ would point south everywhere. Similarly, $\boldsymbol{e}_\phi$ would point east. It's a simple yet instructive exercise to use (8.4) to show that at every point, $\boldsymbol{e}_\theta \cdot \boldsymbol{e}_\phi = 0$, while $\boldsymbol{e}_\theta \times \boldsymbol{e}_\phi$ is parallel to the radial direction—"up" on a spherical Earth. And, of course, this is all as it should be.

Abstracting the Function $\boldsymbol{X}$

So far, the discussion of vectors has used a map $\boldsymbol{X}$ that takes a given set of coordinates, such as θ, ϕ, and *embeds* the resulting surface into a euclidean 3-space. More generally, $\boldsymbol{X}$ could map into a more exotic space, or it might not even be specified at all. Omitting it from expressions such as (8.1) and (8.2) doesn't stop vectors from being defined; we just leave vacant the position that $\boldsymbol{X}$ would normally occupy. The usual interpretation for this procedure is that the vector is now defined to be a derivative operator that acts on any given function $\boldsymbol{X}$ to produce what we have been deriving up until now. In that case, (8.1) and (8.2) will be written as

$$\text{“}\, \boldsymbol{v} = v^\alpha \frac{\partial}{\partial x^\alpha} \,\text{”}, \tag{8.6}$$

with the basis vectors now regarded as operators:

$$\text{“}\, \boldsymbol{e}_\alpha \equiv \frac{\partial}{\partial x^\alpha} \,\text{”}. \tag{8.7}$$

These operators are then available to act on a function $\boldsymbol{X}$ if there is any requirement or freedom to embed the surface in a higher-dimensional space. We will not be so abstract in our discussion. Instead, we take the more pragmatic view that the surface certainly can be embedded in such a space, so that the operating has already been done, so to speak. Our proper vectors are arrows, not operators.

8.1.2 Two Types of Vectors

Confusion often arises from the fact that physicists give the name "vector" to two distinct entities. To reiterate:

Vectors: Arrows or Operators?

It can be disconcerting to find that vectors lose their "arrow" character in texts that treat the subject very abstractly, instead seeming somehow to turn into derivative operators in the way of (8.7). And as every archer knows, an arrow is not an operator. They are certainly different things. Notationally, there *is* a one-to-one correspondence between writing a basis vector $\boldsymbol{e}_\alpha$ as an arrow $\partial \boldsymbol{X}/\partial x^\alpha$ and omitting the embedding function $\boldsymbol{X}$ to write $\boldsymbol{e}_\alpha$ as an operator $\partial/\partial x^\alpha$; and derivative operators *are* elements of a vector space, even though they are not arrows. The identification of arrows and operators is no deeper than that. The fact is, that an arrow is a *picture*, and a picture is something concrete that requires an embedding. In this book, we always assume the existence of an embedding function $\boldsymbol{X}$, which allows vectors to be treated as arrows.

Position vectors, also called relative vectors or radius vectors, are elements of a vector space in the mathematical sense. They typically locate a particle in some reference frame, and are the sort of vector commonly used in physics texts to deal with classical mechanics scenarios, such as the position of a planet relative to the Sun, or the position of a ball on a table. They are also the vectors $\boldsymbol{r}$ being rotated around arbitrary axes in Chap. 4.

Proper vectors are based on the idea of infinitesimal displacements as described in the current chapter. They are also elements of a vector space, but as we shall see, proper vectors are also required to transform appropriately under a specified change of coordinates. Proper vectors, whose components might form, for example, the four-vectors of the Lorentz transform, are the main sort of vector used in mathematical physics. We'll drop the adjective "proper" and just refer to these as vectors unless a distinction must be made. Position vectors are proper vectors under a rotation about the origin, while derivatives of position vectors in cartesian coordinates are proper vectors with respect to translations and rotations. Noninfinitesimal displacements are generally not counted as proper vectors for the reason described at the start of this chapter. As it turns out, they are not guaranteed to transform linearly with nonlinear changes in coordinates, as opposed to infinitesimal displacements, which always will. However, if we restrict ourselves to linear changes in coordinates, then noninfinitesimal displacements are certainly proper vectors. But since they are not generally useful in a world best described by calculus, they are generally avoided in vector analysis.

8.2 Vectors and Coordinate Changes

Suppose we wish to transform coordinates in order to see how the fluid flow vectors of the previous section are quantified in some other frame. In this example, we are using the three dimensions of ordinary space. It would do no good to try to make this four-dimensional by introducing a time parameter, since that would allow a Galilei transform to reduce a constant-velocity vector field to zero—which is contrary to our notion of a vector since a vector is meant to be frame-independent. After all, in previous chapters, if we had a set of vector components v^α that were all zero in some frame, then they would *have* to transform to zero in all other frames, since their values in any other frame are just linear combinations: $v^{\mu'} = \Lambda^{\mu'}_{\alpha} v^\alpha = 0$. Zero vectors are zero in *all* frames! But in the four dimensions of spacetime, we do consider time to stand on an equal footing with the space axes, and the great thing that we have seen emerging in previous chapters is that relativity introduces us to objects that *do* transform as vectors (under the Lorentz transform), even when time is included.

A familiar example of basis vectors is the set $\{\boldsymbol{e}_x, \boldsymbol{e}_y, \boldsymbol{e}_z\}$, each of which points along one of the cartesian axes that describe our everyday three dimensions. These are especially useful because they do not change with position, a fact that we'll often put to good use in this chapter. How can we relate each of these vectors to the more familiar forms, such as $(1, 0, 0)$? Remember from p. 9 that the ordered set of three numbers $(1, 0, 0)$ is a *coordinate vector* with respect to some basis. Trivially then, the coordinate vector of $\boldsymbol{e}_x$ with respect to the usual cartesian basis $\{\boldsymbol{e}_x, \boldsymbol{e}_y, \boldsymbol{e}_z\}$ will be $(1, 0, 0)$ because

$$\boldsymbol{e}_x = 1\,\boldsymbol{e}_x + 0\,\boldsymbol{e}_y + 0\,\boldsymbol{e}_z\,. \tag{8.8}$$

This is an important distinction; $\boldsymbol{e}_x$ does *not* "equal" $(1, 0, 0)$. Rather, the coordinate vector of $\boldsymbol{e}_x$ with respect to the $\{\boldsymbol{e}_x, \boldsymbol{e}_y, \boldsymbol{e}_z\}$ basis is $(1, 0, 0)$. Far from being a pedantic point, this is an important concept to keep in mind. The coordinate vector of $\boldsymbol{e}_x$ with respect to a different basis will be a different set of three numbers, and we are free to use whatever basis is convenient. While tensor analysis was originally constructed to deal with coordinate vectors only ("components" is the more common term), these are only *representatives* of the actual vector with respect to some chosen basis.

Alternatively to (8.8), note that an arbitrary step in space can be written as

$$\begin{aligned} \mathrm{d}\boldsymbol{X} &= \frac{\partial \boldsymbol{X}}{\partial x}\mathrm{d}x + \frac{\partial \boldsymbol{X}}{\partial y}\mathrm{d}y + \frac{\partial \boldsymbol{X}}{\partial z}\mathrm{d}z \\ &= \boldsymbol{e}_x\,\mathrm{d}x + \boldsymbol{e}_y\,\mathrm{d}y + \boldsymbol{e}_z\,\mathrm{d}z\,, \end{aligned} \tag{8.9}$$

whose coordinate vector with respect to $\{\boldsymbol{e}_x, \boldsymbol{e}_y, \boldsymbol{e}_z\}$ is

$$(\mathrm{d}x, \mathrm{d}y, \mathrm{d}z) = \mathrm{d}x\,(1, 0, 0) + \mathrm{d}y\,(0, 1, 0) + \mathrm{d}z\,(0, 0, 1)\,, \tag{8.10}$$

and again we see the familiar forms of $(1, 0, 0)$, etc., appearing.

The definitions of vector components and basis vectors in (8.1) and (8.2) are completely general in any number of dimensions with any type of coordinates; the definition of a vector does not favour cartesian coordinates. For example, if we have a set of vector components $\mathrm{d}x^\beta$, then in a different frame marked by bars over the coordinate labels, the corresponding vector components are

$$\boxed{\mathrm{d}x^{\bar\alpha} = \frac{\partial x^{\bar\alpha}}{\partial x^\beta}\,\mathrm{d}x^\beta = \Lambda^{\bar\alpha}_\beta\,\mathrm{d}x^\beta\,,} \tag{8.11}$$

where we've used the familiar Λ notation introduced in (6.37) and properly generalised in (7.44). The new (barred) basis vectors are

$$\boxed{\boldsymbol{e}_{\bar\alpha} \equiv \frac{\partial \boldsymbol{X}}{\partial x^{\bar\alpha}} = \frac{\partial \boldsymbol{X}}{\partial x^\beta}\frac{\partial x^\beta}{\partial x^{\bar\alpha}} = \Lambda^\beta_{\bar\alpha}\,\boldsymbol{e}_\beta\,.} \tag{8.12}$$

Although infinitesimals $\mathrm{d}x^{\bar\alpha}, \mathrm{d}x^\beta$ were used in (8.11), we know from (8.1) and the discussion following (8.2) that the vector components $v^{\bar\alpha}$ behave in the same way as $\mathrm{d}x^{\bar\alpha}$. So (8.11) can be rewritten for vector components $v^{\bar\alpha}$ as

$$v^{\bar\alpha} = \Lambda^{\bar\alpha}_\beta\,v^\beta\,. \tag{8.13}$$

Equations (8.11) [or (8.13)] and (8.12) ensure that a vector $\boldsymbol{v}$ has the same form in tensor notation, irrespective of coordinate system:

$$\begin{aligned}\boldsymbol{v} = v^{\bar\alpha}\,\boldsymbol{e}_{\bar\alpha} &= \Lambda^{\bar\alpha}_\beta\,v^\beta \times \Lambda^\mu_{\bar\alpha}\,\boldsymbol{e}_\mu \\ &= \delta^\mu_\beta\,v^\beta\,\boldsymbol{e}_\mu = v^\mu\,\boldsymbol{e}_\mu\,.\end{aligned} \tag{8.14}$$

To preserve the up–down index positions, the Kronecker delta function has been written with one index up and one down. It must always be written this way in tensor analysis. Later we'll see what happens if it is written with both indices up or both down.

Equations (8.11) and (8.12) are very basic to any study of tensor analysis, and something quite fundamental is being demonstrated in (8.14). That is, whenever an index appears both up and down (and therefore is summed over), that index can be switched to one of any other set of coordinates, indicating a frame independence that allows us to construct a geometrical picture of vectors. This is called *contracting* over that index. This is perhaps our first indication of the utility of tensor notation. The quantities (and ultimately equations) that we write down have the same form in any set of coordinates—which makes them ideal for describing physical laws, since these laws must be independent of coordinate systems. Physical frame choices (such as inertial or accelerated) are not as arbitrary as coordinate choices within a frame, but we can expect the covariant language of raised and lowered indices to give, or suggest, the new form of a law in a different frame. In Sect. 8.7 we'll draw more attention to the distinction between frames and coordinates.

"Invariance" versus "Covariance"

The words invariance and covariance are sometimes confused. What does each mean? Invariance refers to the desirable notion that a quantity will not change when the coordinate system or frame is changed. The simplest such quantities—pure numbers that everyone agrees on, such as temperature—are called scalars. But more complicated entities *do* change under the same circumstances. However, as outlined in the previous chapters and this one, although tensor components and basis vectors change, they change with coordinates in such a way—through multiplication by matrices of partial derivatives, as in (8.11) and (8.12)—that the more refined concept of a full tensor of components plus basis *is* invariant.

This idea of varying *with* ("co") coordinates is what gives the index notation of tensor components and basis vectors the name "covariant". The word denotes our acknowledgement that although some quantities are not invariant, they are not arbitrary either, and the way that they change is constrained in a certain way. But the overriding concept is invariance—whether of a scalar or of a complete tensor (components *together with* basis vectors). Invariance of any object implies (in fact, defines) covariance of the elements that make it up, which in the case of tensors are the components and basis vectors. Invariance is the key concept; covariance is a description of how the notation of vectors and tensors recognises invariance.

Although the whole index notation is termed covariant, the word is also given a meaning within that context. To ensure the overall invariance of a vector $\boldsymbol{v}$ (components and basis vectors), its components must transform oppositely from the basis vectors. Historically, the index-down set of components has come to be designated as changing "with" the coordinate transformation (*co*variant) (where "with" clearly does not mean "identically to" here!). The other set (index up) has then been designated as changing "against" the coordinate transformation (*contra*variant). So, for example, the term *covariant vector* is traditionally used to denote v_α, the coordinate vector with respect to the cobasis of Sect. 8.4, whereas *contravariant vector* is similarly used to denote v^α, the coordinate vector with respect to a basis.

Fundamental to the idea of frame changing with vectors is that the coordinate vector (the ordered set of components) v^α—such as the four-vector of special relativity—is not frame-independent, but those components *together with a basis*, $v^\alpha \boldsymbol{e}_\alpha$, certainly are frame-independent, as demonstrated by (8.14). The word "vector" is somewhat overused in tensor analysis. The components v^α constitute the coordinate vector with respect to the $\boldsymbol{e}_\alpha$ basis, and are simply called a vector. They depend on frame choice: $\{v^\alpha\} \neq \{v^{\bar\beta}\}$ in general. But the sum $v^\alpha \boldsymbol{e}_\alpha$ is also called a vector, and is *in*dependent of frame choice: $v^\alpha \boldsymbol{e}_\alpha = v^{\bar\beta} \boldsymbol{e}_{\bar\beta}$. Most tensor analysis concentrates on components only; indeed, we'll see that concepts such as the metric and the covariant derivative (both defined later in this chapter) were created to allow us to forget about the basis vectors entirely, by bundling any effects due to them into the component manipulations. Even so, it's wise to remember that the basis vectors do form the other side of the coin and are sometimes too easily forgotten. Explicitly including them in a calculation is often a very useful thing to do.

Two types of indices are sometimes needed, using Greek and Latin letters, to distinguish vectors $v^\alpha \boldsymbol{e}_\alpha$ that exist in some higher-dimensional space, say the four dimensions of spacetime, from those confined to a subspace and written $v^a \boldsymbol{e}_a$, where the subspace might be the familiar three spatial dimensions. In this book, Greek indices indicate a summation over the whole space, while Latin letters indicate a sum over a subspace. So, for example, in spacetime, Greek indices mean 0 to 3 (e.g., t, x, y, z, or t, r, θ, ϕ), while Latin denote 1 to 3 (e.g., x, y, z, or r, θ, ϕ). We have not yet needed to make this distinction but will certainly do so in Chap. 9. In some texts, the roles of Latin and Greek indices are reversed.

We should check that basis vectors $\boldsymbol{e}_\alpha$ are really vectors; i.e., that their components transform as vectors. This requires showing that their components in a barred frame are related to their components in an unbarred frame by the usual relation (8.11): $(\boldsymbol{e}_\alpha)^{\bar\mu} = \Lambda^{\bar\mu}_\beta\,(\boldsymbol{e}_\alpha)^\beta$. First realise that $\boldsymbol{e}_\alpha = \Lambda^{\bar\mu}_\alpha\,\boldsymbol{e}_{\bar\mu}$, so that the components of $\boldsymbol{e}_\alpha$ in the barred frame are $\Lambda^{\bar\mu}_\alpha$. Next, write $\boldsymbol{e}_\alpha = \delta^\beta_\alpha\,\boldsymbol{e}_\beta$, so the components of $\boldsymbol{e}_\alpha$ in the unbarred frame are δ^β_α. Hence,

$$\begin{aligned}\mu^{\text{th}} \text{ component of } \boldsymbol{e}_\alpha \text{ in barred frame} &= (\boldsymbol{e}_\alpha)^{\bar\mu}\\ &= \Lambda^{\bar\mu}_\alpha = \Lambda^{\bar\mu}_\beta\,\delta^\beta_\alpha = \Lambda^{\bar\mu}_\beta\,(\boldsymbol{e}_\alpha)^\beta\\ &= \Lambda^{\bar\mu}_\beta \times \beta^{\text{th}} \text{ component of } \boldsymbol{e}_\alpha \text{ in unbarred frame.} \qquad (8.15)\end{aligned}$$

The components of $\boldsymbol{e}_\alpha$ transform in the way of (8.11) as required, and so $\boldsymbol{e}_\alpha$ is indeed a vector.

We are beginning to see that four-vectors are not really the separate breed that they first appeared to be. They form part of a larger set of proper vectors that includes other transformations. These proper vectors are always defined by the way that their components transform: through the partial derivatives of coordinates, $\Lambda^{\bar\alpha}_\beta$ and $\Lambda^\beta_{\bar\alpha}$, of (8.11) and (8.12). (The basis vectors always transform correctly, by construction.) We cannot call a set of numbers a vector unless we also specify how it transforms. So four-vectors together with their basis are proper vectors under the Lorentz transform and, as such, are also called Lorentz vectors. From now on, we will refer to them simply as vectors, realising that they fit into this much bigger scheme in which a vector transforms via the general equations (8.11) and (8.12).

8.3 Generalising the Idea of Vector Length

In Sect. 5.6, we introduced the idea, originally due to Minkowski, that spacetime could be considered as a unified entity. This was done by showing that all observers agree on the value of a unique number attached to each four-vector, and that this "scalar" is a natural extension to the three-dimensional idea of length. We calculated this length for the four-velocity in Sect. 6.3.2. It appears, then, that just as the familiar dot product in three dimensions allows us to calculate the euclidean length of a vector, so might a similar idea be applicable to higher dimensions. A higher-dimensional dot product or *norm* can be *defined* to give a vector length (not necessarily euclidean) through the metric:

$$|\boldsymbol{v}|^2 \equiv \boldsymbol{v}\cdot\boldsymbol{v} = v^\alpha \boldsymbol{e}_\alpha \cdot v^\beta \boldsymbol{e}_\beta = v^\alpha v^\beta \; \boldsymbol{e}_\alpha \cdot \boldsymbol{e}_\beta \equiv v^\alpha v^\beta g_{\alpha\beta} \,. \tag{8.16}$$

The action of the metric in combining pairs of vector components into an overall norm is encoded in the *metric components* $g_{\alpha\beta}$, which obey the key identity

$$\boxed{g_{\alpha\beta} \equiv \boldsymbol{e}_\alpha \cdot \boldsymbol{e}_\beta \,.} \tag{8.17}$$

Now, since each length is a scalar (i.e., independent of reference frame) and since $\boldsymbol{v}$ is a vector (i.e., having a reality independent of frame), we can infer that the numbers $g_{\alpha\beta}$ are also the components of a new object that also has a reality independent of frame.

Our making this inference is an example of the use of the *quotient theorem* of tensor analysis, which is a rigorous statement involving tensor indices that effectively states what we have just said.

This new object, the metric tensor, is our first generalisation of proper vectors to the larger set of objects that need more than one index to describe their components. These are the *tensors*, and have an existence independent of frame when coupled to an appropriate basis. In general, tensors can have any number of up and down indices, and so rather than have a separate notation for each type, we tend to refer to each by their components. Notice that the metric is symmetric, simply because the dot product is: $g_{\alpha\beta} = \boldsymbol{e}_\alpha \cdot \boldsymbol{e}_\beta = g_{\beta\alpha}$; so we can, and will, swap the indices freely whenever convenient. And just as is done for vectors, the word "tensor" can refer to the whole set of components, such as $g_{\alpha\beta}$, it can refer to that set together with a basis (which we'll consider in Sect. 8.5.1), or it can refer to one of those basis elements.

The metric tensor is the real quantifier of a space's geometry. An example is the familiar euclidean metric for cartesian coordinates. Pythagoras's theorem tells us that $\mathrm{d}\ell^2 = \mathrm{d}x^2 + \mathrm{d}y^2 + \mathrm{d}z^2$, so writing

$$\begin{aligned} \mathrm{d}\ell^2 &= |\mathrm{d}\boldsymbol{x}|^2 = \mathrm{d}x^\alpha \boldsymbol{e}_\alpha \cdot \mathrm{d}x^\beta \boldsymbol{e}_\beta = \mathrm{d}x^\alpha \mathrm{d}x^\beta g_{\alpha\beta} \\ &= \mathrm{d}x^2 + \mathrm{d}y^2 + \mathrm{d}z^2 \end{aligned} \tag{8.18}$$

implies that the euclidean metric has components

$$g_{xx} = g_{yy} = g_{zz} = 1 \,, \quad g_{xy} = g_{xz} = \dots = 0 \,. \tag{8.19}$$

(Strictly speaking, $d\ell^2$ is the line element, while the set $g_{\alpha\beta}$ is the metric, but the terms are often mixed. This is harmless.) Being a quadratic form, the metric components are naturally written as a matrix:

$$d\ell^2 = \begin{bmatrix} dx & dy & dz \end{bmatrix} \begin{bmatrix} 1 & 0 & 0 \\ 0 & 1 & 0 \\ 0 & 0 & 1 \end{bmatrix} \begin{bmatrix} dx \\ dy \\ dz \end{bmatrix} . \tag{8.20}$$

> Having said this, the metric is *not* a matrix; rather, its elements can be written in a matrix to allow them to be manipulated easily. A matrix is purely a set of numbers written in a tableau that enables us to manipulate them according to useful rules. We'll see more of this distinction later in Sect. 8.5.2.

An example of a metric for a different space is that of special relativity with cartesian coordinates. This metric is always written $\eta_{\alpha\beta}$. A look at (6.24) shows that the sixteen entries of $\eta_{\alpha\beta}$ must be

$$\eta_{tt} = 1, \quad \eta_{xx} = \eta_{yy} = \eta_{zz} = -1; \quad \text{all other } \eta_{\alpha\beta} = 0. \tag{8.21}$$

The alternative sign choice $(-{+}{+}{+})$ is commonly used also. Both sign conventions can be accommodated in any one expression by writing the matrix of $\eta_{\alpha\beta}$ as $\eta_{tt}\,\text{diag}(1, -1, -1, -1)$. We'll do this in Chap. 10.

8.3.1 Coordinate Transformation of the Metric

What are the components of the metric tensor in another frame? They follow straightaway from the definition of the metric:

$$g_{\bar{\mu}\bar{\nu}} = \boldsymbol{e}_{\bar{\mu}} \cdot \boldsymbol{e}_{\bar{\nu}} = \Lambda^{\alpha}_{\bar{\mu}}\, \boldsymbol{e}_{\alpha} \cdot \Lambda^{\beta}_{\bar{\nu}}\, \boldsymbol{e}_{\beta} = \Lambda^{\alpha}_{\bar{\mu}}\, \Lambda^{\beta}_{\bar{\nu}}\, g_{\alpha\beta} \,. \tag{8.22}$$

This transformation law is just an extension of the covector law that we first met in Sect. 6.5 and equation (6.50). It generalises to define a tensor with any number of up and down components. For each component, there should be a Λ-factor with appropriate indices, so that all indices not being summed over appear in the same up or down positions on both sides of any tensor equation. When changing frames, the various Λ-factors will ultimately be multiplied together, and when they do, the same sort of cancellation occurs as in (8.14). This ensures frame independence; summing over all up and down indices will always produce a quantity that is independent of frame, be it a proper vector (vector components with basis) in (8.14) or a scalar in (8.16). No other positioning of indices makes sense in tensor analysis.

The Basis and Metric for Polar Coordinates

Polar coordinates in the plane are very useful for developing good sense with the ideas of vectors and metrics. For example, what are the basis vectors and

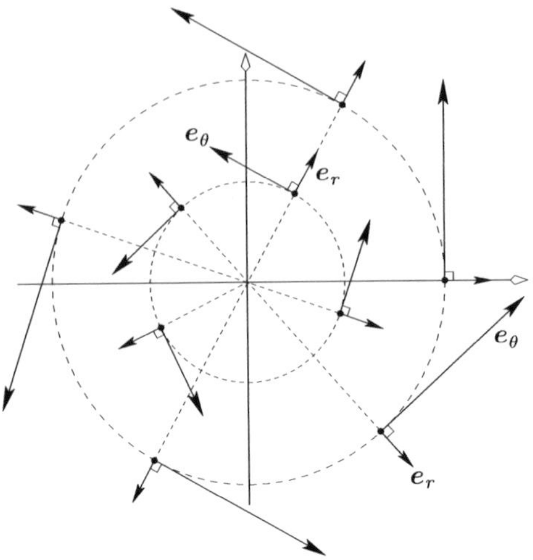

Fig. 8.4. Polar coordinate basis vectors at various points. Not surprisingly, all the $\boldsymbol{e}_r$'s point radially outward and have unit length. Why? Because they are just $\partial \boldsymbol{X}/\partial r$, which means we need only ask for the step $\mathrm{d}\boldsymbol{X}$ taken when r is increased by $\mathrm{d}r > 0$ with θ held constant. This step is radially outward for a distance of $\mathrm{d}r$, so that the ratio $|\mathrm{d}\boldsymbol{X}|/\mathrm{d}r = 1$. Similarly, the step $\mathrm{d}\boldsymbol{X}$ taken when θ is increased by $\mathrm{d}\theta > 0$ and r held constant is tangential to a circle of radius r, and has length $r\,\mathrm{d}\theta$. Thus the ratio $|\mathrm{d}\boldsymbol{X}|/\mathrm{d}\theta = r$, so that $\boldsymbol{e}_\theta$ has length r. Naturally, these lengths and directions agree with those calculated from (8.25). Bear in mind that the units of these vectors are different; see the box on p. 50.

metric in polar coordinates? Start with what we *do* know: the cartesian basis vectors $\boldsymbol{e}_x, \boldsymbol{e}_y$ with their euclidean metric. The polar basis is given by (8.12):

$$\begin{aligned} \boldsymbol{e}_r &= \frac{\partial x}{\partial r}\,\boldsymbol{e}_x + \frac{\partial y}{\partial r}\,\boldsymbol{e}_y\,, \\ \boldsymbol{e}_\theta &= \frac{\partial x}{\partial \theta}\,\boldsymbol{e}_x + \frac{\partial y}{\partial \theta}\,\boldsymbol{e}_y\,. \end{aligned} \tag{8.23}$$

Polar and cartesian bases are related by

$$x = r\cos\theta\,, \quad y = r\sin\theta\,, \tag{8.24}$$

in which case the polar basis vectors become

$$\begin{aligned} \boldsymbol{e}_r &= \cos\theta\,\boldsymbol{e}_x + \sin\theta\,\boldsymbol{e}_y\,, \\ \boldsymbol{e}_\theta &= -r\sin\theta\,\boldsymbol{e}_x + r\cos\theta\,\boldsymbol{e}_y\,. \end{aligned} \tag{8.25}$$

The polar basis vectors at various points are shown in Fig. 8.4. Using these, we can calculate the polar metric to be

$$\begin{aligned} g_{rr} &= \boldsymbol{e}_r \cdot \boldsymbol{e}_r = (\cos\theta\, \boldsymbol{e}_x + \sin\theta\, \boldsymbol{e}_y) \cdot (\text{same}) = 1\,, \\ g_{r\theta} = g_{\theta r} &= \boldsymbol{e}_r \cdot \boldsymbol{e}_\theta = (\cos\theta\, \boldsymbol{e}_x + \sin\theta\, \boldsymbol{e}_y) \cdot (-r\sin\theta\, \boldsymbol{e}_x + r\cos\theta\, \boldsymbol{e}_y) = 0\,, \\ g_{\theta\theta} &= \boldsymbol{e}_\theta \cdot \boldsymbol{e}_\theta = (-r\sin\theta\, \boldsymbol{e}_x + r\cos\theta\, \boldsymbol{e}_y) \cdot (\text{same}) = r^2\,, \end{aligned} \tag{8.26}$$

so that the polar line element is

$$\mathrm{d}\ell^2 = \mathrm{d}r^2 + r^2 \mathrm{d}\theta^2\,. \tag{8.27}$$

Of course, the polar line element is usually calculated in a much more straightforward way through a geometric, pictorial procedure of applying Pythagoras's theorem to the right triangle formed by increases of $\mathrm{d}r$ and $r\,\mathrm{d}\theta$. It can also be calculated by writing

$$\mathrm{d}\ell^2 = \mathrm{d}x^2 + \mathrm{d}y^2 = \left(\frac{\partial x}{\partial r}\,\mathrm{d}r + \frac{\partial x}{\partial \theta}\,\mathrm{d}\theta\right)^2 + \left(\frac{\partial y}{\partial r}\,\mathrm{d}r + \frac{\partial y}{\partial \theta}\,\mathrm{d}\theta\right)^2 \tag{8.28}$$

and then inserting the various partial derivatives. These approaches don't use basis vectors, of course, but following the procedure of (8.26) is a useful exercise in learning how everything fits together.

8.4 A Natural Basis for Covectors

The set of basis vectors $\{\boldsymbol{e}_\alpha\}$ pairs naturally with the single coordinate vector v^α, producing a single vector $\boldsymbol{v} \equiv v^\alpha \boldsymbol{e}_\alpha$. In Sect. 2.3 we introduced another basis vector set $\{\boldsymbol{e}^\alpha\}$ together with its corresponding coordinate vector v_α. These *cobasis* vectors had the property shown in (2.19) of being orthonormal to the basis $\{\boldsymbol{e}_\alpha\}$, and were calculated using a cross product in (2.20). We wish to generalise $\boldsymbol{e}^\alpha$ now to an arbitrary space using tensor notation. The cobasis $\{\boldsymbol{e}^\alpha\}$ is defined by how it relates to the basis $\{\boldsymbol{e}_\alpha\}$:

$$\boldsymbol{e}^\alpha \cdot \boldsymbol{e}_\beta \equiv \delta^\alpha_\beta\,. \tag{8.29}$$

Being vectors, the $\boldsymbol{e}^\alpha$ are expressible as linear combinations of the basis set $\{\boldsymbol{e}_\beta\}$, and the goal is to find these linear combinations. Prepare for this by defining some quantities that will be very useful for the required manipulations, as well as giving meaning to the notation that results. Just as $g_{\alpha\beta} \equiv \boldsymbol{e}_\alpha \cdot \boldsymbol{e}_\beta$, define

$$\boxed{\begin{aligned} g^\alpha{}_\beta &\equiv \boldsymbol{e}^\alpha \cdot \boldsymbol{e}_\beta = \delta^\alpha_\beta = \boldsymbol{e}_\beta \cdot \boldsymbol{e}^\alpha = g_\beta{}^\alpha\,, \\ g^{\alpha\beta} &\equiv \boldsymbol{e}^\alpha \cdot \boldsymbol{e}^\beta = \boldsymbol{e}^\beta \cdot \boldsymbol{e}^\alpha = g^{\beta\alpha}\,. \end{aligned}} \tag{8.30}$$

The spacing in the $g^\alpha{}_\beta$ indices is not really necessary, but we have written it this way only to mimic those indices' positions in $\boldsymbol{e}^\alpha \cdot \boldsymbol{e}_\beta$.

Note that the set of $g^{\alpha\beta}$ is usually defined in such a way that a matrix composed of them is the inverse of the matrix composed of the $g_{\alpha\beta}$. We have not done that because at this stage there is no motivation for doing so. Rather, the definitions in (8.30) are all about symmetry in the notation. But we'll shortly find in (8.35) that matrices composed of $g^{\alpha\beta}$ and $g_{\alpha\beta}$ are indeed inverses of one other.

If the $\boldsymbol{e}^\alpha$ are to be written as linear combinations of the basis set $\{\boldsymbol{e}_\mu\}$, define the coefficients as $G^{\alpha\mu}$:

$$\boldsymbol{e}^\alpha = G^{\alpha\mu}\boldsymbol{e}_\mu \,. \tag{8.31}$$

Now use (8.29) to write

$$\delta^\alpha_\beta = \boldsymbol{e}^\alpha\cdot\boldsymbol{e}_\beta = G^{\alpha\mu}\boldsymbol{e}_\mu\cdot\boldsymbol{e}_\beta = G^{\alpha\mu}g_{\mu\beta}\,, \tag{8.32}$$

in which case this last expression together with (8.30) and (8.31) produces

$$g^{\alpha\beta} = G^{\alpha\mu}\boldsymbol{e}_\mu\cdot G^{\beta\nu}\boldsymbol{e}_\nu = G^{\alpha\mu}G^{\beta\nu}g_{\mu\nu} = G^{\alpha\mu}\delta^\beta_\mu = G^{\alpha\beta}\,. \tag{8.33}$$

So the coefficients $G^{\alpha\beta}$ are just equal to $g^{\alpha\beta}$, allowing us to write

$$\boxed{\boldsymbol{e}^\alpha = g^{\alpha\mu}\boldsymbol{e}_\mu\,,} \tag{8.34}$$

where (8.32) becomes

$$g^{\alpha\mu}g_{\mu\beta} = \delta^\alpha_\beta\,. \tag{8.35}$$

This last relation shows the fundamental result that the matrix with $\alpha\mu^{\text{th}}$ entry $g^{\alpha\mu}$ is the inverse of the matrix with $\mu\beta^{\text{th}}$ entry $g_{\mu\beta}$. Note also that (8.34) inverts easily by multiplying by the metric and summing:

$$g_{\beta\alpha}\boldsymbol{e}^\alpha = g_{\beta\alpha}g^{\alpha\mu}\boldsymbol{e}_\mu = \delta^\mu_\beta\boldsymbol{e}_\mu = \boldsymbol{e}_\beta\,. \tag{8.36}$$

Since $\boldsymbol{e}^\alpha$ is a vector, its components should transform as required of vector components, and indeed they do:

$$\begin{aligned}\mu^{\text{th}}\text{ component of }\boldsymbol{e}^\alpha\text{ in barred frame} &= (\boldsymbol{e}^\alpha)^{\bar\mu}\\ &= \left(g^{\alpha\beta}\Lambda^{\bar\nu}_\beta\,\boldsymbol{e}_{\bar\nu}\right)^{\bar\mu} = g^{\alpha\beta}\Lambda^{\bar\mu}_\beta = \Lambda^{\bar\mu}_\beta\,(g^{\alpha\gamma}\boldsymbol{e}_\gamma)^\beta = \Lambda^{\bar\mu}_\beta\,(\boldsymbol{e}^\alpha)^\beta\\ &= \Lambda^{\bar\mu}_\beta\times\beta^{\text{th}}\text{ component of }\boldsymbol{e}^\alpha\text{ in unbarred frame.}\end{aligned} \tag{8.37}$$

This is the behaviour required of raised components.

Are $g^{\alpha\beta}$ and $g^\alpha{}_\beta$ really tensors just like $g_{\alpha\beta}$? They are, and to prove this, we must show that they transform in the required way. Consider $g^{\alpha\beta}$, beginning with a guess at its transformation in the parentheses in (8.38), where it is certainly true that

$$\left(\Lambda^{\bar\alpha}_\mu\,\Lambda^{\bar\beta}_\nu\,g^{\mu\nu}\right)g_{\bar\beta\bar\gamma} = \Lambda^{\bar\alpha}_\mu\,\Lambda^{\bar\beta}_\nu\,g^{\mu\nu}\,\Lambda^\varrho_{\bar\beta}\,\Lambda^\sigma_{\bar\gamma}\,g_{\varrho\sigma} = \delta^{\bar\alpha}_{\bar\gamma}\,. \tag{8.38}$$

Thus we can infer that the term in parentheses is the inverse of the metric:

$$\Lambda^{\bar\alpha}_{\mu}\,\Lambda^{\bar\beta}_{\nu}\,g^{\mu\nu} = g^{\bar\alpha\bar\beta}\,. \tag{8.39}$$

QED; this is the expected transformation for a tensor with both indices up (known as a (2,0) tensor). A similar argument shows that $g^{\alpha}{}_{\beta}$ also transforms in the appropriate way.

Finally, do the $\boldsymbol{e}^{\alpha}$ really form a basis? They will if they are linearly independent. We can prove that they are by contradiction. Assume there exists a set c_α not all equal to zero such that $c_\alpha\,\boldsymbol{e}^{\alpha} = 0$. Then $c_\alpha\,g^{\alpha\beta}\,\boldsymbol{e}_\beta = 0$. But, because the $\boldsymbol{e}_\beta$ are linearly independent, this means $c_\alpha\,g^{\alpha\beta} = 0$ for all β, and so on multiplying by $g_{\beta\gamma}$, we arrive at $c_\gamma = 0$ for all γ, which is a contradiction. So the $\boldsymbol{e}^{\alpha}$ really do form a basis.

In three-dimensional euclidean space, the basis vectors $\boldsymbol{e}^{\alpha}$ are exactly the cobasis of Sect. 2.3 used in crystallography. We see here the compactness of tensor notation: the expression for the cobasis in (8.34) is in fact equivalent to the more complicated crystallography version, (2.20). The fact that the two are identical follows from the fact that they both express orthonormality between the basis and the cobasis, as in (2.19) and (8.29).

A simple example shows how a cobasis is constructed and what it looks like. Work in the xy-plane with the usual cartesian basis $\boldsymbol{e}_x, \boldsymbol{e}_y$, and consider another basis of two vectors, $\boldsymbol{e}_1, \boldsymbol{e}_2$, with coordinates in the $\boldsymbol{e}_x, \boldsymbol{e}_y$ basis of

$$\boldsymbol{e}_1 = \begin{bmatrix}1 & 0\end{bmatrix}, \quad \boldsymbol{e}_2 = \begin{bmatrix}1 & 1\end{bmatrix}. \tag{8.40}$$

(Whether we write the coordinate vectors as rows or columns in this example is immaterial; here we are only interested in their components.) Refer to Fig. 8.5. The metric in this new basis has components $g_{11} \equiv \boldsymbol{e}_1\cdot\boldsymbol{e}_1 = 1$, etc. Letting α, β stand for any index in the set $\{1, 2\}$, insert these components into a matrix and invert it to find the $g^{\alpha\beta}$:

$$g_{\alpha\beta} = \begin{bmatrix}1 & 1\\ 1 & 2\end{bmatrix}, \quad g^{\alpha\beta} = \begin{bmatrix}2 & -1\\ -1 & 1\end{bmatrix}. \tag{8.41}$$

Thus

$$\begin{aligned}\boldsymbol{e}^1 &= g^{1\alpha}\boldsymbol{e}_\alpha = 2\boldsymbol{e}_1 - \boldsymbol{e}_2 = \begin{bmatrix}1 & -1\end{bmatrix},\\ \boldsymbol{e}^2 &= g^{2\alpha}\boldsymbol{e}_\alpha = -\boldsymbol{e}_1 + \boldsymbol{e}_2 = \begin{bmatrix}0 & 1\end{bmatrix}.\end{aligned} \tag{8.42}$$

As a check, $\boldsymbol{e}^1\cdot\boldsymbol{e}^1 = 2 = g^{11}$ as expected, and similarly for the other elements of the inverse metric matrix. The vectors $\boldsymbol{e}_1, \boldsymbol{e}_2, \boldsymbol{e}^1, \boldsymbol{e}^2$ are drawn in Fig. 8.5. Note that $\boldsymbol{e}_1\cdot\boldsymbol{e}^2 = \boldsymbol{e}_2\cdot\boldsymbol{e}^1 = 0$, as expected from (8.29).

Although the $\boldsymbol{e}^{\alpha}$ are vectors, there is certainly a fundamental difference between them and the set of $\boldsymbol{e}_\alpha$ under a frame change:

$$\begin{aligned}\boldsymbol{e}_{\bar\mu} &= \Lambda^{\alpha}_{\bar\mu}\,\boldsymbol{e}_\alpha\,, \quad \text{while}\\ \boldsymbol{e}^{\bar\mu} &= g^{\bar\mu\bar\nu}\,\boldsymbol{e}_{\bar\nu} = \Lambda^{\bar\mu}_{\alpha}\,\Lambda^{\bar\nu}_{\beta}\,g^{\alpha\beta}\,\Lambda^{\sigma}_{\bar\nu}\,\boldsymbol{e}_\sigma = \Lambda^{\bar\mu}_{\alpha}\,g^{\alpha\sigma}\boldsymbol{e}_\sigma = \Lambda^{\bar\mu}_{\alpha}\,\boldsymbol{e}^{\alpha}.\end{aligned} \tag{8.43}$$

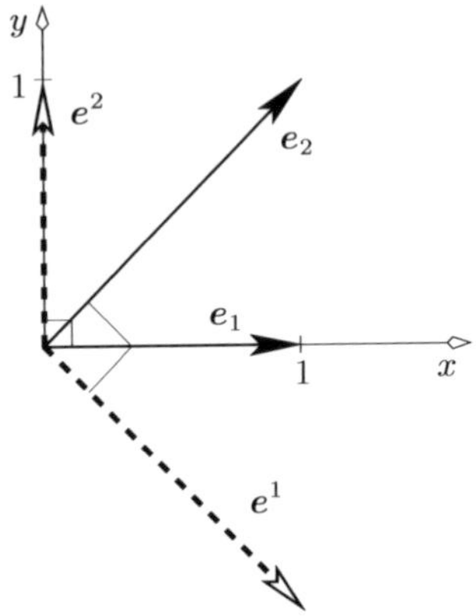

Fig. 8.5. The basis $\boldsymbol{e}_1, \boldsymbol{e}_2$ together with its cobasis $\boldsymbol{e}^1, \boldsymbol{e}^2$, discussed in (8.40)–(8.42). As expected, $\boldsymbol{e}_1 \perp \boldsymbol{e}^2$ and $\boldsymbol{e}^1 \perp \boldsymbol{e}_2$.

Let's look closely to find the essential difference between the two transformations of (8.43). A good magnifying glass is needed here, so we'll take the time to do the calculation slowly and carefully. It can be done very neatly by using a matrix formalism to write the two contractions of (8.43). Remember that matrix multiplication does not distinguish between up and down indices; a matrix is, after all, just a tableau of numbers subject to certain rules. For example, $\Lambda^{\alpha}_{\bar{\beta}}$ is really the $\alpha\beta^{\text{th}}$ component of the jacobian (6.37), so define a jacobian matrix Λ with $\alpha\beta^{\text{th}}$ component $\Lambda^{\alpha}_{\bar{\beta}}$ as follows:

$$\text{matrix } \Lambda \text{ has } \alpha\beta^{\text{th}} \text{ component } (\Lambda)_{\alpha\beta} \equiv \Lambda^{\alpha}_{\bar{\beta}}\,. \tag{8.44}$$

The mixture of barred and unbarred indices is no mistake in the notation. It allows us to write (8.43) as a matrix multiplication by ensuring that the summed indices appear as neighbours, so to speak. This implies that

$$\begin{aligned} &\text{matrix } \Lambda^t \text{ has } \alpha\beta^{\text{th}} \text{ component } (\Lambda^t)_{\alpha\beta} = (\Lambda)_{\beta\alpha} = \Lambda^{\beta}_{\bar{\alpha}}\,, \quad \text{and} \\ &\text{matrix } \Lambda^{-1} \text{ has } \alpha\beta^{\text{th}} \text{ component } (\Lambda^{-1})_{\alpha\beta} = \Lambda^{\bar{\alpha}}_{\beta}\,, \\ &\qquad \text{since then } (\Lambda^{-1}\Lambda)_{\mu\nu} = \sum_{\alpha} (\Lambda^{-1})_{\mu\alpha}(\Lambda)_{\alpha\nu} = \Lambda^{\bar{\mu}}_{\alpha}\,\Lambda^{\alpha}_{\bar{\nu}} = \delta^{\bar{\mu}}_{\bar{\nu}} \\ &\qquad\qquad\qquad\qquad\qquad = (1)_{\mu\nu} \quad \text{as expected,} \end{aligned} \tag{8.45}$$

where "1" is the unit matrix. Also needed are matrices that hold not numbers but basis or cobasis vectors. This is not as strange as it sounds, since all we are doing is arranging all the objects into neat tableaux to allow easy manipulation via matrix formalism:

$$\begin{aligned}
&\text{matrix } e \text{ has } \alpha 1^{\text{th}} \text{ component } (e)_{\alpha 1} \equiv \boldsymbol{e}_\alpha\,,\\
&\text{matrix } \bar{e} \text{ has } \alpha 1^{\text{th}} \text{ component } (\bar{e})_{\alpha 1} \equiv \boldsymbol{e}_{\bar{\alpha}}\,,\\
&\text{matrix } E \text{ has } \alpha 1^{\text{th}} \text{ component } (E)_{\alpha 1} \equiv \boldsymbol{e}^\alpha,\\
&\text{matrix } \bar{E} \text{ has } \alpha 1^{\text{th}} \text{ component } (\bar{E})_{\alpha 1} \equiv \boldsymbol{e}^{\bar{\alpha}}.
\end{aligned} \tag{8.46}$$

Now we can write the basis vector transform in (8.43) as (where repeated matrix indices are assumed to be summed over)

$$(\bar{e})_{\mu 1} = \boldsymbol{e}_{\bar{\mu}} = \Lambda^{\alpha}_{\bar{\mu}}\, \boldsymbol{e}_\alpha = (\Lambda)_{\alpha\mu}\, (e)_{\alpha 1} = (\Lambda^t)_{\mu\alpha}\, (e)_{\alpha 1} = (\Lambda^t e)_{\mu 1}\,, \tag{8.47}$$

or in other words

$$\bar{e} = \Lambda^t e\,. \tag{8.48}$$

Similarly, the cobasis vector transform in (8.43) is written as

$$(\bar{E})_{\mu 1} = \boldsymbol{e}^{\bar{\mu}} = \Lambda^{\bar{\mu}}_{\alpha}\, \boldsymbol{e}^\alpha = (\Lambda^{-1})_{\mu\alpha}\, (E)_{\alpha 1} = (\Lambda^{-1} E)_{\mu 1}\,, \tag{8.49}$$

so that

$$\bar{E} = \Lambda^{-1} E\,. \tag{8.50}$$

Now compare the basis transformation (8.48) with the cobasis transformation (8.50). One uses the matrix Λ^t while the other uses Λ^{-1}. This is the sense in which basis vectors and cobasis vectors differ. When the coordinate transformation is a two-dimensional rotation (4.7), the jacobian is just the rotation matrix, whose inverse and transpose are identical. Thus, under a rotation, the basis and cobasis vectors rotate in exactly the same way, in the manner of arrows.

An important lesson can be seen in this analysis. That is, when writing tensor components such as $A^\alpha{}_{\bar{\beta}}$, we should avoid repeating an index in two different coordinates, so that "$A^\alpha{}_{\bar{\alpha}}$" is avoided. The reason is that $A^\alpha{}_{\bar{\alpha}}$ clashes with the idea of writing these components in a matrix, since an expression like (8.44) then appears to be dealing with the $\alpha\alpha^{\text{th}}$ component of a matrix, which is a diagonal entry only. In contrast, $A^\alpha{}_{\bar{\beta}}$ is unambiguously the $\alpha\beta^{\text{th}}$ component of a matrix, which is not confined to the diagonal.

Generalising Matrix Multiplication?

Matrix multiplication consists of forming euclidean dot products between rows and columns. In fact, a more general inner product could be used instead. The use of such an inner product is equivalent to a choice of metric with, say, the summed index appearing only in a down position. This renders the contraction equivalent to the inner product of two sets of numbers. These sets can be written as the row of a first matrix and a column of a second. However, by absorbing the metric through the raising or lowering of the summation index, the contraction is effectively reduced to the use of the euclidean dot product. Because of this, there is never any need to generalise matrix multiplication to arbitrary inner products. Raising or lowering indices converts those inner products to the euclidean one, which is equivalent to our only using the usual form of matrix multiplication.

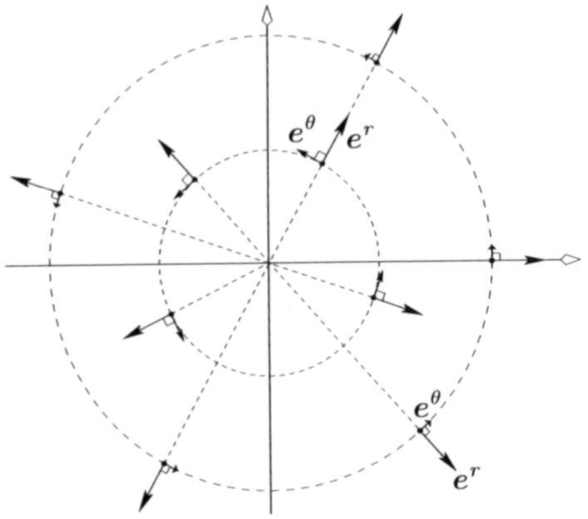

Fig. 8.6. Polar cobasis vectors at various points, drawn as a comparison with the basis vectors in Fig. 8.4. The radial cobasis vectors are identical to the radial basis vectors: $\boldsymbol{e}^r = \boldsymbol{e}_r$. In contrast, the transverse cobasis vectors are parallel to the transverse basis vectors but with different lengths; $\boldsymbol{e}^\theta = \boldsymbol{e}_\theta / r^2$, so their lengths are inversely proportional to r. For a discussion of the length units, see the box on p. 50.

The Cobasis for Polar Coordinates

What is the polar coordinate cobasis? Equation (8.34) gives

$$\boldsymbol{e}^r = g^{r\alpha} \boldsymbol{e}_\alpha \,, \quad \boldsymbol{e}^\theta = g^{\theta\alpha} \boldsymbol{e}_\alpha \,. \tag{8.51}$$

Now use the polar metric given in (8.26) to calculate the inverse matrix of the metric elements:

$$\begin{bmatrix} g^{rr} & g^{r\theta} \\ g^{\theta r} & g^{\theta\theta} \end{bmatrix} = \begin{bmatrix} g_{rr} & g_{r\theta} \\ g_{\theta r} & g_{\theta\theta} \end{bmatrix}^{-1} = \begin{bmatrix} 1 & 0 \\ 0 & r^2 \end{bmatrix}^{-1} = \begin{bmatrix} 1 & 0 \\ 0 & 1/r^2 \end{bmatrix} . \tag{8.52}$$

The inverse metric matrix is also diagonal, which means the cobasis is

$$\boldsymbol{e}^r = g^{rr} \boldsymbol{e}_r = \boldsymbol{e}_r \,, \quad \boldsymbol{e}^\theta = g^{\theta\theta} \boldsymbol{e}_\theta = \boldsymbol{e}_\theta / r^2 . \tag{8.53}$$

A sample of cobasis vectors at various points are shown in Fig. 8.6.

8.4.1 Raising and Lowering Indices

Of great notational importance in the previous few pages is the idea that the metric "raises and lowers indices" on the basis vectors and on itself, with expressions such as

$$\boldsymbol{e}^\alpha = g^{\alpha\beta} \boldsymbol{e}_\beta \,, \quad \boldsymbol{e}_\alpha = g_{\alpha\beta} \boldsymbol{e}^\beta \,, \quad g_{\alpha\mu} g^{\mu\beta} = g_\alpha{}^\beta . \tag{8.54}$$

The same idea turns out also to apply to components. Given a vector $\boldsymbol{v}$, we *defined* the set of numbers v_α to be its coordinate vector over the cobasis:

$$v_\alpha \boldsymbol{e}^\alpha \equiv \boldsymbol{v} = v^\alpha \boldsymbol{e}_\alpha \,. \tag{8.55}$$

But we would like to show that the set of v_α transforms as a covector, to match the previous usage of a lowered index in (6.50). How do the v_α relate to v^α? Use the first identity of (8.54) to write (8.55) as

$$v_\alpha g^{\alpha\beta} \boldsymbol{e}_\beta \overset{\text{req.}}{=\!=} v^\beta \boldsymbol{e}_\beta \,, \tag{8.56}$$

so that the linear independence of the $\boldsymbol{e}_\beta$ gives

$$v_\alpha g^{\alpha\beta} = v^\beta . \tag{8.57}$$

Thus, the metric "raises" component indices too. This last result inverts to give

$$v_\alpha = v^\beta g_{\alpha\beta} \,, \tag{8.58}$$

which means that the metric also "lowers" component indices. So given the metric, not only can we switch freely between basis and cobasis, but we can also switch between a basis coordinate vector and the corresponding cobasis coordinate vector.

As a side note, in (8.16) a vector's length was expressed in an almost euclidean way. The components still appeared quadratically, but the metric $g_{\alpha\beta}$ seems to have complicated the issue. It's more elegant to absorb the metric into one of the components by using it to lower the β index in (8.16), giving a more suggestive expression for the vector length as a contraction:

$$|\boldsymbol{v}|^2 = v^\alpha v_\alpha \,. \tag{8.59}$$

This is a sum over quadratic-like terms, much as we are used to from Pythagoras's theorem. Notation like (8.59) allows the line element to be written in a kind of euclidean way. For instance, the contraction is a sum of products, which is equivalent to using the euclidean metric.

We can now show, as mentioned just after (8.55), that the set of v_α transforms as a covector in the manner of (6.50).

$$v_{\bar\mu} = v^{\bar\nu} g_{\bar\nu\bar\mu} = \Lambda^{\bar\nu}_{\alpha}\, v^\alpha\, \Lambda^{\beta}_{\bar\nu} \Lambda^{\gamma}_{\bar\mu}\, g_{\beta\gamma} = v^\beta\, \Lambda^{\gamma}_{\bar\mu}\, g_{\beta\gamma} = \Lambda^{\gamma}_{\bar\mu}\, v_\gamma \,. \tag{8.60}$$

The names "covector", referring to the coordinate vector v_α over the cobasis, and "cobasis vector", or sometimes "basis covector", referring to $\boldsymbol{e}^\alpha$, were not created to imply that $v_\alpha \boldsymbol{e}^\alpha$ is a new object called a "covector" and somehow different from the vector $v^\alpha \boldsymbol{e}_\alpha$. These are just alternative ways to write the vector $\boldsymbol{v}$, which is why the terms "contravariant component" and "covariant component" of $\boldsymbol{v}$ were originally created for v^α and v_α. The boxes on pages 280 and 293 discuss this in more detail. Remember that the cobasis *is* nothing more than a basis; it is given its own name purely because of its special relationship to a partner basis in (8.29).

It is worth noting that lowered indices are traditionally *defined* in terms of raised ones via the metric, although it might not immediately be clear just why such a definition should be useful. Our approach has been to define raised- and lowered-index notation quite symmetrically from the start in Chap. 2, long before the ideas of a metric or a tensor were encountered.

8.5 Tensor Components with More than Two Indices

Later, we'll meet tensors whose components have more than two indices. By definition, such components written as, e.g., $A_\alpha{}^\beta{}_{\gamma\delta}$ must transform as

$$A_{\bar\mu}{}^{\bar\nu}{}_{\bar\varrho\bar\sigma} = \Lambda^\alpha_{\bar\mu}\,\Lambda^{\bar\nu}_\beta\,\Lambda^\gamma_{\bar\varrho}\,\Lambda^\delta_{\bar\sigma}\,A_\alpha{}^\beta{}_{\gamma\delta} \tag{8.61}$$

in order to be called *tensor* components; this property is then what *defines* the components of a general tensor. The duality between raised and lowered indices guarantees that a new set of tensor components can be built by shifting the components up and down as we see fit, by using the metric. For example, lowering the second index of $A_\alpha{}^\beta{}_{\gamma\delta}$ gives

$$A_{\alpha\mu\gamma\delta} \equiv g_{\mu\beta}\,A_\alpha{}^\beta{}_{\gamma\delta}\,, \tag{8.62}$$

and these new components are guaranteed to transform as they should. The various jacobians and metric components conspire to make sure of that. A set of tensor components such as $A_\alpha{}^\beta{}_{\gamma\delta}$ is generally not the same as the tensor components $A_{\alpha\beta}{}^\gamma{}_\delta$ created from it using the metric. Because of this, we must always be careful to space the indices in such a way that they *can* be raised or lowered unambiguously. Nevertheless, in the next section we show that by including an appropriate basis, the various sets of components $A_\alpha{}^\beta{}_{\gamma\delta}$, $A_{\alpha\beta}{}^\gamma{}_\delta$, etc., can all be regarded as components of the *same* object, $\boldsymbol{A}$, which is simply *the* tensor with these various component sets over various basis choices.

An exception to the rule of keeping indices correctly spaced is the Kronecker delta δ^α_β, which (in tensor use) is always defined with one index up and one down, with no spacing necessary since its meaning is unambiguous. Notationally, when we wear our covariant hats and thus are careful to keep all indices in their correct up and down positions, we must never write the Kronecker delta as $\delta_{\alpha\beta}$. If we do wish to play the raising-lowering game with the Kronecker delta, then we are free to define

$$\delta_{\alpha\beta} \equiv \delta^\gamma_\beta\,g_{\gamma\alpha} = g_{\beta\alpha} = g_{\alpha\beta}\,. \tag{8.63}$$

So beware: the usual Kronecker delta *must* be written with one index up and one down if it's to retain its usual "zero or one" meaning. With both indices down, the Kronecker delta is really just another name for the metric! That is, when using covariant index notation, δ and g mean exactly the same

Our Terminology of Vectors and Their Components

The terms used to describe four-vectors, covectors, and their respective bases can be confusing since the word "vector" is heavily used in physics. Throughout this book, we use the following language. First, position vectors and proper vectors are both denoted with boldface, e.g. $\boldsymbol{v}$; this is standard, and there should never be any real confusion in using the same symbol since the two objects are both "arrows", and are both elements of a vector space. They also tend to be used in different contexts. At the start of Chap. 2 we already referred to the linear algebra term "coordinate vector" as the set of coefficients of $\boldsymbol{v}$ relative to some chosen basis. Sets of vector components are generally known by different but equivalent names:

- The *vector* is the coordinate vector over a basis, or a "basis coordinate vector". Also known as the "contravariant components of $\boldsymbol{v}$" and called a four-vector when we restrict ourselves to the Lorentz transform.
- The *covector* is the coordinate vector over a cobasis, or a "cobasis coordinate vector". Also known as the "covariant components of $\boldsymbol{v}$".

Note also that referring to a basis vector as $\boldsymbol{e}_\alpha$ means just one vector, such as $\boldsymbol{e}_0$ or $\boldsymbol{e}_1$, as opposed to the whole set (in, say, four dimensions) $\{\boldsymbol{e}_0, \boldsymbol{e}_1, \boldsymbol{e}_2, \boldsymbol{e}_3\}$. Referring to a vector as v^α always means the whole coordinate vector: the ordered set $\{v^\alpha\}$, such as (v^0, v^1, v^2, v^3). A similar remark will apply later in this chapter, where a cobasis vector, also sometimes called a "basis covector" $\boldsymbol{e}^\alpha$, will mean just one vector of the cobasis, such as $\boldsymbol{e}^0$, whereas "a covector v_α" will always mean the whole coordinate vector with respect to the cobasis, e.g. (v_0, v_1, v_2, v_3).

Later, these remarks will also apply to tensors since these are just straightforward generalisations of vectors. The whole set of components of a tensor is usually called a tensor, but that set together with a basis is also called a tensor. This practice is common, and no real confusion should result. Occasionally, if we want to stress the spatial part of a four-dimensional vector in relativity such as in Chap. 6, we'll write the coordinate vector as $\vec{v} = (v^0, \boldsymbol{v}) = (v^0, v^1, v^2, v^3)$, with the three-dimensional spatial part bold.

In summary, use the terms

- **vector** or **coordinate vector over a basis** for $\{v^\alpha\}$,
- **covector** or **coordinate vector over a cobasis** for $\{v_\alpha\}$,
- **basis vector** for $\boldsymbol{e}_\alpha$,
- **cobasis vector**, or occasionally **basis covector**, for $\boldsymbol{e}^\alpha$, and
- **vector** or **proper vector** for $\boldsymbol{v} = v^\alpha \boldsymbol{e}_\alpha = v_\alpha \boldsymbol{e}^\alpha$.

thing when taking indices in the same positions. A good example of this can be found in Sect. 10.3.6 when defining the lowered-indices version of the energy–momentum tensor.

Another point worth noting is that the notation of the jacobian matrix elements is very closely related to the metric notation, as is evident when we consider expressions such as

$$\boldsymbol{e}^{\bar{\alpha}}\cdot\boldsymbol{e}_\beta = \Lambda^{\bar{\alpha}}_{\mu}\,\boldsymbol{e}^{\mu}\cdot\boldsymbol{e}_\beta = \Lambda^{\bar{\alpha}}_{\beta} \quad\text{and}\quad \boldsymbol{e}^{\alpha}\cdot\boldsymbol{e}_\beta = \delta^{\alpha}_{\beta} = \frac{\partial x^\alpha}{\partial x^\beta} = \Lambda^{\alpha}_{\beta}\,. \tag{8.64}$$

These can be compared with the definitions of the metric in its various guises, (8.17) and (8.30).

8.5.1 Bases for More General Tensors

Previously, we made the point that along with vector components v^α and v_α, the metric coefficients $g_{\alpha\beta}, g^{\alpha}{}_{\beta}$, and $g^{\alpha\beta}$ are all tensor components because they transform in the appropriate way. In the one-index case, we found that the basis and cobasis could be used to construct a frame-independent object, the proper vector. Now we wish to extend this idea to sets of tensor components having two or more indices. We'll pave the way by reiterating the idea of linear independence of the basis. Write a general vector again as

$$\boldsymbol{v} = v^\alpha \boldsymbol{e}_\alpha = v_\alpha \boldsymbol{e}^\alpha\,, \quad \text{with}\ \boldsymbol{e}^\alpha = g^{\alpha\beta}\boldsymbol{e}_\beta\,. \tag{8.65}$$

Consider (2.25):

$$\boldsymbol{v}\cdot\boldsymbol{e}^\alpha = v^\beta \boldsymbol{e}_\beta\cdot\boldsymbol{e}^\alpha = v^\alpha. \tag{8.66}$$

"Dotting" with $\boldsymbol{e}^\alpha$ picks out v^α, the component of $\boldsymbol{e}_\alpha$. Now suppose that $\boldsymbol{v} = \boldsymbol{0}$. An axiom of linear algebra then allows us to write $\boldsymbol{v}\cdot\boldsymbol{e}^\alpha = 0$. In that case, (8.66) implies that $v^\alpha = 0$, so the zero vector must have all components equal to zero, a desirable property that goes hand in hand with the linear independence of the basis. Similar remarks apply to the cobasis.

This idea of linear independence helps us to construct a basis for a general second-order tensor $\boldsymbol{T}$. First, in analogy with the two alternative bases $\{\boldsymbol{e}_\alpha\}$ and $\{\boldsymbol{e}^\alpha\}$, we simply define four bases using a new operator "$\otimes$" that combines basis vectors as $\{\boldsymbol{e}_\alpha\otimes\boldsymbol{e}_\beta\}$, $\{\boldsymbol{e}_\alpha\otimes\boldsymbol{e}^\beta\}$, $\{\boldsymbol{e}^\alpha\otimes\boldsymbol{e}_\beta\}$, and $\{\boldsymbol{e}^\alpha\otimes\boldsymbol{e}^\beta\}$, so that the analogy with the first part of (8.65) is

$$\begin{aligned}\boldsymbol{T} &= T^{\alpha\beta}\,\boldsymbol{e}_\alpha\otimes\boldsymbol{e}_\beta = T^{\alpha}{}_{\beta}\,\boldsymbol{e}_\alpha\otimes\boldsymbol{e}^\beta \\ &= T_{\alpha}{}^{\beta}\,\boldsymbol{e}^\alpha\otimes\boldsymbol{e}_\beta = T_{\alpha\beta}\,\boldsymbol{e}^\alpha\otimes\boldsymbol{e}^\beta.\end{aligned} \tag{8.67}$$

Let's analyse one of these expressions by asking: just what is $\boldsymbol{e}_\alpha\otimes\boldsymbol{e}_\beta$—how does it behave? First, note that we require linear behaviour from this new basis:

$$\begin{aligned}T^{\alpha\beta}\boldsymbol{e}_\alpha\otimes\boldsymbol{e}_\beta &= T^{\bar{\mu}\bar{\nu}}\Lambda^{\alpha}_{\bar{\mu}}\,\Lambda^{\beta}_{\bar{\nu}}\,\boldsymbol{e}_\alpha\otimes\boldsymbol{e}_\beta \overset{\text{req.}}{=\!=} T^{\bar{\mu}\bar{\nu}}\left(\Lambda^{\alpha}_{\bar{\mu}}\boldsymbol{e}_\alpha\right)\otimes\left(\Lambda^{\beta}_{\bar{\nu}}\boldsymbol{e}_\beta\right) \\ &= T^{\bar{\mu}\bar{\nu}}\boldsymbol{e}_{\bar{\mu}}\otimes\boldsymbol{e}_{\bar{\nu}}\,,\end{aligned} \tag{8.68}$$

in which case it must follow that for any constant c,

$$(c\,\boldsymbol{e}_\alpha)\otimes\boldsymbol{e}_\beta = \boldsymbol{e}_\alpha\otimes\left(c\,\boldsymbol{e}_\beta\right) = c\left(\boldsymbol{e}_\alpha\otimes\boldsymbol{e}_\beta\right). \tag{8.69}$$

It's possible to show that the set of $\boldsymbol{e}_\alpha\otimes\boldsymbol{e}_\beta$ is linearly independent by analogy with the dot product discussion around (8.66). To do so, more than two basis vectors need to be combined at once, so the dot product, being a binary operator, is now insufficient for the job. We will replace it by the functional notation first used in Sect. 2.4.4. Using this notation, the orthonormality of basis and cobasis vectors, written as $\boldsymbol{e}_\alpha\cdot\boldsymbol{e}^\beta = \delta^\beta_\alpha$, could just as well be expressed as

$$\boldsymbol{e}_\alpha\left(\boldsymbol{e}^\beta\right) \equiv \boldsymbol{e}^\beta\left(\boldsymbol{e}_\alpha\right) \equiv \delta^\beta_\alpha\,. \tag{8.70}$$

It should be stressed that we are *not* redefining basis and cobasis vectors to be functions of each other! As was pointed out on p. 26, this sort of functional notation is just a way of combining more than two vectors. (Some texts do use an expression like (8.70) as functional notation, but whether they really mean to be circular is not clear.) For a second-order tensor, then, define

$$\boldsymbol{e}_\alpha\otimes\boldsymbol{e}_\beta\left(\boldsymbol{e}^\mu,\boldsymbol{e}^\nu\right) \equiv \boldsymbol{e}_\alpha\left(\boldsymbol{e}^\mu\right)\;\boldsymbol{e}_\beta\left(\boldsymbol{e}^\nu\right) = \delta^\mu_\alpha\,\delta^\nu_\beta\,. \tag{8.71}$$

This approach enables the tensor $\boldsymbol{T}$ to be treated as a function of the basis vectors that picks out $T^{\mu\nu}$ in analogy with (8.66). Just as (8.66) could have been written as $\boldsymbol{v}\left(\boldsymbol{e}^\alpha\right) = v^\alpha$, more generally we can write

$$\begin{aligned}\boldsymbol{T}\left(\boldsymbol{e}^\mu,\boldsymbol{e}^\nu\right) &= T^{\alpha\beta}\;\boldsymbol{e}_\alpha\otimes\boldsymbol{e}_\beta\left(\boldsymbol{e}^\mu,\boldsymbol{e}^\nu\right)\\ &= T^{\alpha\beta}\;\delta^\mu_\alpha\;\delta^\nu_\beta = T^{\mu\nu}.\end{aligned} \tag{8.72}$$

So far, so good. And now to answer the question: is the set $\left\{\boldsymbol{e}_\alpha\otimes\boldsymbol{e}_\beta\right\}$ linearly independent? As for the vector case, consider the zero tensor $\boldsymbol{T}=\boldsymbol{0}$, and demand that $\boldsymbol{T}\left(\boldsymbol{e}^\mu,\boldsymbol{e}^\nu\right)=0$. But from what we have just done, this is equivalent to demanding that $T^{\mu\nu}=0$, in which case $\left\{\boldsymbol{e}_\alpha\otimes\boldsymbol{e}_\beta\right\}$ is certainly linearly independent.

Similar arguments apply to other basis tensors, such as $\boldsymbol{e}_\alpha\otimes\boldsymbol{e}^\beta$. The symbol $\otimes$ is tedious to write but can be necessary to prevent an equation such as (8.69) from looking like functional notation. Still, it really is nothing more than a kind of spacer, and after serving a pedagogical purpose here, and if there's no ambiguity, it can be dropped in practice—as long as the result is not confused with the geometric product (4.93)! Thus $\boldsymbol{T} = T^{\alpha\beta}\,\boldsymbol{e}_\alpha\,\boldsymbol{e}_\beta$ or even $\boldsymbol{T} = T^{\alpha\beta}\,\boldsymbol{e}_{\alpha\beta}$, where

$$\boldsymbol{e}_{\alpha\beta} \equiv \boldsymbol{e}_\alpha\boldsymbol{e}_\beta \equiv \boldsymbol{e}_\alpha\otimes\boldsymbol{e}_\beta\,. \tag{8.73}$$

Similarly, $\boldsymbol{e}^{\alpha\beta}, \boldsymbol{e}^\alpha{}_\beta$, and $\boldsymbol{e}_\alpha{}^\beta$ can also be defined.

Wedge Products Again. In Sect. 2.4.4, we met with the idea of using a wedge product to define "multivectors", objects that produce signed volumes when combined with cobasis vectors. Based on what has been done in the last few paragraphs, we could equally well consider the signed area of (2.60) to be the raised (1, 2)-component of the bivector $\boldsymbol{a} \wedge \boldsymbol{b}$. More generally, this bivector has a raised (α, β)-component of

$$\boldsymbol{a} \wedge \boldsymbol{b}\left(\mathbf{e}^{\alpha}, \mathbf{e}^{\beta}\right) = \begin{vmatrix} \boldsymbol{a}\cdot\mathbf{e}^{\alpha} & \boldsymbol{a}\cdot\mathbf{e}^{\beta} \\ \boldsymbol{b}\cdot\mathbf{e}^{\alpha} & \boldsymbol{b}\cdot\mathbf{e}^{\beta} \end{vmatrix} = \boldsymbol{a}\otimes\boldsymbol{b}\left(\mathbf{e}^{\alpha}, \mathbf{e}^{\beta}\right) - \boldsymbol{b}\otimes\boldsymbol{a}\left(\mathbf{e}^{\alpha}, \mathbf{e}^{\beta}\right). \tag{8.74}$$

In that case,

$$\boldsymbol{a} \wedge \boldsymbol{b} = \boldsymbol{a}\otimes\boldsymbol{b} - \boldsymbol{b}\otimes\boldsymbol{a}\,. \tag{8.75}$$

Similarly, the signed volume that we met in (2.62) can be considered as the (1, 2, 3)-component of

$$\boldsymbol{a} \wedge \boldsymbol{b} \wedge \boldsymbol{c} = \boldsymbol{a}\otimes\boldsymbol{b}\otimes\boldsymbol{c} - \boldsymbol{b}\otimes\boldsymbol{a}\otimes\boldsymbol{c} + \boldsymbol{b}\otimes\boldsymbol{c}\otimes\boldsymbol{a} - \cdots, \tag{8.76}$$

where the sum is over all permutations of $\boldsymbol{a}, \boldsymbol{b}, \boldsymbol{c}$, with a plus sign for an even permutation and a minus sign for an odd permutation. Since wedge products are intimately tied to volumes in n dimensions, permutations with associated signs appear frequently when tensor analysis is used in geometry.

8.5.2 The Metric Tensor Versus the Metric Matrix

It might seem strange that the metric tensor $\boldsymbol{g} = g_{\alpha\beta}\,\mathbf{e}^{\alpha\beta} = g^{\alpha\beta}\mathbf{e}_{\alpha\beta}$ has components $g_{\alpha\beta}$ and $g^{\alpha\beta}$, depending on the basis chosen. After all, the matrix of $g^{\alpha\beta}$ is the *inverse* of the matrix of $g_{\alpha\beta}$!

But some reflection shows that there is nothing unreasonable about this. Rather, it even underlines the fact that matrices and tensors are two different things. The matrix of tensor components is simply an array of numbers designed for efficient bookkeeping in linear algebra manipulations. The tensor $\boldsymbol{g}$ includes a basis, but the matrix g of components $g_{\alpha\beta}$ does not. Similarly, the inverse matrix g^{-1} of components $g^{\alpha\beta}$ incorporates no basis. Remember that there is no tensor called $\boldsymbol{g}^{-1}$; the matrix g^{-1} simply holds the components of $\boldsymbol{g}$ over the basis $\mathbf{e}_{\alpha\beta}$.

Further insight into this distinction comes by considering an analogy purely within the context of matrix algebra. Write the (symmetric) metric matrix g and its inverse g^{-1} as

$$g = \begin{bmatrix} g_{11} & g_{12} \\ g_{12} & g_{22} \end{bmatrix}, \quad g^{-1} \equiv \begin{bmatrix} g^{11} & g^{12} \\ g^{21} & g^{22} \end{bmatrix} = \frac{1}{\Delta}\begin{bmatrix} g_{22} & -g_{12} \\ -g_{12} & g_{11} \end{bmatrix}, \tag{8.77}$$

where the metric determinant is $\Delta \equiv g_{11}g_{22} - g_{12}^2$, and g^{-1} is also seen to be symmetric. Now suppose we write g in terms of a *matrix* basis as $g = g_{\alpha\beta}\mathrm{e}^{\alpha\beta}$, so that

$$\boldsymbol{e}^{11} \equiv \begin{bmatrix} 1 & 0 \\ 0 & 0 \end{bmatrix}, \boldsymbol{e}^{12} \equiv \begin{bmatrix} 0 & 1 \\ 0 & 0 \end{bmatrix}, \boldsymbol{e}^{21} \equiv \begin{bmatrix} 0 & 0 \\ 1 & 0 \end{bmatrix}, \boldsymbol{e}^{22} \equiv \begin{bmatrix} 0 & 0 \\ 0 & 1 \end{bmatrix}. \tag{8.78}$$

The question is: if a new basis $\boldsymbol{e}_{\alpha\beta}$ is defined by lowering the indices of the $\boldsymbol{e}^{\alpha\beta}$ using the metric, will we be able to write $g = g^{\alpha\beta}\boldsymbol{e}_{\alpha\beta}$? We hope that the answer is yes since this is the way covariant notation has been defined; raising one set of indices and lowering another must involve the metric and its inverse, which cancel internally to give no change. But we can see that the answer will be yes quite explicitly with the basis matrices of (8.78). For example,

$$\boldsymbol{e}_{11} \equiv \boldsymbol{e}_1 \otimes \boldsymbol{e}_1 = g_{1\alpha}\,\boldsymbol{e}^{\alpha} \otimes g_{1\beta}\,\boldsymbol{e}^{\beta} = g_{1\alpha}\,g_{1\beta}\,\boldsymbol{e}^{\alpha\beta} = \begin{bmatrix} g_{11}^2 & g_{11}\,g_{12} \\ g_{11}\,g_{12} & g_{12}^2 \end{bmatrix}. \tag{8.79}$$

The other basis matrices follow similarly. If we then form the product $g^{\alpha\beta}\boldsymbol{e}_{\alpha\beta}$ of the inverse elements from (8.77) and the matrix basis from (8.79), the result is indeed the matrix g, as expected. So not only can g be expressed as a linear combination of basis matrices using its elements $g_{\alpha\beta}$ as the coefficients, but the same can also be done using the elements $g^{\alpha\beta}$ of its inverse. To reiterate, there is no such tensor as $\boldsymbol{g}^{-1}$.

In the last example, the bases were matrices. But in a tensor space, the bases are sets of more abstract entities such as $\boldsymbol{e}_{\alpha\beta}$, and so matrices of tensor *components* do not themselves constitute the whole tensor. Thus the metric tensor $\boldsymbol{g}$ has components $g_{\alpha\beta}$ that comprise a matrix called g, and components $g^{\alpha\beta}$ that comprise a matrix called g^{-1}. It even has the unit matrix components δ^{α}_{β} over the $\boldsymbol{e}_{\alpha}{}^{\beta}$ and $\boldsymbol{e}^{\beta}{}_{\alpha}$ bases! That is,

$$\boldsymbol{g} = g^{\alpha}{}_{\beta}\,\boldsymbol{e}_{\alpha}{}^{\beta} = \delta^{\alpha}_{\beta}\,\boldsymbol{e}_{\alpha}{}^{\beta} = \boldsymbol{e}_{\alpha}{}^{\alpha}\,, \quad \text{and similarly } \boldsymbol{g} = \boldsymbol{e}^{\alpha}{}_{\alpha}\,. \tag{8.80}$$

For general tensors, as long as we know the positions of the indices (whether raised or lowered), any of the matrices of tensor components contain exactly the same information that the complete tensor holds.

8.6 The Gradient Operator and the Cobasis

Some of the real power of tensor notation can be seen when we begin to do differential calculus in general coordinates. Fundamental here is the notion of the *gradient* of a function. Let's begin by calculating the gradient operator ∇ in polar coordinates but using *non*covariant notation, since this is a good way to make a comparison with the covariant version and ultimately to motivate the use of covariant language by demonstrating its power and elegance. We'll define the gradient more carefully a few pages hence, but for now we simply use the well-known expression for it in terms of cartesian coordinates.

The symbol ∇ used as part of the gradient, divergence, and curl operators was originally called "nabla" (a type of ancient Near-Eastern harp), and still is, but is also often read as "del". Usage varies, but in the author's opinion del might perhaps better be left to refer to the partial derivative operator ∂, which itself is often called "partial".

To save notation in what follows, partial derivatives of a function of all the coordinates $f(x^\alpha)$ in tensor theory are usually written in the shortened form

$$f_{,\alpha} \equiv \partial_\alpha f \equiv \frac{\partial f}{\partial x^\alpha}, \tag{8.81}$$

each having the α in a down position. Now, in cartesian coordinates, we know that the gradient operator gives derivatives

$$f(x + \mathrm{d}x, y + \mathrm{d}y) - f(x,y) = \nabla f \cdot (\mathrm{d}x, \mathrm{d}y) = f_{,x}\,\mathrm{d}x + f_{,y}\,\mathrm{d}y\,, \tag{8.82}$$

so that

$$\nabla f = f_{,x}\,\boldsymbol{e}_x + f_{,y}\,\boldsymbol{e}_y \quad \text{(noncovariant!)} \tag{8.83}$$

or

$$\nabla = \boldsymbol{e}_x\,\partial_x + \boldsymbol{e}_y\,\partial_y \quad \text{(also noncovariant)}. \tag{8.84}$$

Expressions (8.83) and (8.84) are not covariant because the summed indices are everywhere down. Convert all terms to polar by using the chain rule for partial derivatives:

$$\begin{aligned} \boldsymbol{e}_x &= \frac{\partial r}{\partial x}\,\boldsymbol{e}_r + \frac{\partial \theta}{\partial x}\,\boldsymbol{e}_\theta\,, \\ \frac{\partial}{\partial x} &= \frac{\partial r}{\partial x}\frac{\partial}{\partial r} + \frac{\partial \theta}{\partial x}\frac{\partial}{\partial \theta}\,, \end{aligned} \tag{8.85}$$

and similarly for the y-components. Notice that the basis vectors transform in exactly the same way as the partial derivative operators, which is reasonable because that is just how basis vectors were defined in (8.2). The presence of terms such as $\partial r/\partial x$ brings up an interesting point. The cartesian polar coordinate transform is

$$x = r\cos\theta\,, \quad y = r\sin\theta\,, \tag{8.86}$$

so we can easily calculate $\partial x/\partial r$ and so on. But what is $\partial r/\partial x$? (The common first guess that it's just the reciprocal of $\partial x/\partial r$ is *not* true.) In this particular example, (8.86) can be solved for r and θ, which are then simple to differentiate; but in general the defining equations might be difficult to invert. Is there some other way?

Equation (8.86) is not difficult to invert, although the inverse tangent needs some thought to map θ onto a complete 2π radians. Perhaps the simplest solution is

$$r = \sqrt{x^2 + y^2}\,, \quad \theta = \tan^{-1}\frac{y}{x} + \{\pi \text{ if } x < 0\}\,, \tag{8.87}$$

which puts θ into the interval $(-\pi/2, 3\pi/2)$. These can be differentiated to obtain the correct answers for $\partial r/\partial x$ and so on. But beware of a subtle trap. A common way of inverting (8.86) omits the constant in θ to write

$$r = \sqrt{x^2 + y^2}\,, \quad \theta = \tan^{-1}\frac{y}{x}\,. \tag{8.88}$$

Of course, the forgotten constant doesn't affect any differentiation of θ, which is why this naïve expression gives the right answer. But if used for numerical work in trigonometry, (8.88) can place θ into the wrong quadrant and so is liable to fail—sometimes catastrophically.

In elementary calculus, we are certainly aware that for a function of one variable, $y = f(x)$, an inverse relationship holds for the derivatives:

$$\frac{\mathrm{d}x}{\mathrm{d}y} = \frac{1}{\mathrm{d}y/\mathrm{d}x}\,. \tag{8.89}$$

In fact, the multivariate problem also involves inverses—matrix inverses. Why? Looking back to (6.37), it's not hard to see that the product of two jacobian matrices (partial derivatives from one coordinate system to another) is the identity

$$\begin{bmatrix} \dfrac{\partial r}{\partial x} & \dfrac{\partial r}{\partial y} \\[2ex] \dfrac{\partial \theta}{\partial x} & \dfrac{\partial \theta}{\partial y} \end{bmatrix} \begin{bmatrix} \dfrac{\partial x}{\partial r} & \dfrac{\partial x}{\partial \theta} \\[2ex] \dfrac{\partial y}{\partial r} & \dfrac{\partial y}{\partial \theta} \end{bmatrix} = \begin{bmatrix} 1 & 0 \\ 0 & 1 \end{bmatrix}. \tag{8.90}$$

This matrix multiplication is transparent enough to be obviously true in general when applied to the jacobian matrices relating any two different coordinate systems in any number of dimensions. This is a very useful and often overlooked fact. Its covariant form uses the $\Lambda^{\alpha}_{\bar{\beta}}$ notation, as discussed back on p. 228. Writing polar coordinates unbarred and cartesian coordinates barred, (8.90) is equivalent to

$$\Lambda^{\alpha}_{\bar{\beta}}\, \Lambda^{\bar{\beta}}_{\gamma} = \delta^{\alpha}_{\gamma}\,, \tag{8.91}$$

which is just the usual chain rule, but written covariantly. Returning to the task of calculating partial derivatives such as $\partial r/\partial x$, the first step is to use the basic defining equations for polar coordinates (8.86) to write

$$\begin{bmatrix} \dfrac{\partial x}{\partial r} & \dfrac{\partial x}{\partial \theta} \\[2ex] \dfrac{\partial y}{\partial r} & \dfrac{\partial y}{\partial \theta} \end{bmatrix} = \begin{bmatrix} \cos\theta & -r\sin\theta \\ \sin\theta & r\cos\theta \end{bmatrix}. \tag{8.92}$$

Then, by (8.90) we need only invert to find the other jacobian matrix:

$$\begin{bmatrix} \dfrac{\partial r}{\partial x} & \dfrac{\partial r}{\partial y} \\[2ex] \dfrac{\partial \theta}{\partial x} & \dfrac{\partial \theta}{\partial y} \end{bmatrix} = \begin{bmatrix} \cos\theta & \sin\theta \\[1ex] \dfrac{-\sin\theta}{r} & \dfrac{\cos\theta}{r} \end{bmatrix}. \tag{8.93}$$

So, in particular, $\partial r/\partial x = \partial x/\partial r = \cos\theta$, and this inversion of partial derivatives highlights the fact that the numerator and denominator of an expression such as $\partial r/\partial x$ do not behave as parts of a fraction—as opposed to the $\mathrm{d}y/\mathrm{d}x$ of single-variable calculus, where $\mathrm{d}y$ and $\mathrm{d}x$ *do* behave as parts of a fraction.

We now have what is needed to express ∇ in polar coordinates. Incorporate (8.85) into (8.84) using these partial derivatives:

$$\begin{aligned}\nabla &= \left(\cos\theta\,\boldsymbol{e}_r - \frac{\sin\theta}{r}\,\boldsymbol{e}_\theta\right)\left(\cos\theta\,\partial_r - \frac{\sin\theta}{r}\,\partial_\theta\right)\\ &\quad+\left(\sin\theta\,\boldsymbol{e}_r + \frac{\cos\theta}{r}\,\boldsymbol{e}_\theta\right)\left(\sin\theta\,\partial_r + \frac{\cos\theta}{r}\,\partial_\theta\right)\\ &= \boldsymbol{e}_r\,\partial_r + \frac{\boldsymbol{e}_\theta}{r^2}\partial_\theta\,. \end{aligned} \tag{8.94}$$

In practice, the polar basis vectors are usually normalised; this has its advantages and is physically meaningful, as we'll see shortly. To convert $\boldsymbol{e}_r, \boldsymbol{e}_\theta$ to unit basis vectors $\boldsymbol{e}_{\widehat{r}}, \boldsymbol{e}_{\widehat{\theta}}$, we need only divide $\boldsymbol{e}_r, \boldsymbol{e}_\theta$ by their own lengths. A very direct way of calculating these lengths is via a dot product in cartesian coordinates. We did this before in (8.26):

$$\begin{aligned}|\boldsymbol{e}_r|^2 &= \boldsymbol{e}_r\cdot\boldsymbol{e}_r = g_{rr} = 1\,,\\ |\boldsymbol{e}_\theta|^2 &= \boldsymbol{e}_\theta\cdot\boldsymbol{e}_\theta = g_{\theta\theta} = r^2.\end{aligned} \tag{8.95}$$

In this simple case, we can also opt for what is perhaps a more intuitive approach, as pointed out in Fig. 8.4. Since $\boldsymbol{e}_r = \partial\boldsymbol{X}/\partial r$ and similarly for $\boldsymbol{e}_\theta$, it must be true that

$$\begin{aligned}|\boldsymbol{e}_r| &= \frac{|\mathrm{d}\boldsymbol{X}|_{\theta\text{ const.}}}{|\mathrm{d}r|} = \frac{|\mathrm{d}r|}{|\mathrm{d}r|} = 1\,,\\ |\boldsymbol{e}_\theta| &= \frac{|\mathrm{d}\boldsymbol{X}|_{r\text{ const.}}}{|\mathrm{d}\theta|} = \frac{r|\mathrm{d}\theta|}{|\mathrm{d}\theta|} = r\,,\end{aligned} \tag{8.96}$$

in agreement with (8.95).

The basis vectors now can be normalised:

$$\boldsymbol{e}_{\widehat{r}} = \frac{\boldsymbol{e}_r}{|\boldsymbol{e}_r|} = \boldsymbol{e}_r\,,\quad \boldsymbol{e}_{\widehat{\theta}} = \frac{\boldsymbol{e}_\theta}{|\boldsymbol{e}_\theta|} = \frac{\boldsymbol{e}_\theta}{r}\,. \tag{8.97}$$

The normalised form of the gradient operator in polar coordinates then becomes, from (8.94),

$$\nabla = \boldsymbol{e}_{\widehat{r}}\,\partial_r + \frac{\boldsymbol{e}_{\widehat{\theta}}}{r}\partial_\theta\,. \tag{8.98}$$

Contrast this with (8.94). Equation (8.98) tends to be the preferred form in reference books, but in fact few will write it covariantly. Instead, for an arbitrary function f, they will probably write it in the following noncovariant way, with lowered indices that have nothing to do with *co*vectors, and also no carets:

$$\text{``}\,(\nabla f)_r = \frac{\partial f}{\partial r}\,, \quad (\nabla f)_\theta = \frac{1}{r}\frac{\partial f}{\partial \theta}\,\text{''}\,. \tag{8.99}$$

Clearly, it's important to be aware of the convention being used in tabled formulae such as these.

8.6.1 The Gradient Operator in Fully Covariant Notation

Looking back, this whole gradient calculation was a little cumbersome and apparently not generalisable to arbitrary coordinates. For one thing, there were lots of sums over the coordinates to be made: a situation ripe for the summation convention, which we were not able to make use of because the notation was not truly covariant. But we also cannot fail to be reminded of the frequency–wavenumber in Sect. 6.4 which, like ∇, was also defined using partial derivatives. There we found that a more elegant, natural, and generalisable treatment was to use a covector approach by making sure that k_α was specified with *lowered* indices, just like the $f_{,\alpha}$ in (8.81). After all, like the gradient, the frequency–wavenumber was defined using partial derivatives in (2.147) or (6.31).

But what exactly *is* the gradient? We have democratised the basis vectors as regards up/down indices, so consider again the arbitrary function $f(x^\alpha)$ introduced at the start of this section. A small increase in f is $\mathrm{d}f = f_{,\alpha}\mathrm{d}x^\alpha$. This must be a scalar, independent of whatever coordinates are used. Why? Because we are inducing an increase, $\mathrm{d}f$, in the function by taking an infinitesimal step to a neighbouring point in its domain. Coordinates have no bearing on how much the function's value changes, which is well defined for the step that we took. (For example, the temperature difference between two points in a room must be independent of whether we choose cartesian or polar coordinates to locate those points.) So $\mathrm{d}f = f_{,\alpha}\mathrm{d}x^\alpha$ is a scalar. If we *define* the gradient ∇f via

$$\mathrm{d}f = \nabla f \cdot \mathrm{d}\boldsymbol{x}\,, \tag{8.100}$$

then we can rewrite (8.100) as

$$\begin{aligned} \underbrace{f_{,\alpha}\mathrm{d}x^\alpha}_{\mathrm{d}f} &= f_{,\alpha}\mathrm{d}x^\beta\, \delta^\alpha_\beta = f_{,\alpha}\mathrm{d}x^\beta\, \boldsymbol{e}^\alpha\cdot\boldsymbol{e}_\beta \\ &= \underbrace{f_{,\alpha}\boldsymbol{e}^\alpha}_{\therefore\,\equiv\,\nabla f} \cdot \underbrace{\mathrm{d}x^\beta\boldsymbol{e}_\beta}_{\mathrm{d}\boldsymbol{x}}\,. \end{aligned} \tag{8.101}$$

This shows that the gradient is in fact $\nabla f = f_{,\alpha}\boldsymbol{e}^\alpha$. But the set of derivatives $f_{,\alpha}$ forms a covector because they transform in the required way, as can be seen by simply applying the chain rule for partial derivatives:

$$f_{,\bar\beta} = \Lambda^\alpha_{\bar\beta}\, f_{,\alpha}\,. \tag{8.102}$$

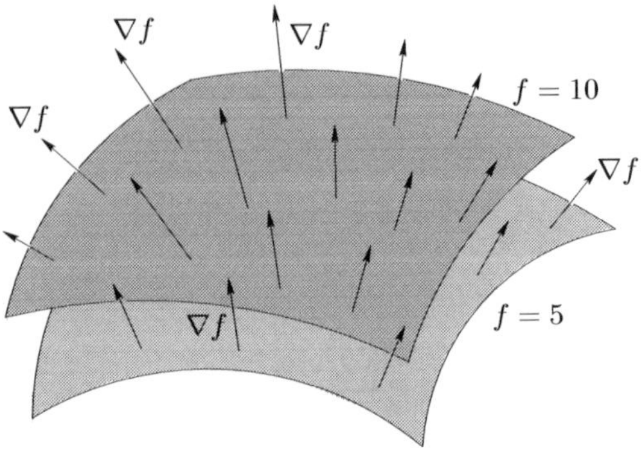

Fig. 8.7. The gradient vector of some function f is always perpendicular to the surfaces of constant f, pointing in the direction where f is increasing most rapidly. The reason comes from $\mathrm{d}f = \nabla f \cdot \mathrm{d}\boldsymbol{x}$. First, the perpendicularity is ensured because a step $\mathrm{d}\boldsymbol{x}$ taken along the constant-f surface must give $\mathrm{d}f = 0$, which will be the case if this step is perpendicular to ∇f, so that ∇f must be perpendicular to the surface. Second, that f is increasing most rapidly in the direction of ∇f is ensured because a step $\mathrm{d}\boldsymbol{x}$ taken in the direction of ∇f will maximise $\nabla f \cdot \mathrm{d}\boldsymbol{x}$, which implies $\mathrm{d}f$ is maximised in this direction.

Thus the gradient ∇f is a proper vector, and is most naturally expressed over a *cobasis* as in (8.101). The use of the cobasis shows that the gradient operator itself is quite naturally written as

$$\boxed{\nabla = \boldsymbol{e}^\alpha \partial_\alpha\,.} \tag{8.103}$$

This is a very important and useful expression that we'll use extensively later in this chapter, as well as in Chap. 12. It is the central identity connecting vector notation with tensor notation in *any* coordinates. Contrast it with the less useful noncovariant form (8.84). The difference is academic in the case of cartesians (where raised indices are no different from lowered ones), but assumes its full power when other bases are considered.

On the strength of the fact that the set of components $f_{,\alpha}$ of ∇f is a covector, the gradient is often referred to as a covector and is said to be something quite different from a vector. But the discussion of the previous pages, and in particular (8.55), shows that ∇f is certainly a vector, as has been drawn in Fig. 8.7. In fact, there is nothing to stop us from raising the derivative index along with its comma by defining a new set of numbers called $f^{,\alpha}$, also written $\partial^\alpha f$:

$$\nabla f = f_{,\alpha}\boldsymbol{e}^\alpha = g_{\alpha\mu} f^{,\mu} g^{\alpha\nu} \boldsymbol{e}_\nu = f^{,\mu}\boldsymbol{e}_\mu\,. \tag{8.104}$$

So $f_{,\alpha}\boldsymbol{e}^\alpha$ is identical to $f^{,\alpha}\boldsymbol{e}_\alpha$, and indeed it's trivial to show that the $f^{,\alpha}$ transform as vector components should.

As is shown in Fig. 8.7, the gradient of a function f always points in the direction in which f is increasing the most rapidly. A simple example of this is when $f(r,\theta) = r$, in which case

$$\nabla r = r_{,r}\boldsymbol{e}^r + r_{,\theta}\boldsymbol{e}^\theta = \boldsymbol{e}^r = \boldsymbol{e}_r = \boldsymbol{e}_{\widehat{r}}\,, \tag{8.105}$$

which is no different in three dimensions:

$$\nabla r = r_{,r}\boldsymbol{e}^r + r_{,\theta}\boldsymbol{e}^\theta + r_{,\phi}\boldsymbol{e}^\phi = \boldsymbol{e}^r = \boldsymbol{e}_r = \boldsymbol{e}_{\widehat{r}}\,. \tag{8.106}$$

Indeed, $\boldsymbol{e}_r$ always points radially away from the origin. No more complicated is the case of $f(r,\theta) = \theta$:

$$\nabla\theta = \theta_{,\alpha}\boldsymbol{e}^\alpha = \boldsymbol{e}^\theta = \boldsymbol{e}_\theta/r^2 = \boldsymbol{e}_{\widehat{\theta}}/r\,. \tag{8.107}$$

Note that $\nabla r = \boldsymbol{e}^r$ and $\nabla\theta = \boldsymbol{e}^\theta$. This is true for any general coordinate x^α:

$$\nabla x^\alpha = x^\alpha_{\ ,\beta}\,\boldsymbol{e}^\beta = \delta^\alpha_\beta\,\boldsymbol{e}^\beta = \boldsymbol{e}^\alpha. \tag{8.108}$$

On the strength of this, the cobasis vector $\boldsymbol{e}^\alpha$ could be written ∇x^α. We'll meet this expression again on p. 307 and at the end of this chapter.

The Inverse-Square Force in Polar Coordinates

A common example of a polar coordinate gradient is encountered in field theory, where the gradient of the potential function $1/r$ must be calculated. While the calculation in cartesian coordinates is more lengthy (though not difficult), it becomes quite trivial in polar coordinates:

$$\nabla\frac{1}{r} = \left(\frac{1}{r}\right)_{,\alpha}\boldsymbol{e}^\alpha = \frac{-1}{r^2}\,\boldsymbol{e}^r = \frac{-1}{r^2}\,\boldsymbol{e}_r = \frac{-1}{r^2}\,\boldsymbol{e}_{\widehat{r}}\,. \tag{8.109}$$

This expression is related to the concept of an inverse-square force law, which we'll have occasion to employ in (10.80).

Finally, the nabla operator can be applied to tensors of any order at all. For example, for a second-order tensor,

$$\nabla\boldsymbol{T} = \boldsymbol{e}^\alpha\partial_\alpha\left(T^{\mu\nu}\boldsymbol{e}_{\mu\nu}\right) \equiv \boldsymbol{e}^\alpha\otimes\partial_\alpha\left(T^{\mu\nu}\boldsymbol{e}_{\mu\nu}\right). \tag{8.110}$$

The techniques to come in this chapter will show how to calculate this tensor derivative. In fact, in Sects 8.10 and 12.7 we'll concentrate heavily on a more advanced idea: the very useful marriage of nabla with a wedge product to give the operator $\nabla\wedge$, known as the exterior derivative.

Many Sign Changes Make Hard Work

The expression for the nabla operator, $\nabla = \boldsymbol{e}^\alpha \partial_\alpha$, is valid for any coordinate system in any number of dimensions. What results when it's expressed in the $txyz$-coordinates of Minkowski spacetime using basis vectors?

$$\begin{aligned}\nabla &= \boldsymbol{e}^t\partial_t + \boldsymbol{e}^x\partial_x + \boldsymbol{e}^y\partial_y + \boldsymbol{e}^z\partial_z \\ &= \eta^{tt}\boldsymbol{e}_t\partial_t + \eta^{xx}\boldsymbol{e}_x\partial_x + \eta^{yy}\boldsymbol{e}_y\partial_y + \eta^{zz}\boldsymbol{e}_z\partial_z \\ &= \eta_{tt}\left(\boldsymbol{e}_t\partial_t - \boldsymbol{e}_x\partial_x - \boldsymbol{e}_y\partial_y - \boldsymbol{e}_z\partial_z\right). \end{aligned} \tag{8.111}$$

There is now a mixture of signs. Unfortunately, the four-dimensional nabla is often *defined* as the last line of (8.111) (and written as a box $\Box$ in analogy to the three sides of ∇, which was originally a three-dimensional operator only). This definition is really as nongeneralisable as the noncovariant euclidean 3-space nabla (8.84). Neither expression lends itself to being converted to arbitrary coordinates, so these noncovariant definitions should be seen for what they are: useful for cartesian coordinates only—and provided we don't trip up on the sign changes.

A similar noncovariance happens when considering the *Minkowski dot product*. The correct expression is

$$\begin{aligned}\boldsymbol{A}\cdot\boldsymbol{B} &= A^\alpha \boldsymbol{e}_\alpha \cdot B^\beta \boldsymbol{e}_\beta = \eta_{\alpha\beta}A^\alpha B^\beta = \sum_\alpha \eta_{\alpha\alpha}A^\alpha B^\alpha \\ &= \eta_{tt}A^tB^t + \eta_{xx}A^xB^x + \eta_{yy}A^yB^y + \eta_{zz}A^zB^z \\ &= \eta_{tt}\left(A^tB^t - A^xB^x - A^yB^y - A^zB^z\right). \end{aligned} \tag{8.112}$$

Again, unfortunately, the Minkowski dot product is often *defined* noncovariantly as the last line in (8.112) (with $\eta_{tt} = +1$ and -1 both used), which makes for difficulty in seeing how it might be written in general coordinates; and there are more mixed signs to watch out for. Perhaps not surprisingly, these noncovariant expressions for the Minkowski nabla and dot product find little use in practice; tensor notation tends to be the preferred approach. We can always use nabla and the dot product with impunity by writing them covariantly. Only then are we guaranteed to get everything right for any metric at all. A good example is the divergence calculation in (12.41).

Taylor's Theorem in Various Guises

On the subject of the gradient, it's useful to write down Taylor's theorem using both tensor and nontensor notation. Taylor's theorem is used extensively in mathematical physics, but its n-dimensional form is seldom written explicitly. We saw it briefly for one dimension in (2.221) and for three dimensions (really n dimensions) in (2.236). Here we show that this n-dimensional form is compact and elegant using both vector and tensor notation.

That the basic first-order increase in a scalar function $f(x^\alpha)$ is $\mathrm{d}f = f_{,\alpha}\,\mathrm{d}x^\alpha$ can be seen heuristically, by using cartesian coordinates with the analogy of climbing a staircase with 90° turns at its corners on a ground defined by the xy-plane. Starting at (x, y), climbing in the x-direction is akin to increasing f in the x-direction, holding y constant, so the height we gain is $f_{,x}\,\mathrm{d}x$. We turn the corner and gain more height by increasing f in the y-direction with x held constant, further increasing the height by $f_{,y}\,\mathrm{d}y$.

We are now standing right above $(x + \mathrm{d}x, y + \mathrm{d}y)$, and the total height gained has been $f_{,x}\,\mathrm{d}x + f_{,y}\,\mathrm{d}y$. But each factor in this expression $\mathrm{d}f = f_{,\alpha}\,\mathrm{d}x^\alpha$ in cartesian coordinates transforms to any other coordinates by way of a jacobian matrix, and so $\mathrm{d}f = f_{,\alpha}\,\mathrm{d}x^\alpha$ holds in *all* coordinates. The same expression holds in an arbitrary number of dimensions, which is not hard to prove by applying a succession of one-dimensional Taylor expansions to $f(x^1 + \mathrm{d}x^1, x^2 + \mathrm{d}x^2, \ldots)$.

In vector form, the increase $\mathrm{d}f$ is, of course,

$$\mathrm{d}f = f_{,\alpha}\,\mathrm{d}x^\alpha = (\nabla f)_\alpha\,\mathrm{d}x^\alpha = \nabla f \cdot \mathrm{d}\boldsymbol{x}\,. \tag{8.113}$$

Taylor's theorem extends this result to the case of a noninfinitesimal step. For appropriately well-behaved functions, it can be written in the following ways, where $\boldsymbol{x} \equiv (x^1, \ldots, x^n)$ and f is always evaluated at $\boldsymbol{x}$. First,

$$\begin{aligned} f(\boldsymbol{x} + \Delta\boldsymbol{x}) = f(\boldsymbol{x}) &+ f_{,\alpha}\,\Delta x^\alpha + \frac{1}{2!} f_{,\alpha\beta}\,\Delta x^\alpha\,\Delta x^\beta \\ &+ \frac{1}{3!} f_{,\alpha\beta\gamma}\,\Delta x^\alpha\,\Delta x^\beta\,\Delta x^\gamma + \cdots . \end{aligned} \tag{8.114}$$

All terms in this expansion are scalars. The zeroth-order term $f(\boldsymbol{x})$ is obviously so, the first-order term is a dot product between two vectors, and the second-order term is a quadratic form, and so can be written as a matrix multiplication. The matrix of second derivatives is called the *hessian* of f. The higher-order terms are not easily expressed using vectors or matrices. Because of this, the terms of the Taylor series are often described as being of "scalar, vector, matrix, and tensor" nature, but this is a little overdone and gives the wrong impression that there is some dramatic increase in abstraction in the terms as the sequence progresses. The point is only that there is no commonly used notation for the third- and higher-order terms that avoids indices; but those terms are not fundamentally any more complicated than the first three terms.

With gradient notation, the theorem has a different sort of symmetry. Again f is understood to be evaluated at $\boldsymbol{x}$:

$$\begin{aligned} f(\boldsymbol{x} + \Delta\boldsymbol{x}) = f(\boldsymbol{x}) &+ \nabla f \cdot \Delta\boldsymbol{x} + \frac{1}{2!} \nabla\left[\nabla f \cdot \Delta\boldsymbol{x}\right] \cdot \Delta\boldsymbol{x} \\ &+ \frac{1}{3!} \nabla\left[\nabla\left[\nabla f \cdot \Delta\boldsymbol{x}\right] \cdot \Delta\boldsymbol{x}\right] \cdot \Delta\boldsymbol{x} + \cdots . \end{aligned} \tag{8.115}$$

As a novelty, this can also be written as in (8.116), with no factorial signs, and where the gradient is taken *before* the "previous term" is numerically evaluated, of course!

$$f(\boldsymbol{x}+\Delta\boldsymbol{x}) = f(\boldsymbol{x}) + \nabla(\text{previous term})\cdot\Delta\boldsymbol{x} + \frac{1}{2}\nabla(\text{previous term})\cdot\Delta\boldsymbol{x} + \frac{1}{3}\nabla(\text{previous term})\cdot\Delta\boldsymbol{x} + \cdots . \quad (8.116)$$

8.6.2 Is a Metric Needed?

Equation (8.113) calculates the infinitesimal increase $\mathrm{d}f$ in a function f along a step $\mathrm{d}\boldsymbol{x}$ by using the dot product: $\mathrm{d}f = \nabla f\cdot\mathrm{d}\boldsymbol{x}$. The dot product makes use of the metric, and yet the increase $\mathrm{d}f$ is independent of any metric since $\mathrm{d}f = f_{,\alpha}\mathrm{d}x^\alpha$. That might imply that involving a metric in any discussion of ∇, via the use of the cobasis, is unnecessary.

Traditionally, the lack of need for a metric to calculate $\mathrm{d}f$ has been signalled by regarding the gradient not as a vector but as a new object called a *one-form.* A one-form is defined to be a function mapping vectors to real numbers. Thus ∇f might be regarded as a one-form function that takes a single vector argument $\mathrm{d}\boldsymbol{x}$, producing the real number $\mathrm{d}f$.

The one-form function tends to be considered as a separate object (more usually denoted ω^α instead of $\boldsymbol{e}^\alpha$), visualised not as an arrow but as a set of parallel planes. These planes need labelling to distinguish the one-form from its negative, but this appears to be seldom, if ever, done. The operation of the one-form on a vector is then imagined as the arrow piercing the planes; the number of planes pierced denotes the real number output by the one-form function. We saw the appearance of sets of planes in Sect. 2.3 and made the point there that a cobasis *vector*, while not a set of planes, does help describe the orientation of lattice planes in a crystal.

Users of the one-form as a function would regard (8.29) not as a dot product between two vectors $\boldsymbol{e}^\alpha$ and $\boldsymbol{e}_\beta$ but rather as a functional relationship: $\omega^\alpha(\boldsymbol{e}_\beta) \equiv \delta^\alpha_\beta$. Some will write it as an inner product expression, $\langle\omega^\alpha|\boldsymbol{e}_\beta\rangle \equiv \delta^\alpha_\beta$. The use of the inner product is perhaps meant to emphasise the one-form as a separate object, just as a vector is an object. But certainly this use of the inner product does come very close to the dot product expression of (8.29).

Our view is that introducing a new set of objects ω^α, in order to omit the metric, is not especially useful in physics. A good analogy lies in computing the length of a line segment in elementary euclidean geometry. While this length can be calculated by introducing a cartesian set of coordinates and applying Pythagoras's theorem, its value is nevertheless independent of that coordinate choice. Even so, that does not mean we should make a point of discarding coordinates and devising a theory that doesn't use them or replaces them with something new that ultimately is only equivalent to their use. It is perfectly understood that coordinates are very useful for calculating things that are independent of the coordinate choice; this is done all the time in

Is Force a Vector or a One-Form?

One-forms are not absolutely necessary in physics, and insisting on their use when a metric is present can lead to problems of interpretation. An example lies in calculating the infinitesimal work $\mathrm{d}W$ done by a force $\boldsymbol{F}$ that accelerates a mass through some infinitesimal displacement $\mathrm{d}\boldsymbol{x}$. Written as $\mathrm{d}W = \boldsymbol{F} \cdot \mathrm{d}\boldsymbol{x}$, it is sometimes interpreted that force is a one-form that combines with, or acts on, the vector $\mathrm{d}\boldsymbol{x}$ to produce a scalar $\mathrm{d}W$. But then, since Newton tells us that $\boldsymbol{F} = m\boldsymbol{a}$, and acceleration is a vector (being derived from the vector $\mathrm{d}\boldsymbol{x}$), we might conclude that force is a vector. So is force a vector or a one-form?

For us the question does not arise, since we always take the pragmatic view that a metric exists, and so there are only vectors. After all, every laboratory has at least one ruler! Physicists measure things, so a metric is a natural place from which to start. Furthermore, a metric is *always* eventually introduced in any textbook on tensor theory in physics, and this metric then has the effect of rendering the idea of one-forms superfluous.

physics. Likewise, the cobasis and metric can do everything that one-forms can do.

The cobasis was defined more generally by (8.29), and because we have always used a metric in all of our tensor discussions, we have been free to generate the cobasis vectors from the basis vectors by way of the important expression (8.34). Users of one-forms cannot benefit from this expression and so need to find alternative paths through the calculations of the next few sections, calculations that are here rendered very transparent through our being able to use (8.34) to switch from basis to cobasis at will.

A major difficulty with using a one-form basis ω^α, visualised as sets of parallel planes, along with a never-specified definition of just what it means to add these sets of planes, is that the object $v_\alpha \omega^\alpha$ is necessarily distinct from $v^\alpha \boldsymbol{e}_\alpha$. The former is a linear combination of sets of parallel planes, while the latter is a linear combination of arrows. Such an arbitrary distinction between basis and cobasis destroys all of the enormous power gained by relating the two.

In (8.108), we saw that the cobasis vector $\boldsymbol{e}^\alpha$ is equal to ∇x^α. This resembles the established notation for one-forms, which takes a synonymous name for ω^α as $\mathrm{d}x^\alpha$, which is (apparently) supposed to define a rigorous notion of an infinitesimal. This sort of identification would presumably demand that a simple derivative such as $\mathrm{d}y/\mathrm{d}x$ be considered a ratio of sets of parallel lines, although it is entirely unclear what that's supposed to mean. Needless to say, we do not view things in this way, and we certainly only use infinitesimal notation to mean infinitesimals; it has nothing to do with one-forms. And, of course, ∇x^α does not equal $\mathrm{d}x^\alpha$; they are related through the well-known identity $\mathrm{d}x^\alpha = \nabla x^\alpha \cdot \mathrm{d}\boldsymbol{x}$. The "d" notation is discussed further at the end of this chapter.

Typically, the infinitesimal notation $\mathrm{d}x^\alpha$ for a one-form may or may not be normalised, carry a tilde, or be written in a different font. This plethora of symbols makes for fantastic typesetting in some books on tensor analysis, but the main effect it seems to have is to engender a discomfort and vagueness about what one-forms really are, as well as giving rise to the traditional view that tensor analysis is a difficult subject.

Finally, we reiterate the point made on p. 232, that although many books use the words covector and one-form interchangeably, for us they are separate objects. Our covector is an ordered set of numbers $\{v_\alpha\}$, a coordinate vector over a particular set of basis vectors called the cobasis.

8.7 Normalised Basis Vectors

Basis vectors that have been normalised to unit length are very useful and tend to be *de rigueur* in texts covering vector calculus in euclidean space with different coordinate systems. As we noted in (8.99), this sometime lack of indicating the normalisation is a trap for young players when referring to tables of vector identities using, for example, polar coordinates. Just as we did for the polar basis, normalise a general basis vector by dividing by its length, indicated by placing a caret over the index:

$$\boldsymbol{e}_{\widehat{\alpha}} \equiv \frac{\boldsymbol{e}_\alpha}{|\boldsymbol{e}_\alpha|} \quad \text{(no sum over } \alpha\text{)}. \tag{8.117}$$

(A lowered index in a denominator counts as a raised index overall and so would normally be summed over in (8.117). But we are not summing in this equation.) In order to preserve the usual covariant notation that writes $\boldsymbol{v} = v^{\widehat{\alpha}} \boldsymbol{e}_{\widehat{\alpha}} = v^\alpha \boldsymbol{e}_\alpha$, it must follow that the components $v^{\widehat{\alpha}}$ of a vector $\boldsymbol{v}$ are

$$v^{\widehat{\alpha}} = v^\alpha \, |\boldsymbol{e}_\alpha| \quad \text{(no sum)}. \tag{8.118}$$

Similarly,

$$\boldsymbol{e}^{\widehat{\alpha}} \equiv \frac{\boldsymbol{e}^\alpha}{|\boldsymbol{e}^\alpha|}, \quad v_{\widehat{\alpha}} = v_\alpha \, |\boldsymbol{e}^\alpha| \quad \text{(no sums)}. \tag{8.119}$$

Expressions such as these show that a careted index is not quite covariant notation—and it will not obey rules such as raising and lowering of indices. (We can certainly define careted quantities that do obey such rules by definition, as will be done in Sect. 12.7.) In texts that don't use covariant notation, these normalised basis vectors might be called, in the case of polar coordinates, $\widehat{\boldsymbol{r}}$ for $\boldsymbol{e}_{\widehat{r}}$ and $\widehat{\boldsymbol{\theta}}$ for $\boldsymbol{e}_{\widehat{\theta}}$. They are generally the basis vectors of choice, and the caretless *sub*scripts r, θ will be written on the components, in contrast to true covariant notation, which would demand $\widehat{r}, \widehat{\theta}$ *super*scripts on the components.

The fact that the careted indices are not fully covariant gives a hint that normalised basis vectors are a slightly different breed from the usual sort. The lack of covariance suggests that there might not exist new coordinates called, in the polar case, $\widehat{r}$ and $\widehat{\theta}$. We can show this explicitly as follows. If there really were such coordinates $\widehat{r}, \widehat{\theta}$, then we could write basis-changing expressions involving $\Lambda^{\widehat{r}}_{x}$, etc. Could this really be done? If so, then

$$\begin{aligned} \boldsymbol{e}^{\widehat{r}} = \boldsymbol{e}^{r} \text{ would imply that } \Lambda^{\widehat{r}}_{r} = 1\,,\ \Lambda^{\widehat{r}}_{\theta} = 0\,, \\ \text{and } \boldsymbol{e}^{\widehat{\theta}} = r\boldsymbol{e}^{\theta} \text{ would imply that } \Lambda^{\widehat{\theta}}_{r} = 0\,,\ \Lambda^{\widehat{\theta}}_{\theta} = r\,. \end{aligned} \tag{8.120}$$

But the equality of mixed partial derivatives implies that, in particular, $\Lambda^{\widehat{\theta}}_{r,\theta} = 0$ would have to be identical to $\Lambda^{\widehat{\theta}}_{\theta,r} = 1$—which clearly is not the case. So there really do not exist coordinates called $\widehat{r}$ and $\widehat{\theta}$. The set $\{\boldsymbol{e}_{\widehat{r}}, \boldsymbol{e}_{\widehat{\theta}}\}$ is called a *noncoordinate basis*. It is in fact quite meaningful physically because, after all, we make our day-to-day measurements using normalised basis vectors; the very ruler we use to measure the dimensions of this book is part of a normalised basis. So although up until now we have not been overly careful to distinguish between frames and coordinates, we now make the distinction that a frame is a physical system within which measurements are made. It can be normalised, having a set of normalised basis vectors at each point in space (or spacetime). Coordinates, on the other hand, are a set of numbers at each point in spacetime; they might cover the whole of spacetime or just parts of it, so that several coordinate systems might be required to cover all of spacetime, depending on how they are defined.

Although the normalised basis vectors $\boldsymbol{e}_{\widehat{\alpha}}$ described in (8.117) were parallel to $\boldsymbol{e}_{\alpha}$, they certainly don't have to be, and if the coordinate basis is not orthogonal, then we might wish to orthogonalise it before normalising. Call the set of orthonormalised basis vectors $\{\boldsymbol{e}_{\widehat{\alpha}}\}$, where $\widehat{\alpha}$ is any label that denotes a basis vector. In that case, every vector of either set is a linear combination of the vectors of the other set:

$$\boldsymbol{e}_{\widehat{\alpha}} = \Lambda^{\mu}_{\widehat{\alpha}}\,\boldsymbol{e}_{\mu}\,, \quad \boldsymbol{e}_{\mu} = \Lambda^{\widehat{\alpha}}_{\mu}\,\boldsymbol{e}_{\widehat{\alpha}}\,. \tag{8.121}$$

Note that $\Lambda^{\mu}_{\widehat{\alpha}}$ and $\Lambda^{\widehat{\alpha}}_{\mu}$ will only be partial derivatives as used in (7.44) if both indices refer to coordinate bases, which in general will not be the case. So, e.g., here $\Lambda^{\mu}_{\widehat{\alpha}}$ just denotes the μ-component of $\boldsymbol{e}_{\widehat{\alpha}}$ over the $\{\boldsymbol{e}_{\mu}\}$ basis, and we could equally well have written

$$\boldsymbol{e}_{\widehat{\alpha}} = (\boldsymbol{e}_{\widehat{\alpha}})^{\mu}\,\boldsymbol{e}_{\mu}\,, \quad \boldsymbol{e}_{\mu} = (\boldsymbol{e}_{\mu})^{\widehat{\alpha}}\,\boldsymbol{e}_{\widehat{\alpha}}\,. \tag{8.122}$$

This notation is common, but certainly quite awkward.

The set of vectors $\{\boldsymbol{e}_{\widehat{\alpha}}\}$ that can be constructed at each point, whether or not it constitutes a coordinate basis, is really what is meant by a frame. In particular, the metric of the orthonormal frame is

$$\eta_{\widehat{\alpha}\widehat{\beta}} = \boldsymbol{e}_{\widehat{\alpha}} \cdot \boldsymbol{e}_{\widehat{\beta}} = \Lambda^{\mu}_{\widehat{\alpha}}\,\Lambda^{\nu}_{\widehat{\beta}}\,\boldsymbol{e}_{\mu}\cdot\boldsymbol{e}_{\nu} = \Lambda^{\mu}_{\widehat{\alpha}}\,\Lambda^{\nu}_{\widehat{\beta}}\,g_{\mu\nu}\,. \tag{8.123}$$

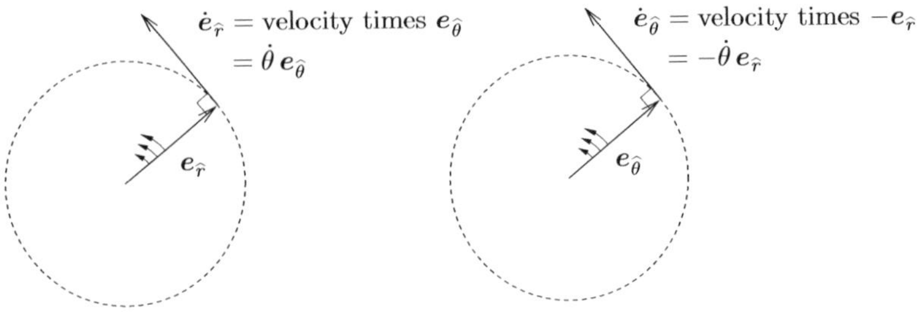

Fig. 8.8. Left: The head of a normalised radial basis vector $\boldsymbol{e}_{\hat{r}}$ turns with velocity $\dot{\theta}$, so that the vector's tangent is $\dot{\theta}\,\boldsymbol{e}_{\hat{\theta}}$. The word "velocity" here is used in its one-dimensional sense, a signed speed as opposed to a vector, so that the multiplication is the ordinary "number × vector" type. **Right**: Similarly, the head of a normalised angular basis vector $\boldsymbol{e}_{\hat{\theta}}$ turns with the same velocity $\dot{\theta}$, producing a tangent $-\dot{\theta}\,\boldsymbol{e}_{\hat{r}}$.

We'll see more of these orthonormal basis vectors in a general relativity context in Chap. 12.

8.7.1 The Normalised Polar Basis in Celestial Mechanics

A great example of the utility of normalised basis vectors arises when studying planetary motion. Here we wish to find the motion of a small mass m subject to a central gravity force produced by a large mass M using $\boldsymbol{F} = m\boldsymbol{a}$. Polar coordinates are most useful since the force $\boldsymbol{F}$ is radial only, in which case the acceleration $\boldsymbol{a}$ must be expressed in terms of polar coordinates. This can be done in quite a long-winded way by changing to a cartesian basis, time-differentiating the position vector twice (easily done since the cartesians have zero time derivatives), and then converting back to polar coordinates.

But the same result is much more simply achieved by using a *normalised* polar basis. To see why, realise that no matter what motion a planet or comet might have, the normalised polar basis vectors always turn in circles—they have constant length. We need only know how fast their heads turn, which is easy. The velocity of these equals the vectors' (unit) length multiplied by their angular velocity $\dot{\theta} \equiv \mathrm{d}\theta/\mathrm{d}t$, as shown in Fig. 8.8.

Now, from the last section, we know that $\boldsymbol{e}_{\hat{r}}$ and $\boldsymbol{e}_{\hat{\theta}}$ are orthogonal. Thus the time derivative of $\boldsymbol{e}_{\hat{r}}$ is proportional to $\boldsymbol{e}_{\hat{\theta}}$ since the tangent to a circle is always orthogonal to its radius vector. (This same idea of a tangent to a unit circle is used frequently in differential geometry, as we'll see in Chap. 9.)

The constant of proportionality is just the velocity with which the head of $\boldsymbol{e}_{\hat{r}}$ moves, which is $\dot{\theta}$ for a unit circle, giving $\dot{\boldsymbol{e}}_{\hat{r}} = \dot{\theta}\,\boldsymbol{e}_{\hat{\theta}}$. A similar argument produces $\dot{\boldsymbol{e}}_{\hat{\theta}} = -\dot{\theta}\,\boldsymbol{e}_{\hat{r}}$, the only difference being the minus sign that was introduced because we effectively turned two right angles to the original $\boldsymbol{e}_{\hat{r}}$.

The acceleration is the second time derivative of the position vector $\boldsymbol{r}$,

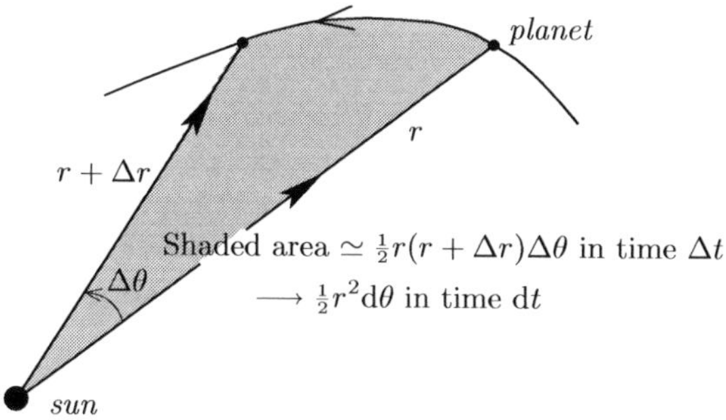

Fig. 8.9. Kepler's second law. As a planet moves around its sun, its position vector sweeps out some area in a given time interval. The shaded region is approximately a triangle of base r and height $(r+\Delta r)\Delta\theta$, with area half × base × height. Hence, in a time $\mathrm{d}t$, the area swept out by a planet's position vector is $\frac{1}{2}r^2\mathrm{d}\theta$. So the area swept per unit time equals $\frac{1}{2}r^2\dot{\theta}$. But (8.127) says that this quantity is a constant. Thus a planet's position vector sweeps out equal areas in equal times.

$$\begin{aligned} \boldsymbol{a} \equiv \ddot{\boldsymbol{r}} &= \mathrm{d}^2(r\boldsymbol{e}_{\widehat{r}})/\mathrm{d}t^2 \\ &= \ddot{r}\,\boldsymbol{e}_{\widehat{r}} + 2\,\dot{r}\,\dot{\boldsymbol{e}}_{\widehat{r}} + r\,\ddot{\boldsymbol{e}}_{\widehat{r}}\,, \end{aligned} \tag{8.124}$$

so that using the basis vector derivatives in Fig. 8.8, we can write

$$\begin{aligned} \dot{\boldsymbol{e}}_{\widehat{r}} &= \dot{\theta}\,\boldsymbol{e}_{\widehat{\theta}}\,, \\ \ddot{\boldsymbol{e}}_{\widehat{r}} &= \ddot{\theta}\,\boldsymbol{e}_{\widehat{\theta}} + \dot{\theta}\,\dot{\boldsymbol{e}}_{\widehat{\theta}} = \ddot{\theta}\,\boldsymbol{e}_{\widehat{\theta}} - \dot{\theta}^2\,\dot{\boldsymbol{e}}_{\widehat{r}}\,. \end{aligned} \tag{8.125}$$

The acceleration then becomes

$$\boldsymbol{a} = \left(\ddot{r} - r\,\dot{\theta}^2\right)\boldsymbol{e}_{\widehat{r}} + \left(2\,\dot{r}\,\dot{\theta} + r\,\ddot{\theta}\right)\boldsymbol{e}_{\widehat{\theta}}\,, \tag{8.126}$$

without any hard work. Newton tells us that this acceleration vector equals the force per unit mass, $-GM/r^2\,\boldsymbol{e}_{\widehat{r}}$, in which case equating components of this with those of $\boldsymbol{a}$ in (8.126) gives

$$\begin{aligned} &\ddot{r} - r\,\dot{\theta}^2 = -GM/r^2\,, \\ &2\,\dot{r}\,\dot{\theta} + r\,\ddot{\theta} = 0\,, \quad \text{i.e.}\quad \frac{\mathrm{d}(r^2\dot{\theta})}{\mathrm{d}t} = 0\,, \end{aligned} \tag{8.127}$$

from which the standard analysis of orbits can be deduced. For example, the last equation of (8.127) says that $r^2\dot{\theta}$ is constant in time, so that $r^2\mathrm{d}\theta = \text{constant} \times \mathrm{d}t$. Figure 8.9 shows how this implies that the infinitesimal area, $^1\!/_2\, r^2\mathrm{d}\theta$, swept out by a planet's position vector (extending from its sun) in a time $\mathrm{d}t$ is proportional to $\mathrm{d}t$. This is, of course, Kepler's law of equal areas swept in equal times.

These normalised polar vectors also make it very easy to see how angular momentum is related to orbital motion. With no basis-changing fuss, we can straightaway write

$$\begin{aligned}\boldsymbol{L} &\equiv \boldsymbol{r} \times \boldsymbol{p} = \boldsymbol{r} \times m\dot{\boldsymbol{r}} \\ &= r\,\boldsymbol{e}_{\widehat{r}} \times m\left(\dot{r}\,\boldsymbol{e}_{\widehat{r}} + r\,\dot{\theta}\,\boldsymbol{e}_{\widehat{\theta}}\right) \\ &= mr^2\dot{\theta}\,\boldsymbol{e}_{\widehat{r}} \times \boldsymbol{e}_{\widehat{\theta}} = mr^2\dot{\theta}\,\boldsymbol{e}_z\,, \end{aligned} \tag{8.128}$$

where $\boldsymbol{e}_z$ points perpendicular to the plane of the orbit and has unit length. Thus the angular momentum has magnitude $L = mr^2\dot{\theta}$. This, together with the second equation of (8.127), shows that orbital angular momentum is conserved over time. The calculation has been quite simple; in contrast, working with the cross product $\boldsymbol{r} \times \boldsymbol{p}$ in cartesians is much more laborious.

8.7.2 An Example of Using Vectors to Calculate an Effective Potential

A good example of some of the important ideas covered so far in this chapter lies in answering the following question. Chapter 4 derived expressions for the Coriolis and centrifugal forces felt by a mass on the surface of a rotating body. Excluding the Coriolis force with its velocity dependence, what is the effective potential felt by someone at rest on Earth's surface (assumed to be a sphere)? We will investigate the idea of a potential more fully in Chap. 10, but for now will just quote the main result that defines the potential in terms of the force felt.

Call this potential Φ. It relates to the force $\boldsymbol{F}$ experienced by a mass m via $-\nabla\Phi = \boldsymbol{F}/m$. In (4.48) we saw an expression for the centrifugal force in terms of Earth's angular velocity $\boldsymbol{\omega}$ and the position vector $\boldsymbol{r}$ of the mass, and by adding the force due to gravity and using spherical polar coordinates, the force per unit mass experienced at any point on Earth's surface can be written as

$$-\nabla\Phi = \frac{-GM}{r^2}\boldsymbol{e}_{\widehat{r}} - \boldsymbol{\omega} \times (\boldsymbol{\omega} \times \boldsymbol{r})\,. \tag{8.129}$$

This equation must be solved for the scalar Φ as a function of position. The idea is to write (8.129) in terms of any basis (or cobasis, which is really just a basis anyway!) and then equate the coefficients of each basis vector. A useful set might be the spherical polar basis since (8.129) already uses these coordinates, so we'll express both sides of (8.129) in terms of the basis $\boldsymbol{e}_r, \boldsymbol{e}_\theta, \boldsymbol{e}_\phi$. These are easy to use because at any point on the surface of a spherical Earth, $\boldsymbol{e}_r$ points up (i.e, radially outward from Earth's centre), $\boldsymbol{e}_\theta$ points south, and $\boldsymbol{e}_\phi$ points east. The last two of these are shown in Fig. 8.3.

Since the gradient $\nabla\Phi$ is naturally expressed in terms of the *co*basis, we require the spherical polar metric to convert cobasis to basis. Spherical polar coordinates are related to cartesians by

$$\begin{aligned} x &= r\sin\theta\,\cos\phi\,, \\ y &= r\sin\theta\,\sin\phi\,, \\ z &= r\cos\theta\,, \end{aligned} \tag{8.130}$$

so that the line element is $\mathrm{d}x^2 + \mathrm{d}y^2 + \mathrm{d}z^2 = \mathrm{d}r^2 + r^2\mathrm{d}\theta^2 + r^2\sin^2\theta\,\mathrm{d}\phi^2$. If the coordinates are ordered as r, θ, ϕ, the metric and its inverse will have elements

$$\begin{aligned} (g_{\alpha\beta}) &= \mathrm{diag}\left(1, r^2, r^2\sin^2\theta\right), \\ (g^{\alpha\beta}) &= \mathrm{diag}\left(1, \frac{1}{r^2}, \frac{1}{r^2\sin^2\theta}\right). \end{aligned} \tag{8.131}$$

These allow the cobasis and basis to be related:

$$\boldsymbol{e}^r = \boldsymbol{e}_r\,, \qquad \boldsymbol{e}^\theta = \frac{\boldsymbol{e}_\theta}{r^2}\,, \qquad \boldsymbol{e}^\phi = \frac{\boldsymbol{e}_\phi}{r^2\sin^2\theta}\,, \tag{8.132}$$

which allows the fundamental gradient expression (8.103) to be converted from a cobasis to a basis:

$$\begin{aligned} \nabla\Phi &= \Phi_{,r}\,\boldsymbol{e}^r + \Phi_{,\theta}\,\boldsymbol{e}^\theta + \Phi_{,\phi}\,\boldsymbol{e}^\phi \\ &= \Phi_{,r}\,\boldsymbol{e}_r + \frac{\Phi_{,\theta}}{r^2}\,\boldsymbol{e}_\theta + \frac{\Phi_{,\phi}}{r^2\sin^2\theta}\,\boldsymbol{e}_\phi\,. \end{aligned} \tag{8.133}$$

We could of course have avoided a basis entirely and left $\nabla\Phi$ written over a cobasis, although then the right-hand side of (8.129) would have to be converted to a cobasis. The choice between using the basis or cobasis is quite arbitrary.

This takes care of the left-hand side of (8.129). For its right-hand side, we are required to calculate

$$\boldsymbol{\omega}\times(\boldsymbol{\omega}\times\boldsymbol{r}) = \omega\boldsymbol{e}_z\times(\omega\boldsymbol{e}_z\times r\boldsymbol{e}_{\widehat{r}}) = \omega^2 r\sin\theta\,\boldsymbol{e}_z\times\boldsymbol{e}_{\widehat{\phi}}\,. \tag{8.134}$$

Note that normalised basis vectors have automatically appeared here. We can easily convert back and forth between them using the metric, giving

$$\boldsymbol{e}_{\widehat{r}} = \boldsymbol{e}_r\,, \qquad \boldsymbol{e}_{\widehat{\theta}} = \frac{\boldsymbol{e}_\theta}{r}\,, \qquad \boldsymbol{e}_{\widehat{\phi}} = \frac{\boldsymbol{e}_\phi}{r\sin\theta}\,, \tag{8.135}$$

but there is no need to do so until it becomes really necessary.

The last cross product in (8.134) is a mixture of a cartesian basis vector and a polar basis vector, but it will be easier to calculate if we can convert it either to all cartesian or all polar. Let's choose polar, so that we must express $\boldsymbol{e}_z$ in the polar basis:

$$\boldsymbol{e}_z = \Lambda_z^r\,\boldsymbol{e}_r + \Lambda_z^\theta\,\boldsymbol{e}_\theta + \Lambda_z^\phi\,\boldsymbol{e}_\phi\,. \tag{8.136}$$

The partial derivatives can be calculated either by expressing r, θ, ϕ in terms of cartesians, which is actually not difficult, or by the more general route that was outlined in Sect. 8.6, which writes

$$\begin{bmatrix} \Lambda^r_x & \Lambda^r_y & \Lambda^r_z \\ \Lambda^\theta_x & \Lambda^\theta_y & \Lambda^\theta_z \\ \Lambda^\phi_x & \Lambda^\phi_y & \Lambda^\phi_z \end{bmatrix} = \begin{bmatrix} \Lambda^x_r & \Lambda^x_\theta & \Lambda^x_\phi \\ \Lambda^y_r & \Lambda^y_\theta & \Lambda^y_\phi \\ \Lambda^z_r & \Lambda^z_\theta & \Lambda^z_\phi \end{bmatrix}^{-1}$$

$$= \begin{bmatrix} \sin\theta\cos\phi & r\cos\theta\cos\phi & -r\sin\theta\sin\phi \\ \sin\theta\sin\phi & r\cos\theta\sin\phi & r\sin\theta\cos\phi \\ \cos\theta & -r\sin\theta & 0 \end{bmatrix}^{-1}$$

$$= \begin{bmatrix} \sin\theta\cos\phi & \sin\theta\sin\phi & \cos\theta \\ \frac{1}{r}\cos\theta\cos\phi & \frac{1}{r}\cos\theta\sin\phi & \frac{-1}{r}\sin\theta \\ \frac{-\sin\phi}{r\sin\theta} & \frac{\cos\phi}{r\sin\theta} & 0 \end{bmatrix}. \tag{8.137}$$

Hence (8.136) gives

$$\boldsymbol{e}_z = \cos\theta\,\boldsymbol{e}_r - \frac{\sin\theta}{r}\,\boldsymbol{e}_\theta\,, \tag{8.138}$$

in which case (8.134) becomes

$$\begin{aligned} \boldsymbol{\omega} \times (\boldsymbol{\omega} \times \boldsymbol{r}) &= \omega^2 r \sin\theta \left(\cos\theta\,\boldsymbol{e}_r - \frac{\sin\theta}{r}\,\boldsymbol{e}_\theta \right) \times \boldsymbol{e}_{\widehat{\phi}} \\ &= \omega^2 r \sin\theta \left(-\cos\theta\,\boldsymbol{e}_{\widehat{\theta}} - \sin\theta\,\boldsymbol{e}_{\widehat{r}} \right) \\ &= \omega^2 r \sin\theta \left(\frac{-\cos\theta}{r}\,\boldsymbol{e}_\theta - \sin\theta\,\boldsymbol{e}_r \right). \end{aligned} \tag{8.139}$$

We now have everything needed to rewrite (8.129) in terms of a single basis. Equations (8.133) and (8.139) together give

$$-\Phi_{,r}\,\boldsymbol{e}_r - \frac{\Phi_{,\theta}}{r^2}\,\boldsymbol{e}_\theta - \frac{\Phi_{,\phi}}{r^2\sin^2\theta}\,\boldsymbol{e}_\phi = \left(\frac{-GM}{r^2} + \omega^2 r \sin^2\theta \right) \boldsymbol{e}_r + \omega^2 \sin\theta\cos\theta\,\boldsymbol{e}_\theta\,. \tag{8.140}$$

This equation is really three separate ones:

$$\Phi_{,r} = \frac{GM}{r^2} - \omega^2 r \sin^2\theta\,,$$

$$\Phi_{,\theta} = -\omega^2 r^2 \sin\theta \cos\theta\,, \quad \Phi_{,\phi} = 0\,. \tag{8.141}$$

These are readily integrated, giving the required effective potential:

$$\Phi = \frac{-GM}{r} - \frac{1}{2}\omega^2 r^2 \sin^2\theta + \text{constant}. \tag{8.142}$$

Although somewhat lengthy, the calculation has been straightforward. We made liberal use of the metric to convert to and from normalised basis vectors when needed, which made the cross products trivial. The calculation of the required jacobian matrix took some effort, which could have been avoided had we expressed the polar coordinates in terms of cartesians; but for the sake of the example, it was assumed here that, in general, the relevant coordinate transformation might be difficult to invert. There are other approaches to calculating Φ, but the procedure we have followed here is certainly broadly useful and ties together several ideas.

8.7.3 Some Final Remarks on Vector Terminology

In the previous sections, such as the box on p. 307, we have taken the approach that physicists make measurements, so that it's completely natural to begin tensor analysis with the idea of a metric already in place. Because of this, we have not needed the idea of a one-form at all. It should be noted that the conventional use of one-forms does *not* simply take our $\boldsymbol{e}^\alpha$ and rename it a one-form ω^α. After all, one-forms are not supposed to be related to basis vectors, whereas our $\boldsymbol{e}^\alpha$ certainly is: $\boldsymbol{e}^\alpha = g^{\alpha\beta}\boldsymbol{e}_\beta$. Divorcing one-forms from basis vectors loses the power of equations such as (8.132), which can make vector analysis more difficult.

Although we have been careful to define the terminology used in this chapter, the various terms collected on p. 293 are not meant to give any impression that vector analysis is complicated by terminology. In the end, one of the important reasons why linear algebra introduces the idea of a basis is that it *does* give us the freedom to choose any basis that makes a task simpler. The cobasis is just another basis. The main point to remember is that vectors and tensors are invariant objects, and they are expressible as components with a basis, irrespective of any terminology used. There is just one entity $\boldsymbol{v}$, and whether it's expressed as $v^\alpha \boldsymbol{e}_\alpha$ or as $v_\alpha \boldsymbol{e}^\alpha$ is entirely our choice. One version might be more useful than the other in rendering a given problem more tractable, and certainly the ability to switch between them lends great power to the analyses we make.

8.8 Volume Elements, Determinants, and Cross Products Again

Having begun to establish derivative formalism in tensor language, this is a good opportunity to take a side trip to lay the foundation for some ideas we

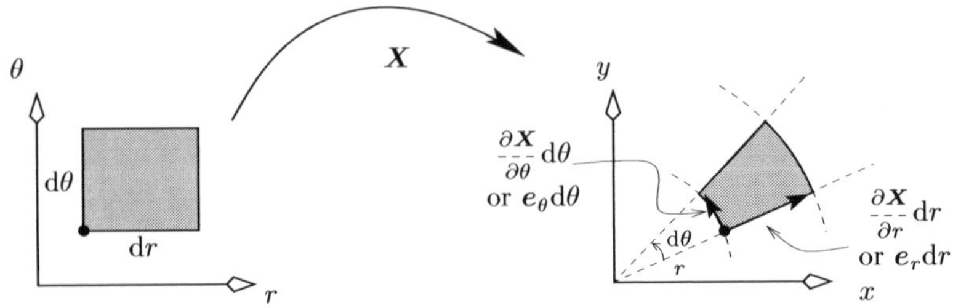

Fig. 8.10. A rectangular tiling of the $r\theta$-plane induces a corresponding tiling of the xy-plane. The area of this new tile, in the infinitesimal limit, is the area of a rectangle with sides $\mathrm{d}r$ and $r\,\mathrm{d}\theta$; *or* it can be calculated from a cross product of the vectors shown.

use later when treating tensor derivatives in more detail, and to build on the theory of volumes and determinants from Chap. 2.

We wish to investigate changing variables in a multidimensional integration using the idea of vectors as arrows that delineate a volume of integration in any number of dimensions. So consider an integral

$$I = \iint f(x,y)\,\mathrm{d}x\,\mathrm{d}y\,, \tag{8.143}$$

which is required, for example, to be converted to polar coordinates. The task is more involved than simply expressing the product $\mathrm{d}x\,\mathrm{d}y$ in terms of $\mathrm{d}r$ and $\mathrm{d}\theta$ using $x = r\cos\theta, y = r\sin\theta$, since such a product will produce a sum of $\mathrm{d}r^2, \mathrm{d}r\,\mathrm{d}\theta$, and $\mathrm{d}\theta^2$ terms, which is not appropriate for an integration in polar coordinates. Rather, we wish to tile the plane in a way to suit the r, θ integration. So we need to find an expression only involving $\mathrm{d}r\,\mathrm{d}\theta$, and must tile the $r\theta$-plane with $\mathrm{d}r\,\mathrm{d}\theta$ rectangles in polar coordinates, asking what these look like when transformed to x, y-coordinates. Only in this way can we be assured of not "over integrating", i.e. not over counting the infinitesimal integration measures over the xy-plane; there must be a one-to-one correspondence between the integration areas in each set of coordinates. For small increases $\mathrm{d}r$ and $\mathrm{d}\theta$, the resulting tesselation is shown in Fig. 8.10, which shows a map

$$(x,y) = \boldsymbol{X}(r,\theta) = (r\cos\theta, r\sin\theta)\,. \tag{8.144}$$

The resulting tile on the cartesian plane has sides of lengths $\mathrm{d}r$ and $r\,\mathrm{d}\theta$, giving a total area of $r\,\mathrm{d}r\,\mathrm{d}\theta$, which becomes the corresponding measure for the polar integration:

$$I = \iint f(r\cos\theta, r\sin\theta)\, r\,\mathrm{d}r\,\mathrm{d}\theta\,. \tag{8.145}$$

The area of this tile was calculated by treating it as an infinitesimal rectangle of side lengths $\mathrm{d}r$ and $r\,\mathrm{d}\theta$. But we could also have treated it as a small parallelogram with side vectors $\partial\boldsymbol{X}/\partial r\,\mathrm{d}r$ and $\partial\boldsymbol{X}/\partial\theta\,\mathrm{d}\theta$, and then used the determinant idea of Chap. 2 to calculate the area of this parallelogram. Specifically, (2.46) gives

$$\text{area} = \text{abs}\begin{vmatrix} — & \dfrac{\partial \boldsymbol{X}}{\partial r}\,\mathrm{d}r & — \\ & \vdots & \\ — & \dfrac{\partial \boldsymbol{X}}{\partial \theta}\,\mathrm{d}\theta & — \end{vmatrix} = \text{abs}\begin{vmatrix} \Lambda^x_r & \Lambda^y_r \\ \Lambda^x_\theta & \Lambda^y_\theta \end{vmatrix}\mathrm{d}r\,\mathrm{d}\theta$$

$$= \text{abs}\begin{vmatrix} \cos\theta & \sin\theta \\ -r\sin\theta & r\cos\theta \end{vmatrix}\mathrm{d}r\,\mathrm{d}\theta = r\,\mathrm{d}r\,\mathrm{d}\theta\,. \tag{8.146}$$

The absolute value of the determinant must be used in case we had swapped the order of r and θ, which would have swapped the matrix rows and changed the sign of its determinant.

How is this extended to higher dimensions? The clue lies in the fact that the xy-tile has sides given by the vectors $\partial\boldsymbol{X}/\partial r\ \mathrm{d}r$ and $\partial\boldsymbol{X}/\partial\theta\ \mathrm{d}\theta$. In n dimensions, use the generic variables

$$\begin{aligned} x^{1'},\dots,x^{n'} &\text{ instead of } r,\theta\,, \\ x^1,\dots,x^n &\text{ instead of } x,y\,, \end{aligned} \tag{8.147}$$

where the set $x^1,\dots,x^n$ still uses a euclidean metric, since the theory of determinants from Chap. 2 was only described for that metric. Suppose that an integral

$$I = \int f(x^1,\dots,x^n)\,\mathrm{d}x^1\dots\mathrm{d}x^n \tag{8.148}$$

must be transformed to the new variables $x^{1'}\dots x^{n'}$ related by

$$\boldsymbol{X}(x^{1'},\dots,x^{n'}) = (x^1,\dots,x^n)\,. \tag{8.149}$$

By analogy with the polar coordinate case in Fig. 8.10, the correct integration measure must be the volume of the n-sided parallelepiped with sides in euclidean space of

$$\frac{\partial\boldsymbol{X}}{\partial x^{1'}}\,\mathrm{d}x^{1'}\,,\ \dots,\ \frac{\partial\boldsymbol{X}}{\partial x^{n'}}\,\mathrm{d}x^{n'}. \tag{8.150}$$

Again, by (2.46), the volume of this parallelepiped is

$$\text{abs}\begin{vmatrix} — & \dfrac{\partial \boldsymbol{X}}{\partial x^{1'}}\,\mathrm{d}x^{1'} & — \\ & \vdots & \\ — & \dfrac{\partial \boldsymbol{X}}{\partial x^{n'}}\,\mathrm{d}x^{n'} & — \end{vmatrix} = \text{abs}\begin{vmatrix} \Lambda^1_{1'} & \dots & \Lambda^n_{1'} \\ & \vdots & \\ \Lambda^1_{n'} & \dots & \Lambda^n_{n'} \end{vmatrix}\mathrm{d}x^{1'}\dots\mathrm{d}x^{n'}, \tag{8.151}$$

where the integration is such that the $\mathrm{d}x^{\alpha'}$ are always positive—just as the $\mathrm{d}r, \mathrm{d}\theta$ were in Fig. 8.10. The transpose of the last matrix in (8.151) is the jacobian matrix for this coordinate transformation. We encountered these matrices earlier in (6.38) and (8.90). The determinant of the jacobian matrix is the *jacobian determinant*:

$$\frac{\partial(x^1,\ldots,x^n)}{\partial(x^{1'},\ldots,x^{n'})} \equiv \begin{vmatrix} \Lambda^1_{1'} & \ldots & \Lambda^1_{n'} \\ & \vdots & \\ \Lambda^n_{1'} & \ldots & \Lambda^n_{n'} \end{vmatrix} = \begin{vmatrix} \Lambda^1_{1'} & \ldots & \Lambda^n_{1'} \\ & \vdots & \\ \Lambda^1_{n'} & \ldots & \Lambda^n_{n'} \end{vmatrix} . \tag{8.152}$$

(The fact that the *transpose* of the last matrix in (8.151) is used in the definition of the jacobian matrix is not important; we are only interested in its determinant, and any nonsingular matrix and its transpose have the same determinant anyway.)

So the volume element required in the integration is

$$\left|\frac{\partial(x^1,\ldots,x^n)}{\partial(x^{1'},\ldots,x^{n'})}\right| \mathrm{d}x^{1'}\ldots\mathrm{d}x^{n'}, \tag{8.153}$$

and the integral in (8.148) becomes

$$\boxed{\int f(x^1,\ldots,x^n)\,\mathrm{d}x^1\ldots\mathrm{d}x^n = \int f(x^{1'},\ldots,x^{n'})\left|\frac{\partial(x^1,\ldots,x^n)}{\partial(x^{1'},\ldots,x^{n'})}\right| \mathrm{d}x^{1'}\ldots\mathrm{d}x^{n'},} \tag{8.154}$$

where the notation $f(x^{1'},\ldots,x^{n'})$ is shorthand for what is really a new function, equivalent to writing the $f(r\cos\theta, r\sin\theta)$ of (8.145) as $f(r,\theta)$. (See further the discussion at the start of Sect. 2.9.)

Equation (8.154) was developed using a euclidean metric for the variables $x^1,\ldots,x^n$. But, in fact, it holds quite generally even when these variables do not have a euclidean metric, as can be seen from the behaviour of the jacobian determinant. To prove this, it is first of all straightforward to show that the chain rule of partial derivatives can be written as a product of the relevant jacobian matrices. The reason is simply that the jacobian matrix relating two sets of variables $x^1,\ldots,x^n$ and $x^{1'},\ldots,x^{n'}$ is

$$\left(\Lambda^\alpha_{\beta'}\right) \equiv \begin{bmatrix} \Lambda^1_{1'} & \ldots & \Lambda^1_{n'} \\ & \vdots & \\ \Lambda^n_{1'} & \ldots & \Lambda^n_{n'} \end{bmatrix}, \tag{8.155}$$

so that a multiplication of jacobian matrices for general coordinates $x^\alpha, x^{\beta'}, x^{\mu''}$ can be written as

$$\left(\Lambda^\alpha_{\mu''}\right) = \left(\Lambda^\alpha_{\beta'}\right)\left(\Lambda^{\beta'}_{\mu''}\right), \tag{8.156}$$

from which the chain rule is apparent. In that case, taking the determinant of both sides of this last equation gives the following relation for any completely general coordinates (i.e., none of them need be cartesian):

$$\frac{\partial(x^1,\dots,x^n)}{\partial(x^{1''},\dots,x^{n''})} = \frac{\partial(x^1,\dots,x^n)}{\partial(x^{1'},\dots,x^{n'})}\,\frac{\partial(x^{1'},\dots,x^{n'})}{\partial(x^{1''},\dots,x^{n''})}\,. \tag{8.157}$$

Now, to show that (8.154) holds for any general coordinates, let's use a two-dimensional case as an example, purely to avoid a cumbersome discussion regarding the notation. Suppose an integration over two variables α, β is to be converted to two other variables μ, ν. We can make use of (8.154) and (8.157) by employing cartesians as a kind of guide rail, eventually to be let go. For arbitrary f,

$$\begin{aligned}\int f\,\mathrm{d}\alpha\,\mathrm{d}\beta &\overset{(8.157)}{=\!=} \int f \left|\frac{\partial(\alpha,\beta)}{\partial(x,y)}\right| \left|\frac{\partial(x,y)}{\partial(\alpha,\beta)}\right| \mathrm{d}\alpha\,\mathrm{d}\beta \overset{(8.154)}{=\!=} \int f \left|\frac{\partial(\alpha,\beta)}{\partial(x,y)}\right| \mathrm{d}x\,\mathrm{d}y \\ &\overset{(8.154)}{=\!=} \int f \left|\frac{\partial(\alpha,\beta)}{\partial(x,y)}\right| \left|\frac{\partial(x,y)}{\partial(\mu,\nu)}\right| \mathrm{d}\mu\,\mathrm{d}\nu \overset{(8.157)}{=\!=} \int f \left|\frac{\partial(\alpha,\beta)}{\partial(\mu,\nu)}\right| \mathrm{d}\mu\,\mathrm{d}\nu\,.\end{aligned} \tag{8.158}$$

All reference to x, y has disappeared entirely, so we conclude that, in general, (8.154) holds for any coordinates.

The jacobian determinant $\partial(\cdot)/\partial(\cdot)$ can be written in an alternative and very useful way. As it's related to volumes, we expect it should also be related to the relevant metrics. They can be introduced via

$$g_{\alpha'\beta'} = \Lambda^{\mu}_{\alpha'}\Lambda^{\nu}_{\beta'}\,g_{\mu\nu}\,. \tag{8.159}$$

Convert this expression to a matrix multiplication, which will then directly allow the jacobian determinant to be calculated and related to the metrics. The mixture of raised and lowered indices requires some care, but we have encountered this before in (8.44) and (8.45). So define the following three matrices, remembering that $g_{\alpha'\beta'}$ is really the $\alpha\beta^{\text{th}}$ component of the metric matrix g', and notice that the frame-changing Λ matrix is just the jacobian matrix of (8.152):

$$\begin{aligned}&\text{matrix } g \text{ has } \alpha\beta^{\text{th}} \text{ component } (g)_{\alpha\beta} \equiv g_{\alpha\beta}\,,\\ &\text{matrix } g' \text{ has } \alpha\beta^{\text{th}} \text{ component } (g')_{\alpha\beta} \equiv g_{\alpha'\beta'}\,,\\ &\text{matrix } \Lambda \text{ has } \alpha\beta^{\text{th}} \text{ component } (\Lambda)_{\alpha\beta} \equiv \Lambda^{\alpha}_{\beta'}\,.\end{aligned} \tag{8.160}$$

These allow (8.159) to be written as a matrix multiplication:

$$\begin{aligned}(g')_{\alpha\beta} &= \sum_{\mu\nu}(\Lambda)_{\mu\alpha}\,(\Lambda)_{\nu\beta}\,(g)_{\mu\nu} = \sum_{\mu\nu}(\Lambda^t)_{\alpha\mu}\,(g)_{\mu\nu}\,(\Lambda)_{\nu\beta}\\ &= (\Lambda^t\,g\,\Lambda)_{\alpha\beta}\,,\end{aligned} \tag{8.161}$$

or in other words,

$$g' = \Lambda^t\,g\,\Lambda\,. \tag{8.162}$$

This, the matrix version of (8.159), is very useful because forming its determinant and taking an absolute value produces

$$\left|\frac{\partial(x^1,\ldots,x^n)}{\partial(x^{1'},\ldots,x^{n'})}\right| = |\det \Lambda| = \sqrt{\frac{\det g'}{\det g}}\,. \tag{8.163}$$

This is just what is needed to write the change of variables in (8.154). Also, there is no real ambiguity in writing g to mean the determinant of the metric matrix g; this is universally done. In that case, the integral (8.154) can be written as

$$\boxed{\int f(x^1,\ldots,x^n)\,\mathrm{d}x^1\ldots\mathrm{d}x^n = \int f(x^{1'},\ldots,x^{n'})\sqrt{g'/g}\,\mathrm{d}x^{1'}\ldots\mathrm{d}x^{n'}.} \tag{8.164}$$

We'll use this expression in Sect. 12.8 when integrating over spacetime. An example of it here is when the unprimed coordinates are x, y and the primed coordinates are r, θ. In that case, $g = 1$ and $g' = r^2$, so that $\sqrt{g'/g} = r$, and the polar volume measure is $r\,\mathrm{d}r\,\mathrm{d}\theta$, as expected.

Note that although the volume elements are related by

$$\mathrm{d}x^1\ldots\mathrm{d}x^n \longleftrightarrow \sqrt{g'/g}\,\mathrm{d}x^{1'}\ldots\mathrm{d}x^{n'}, \tag{8.165}$$

we must remember that because they are integration measures, they relate to limits on different coordinates and so need not be equal. This can be seen easily in the polar case: it makes no sense to equate $r\,\mathrm{d}r\,\mathrm{d}\theta$ with $\mathrm{d}x\,\mathrm{d}y$, since the tile of area $r\,\mathrm{d}r\,\mathrm{d}\theta$ in Fig. 8.10 bears no relation to a rectangle in the xy-plane of area $\mathrm{d}x\,\mathrm{d}y$. The tiles correspond with each other, but their areas are not required to be equal and in general won't be, even though some books will treat (8.165) as an equality.

That the square root of the metric appears in relation to higher-dimensional volumes might not be surprising here: we first encountered it back in Sect. 2.4.3, where the Gram matrix was introduced. The Gram matrix as constructed from basis vectors is none other than the metric matrix.

8.8.1 A Final Word: The Cross Product in General Coordinates

In Chap. 2, we saw the Levi-Civita way of writing the cross product in (2.51), where cartesian coordinates were used. But cartesian coordinates in euclidean space have identical basis and cobasis vectors. To begin to make a form of the cross product that is valid in all coordinates, write (2.51) so that it obeys the Einstein summation convention, which just means changing the basis vectors in the sum to cobasis vectors:

$$\operatorname{cross}(\boldsymbol{\alpha}_1,\ldots,\boldsymbol{\alpha}_{n-1}) = \underbrace{\varepsilon_{\mu\ldots\omega}}_{n\text{ indices}}\,\alpha_1^\mu\,\alpha_2^\nu\,\ldots\,\alpha_{n-1}^\psi\,\mathbf{e}^\omega. \tag{8.166}$$

It now becomes apparent that, for general coordinates, we actually might better focus on the *co*basis $\{e^1 \ldots e^n\}$, and of course in general coordinates the basis and cobasis will not be the same. Consider again the $n-1$ linearly independent vectors $\{\boldsymbol{\alpha}_1, \ldots, \boldsymbol{\alpha}_{n-1}\}$ of (2.48) in $\mathbb{E}^n$. This time we will be using unprimed and primed coordinates, so write their coordinate vectors as, e.g., $[\boldsymbol{\alpha}_1], [\boldsymbol{\alpha}_1]'$ in unprimed and primed coordinates (cf. the more relaxed notation of Sect. 2.4, which only used one set of coordinates). Let the unprimed coordinates be cartesian. Because using the cobasis appears to be more reasonable in general coordinates, write

$$\operatorname{cross}(\boldsymbol{\alpha}_1, \ldots, \boldsymbol{\alpha}_{n-1}) \equiv \begin{vmatrix} — & [\boldsymbol{\alpha}_1] & — \\ & \vdots & \\ — & [\boldsymbol{\alpha}_{n-1}] & — \\ e^1 & \ldots & e^n \end{vmatrix}$$

$$= \begin{vmatrix} \alpha_1^1 & \ldots & \alpha_1^n \\ & \vdots & \\ \alpha_{n-1}^1 & \ldots & \alpha_{n-1}^n \\ e^1 & \ldots & e^n \end{vmatrix} = \begin{vmatrix} \Lambda^1_{\mu'}\alpha_1^{\mu'} & \ldots & \Lambda^n_{\mu'}\alpha_1^{\mu'} \\ & \vdots & \\ \Lambda^1_{\mu'}\alpha_{n-1}^{\mu'} & \ldots & \Lambda^n_{\mu'}\alpha_{n-1}^{\mu'} \\ \Lambda^1_{\mu'}e^{\mu'} & \ldots & \Lambda^n_{\mu'}e^{\mu'} \end{vmatrix} . \tag{8.167}$$

We get a hint now of just why a cobasis should be useful in the last row of (8.167): because that way the basis vectors transform in exactly the same way as the vector components in the other rows, which gives the matrices a uniformity without which the analysis would not get very far. The last determinant of (8.167) can now be factored to give

$$\operatorname{cross}(\boldsymbol{\alpha}_1, \ldots, \boldsymbol{\alpha}_{n-1}) \equiv \begin{vmatrix} — & [\boldsymbol{\alpha}_1]' & — \\ & \vdots & \\ — & [\boldsymbol{\alpha}_{n-1}]' & — \\ e^{1'} & \ldots & e^{n'} \end{vmatrix} \begin{vmatrix} \Lambda^1_{1'} & \ldots & \Lambda^n_{1'} \\ & \vdots & \\ \Lambda^1_{n'} & \ldots & \Lambda^n_{n'} \end{vmatrix}$$

$$= \frac{\partial(x^1, \ldots, x^n)}{\partial(x^{1'}, \ldots, x^{n'})} \begin{vmatrix} — & [\boldsymbol{\alpha}_1]' & — \\ & \vdots & \\ — & [\boldsymbol{\alpha}_{n-1}]' & — \\ e^{1'} & \ldots & e^{n'} \end{vmatrix} = (\pm)\sqrt{\frac{g'}{g}} \begin{vmatrix} — & [\boldsymbol{\alpha}_1]' & — \\ & \vdots & \\ — & [\boldsymbol{\alpha}_{n-1}]' & — \\ e^{1'} & \ldots & e^{n'} \end{vmatrix} , \tag{8.168}$$

where the plus sign in front of the last square root is used if the handedness is unchanged by the coordinate transformation (which is usually the case). The unprimed indices are cartesian: $g = 1$, and so the general expression for the n-dimensional cross product becomes

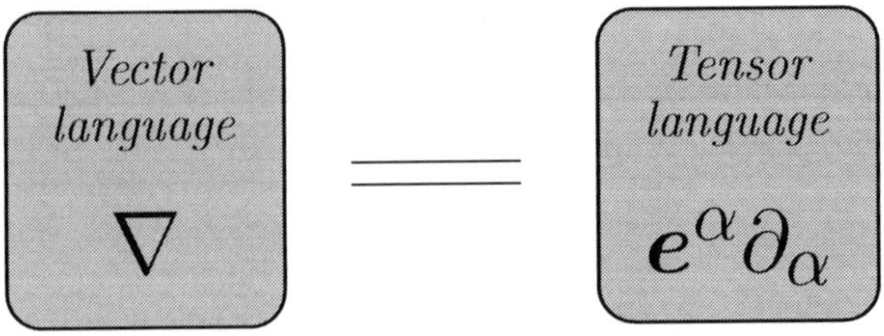

Fig. 8.11. Expressing ∇ over a vector cobasis as $\boldsymbol{e}^\alpha \partial_\alpha$ enables sometimes-convoluted vector expressions to be written in streamlined tensor language.

$$\operatorname{cross}(\boldsymbol{\alpha}_1, \dots, \boldsymbol{\alpha}_{n-1}) = (\pm)\sqrt{g'} \begin{vmatrix} — & [\boldsymbol{\alpha}_1]' & — \\ & \vdots & \\ — & [\boldsymbol{\alpha}_{n-1}]' & — \\ \boldsymbol{e}^{1'} & \dots & \boldsymbol{e}^{n'} \end{vmatrix}$$

$$\overset{(2.33)}{=\!=} (\pm)\sqrt{g'}\, \underbrace{\varepsilon_{\mu' \dots \omega'}}_{n \text{ indices}} \alpha_1^{\mu'} \alpha_2^{\nu'} \dots \alpha_{n-1}^{\psi'} \boldsymbol{e}^{\omega'} . \tag{8.169}$$

Comparing this with the expressions (8.166) and (8.167) for the cartesian case, we see that the Levi-Civita symbol has become (dropping the primes) $\sqrt{g}\, \varepsilon_{\mu\nu\dots\omega}$ in general coordinates. The matrix part of (8.169) will be used in Sect. 8.9.5 when writing the general expression for a curl.

8.9 From Vector Calculus to Tensor Calculus

The standard, sometimes arcane "div-grad-curl" identities of vector calculus can be derived very elegantly and quickly from (8.103) (see Fig. 8.11). This covariant form makes the nabla operator far more useful than the noncovariant cartesian-specific form of (8.84). To convert (8.103) to the more common noncovariant notation that uses normalised basis vectors, we need only write

$$\nabla f = \boldsymbol{e}^\alpha \partial_\alpha f = f_{,\alpha}\, \boldsymbol{e}^\alpha = f_{,\alpha}\, g^{\alpha\beta} \boldsymbol{e}_\beta \,. \tag{8.170}$$

The metric is often diagonal, so that $g^{\alpha\alpha} = 1/g_{\alpha\alpha}$ with all other metric coefficients zero. In that case write, without any implied summation,

$$\begin{aligned} \nabla f &= f_{,\alpha}\, g^{\alpha\alpha} \boldsymbol{e}_\alpha = \frac{f_{,\alpha}}{g_{\alpha\alpha}} \boldsymbol{e}_\alpha \quad \text{(no sum)} \\ &= \frac{f_{,\alpha}}{\sqrt{g_{\alpha\alpha}}} \boldsymbol{e}_{\widehat{\alpha}} \quad \text{(no sum)}. \end{aligned} \tag{8.171}$$

For general orthogonal coordinate systems (i.e. those with a diagonal metric), (8.171) is usually the form appearing in reference books—although as we said earlier, the basis is seldom written, let alone including a caret.

8.9.1 The Divergence in Tensor Notation

Once we are accustomed to the gradient operator ∇, the next such operator to focus on is the divergence:

$$\nabla\cdot\boldsymbol{A} = (\boldsymbol{e}^{\alpha}\,\partial_{\alpha})\cdot(A^{\mu}\,\boldsymbol{e}_{\mu}) = \boldsymbol{e}^{\alpha}\cdot\partial_{\alpha}(A^{\mu}\,\boldsymbol{e}_{\mu}) = \boldsymbol{e}^{\alpha}\cdot(A^{\mu}{}_{,\alpha}\,\boldsymbol{e}_{\mu} + A^{\mu}\,\boldsymbol{e}_{\mu,\alpha})\,. \quad (8.172)$$

At this point, we are faced with the challenge of calculating changes in basis vectors over the coordinate system itself, $\boldsymbol{e}_{\mu,\alpha}$. This is never encountered when using cartesian coordinates due to the fact that $\boldsymbol{e}_x, \boldsymbol{e}_y, \boldsymbol{e}_z$ don't change with position. Changing basis vectors are encountered when Newton's laws are invoked to describe planetary motion. Orbital mechanics is far more easily worked out in polar coordinates and hence uses nonvanishing expressions such as $\boldsymbol{e}_{\theta,r}$. These polar coordinate expressions are particularly straightforward to manipulate because we have the advantage of being able to draw a picture and think geometrically, which is what we did on p. 310 in applying Newton's laws to orbits. The tensor approach comes into its own for more general coordinates.

Let's explore the task of calculating $\boldsymbol{e}_{\mu,\alpha}$ using polar coordinates as a specific example, following two different paths:

1. If the general coordinates can be related to cartesians, then we are in luck; we can express everything in terms of $\boldsymbol{e}_x, \boldsymbol{e}_y, \boldsymbol{e}_z$ since these are constant.
2. But if only the metric is known (as is the case in general relativity), then a more cunning approach will be required.

First Method of Calculating $\boldsymbol{e}_{\mu,\alpha}$: If Related to Cartesians

Focus on calculating $\boldsymbol{e}_{\theta,r}$ in polar coordinates in the plane by referring the problem back to cartesian coordinates with their zero derivatives. Let polar coordinates be unprimed, with cartesian coordinates primed. We have

$$\boldsymbol{e}_{\theta} = \Lambda^{\kappa'}_{\theta}\,\boldsymbol{e}_{\kappa'}\,. \quad (8.173)$$

It's now straightforward to differentiate both sides with respect to r:

$$\boldsymbol{e}_{\theta,r} = \Lambda^{\kappa'}_{\theta,r}\,\boldsymbol{e}_{\kappa'}\,. \quad (8.174)$$

Now that the cartesian basis vectors have served their purpose of being constant in the differentiation, revert to the unprimed indices of the polar basis:

$$\boldsymbol{e}_{\theta,r} = \Lambda^{\kappa'}_{\theta,r}\,\Lambda^{\mu}_{\kappa'}\,\boldsymbol{e}_{\mu} \quad (\kappa' \text{ must be cartesian}). \quad (8.175)$$

It's instructive to write out all the components for this exercise, although we won't do that here. The various partial derivatives were calculated in (8.92) and (8.93). Using these, (8.175) becomes

$$\boldsymbol{e}_{\theta,r} = \boldsymbol{e}_{\theta}/r\,. \quad (8.176)$$

Second Method of Calculating $\boldsymbol{e}_{\mu,\alpha}$: When Only the Metric Is Known

The second approach to calculating $\boldsymbol{e}_{\theta,r}$ doesn't depend on cartesian coordinates, although they will turn up here in the final stage. First, note that (8.175) writes the derivatives of basis vectors as linear combinations over the same basis.

This can be done because the function $\boldsymbol{X}$ maps a space into itself, which is really what we are doing when changing coordinates. In this case, a coordinate change like (8.86) is represented by a function $\boldsymbol{X}\colon\{r,\theta\} \to \{x,y\}$ or $\boldsymbol{X}\colon\mathbb{R}^2 \to \mathbb{E}^2$. In the next chapter, we'll consider cases such as $\boldsymbol{X}\colon\mathbb{R}^2 \to \mathbb{E}^3$ that embed a surface into a euclidean 3-space. In that case, the original two basis vectors will no longer be sufficient to span the 3-space. See (9.18) for the extended version of (8.177).

Writing the derivative of a basis vector as a linear combination over the same basis (and not just cartesian or polar) defines *Christoffel symbols* $\Gamma^{\mu}{}_{\alpha\beta}$:

$$\boldsymbol{e}_{\alpha,\beta} \equiv \Gamma^{\mu}{}_{\alpha\beta}\, \boldsymbol{e}_{\mu}\,. \tag{8.177}$$

In particular, the same analysis that led to (8.175) allows us to write the Christoffel symbols immediately by using cartesian coordinates:

$$\Gamma^{\mu}{}_{\alpha\beta} = \Lambda^{\kappa'}_{\alpha,\beta}\, \Lambda^{\mu}_{\kappa'} \quad (\kappa' \text{ must be cartesian}). \tag{8.178}$$

But, in general, we might not be able to use cartesian coordinates to calculate the Christoffel symbols. By exploiting a particular symmetry, the Christoffel symbols can be expressed in terms of the metric in the following way. First consider the dot product

$$\boldsymbol{e}_{\mu}\cdot\boldsymbol{e}_{\alpha,\beta} = \boldsymbol{e}_{\mu}\cdot\left(\Gamma^{\nu}{}_{\alpha\beta}\,\boldsymbol{e}_{\nu}\right) = \Gamma^{\nu}{}_{\alpha\beta}\, g_{\mu\nu} \equiv \Gamma_{\mu\alpha\beta}\,, \tag{8.179}$$

where as a notational convenience we have lowered the first Christoffel index.

It will be shown in (8.195) that the Christoffel symbols are not tensors: they do not transform under a change of coordinates in the way that a tensor must. Even so, there is nothing to stop us from raising or lowering their indices if that suits us notationally. The $\Gamma_{\mu\alpha\beta}$ are known as *Christoffel symbols of the first kind*, while $\Gamma^{\mu}{}_{\alpha\beta}$ are known as *Christoffel symbols of the second kind*.

Now notice that the order of the α, β indices in (8.179) doesn't matter since

$$\boldsymbol{e}_{\alpha,\beta} = \frac{\partial^2 \boldsymbol{X}}{\partial x^{\beta}\,\partial x^{\alpha}} = \frac{\partial^2 \boldsymbol{X}}{\partial x^{\alpha}\,\partial x^{\beta}} = \boldsymbol{e}_{\beta,\alpha}\,. \tag{8.180}$$

This yields two useful symmetries:

$$\Gamma_{\mu\alpha\beta} = \Gamma_{\mu\beta\alpha}\,, \quad \Gamma^{\mu}{}_{\alpha\beta} = \Gamma^{\mu}{}_{\beta\alpha}\,. \tag{8.181}$$

Now relate the Christoffel symbols to the metric by writing

$$\begin{aligned} g_{\mu\beta,\gamma} &= \partial_\gamma(\boldsymbol{e}_\mu \cdot \boldsymbol{e}_\beta) = \boldsymbol{e}_{\mu,\gamma} \cdot \boldsymbol{e}_\beta + \boldsymbol{e}_\mu \cdot \boldsymbol{e}_{\beta,\gamma} \\ &= \Gamma_{\beta\mu\gamma} + \Gamma_{\mu\beta\gamma} \,. \end{aligned} \tag{8.182}$$

The swapping of indices occurring here gives a hint that if we write out all permutations, we just might arrive at a set of equations that has just enough cancellation to isolate the Christoffel symbols. And that is precisely what happens:

$$g_{\mu\beta,\gamma} - g_{\beta\gamma,\mu} + g_{\gamma\mu,\beta} = 2\Gamma_{\mu\beta\gamma} \,. \tag{8.183}$$

Raising the first Christoffel index produces the required expression

$$\boxed{\Gamma^\alpha{}_{\beta\gamma} = \frac{g^{\alpha\mu}}{2}\left(g_{\mu\beta,\gamma} - g_{\beta\gamma,\mu} + g_{\gamma\mu,\beta}\right),} \tag{8.184}$$

which is highly useful, since it gives the Christoffel symbols for any metric without the need to refer to a cartesian basis. It's worth committing to memory, which is made easier by the cyclic subscripts and symmetrical $+ - +$ signs.

Returning to the original task of calculating $\boldsymbol{e}_{\theta,r}$ requires that we write

$$\boldsymbol{e}_{\theta,r} = \Gamma^\alpha{}_{\theta r}\, \boldsymbol{e}_\alpha \quad (\alpha \text{ is a polar index}). \tag{8.185}$$

Calculating the necessary Christoffel symbols $\Gamma^r{}_{\theta r}, \Gamma^\theta{}_{\theta r}$ via (8.184) requires the polar metric. We did this on p. 284, but will cover it again here. From the polar line element

$$\mathrm{d}\ell^2 = \mathrm{d}r^2 + r^2\mathrm{d}\theta^2, \tag{8.186}$$

we know immediately that

$$g_{rr} = 1\,, \quad g_{r\theta} = g_{\theta r} = 0\,, \quad g_{\theta\theta} = r^2. \tag{8.187}$$

But it's instructive to be more pedestrian about calculating this metric. At this point, because the polar coordinates are defined in terms of cartesians, we *will* need to refer back to x and y. For example,

$$g_{rr} = \boldsymbol{e}_r \cdot \boldsymbol{e}_r = \left(\Lambda^x_r\, \boldsymbol{e}_x + \Lambda^y_r\, \boldsymbol{e}_y\right) \cdot (\text{same}) = 1\,, \tag{8.188}$$

just as we found in (8.26). But (8.188) is none other than the metric tensor transformation:

$$\begin{aligned} g_{rr} &= \Lambda^\alpha_r\, \Lambda^\beta_r\, g_{\alpha\beta} \\ &= (\Lambda^x_r)^2 + (\Lambda^y_r)^2 = 1\,, \quad \text{using (8.92)}. \end{aligned} \tag{8.189}$$

The same approach gives the other metric components in (8.187). Equation (8.184) also requires the inverse of the metric matrix. For the index ordering r, θ, we have

$$g \equiv (g_{\alpha\beta}) = \begin{bmatrix} 1 & 0 \\ 0 & r^2 \end{bmatrix} \quad \text{so} \quad \left(g^{\alpha\beta}\right) \equiv g^{-1} = \begin{bmatrix} 1 & 0 \\ 0 & 1/r^2 \end{bmatrix}, \tag{8.190}$$

so that

$$g^{rr} = 1\,, \quad g^{r\theta} = g^{\theta r} = 0\,, \quad g^{\theta\theta} = 1/r^2\,. \tag{8.191}$$

It now becomes a simple exercise to apply (8.184), giving

$$\Gamma^r{}_{\theta r} = 0\,, \quad \Gamma^\theta{}_{\theta r} = 1/r\,, \tag{8.192}$$

in which case

$$\boldsymbol{e}_{\theta,r} = \Gamma^r{}_{\theta r}\boldsymbol{e}_r + \Gamma^\theta{}_{\theta r}\boldsymbol{e}_\theta = \boldsymbol{e}_\theta/r\,. \tag{8.193}$$

This agrees of course with the first approach in (8.176). This second method of using Christoffel symbols is necessary to know because, in general, there might not be a cartesian basis that can enable the first method on p. 323 to be used. The Christoffel symbols are of fundamental importance to differential geometry in that they encapsulate how basis vectors change over a surface. In the next chapter, we'll see how this enables us to calculate the curvature of that surface, as well as giving meaning to curvature in higher dimensions, where cartesian coordinates might play no role at all. And because gravity is associated with spacetime curvature in general relativity, the Christoffel symbols are ultimately the things that describe the gravitational field in that theory, where again we don't have the luxury of a natural cartesian frame.

8.9.2 Christoffel Symbols for Cartesian Coordinates

What are the Christoffel symbols for cartesian coordinates? Because the basis vectors are constant in space, their derivatives are zero, and so (8.177) implies that the Christoffel symbols must vanish. Alternatively, the cartesian metric tensor is constant and so has zero derivatives, again leading to zero Christoffel symbols from (8.184). We see then that the Christoffel symbols cannot be tensor components because if they were, then being zero in one coordinate set would imply that they were zero in all others (after all, the tensor transformation rules write the components of any tensor as a linear combination of components in another set of coordinates). But we know they are *not* necessarily zero in other coordinates, so they cannot form a tensor.

This nontensorial nature can be quantified by expressing $\boldsymbol{e}_{\alpha,\beta}$ in terms of a general set of (primed) coordinates. Writing

$$\begin{aligned} \Gamma^\varrho{}_{\alpha\beta}\, \boldsymbol{e}_\varrho = \boldsymbol{e}_{\alpha,\beta} &= \left(\Lambda^{\mu'}_\alpha\, \boldsymbol{e}_{\mu'}\right)_{,\beta} \\ &= \left(\Lambda^{\mu'}_{\alpha,\beta} + \Lambda^{\lambda'}_\alpha\, \Lambda^{\gamma'}_\beta\, \Gamma^{\mu'}{}_{\lambda'\gamma'}\right) \boldsymbol{e}_\varrho\, \Lambda^\varrho_{\mu'} \end{aligned} \tag{8.194}$$

means it must follow that

$$\Gamma^\varrho{}_{\alpha\beta} = \Lambda^\varrho_{\mu'}\, \Lambda^{\mu'}_{\alpha,\beta} + \Lambda^\varrho_{\mu'}\, \Lambda^{\lambda'}_\alpha\, \Lambda^{\gamma'}_\beta\, \Gamma^{\mu'}{}_{\lambda'\gamma'}\,. \tag{8.195}$$

The second term in (8.195) is the usual tensor transformation that we would expect if Christoffel symbols were tensors; but the first term in (8.195) is the extra part that muddies the water, ensuring they are not. And indeed the first term is the only one that survives when the unprimed coordinates are cartesian, just as we saw in (8.178). Later, in Sect. 12.3.2, we'll encounter the idea of forcing the Christoffel symbols to be zero by way of a suitable coordinate transformation, an idea that corresponds to a freely falling laboratory. It will bring us as close as possible to "removing" gravity from that laboratory.

Following from (8.177), what is the Christoffel expression for $\boldsymbol{e}^{\alpha}{}_{,\beta}$? It can be calculated by writing $\boldsymbol{e}^{\alpha}{}_{,\beta} = \left(g^{\alpha\mu}\boldsymbol{e}_{\mu}\right)_{,\beta}$ and then expanding by the product rule. But a simpler approach starts with the orthonormality of basis with cobasis to write $\left(\boldsymbol{e}^{\alpha}\cdot\boldsymbol{e}_{\mu}\right)_{,\beta} = 0$. Differentiation of the dot product then gives

$$\boldsymbol{e}^{\alpha}{}_{,\beta}\cdot\boldsymbol{e}_{\mu} = -\boldsymbol{e}^{\alpha}\cdot\boldsymbol{e}_{\mu,\beta} = -\boldsymbol{e}^{\alpha}\cdot\Gamma^{\nu}{}_{\mu\beta}\,\boldsymbol{e}_{\nu} = -\Gamma^{\alpha}{}_{\mu\beta}\,. \tag{8.196}$$

Remembering (2.24) then allows us to write

$$\boldsymbol{e}^{\alpha}{}_{,\beta} = \left(\boldsymbol{e}^{\alpha}{}_{,\beta}\cdot\boldsymbol{e}_{\mu}\right)\boldsymbol{e}^{\mu} = -\Gamma^{\alpha}{}_{\beta\mu}\,\boldsymbol{e}^{\mu}. \tag{8.197}$$

We'll have need of this expression in the next few pages.

We have yet to produce a tensor expression for the divergence (8.172). While developing that equation, we encountered the vector component derivative $A^{\mu}{}_{,\alpha}$. This derivative can be shown not to be a tensor by using just the same sort of direct approach that was used for the Christoffel symbols in (8.194) and (8.195). We also saw the need in the divergence expression (8.172) for differentiating basis vectors. Nonconstant basis vectors are so common in differential geometry and general relativity that the language has been arranged to enable us to forget about them—in a sense it allows us to pretend that the basis vectors are constant, and so carry out calculations with the same sense of abandon as if we were simply using cartesian coordinates. For example, in an elementary physics course we might encounter cartesian expressions such as

$$\frac{\mathrm{d}}{\mathrm{d}t}(t,t^2) = (1,2t)\,. \tag{8.198}$$

In such a course it's not always appropriate to stress that the basis vectors should also be differentiated, and indeed have been—but are constant. Just as a comma subscript denotes partial differentiation, a semicolon subscript is defined to allow us to forget about basis vector derivatives, by defining a *covariant derivative*:

$$\begin{aligned}\frac{\partial \boldsymbol{A}}{\partial x^{\nu}} &= (A^{\mu}\,\boldsymbol{e}_{\mu})_{,\nu} \equiv A^{\mu}{}_{;\nu}\,\boldsymbol{e}_{\mu}\\ &= (A_{\mu}\,\boldsymbol{e}^{\mu})_{,\nu} \equiv A_{\mu;\nu}\,\boldsymbol{e}^{\mu}.\end{aligned} \tag{8.199}$$

These semicoloned components are elements of a tensor. To see this, consider, for example,

$$A^{\alpha'}{}_{;\beta'}\,\boldsymbol{e}_{\alpha'} = \frac{\partial \boldsymbol{A}}{\partial x^{\beta'}} = \Lambda^{\mu}_{\beta'}\,\frac{\partial \boldsymbol{A}}{\partial x^{\mu}} = \Lambda^{\mu}_{\beta'}\,A^{\nu}{}_{;\mu}\,\boldsymbol{e}_{\nu} = \Lambda^{\mu}_{\beta'}\,\Lambda^{\alpha'}_{\nu}\,A^{\nu}{}_{;\mu}\,\boldsymbol{e}_{\alpha'}\,, \tag{8.200}$$

so that

$$A^{\alpha'}{}_{;\beta'} = \Lambda^{\mu}_{\beta'}\,\Lambda^{\alpha'}_{\nu}\,A^{\nu}{}_{;\mu}\,, \tag{8.201}$$

which is, of course, just the transformation that tensor components must obey. The covariant derivative is the key to generalising any derivative expression from cartesian coordinates to arbitrary coordinates. Because the Christoffel symbols are zero in cartesian coordinates, covariant differentiation (semicolons) is idential to partial differentiation (commas) in these coordinates. So, in cartesian coordinates only, we can always change all commas to semicolons. However, such an expression will then become a true tensor expression, and so will be valid in *any* coordinates. Thus, we can convert any derivative expression in cartesian coordinates to general coordinates simply by changing all commas to semicolons. This is often called the *comma-goes-to-semicolon rule.*

The $A^{\mu}{}_{;\nu}$ can be viewed as the components of the derivative of the vector A^{μ}, and we have not needed the basis at all. Naturally, this simplicity comes at a price—the covariant derivative is going to be more complicated than the usual sort. It must incorporate Christoffel symbols to take account of the changing basis:

$$\begin{aligned} A^{\mu}{}_{;\nu}\,\boldsymbol{e}_{\mu} &= A^{\mu}{}_{,\nu}\,\boldsymbol{e}_{\mu} + A^{\mu}\,\boldsymbol{e}_{\mu,\nu} \\ &= A^{\mu}{}_{,\nu}\,\boldsymbol{e}_{\mu} + A^{\mu}\,\Gamma^{\alpha}{}_{\mu\nu}\,\boldsymbol{e}_{\alpha} \\ &= (A^{\mu}{}_{,\nu} + \Gamma^{\mu}{}_{\alpha\nu}A^{\alpha})\,\boldsymbol{e}_{\mu}\,. \end{aligned} \tag{8.202}$$

(The need to swap α and μ indices in this expression is very common in tensor algebra, especially when moving common factors outside parentheses.) Thus we have

$$A^{\mu}{}_{;\nu} = A^{\mu}{}_{,\nu} + \Gamma^{\mu}{}_{\alpha\nu}A^{\alpha}, \tag{8.203}$$

and similarly

$$A_{\mu;\nu} = A_{\mu,\nu} - \Gamma^{\alpha}{}_{\mu\nu}A_{\alpha}\,. \tag{8.204}$$

This more complicated derivative reminds us of what was called "inertial differentiation" in Chap. 4, where we modified the time derivative in (4.43) and (4.44) to take Earth's rotation into account, so that the rotation was automatically and seamlessly included in expressions such as (4.47).

Covariant differentiation can be defined for tensor indices of any order. For scalars it's simply defined to be partial differentiation, since no basis vectors are involved. It must therefore obey a product rule when operating on scalars. Begin with this to write

$$\begin{aligned} (A^{\mu}B_{\mu})_{;\alpha} &\equiv (A^{\mu}B_{\mu})_{,\alpha} \\ &= A^{\mu}{}_{,\alpha}B_{\mu} + A^{\mu}B_{\mu,\alpha}\,. \end{aligned} \tag{8.205}$$

Now add and subtract equal Christoffel terms to write (while taking care to interchange μ and β dummy indices as necessary)

$$\begin{aligned}(A^\mu B_\mu)_{;\alpha} &= \left(A^\mu{}_{,\alpha} + \Gamma^\mu{}_{\alpha\beta} A^\beta\right) B_\mu + A^\beta \left(B_{\beta,\alpha} - \Gamma^\mu{}_{\beta\alpha} B_\mu\right) \\ &= A^\mu{}_{;\alpha} B_\mu + A^\mu B_{\mu;\alpha}\,, \end{aligned} \tag{8.206}$$

so that the covariant derivative also obeys a product rule when operating on vectors.

Covariant derivatives of higher-order tensor components are built upon these few expressions. For example, what is $A^\mu{}_{\nu;\alpha}$? Introduce a tensor basis $\boldsymbol{e}_\mu \boldsymbol{e}^\nu$, and from the definition of covariant differentiation write

$$\begin{aligned} A^\mu{}_{\nu;\alpha}\, \boldsymbol{e}_\mu\, \boldsymbol{e}^\nu &\equiv \left(A^\mu{}_\nu\, \boldsymbol{e}_\mu\, \boldsymbol{e}^\nu\right)_{,\alpha} \\ &= A^\mu{}_{\nu,\alpha}\, \boldsymbol{e}_\mu\, \boldsymbol{e}^\nu + A^\mu{}_\nu \underbrace{\boldsymbol{e}_{\mu,\alpha}}_{\equiv \Gamma^\beta{}_{\mu\alpha} \boldsymbol{e}_\beta} \boldsymbol{e}^\nu + A^\mu{}_\nu\, \boldsymbol{e}_\mu \underbrace{\boldsymbol{e}^\nu{}_{,\alpha}}_{= -\Gamma^\nu{}_{\alpha\mu} \boldsymbol{e}^\mu}\,. \end{aligned} \tag{8.207}$$

Substituting in the Christoffel expressions and factoring out the basis vectors leads to

$$A^\mu{}_{\nu;\alpha} = A^\mu{}_{\nu,\alpha} + \Gamma^\mu{}_{\beta\alpha} A^\beta{}_\nu - \Gamma^\beta{}_{\nu\alpha} A^\mu{}_\beta\,. \tag{8.208}$$

The general rule extends to tensors of any order: every "up" index produces an added Christoffel term that replaces that index with a summation index, and every "down" index calls for a subtracted Christoffel term in the same way.

Covariant derivatives of the metric are always zero:

$$\begin{aligned} g_{\alpha\beta;\gamma} &= g_{\alpha\beta,\gamma} - \Gamma^\mu{}_{\alpha\gamma}\, g_{\mu\beta} - \Gamma^\mu{}_{\beta\gamma}\, g_{\alpha\mu} \\ &\overset{(8.182)}{=\!=} \Gamma_{\alpha\beta\gamma} + \Gamma_{\beta\alpha\gamma} - \Gamma_{\beta\alpha\gamma} - \Gamma_{\alpha\beta\gamma} \\ &= 0\,. \end{aligned} \tag{8.209}$$

Likewise, $g^{\alpha\beta}{}_{;\gamma} = g^\alpha{}_{\beta;\gamma} = 0$. These expressions are useful because they imply that indices can be raised or lowered "within a semicolon", reinforcing the idea that unlike partial differentiation, covariant differentiation produces tensor components:

$$g_{\alpha\mu}\, A^\alpha{}_{;\beta} = \left(g_{\alpha\mu}\, A^\alpha\right)_{;\beta} = A_{\mu;\beta}\,. \tag{8.210}$$

8.9.3 Preparing to Make the Divergence Covariant

Finally, we are in a position to express (8.172) in a covariant form. The covariant derivative simplifies the expressions heavily:

$$\begin{aligned} \nabla\cdot \boldsymbol{A} &= (\boldsymbol{e}^\alpha\, \partial_\alpha)\cdot(A^\mu\, \boldsymbol{e}_\mu) \\ &= A^\mu{}_{;\alpha}\, \boldsymbol{e}_\mu\cdot \boldsymbol{e}^\alpha = A^\mu{}_{;\alpha}\, \delta^\alpha_\mu \\ &= A^\alpha{}_{;\alpha}\,. \end{aligned} \tag{8.211}$$

Why the Covariant Derivative is Useful

The covariant derivative is highly useful in tensor analysis. Two reasons for this are as follows. Firstly, it invests the differentiation process completely in the tensor components, such as in (8.211), thus enabling us to treat basis vectors in a differentiation as if they were constant, just as we are accustomed to doing with the cartesian basis. Secondly, the covariant derivative enables any partial derivative expression calculated in cartesian coordinates to be written in arbitrary coordinates simply by replacing all commas with semicolons. These two ideas are interrelated in expressions such as (8.199).

Has anything been gained by writing $\nabla \cdot \boldsymbol{A}$ in this way, or has the work of calculating the divergence merely been condensed into a slick notation that just has to be "unpacked" whenever we want to do anything with it? It turns out that this notation really has simplified the calculation, due to the following two theorems involving the metric determinant g:

For all vectors A^α:

$$\sqrt{|g|}\, A^\alpha{}_{;\alpha} = \left(\sqrt{|g|}\, A^\alpha\right)_{,\alpha} . \tag{8.212}$$

For all antisymmetric tensors $F^{\alpha\beta}$ *(i.e.* $F^{\alpha\beta} = -F^{\beta\alpha}$*):*

$$\sqrt{|g|}\, F^{\alpha\beta}{}_{;\beta} = \left(\sqrt{|g|}\, F^{\alpha\beta}\right)_{,\beta} . \tag{8.213}$$

Proving these involves the following useful identity:

$$\Gamma^\alpha{}_{\alpha\beta} = \partial_\beta \ln \sqrt{|g|} . \tag{8.214}$$

We prove (8.214) using some index manipulation. Use (8.184) to write the left-hand side as

$$\Gamma^\alpha{}_{\alpha\beta} = \frac{1}{2} g^{\alpha\mu} g_{\alpha\mu,\beta} . \tag{8.215}$$

Differentiating $\ln \sqrt{|g|}$ involves more work. In particular, we need to know $g_{,\beta}$. The answer to this builds usefully on work we've done previously, since the differentiation is accomplished with the aid of some determinant theory. Suppose we have a matrix $A = (A_{\alpha\beta})$ with determinant $|A|$, for which we must calculate $\partial|A|/\partial A_{\alpha\beta}$. Of use is the *adjugate* of A, adj A, a low-profile linear algebra concept seldom found in a physics context. But here we have a perfect use for it. The adjugate is the transposed matrix of the cofactors of A. As an example, the adjugate of a general 2×2 matrix (A_{ij}) is easily written down:

$$\operatorname{adj} \begin{bmatrix} A_{11} & A_{12} \\ A_{21} & A_{22} \end{bmatrix} = \begin{bmatrix} A_{22} & -A_{12} \\ -A_{21} & A_{11} \end{bmatrix} . \tag{8.216}$$

In general, the following result holds for any square matrix A, where I is the same-sized identity:

$$|A|\,I = A \text{ adj } A\,. \tag{8.217}$$

(This is the second important theorem involving cofactors referred to on p. 19.) The proof involves expanding A adj A about a general row, as well as examining what happens when one row of A is replaced by another. Although not difficult, it's omitted as being more suited to a linear algebra course. The elements on the diagonal of $|A|I$ are then

$$|A| = (A \,\text{adj}\, A)_{\alpha\alpha} = \sum_\beta A_{\alpha\beta}\,(\text{adj}\, A)_{\beta\alpha}\,. \tag{8.218}$$

The utility of using the adjugate is that $(\text{adj}\, A)_{\beta\alpha}$ is a function of all but one of the elements of A. Conspicuously, it's *not* a function of $A_{\alpha\beta}$ since, after all, $(\text{adj}\, A)_{\beta\alpha}$ was obtained by crossing $A_{\alpha\beta}$ out from A. With this in mind, we can immediately use (8.218) to write

$$\frac{\partial|A|}{\partial A_{\alpha\beta}} = (\text{adj}\, A)_{\beta\alpha} \overset{(8.217)}{=\!=\!=} |A|\left(A^{-1}\right)_{\beta\alpha}\,. \tag{8.219}$$

This is true for any square matrix at all. In particular, it holds for A^{-1}, leading to

$$\frac{\partial|A|}{\partial(A^{-1})_{\alpha\beta}} = -|A|\,A_{\beta\alpha}\,. \tag{8.220}$$

Specifically, choose A to be the metric $(g_{\alpha\beta})$, with determinant g. In that case (8.219) and (8.220) produce the following covariant results—remembering that the metric and its inverse are symmetric in their indices:

$$\frac{\partial g}{\partial g_{\alpha\beta}} = g\,g^{\alpha\beta} \quad\text{and}\quad \frac{\partial g}{\partial g^{\alpha\beta}} = -g\,g_{\alpha\beta}\,. \tag{8.221}$$

(Notice that because of the useful way in which $g^{\alpha\beta}$ was defined, these equations are covariant for the metric. That is, the up/down position of the indices is the same on both sides of each equation.)

With the required derivatives, we can now evaluate the right-hand side of (8.214). The chain rule, using $\mathrm{d}|x|/\mathrm{d}x = |x|/x$, gives

$$\begin{aligned}\partial_\beta \ln\sqrt{|g|} &= \frac{1}{\sqrt{|g|}}\,\frac{1}{2\sqrt{|g|}}\,\frac{|g|}{g}\,\frac{\partial g}{\partial g_{\alpha\mu}}\,g_{\alpha\mu,\beta}\\ &\overset{(8.221)}{=\!=\!=} \frac{1}{2}\,g^{\alpha\mu}\,g_{\alpha\mu,\beta}\\ &\overset{(8.215)}{=\!=\!=} \Gamma^\alpha{}_{\alpha\beta}\,.\end{aligned} \tag{8.222}$$

This proves (8.214). Equations (8.212) and (8.213) are then proved by expanding the left-hand side "semicolon" part of each in terms of commas and

Christoffel symbols. For the case of $F^{\alpha\beta}$, we need to invoke the antisymmetry of $F^{\alpha\beta}$ and the symmetry of the Christoffel terms and swap two indices; the work is left as a straightforward exercise for the reader. Here we concentrate on giving an example of (8.211) and (8.212) to calculate the divergence for polar coordinates in the plane. The metric for $(x^1, x^2) \equiv (r, \theta)$ is

$$g_{\alpha\beta} = \begin{bmatrix} 1 & 0 \\ 0 & r^2 \end{bmatrix}, \tag{8.223}$$

so that $g = r^2$, and

$$\begin{aligned} \nabla\cdot\boldsymbol{A} = A^{\alpha}{}_{;\alpha} &= \frac{1}{\sqrt{g}}\left(\sqrt{g}\,A^{\alpha}\right)_{,\alpha} = \frac{1}{r}\left(rA^{\alpha}\right)_{,\alpha} \\ &= \frac{1}{r}\left(rA^{r}\right)_{,r} + A^{\theta}{}_{,\theta} \\ &= \frac{1}{r}\left(rA^{\hat{r}}\right)_{,r} + \frac{1}{r}A^{\hat{\theta}}{}_{,\theta}\,, \end{aligned} \tag{8.224}$$

where the last line is written using (8.118) by way of (8.96), since normalised components are commonly used in practice.

8.9.4 The Covariant Laplacian

Much of the work of the last few pages comes together when we require the covariant form of the laplacian ∇^2. Again, the fundamental identity (8.103) gives

$$\begin{aligned} \nabla^2\Phi = \nabla\cdot\nabla\Phi &= (\boldsymbol{e}^{\alpha}\partial_{\alpha})\cdot(\boldsymbol{e}^{\mu}\Phi_{,\mu}) \\ &= \boldsymbol{e}^{\alpha}\cdot\boldsymbol{e}^{\mu}\,\Phi_{,\mu;\alpha} = g^{\alpha\mu}\,\Phi_{,\mu;\alpha} = \Phi^{,\alpha}{}_{;\alpha}\,, \end{aligned} \tag{8.225}$$

which can also be written as $\Phi^{;\alpha}{}_{;\alpha}$ since Φ is a scalar. We infer that $\nabla^2\Phi$ is also a scalar.

As an example of the use of (8.225), calculate the laplacian in spherical polar coordinates r, θ, ϕ. Here

$$g_{\alpha\beta} = \mathrm{diag}(1, r^2, r^2\sin^2\theta)\,, \tag{8.226}$$

so that $g = r^4\sin^2\theta$, and

$$\begin{aligned} \nabla^2\Phi = \Phi^{,\alpha}{}_{;\alpha} &= \frac{1}{\sqrt{g}}\,\sqrt{g}\,\Phi^{,\alpha}{}_{;\alpha} \overset{(8.212)}{=\!=\!=} \frac{1}{\sqrt{g}}\left(\sqrt{g}\,\Phi^{,\alpha}\right)_{,\alpha} \\ &= \frac{1}{\sqrt{g}}\left(\sqrt{g}\,g^{\alpha\beta}\,\Phi_{,\beta}\right)_{,\alpha} \xrightarrow[\text{diagonal metric}]{\text{sum over }\beta,} \sum_{\alpha}\frac{1}{\sqrt{g}}\left(\sqrt{g}\,g^{\alpha\alpha}\,\Phi_{,\alpha}\right)_{,\alpha} \\ &= \sum_{\alpha}\frac{1}{r^2\sin\theta}\left(r^2\sin\theta\,g^{\alpha\alpha}\,\Phi_{,\alpha}\right)_{,\alpha} \\ &= \frac{1}{r^2}\left(r^2\,\Phi_{,r}\right)_{,r} + \frac{(\sin\theta\,\Phi_{,\theta})_{,\theta}}{r^2\sin\theta} + \frac{\Phi_{,\phi\phi}}{r^2\sin^2\theta}\,. \end{aligned} \tag{8.227}$$

It's pleasing that such a complicated result can be derived by the various tensor manipulation theorems—and no Christoffel symbols were required!

8.9.5 The Covariant Curl

Writing the curl covariantly requires some care, but is not difficult based on what we have done up until now. One approach is to apply (8.169) by considering the curl to be the cross product of ∇ and a vector $\boldsymbol{A}$. First, in *any* coordinates,

$$\nabla\times\boldsymbol{A} = \boldsymbol{e}^\alpha\partial_\alpha \times (A^\beta \boldsymbol{e}_\beta) = \boldsymbol{e}_\alpha\partial^\alpha \times (A^\beta \boldsymbol{e}_\beta)\,. \tag{8.228}$$

As we have seen in the last few pages, the basis vectors can be lifted outside the parentheses provided the derivative becomes covariant. A simple way of writing this uses the notation

$$\mathrm{D}_\alpha f \equiv f_{;\alpha}\,, \quad \mathrm{D}^\alpha f \equiv f^{;\alpha}\,. \tag{8.229}$$

Using this, (8.228) becomes

$$\nabla\times\boldsymbol{A} = \boldsymbol{e}_\alpha\partial^\alpha \times (A^\beta \boldsymbol{e}_\beta) = \boldsymbol{e}_\alpha \mathrm{D}^\alpha \times A^\beta \boldsymbol{e}_\beta\,, \tag{8.230}$$

where D^α only acts on A^β. Equation (8.169) then writes this as (with the sign chosen as positive to make the curl right handed)

$$\nabla\times\boldsymbol{A} = \sqrt{g}\begin{vmatrix} \mathrm{D}^1 & \mathrm{D}^2 & \mathrm{D}^3 \\ A^1 & A^2 & A^3 \\ \boldsymbol{e}^1 & \boldsymbol{e}^2 & \boldsymbol{e}^3 \end{vmatrix}. \tag{8.231}$$

It is perhaps more natural to lower the indices on the covariant derivatives, although the resulting expression is more symmetrical if *all* indices are lowered. Do this by realising that metric coefficients can slip in and out of the covariant derivatives with impunity, enabling (8.231) to be written as

$$\nabla\times\boldsymbol{A} = \sqrt{g}\begin{vmatrix} g^{1\alpha}\mathrm{D}_\alpha & g^{2\alpha}\mathrm{D}_\alpha & g^{3\alpha}\mathrm{D}_\alpha \\ g^{1\alpha}A_\alpha & g^{2\alpha}A_\alpha & g^{3\alpha}A_\alpha \\ g^{1\alpha}\boldsymbol{e}_\alpha & g^{2\alpha}\boldsymbol{e}_\alpha & g^{3\alpha}\boldsymbol{e}_\alpha \end{vmatrix} = \sqrt{g}\begin{vmatrix} \mathrm{D}_1 & \mathrm{D}_2 & \mathrm{D}_3 \\ A_1 & A_2 & A_3 \\ \boldsymbol{e}_1 & \boldsymbol{e}_2 & \boldsymbol{e}_3 \end{vmatrix}\begin{vmatrix} g^{11} & g^{21} & g^{31} \\ g^{12} & g^{22} & g^{32} \\ g^{13} & g^{23} & g^{33} \end{vmatrix}$$

$$= \frac{1}{\sqrt{g}}\begin{vmatrix} \mathrm{D}_1 & \mathrm{D}_2 & \mathrm{D}_3 \\ A_1 & A_2 & A_3 \\ \boldsymbol{e}_1 & \boldsymbol{e}_2 & \boldsymbol{e}_3 \end{vmatrix}. \tag{8.232}$$

Equations (8.231) and (8.232) are expressions for the curl in general coordinates. But note that in (8.232) the components are of the form $A_{\alpha;\beta} - A_{\beta;\alpha}$. This expression equals $A_{\alpha,\beta} - A_{\beta,\alpha}$ (because the relevant Christoffel terms cancel due to the symmetry of their lower two indices), so that the expression for the curl with lowered indices is more easily written as

$$\nabla\times\boldsymbol{A} = \frac{1}{\sqrt{g}}\begin{vmatrix} \partial_1 & \partial_2 & \partial_3 \\ A_1 & A_2 & A_3 \\ \boldsymbol{e}_1 & \boldsymbol{e}_2 & \boldsymbol{e}_3 \end{vmatrix}. \tag{8.233}$$

The cancelling of Christoffel terms would not have occurred had we dealt with the $A^{\alpha;\beta} - A^{\beta;\alpha}$ of (8.231) instead; this does not equal $A^{\alpha,\beta} - A^{\beta,\alpha}$ in general. Thus (8.231) is not so simply written using partials $\partial^1, \partial^2, \partial^3$. (We cannot simply raise the indices in (8.233) using the reverse procedure of (8.231) and (8.232) because in doing so, the partials would also act on the metric coefficients, leading to complications.) The antisymmetric expression $A_{\alpha,\beta} - A_{\beta,\alpha}$ is the hallmark of a curl in tensor notation.

A slightly different approach offers further insight into covariant notation for the curl. As in (8.228), start by exploring some of the up/down index combinations:

$$\nabla\times\boldsymbol{A} = \boldsymbol{e}^\alpha\partial_\alpha \times (A^\beta\boldsymbol{e}_\beta) = \boldsymbol{e}^\alpha\times\boldsymbol{e}_\beta\, A^\beta{}_{;\alpha} = \boldsymbol{e}_\alpha\times\boldsymbol{e}_\beta\, A^{\beta;\alpha} = \boldsymbol{e}^\alpha\times\boldsymbol{e}^\beta\, A_{\beta;\alpha}\,. \tag{8.234}$$

The cross product of $\boldsymbol{e}_\alpha$ and $\boldsymbol{e}_\beta$ produces a vector orthogonal to both, so for simplicity we'll restrict the following discussion to orthogonal coordinates (i.e. those with a diagonal metric), specifically those that are right handed, for which $\boldsymbol{e}_{\hat{1}}\times\boldsymbol{e}_{\hat{2}} = \boldsymbol{e}_{\hat{3}}$ with cyclic permutations. But since, a moment ago, we saw the Christoffel symbols cancel for lowered indices, it might be more useful to make the cobasis-vector choice $\boldsymbol{e}^\alpha\times\boldsymbol{e}^\beta$ in (8.234). So translate between basis and cobasis with

$$\boldsymbol{e}^\alpha = g^{\alpha\alpha}\boldsymbol{e}_\alpha \text{ (no sum)} = \frac{1}{g_{\alpha\alpha}}\sqrt{g_{\alpha\alpha}}\,\boldsymbol{e}_{\hat{\alpha}} = \frac{\boldsymbol{e}_{\hat{\alpha}}}{\sqrt{g_{\alpha\alpha}}}\,, \tag{8.235}$$

in which case

$$\boldsymbol{e}^1\times\boldsymbol{e}^2 = \frac{\boldsymbol{e}_3}{\sqrt{g}} \quad \text{(and cyclic permutations)}, \tag{8.236}$$

as well as the usual $\boldsymbol{e}^\alpha\times\boldsymbol{e}^\alpha = \boldsymbol{0}$. The result in (8.236) is written as a basis vector, as opposed to a cobasis vector, since we wish to express the curl here using the basis. (That need not be confusing. All we are doing is choosing to work with either a basis or a cobasis, depending on which is more suitable for whatever we need to calculate.)

In order to collect components together, write the curl in (8.234) as

$$\begin{aligned}\nabla\times\boldsymbol{A} &= (A_{\beta;\alpha} - A_{\alpha;\beta})\,\boldsymbol{e}^\alpha\times\boldsymbol{e}^\beta \quad (\alpha<\beta)\\ &= (A_{\beta,\alpha} - A_{\alpha,\beta})\,\boldsymbol{e}^\alpha\times\boldsymbol{e}^\beta \quad (\alpha<\beta)\,.\end{aligned} \tag{8.237}$$

Rewrite this last expression using (8.236):

$$\begin{aligned}\nabla\times\boldsymbol{A} &= (A_{2,1}-A_{1,2})\,\boldsymbol{e}^1\times\boldsymbol{e}^2 + (A_{3,1}-A_{1,3})\,\boldsymbol{e}^1\times\boldsymbol{e}^3 + (A_{3,2}-A_{2,3})\,\boldsymbol{e}^2\times\boldsymbol{e}^3 \\ &= \frac{1}{\sqrt{g}}\left[(A_{2,1}-A_{1,2})\,\boldsymbol{e}_3 + (A_{3,1}-A_{1,3})\,\boldsymbol{e}_2 + (A_{3,2}-A_{2,3})\,\boldsymbol{e}_1\right] \\ &= \frac{1}{\sqrt{g}}\begin{vmatrix}\partial_1 & \partial_2 & \partial_3 \\ A_1 & A_2 & A_3 \\ \boldsymbol{e}_1 & \boldsymbol{e}_2 & \boldsymbol{e}_3\end{vmatrix}, \quad \text{which is (8.233), as expected.}\end{aligned} \tag{8.238}$$

The use of both basis and cobasis vectors has enabled us to specify different forms for the curl, each having a symmetry of its own, which is all valuable pedagogically.

In Chap. 10, we will find that the electromagnetic potential can be written as a vector A^α, and it will also be convenient to define the *Faraday tensor* $F_{\alpha\beta}$,

$$F_{\alpha\beta} \equiv A_{\beta;\alpha} - A_{\alpha;\beta} = A_{\beta,\alpha} - A_{\alpha,\beta}\,, \tag{8.239}$$

whose components will turn out to be just the electric and magnetic fields. As might be expected, the Faraday tensor appears in the covariant approach to the Lorentz force and Maxwell's equations, a subject we'll explore further in that chapter.

8.10 Exterior Calculus and the Theorems of Stokes and Gauss in Higher Dimensions

Stokes' and Gauss's theorems are the main tools that aid us in developing a physical intuition for Maxwell's equations, since these two theorems enable integrals over lines or surfaces to be converted into integrals over surfaces and volumes, respectively. They are highly adapted to the concepts of *circulation* and *flux*. We met flux in Chap. 6 in the context of rainfall. The flux of rain passing through a surface was defined to be the amount of rain passing through the surface in a given time interval. The flux through a surface is defined more generally for any vector field:

$$\text{flux of } \boldsymbol{A} \text{ through surface } S \equiv \int_S \boldsymbol{A}\cdot\boldsymbol{n}\,\mathrm{d}S\,, \tag{8.240}$$

where $\boldsymbol{n}$ is a unit vector normal to the infinitesimal surface element, which itself has area $\mathrm{d}S$. In contrast, the circulation is a concept we have not needed up until now, but it, too, pervades electromagnetic theory. For any vector field $\boldsymbol{A}$, the circulation around a closed, not necessarily planar loop ℓ in three dimensions is defined to be

$$\text{circulation of } \boldsymbol{A} \text{ around loop } \ell \equiv \oint \boldsymbol{A}\cdot\mathrm{d}\boldsymbol{\ell}\,, \tag{8.241}$$

where $\mathrm{d}\boldsymbol{\ell}$ measures an infinitesimal segment of the loop (using a bold font to show that it's a vector).

We start the discussion of the theorems' generalisation by stating each of them as they are used in electromagnetism. Pictured in Fig. 8.12 on p. 339, Stokes' theorem (which in fact was originally formulated by Kelvin) says that the circulation of $\boldsymbol{A}$ around a loop in three dimensions is equal to the flux of $\nabla\times\boldsymbol{A}$ through *any* surface that the loop encloses:

$$\oint \boldsymbol{A}\cdot\mathrm{d}\boldsymbol{\ell} = \iint_S \nabla\times\boldsymbol{A}\cdot\boldsymbol{n}\,\mathrm{d}S\,. \tag{8.242}$$

Incidentally, this allows a picture of the curl to be formed. If the loop is chosen to be of infinitesimal size and planar, then (8.242) says that

$$\frac{\text{circulation}}{\text{area of loop}} = \text{normal component of } \nabla\times\boldsymbol{A}\,, \tag{8.243}$$

where the normal component is defined in a right-handed sense; that is, if the fingers of the right hand curl around the loop in the direction of increasing ℓ, then the thumb points in the direction of $\boldsymbol{n}$. So the curl gives the circulation per unit area, and it points in the direction normal to the plane in which the circulation is a maximum.

Whereas Stokes' theorem is concerned with circulation around a closed loop, Gauss's theorem says that the flux of $\boldsymbol{A}$ through a closed surface S is equal to the divergence of $\boldsymbol{A}$ integrated over the enclosed volume V:

$$\iint_S \boldsymbol{A}\cdot\boldsymbol{n}\,\mathrm{d}S = \iiint_V \nabla\cdot\boldsymbol{A}\,\mathrm{d}V\,. \tag{8.244}$$

We can picture the divergence by considering an infinitesimal volume. For this, Gauss's theorem says

$$\frac{\text{total flux out of surface}}{\text{volume enclosed}} = \nabla\cdot\boldsymbol{A}\,, \tag{8.245}$$

so that $\nabla\cdot\boldsymbol{A}$ gives the amount of flux emanating from a closed surface per unit volume. For example, one of Maxwell's equations says that the divergence of the magnetic field is $\nabla\cdot\boldsymbol{B}=0$, which means, from (8.245), that the magnetic flux coming out of any closed surface is equal to that going in, since the different directions contribute opposite signs to the total, and the total emanating from the volume enclosed is zero. Thus there are no sources of a magnetic field in Maxwell's theory, and this gives rise to a picture of unbroken curves representing that field. Of course, the remaining equations of Maxwell are interpreted in similar ways.

Stokes' and Gauss's theorems are actually two cases of a more general theorem that applies to higher dimensions. We will state the theorem, while leaving its proof to advanced texts and instead focussing on showing how it reduces to Stokes' and Gauss's theorems as special cases. The theorem is not so much quoted for its own sake; instead, by showing how Stokes' and

The Magic of Electromagnetic Field Lines

A common tutorial question in introductory electromagnetism asks the student to show that the electric field due to a uniformly charged flat plate of infinite extent is independent of the distance from the plate. The student draws electric field lines emanating from the plate and reasons that, due to symmetry, the only way the lines can "flow" is perpendicular to the plate. Since the strength of the field equals the areal density of the lines, the field must be a constant everywhere outside the plate.

It's the right answer, but what an answer! Apparently based on nothing more than a simple sketch, something quite astounding has been inferred about the electric field. Where did the magic happen?

The magic had happened long before the student ever sat down. The only way the usual idea of field lines can be applied to a field in the sense of emanating from a point is if the field falls off as an inverse square. That happens to be the case for the electric field, of course. Close to the infinite flat plate, the forces on a test charge caused by charges in the plate are strong but are mostly cancelled by being set against each other. Far from the plate, the forces are weaker but mostly act in the same direction. It's a marvellous special property of an inverse-square law that these things are exactly balanced to the effect that the field is the same both near and far from the plate. And the language of electromagnetism, with its pictures of field lines and vectors, embodies this magic property so very simply that we don't even know it's happening. (The box on p. 498 discusses another example of how special the inverse-square law is.)

In contrast, field lines cannot quite be drawn for a strong gravitational field because the field is stronger than inverse square near the source of the gravity, as we'll see in Chap. 12. (On a newtonian level, inverse-square forces give rise to nonprecessing orbits, but the tiny amount of "extra" gravity close to the Sun is what slowly pulls Mercury's orbit around to give its well-studied precession.) If we insist on drawing field lines for the Sun's gravity, we might have to draw just a few more of them closer to the Sun (but then how do they terminate?), or perhaps modify their meaning in some way by giving them some thickness or some structure that embodies the right strength of gravity. It wouldn't be easy, but it *might* just be able to be done with the right imagination.

Gauss's theorems result, we will gain insight into the techniques of this area of mathematics, called *exterior calculus*.

Before doing so, two important points must be noted. The first is that, apart from scalars and vectors, tensors of higher order used by the theorem must be antisymmetric. Why is antisymmetry so important? Because when dealing with volumes and areas, antisymmetry appears from the very beginning, in the properties of the determinant that we saw in Chap. 2.

We've come across antisymmetry already in the previous section, in the case of a curl, where $F_{\alpha\beta} = -F_{\beta\alpha}$. Similarly, antisymmetry for the lowered components of higher-order tensors means that all index permutations introduce a minus sign. So, for example, a third-order tensor $\boldsymbol{F}$ is said to be antisymmetric if

$$F_{\alpha\beta\gamma} = -F_{\alpha\gamma\beta} = +F_{\beta\gamma\alpha} = \dots \,. \tag{8.246}$$

Although any tensor $\boldsymbol{T}$ (not necessarily antisymmetric) can always be written over a cobasis as $\boldsymbol{T} = T_{\alpha\beta\gamma}\, \mathbf{e}^\alpha \mathbf{e}^\beta \mathbf{e}^\gamma$, for antisymmetric tensors, we can make use of expressions such as (8.75) and (8.76) to use wedge products of cobasis vectors as a basis—and these are just the multivectors of Sect. 2.4.4. For example, in the case of $\boldsymbol{F}$ in (8.246), write

$$\begin{aligned}
\boldsymbol{F} &= F_{\alpha\beta\gamma}\, \mathbf{e}^\alpha \mathbf{e}^\beta \mathbf{e}^\gamma \quad \text{(true for any tensor)} \\
&= {}^1\!/_{3!} \left(F_{\alpha\beta\gamma}\, \mathbf{e}^\alpha \mathbf{e}^\beta \mathbf{e}^\gamma + F_{\alpha\gamma\beta}\, \mathbf{e}^\alpha \mathbf{e}^\gamma \mathbf{e}^\beta + \cdots \right) \quad \text{(just relabelling indices)} \\
&= {}^1\!/_{3!} \left(F_{\alpha\beta\gamma}\, \mathbf{e}^\alpha \mathbf{e}^\beta \mathbf{e}^\gamma - F_{\alpha\beta\gamma}\, \mathbf{e}^\alpha \mathbf{e}^\gamma \mathbf{e}^\beta + \cdots \right) \quad \text{(using antisymmetry)} \\
&= {}^1\!/_{3!}\, F_{\alpha\beta\gamma}\, \mathbf{e}^\alpha \wedge \mathbf{e}^\beta \wedge \mathbf{e}^\gamma \,.
\end{aligned} \tag{8.247}$$

That is, $\boldsymbol{F}$ can be expanded over a trivector basis $\mathbf{e}^\alpha \wedge \mathbf{e}^\beta \wedge \mathbf{e}^\gamma$. This idea allows a general antisymmetric tensor to be written as a multivector. They are the same thing.[1]

In contrast, users of one-forms will expand $\boldsymbol{F}$ over a basis of one-forms, writing each line of (8.247) using one-forms ω^α instead of cobasis vectors $\mathbf{e}^\alpha$. Thus they consider wedge products $\omega^\alpha \wedge \omega^\beta \wedge \omega^\gamma$, calling these *three-forms*. In the language of forms, the curl $F_{\alpha\beta}$ of the last section, or really the associated tensor $\boldsymbol{F}$, is referred to as a two-form. (Unfortunately, the word "form" can be misleading; the first and second "fundamental forms" of differential geometry in the next chapter are both symmetric tensors. But when the word "form" is used in the current context, it signifies an antisymmetric tensor.)

The second important point to note is that the generalised Stokes–Gauss theorem involves a more sophisticated version of the curl that uses a wedge product instead of a cross product. In Sect. 2.4.4, we saw the similarity between $\boldsymbol{a} \wedge \boldsymbol{b}$ and $\boldsymbol{a} \times \boldsymbol{b}$, noting that $\boldsymbol{a} \wedge \boldsymbol{b}$ combines with two vectors, while $\boldsymbol{a} \times \boldsymbol{b}$ can only be dotted with one. Although the wedge and cross products are essentially the same, the cross product is conventionally reserved for vectors only. Likewise, here we'll write the curl as $\nabla \wedge \boldsymbol{A}$ instead of $\nabla \times \boldsymbol{A}$. The generalised curl $\nabla \wedge$ is so important in forming a geometrical view of tensors that it has been given the name *exterior derivative*. It is, in fact, nothing more than a higher-dimensional version of the curl and divergence operators.

[1] Note that while a four-component antisymmetric tensor might be called a quadrivector, it is definitely not a four-vector, since a four-vector is described by just one index, not four! Similarly, an n-component antisymmetric tensor is more meaningfully called an n-multivector rather than an n-vector.

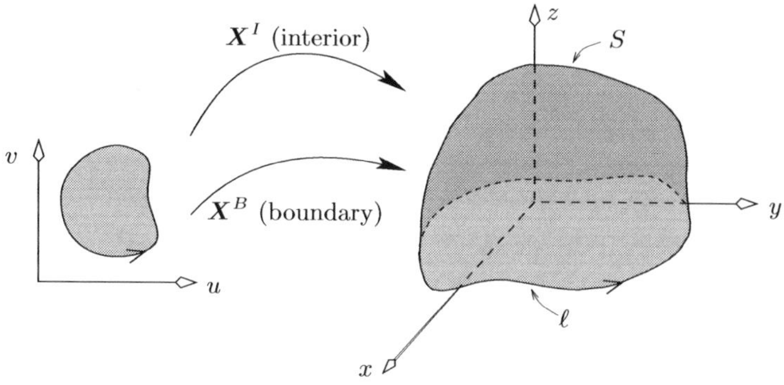

Fig. 8.12. Stokes' theorem is concerned with equating the circulation of a vector field around the loop ℓ with the flux of the field through any surface S enclosed by the loop. A real calculation requires us to construct the boundary curve and its interior (the surface) by way of some parametrisation u, v using two maps, one determining the boundary and the other the interior.

The generalised Stokes–Gauss theorem rests on our being aware of the maps that specify the various contours, surfaces, and volumes being integrated over. An example for Stokes' theorem is shown in Fig. 8.12. The boundary-creating map $\boldsymbol{X}^B$ takes a loop parametrised by u into a not necessarily planar boundary loop in euclidean 3-space. The domain loop (in the uv-plane) needs just one parameter to specify it, since v is a function of u in this case. The interior-creating map, $\boldsymbol{X}^I$, uses u, v as independent parameters and takes the encircled region in the uv-plane into a 2-surface interior to the loop in the same euclidean 3-space. The maps are related by

$$\boldsymbol{X}^B(u) \equiv \boldsymbol{X}^I\Big(u, v(u)\Big)\,, \tag{8.248}$$

where $v = v(u)$ describes the boundary in the uv-plane. The 2-surface S can be any surface bounded by the loop; mathematically, this means that the surface integral (i.e. the flux) is independent of the parametrisation $\boldsymbol{X}^I$. The loop and surface in the euclidean 3-space are integrated over in (8.242).

Suppose we are given an n-component antisymmetric tensor $\boldsymbol{F}_n$ along with the following maps that take $\boldsymbol{u}$-space into $\boldsymbol{x}$-space:

$$\begin{aligned} \boldsymbol{X}^B\colon u^1, \dots, u^n &\longrightarrow x^1, \dots, x^{m \geqslant n+1}\,, \\ \boldsymbol{X}^I\colon u^1, \dots, u^{n+1} &\longrightarrow x^1, \dots, x^m\,. \end{aligned} \tag{8.249}$$

These maps, along with the parameters $u^1, \dots, u^{n+1}$, define basis vectors in $\boldsymbol{x}$-space:

$$\boldsymbol{e}_i^B \equiv \frac{\partial \boldsymbol{X}^B}{\partial u^i}\ (i = 1 \dots n)\,, \quad \boldsymbol{e}_i^I \equiv \frac{\partial \boldsymbol{X}^I}{\partial u^i}\ (i = 1 \dots n+1)\,. \tag{8.250}$$

So, $\boldsymbol{X}^B$ maps an n-dimensional boundary from $\boldsymbol{u}$-space into $\boldsymbol{x}$-space, and $\boldsymbol{X}^I$ maps its $n+1$-dimensional interior into $\boldsymbol{x}$-space. (We'll see examples of these in a moment.) The generalised Stokes–Gauss theorem states that

$$\boxed{\int\limits_{n\text{-dim. boundary}} \boldsymbol{F}_n\left(\boldsymbol{e}_1^B, \ldots, \boldsymbol{e}_n^B\right) \mathrm{d}u^1 \ldots \mathrm{d}u^n = \int\limits_{n+1\text{-dim. interior}} (\nabla \wedge \boldsymbol{F}_n)\left(\boldsymbol{e}_1^I, \ldots, \boldsymbol{e}_{n+1}^I\right) \mathrm{d}u^1 \ldots \mathrm{d}u^{n+1},} \tag{8.251}$$

in which the left-hand side really has n integral signs and the right-hand side has $n+1$ integral signs. This can be rewritten by using the idea of the dot product for wedges in (2.65) and (2.66):

$$\int\limits_{n\text{-dim. boundary}} \boldsymbol{F}_n \cdot \left(\boldsymbol{e}_1^B \wedge \cdots \wedge \boldsymbol{e}_n^B\right) \mathrm{d}u^1 \ldots \mathrm{d}u^n = \int\limits_{n+1\text{-dim. interior}} (\nabla \wedge \boldsymbol{F}_n) \cdot \left(\boldsymbol{e}_1^I \wedge \cdots \wedge \boldsymbol{e}_{n+1}^I\right) \mathrm{d}u^1 \ldots \mathrm{d}u^{n+1}. \tag{8.252}$$

Let's see how it all works by studying the cases of $n = 0, 1$, and 2.

$n = 0$: *The Line Integral.* This is a special case since (8.249) demands $n > 0$. But the generalised Stokes–Gauss theorem is still meaningful here. It equates the "integral" over a 0-dimensional boundary (really no integral at all!) with the line integral along a curve. The 0-order tensor $\boldsymbol{F}_0$ is simply a scalar function, which we'll call f; the notion of antisymmetry is irrelevant here. Since the boundary is 0-dimensional, there is no map $\boldsymbol{X}^B$ and the left-hand side of (8.251) has no integral signs. It is to be interpreted as the increase in the function over the 0-dimensional boundary, which is the two end points. The map $\boldsymbol{X}^I$: $u \to x, y, z$ creates a 1-dimensional "interior" (a curve), so that the right-hand side of (8.251) or (8.252) is an integral along this curve, and the theorem becomes

$$\Delta f = \int\limits_{\text{curve}} (\nabla \wedge f) \cdot \boldsymbol{e}_u^I \, \mathrm{d}u \, . \tag{8.253}$$

What is $\nabla \wedge f$? Since f is a scalar, the wedge is superfluous. Write $\nabla \wedge f = \nabla f$, while $\boldsymbol{e}_u^I \, du$ is more commonly called $\mathrm{d}\boldsymbol{\ell}$, a step along the curve in xyz-space. Thus

$$(\nabla \wedge f) \cdot \boldsymbol{e}_u^I \, \mathrm{d}u = \nabla f \cdot \mathrm{d}\boldsymbol{\ell} \, , \tag{8.254}$$

and finally, (8.253) becomes

$$\Delta f = \int\limits_{\text{curve}} \nabla f \cdot \mathrm{d}\boldsymbol{\ell} \, , \tag{8.255}$$

which is the familiar expression for a line integral.

$n = 1$: *Stokes' Theorem.* The first-order tensor $\boldsymbol{F}_1$ is now a vector, so call it $\boldsymbol{A}$. The maps are $\boldsymbol{X}^B\colon u \to x, y, z$, creating a 1-dimensional boundary, and $\boldsymbol{X}^I\colon u, v \to x, y, z$, giving the 2-surface interior to the boundary. This surface can be embedded in two or three dimensions in xyz-space. Here and in Fig. 8.12, we have chosen three dimensions for the embedding, but if the boundary and surface are embedded in the plane, then *Green's theorem* results, which is thus a reduced version of Stokes' theorem.

We have seen the left-hand side of (8.251) already in the $n = 0$ case: it's the circulation of $\boldsymbol{A}$ around the *closed* loop boundary. Thus, (8.251) says

$$\oint \boldsymbol{A}\cdot \mathrm{d}\boldsymbol{\ell} = \iint (\nabla \wedge \boldsymbol{A}) \left(\boldsymbol{e}_u^I, \boldsymbol{e}_v^I\right) \mathrm{d}u\, \mathrm{d}v\,. \tag{8.256}$$

As for $\nabla \wedge \boldsymbol{A}$, this combines with two basis vectors and so is better expressed in terms of a cobasis. This eliminates the need to deal with the metric (which would happen if a basis were used exclusively, since then we'd have expressions such as $\boldsymbol{e}_\alpha \cdot \boldsymbol{e}_\beta = g_{\alpha\beta}$). So write $\boldsymbol{A}$ over a cobasis, using Greek indices for $\boldsymbol{x}$-space:

$$\begin{aligned}\nabla \wedge \boldsymbol{A} &= \boldsymbol{e}^\alpha \partial_\alpha \wedge \left(A_\beta\, \boldsymbol{e}^\beta\right) = A_{\beta;\alpha}\, \boldsymbol{e}^\alpha \wedge \boldsymbol{e}^\beta \\ &= A_{\beta;\alpha} \left(\boldsymbol{e}^\alpha \boldsymbol{e}^\beta - \boldsymbol{e}^\beta \boldsymbol{e}^\alpha\right) = \left(A_{\beta;\alpha} - A_{\alpha;\beta}\right) \boldsymbol{e}^\alpha \boldsymbol{e}^\beta \\ &= \left(A_{\beta,\alpha} - A_{\alpha,\beta}\right) \boldsymbol{e}^\alpha \boldsymbol{e}^\beta = A_{\beta,\alpha}\, \boldsymbol{e}^\alpha \wedge \boldsymbol{e}^\beta.\end{aligned} \tag{8.257}$$

In other words,

$$\begin{aligned}\nabla \wedge \boldsymbol{A} &= \left(A_{z,y} - A_{y,z}\right) \boldsymbol{e}^y \wedge \boldsymbol{e}^z + \left(A_{x,z} - A_{z,x}\right) \boldsymbol{e}^z \wedge \boldsymbol{e}^x \\ &\quad + \left(A_{y,x} - A_{x,y}\right) \boldsymbol{e}^x \wedge \boldsymbol{e}^y \\ &\equiv (\nabla \times \boldsymbol{A}) \cdot \left(\boldsymbol{e}^y \wedge \boldsymbol{e}^z,\ \boldsymbol{e}^z \wedge \boldsymbol{e}^x,\ \boldsymbol{e}^x \wedge \boldsymbol{e}^y\right),\end{aligned} \tag{8.258}$$

where the last line is written for notational convenience using the dot product notation, although the three wedge products certainly do not form a real vector. To calculate $(\nabla \wedge \boldsymbol{A}) \left(\boldsymbol{e}_u^I, \boldsymbol{e}_v^I\right)$, we need, e.g.,

$$\boldsymbol{e}^y \wedge \boldsymbol{e}^z \left(\boldsymbol{e}_u^I, \boldsymbol{e}_v^I\right) = \begin{vmatrix} \boldsymbol{e}^y \cdot \boldsymbol{e}_u^I & \boldsymbol{e}^y \cdot \boldsymbol{e}_v^I \\ \boldsymbol{e}^z \cdot \boldsymbol{e}_u^I & \boldsymbol{e}^z \cdot \boldsymbol{e}_v^I \end{vmatrix} = \begin{vmatrix} \left(\boldsymbol{e}_u^I\right)^y & \left(\boldsymbol{e}_v^I\right)^y \\ \left(\boldsymbol{e}_u^I\right)^z & \left(\boldsymbol{e}_v^I\right)^z \end{vmatrix} = \left(\boldsymbol{e}_u^I \times \boldsymbol{e}_v^I\right)^x, \tag{8.259}$$

and similarly for cyclic permutations of x, y, z. Hence

$$(\nabla \wedge \boldsymbol{A}) \left(\boldsymbol{e}_u^I, \boldsymbol{e}_v^I\right) = (\nabla \times \boldsymbol{A}) \cdot \left(\boldsymbol{e}_u^I \times \boldsymbol{e}_v^I\right), \tag{8.260}$$

which we might also have written down straightaway by recalling (2.63). Now, remember that

$$\boldsymbol{e}_u^I = \frac{\partial \boldsymbol{X}^I}{\partial u} = \left(\frac{\partial x}{\partial u}, \frac{\partial y}{\partial u}, \frac{\partial z}{\partial u}\right) \tag{8.261}$$

and similarly for $\boldsymbol{e}_v^I$, so

$$\boldsymbol{e}_u^I \times \boldsymbol{e}_v^I \,\mathrm{d}u\,\mathrm{d}v = \left(\frac{\partial x}{\partial u}, \frac{\partial y}{\partial u}, \frac{\partial z}{\partial u}\right) \times \left(\frac{\partial x}{\partial v}, \frac{\partial y}{\partial v}, \frac{\partial z}{\partial v}\right) \mathrm{d}u\,\mathrm{d}v$$
$$= \left(\frac{\partial(y,z)}{\partial(u,v)}, \frac{\partial(z,x)}{\partial(u,v)}, \frac{\partial(x,y)}{\partial(u,v)}\right) \mathrm{d}u\,\mathrm{d}v\,. \tag{8.262}$$

But notice that this is precisely equal to the product of the normal vector $\boldsymbol{n}$ to the surface in $\boldsymbol{x}$-space and an infinitesimal area. To see this, make $\boldsymbol{e}_u^I, \boldsymbol{e}_v^I, \boldsymbol{n}$ a right-handed set, and write such a product as

$$\boldsymbol{n}\,\mathrm{d}S \equiv \frac{\boldsymbol{e}_u^I \times \boldsymbol{e}_v^I}{\left|\boldsymbol{e}_u^I \times \boldsymbol{e}_v^I\right|} \left|\boldsymbol{e}_u^I\,\mathrm{d}u \times \boldsymbol{e}_v^I\,\mathrm{d}v\right| = \boldsymbol{e}_u^I \times \boldsymbol{e}_v^I \,\mathrm{d}u\,\mathrm{d}v\,. \tag{8.263}$$

In that case, (8.260) becomes

$$(\nabla \wedge \boldsymbol{A})\left(\boldsymbol{e}_u^I, \boldsymbol{e}_v^I\right) \mathrm{d}u\,\mathrm{d}v = \nabla \times \boldsymbol{A} \cdot \boldsymbol{n}\,\mathrm{d}S\,, \tag{8.264}$$

allowing (8.256) to be written more recognisably as Stokes' theorem (8.242).

In passing, note that, for example, (8.259) and (8.262) give

$$\boldsymbol{e}^y \wedge \boldsymbol{e}^z \left(\boldsymbol{e}_u^I, \boldsymbol{e}_v^I\right) \mathrm{d}u\,\mathrm{d}v = \frac{\partial(y,z)}{\partial(u,v)}\,\mathrm{d}u\,\mathrm{d}v = \text{a signed area}, \tag{8.265}$$

with a handedness-dependent sign. This signed area is independent of any parametrisation when integrated, and so is usually simply called $\mathrm{d}y \wedge \mathrm{d}z$. Also, if x, y, z and u, v have the same handedness (which will almost always be the case), then the sign of this area is $+1$, and it can be written simply as $\mathrm{d}y\,\mathrm{d}z$.

$n = 2$: Gauss's Theorem. Here we are integrating a second-order tensor $\boldsymbol{F}_2$ and now must pay attention to the required antisymmetry. Expand over a cobasis, writing $\boldsymbol{F}_2 = F_{\alpha\beta}\,\boldsymbol{e}^\alpha \boldsymbol{e}^\beta$ with $F_{\alpha\beta} = -F_{\beta\alpha}$. A cobasis has been used to make combining with a basis easier, just as we did for Stokes' theorem.

The maps $\boldsymbol{X}^B, \boldsymbol{X}^I$ now deal with a closed 2-dimensional boundary enclosing a 3-surface in uvw-space, and this surface can be embedded in three or more dimensions in $\boldsymbol{x}$-space. We will stay with three dimensions in $\boldsymbol{x}$-space to produce Gauss's theorem. The maps are

$$\boldsymbol{X}^B\colon u, v \longrightarrow x, y, z\,, \quad \boldsymbol{X}^I\colon u, v, w \longrightarrow x, y, z\,, \tag{8.266}$$

with $\boldsymbol{X}^B(u,v) \equiv \boldsymbol{X}^I\big(u, v, w(u,v)\big)$, since w is constrained to be a function of u, v on the boundary. The Stokes–Gauss theorem (8.251) becomes

$$\iint \boldsymbol{F}_2\left(\boldsymbol{e}_u^B, \boldsymbol{e}_v^B\right) \mathrm{d}u\,\mathrm{d}v = \iiint (\nabla \wedge \boldsymbol{F}_2)\left(\boldsymbol{e}_u^I, \boldsymbol{e}_v^I, \boldsymbol{e}_w^I\right) \mathrm{d}u\,\mathrm{d}v\,\mathrm{d}w\,. \tag{8.267}$$

All of this requires careful unravelling. Start with the left-hand side of (8.267). Again using Greek indices for $\boldsymbol{x}$-space, write

$$\begin{aligned}\boldsymbol{F}_2 &= F_{\alpha\beta}\,\mathbf{e}^\alpha\mathbf{e}^\beta = {}^1\!/_2\,F_{\alpha\beta}\,\mathbf{e}^\alpha\mathbf{e}^\beta + {}^1\!/_2\,F_{\beta\alpha}\,\mathbf{e}^\beta\mathbf{e}^\alpha = {}^1\!/_2\,F_{\alpha\beta}\left(\mathbf{e}^\alpha\mathbf{e}^\beta - \mathbf{e}^\beta\mathbf{e}^\alpha\right)\\ &= {}^1\!/_2\,F_{\alpha\beta}\,\mathbf{e}^\alpha\wedge\mathbf{e}^\beta = {}^1\!/_2\,(F_{xy}\,\mathbf{e}^x\wedge\mathbf{e}^y + F_{yx}\,\mathbf{e}^y\wedge\mathbf{e}^x + \cdots)\\ &= \underbrace{(F_{yz},\, F_{zx},\, F_{xy})}_{\equiv\,\boldsymbol{G}}\cdot(\mathbf{e}^y\wedge\mathbf{e}^z,\, \mathbf{e}^z\wedge\mathbf{e}^x,\, \mathbf{e}^x\wedge\mathbf{e}^y)\,,\end{aligned} \tag{8.268}$$

where we have used the antisymmetry of $F_{\alpha\beta}$ as necessary. Repeating the calculations we did for Stokes' theorem allows the left-hand side of (8.267) to be written as

$$\iint \boldsymbol{F}_2\left(\mathbf{e}_u^B, \mathbf{e}_v^B\right) \mathrm{d}u\,\mathrm{d}v = \iint \boldsymbol{G}\cdot\boldsymbol{n}\,\mathrm{d}S\,. \tag{8.269}$$

Although it seems that only three components of $F_{\alpha\beta}$ have been singled out to make $\boldsymbol{G}$, they do in fact encapsulate all of the information in $F_{\alpha\beta}$ owing to its antisymmetry. The integral in (8.269) is now a flux through a closed surface.

Unravelling the right-hand side of (8.267) presents a little more work. It follows much the same ideas as in Stokes' theorem except there are now more indices to deal with, so we'll search for a more general method of handling the manipulations. Write

$$\nabla\wedge\boldsymbol{F}_2 = \mathbf{e}^\gamma\partial_\gamma\wedge\left({}^1\!/_2\,F_{\alpha\beta}\,\mathbf{e}^\alpha\wedge\mathbf{e}^\beta\right) = {}^1\!/_2\,F_{\alpha\beta;\gamma}\,\mathbf{e}^\gamma\wedge\mathbf{e}^\alpha\wedge\mathbf{e}^\beta. \tag{8.270}$$

As occurred in (8.257), the semicolon here also can be replaced with a comma. This is a very important fact that we'll take the time to prove, since it sheds light on the index manipulations that are typical of exterior calculus. First, define, for any tensor (say third order as an example),

$$T_{[abc]} \equiv T_{abc} - T_{bac} + T_{bca} - \cdots\,, \tag{8.271}$$

where the sum is over all signed permutations. We will prove the following:

$$T_{[123\ldots n;\,n+1]} = T_{[123\ldots n,\,n+1]}\,. \tag{8.272}$$

Use the Levi-Civita permutation symbol ε from Chap. 2, but write it with indices raised in order to use the summation convention. (They have not been raised with any metric: it's still the case that $\varepsilon^{123\ldots} \equiv 1$.) The left-hand side of (8.272) is

$$\varepsilon^{\alpha\beta\ldots\omega}\,T_{\alpha\beta\ldots;\,\omega} = \varepsilon^{\alpha\beta\ldots\omega}\left(T_{\alpha\beta\ldots,\,\omega} - \Gamma^\mu{}_{\alpha\omega}T_{\mu\beta\ldots} - \Gamma^\mu{}_{\beta\omega}T_{\alpha\mu\ldots} - \cdots\right). \tag{8.273}$$

The antisymmetry of the Levi-Civita symbol combined with the symmetry of the Christoffel terms ensures that each of the ΓT terms in the parentheses vanishes, so that

$$\text{LHS of (8.272)} = \varepsilon^{\alpha\beta\ldots\omega}\, T_{\alpha\beta\ldots;\,\omega} = \varepsilon^{\alpha\beta\ldots\omega}\, T_{\alpha\beta\ldots,\,\omega} = \text{RHS of (8.272)}. \tag{8.274}$$

Now, (8.272) enables us to calculate the general exterior derivative

$$\nabla \wedge \left(T_{\alpha\beta\ldots\zeta}\, \boldsymbol{e}^\alpha \wedge \boldsymbol{e}^\beta \wedge \cdots \wedge \boldsymbol{e}^\zeta\right). \tag{8.275}$$

To see how, note that using the same permutation manipulation that appeared in the discussion around (2.36) allows the wedge product in (8.275) to be written as

$$\begin{aligned}
\boldsymbol{e}^\omega \partial_\omega \wedge \left(T_{\alpha\beta\ldots\zeta}\, \boldsymbol{e}^\alpha \wedge \boldsymbol{e}^\beta \wedge \cdots \wedge \boldsymbol{e}^\zeta\right) &= T_{\alpha\beta\ldots\zeta;\,\omega}\, \boldsymbol{e}^\omega \wedge \boldsymbol{e}^\alpha \wedge \cdots \wedge \boldsymbol{e}^\zeta \\
&= T_{123\ldots n;\, n+1}\, \boldsymbol{e}^{n+1} \wedge \boldsymbol{e}^1 \wedge \boldsymbol{e}^2 \wedge \cdots \wedge \boldsymbol{e}^n \\
&\quad + T_{213\ldots n;\, n+1}\, \boldsymbol{e}^{n+1} \wedge \boldsymbol{e}^2 \wedge \boldsymbol{e}^1 \wedge \cdots \wedge \boldsymbol{e}^n + \cdots \\
&= T_{123\ldots n;\, n+1}\, \boldsymbol{e}^{n+1} \wedge \boldsymbol{e}^1 \wedge \boldsymbol{e}^2 \wedge \cdots \wedge \boldsymbol{e}^n \\
&\quad - T_{213\ldots n;\, n+1}\, \boldsymbol{e}^{n+1} \wedge \boldsymbol{e}^1 \wedge \boldsymbol{e}^2 \wedge \cdots \wedge \boldsymbol{e}^n + \cdots \\
&= T_{[123\ldots n;\, n+1]}\, \boldsymbol{e}^{n+1} \wedge \boldsymbol{e}^1 \wedge \boldsymbol{e}^2 \wedge \cdots \wedge \boldsymbol{e}^n \\
&= T_{[123\ldots n,\, n+1]}\, \boldsymbol{e}^{n+1} \wedge \boldsymbol{e}^1 \wedge \boldsymbol{e}^2 \wedge \cdots \wedge \boldsymbol{e}^n \text{ (and now work backward)} \\
&= T_{\alpha\beta\ldots\zeta,\,\omega}\, \boldsymbol{e}^\omega \wedge \boldsymbol{e}^\alpha \wedge \cdots \wedge \boldsymbol{e}^\zeta.
\end{aligned} \tag{8.276}$$

So we have shown that

$$\nabla \wedge \left(T_{\alpha\beta\ldots\zeta}\, \boldsymbol{e}^\alpha \wedge \boldsymbol{e}^\beta \wedge \cdots \wedge \boldsymbol{e}^\zeta\right) = T_{\alpha\beta\ldots\zeta,\,\omega}\, \boldsymbol{e}^\omega \wedge \boldsymbol{e}^\alpha \wedge \cdots \wedge \boldsymbol{e}^\zeta. \tag{8.277}$$

This proof might appear to be long but really only involves permutations. It is certainly useful to understand because some texts take (8.277) as a definition—which is certainly not an obvious thing to do, not only because the comma does not guarantee a tensor but also because a different notation is typically used (the "d" notation, which we'll encounter soon). However, the antisymmetry guarantees that the comma is allowed, as we have seen.

Now we are able to simplify the right-hand side of (8.267) by way of (8.270). Converting the semicolon to a comma and using the antisymmetry of $F_{\alpha\beta}$ gives

$$\begin{aligned}
\nabla \wedge \boldsymbol{F}_2 = {}^1\!/\!_2\, F_{\alpha\beta;\gamma}\, \boldsymbol{e}^\gamma \wedge \boldsymbol{e}^\alpha \wedge \boldsymbol{e}^\beta &= {}^1\!/\!_2\, F_{\alpha\beta,\gamma}\, \boldsymbol{e}^\gamma \wedge \boldsymbol{e}^\alpha \wedge \boldsymbol{e}^\beta \\
&= {}^1\!/\!_2 \left(F_{yz,x} - F_{zy,x} + \cdots\right) \boldsymbol{e}^x \wedge \boldsymbol{e}^y \wedge \boldsymbol{e}^z \\
&= \left(F_{yz,x} + F_{zx,y} + F_{xy,z}\right) \boldsymbol{e}^x \wedge \boldsymbol{e}^y \wedge \boldsymbol{e}^z \\
&= \nabla\cdot\boldsymbol{G}\, \boldsymbol{e}^x \wedge \boldsymbol{e}^y \wedge \boldsymbol{e}^z.
\end{aligned} \tag{8.278}$$

So the right-hand side of (8.267) is

$$\iiint \nabla\cdot\boldsymbol{G}\; \mathbf{e}^x \wedge \mathbf{e}^y \wedge \mathbf{e}^z \left(\mathbf{e}^I_u, \mathbf{e}^I_v, \mathbf{e}^I_w\right) \mathrm{d}u\, \mathrm{d}v\, \mathrm{d}w\,. \tag{8.279}$$

Also,

$$\mathbf{e}^x \wedge \mathbf{e}^y \wedge \mathbf{e}^z \left(\mathbf{e}^I_u, \mathbf{e}^I_v, \mathbf{e}^I_w\right) \mathrm{d}u\, \mathrm{d}v\, \mathrm{d}w = \begin{vmatrix} \mathbf{e}^x\cdot\mathbf{e}^I_u & \dots & \mathbf{e}^x\cdot\mathbf{e}^I_w \\ & \vdots & \\ \mathbf{e}^z\cdot\mathbf{e}^I_u & \dots & \mathbf{e}^z\cdot\mathbf{e}^I_w \end{vmatrix} \mathrm{d}u\, \mathrm{d}v\, \mathrm{d}w$$

$$= \begin{vmatrix} \Lambda^x_u & \dots & \Lambda^x_w \\ & \vdots & \\ \Lambda^z_u & \dots & \Lambda^z_w \end{vmatrix} \mathrm{d}u\, \mathrm{d}v\, \mathrm{d}w = \frac{\partial(x,y,z)}{\partial(u,v,w)}\, \mathrm{d}u\, \mathrm{d}v\, \mathrm{d}w\,. \tag{8.280}$$

This is a signed volume and is independent of parametrisation when integrated, so is usually denoted $\mathrm{d}x \wedge \mathrm{d}y \wedge \mathrm{d}z$. Again, x, y, z and u, v, w will usually have the same handedness, so that the sign of this area is $+1$ and it can be written simply as $\mathrm{d}x\, \mathrm{d}y\, \mathrm{d}z$.

Finally, combining (8.267), (8.269), (8.279), and (8.280) gives

$$\iint \boldsymbol{G}\cdot\boldsymbol{n}\, \mathrm{d}S = \iiint \nabla\cdot\boldsymbol{G}\, \mathrm{d}V\,, \tag{8.281}$$

which is Gauss's theorem.

Alternative Notation for the Exterior Derivative

Exterior calculus texts generally abbreviate the manipulations involved with the generalised Stokes–Gauss theorem. Consider, for example, (8.259) together with (8.262), writing

$$\mathbf{e}^y \wedge \mathbf{e}^z \left(\mathbf{e}^I_u, \mathbf{e}^I_v\right) \mathrm{d}u\, \mathrm{d}v = \frac{\partial(y,z)}{\partial(u,v)}\, \mathrm{d}u\, \mathrm{d}v \equiv \mathrm{d}y \wedge \mathrm{d}z\,. \tag{8.282}$$

The abbreviation $\mathrm{d}y \wedge \mathrm{d}z$ means that effectively the terms involving u and v have vanished, and the cobasis vectors $\mathbf{e}^y, \mathbf{e}^z$ have metamorphosed into $\mathrm{d}y, \mathrm{d}z$. This is true, but pedagogy can suffer heavily when the entire calculation is shortened considerably in most exterior calculus texts, which essentially *begin* by treating their cobasis vectors as a new breed of object, a one-form, and calling them $\mathrm{d}y, \mathrm{d}z$ (even though these are *not* infinitesimals). In the light of our previous discussion of one-forms in Sect. 8.6.2, we suggest that this presents a great deal of needless confusion.

Unfortunately, from there, in a sort of chain reaction, several increasingly obscure things tend to be written. First, the vector $\boldsymbol{A}$, which would normally be written over a cobasis as $A_\alpha \mathbf{e}^\alpha$, is instead written as $A_\alpha \mathrm{d}x^\alpha$, even though $\boldsymbol{A}$ is *not* an infinitesimal. Next, the curl operator $\nabla\wedge$ needs to act on the $A_\alpha \mathrm{d}x^\alpha$

to produce $A_{\beta,\alpha}\mathrm{d}x^\alpha \wedge \mathrm{d}x^\beta$. This tends to be arranged by simply defining it to be so; the notion that a covariant derivative might need to be used is not always mentioned, although, as we showed in (8.276), it's true that partial differentiation does suffice. But this means that the operator $\nabla\wedge \equiv \boldsymbol{e}^\alpha\partial_\alpha\wedge$ is replaced by $\mathrm{d}x^\alpha\partial_\alpha\wedge$. The $\mathrm{d}x^\alpha\partial_\alpha$ is then reinterpreted as almost the usual infinitesimal operator "d", based on the identity $\mathrm{d}f = \mathrm{d}x^\alpha\partial_\alpha f$, while at the same time not really being an infinitesimal at all.

At the end of this convoluted path, the exterior product $\nabla\wedge$ becomes written as "d", and (8.277) is typically written as a definition:

$$\text{“}\,\mathrm{d}\left(T_{\alpha\beta\ldots\zeta}\,\boldsymbol{e}^\alpha\wedge\boldsymbol{e}^\beta\wedge\cdots\wedge\boldsymbol{e}^\zeta\right)\equiv T_{\alpha\beta\ldots\zeta,\,\omega}\,\boldsymbol{e}^\omega\wedge\boldsymbol{e}^\alpha\wedge\cdots\wedge\boldsymbol{e}^\zeta\,\text{”}. \tag{8.283}$$

We can see a correspondence with the notation of this chapter in that, while we might write (for no particular reason)

$$\nabla\wedge x^\alpha = \boldsymbol{e}^\beta\partial_\beta\wedge x^\alpha \equiv \boldsymbol{e}^\beta\partial_\beta x^\alpha = \boldsymbol{e}^\alpha, \tag{8.284}$$

users of the "d", in contrast, would simply replace the $\nabla\wedge x^\alpha$ with $\mathrm{d}x^\alpha$. If they were using one-forms ω^α, instead of our cobasis vectors $\boldsymbol{e}^\alpha$, this replacement would have the effect of either converting (8.284) to the odd-looking identity "$\mathrm{d}x^\alpha = \omega^\alpha$" as mentioned earlier on p. 307, or else reducing (8.284) to a triviality, since the cobasis vectors $\boldsymbol{e}^\alpha$, or rather the one-forms ω^α, would have been written as $\mathrm{d}x^\alpha$ from the very beginning anyway. None of this renders "d" a good notation for the exterior derivative.

Another example of the use of "d" is that our fundamental expression $\boldsymbol{e}^\alpha\cdot\boldsymbol{e}_\beta \equiv \delta^\alpha_\beta$ is written in "d" notation—and referring to the discussion around (8.7)—as "$\langle\mathrm{d}x^\alpha|\,\partial_\beta\rangle \equiv \delta^\alpha_\beta$", which has none of the symmetry of the basis-cobasis expression. With the wisdom of hindsight, we can see what is happening; but without that wisdom, using the "d" in this way is nothing if not obtuse. We saw this sort of notation earlier in (8.108) together with the discussion following that equation.

Further obscurity follows when the d is used to relate a tensor to an exterior derivative (which, remember, is just a generalised curl). For example, (8.257) shows that the Faraday tensor $\boldsymbol{F}$ can be written as a generalised curl of the electromagnetic vector potential $\boldsymbol{A}$:

$$\nabla\wedge\boldsymbol{A} = (A_{\beta,\alpha} - A_{\alpha,\beta})\,\boldsymbol{e}^\alpha\boldsymbol{e}^\beta = F_{\alpha\beta}\,\boldsymbol{e}^\alpha\boldsymbol{e}^\beta = \boldsymbol{F}. \tag{8.285}$$

Texts using the d notation for a generalised curl will write this as $\boldsymbol{F} = \mathrm{d}\boldsymbol{A}$ (even though there are no u, v terms present). As far as notation is concerned, this has an odd look about it, seeming to mix a noninfinitesimal with what appears to be an infinitesimal (but is not), and so has quite a potential to mislead. It tends to be justified by describing the d as supplying the underlying rigor when using infinitesimals—although what this really means is not at all clear. Infinitesimals pervade both mathematical physics and pure mathematics quite deeply, and physicists and mathematicians use them all the time.

Although the idea of a rigorous way to treat infinitesimals is interesting and has a place within mathematics, the idea that they are something very small is never misleading and always highly useful—especially when they appear alone in an integral, as opposed to occurring in pairs for derivatives, where there is less need to think of them as having infinitesimal size.

So, we choose not to redefine d to mean an exterior derivative. The d in (8.283) is not denoting an infinitesimal, and the a priori use of the comma in that *definition* has a very untensorial look about it, which is not satisfying from a notational point of view. And while it all works for the Stokes–Gauss theorem—since this has the required $(\boldsymbol{e}_u, \boldsymbol{e}_v)\, \mathrm{d}u\, \mathrm{d}v$ term in it from the outset, which was the necessary ingredient to produce expressions such as $\mathrm{d}y \wedge \mathrm{d}z$—there is no reason that it should be more generally useful.

Of course, the use of d means that the expressions that we have calculated in this section via determinants can appear top-heavy because, while we write (8.282) as it stands, users of the d notation will simply omit the $(\boldsymbol{e}^I_u, \boldsymbol{e}^I_v)\, \mathrm{d}u\, \mathrm{d}v$ and write $\mathrm{d}y \wedge \mathrm{d}z$ straightaway since they already write $\boldsymbol{e}^y, \boldsymbol{e}^z$ as $\mathrm{d}y, \mathrm{d}z$ from the outset. This is certainly concise, and abbreviating notation in meaningful and useful ways is of course an important concept in mathematical physics. But pedagogically, abbreviating the language too soon, before the concepts have been understood, does not give the student any opportunity to understand why the subject has been built in the way that it has. It's true that $\boldsymbol{F} = \nabla \wedge \boldsymbol{A}$ can be cumbersome to write over and over, and shortening it to, e.g., "$\boldsymbol{F} = \partial \boldsymbol{A}$" can be useful, especially if it suggests that partial differentiations are happening without looking too much like an infinitesimal (cf. $\boldsymbol{F} = \mathrm{d}\boldsymbol{A}$ on the previous page). Notation should not confuse or mislead.

Finally, the "d" notation is conventionally used to write the generalised Stokes–Gauss theorem (8.251) or (8.252) as

$$\text{“} \int_{\text{boundary}} \boldsymbol{F} = \int_{\text{interior}} \mathrm{d}\boldsymbol{F} \text{”}. \tag{8.286}$$

This has a deceptively simple form that requires a lot of work to decrypt, as we have seen in this section.

Divs of Curls, and Curls of Grads

The result of operating twice with the exterior derivative $\nabla\wedge$ on any multivector is zero because each operator brings in a partial derivative, and the two partial derivatives commute while the wedges anticommute. That is, for any n-index antisymmetric tensor $\boldsymbol{F}$,

$$\boldsymbol{F} \text{ (being antisymmetric)} = {}^1\!/\!_{n!}\, F_{\alpha\ldots\varsigma}\, \boldsymbol{e}^\alpha \wedge \cdots \wedge \boldsymbol{e}^\varsigma\,, \tag{8.287}$$

so that

$$\begin{aligned} \nabla \wedge \boldsymbol{F} &= {}^1\!/\!_{n!}\, F_{\alpha\ldots\varsigma,\,\omega}\, \boldsymbol{e}^\omega \wedge \boldsymbol{e}^\alpha \wedge \cdots \wedge \boldsymbol{e}^\varsigma\,, \\ \nabla \wedge (\nabla \wedge \boldsymbol{F}) &= {}^1\!/\!_{n!}\, F_{\alpha\ldots\varsigma,\,\omega\eta}\, \boldsymbol{e}^\eta \wedge \boldsymbol{e}^\omega \wedge \boldsymbol{e}^\alpha \wedge \cdots \wedge \boldsymbol{e}^\varsigma = \boldsymbol{0}\,. \end{aligned} \tag{8.288}$$

It is straightforward to show that the identity $\nabla \wedge (\nabla \wedge \boldsymbol{F}) = \boldsymbol{0}$ also holds when $\boldsymbol{F}$ is a scalar or vector function, even though the notion of antisymmetry does not apply to these. For a scalar function, it reduces to the very well-known "curl(grad) = **0**", while if $\boldsymbol{F}$ is a vector function, we obtain "div(curl) = 0". So if we consider scalars and vectors to be the simplest types of multivectors, then the following general identity holds:

$$\nabla \wedge (\nabla \wedge \text{multivector}) = \boldsymbol{0}\,. \tag{8.289}$$

In hindsight this might well seem reasonable (or at least easily memorised) since we already know that the wedge product vanishes when two of its factors are identical, for the same reason that a determinant vanishes when two of its rows (or columns) are identical. Even so, we must remember that beyond scalars and vectors, (8.289) requires antisymmetry in the tensor indices. It becomes shortened to "$\mathrm{d}^2 = 0$" in exterior calculus texts (where the d^2 acts on an n-form), which is certainly obscure.

The various theorems and index manipulations of this section are part of the subject of exterior calculus, which seeks to bring a geometrical view to bear on tensor analysis, thereby shedding light on subjects such as electromagnetism and relativity. We have already seen in (8.285) that the Faraday tensor is the exterior derivative—that is, the generalised curl—of the electromagnetic vector potential: $\boldsymbol{F} = \nabla \wedge \boldsymbol{A}$. And in Chap. 12 we'll encounter the exterior derivative again, using it as a shortcut in deriving the quantities needed to calculate gravitational fields.

9 Curvature and Differential Geometry

We spent the last chapter building up the notation of tensor calculus and developing a feel for covariant notation. Now we wish to look at some of the basic ideas behind *differential geometry*, the detailed study of curvature, since these will prove valuable in paving the way to general relativity.

9.1 Curvature in the Plane

A natural place to begin to think of curvature is to ask how much we need to turn the steering wheel of a car in order to drive along a curved road on a flat plane, shown on the left in Fig. 9.1. For a given distance travelled, Δs, the greater the angle $\Delta\theta$ through which we must turn the steering wheel, then intuitively the more curved the road must be. The curvature at a single point is defined as the limit of this angle turned through per unit of distance travelled:

$$\text{curvature } k \equiv \frac{d\theta}{ds} \,. \tag{9.1}$$

Being one-dimensional, the curve has just one parameter to quantify it, and this is arc length s or any parameter related to s. Parametrisations using arc length turn up again and again in differential geometry. Their analogy in the geometrical view of relativity is the proper time used in spacetime

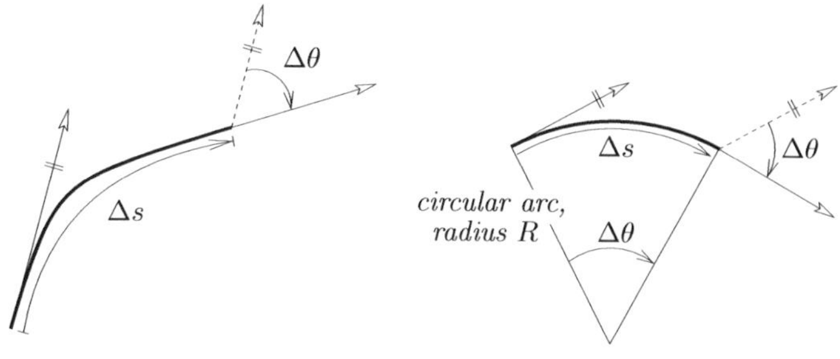

Fig. 9.1. Left: Curvature is defined by how far we must turn a steering wheel to get around a corner. **Right**: In the case of a circle, the curvature is constant and equal to the reciprocal of its radius.

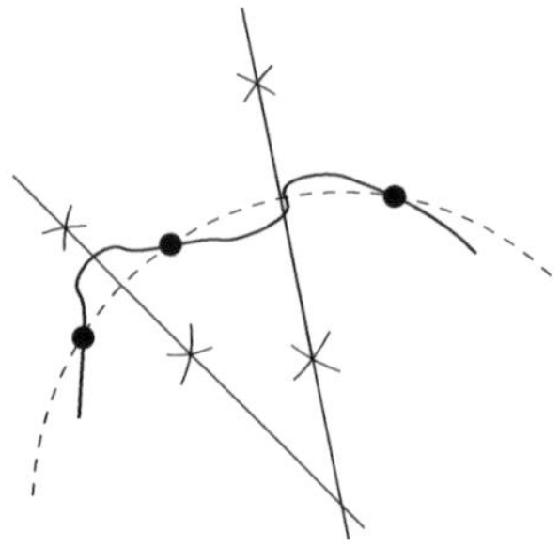

Fig. 9.2. Any three noncollinear points can always be joined by a circle. The proof is by construction. Use a compass and ruler to draw perpendicular bisectors through pairs of points. These lines must intersect at the circle's centre.

diagrams, which we know is the invariant interval between events. An element of arc length for a general curve in the xy-plane is given by Pythagoras's theorem as $\mathrm{d}s = \sqrt{\mathrm{d}x^2 + \mathrm{d}y^2} = \sqrt{1 + (\mathrm{d}y/\mathrm{d}x)^2}\,\mathrm{d}x$. In general, this is difficult or impossible to integrate analytically in terms of simple functions, even for the most basic curves, but we'll have no need to do so in this chapter.

On the right-hand side in Fig. 9.1 is shown the case of driving around a circle of radius R. Here $\Delta\theta/\Delta s = 1/R$, and this doesn't change in the limit as $\Delta s \to 0$, which matches our intuition that a circular road requires a constant turn on the steering wheel. So a good quantity to help develop a feel for curvature is

$$\text{curvature } k \text{ of a circle } = 1/\text{radius}. \tag{9.2}$$

The idea that a circle has a constant curvature can be applied to any point of an arbitrary curve in the plane to give the concept of the *osculating circle* at that point. Choose any three points on the curve. As shown in Fig. 9.2, a circle can always be drawn through them since its centre must lie at the intersection of the perpendicular bisectors of any two pairs of the points. Fig. 9.2 shows this construction using a compass and ruler. By letting all three points approach the point at which we wish to know the curvature, this circle drawn through them tends toward a unique circle at that point, known as the osculating circle (from the Latin "to kiss"). We'll omit the long but straightforward exercise of proving that the curvature $k = \mathrm{d}\theta/\mathrm{d}s$ is indeed the reciprocal of the osculating circle's radius.

If the tangent vectors are all drawn with unit length (a procedure we'll quantify shortly), they can be collected along a curve and placed with their tails all at one point, shown in Fig. 9.3. This is a *circle map*, which allows us to visualise the swept angle much as was done in Fig. 9.1 for the circle. The angle swept through is just the length of the arc, with no units since the circle is an abstract entity:

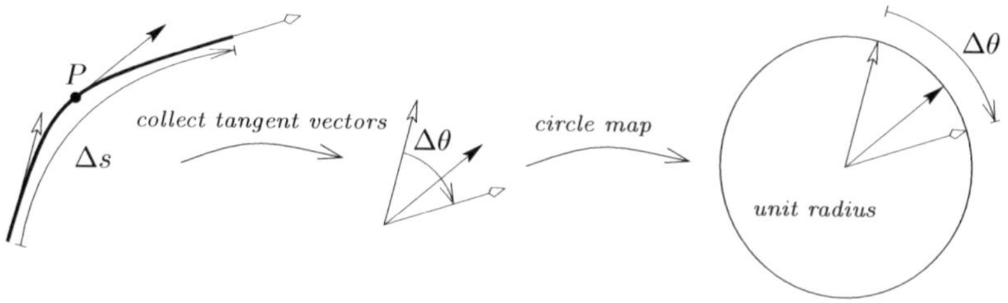

Fig. 9.3. Collecting the tangent vectors (arranged to have unit length) along the curve in the vicinity of a point P makes a *circle map* of that curve to a unit circle. For clarity, the three representative vectors drawn have been given three different arrowheads to show where each lies in the circle map. (Thus, all vectors with any one arrowhead are parallel across the three pictures above.) The angle $\Delta\theta$ is also just the length of the arc that the vectors encompass on this circle.

$$k = \frac{\mathrm{d}\theta}{\mathrm{d}s} = \lim_{\Delta s \to 0} \frac{\begin{bmatrix}\text{dimensionless arc length swept out on unit} \\ \text{circle by tangents or normals}\end{bmatrix}}{\text{corresponding arc length } \Delta s \text{ of actual curve}}\,. \tag{9.3}$$

For nonplanar curves in euclidean 3-space, this notion of curvature still applies. But the plane containing the osculating circle at any point of the curve, known as the *osculating plane*, will now twist about as it follows the curve's meanderings through space. To quantify this further, we'll take our basic curve to be a one-parameter map, embedded in any number of higher dimensions (three for the sake of a picture), as shown in Fig. 9.4. The vector $\Delta\alpha$ might have any length, depending on the parameter s describing the curve. But there are advantages to having s measure the curve length, or at least something proportional to it, and this choice is the main one in the

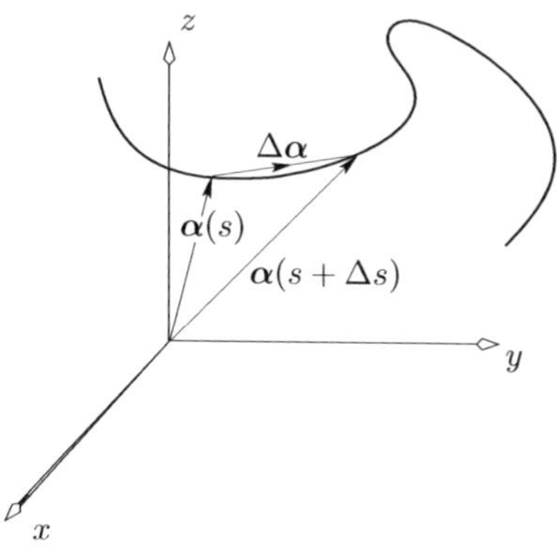

Fig. 9.4. A particle moving along a curve has a velocity vector that's independent of any coordinates. Dispensing with the particle leaves us with a tangent vector.

study of differential geometry. The analogous parameter in the spacetime of relativity is proper time—the time measured by a clock traversing a given worldline. But this, too, is a "length" for a particle's worldline, as was discussed on p. 201.

So the parameter s might be geometrical length or proper time. That means that in the limit $\Delta s \to 0$, the velocity vector has a length that is just the length of the curve element. Thus scaling the vector by $\mathrm{d}s$ ensures it will always have unit length, again a desirable property. And sure enough, this is exactly what we did when defining the four-velocity in (6.13), where proper time $\mathrm{d}\tau$ had the same role as $\mathrm{d}s$.

To eliminate the need for drawing an osculating circle at a point of interest on a given curve, we can approach curvature slightly differently by way of the curve's tangent vector. The special, and useful, property of a curve parametrised by arc length is that its tangent vector always has unit length, which follows from Fig. 9.4:

$$|\boldsymbol{\alpha}'(s)| = \lim_{\Delta s \to 0} \frac{|\boldsymbol{\alpha}(s+\Delta s) - \boldsymbol{\alpha}(s)|}{\Delta s} = \frac{|\mathrm{d}\boldsymbol{\alpha}|}{\mathrm{d}s} = \frac{\mathrm{d}s}{\mathrm{d}s} = 1\,. \tag{9.4}$$

(Compare this with the four-velocity of relativity, which for the same reason always has unit length.) For a general parametrisation $\boldsymbol{\alpha}(t)$, where t is not necessarily arc length, the tangent vector is $\boldsymbol{\alpha}'(t)$, and in general the tangent vector's length depends on this parametrisation since

$$\boldsymbol{\alpha}'(t) = \boldsymbol{\alpha}'(s)\,\frac{\mathrm{d}s}{\mathrm{d}t}\,. \tag{9.5}$$

In that case, the length of the tangent for an arbitrary parametrisation is $|\boldsymbol{\alpha}'(t)| = |\mathrm{d}s/\mathrm{d}t|$. If t is thought of as time, then $|\mathrm{d}s/\mathrm{d}t|$ is the speed that a particle traversing the curve would have. In essence we're saying that if we measure time in terms of distance travelled, then no matter how fast we go, our "speed" is always one—since it's the distance travelled divided by itself.

It certainly makes sense to concentrate on a parametrisation where the speed is at least constant (if not unity), since that endows higher derivatives with geometric significance. Why? Because if $|\boldsymbol{\alpha}'(t)| = \text{constant}$, then

$$\begin{aligned} &\boldsymbol{\alpha}'(t)\cdot\boldsymbol{\alpha}'(t) = \text{constant},\\ \text{i.e.}\quad &\boldsymbol{\alpha}'(t)\cdot\boldsymbol{\alpha}''(t) = 0 \text{ (by differentiating)}\\ \text{or}\quad &\boldsymbol{\alpha}'(t) \perp \boldsymbol{\alpha}''(t)\,, \end{aligned} \tag{9.6}$$

so that for such a parameter t, $\boldsymbol{\alpha}''(t)$ is normal to the curve. (We've seen this procedure before when relating four-velocity to four-acceleration in the box on p. 245.) Thus $\boldsymbol{\alpha}''(t)$ points toward the centre of the osculating circle. For a particle tracing out the curve with constant speed, this vector is none other than the centripetal acceleration familiar from newtonian kinematics. So, for example, if a particle moves in a circle at constant speed,

$$\boldsymbol{\alpha}(t) = \Big(r\cos(\omega t + \theta_0), r\sin(\omega t + \theta_0)\Big), \tag{9.7}$$

then its centripetal acceleration is

$$\boldsymbol{\alpha}''(t) = -\omega^2\,\boldsymbol{\alpha}(t), \tag{9.8}$$

which clearly does point toward the centre of the circle.

We could equally well consider other constant speeds. For these, t can be written in terms of arc length s using two constants c_1, c_2, as

$$t = c_1 s + c_2\,. \tag{9.9}$$

Here c_1 is really a measure of the units used and c_2 sets a starting time. The relationship (9.9) sets t to be an *affine* parameter. Neither of the constants c_1, c_2 is relevant physically, so we can restrict ourselves to the special parametrisation of arc length s, corresponding to the particle traversing the curve with unit speed. (Many of the results in this chapter hold, or are easily extended to hold, for affine parameters since the derivative $\mathrm{d}s/\mathrm{d}t$ is constant and so changes the equations only trivially.)

To summarise, the unit tangent vector is

$$\boldsymbol{T}(s) \equiv \boldsymbol{\alpha}'(s), \tag{9.10}$$

and the *principal normal vector* (also of unit length) is defined as

$$\boldsymbol{N}(s) \equiv \frac{\boldsymbol{\alpha}''(s)}{|\boldsymbol{\alpha}''(s)|} = \frac{\boldsymbol{T}'(s)}{|\boldsymbol{T}'(s)|}, \tag{9.11}$$

which lies in the osculating plane. (The basis set is completed by defining the *binormal vector* as $\boldsymbol{B} \equiv \boldsymbol{T} \times \boldsymbol{N}$, which is normal to the osculating plane.) We defined the curvature as the rate of increase of the tangent vector angle with respect to arc length; but for a *unit* tangent vector $\boldsymbol{T}(s)$, this angular increase is just the arc length that $\boldsymbol{T}(s)$ traces out as it turns, or $|\mathrm{d}\boldsymbol{T}(s)|$. Thus

$$k = \frac{1}{r} = \frac{\mathrm{d}\theta}{\mathrm{d}s} = \frac{|\mathrm{d}\boldsymbol{T}(s)|}{\mathrm{d}s} = |\boldsymbol{T}'(s)| = |\boldsymbol{\alpha}''(s)|, \tag{9.12}$$

and this general expression for the curvature no longer depends on constructing an osculating circle. The vector $\boldsymbol{\alpha}''(s)$ is called the *curvature vector* and always points toward the centre of curvature. Sometimes it's called the *acceleration vector*, which is misleading since it is really only the acceleration when a particle traversing the curve has unit speed. The vector $\boldsymbol{\alpha}''(s)$ will not always point toward the centre of curvature if the particle's speed varies.

On that note, the usual expressions for velocity and acceleration can now easily be produced in the following way. The position vector $\boldsymbol{\alpha}(t)$ differentiates to give the proper vectors $\boldsymbol{\alpha}'(t)$ and $\boldsymbol{\alpha}''(t)$:

$$\boldsymbol{\alpha}'(t) = \boldsymbol{\alpha}'(s)\frac{\mathrm{d}s}{\mathrm{d}t} = \frac{\mathrm{d}s}{\mathrm{d}t}\boldsymbol{T} \equiv v\boldsymbol{T}; \tag{9.13}$$

i.e., the velocity is tangential, while the acceleration is

$$\begin{aligned}\boldsymbol{\alpha}''(t) &= \frac{\mathrm{d}v}{\mathrm{d}t}\boldsymbol{T} + v^2\,\boldsymbol{T}'(s)\\ &= \frac{\mathrm{d}v}{\mathrm{d}t}\boldsymbol{T} + \frac{v^2}{r}\boldsymbol{N}\,,\end{aligned} \tag{9.14}$$

where r is the radius of curvature. So the acceleration is a mixture of centripetal and tangential components, a concept familiar from classical mechanics.

9.1.1 Curves on Surfaces

Up until now, our discussion has been restricted to curves in euclidean 3-space. But suppose that such a curve actually resides on a surface within that space. Because such a surface needs only two parameters to describe it, it's called a 2-surface, and is produced by a map $\boldsymbol{X}\colon\mathbb{R}^2 \to \mathbb{E}^3$ such as shown in Figs 8.2, 8.3, and 9.5. We will here label the two parameters as u^1, u^2, so that the map is $\boldsymbol{X}\colon\{u^1, u^2\} \to \{x, y, z\}$, giving the surface two basis vectors $\boldsymbol{e}_1, \boldsymbol{e}_2$. As is fairly conventional, Latin subscripts are used throughout the following calculations to emphasise that we are only using the variables u^1, u^2; that is, a subset of the surrounding 3-space:

$$\boldsymbol{e}_a \equiv \frac{\partial \boldsymbol{X}}{\partial u^a}\,. \tag{9.15}$$

These basis vectors are tangential to the surface. For the *2-sphere* of Fig. 8.3, $\boldsymbol{e}_\theta$ points everywhere south and $\boldsymbol{e}_\phi$ points everywhere east, but in general they can point anywhere tangentially and need not be orthogonal. The euclidean 3-space itself is spanned by the two basis vectors of the embedded 2-surface along with one more vector (set to have unit length): the vector $\boldsymbol{U}$ normal to the surface:

$$\boldsymbol{U} \equiv \frac{\boldsymbol{e}_1 \times \boldsymbol{e}_2}{|\boldsymbol{e}_1 \times \boldsymbol{e}_2|}\,. \tag{9.16}$$

The curve $\boldsymbol{\alpha}$ is now the image of a curve in the u^1-u^2 space and is shown in Fig. 9.5.

We can investigate how the tangent and curvature vectors relate to curvature by writing[1]

$$\begin{aligned}\boldsymbol{\alpha}(s) &= \boldsymbol{X}\left(u^1(s), u^2(s)\right),\\ \boldsymbol{\alpha}'(s) &= u^{a\prime}(s)\,\frac{\partial \boldsymbol{X}}{\partial u^a} = u^{a\prime}\,\boldsymbol{e}_a\,, \quad \text{where } u^{a\prime} \equiv u^{a\prime}(s) \equiv \mathrm{d}u^a/\mathrm{d}s\,,\\ \boldsymbol{\alpha}''(s) &= u^{a\prime\prime}\,\boldsymbol{e}_a + u^{a\prime}\,u^{b\prime}\,\boldsymbol{e}_{a,b}\,.\end{aligned} \tag{9.17}$$

[1] Don't confuse the derivatives $u^{a\prime}$ with $u^{a'}$, which are components in a primed basis! We won't be using the latter.

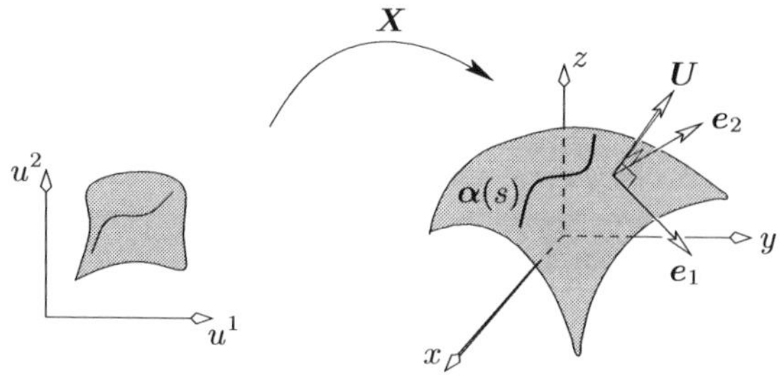

Fig. 9.5. The function $\boldsymbol{X}: \{u^1, u^2\} \to \{x, y, z\}$ maps a curve $\big(u^1(s), u^2(s)\big)$ in $\mathbb{R}^2$ into another curve $\boldsymbol{\alpha}(s)$ in $\mathbb{E}^3$ on the 2-surface embedded in euclidean 3-space.

(Notice how different the expressions are for the position vector $\boldsymbol{\alpha}$ as opposed to the proper vectors $\boldsymbol{\alpha}'$ and $\boldsymbol{\alpha}''$, which make use of components, a basis, and the summation convention.) We have seen the expression $\boldsymbol{e}_{a,b}$ previously in (8.177), except that there it was introduced without a context of embedding. Now we must alter it to ensure that the entire 3-space is spanned, by adding a normal contribution, giving what are known as *Gauss's formulae*:

$$\boldsymbol{e}_{a,b} \equiv \Gamma^c{}_{ab}\, \boldsymbol{e}_c + L_{ab}\, \boldsymbol{U}\,. \tag{9.18}$$

The set of new coefficients L_{ab} is known as the *second fundamental form.* (The metric is sometimes called the first fundamental form; see the comment on p. 338.) The second fundamental form is easily determined from (9.18) by dotting each side with $\boldsymbol{U}$ to give

$$L_{ab} = \boldsymbol{e}_{a,b} \cdot \boldsymbol{U}\,. \tag{9.19}$$

Like the Christoffel symbols, L_{ab} is symmetric since $\boldsymbol{e}_{a,b} = \boldsymbol{e}_{b,a}$.

Figure 9.6 shows the relationships among the derivatives of $\boldsymbol{\alpha}(s)$ and their components. We can see in this figure that the curvature vector $\boldsymbol{\alpha}''(s)$ resolves into two components, tangential and normal to the surface. Evidently, the tangential component $\boldsymbol{\alpha}''_{\text{tan}}(s)$ measures how the curve bends within the surface, while the normal component $\boldsymbol{\alpha}''_{\text{nor}}(s)$ is due to the bending of the surface itself within the ambient euclidean 3-space. So write the curvature vector in (9.17) as

$$\boldsymbol{\alpha}''(s) = \underbrace{\left(u^{c\prime\prime} + \Gamma^c{}_{ab}\, u^{a\prime}\, u^{b\prime}\right) \boldsymbol{e}_c}_{\equiv\, \boldsymbol{\alpha}''_{\text{tan}}} + \underbrace{L_{ab}\, u^{a\prime}\, u^{b\prime}\, \boldsymbol{U}}_{\equiv\, \boldsymbol{\alpha}''_{\text{nor}}}. \tag{9.20}$$

The two components of the curvature vector can now be seen explicitly. First is the **geodesic curvature of the curve,** k_g:

$$\boldsymbol{\alpha}''_{\text{tan}} = \left(u^{c\prime\prime} + \Gamma^c{}_{ab}\, u^{a\prime}\, u^{b\prime}\right) \boldsymbol{e}_c \equiv k_g\, \underbrace{\boldsymbol{U} \times \boldsymbol{\alpha}'}_{\text{“geodesic normal vector”}}. \tag{9.21}$$

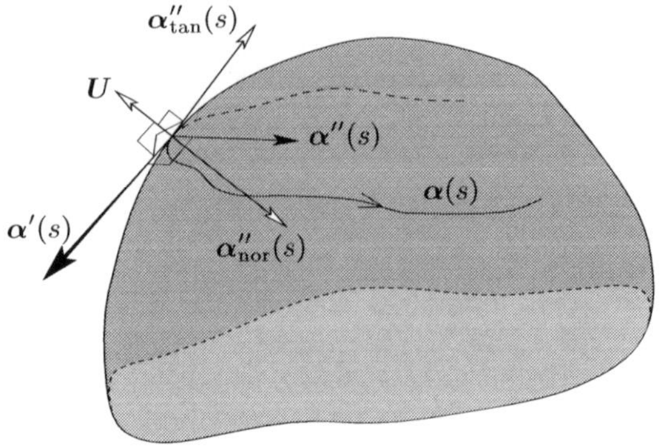

Fig. 9.6. Relationships of the various derivatives along a curve $\boldsymbol{\alpha}(s)$, with reference to (9.20). The tangent or velocity vector $\boldsymbol{\alpha}'(s)$ points out of the page. The curvature or acceleration vector $\boldsymbol{\alpha}''(s)$ points toward the centre of the osculating circle of $\boldsymbol{\alpha}$ at the point of interest. The curvature vector $\boldsymbol{\alpha}''(s)$ resolves into components $\boldsymbol{\alpha}''_{\rm nor}$ normal to, and $\boldsymbol{\alpha}''_{\rm tan}$ tangential to, the surface. The normal vector $\boldsymbol{U}$ is shown, while $\boldsymbol{e}_1, \boldsymbol{e}_2$ are omitted for clarity. They span the tangent plane, as do $\boldsymbol{\alpha}'$ and $\boldsymbol{\alpha}''_{\rm tan}$. It's evident that $\boldsymbol{\alpha}''_{\rm nor} \propto \boldsymbol{U}$, while $\boldsymbol{\alpha}''_{\rm tan} \perp \boldsymbol{U}$. Also, it must be true that $\boldsymbol{\alpha}''_{\rm tan} \perp \boldsymbol{\alpha}'$ since $\boldsymbol{\alpha}''_{\rm tan} \cdot \boldsymbol{\alpha}' = (\boldsymbol{\alpha}''_{\rm tan} + \boldsymbol{\alpha}''_{\rm nor}) \cdot \boldsymbol{\alpha}' = \boldsymbol{\alpha}'' \cdot \boldsymbol{\alpha}' = 0$. Thus, because $\boldsymbol{\alpha}''_{\rm tan}$ is perpendicular to both $\boldsymbol{U}$ and $\boldsymbol{\alpha}'$, we conclude that $\boldsymbol{\alpha}''_{\rm tan} \propto \boldsymbol{U} \times \boldsymbol{\alpha}'$. This relationship is used in (9.21).

The geodesic curvature quantifies how the curve bends within the surface and so is known as *intrinsic* to the surface.

The second component of the curvature vector is the **normal curvature of the surface in the $\boldsymbol{\alpha}'(s)$ direction**, defined as $k_n(\boldsymbol{\alpha}')$ (which we'll write as k_n):

$$\boldsymbol{\alpha}''_{\rm nor} = L_{ab}\, u^{a\prime}\, u^{b\prime}\, \boldsymbol{U} \equiv k_n \boldsymbol{U}. \tag{9.22}$$

The normal curvature of the surface as a function of direction has nothing to do with the curve $\boldsymbol{\alpha}(s)$: it's a property of the surface alone, measuring how the surface bends within the surrounding euclidean 3-space. This is termed *extrinsic* curvature. The study of intrinsic and extrinsic curvatures is fundamental to gaining an understanding of the curved spacetime of general relativity.

We have now defined three curvatures: k from (9.12) and earlier equations, k_g from (9.21), and k_n from (9.22). Useful for visualising these is the fact that they are related via Pythagoras's theorem. To see this, note that (9.20) becomes

$$\boldsymbol{\alpha}''(s) = k_g\, \boldsymbol{U} \times \boldsymbol{\alpha}' + k_n\, \boldsymbol{U}. \tag{9.23}$$

But the vectors $\boldsymbol{U}$ and $\boldsymbol{U} \times \boldsymbol{\alpha}'$ are orthogonal, so we can apply Pythagoras's theorem to write

$$k^2 = k_g^2 + k_n^2\,, \tag{9.24}$$

which is useful to keep in mind in developing an intuitive feel for the various curvatures, especially as we are usually only interested in their absolute values.

The two curvature components of (9.23) can be extracted simply by taking the dot product of its left-hand side with each of its right-hand side's unit vectors (and permuting the mixed dot/cross product that results):

$$\boxed{\begin{aligned} k_n &= \boldsymbol{U} \cdot \boldsymbol{\alpha}'' , \\ k_g &= \boldsymbol{U} \cdot \boldsymbol{\alpha}' \times \boldsymbol{\alpha}'' . \end{aligned}} \tag{9.25}$$

So the normal curvature of the surface in the direction of the curve's tangent is the normal component of the curvature vector $\boldsymbol{\alpha}''$, while the geodesic curvature of the curve itself is the normal component of the crossed tangent and curvature vectors $\boldsymbol{\alpha}' \times \boldsymbol{\alpha}''$. For example, if the surface is a plane, then $\boldsymbol{\alpha}''$ must lie within it and be everywhere orthogonal to $\boldsymbol{U}$, so that the normal curvature must vanish in all directions at every point. All of the curvature of a plane curve must therefore be geodesic curvature.

The normal curvature in some direction has another interpretation. Suppose that, at some point P of interest, we draw a new curve $\boldsymbol{\beta}$ whose osculating plane is normal to the surface. In that case, $\boldsymbol{\beta}''$ is parallel to $\boldsymbol{U}$ so

$$k_n = \boldsymbol{U} \cdot \boldsymbol{\beta}'' = \boldsymbol{U} \cdot \left(\pm \left|\boldsymbol{\beta}''\right| \boldsymbol{U}\right) = \pm |\boldsymbol{\beta}''| , \tag{9.26}$$

where the plus sign indicates that the centripetal acceleration of a point traversing the curve at constant speed is parallel to $\boldsymbol{U}$, and a minus sign indicates it is antiparallel. This gives a more intuitive feel for the normal curvature of the surface in a given direction at P. We merely draw a new curve on the surface in a normal plane, with its tangent pointing in the given direction, and measure the curvature k of the curve at P. This will equal k_n (up to a sign that doesn't concern us). This new curve has no geodesic curvature ($k_g = 0$).

To illustrate these points, let's calculate k, k_n, k_g for a circle of constant latitude on Earth, known as a *small circle*. Refer to Fig. 9.7 for the map, where Earth is modelled as a sphere of radius R:

$$\boldsymbol{X}(\theta, \phi) = (R \sin\theta \cos\phi, \quad R \sin\theta \sin\phi, \quad R \cos\theta) . \tag{9.27}$$

The small circle runs around from $\phi = 0 \to 2\pi$, where ϕ itself is a function of the curve length s:

$$\boldsymbol{\alpha}(s) \equiv \boldsymbol{X}\Big(\theta_0, \phi(s)\Big) = (R \sin\theta_0 \cos\phi(s), \quad R \sin\theta_0 \sin\phi(s), \quad R \cos\theta_0) , \tag{9.28}$$

with $\mathrm{d}s = R \sin\theta_0 \, \mathrm{d}\phi$. The derivatives needed for (9.25) are easily calculated:

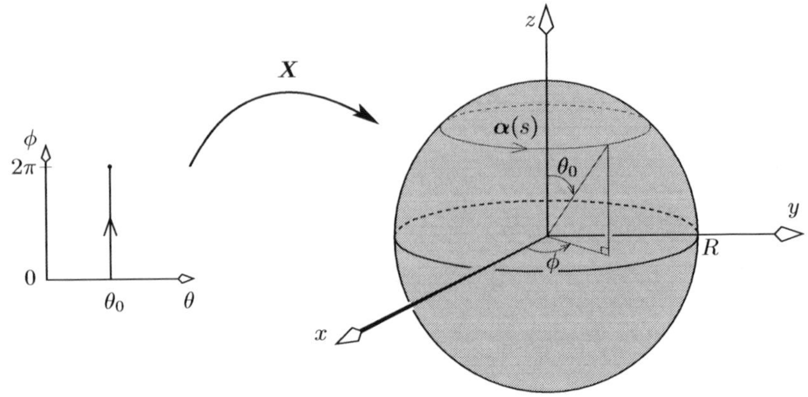

Fig. 9.7. Defining a path following a circle of constant latitude on a sphere enables us to calculate the normal curvature k_n of the sphere along the path, along with the path's geodesic curvature k_g, as well as verifying that $k^2 = k_n^2 + k_g^2$.

$$\begin{aligned}\boldsymbol{\alpha}'(s) &= (-\sin\phi, \cos\phi, 0)\,,\\ \boldsymbol{\alpha}''(s) &= \frac{1}{R\sin\theta_0}(-\cos\phi, -\sin\phi, 0)\,,\\ \boldsymbol{\alpha}'\times\boldsymbol{\alpha}'' &= \frac{1}{R\sin\theta_0}(0,0,1)\,.\end{aligned} \tag{9.29}$$

The normal vector to the surface at $\theta = \theta_0$ is

$$\boldsymbol{U} = \frac{\boldsymbol{e}_\theta \times \boldsymbol{e}_\phi}{|\boldsymbol{e}_\theta \times \boldsymbol{e}_\phi|} \overset{(8.4)}{=\!=} (\sin\theta_0\cos\phi,\ \sin\theta_0\sin\phi,\ \cos\theta_0)\,. \tag{9.30}$$

Hence, (9.25) gives

$$k_n = \boldsymbol{U}\cdot\boldsymbol{\alpha}'' = \frac{-1}{R}\,, \quad k_g = \boldsymbol{U}\cdot\boldsymbol{\alpha}'\times\boldsymbol{\alpha}'' = \frac{\cot\theta_0}{R}\,. \tag{9.31}$$

The curvature of the small circle in the euclidean 3-space is the positive quantity $k = 1/(R\sin\theta_0)$. The signs of k_n and k_g are sensitive to the choices of the sign of $\boldsymbol{U}$ and the direction of the path, but as can easily be verified, the relation $k^2 = k_n^2 + k_g^2$ does certainly hold.

Notice that the normal curvature in the direction of the path, k_n, is constant regardless of the latitude of the small circle, as expected from our discussion immediately following (9.26). We expect that $|k_n| = 1/R$ because k_n measures the curvature of a great circle, which we know has $k = 1/R$ on a sphere. On the other hand, the geodesic curvature k_g certainly changes sign as we move from the Northern Hemisphere to the Southern Hemisphere. For our choice of normal $\boldsymbol{U}$ and path direction, k_g tends to $+\infty$ near the North Pole, reduces to zero at the equator, then switches sign in the Southern Hemisphere, tending to $-\infty$ near the South Pole. It measures the amount by which

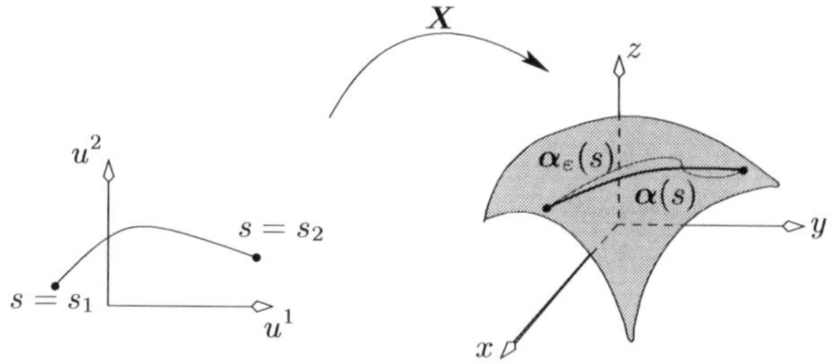

Fig. 9.8. A curve parametrised by length s in 2-space, mapped to an image $\boldsymbol{\alpha}(s)$ on a surface in 3-space. The lighter curve is a perturbation $\boldsymbol{\alpha}_\varepsilon(s)$.

we would have to turn the steering wheel in a car travelling east along the small circle: turning constantly to the left in the Northern Hemisphere, travelling straight ahead at the equator, and turning constantly to the right in the Southern Hemisphere.

9.2 Geodesics: Curves with No Geodesic Curvature

The geodesic and normal curvatures serve to introduce two important concepts that will become useful for thinking about general relativity. Geodesic curvature k_g is a property of the curve alone, and the lack of it defines a *geodesic* curve. On the other hand, normal curvature k_n is a property of the surface alone and appears when we define the curvature of the surface itself, which we'll do shortly.

As a curve meanders over a surface, its curvature k might vary between being mostly normal and mostly geodesic. A curve with no geodesic curvature is called a geodesic; its curvature is all normal, being completely inherited from the way the surface curves in its ambient 3-space. (Perhaps this is confusing terminology, but we can think of geodesic curvature as being the amount by which a curve veers away from being geodesic.) This definition means that either of the following equations determines a geodesic. All derivatives are taken, as usual, with respect to arc length s:

$$u^{c\prime\prime} + \Gamma^c{}_{ab}\, u^{a\prime}\, u^{b\prime} = 0 \text{ for all } c\,, \tag{9.32}$$

$$\boldsymbol{U} \cdot \boldsymbol{\alpha}' \times \boldsymbol{\alpha}'' = 0\,. \tag{9.33}$$

We'll concentrate on the first expression since it serves to define a geodesic in any number of dimensions, unlike the second. The first expression can be used to show the most important property of a geodesic: *a curve is a geodesic if and only if it has stationary length* with respect to perturbations of the curve when the two end points are held fixed, such as shown in Fig. 9.8.

To prove this assertion, first set up the scenario. The curve $\boldsymbol{\alpha}(s)$ results from the usual map $\boldsymbol{X}\colon \{u^1, u^2\} \to \{x, y, z\}$:

$$\boldsymbol{\alpha}(s) \equiv \boldsymbol{X}\left(u^a(s)\right). \tag{9.34}$$

Perturb $\boldsymbol{\alpha}(s)$ by constructing another curve anchored to the end points,

$$\boldsymbol{\alpha}_\varepsilon(s) \equiv \boldsymbol{X}\left(\xi^a(s,\varepsilon)\right), \tag{9.35}$$

which adds a small amount ε of a perturbing function $\eta^a(s)$ to the main curve $\boldsymbol{\alpha}(s)$. Describe this curve using perturbed parameters ξ^a:

$$\xi^a(s,\varepsilon) \equiv u^a(s) + \varepsilon\,\eta^a(s), \tag{9.36}$$

$$\eta^a(s_1) = \eta^a(s_2) = 0 \quad \text{(end-point anchoring).} \tag{9.37}$$

Suppose $L(\varepsilon)$ is the length of $\boldsymbol{\alpha}_\varepsilon(s)$ from $s = s_1 \to s_2$. Then

$$\mathrm{d}L(\varepsilon) = |\mathrm{d}\boldsymbol{\alpha}_\varepsilon| = \left|\frac{\partial \boldsymbol{X}}{\partial \xi^a}\,\mathrm{d}\xi^a\right|. \tag{9.38}$$

But (9.36) implies that

$$\frac{\partial \boldsymbol{X}}{\partial \xi^a} = \frac{\partial \boldsymbol{X}}{\partial u^a} \equiv \boldsymbol{e}_a\,, \tag{9.39}$$

so that

$$\mathrm{d}L(\varepsilon) = |\mathrm{d}\xi^a\,\boldsymbol{e}_a| = \sqrt{g_{ab}\,\mathrm{d}\xi^a\,\mathrm{d}\xi^b}\,, \tag{9.40}$$

where $g_{ab} \equiv \boldsymbol{e}_a\!\cdot\!\boldsymbol{e}_b$ is the metric on the surface. Thus

$$L(\varepsilon) = \int_{s_1}^{s_2} \left[g_{ab}\,\frac{\partial \xi^a}{\partial s}\,\frac{\partial \xi^b}{\partial s}\right]^{1/2} \mathrm{d}s \equiv \int_{s_1}^{s_2} \lambda(s,\varepsilon)\,\mathrm{d}s\,. \tag{9.41}$$

Requiring the geodesic length to be stationary means requiring $L'(0) = 0$. What is $L'(0)$? Since λ and $\partial\lambda/\partial\varepsilon$ are continuous, we can differentiate under the integral sign to give

$$L'(0) = \int_{s_1}^{s_2} \frac{\partial \lambda(s,0)}{\partial \varepsilon}\,\mathrm{d}s\,, \tag{9.42}$$

where (with a prime meaning $\mathrm{d}/\mathrm{d}s$ as usual)

$$\frac{\partial \lambda(s,\varepsilon)}{\partial \varepsilon} = \frac{1}{2\lambda(s,\varepsilon)}\left[\frac{\partial g_{ab}}{\partial \xi^c}\,\eta^c\,\frac{\partial \xi^a}{\partial s}\,\frac{\partial \xi^b}{\partial s} + 2\,g_{ab}\,\eta^{a\prime}\,\frac{\partial \xi^b}{\partial s}\right]. \tag{9.43}$$

Setting $\varepsilon = 0$ converts the ξ^a to u^a, while $\lambda(s,0) = 1$ since s is, after all, the length parameter:

$$\frac{\partial \lambda(s,0)}{\partial \varepsilon} = \frac{1}{2}\,g_{ab,c}\,\eta^c\,u^{a\prime}\,u^{b\prime} + g_{ab}\,\eta^{a\prime}\,u^{b\prime}, \quad \text{where } g_{ab,c} \equiv \partial g_{ab}/\partial u^c\,. \tag{9.44}$$

Thus

$$L'(0) = \int_{s_1}^{s_2} \left[\frac{1}{2}\, g_{ab,c}\, \eta^c\, u^{a\prime}\, u^{b\prime} + g_{ab}\, \eta^{a\prime}\, u^{b\prime} \right] \mathrm{d}s\,. \tag{9.45}$$

The second term in the brackets integrates by parts:

$$L'(0) = \int_{s_1}^{s_2} \frac{1}{2}\, g_{ab,c}\, \eta^c\, u^{a\prime}\, u^{b\prime}\, \mathrm{d}s + \underbrace{\left[g_{ab}\, \eta^a\, u^{b\prime} \right]_{s_1}^{s_2}}_{\Uparrow} - \int_{s_1}^{s_2} \frac{\mathrm{d}}{\mathrm{d}s} \left(g_{ab}\, u^{b\prime} \right) \eta^a\, \mathrm{d}s$$

equals zero because $\eta^a(s_1) = \eta^a(s_2) = 0$

$$= \int_{s_1}^{s_2} \left[\frac{1}{2}\, g_{ab,c}\, u^{a\prime}\, u^{b\prime} - \frac{\mathrm{d}}{\mathrm{d}s} \left(g_{cb}\, u^{b\prime} \right) \right] \eta^c\, \mathrm{d}s$$

$$= \int_{s_1}^{s_2} \left[\frac{1}{2}\, g_{ab,c}\, u^{a\prime}\, u^{b\prime} - g_{cb,a}\, u^{a\prime}\, u^{b\prime} - g_{cb}\, u^{b\prime\prime} \right] \eta^c\, \mathrm{d}s\,. \tag{9.46}$$

The expression in the brackets can be further simplified. The metric derivatives recall the Christoffel symbol of (8.182) together with its symmetries:

$$g_{ab,c} = \Gamma_{abc} + \Gamma_{bac}\,, \qquad \Gamma_{abc} = \Gamma_{acb}\,. \tag{9.47}$$

Using these, swapping some dummy indices, and finally raising the c subscript on the Christoffel symbols—as well as lowering it on the perturbing function—produces

$$L'(0) = -\int_{s_1}^{s_2} \left[\Gamma^c{}_{ab}\, u^{a\prime}\, u^{b\prime} + u^{c\prime\prime} \right] \eta_c\, \mathrm{d}s\,. \tag{9.48}$$

Now the hard work is done and we're in a position to prove the original assertion. If a curve is a geodesic, then (9.32) holds, so that (9.48) trivially produces $L'(0) = 0$, and the curve $\boldsymbol{\alpha}(s)$ has stationary length. Conversely, if $\boldsymbol{\alpha}(s)$ has stationary length, then $L'(0)$ must equal zero for any perturbation η_c in (9.48). That implies that the brackets in the integrand of (9.48) must be zero, which means that (9.32) holds, so $\boldsymbol{\alpha}(s)$ is a geodesic and the assertion is proved. (Equation (9.32) will also hold for an affine parameter, since such a parameter only introduces a multiplicative constant that can be divided away.)

So it is that a curve with stationary length on a surface must have no geodesic curvature—in essence it has no curvature of its own. All of its curvature is normal curvature, inherited from the surface itself.

It's not difficult to prove that, given a point P on a surface together with a unit tangent vector $\boldsymbol{v}$ at that point, it is always possible to draw a unique geodesic through P whose tangent equals $\boldsymbol{v}$. That is, a point and a direction suffice to determine a unique geodesic. Why? Referring to (9.17), the geodesic will be $\boldsymbol{\alpha}(s) \equiv \boldsymbol{X}\left(u^1(s), u^2(s)\right)$ with $\boldsymbol{\alpha}'(0) = \boldsymbol{v}$, so we wish to show that there exist functions $u^a(s)$ with a length parameter s, such that for the initial condition parameters u_0^1, u_0^2, v^1, v^2 satisfying

$$P \equiv \boldsymbol{X}(u_0^1, u_0^2)\,, \quad \text{and } \boldsymbol{v} \equiv v^a \boldsymbol{e}_a(u_0^1, u_0^2)\,, \tag{9.49}$$

the functions will satisfy

$$\begin{gathered} u^{a\prime\prime} + \Gamma^a_{\ ij}\, u^{i\prime}\, u^{j\prime} = 0\,, \\ u^a(0) = u_0^a\,, \quad u^{a\prime}(0) = v^a\,. \end{gathered} \tag{9.50}$$

The theory of differential equations guarantees that unique functions $u^a(s)$ satisfying (9.50) do exist in a neighbourhood of P, so we need only show that s is indeed a length parameter. To do this, consider the squared length of the tangent vector:

$$|\boldsymbol{\alpha}'(s)|^2 = g_{ab}\, u^{a\prime}(s)\, u^{b\prime}(s)\,. \tag{9.51}$$

First, $|\boldsymbol{\alpha}'(0)| = |\boldsymbol{v}| = 1$. Second, the derivative of $g_{ab}\, u^{a\prime}(s)\, u^{b\prime}(s)$ with respect to s, using (9.50), is zero, which shows that $|\boldsymbol{\alpha}'(s)|$ is a constant for all s. Thus $|\boldsymbol{\alpha}'(s)|$ always equals one, or $|\mathrm{d}\boldsymbol{\alpha}| = \mathrm{d}s$, and so s is indeed arc length and the theorem is proved. We'll use this theorem in the next section.

9.3 The Curvature of a Surface

In the discussion around Fig. 9.6 and equation (9.22), we discussed the idea of measuring the normal curvature of a surface in a given direction at some point P, since this was just the curvature of a curve drawn on the surface in a plane containing the normal vector $\boldsymbol{U}$, with that curve's tangent pointing in the given direction. We see now that, locally at least, such a curve is a geodesic, and it can always be drawn, by the theorem we proved at the end of the last section. Its curvature k at P is equal to k_n since $k_g = 0$.

How does the normal curvature at P depend on direction? Since it is the curvature k of a locally geodesic curve, we can imagine the infinite number of such curves through P, each lying in a different normal plane. If we slowly rotate this plane through 360° around the normal through P, the sequence of curvatures k of the local geodesics varies smoothly as a function of angle, and so must attain maximal and minimal values in what are known as *principal directions*. (These, we will show later, are happily always orthogonal.) These extremal values of k are known as the *principal normal curvatures* at P.

It turns out that the whole set of normal curvatures can be condensed in a simple way to define just one number quantifying the extrinsic curvature of the surface at P. Before doing this, however, we'll follow a different approach by extending the notion of the curvature of a one-parameter curve to a two-parameter surface.

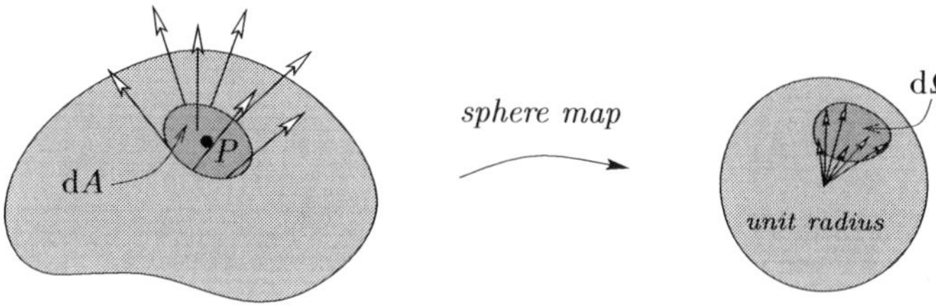

Fig. 9.9. The 2-surface analogy of Fig. 9.3. The normals for all points in a small area $\mathrm{d}A$ on a surface encompass a solid angle $\mathrm{d}\Omega$, which equals the normals' swept area on a unit sphere. This is called a *sphere map* or a *Gauss map*.

Here is how that's done. In Fig. 9.3, we pictured a curve in the neighbourhood of some point P as having a set of tangent vectors along it (or, equivalently, principal normal vectors), fanning out as it were. Its curvature k was defined in (9.3) to be the angle made by the fan per unit arc length: thus $k = \mathrm{d}\theta/\mathrm{d}s$. Equivalently, since any angle equals its corresponding arc length on a unit-radius circle, the curvature k was equal to the dimensionless arc length swept out on a unit-radius circle by the curve's tangent (or normal) vector per unit real arc length of the curve itself. A circle turned out to have a constant curvature of 1/radius.

The same idea applies to a 2-surface. Referring to Fig. 9.9, define the *Gauss curvature* $K(P)$ of the surface at the point P to be the solid angle swept out by the normals in a vanishing neighbourhood of P per unit real area of the surface. (Remember that the solid angle subtended by a set of rays is defined to be equal to the dimensionless area that those rays sweep out over the surface of a unit-radius sphere—or equivalently, the area swept out on *any* sphere divided by the square of that sphere's radius.)

$$K(P) \equiv \frac{\mathrm{d}\Omega}{\mathrm{d}A} = \lim_{\Delta A \to 0} \frac{\begin{bmatrix}\text{dimensionless area swept out}\\ \text{on unit sphere by normals}\end{bmatrix}}{\text{corresponding area } \Delta A \text{ of actual surface}}\,. \tag{9.52}$$

Figure 9.9 defines the *sphere map* as the generalisation of Fig. 9.3's circle map. We can see immediately that a sphere of radius r has a Gauss curvature of $4\pi/(4\pi r^2) = 1/r^2$, which is a reasonable extension of the one-parameter case of a circle (which has $k = 1/r$).

This is all intuitive but needs to be further quantified. To do so, we need to know how to measure an arbitrary surface area. Remember that the surface area is given by the cross product in (2.52). In Fig. 9.5, an infinitesimal parallelogram drawn in the curved surface along directions of constant u^a has edge vectors $\partial \boldsymbol{X}/\partial u^1\, \mathrm{d}u^1$ and $\partial \boldsymbol{X}/\partial u^2\, \mathrm{d}u^2$, written as $\boldsymbol{X}_{,1}\, \mathrm{d}u^1$ and $\boldsymbol{X}_{,2}\, \mathrm{d}u^2$. The area of the surface element on the left in Fig. 9.9 is thus

$$\mathrm{d}A = |\boldsymbol{X}_{,1} \times \boldsymbol{X}_{,2}|\, \mathrm{d}u^1\, \mathrm{d}u^2 = |\boldsymbol{e}_1 \times \boldsymbol{e}_2|\, \mathrm{d}u^1\, \mathrm{d}u^2. \tag{9.53}$$

What about the area $d\Omega$ of the corresponding element in the spherical image in Fig. 9.9? Vectors that bound an infinitesimal parallelogram in the image's spherical surface are always perpendicular to its radius vectors. But these radius vectors are just the unit normals $\boldsymbol{U}(u^1, u^2)$. The parallelogram must therefore have edge vectors $\partial\boldsymbol{U}/\partial u^1\, du^1$ and $\partial\boldsymbol{U}/\partial u^2\, du^2$, written as $\boldsymbol{U}_{,1}\, du^1$ and $\boldsymbol{U}_{,2}\, du^2$. The area of the surface element will be the magnitude of their cross product, or $|\boldsymbol{U}_{,1}\, du^1 \times \boldsymbol{U}_{,2}\, du^2|$. The cross product itself must be proportional to $\boldsymbol{U}$ since the normal to the sphere's surface is everywhere proportional to its radius vector $\boldsymbol{U}$. Thus $\boldsymbol{U}_{,1} \times \boldsymbol{U}_{,2} \propto \boldsymbol{U}$.

Actually, we can improve on this result by introducing a signed area corresponding to whether the ordered set $\{\boldsymbol{U}_{,1}, \boldsymbol{U}_{,2}, \boldsymbol{U}\}$ is right- or left handed. A change in handedness corresponds to the left-hand surface in Fig. 9.9 changing its curvature in such a way that its normal vectors fan in instead of out. These normals still trace out an area on the image sphere but their ordering has changed. (We could have done this in (9.3) for the one-dimensional case but decided against it to keep the discussion simple.) This all means that

$$\boldsymbol{U} = \pm \frac{\boldsymbol{U}_{,1} \times \boldsymbol{U}_{,2}}{|\boldsymbol{U}_{,1} \times \boldsymbol{U}_{,2}|}, \tag{9.54}$$

with $\pm$ corresponding to whether $\{\boldsymbol{U}_{,1}, \boldsymbol{U}_{,2}, \boldsymbol{U}\}$ is right- or left handed, respectively. In that case, define the signed area as

$$\begin{aligned} d\Omega &\equiv \pm|\boldsymbol{U}_{,1} \times \boldsymbol{U}_{,2}|\, du^1\, du^2 \\ &= \boldsymbol{U}\cdot(\boldsymbol{U}_{,1} \times \boldsymbol{U}_{,2})\, du^1\, du^2, \end{aligned} \tag{9.55}$$

so that the Gauss curvature becomes

$$K = \frac{d\Omega}{dA} = \frac{\boldsymbol{U}\cdot(\boldsymbol{U}_{,1} \times \boldsymbol{U}_{,2})}{|\boldsymbol{e}_1 \times \boldsymbol{e}_2|}. \tag{9.56}$$

What is $\boldsymbol{U}_{,i}$ in terms of the basis $\boldsymbol{e}_1, \boldsymbol{e}_2$? A clue comes from (9.19), which has all of the ingredients. So begin with the product rule for differentiation:

$$(\boldsymbol{e}_a\cdot\boldsymbol{U})_{,b} = \boldsymbol{e}_{a,b}\cdot\boldsymbol{U} + \boldsymbol{e}_a\cdot\boldsymbol{U}_{,b}\,. \tag{9.57}$$

But $\boldsymbol{U}$ is perpendicular to $\boldsymbol{e}_1, \boldsymbol{e}_2$, so this last equation combined with (9.19) gives

$$\boldsymbol{e}_a\cdot\boldsymbol{U}_{,b} = -L_{ab}\,. \tag{9.58}$$

Now introduce coefficients $\gamma^c{}_b$ to write $\boldsymbol{U}_{,b}$ as a linear combination of basis vectors:

$$\boldsymbol{U}_{,b} = \gamma^c{}_b\, \boldsymbol{e}_c\,. \tag{9.59}$$

Dotting this with $\boldsymbol{e}_a$ then leads to $\gamma_{ab} = -L_{ab}$, bringing us to the *equations of Weingarten*:

$$\boldsymbol{U}_{,b} = -L^c{}_b\, \boldsymbol{e}_c\,. \tag{9.60}$$

In that case, writing $\left(L^i{}_j\right)$ as a matrix whose ij^{th} element is $L^i{}_j$,

$$\begin{aligned} \boldsymbol{U}_{,1} \times \boldsymbol{U}_{,2} &= L^a{}_1\, \boldsymbol{e}_a \times L^b{}_2\, \boldsymbol{e}_b \\ &= \det\left(L^i{}_j\right)\, \boldsymbol{e}_1 \times \boldsymbol{e}_2\,. \end{aligned} \tag{9.61}$$

But, since $L^i{}_j = g^{ia} L_{aj}$, it must be that

$$\det\left(L^i{}_j\right) = L/g\,, \quad \text{where } L \equiv \det\left(L_{ij}\right) \tag{9.62}$$

and g is the usual surface metric determinant, $\det\left(g_{ij}\right)$. The Gauss curvature becomes

$$K = \frac{\frac{L}{g}\, \boldsymbol{U} \cdot (\boldsymbol{e}_1 \times \boldsymbol{e}_2)}{|\boldsymbol{e}_1 \times \boldsymbol{e}_2|} = \frac{L}{g} \tag{9.63}$$

using (9.16). So we see that the Gauss curvature is related to the first and second fundamental forms.

The Gauss curvature, being a measure of how the surface bends within the surrounding 3-space, should be related to the set of normal curvatures referred to at the start of this section on p. 362. (Remember that the normal curvature $k_n(\boldsymbol{v})$ was defined in a given tangent direction $\boldsymbol{v}$.) With the benefit of hindsight, this relationship can be found by determining the extremal values of the normal curvature. We know that because the surface is smooth, the normal curvature must take on a continuum of values as the tangent vector $\boldsymbol{v}$ is rotated through a full 360°, and so k_n must attain a maximum and a minimum. How do we do this? Earlier, in (9.22), we defined the normal curvature in a direction $\boldsymbol{v}$ as

$$k_n(\boldsymbol{v}) = L_{ab}\, v^a\, v^b\,, \tag{9.64}$$

where $\boldsymbol{v}$ is a unit vector tangent to the surface. We must extremise $k_n(\boldsymbol{v})$ subject to $g_{ab}\, v^a\, v^b = 1$. The calculation actually becomes difficult if we simply write

$$k_n = L_{11} \left(v^1\right)^2 + 2L_{12}\, v^1\, v^2 + L_{22} \left(v^2\right)^2 \tag{9.65}$$

and then express, say, v^2 in terms of v^1 to convert the extremisation to an exercise in single-variable calculus. An alternative approach uses the method of Lagrange multipliers, so we'll digress for a moment to discuss this.

9.3.1 The Method of Lagrange Multipliers

As a simple example of the method, suppose we wish to extremise the value of $f(x, y) = x + y$ subject to a constraint $x^2 + y^2 = 1$. In Fig. 9.10 are plotted contours of $f =$ *constant*, together with the constraint $x^2 + y^2 = 1$. Our only interest is the contours that intersect the constraint; in particular, we wish to find the extremal contour that just *touches* the constraint. Continuity of the functions involved guarantees that such a contour exists.

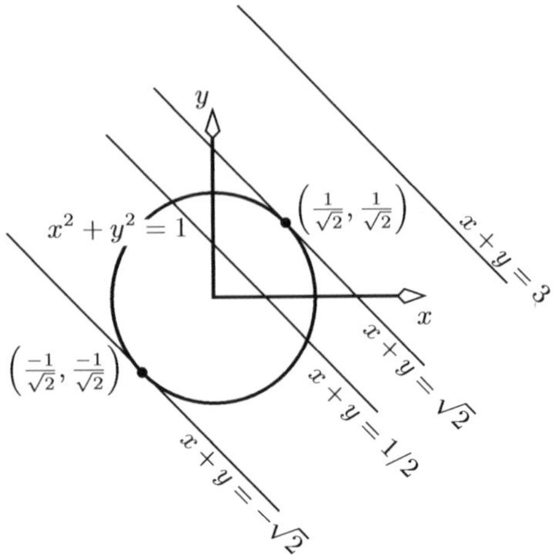

Fig. 9.10. A simple exercise in Lagrange multipliers. Extremising $f(x, y) = x + y$ subject to the constraint $x^2 + y^2 = 1$ involves choosing a contour of f that not only touches the constraint circle but is tangential to it. Thus, at the point(s) of interest, normal vectors of the function and the constraint must point in the same direction.

How do we find this tangential contour? A small step in any direction $\mathrm{d}\boldsymbol{x}$ gives rise to an increase in f of $\mathrm{d}f = \nabla f \cdot \mathrm{d}\boldsymbol{x}$. If we step along a contour of constant f, it must be true that $\mathrm{d}f = 0$, so that $\nabla f \perp \mathrm{d}\boldsymbol{x}$. So the gradient ∇f always points perpendicular to the contours of constant f, precisely as we saw in Fig. 8.7. Of course, this orthogonality of the gradient applies to any constraint as well, in which case at the extremal points the gradients of f and the constraint must line up—they must be proportional. So, in more formal terms, to extremise the value of $f(x, y)$ subject to a constraint $g(x, y) = 0$, we must solve

$$\begin{aligned} \nabla f(x, y) &= \lambda \nabla g(x, y)\,, \\ g(x, y) &= 0\,. \end{aligned} \tag{9.66}$$

This is the method of Lagrange multipliers. The parameter λ is called the Lagrange multiplier and embodies the proportionality of the gradient vectors, but apart from that we're not interested in its value. In the case of Fig. 9.10, the function to extremise is $f(x, y) = x + y$, while the constraint is $g(x, y) = x^2 + y^2 - 1$, so (9.66) becomes

$$\begin{aligned} (1, 1) &= \lambda(2x, 2y)\,, \\ x^2 + y^2 &= 1\,. \end{aligned} \tag{9.67}$$

These give $(x, y) = \pm(1/\sqrt{2}, 1/\sqrt{2})$, together with the extremal values of $\pm\sqrt{2}$ for f. Details of solving nonlinear simultaneous equations are in the box on the following page.

Solving Nonlinear Simultaneous Equations

How do we solve a set of three nonlinear simultaneous equations such as (9.67)? First, write them as

$$\begin{aligned} f_1(x, y, \lambda) &= 0\,, \\ f_2(x, y, \lambda) &= 0\,, \\ f_3(x, y, \lambda) &= 0\,. \end{aligned} \tag{9.68}$$

There are three equations in three unknowns here. Set $\boldsymbol{x} \equiv [x\ y\ \lambda]^t$ (a column vector), and write them as $f_1(\boldsymbol{x}) = 0$, etc. For some $\boldsymbol{x}_0$ not too far from $\boldsymbol{x}$, we can Taylor-expand the functions to first order, writing (and remembering that $\nabla f_1, \dots \nabla f_3$ are row vectors)

$$\begin{bmatrix} f_1(\boldsymbol{x}) \\ f_2(\boldsymbol{x}) \\ f_3(\boldsymbol{x}) \end{bmatrix} \simeq \begin{bmatrix} f_1(\boldsymbol{x}_0) \\ f_2(\boldsymbol{x}_0) \\ f_3(\boldsymbol{x}_0) \end{bmatrix} + \underbrace{\begin{bmatrix} \nabla f_1(\boldsymbol{x}_0) \\ \nabla f_2(\boldsymbol{x}_0) \\ \nabla f_3(\boldsymbol{x}_0) \end{bmatrix}}_{\equiv J(\boldsymbol{x}_0)} (\boldsymbol{x} - \boldsymbol{x}_0)\,. \tag{9.69}$$

But the left-hand side of this equation is the zero vector, in which case

$$0 \simeq \begin{bmatrix} f_1(\boldsymbol{x}_0) \\ f_2(\boldsymbol{x}_0) \\ f_3(\boldsymbol{x}_0) \end{bmatrix} + J(\boldsymbol{x}_0)\,(\boldsymbol{x} - \boldsymbol{x}_0)\,, \tag{9.70}$$

which rearranges to give

$$\boldsymbol{x} \simeq \boldsymbol{x}_0 - J^{-1}(\boldsymbol{x}_0) \begin{bmatrix} f_1(\boldsymbol{x}_0) \\ f_2(\boldsymbol{x}_0) \\ f_3(\boldsymbol{x}_0) \end{bmatrix}. \tag{9.71}$$

This last equation can be used iteratively, hopefully converging to the solution: the initial estimate of the parameters x, y, λ is $\boldsymbol{x}_0$, and the new estimate is given by the right-hand side of (9.71). The linearisation used here is reminiscent of the approach taken for nonlinear least squares fitting in the box on p. 101.

For the case of three variables, suppose we must extremise $f(x, y, z)$ subject to constraints $g_1(x, y, z) = 0$ and $g_2(x, y, z) = 0$. The constraints now describe surfaces in the xyz-space that intersect in a curve. Any surface of constant f must intersect this curve, and the extremal one must touch it tangentially. This implies that ∇f must lie in the plane formed by ∇g_1 and ∇g_2. So we must solve

$$\begin{aligned} \nabla f(x, y, z) &= \lambda_1 \nabla g_1(x, y, z) + \lambda_2 \nabla g_2(x, y, z)\,, \\ g_1(x, y, z) &= 0\,, \\ g_2(x, y, z) &= 0\,. \end{aligned} \tag{9.72}$$

With more variables, the method extends quite easily: ∇f must be a linear combination of $\nabla g_1, \ldots, \nabla g_n$, and of course the constraints themselves must always be included. A rigorous proof in higher dimensions can be found in advanced calculus texts.

For the case at hand, we must extremise the normal curvature in a direction given by a unit vector $\boldsymbol{v}$:

$$\text{Extremise } k_n = L_{ab}\, v^a\, v^b \quad \text{subject to } g_{ab}\, v^a\, v^b = 1\,. \tag{9.73}$$

Write these as

$$\begin{aligned} k_n &= L_{11}\left(v^1\right)^2 + 2L_{12}\, v^1\, v^2 + L_{22}\left(v^2\right)^2, \\ g_{11}\left(v^1\right)^2 &+ 2g_{12}\, v^1\, v^2 + g_{22}\left(v^2\right)^2 = 1\,. \end{aligned} \tag{9.74}$$

The gradient part of Lagrange's method is

$$\begin{bmatrix} 2L_{11}\, v^1 + 2L_{12}\, v^2 \\ 2L_{12}\, v^1 + 2L_{22}\, v^2 \end{bmatrix} = \lambda \begin{bmatrix} 2g_{11}\, v^1 + 2g_{12}\, v^2 \\ 2g_{12}\, v^1 + 2g_{22}\, v^2 \end{bmatrix}. \tag{9.75}$$

Writing the matrices (L_{ab}) and (g_{ab}) as $\widehat{L}$ and $\widehat{g}$ allows us to write (9.75) more tidily as

$$\widehat{L}\, \boldsymbol{v} = \lambda \widehat{g}\, \boldsymbol{v}\,, \tag{9.76}$$

so that $(\widehat{L} - \lambda\widehat{g})\boldsymbol{v} = 0$, which implies that the determinant $\left|\widehat{L} - \lambda\widehat{g}\right| = 0$ since $\boldsymbol{v} \neq 0$:

$$\begin{vmatrix} L_{11} - \lambda g_{11} & L_{12} - \lambda g_{12} \\ L_{12} - \lambda g_{12} & L_{22} - \lambda g_{22} \end{vmatrix} = 0\,. \tag{9.77}$$

(Remember that both $\widehat{L}$ and $\widehat{g}$ are symmetric.) Writing (as usual) the determinants of $\widehat{L}$ and $\widehat{g}$ as L and g, this last equation becomes

$$\lambda^2 - \frac{\lambda}{g}\left(L_{11}\, g_{22} + 2L_{12}\, g_{12} + L_{22}\, g_{11}\right) + \frac{L}{g} = 0\,. \tag{9.78}$$

There are two Lagrange multiplier roots to this quadratic, but at this point we notice something: their product must be L/g, which is just the Gauss curvature. But since $k_n = L_{ab}\, v^a\, v^b$, it must follow that

$$k_n = \boldsymbol{v}^t\, \widehat{L}\, \boldsymbol{v} = \boldsymbol{v}^t\, \lambda\widehat{g}\, \boldsymbol{v} = \lambda\, g_{ab}\, v^a\, v^b = \lambda\,. \tag{9.79}$$

Thus, the Lagrange multipliers happen to equal the corresponding maximal and minimal normal curvatures (call these k_1 and k_2), so their product is therefore

$$k_1\, k_2 = L/g\,. \tag{9.80}$$

This is an elegant result, so we decide not to carry through the solution of (9.78); we're content to note that the product of the maximal and minimal normal curvatures is the Gauss curvature.

What about the two principal directions, vectors $\boldsymbol{v}_1$ and $\boldsymbol{v}_2$ that correspond to k_1, k_2? We can show these are orthogonal by a commonly used technique in this type of analysis. First, it's certainly true that

$$\boldsymbol{v}_1^t \, \widehat{L} \, \boldsymbol{v}_2 = \boldsymbol{v}_2^t \, \widehat{L} \, \boldsymbol{v}_1 \,. \tag{9.81}$$

The reason is that these expressions are scalars, so the left-hand side must equal its transpose, which, because $\widehat{L}$ is symmetric, equals the right-hand side. However, we can also eliminate $\widehat{L}$ using (9.76) with $\lambda \to k$:

$$\begin{aligned} \boldsymbol{v}_1^t \, \widehat{L} \, \boldsymbol{v}_2 &= \boldsymbol{v}_1^t \, k_2 \, \widehat{g} \, \boldsymbol{v}_2 = k_2 \, \boldsymbol{v}_1 \cdot \boldsymbol{v}_2 \\ \text{and} \quad \boldsymbol{v}_2^t \, \widehat{L} \, \boldsymbol{v}_1 &= \boldsymbol{v}_2^t \, k_1 \, \widehat{g} \, \boldsymbol{v}_1 = k_1 \boldsymbol{v}_2 \cdot \boldsymbol{v}_1 \,. \end{aligned} \tag{9.82}$$

So $k_2 \, \boldsymbol{v}_1 \cdot \boldsymbol{v}_2 = k_1 \, \boldsymbol{v}_2 \cdot \boldsymbol{v}_1$. If k_1, k_2 differ, then necessarily $\boldsymbol{v}_1 \perp \boldsymbol{v}_2$, which means that the principal directions will be orthogonal. If $k_1 = k_2$, the normal curvatures are everywhere equal, so any two orthogonal directions serve as principal directions.

In summary, the Gauss curvature can be visualised in a simple way. We consider all of the normal plane sections of our surface at the point of interest, each one producing a curve with curvature equal to 1/radius of its osculating circle. By rotating the normal plane through 360° around the point of interest, the curvatures attain maximal and minimal values in the principal directions, which are orthogonal if the curvature changes throughout the 360°. The product of these two curvatures is the Gauss curvature. Examples are shown in Fig. 9.11.

It might be thought that somehow the Gauss curvature favours the principal directions together with their curvatures. After all, as we vary the unit vector $\boldsymbol{v}$ along which we measure normal curvature, there is a continuum of curvature values in between the principal curvatures $\boldsymbol{v}_1$ and $\boldsymbol{v}_2$. Shouldn't these intermediate normal curvatures also have a say in quantifying the Gauss curvature? In a sense, they do—because they can all be derived from the principal curvatures! If we call $(\boldsymbol{v}, \boldsymbol{v}_1)$ the angle between $\boldsymbol{v}$ and $\boldsymbol{v}_1$ and likewise $(\boldsymbol{v}, \boldsymbol{v}_2)$ the angle between $\boldsymbol{v}$ and $\boldsymbol{v}_2$ (so that $(\boldsymbol{v}, \boldsymbol{v}_1) + (\boldsymbol{v}, \boldsymbol{v}_2) = 90°$), then (9.64) together with some arguments from this section can be used to show that all intermediate normal curvatures $k_n(\boldsymbol{v})$ are given by *Euler's formula*:

$$k_n(\boldsymbol{v}) = k_1 \cos^2(\boldsymbol{v}, \boldsymbol{v}_1) + k_2 \cos^2(\boldsymbol{v}, \boldsymbol{v}_2) \,. \tag{9.83}$$

This has a pleasing symmetry and shows that, despite appearances, the Gauss curvature does not really "favour" the principal normal curvatures over their intermediate values.

9.4 Gauss's Extraordinary Theorem

So far, the Gauss curvature $K = L/g$ looks to be a mixture of extrinsic curvature (characterised by L) and intrinsic curvature (characterised by g). But

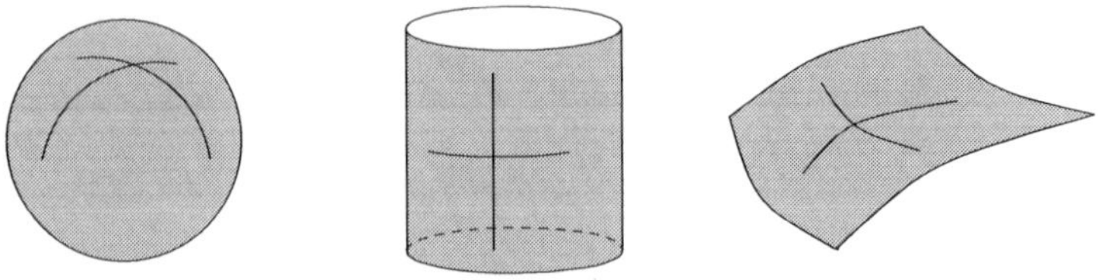

Fig. 9.11. Pairs of curves having the principal normal curvatures at their point of intersection and thus showing the principal directions at that point. The Gauss curvature at the intersection is the product of the two principal normal curvatures. Shown are surfaces of everywhere positive, zero, and negative Gauss curvature. **Left**: All slices of the sphere through any normal have constant normal curvature equal to $1/R$, where R is the sphere's radius. The two principal normal curvatures are thus everywhere $1/R$, so the sphere's Gauss curvature is $1/R^2$ at every point. **Middle**: The cylinder of radius R has its principal directions horizontal and vertical, with principal normal curvatures $1/R$ and zero, respectively. Thus its Gauss curvature is zero at every point. This indicates it can be unrolled into a plane without being deformed (see p. 372). **Right**: The saddle surface has principal normal curvatures of opposite sign, so its Gauss curvature at any point must be negative.

something quite remarkable soon appears; Gauss found it so remarkable that he christened it his *Theorema Egregium*, or Extraordinary Theorem.

To see what so impressed Gauss, consider the first derivatives of basis vectors, $\boldsymbol{e}_{a,b}$ and $\boldsymbol{e}_{b,a}$. These must be equal since each is really the second derivative of the map $\boldsymbol{X}$, and we know that the order of differentiation is immaterial. We originally used this, together with Gauss's formulae (9.18),

$$\boldsymbol{e}_{a,b} = \Gamma^c{}_{ab}\, \boldsymbol{e}_c + L_{ab}\, \boldsymbol{U}\,, \tag{9.84}$$

to infer that $\Gamma^c{}_{ab}$ and L_{ab} are symmetric in a and b. Now, what happens if we consider the next higher order of derivative? The same reasoning gives

$$\boldsymbol{e}_{a,mn} - \boldsymbol{e}_{a,nm} = 0\,, \tag{9.85}$$

which we shorten to

$$\boldsymbol{e}_{a,\underbracket{mn}} = 0\,, \tag{9.86}$$

where the underbracket points to the two indices that are swapped to make a new expression, which is then subtracted from the first. (This notation is useful because, for example, it encapsulates linearity: if $A_{\alpha\beta} = bB_{\alpha\beta} + cC_{\alpha\beta}$, then $A_{\underbracket{\alpha\beta}} = bB_{\underbracket{\alpha\beta}} + cC_{\underbracket{\alpha\beta}}$.) Expanding this using Gauss's equations (9.84) and Weingarten's equations (9.60) produces

$$\left[\Gamma^b{}_{a\underbracket{m,n}} + \Gamma^c{}_{a\underbracket{m}\,\Gamma^b{}_{c\,n}} - L_{a\underbracket{m}\,L^b{}_{n}}\right] \boldsymbol{e}_b + \left[\Gamma^b{}_{a\underbracket{m}\,L_{bn}} + L_{a\underbracket{m,n}}\right] \boldsymbol{U} = 0\,. \tag{9.87}$$

We conclude that each of the brackets equals zero. In particular, the first one is the most interesting since it separates the Γ and the L. Setting it equal to zero results in

$$L_{am}L^b{}_n = \Gamma^b{}_{am,n} + \Gamma^c{}_{am}\Gamma^b{}_{cn} \equiv R^b{}_{anm}\,. \tag{9.88}$$

Expression (9.88) will soon be shown to be a tensor and is called the *Riemann tensor*. A more convenient ordering of the indices makes it easier to memorise, albeit now with minus signs:

$$\boxed{R^a{}_{bcd} = -\Gamma^a{}_{bc,d} - \Gamma^z{}_{bc}\Gamma^a{}_{dz} = -L_{bc}L^a{}_d\,.} \tag{9.89}$$

Upon lowering the first index, the Riemann tensor becomes

$$R_{abcd} = \begin{vmatrix} L_{ac} & L_{ad} \\ L_{bc} & L_{bd} \end{vmatrix}, \tag{9.90}$$

which is a useful expression since it indicates the tensor's three symmetries, distinguishing its first pair of indices from its last pair:

$$\begin{aligned} R_{bacd} &= -R_{abcd}\,, \\ R_{abdc} &= -R_{abcd}\,, \\ R_{cdab} &= R_{abcd}\,. \end{aligned} \tag{9.91}$$

In fact, for 2-surfaces these symmetries mean that most of the Riemann components vanish: the only nonzero ones are R_{1212} and three others with permuted indices, which all have the same magnitude. In particular,

$$R_{1212} = \begin{vmatrix} L_{11} & L_{12} \\ L_{21} & L_{22} \end{vmatrix} = L\,, \tag{9.92}$$

which means the Gauss curvature is

$$K = R_{1212}/g\,. \tag{9.93}$$

But the Riemann tensor is a function of the metric only, from (9.88), implying that the Gauss curvature is also a function of the metric alone. So this curvature that we defined as a measure of how the surface bends within its exterior space, the Gauss curvature, is *intrinsic*—it's purely a function of the metric and its derivatives! This is the content of the *Theorema Egregium*, and it enables us to dispense with imagining the surface as curving in a higher-dimensional space.

Dispensing with the higher embedding space brings about a change in philosophy. A nonexistent normal vector $\boldsymbol{U}$ can give no contribution to (9.18), which then becomes

$$\boldsymbol{e}_{a,b} = \Gamma^c{}_{ab}\,\boldsymbol{e}_c\,. \tag{9.94}$$

(Equivalently, (9.19) gives a zero second fundamental form L_{ab}.) Because we have always reserved Latin indices for an index subset, we signal the dispensing of the embedding space by using Greek indices to write (9.94) as

$$\boldsymbol{e}_{\alpha,\beta} = \Gamma^{\mu}{}_{\alpha\beta}\,\boldsymbol{e}_{\mu}\,, \tag{9.95}$$

which of course was introduced in (8.177) and used extensively in the last chapter. The notion of covariant differentiation, since it uses (9.95), enables us to avoid constantly having to account for the basis vectors in derivatives because covariant differentiation allows the basis to be treated as constant. So in practice the basis is often ignored or even forgotten, which is one reason that vectors are usually held to be synonymous with their coordinate vectors (i.e. components) in physics texts. They are not really synonymous, and the identification is a subtlety that can easily be missed, but some of its roots lie in the *Theorema Egregium* as we've seen here. We'll use this idea in the next section when we further examine the covariant derivative using the form $A^{\alpha}{}_{;\beta} = A^{\alpha}{}_{,\beta} + \Gamma^{\alpha}{}_{\beta\mu}\,A^{\mu}$ that was produced when (9.95) was included in (8.202).

Because the curvature of a surface is intrinsic, any operation that preserves the metric must also preserve the curvature. Bending a sheet of paper alters no distances between infinitesimally separated points, so does not change the metric on its surface, and thus we can see that the intrinsic curvature of the resulting cylinder should still be zero, as shown in Fig. 9.11. Of course, the cylinder does have extrinsic curvature—it curves within the surrounding 3-space after all. So, when standing on Earth's surface, a measurement of the amount by which the horizon drops below eye level in one direction is enough to detect an extrinsic curvature: perhaps Earth is a cylinder. But to show that it has intrinsic curvature (such as a spherical surface), we need to measure a drop below eye level in two directions, as of course we do.

The Riemann Tensor from Anticommuting Covariant Derivatives

The swapping of indices that produced the Riemann tensor brings up another way of seeing how it arises: it measures the degree to which two successive covariant derivatives don't commute. That is, if we write

$$A^{\alpha}{}_{;\mu\nu} = A^{\alpha}{}_{;\mu,\nu} + \Gamma^{\alpha}{}_{\beta\nu}A^{\beta}{}_{;\mu} - \Gamma^{\beta}{}_{\mu\nu}A^{\alpha}{}_{;\beta}\,, \tag{9.96}$$

then expand the covariant derivatives that remain and anticommute over μ and ν, we find that

$$A^{\alpha}{}_{;\underline{\mu\nu}} = R^{\alpha}{}_{\beta\nu\mu}A^{\beta}. \tag{9.97}$$

This shows that the Riemann tensor really is a tensor, because it appears here in an equality whose left-hand side is a tensor, and we know that A^{β} is a tensor. This is another example of the tensor quotient theorem referred to on p. 282.

In fact, expressions like (9.97) hold for any tensor, just as (8.208) uses Christoffel symbols to convert covariant differentiation to normal differentiation, with a separate Christoffel term for each index, using a plus sign for raised indices and a minus sign for lowered indices. That is, for the case of

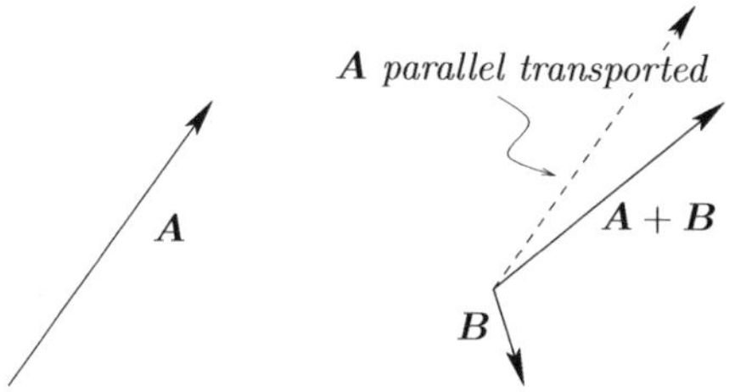

Fig. 9.12. To add vectors at different points in a flat space, we think nothing of translating one of them—"parallel transporting it"—to where the other is located.

anticommuting covariant derivatives, there is a separate Riemann term for each index, again added for raised indices and subtracted for lowered ones. For example,

$$A^\alpha{}_{\beta\gamma;\underset{\sqcup}{\mu\nu}} = R^\alpha{}_{\lambda\nu\mu}A^\lambda{}_{\beta\gamma} - R^\lambda{}_{\beta\nu\mu}A^\alpha{}_{\lambda\gamma} - R^\lambda{}_{\gamma\nu\mu}A^\alpha{}_{\beta\lambda}\,. \tag{9.98}$$

9.5 Translating Vectors by Parallel Transport

In general, vectors cannot be added at different points. This is easy to see by checking whether the sum of the components transforms in the required way:

$$\begin{aligned} A^\alpha(P) + B^\alpha(Q) &= \Lambda^\alpha_{\bar\beta}(P)A^{\bar\beta}(P) + \Lambda^\alpha_{\bar\beta}(Q)B^{\bar\beta}(Q) \\ &\overset{?}{=} \Lambda^\alpha_{\bar\beta}(?)\left(A^{\bar\beta}(P) + B^{\bar\beta}(Q)\right). \end{aligned} \tag{9.99}$$

The second line in (9.99) with its question-marked argument of $\Lambda^\alpha_{\bar\beta}$ cannot be precisely defined. Thus the sum does not obey the basic tensor transformation law—unless $\Lambda^\alpha_{\bar\beta}$ is a constant—so we must conclude that the sum is not a vector.

In a flat space, we're used to adding vectors at different points by translating them to where we need them, as shown in Fig. 9.12. Is it possible to mimic this on a curved surface and so define some kind of addition for vectors there? In other words, given A^α at some point, can we shift it around so that in some sense it remains constant?

Changing A^α continuously requires a knowledge of how vectors can be differentiated. Consider a curve $x^\alpha(s)$ parametrised by its length s, together with a vector field $\boldsymbol{A}$. How do the components A^α change along the curve? They change as

$$\frac{\mathrm{d}A^\alpha}{\mathrm{d}s} = \underbrace{A^\alpha{}_{,\beta}}_{\text{Is not a tensor}} \times \underbrace{\frac{\mathrm{d}x^\beta}{\mathrm{d}s}}_{\text{Is a tensor}}\,. \tag{9.100}$$

We must conclude that $\mathrm{d}A^\alpha/\mathrm{d}s$ is *not* a vector (that is, $\mathrm{d}A^\alpha/\mathrm{d}s$ are not the components of any vector). So how does the vector $\boldsymbol{A}$ itself change?

$$\frac{\mathrm{d}\boldsymbol{A}}{\mathrm{d}s} = \frac{\mathrm{d}\,(A^\alpha \boldsymbol{e}_\alpha)}{\mathrm{d}s} = (A^\alpha \boldsymbol{e}_\alpha)_{,\beta}\,\frac{\mathrm{d}x^\beta}{\mathrm{d}s} = A^\alpha{}_{;\beta}\,\boldsymbol{e}_\alpha\,\frac{\mathrm{d}x^\beta}{\mathrm{d}s}\,. \tag{9.101}$$

This prompts us to define a new set of numbers in analogy to (9.100):

$$\boxed{\frac{\mathrm{D}A^\alpha}{\mathrm{d}s} \equiv \underbrace{A^\alpha{}_{;\beta}\,\frac{\mathrm{d}x^\beta}{\mathrm{d}s}}_{\text{Both tensors}}\,.} \tag{9.102}$$

Since the right-hand side of (9.102) is a tensor, then so also is $\mathrm{D}A^\alpha/\mathrm{d}s$. The vector field derivative (9.101) becomes

$$\frac{\mathrm{d}\boldsymbol{A}}{\mathrm{d}s} = \frac{\mathrm{D}A^\alpha}{\mathrm{d}s}\,\boldsymbol{e}_\alpha\,, \quad \text{or} \quad \mathrm{d}\boldsymbol{A} = \mathrm{D}A^\alpha\,\boldsymbol{e}_\alpha\,. \tag{9.103}$$

In this way we're able to treat the changing basis vectors as if they were constant by investing all of the change of $\boldsymbol{A}$ in the $\mathrm{D}A^\alpha/\mathrm{d}s$ components. The components of this change are

$$\begin{aligned}\mathrm{D}A^\alpha &= A^\alpha{}_{;\gamma}\,\mathrm{d}x^\gamma \qquad (\text{cf. } \mathrm{d}A^\alpha = A^\alpha{}_{,\gamma}\,\mathrm{d}x^\gamma)\\ &= \left(A^\alpha{}_{,\gamma} + \Gamma^\alpha{}_{\beta\gamma}\,A^\beta\right)\mathrm{d}x^\gamma\\ &= \mathrm{d}A^\alpha + \Gamma^\alpha{}_{\beta\gamma}\,A^\beta\,\mathrm{d}x^\gamma\,.\end{aligned} \tag{9.104}$$

Note that in the interest of symmetry in the first line of (9.104), it might be useful to define $\mathrm{D}A^\alpha/\mathrm{d}x^\gamma \equiv A^\alpha{}_{;\gamma}$ [also written $\mathrm{D}_\gamma A^\alpha$ in (8.229)]. That way, a shift of the $\mathrm{d}x^\gamma$ from one side to the other in "$\mathrm{D}A^\alpha/\mathrm{d}x^\gamma = A^\alpha{}_{;\gamma}$" will return the correct expression $\mathrm{D}A^\alpha = A^\alpha{}_{;\gamma}\,\mathrm{d}x^\gamma$. But we must remember that the last of these two expressions contains a sum over γ, while the first does not. And, of course, writing the correct $\mathrm{d}A^\alpha = A^\alpha{}_{,\gamma}\,\mathrm{d}x^\gamma$ certainly does *not* lead to $\mathrm{d}A^\alpha/\mathrm{d}x^\gamma$ being defined as equal to $A^\alpha{}_{,\gamma}$. Rather, the correct expression here is $\partial A^\alpha/\partial x^\gamma = A^\alpha{}_{,\gamma}$.

As long as this is understood, defining $\mathrm{D}A^\alpha/\mathrm{d}x^\gamma \equiv A^\alpha{}_{;\gamma}$ is useful. In fact, we'll see a similar idea in (12.110) and (12.111) when we discuss variational calculus in general relativity. See also the note on p. 518.

To define a notion of *parallel transport*, we asked at the start of this section whether a vector A^α could be moved about in such a way that it remains as constant as possible. So set $\mathrm{d}\boldsymbol{A} = \mathrm{D}A^\alpha\,\boldsymbol{e}_\alpha = 0$; i.e., $\mathrm{D}A^\alpha = 0$ in (9.104). We obtain

$$\mathrm{d}A^\alpha = -\Gamma^\alpha{}_{\beta\gamma}\,A^\beta\,\mathrm{d}x^\gamma\,. \tag{9.105}$$

This is a prescription for how A^α is required to change when stepping from x^α to $x^\alpha + \mathrm{d}x^\alpha$ in order to have a notion of parallelness. But in general the Christoffel symbols are functions of position, so a parallel transport from P to a distant point Q demands that

$$A^\alpha(Q) = A^\alpha(P) + \int_P^Q \mathrm{d}A^\alpha$$
$$= A^\alpha(P) - \underbrace{\int_P^Q \Gamma^\alpha{}_{\beta\gamma}\, A^\beta\, \mathrm{d}x^\gamma}_{\text{Depends on the path!}}. \tag{9.106}$$

The difference in field values at points P and Q is path dependent, so we cannot in general define a unique vector field composed of lots of copies of A^α, parallel transported to every point. The best we can do is parallel transport $\boldsymbol{A}$ from P to Q along some path; the result depends on the path chosen.

If coordinates can be found such that the Christoffel symbols are zero everywhere (such as in euclidean geometry), then the result is not path dependent: it's just a simple translation. This can be used to define *flatness*: a path-independent result means that the space is flat.

The Christoffel symbols and infinitesimals involved in parallel transport might recall to mind the equation for a geodesic, (9.32). To recap, a geodesic with affine parameter t and tangent vector $u^\alpha \equiv \mathrm{d}x^\alpha/\mathrm{d}t$ satisfies[2]

$$\frac{\mathrm{d}u^\alpha}{\mathrm{d}t} + \Gamma^\alpha{}_{\beta\gamma}\, u^\beta\, u^\gamma = 0\,, \tag{9.107}$$

or in other words

$$\mathrm{d}u^\alpha = -\Gamma^\alpha{}_{\beta\gamma}\, u^\beta\, \mathrm{d}x^\gamma, \tag{9.108}$$

so that comparing this with (9.105), we see that a geodesic parallel transports its own tangent vector. (It's also not hard to show the converse: any curve that parallel transports its own tangent vector must be a geodesic.) This makes geodesics special on curved surfaces. They occupy a privileged role in giving a useful meaning to adding vectors at widely separated points.

Picturing Parallel Transport

We can parallel transport a vector along a given path exactly by using (9.106). But what does it all look like—what intuition can be gained? Two theorems are useful in helping us to build an intuitive picture.

Theorem 9.1. *If vectors* $\boldsymbol{A}$ *and* $\boldsymbol{B}$ *are parallel transported along any curve, then their lengths, as well as the angle between them, remain constant.*

Prove this by writing down the parallel-transport increment in $\boldsymbol{A}\cdot\boldsymbol{B}$:

$$\mathrm{d}\,(\boldsymbol{A}\cdot\boldsymbol{B}) \equiv \mathrm{d}\,(A^\alpha B_\alpha) = \mathrm{d}\left(A^\alpha\, B^\beta\, g_{\alpha\beta}\right). \tag{9.109}$$

[2] To keep with accepted formalism in both differential geometry and general relativity, we have run into a minor clash of symbols here. The coordinates u^c in (9.32) are here replaced by x^α.

Differentiating, replacing infinitesimals with (9.105), and also using (8.182) produces a sum of Christoffel terms that equates to zero. Thus $\mathrm{d}(\boldsymbol{A}\cdot\boldsymbol{B}) = 0$, so that $\boldsymbol{A}\cdot\boldsymbol{B}$ does not change along the path. In particular, this means that $|\boldsymbol{A}|^2$ does not change along the path. That is, a parallel-transported vector keeps its length constant. Also, since the angle between the vectors is

$$\cos^{-1} \frac{\boldsymbol{A}\cdot\boldsymbol{B}}{|\boldsymbol{A}|\,|\boldsymbol{B}|}\,, \tag{9.110}$$

each of whose terms is constant along the path, then the angle remains constant along the path, too.

Theorem 9.2. *If a vector* $\boldsymbol{A}$ *is parallel transported along a geodesic, then it keeps a constant angle to the geodesic's tangent vector.*

This is easy to see by setting $\boldsymbol{B}$ in Theorem 9.1 to be the geodesic's tangent vector, which is already being parallel transported by the geodesic. (That is, the geodesic's tangent vector, when parallel transported, remains the geodesic's tangent vector.) Hence, $\boldsymbol{A}$ keeps a constant angle to $\boldsymbol{B}$, and the theorem is proved.

Now picture the transport as follows. To parallel transport a vector along an arbitrary curve, approximate the curve with infinitesimal geodesic segments; then parallel transport along each segment by keeping the vector in the surface and holding a fixed angle between the vector and the segment. We still have the problem of parallel transporting along a segment (unless it's infinitesimally short), but we can do this by drawing a small parallelogram with geodesic diagonals. The construction is shown in Fig. 9.13, and is known as *Schild's ladder*. It enables us to form a very intuitive idea of parallel transport.

As a concrete example, envisage parallel transporting a vector on Earth's curved surface over some specified arc. We'll assume Earth has no oceans, and we will drive a car to transport the vector. The vector is represented by an arrow attached to the floor of the car in some way that will be made explicit in what follows.

First, imagine parallel transporting the vector over the arc of a great circle from Adelaide to Amsterdam. We simply lay the arrow on the car floor and drive from Adelaide to Amsterdam following that arc, which, because it's a geodesic, means we never have to turn the car's steering wheel. Thus the car never tries to turn the arrow in the plane of the floor, and on arriving in Amsterdam, the arrow is the parallel-transported vector.

Next we decide to drive from Amsterdam back to Adelaide, but this time following a more convoluted route with all manner of twists and turns. Now we'll certainly have to turn the steering wheel, and will need to float the arrow in some way on the car floor to decouple it from the left and right turns the car will be making—although it's always constrained to lie in the plane of the floor. We can approximate the route by driving along short geodesic

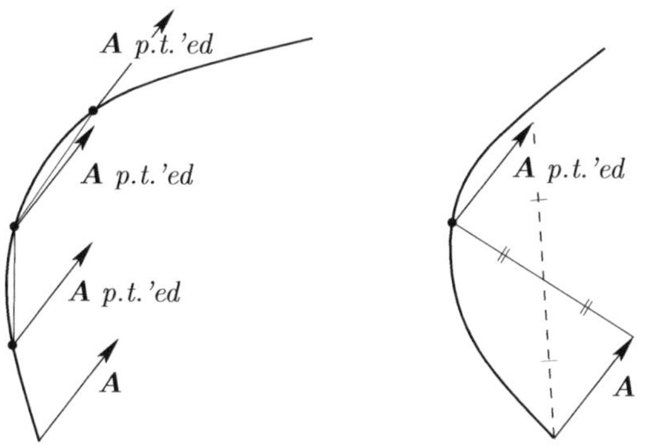

Fig. 9.13. On the left, we parallel transport $\boldsymbol{A}$ along an arbitrary curve by parallel transporting it along the short geodesic segments (drawn straight) connecting close neighbouring points. Do this by maintaining a constant angle between $\boldsymbol{A}$ and each geodesic segment. The construction here uses a flat surface and straight geodesics, but only to show the correspondence with our intuition. On the right, we see further under the microscope in a construction known as Schild's ladder. The actual transport along a segment can be accomplished by drawing a parallelogram with geodesic diagonals. First draw the solid geodesic, then bisect it, and then draw the dashed geodesic through the bisection point. Extending this an equal distance farther results in the new parallel-transported vector's end point.

segments, stopping at the corners to aim the car in the new direction while taking care that the floating arrow does not inadvertently get turned in the process. In fact, with this arrow being completely decoupled from the turns the car makes (but always constrained to lie in the floor plane), we need not even be so careful: we just drive along the designated route, and now, on arriving in Adelaide, the arrow, or vector, has been parallel transported in the required way.

A Picture of DA^α

While $\mathrm{d}A^\alpha\, \boldsymbol{e}_\alpha$ is not a vector, $\mathrm{D}A^\alpha\, \boldsymbol{e}_\alpha$ certainly is and so should be able to be drawn in the usual way as an arrow. Figure 9.14 shows how this is done. First parallel transport $A^\alpha(P)$ an infinitesimal distance from P to Q, giving $A^\alpha_{/\!/}(Q)$. This new vector forms a kind of baseline against which the actual vector $A^\alpha(Q)$ can be compared—it soaks up the extrinsic curvature, as it were. What is left over is the actual increase in the vector, which is $\mathrm{D}A^\alpha$. So $\mathrm{D}A^\alpha$ is the real measure of how A^α is changing along the curve.

Finally, since parallel transport along a curve is accomplished by breaking up the curve into infinitesimal segments, then for the case where P and Q are widely separated, they must be connected by a curve C, and the difference between A^α at Q and its parallel-transported version there is just determined by how much A^α is "really" changing along C:

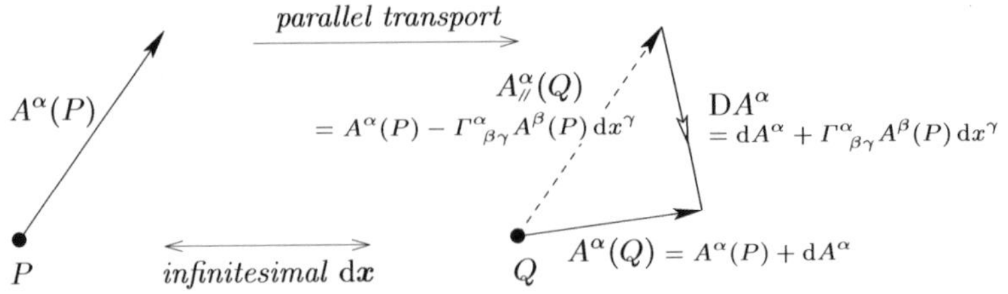

Fig. 9.14. Visualising the vector components $\mathrm{D}A^\alpha$ that carry the real change in $A^\alpha(P)$. Remember that the parallel-transported vector $A^\alpha_{/\!/}(Q)$ only looks as much like $A^\alpha(P)$ as we have drawn when the surface is flat!

$$A^\alpha(Q) - A^\alpha_{/\!/}(Q) = A^\alpha(P) + \int_C \mathrm{d}A^\alpha - \left[A^\alpha(P) + \int_C -\Gamma^\alpha{}_{\beta\gamma}\, A^\beta\, \mathrm{d}x^\gamma\right]$$
$$= \int_C \mathrm{d}A^\alpha + \Gamma^\alpha{}_{\beta\gamma}\, A^\beta\, \mathrm{d}x^\gamma = \int_C \mathrm{D}A^\alpha. \tag{9.111}$$

9.6 Relating Parallel Transport to Curvature

While a difference in vectors might not be well defined over C, a difference in vector *components* always is. If a vector's components change as it's parallel transported along a curve, how much do they change on a round trip? To answer this important question, choose the curved surface's version of a rectangle for simplicity, where each of the coordinates in turn is constant along each side, and the rectangle joins points $0 \to 1 \to 2 \to 3 \to 4$. The end point 4 coincides with the start point 0, shown in Fig. 9.15. We will parallel transport a vector A^α around this loop by calculating the line integral $\oint \mathrm{d}A^\alpha$, where $\mathrm{d}A^\alpha$ is given by (9.105). The task will be made easier if we consider each segment separately, as shown on the next page.

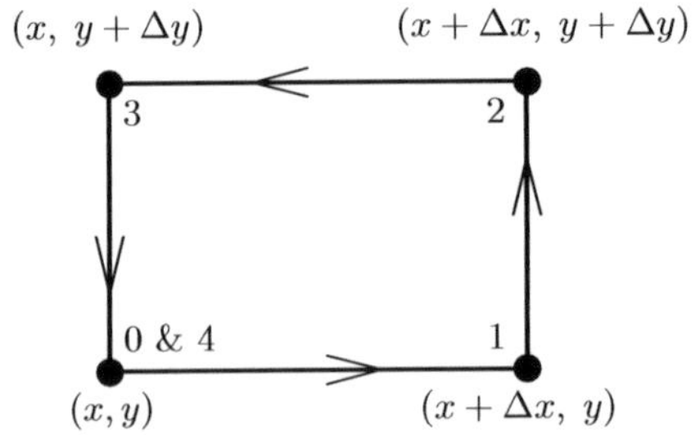

Fig. 9.15. A path along which we wish to parallel transport a vector.

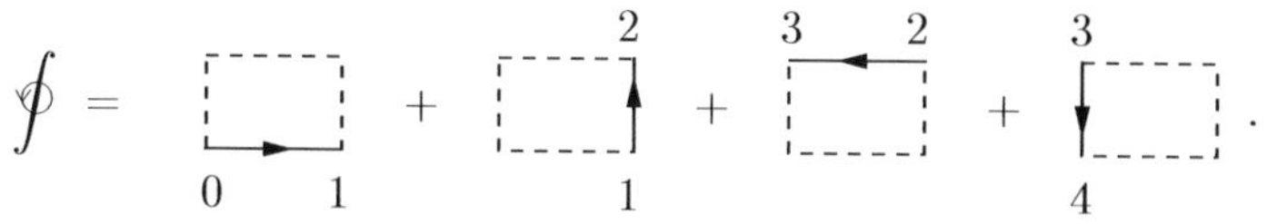

The integral is given by the increase in A^α from the start to the end, or point 0 to point 4. Write it as

$$\oint \mathrm{d}A^\alpha = A^\alpha(\text{point } 4) - A^\alpha(\text{point } 0)\,. \tag{9.112}$$

It might be thought that we could consider the four path legs in two pairs, where each pair consists of opposite legs of the path; then perhaps A^α could be Taylor-expanded "across the void" of the path's interior, so to speak, and various cancellations might occur. In fact, this cannot be done because by its very nature A^α is not a function of position, since its value at any point is dependent on the history of how we got to that point. So A^α cannot just be differentiated with abandon; it can only be differentiated along a path that forms part of the parallel transport. If we do try this approach of differentiating across the void where no path ever went, the calculation will be quite meaningless. It will give results that depend on the order in which the parts of the calculation are assembled, and it can even be arranged to give the right result, with hindsight! But, of course, in no way does that validate the method.

Done correctly, the integral (9.112) can be calculated by stepping A^α along the path from point 0 to point 4 using (9.105) to calculate the increase $\mathrm{d}A^\alpha$ at each step. But with hindsight it proves more economical to start at point 4 and work backward, calculating the contribution to the integral along each of the four legs of the path. We'll expand A^α along each leg from that leg's start point by using a Taylor expansion with terms up to second order in Δx and Δy. (That is, the rectangle in Fig. 9.15 should be *small.*) Each time we encounter a derivative, we'll use (9.105) to replace it with Christoffel terms. For clarity, as the next few expressions will be cluttered, we'll indicate evaluation at, e.g., point 3 by a superscript such as $A^{\alpha(3)}$. So begin with the Taylor expansion of $A^{\alpha(4)}$ about $A^{\alpha(3)}$:

$$A^{\alpha(4)} = A^{\alpha(3)} + A^{\alpha(3)}_{,y}(-\Delta y) + A^{\alpha(3)}_{,yy}\Delta y^2/2 + \cdots\,, \tag{9.113}$$

where the dots indicate terms of third order and higher. But (9.105) specifies that, in parallel transport, the required derivatives are given by

$$A^\alpha_{,y} = -\Gamma^\alpha{}_{y\beta}\, A^\beta\,, \quad A^\alpha_{,yy} = (-\Gamma^\alpha{}_{y\beta,y} + \Gamma^\alpha{}_{y\gamma}\Gamma^\gamma{}_{y\beta})\, A^\beta\,, \tag{9.114}$$

although we won't need the full expression for $A^\alpha_{,yy}$. That's because it only appears in (9.113) to order Δy^2 anyway, and so can be expanded to zeroth order, meaning that its value can be successively changed from $A^{\alpha(3)}_{,yy}$ to $A^{\alpha(2)}_{,yy}$ and so on as we go around the path. Substituting $A^{\alpha}_{,y}$ from (9.114) into (9.113) gives

$$A^{\alpha(4)} = A^{\alpha(3)} + \Gamma^{\alpha(3)}_{y\beta} A^{\beta(3)} \Delta y + A^{\alpha(3)}_{,yy} \Delta y^2/2 + \cdots . \tag{9.115}$$

Now that $A^{\alpha(4)}$ is expressed using terms all evaluated at point 3 with no derivatives, we again step backward one leg, this time Taylor-expanding all terms from point 2. The calculation becomes more involved because we also need to include derivatives of the Christoffel symbols.

$$\begin{aligned} A^{\alpha(4)} &= A^{\alpha(2)} + A^{\alpha(2)}_{,x}(-\Delta x) + A^{\alpha(2)}_{,xx} \Delta x^2/2 + \Gamma^{\alpha(2)}_{y\beta} A^{\beta(2)} \Delta y \\ &\quad + \left[\Gamma^{\alpha(2)}_{y\beta,x} A^{\beta(2)} - \Gamma^{\alpha(2)}_{y\beta} \Gamma^{\beta(2)}_{x\gamma} A^{\gamma(2)}\right](-\Delta x)\Delta y + A^{\alpha(2)}_{,yy} \Delta y^2/2 + \cdots \\ &= A^{\alpha(2)} + \Gamma^{\alpha(2)}_{x\beta} A^{\beta(2)} \Delta x + A^{\alpha(2)}_{,xx} \Delta x^2/2 + \Gamma^{\alpha(2)}_{y\beta} A^{\beta(2)} \Delta y \\ &\quad - \left[\Gamma^{\alpha(2)}_{y\beta,x} - \Gamma^{\alpha(2)}_{y\gamma} \Gamma^{\gamma(2)}_{x\beta}\right] A^{\beta(2)} \Delta x\, \Delta y + A^{\alpha(2)}_{,yy} \Delta y^2/2 + \cdots . \end{aligned} \tag{9.116}$$

Remember that all we are doing is Taylor-expanding to at most second order.

On stepping back to point 1, some cancellation begins to occur, and terms begin to appear in the coefficient of $\Delta x \Delta y$ that form precisely the Riemann tensor. Omitting the laborious details (which are straightforward to fill in), we obtain

$$A^{\alpha(4)} = A^{\alpha(1)} + \Gamma^{\alpha(1)}_{x\beta} A^{\beta(1)} \Delta x + R^{\alpha(1)}_{\ \ \beta yx} A^{\beta(1)} \Delta x\, \Delta y + A^{\alpha(1)}_{,xx} \Delta x^2/2 + \cdots . \tag{9.117}$$

On stepping back to the start point 0, we omit further straightforward details to write

$$A^{\alpha(4)} = A^{\alpha(0)} + R^{\alpha(0)}_{\ \ \beta yx} A^{\beta(0)} \Delta x\, \Delta y + \cdots . \tag{9.118}$$

Finally, the line integral becomes

$$\oint \mathrm{d}A^{\alpha} = A^{\alpha(4)} - A^{\alpha(0)} = R^{\alpha(0)}_{\ \ \beta yx} A^{\beta(0)} \Delta x\, \Delta y + \cdots . \tag{9.119}$$

The contraction of the Riemann tensor with vector components measures the increase in those components when parallel transported around a loop per unit area of the loop. Of course, the third-order terms in (9.119) can be neglected for small loops. But we cannot break a big loop into lots of small ones that share common boundaries in the usual way that's done when proving Stokes' theorem. This is because the value of A^{α} is not a single-valued function of position (unlike the Stokes' theorem case), so integrals along common boundaries between neighbouring cells cannot be expected to cancel. After all, the whole point of parallel transport is that it gives a sort of hysteresis, or memory, to the vector, making its value path-dependent.

More generally, in the x^γ-x^δ plane we can write

$$\oint \mathrm{d}A^\alpha = R^\alpha{}_{\beta\gamma\delta} A^\beta \, \Delta x^\gamma \, \Delta x^\delta + 3^{\text{rd}} \text{ order terms, with no sum on } \gamma, \delta. \tag{9.120}$$

A surface will be flat if it has a zero Riemann tensor, since the zero tensor implies no change to the vector components around the loop—which was, after all, our starting point for beginning to think of parallel transport. So a precise definition of flatness states that a surface of arbitrary dimension is flat if $\oint \mathrm{d}A^\alpha = 0$ for all values of α, independently of the A^α and the plane chosen for the parallel transport. In that case, write

$$\oint \mathrm{d}A^\alpha = 0 \Longleftrightarrow R^\alpha{}_{\beta\gamma\delta} = 0 \quad \forall \alpha, \beta, \gamma, \delta\,. \tag{9.121}$$

Because the Riemann tensor *is* a tensor, being zero in one coordinate system implies that it must be zero in all others, making flatness an absolute property of a surface.

Parallel transport is an interesting procedure, but it's not an absolute necessity for beginning a study of curved spaces. The key players of differential geometry, such as geodesics and curvature, were all defined and derived in this chapter without a notion of parallel transport.

The Ricci Tensor and Ricci Scalar

A contraction of the Riemann tensor is the Ricci tensor $R_{\alpha\beta}$, which forms the basic geometrical part of Einstein's equations that relate the curvature of spacetime to its matter content. The choice of which indices to contract over varies, but the following is commonly used:

$$R_{\alpha\beta} \equiv R^\lambda{}_{\alpha\lambda\beta}\,. \tag{9.122}$$

No matter what the convention, the Ricci tensor can be contracted to produce the *Ricci scalar*:

$$R \equiv R^\alpha{}_\alpha\,. \tag{9.123}$$

Using (9.93), it's a straightforward exercise to show that, for a 2-surface, the Ricci scalar and Gauss curvature are simply related:

$$R = 2K\,. \tag{9.124}$$

So, for example, we know immediately that the Ricci scalar at any point on the surface of a standard sphere (a 2-sphere) of radius r is $2/r^2$. The Ricci tensor $R_{\alpha\beta}$ is sometimes defined as $R^\lambda{}_{\alpha\beta\lambda}$ by contracting over the last Riemann index. Not surprisingly, this just changes signs in various places; for example, (9.124) becomes $R = -2K$.

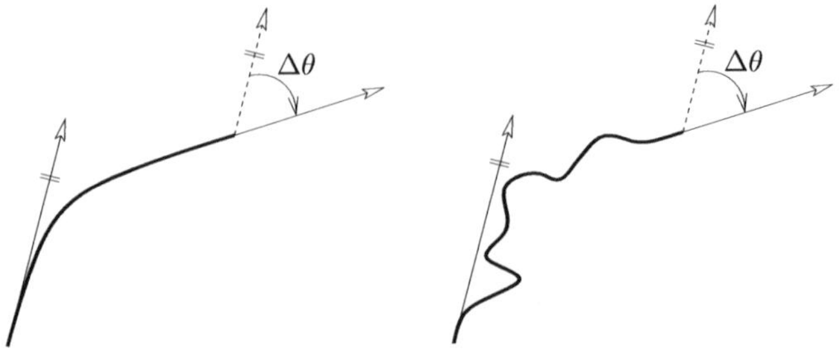

Fig. 9.16. Integrating the curvature k along a curve gives the total angle turned through by its tangent vector, which will be the same for each curve above if the beginning and ending slopes are arranged to be the same. That is, the integral of the curvature is insensitive to the local geometry of each curve.

Although in two dimensions the Gauss curvature relates so simply to the Riemann tensor (9.93), and hence to the Ricci tensor and scalar, in general the Ricci tensor and scalar do not represent a sort of rendered-down version of the curvature. For example, a zero Ricci tensor does not guarantee a zero Riemann tensor, so a zero Ricci tensor does not imply zero curvature. We'll see a good example of this in general relativity in Chap. 12. There, empty space turns out to have a zero Ricci tensor and scalar, but its Riemann tensor (and hence curvature) need not be zero.

9.7 From Geometry to Topology: The Gauss–Bonnet Theorem in Euclidean 3-Space

At the start of this chapter, the curvature of a curve in the plane was defined to be $k = \mathrm{d}\theta/\mathrm{d}s$. Because this is a rate of angular increase of the curve's tangent vector, integrating it along the curve will give the total angle turned through by that vector:

$$\int k \, \mathrm{d}s = \Delta\theta \,. \tag{9.125}$$

At first sight, this might seem to be a trivial result. The curvature that proved so fruitful when analysed in higher dimensions reduces to a single number that simply tells how much the curve turned in total. But a moment's thought shows that actually we can use this simplicity to our advantage. As seen in Fig. 9.16, if we integrate the curvature along a different curve (possibly with different length), then provided that the difference between the beginning and ending slopes is the same for both curves, the total angle turned through will also be the same, so that $\int k \, \mathrm{d}s$ will be identical for both curves. So the curvature integral does not pay heed to any wiggles placed in a curve.

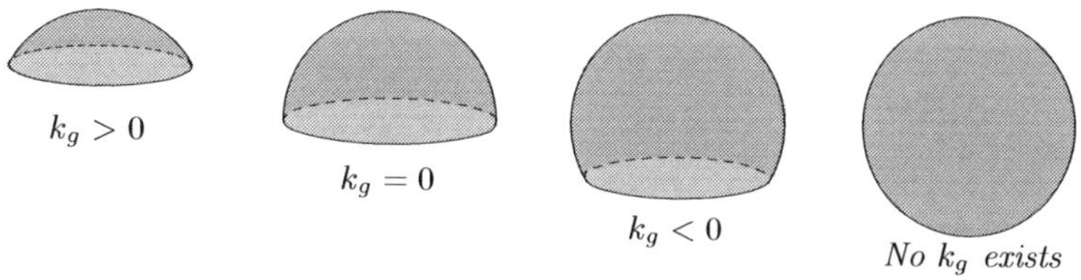

Fig. 9.17. Integrating the Gauss curvature $K = 1/R^2$ for a sphere of radius R over larger and larger areas. Although $\int K\,\mathrm{d}A$ grows from left to right, the geodesic curvature is becoming larger negatively at just the right rate to offset the increasing value of $\int K\,\mathrm{d}A$. That is, as the edge is shifted from the northern to southern hemispheres, we who drive along it from west to east in a car will need to change the direction of turn of our steering wheel from left to right. But a topological change occurs when we integrate over the *entire* sphere.

We can expect the same result from thinking about the circle map. Changing the wiggliness of the curve being integrated along will fan the unit vectors of the circle map out and in, but if the slopes of both ends of the curve are fixed, then the start and end unit vectors won't change their positions on the unit circle, which means the length of the subtended arc of the unit circle won't change either. And, by (9.3), this length is just $\int k\,\mathrm{d}s$.

The same sort of idea also applies to the Gauss curvature K. If we integrate K over a 2-surface, then the result is independent of the surface's local geometry, as long as its edges are held at an unchanging slope. This result follows from the sphere map in the same way that we applied the circle map in the previous paragraph. The unit vectors extending from the centre of the sphere will fan out and in depending on how the surface is deformed, but if the edge slope is everywhere held fixed, then a constant area of the unit sphere will be mapped to, and so we expect $\int K\,\mathrm{d}A$ to be a constant independent of the deformations.

But what happens if the edges are not held fixed? Consider, for example, integrating $K = 1/R^2$ over larger and larger areas of a sphere of radius R, where the edge of the integration is taken for simplicity as a small circle of constant latitude, as in Fig. 9.17. Parametrising the sphere using polar coordinates such that the small circles that bound it are at constant θ, the area of a surface element is given by (9.53), or

$$\mathrm{d}A = |\boldsymbol{e}_\theta \times \boldsymbol{e}_\phi|\,\mathrm{d}\theta\,\mathrm{d}\phi = R^2 \sin\theta\,\mathrm{d}\theta\,\mathrm{d}\phi\,, \tag{9.126}$$

where we have used the parametrisation of (9.27), and will ensure that $\mathrm{d}\theta$ and $\mathrm{d}\phi$ are always positive by integrating in the increasing direction of each.

That is, the surface element is a parallelogram (in this case a rectangle) with side vectors $\mathrm{d}\boldsymbol{X}|_{\phi\text{ const.}}$ and $\mathrm{d}\boldsymbol{X}|_{\theta\text{ const.}}$. These are, respectively, $\partial\boldsymbol{X}/\partial\theta\,\mathrm{d}\theta$ and $\partial\boldsymbol{X}/\partial\phi\,\mathrm{d}\phi$, or, in other words, $\boldsymbol{e}_\theta\,\mathrm{d}\theta$ and $\boldsymbol{e}_\phi\,\mathrm{d}\phi$. Its area $\mathrm{d}A$ is the norm of the cross product of these.

Thus the area of the surface out to some θ_0 is

$$A = \int_0^{2\pi} \int_0^{\theta_0} R^2 \sin\theta \, \mathrm{d}\theta \, \mathrm{d}\phi = 2\pi R^2 \left(1 - \cos\theta_0\right), \tag{9.127}$$

in which case

$$\int K \, \mathrm{d}A = 2\pi \left(1 - \cos\theta_0\right). \tag{9.128}$$

This integral grows with θ_0. However, the fact that the edges are not being held at a constant slope goes hand in hand with a change in their *geodesic curvature* k_g. Referring to the calculation at the end of Sect. 9.1.1, we see that, indeed, around a small circle at constant latitude,

$$\oint k_g \, \mathrm{d}s = \frac{\cot\theta_0}{R} \, 2\pi R \sin\theta_0 = 2\pi \cos\theta_0 . \tag{9.129}$$

Comparing (9.128) with (9.129) shows that, on the sphere, the two integrals add to give a constant:

$$\int K \, \mathrm{d}A + \oint k_g \, \mathrm{d}s = 2\pi . \tag{9.130}$$

That is, the fact that the area is growing while holding a constant Gauss curvature has been offset by the fact that the geodesic curvature along the boundary is becoming more negative. If we were to drive along the edge heading east, the turn required of our steering wheel would change from left in the northern hemisphere, to straight ahead at the equator, to right in the southern hemisphere.

In fact, this idea of the two curvature integrals changing in tandem to sum to a constant turns out to be more generally true, and is known as the *Gauss–Bonnet theorem.* For a more general 2-surface bounded by a more general curve, the Gauss–Bonnet theorem states that

$$\boxed{\int K \, \mathrm{d}A + \oint k_g \, \mathrm{d}s = 2\pi\chi ,} \tag{9.131}$$

where the curve is traversed in the appropriate direction, as we have done here, and χ is the *Euler characteristic* of the surface: a topological invariant. For a polygon or polyhedron, the Euler characteristic is

$$\chi \equiv F - E + V , \tag{9.132}$$

where F, E, and V are the number of faces, edges, and vertices, respectively. For a nonpolyhedral surface, the value we calculate for χ is independent of the way in which we *triangulate* the surface; that is, the way we deform it into a polyhedron in order to calculate F, E, and V.

The sum of any "jump angles"—discontinuous changes in the boundary's tangent vector—can be added to the left-hand side of (9.131), but alternatively, these angles can just be considered as limiting cases of $\oint k_g \, ds$ and so it is not necessary to write them explicitly.

The incomplete circle is topologically equivalent to a disk, or any polygon, and so has one face ($F = 1$) and one vertex for every edge ($E = V$), so that $\chi = 1$ and the Gauss–Bonnet theorem (9.131) reduces to (9.130), as expected.

A common textbook example of the Gauss–Bonnet theorem on the surface of a sphere connects three geodesic segments, each a quarter of the circumference in length, to make a *spherical triangle* that covers one eighth of the sphere's surface, with all three internal angles being 90°. Since the segments are geodesics, $k_g = 0$ along them. The three jump angles at the corners each contribute $\pi/2$ to the integration of (9.131). Thus the Gauss–Bonnet theorem becomes

$$\frac{1}{R^2}\frac{4\pi R^2}{8} + \frac{3\pi}{2} = 2\pi \,, \tag{9.133}$$

which is certainly true.

What happens when we integrate the Gauss curvature over the entire sphere? Now there is no boundary to supply the k_g that would normally act to offset the growing area integral. However, a full sphere is topologically equivalent to *two* disks joined at their edges. In that case, $F = 2$ and $E = V$, giving $\chi = 2$, and the Gauss–Bonnet theorem gives

$$\int K \, dA = 4\pi \,, \tag{9.134}$$

which of course is true, since $K = 1/R^2$ and the surface area is $4\pi R^2$. So the Euler characteristic accounts for the change in topology as the open spherical surfaces in Fig. 9.17 close up to become the rightmost full sphere, with the small-circle boundary shrinking and finally vanishing in the process.

That $\chi = 2$ for the sphere, and hence for anything topologically equivalent to a sphere, is none other than the well-known rule $F - E + V = 2$ for polyhedra.

The Gauss–Bonnet theorem (9.131) is a staple of differential geometry, linking local geometry with global topology, which is an elegant identification that is sometimes described as unintuitive or perhaps startling. But the theorem is not as unintuitive as it might at first seem. After all, the fluctuations of the local geometry are being integrated over, and it should not be too surprising to find that the result is a topological invariant, insensitive to the local geometry of the surface. This is not really different from finding the average of a set of noisy measurements. The average is a kind of topological invariant, too, in the sense that lots of measurements of some fluctuating quantity will tend always to have roughly the same average. The whole point of adding them up (or, in a sense, integrating over them) to produce the average is to allow randomly alternating signs for the noise to cancel. Likewise for the Gauss–Bonnet theorem, integrating over local geometry gives a result that's topologically invariant.

The techniques of differential geometry can take a metric of some space and paint a picture of it as a surface by giving us a feel for whether, where, and how much it is curved. The subject builds heavily on our intuitive ideas of what it means to be curved, and so provides some elegant insight into physical theories. We'll meet some of these theories in the chapters to come. At the head of the list is general relativity, with its key player being a curved spacetime, but any theory that uses a metric can also be analysed in terms of curvature. Two examples that we'll encounter are field theory, with its lagrangians that are closely related to a spacetime metric, and gauge theory, with its covariant derivatives that also admit structure and concepts such as parallel transport. All of these demonstrate just how great a part geometry plays in fundamental physics.

10 Variational Calculus and Field Theory

10.1 The Story of the Fly and the Train

Picture this: a train is steaming along at full tilt along its track when suddenly it encounters, head-on, a fly that happens to be buzzing in the opposite direction. The fly is quite naturally bounced backward by the train, which continues on as if nothing had happened. Now the question often put runs as follows: "Since the fly has had its forward motion changed to backward motion, there must have been a moment in time when it was at rest, when it was in contact with the train. Surely if the fly was at rest then, and yet in contact with the train, then mustn't the train necessarily have been at rest, too, if only for a moment?"

It's a good question and is sometimes discussed in terms of the make-up of the fly's body, the issue of conservation of momentum, the thicknesses of the layers of the fly and train that get compressed during the collision, the atomic theory of what atoms are moving where, binding forces between atoms, and so on. But none of these address the key issue: two particles have a head-on collision using pure newtonian mechanics, and one of them bounces back. Shouldn't both be at rest at some instant?

The problem posed runs deeper than any kind of many-particle treatment, so we will imagine the train to be a single particle that obeys Newton's laws and likewise for the fly. And we lose nothing by doing this because that's what the original question is really all about: the validity and intuitiveness of Newton's laws as applied to structureless particles.

The train moves from the left to right along its track, while the fly moves in the opposite direction. We can plot this on the distance-time graphs of Figs 10.1 and 10.2. There are two basic choices we have in considering how the train and fly actually interact. The first is that the worldlines of each on the distance-time graph will suffer a kink, a point of nondifferentiability, at the moment of impact, as shown in Fig. 10.1. Each has a certain velocity just before impact, and suddenly each has a different velocity. The infinitesimal time it takes for the impact to occur manifests in the fact that each worldline has a kink instead of curving smoothly from its initial to its final slope.

In this scenario, the conundrum is resolved because the fly does not stop the train. Just before the impact, the fly is moving with negative velocity;

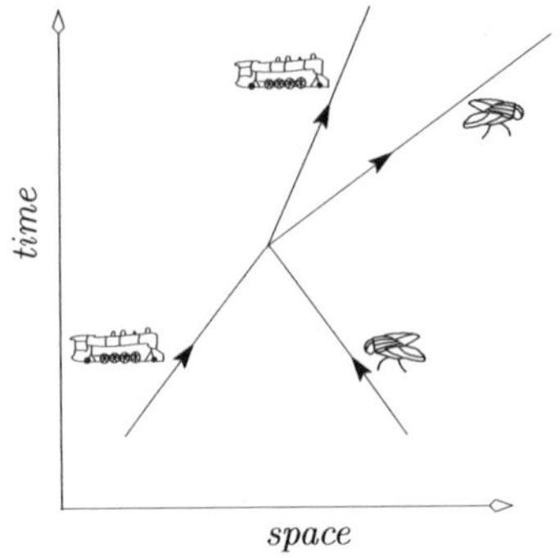

Fig. 10.1. When the fly and train interact instantaneously, they never have zero velocity since neither worldline is ever vertical.

just after impact, it moves with positive velocity. But at no time does its velocity pass through zero: at no time is its worldline vertical!

But there is now a problem of a different kind. The velocities of both fly and train change in an instant, so that for an infinitesimal period of time they undergo infinite accelerations. So we have resolved the conundrum—but at the expense of introducing an infinite acceleration (and so an infinite force), albeit for an infinitesimal time interval. There is a sense of something not right here with really only one way to back out of it, which is to make the worldlines *curved*, as shown in Fig. 10.2.

Now that the worldlines of both fly and train are curved, we have cleared both the conundrum and the acceleration problem. On the left-hand side in Fig. 10.2 we see that although the fly's velocity does become zero at some moment—since its worldline is then vertical—it's not in contact with the train at that moment, and the train's velocity need never be zero. In fact, the fly and train can even touch, as shown on the right-hand side in Fig. 10.2. When they do, they will share the same velocity since the two curves are tangential at the point of contact. And because there are never any kinks in the worldlines of both fly and train, neither ever suffers an infinite acceleration.

So we have resolved the Fly and Train Conundrum, with the benefit of learning something deep and new about Nature; that is, that if Nature wishes to avoid the oddness of an infinite acceleration, then even in a newtonian world of particles interacting with other particles, those interactions must take place via a ghostly action at a distance—or, in modern parlance, a *field.*

The very successful theories of gravity and electromagnetism have already introduced us to the idea of a field, and it's a fairly easy thing to discard the notion of an infinite acceleration, concentrating instead on the idea that forces are mediated by fields alone. Of course, what a field actually *is* is quite another story, one that we won't attempt to answer.

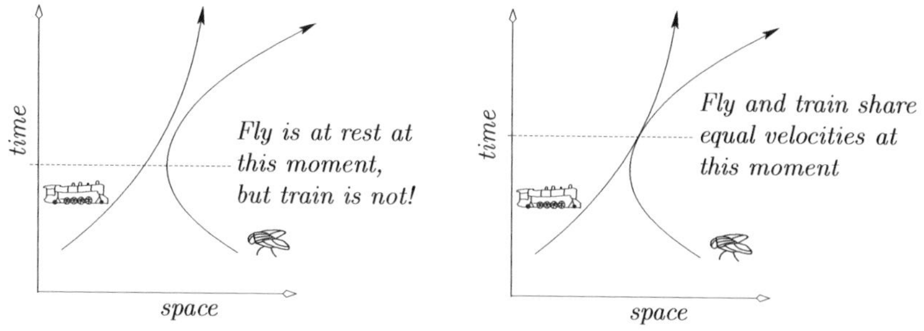

Fig. 10.2. Left: Now the fly does come to rest, but the train does not. **Right**: The fly and train can even touch, sharing the same velocity when they do.

10.2 The Concept of a Field

While the interaction between a fly and a train still has something in the way of contact about it at a macroscopic level, this is not always the case. When an object falls, nothing seems to push it, yet it accelerates to the ground nonetheless. In his analysis of gravity, Newton recognised the need for the gravitational pull of one body on another as not so much mediated by any contact of the bodies but rather as a property of the space between them. He understood that the gravitational force exerted on the Moon by Earth was nevertheless a property of the Moon's position alone, and did not seem to be communicated to it by anything coming out of Earth. And so he was led to the field concept, an invisible "thing" permeating space. Newton remarked that the idea seemed preposterous, but he needed it as part of the framework he had built up to make calculations and predictions using his supremely successful theory of gravity.

Gravity and electromagnetism are always cited as the two early examples of fields. But as we have seen, the field idea enters newtonian mechanics at a very basic level even prior to these.

10.2.1 The Idea of a Potential

Our experience of fields of different types is that they single out distinct quantities on which to act. Gravity only acts on mass (or really energy, which we can view as mass in another guise), while the electromagnetic field acts on charge. What mass and charge are is quite mysterious, but they are the properties that quantify how strongly the appropriate field acts on the object. And to be even more precise we might say that gravity acts on *gravitational mass*, as opposed to the *inertial mass* that is the property of a body that resists acceleration. (Einstein's Equivalence Principle postulates that these two masses are identical, and experiments have verified this to around one part in 10^{12}.)

Fields are defined by our quantifying the force they produce. Experiments indicate that the gravitational force on a body is proportional to its mass, and likewise the electromagnetic force is proportional to charge—although the definitions of mass and charge are tightly bound up in these observations. So focus for now on the gravitational field, and define its strength at any point in space to be the force per unit mass for a mass placed at that point:

$$\boldsymbol{g} \equiv \frac{\boldsymbol{F}}{m} . \tag{10.1}$$

The field is that ghostly thing that produces a force $m\boldsymbol{g}$ on a mass m and as such is a vector field. In our effort to understand just what the field actually is, it's useful to relate it to another quantity: we think of space being filled with another field, this time a scalar field called the *gravitational potential* Φ. The potential is defined such that the resulting force points in the direction in which Φ decreases most rapidly.

Given Φ, what is this direction? We saw the answer previously in Fig. 8.7 but will review the argument here. When stepping along a vector $\mathrm{d}\boldsymbol{x}$, the loss in the potential is

$$\begin{aligned} -\mathrm{d}\Phi &= -\nabla\Phi\cdot\mathrm{d}\boldsymbol{x} \\ &= |\nabla\Phi|\,|\mathrm{d}\boldsymbol{x}|\cos(-\nabla\Phi, \mathrm{d}\boldsymbol{x}) , \end{aligned} \tag{10.2}$$

where "$(-\nabla\Phi, \mathrm{d}\boldsymbol{x})$" denotes the angle between the vectors $-\nabla\Phi$ and $\mathrm{d}\boldsymbol{x}$. Thus, the loss in potential is maximised for a step $\mathrm{d}\boldsymbol{x}$ parallel to $-\nabla\Phi$; that is, the potential decreases most rapidly in the direction of $-\nabla\Phi$. So the field is proportional to $-\nabla\Phi$, and by absorbing the proportionality constant into Φ, we have the freedom to make this an equality:

$$\text{gravitational field} = \boldsymbol{g} = -\nabla\Phi_{\text{gravity}} . \tag{10.3}$$

The same is true for a static electric field:

$$\boldsymbol{E} = -\nabla\Phi_{\text{electric}} . \tag{10.4}$$

It is not, however, true for more complicated fields such as a time-varying electric field; nor for the magnetic field, since the force applied by this depends on the velocity of the charge. Nevertheless, both of these can be accommodated by introducing the *magnetic vector potential* $\boldsymbol{A}$, via

$$\boldsymbol{E} = -\nabla\Phi_{\text{electric}} - \partial\boldsymbol{A}/\partial t , \quad \boldsymbol{B} = \nabla\times\boldsymbol{A} , \tag{10.5}$$

which we'll meet again later.

Note that (10.1) and (10.3) give the force on a mass as

$$\boldsymbol{F} = -\nabla(m\Phi_{\text{gravity}}) \equiv -\nabla V , \quad \text{where } V \equiv m\Phi_{\text{gravity}} , \tag{10.6}$$

with V the *potential energy* of the mass. It's important to bear in mind that since the potential was only defined by the way it changes spatially, it is only defined up to some arbitrary additive constant of integration, and the same is true for potential energy.

We can define the infinitesimal work done by a field when it accelerates a particle as the loss in the associated potential energy:

$$\mathrm{d}W \equiv -\mathrm{d}V = -\nabla V \cdot \mathrm{d}\boldsymbol{x} = \boldsymbol{F} \cdot \mathrm{d}\boldsymbol{x} \,. \tag{10.7}$$

This is an example of the boxed point made on page 48: the infinitesimal "d" means an increase, so that the *loss* in potential energy is $-\mathrm{d}V$.

10.3 The Lagrangian Formalism

Much of modern physics leans on the idea that Nature is constantly extremising various interesting quantities. This might just be something as simple as minimising potential energy: we know that things tend to settle as far "down" as they can. In the case of light travelling between two given points, the ray's path is such that its travel time is minimised. We humans also tend to use this principle in our everyday lives. The ways in which we carry out daily tasks might be many and varied, but they will tend to maximise reward, or perhaps minimise discomfort or cost. Minimising the one does not necessarily maximise the other; perhaps there will be one overriding quantity that is a complicated mixture of the two.

The search for such a *variational principle* has become very fruitful in mathematical physics. It was applied originally to light by Fermat, coming to be called Fermat's Principle, and then to mechanics by Lagrange, although it was Hamilton who really appreciated the underlying principle involved. Finally, it has been applied to theoretical physics under the famous name of the Principle of Least Action.

As an example of a variational principle, consider a simple geometrical problem: what is the equation of the shortest curve joining two given points? Of course it's a straight line, but considering what we mean by the "shortest curve" will be useful in piecing together the salient ideas. So consider a curve joining two points (x_1, y_1) and (x_2, y_2). We know that an element of path length $\mathrm{d}\ell$ along the curve obeys Pythagoras's theorem, $\mathrm{d}\ell^2 = \mathrm{d}x^2 + \mathrm{d}y^2$, or

$$\mathrm{d}\ell = \sqrt{1 + y'^2(x)}\, \mathrm{d}x \,, \tag{10.8}$$

so that the total length of the curve is

$$S \equiv \int \mathrm{d}\ell = \int_{x_1}^{x_2} \sqrt{1 + y'^2}\, \mathrm{d}x \,. \tag{10.9}$$

The question is, given that S must be a minimum, what is y as a function of x? This is not simply an exercise in one-dimensional calculus: instead of the usual calculus approach of changing x and watching how y changes, we

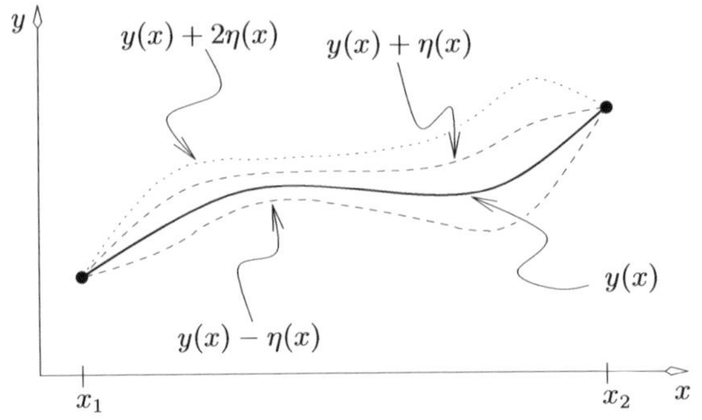

Fig. 10.3. Begin with a function $y(x)$ whose precise form is to be determined, and add any amount of some arbitrary perturbing function $\eta(x)$ to it. The scheme of variational calculus is to demand that some scalar S of interest that is calculated for each of these new functions should have a stationary value for the actual function $y(x)$—irrespective of the form of $\eta(x)$.

now wish to vary the entire function $y(x)$ and watch how S changes. How can this be done?

Take our cue from basic calculus: if a function $y(x)$ has a stationary point for some x, then the usual way of describing this is to say that the slope of y is zero at that value of x. But this really means that if we move away from the stationary point by any amount Δx, then the corresponding y-increase of Δy goes to zero much faster than Δx, as $\Delta x \to 0$. The ratio $\Delta y/\Delta x$ finally becomes zero at the stationary point. Given that the increase in y arbitrarily close to the stationary point is

$$dy = y'(x)\, dx\,, \tag{10.10}$$

it must be that at the stationary point, $y'(x)$ has become zero.

Apply this way of thinking to this new "variational calculus" problem of finding the right curve by reasoning as follows. The function $y(x)$ can be changed by adding some amount λ of an arbitrary perturbing function $\eta(x)$, as shown in Fig. 10.3. This perturbation can be anything, with the only proviso that it be smooth and equal to zero at the end points x_1 and x_2 since the requisite curve is anchored to those points. So consider $S(\lambda)$, the length of the new curve that results when $y(x)$ is replaced by $y(x) + \lambda\eta(x)$:

$$S(\lambda) \equiv \int_{x_1}^{x_2} \sqrt{1 + (y + \lambda\eta)'^2}\, dx\,. \tag{10.11}$$

The original curve length S is now more fully written as $S(0)$. Require that as $\lambda \to 0$, the value of $S(\lambda)$ changes less and less for all functions η:

$$S'(0) = 0 \quad \forall\, \eta(x)\,. \tag{10.12}$$

Now the task has been reduced to the application of some straightforward calculus—although we will take as given that we can differentiate under the integral sign in the next few lines.

$$S'(\lambda) = \int_{x_1}^{x_2} \frac{(y' + \lambda\eta')\,\eta'}{\sqrt{1 + (y + \lambda\eta)'^2}}\, \mathrm{d}x\,, \tag{10.13}$$

so that $S'(0)$ is calculated via an integration by parts:

$$\begin{aligned} S'(0) &= \int \frac{y'}{\sqrt{1 + y'^2}}\eta'\, \mathrm{d}x \\ &= \frac{y'}{\sqrt{1 + y'^2}}\eta(x)\Bigg|_{x_1}^{x_2} - \int_{x_1}^{x_2} \left[\frac{y''}{\sqrt{1 + y'^2}} - \frac{y'^2 y''}{(1 + y'^2)^{3/2}}\right] \eta(x)\, \mathrm{d}x\,. \end{aligned} \tag{10.14}$$

The first term in the last line of (10.14) is zero since $\eta(x_1) \equiv \eta(x_2) \equiv 0$. If $S'(0)$ is to vanish for all choices of η, then the integrand in the second term in the last line of (10.14) must also vanish:

$$\frac{y''}{\sqrt{1 + y'^2}} - \frac{y'^2 y''}{(1 + y'^2)^{3/2}} = 0\,. \tag{10.15}$$

This simplifies neatly to

$$y''(x) = 0\,, \tag{10.16}$$

which is, of course, just the differential equation for a straight line. The two arbitrary coefficients are now fixed by applying the boundary conditions $y(x_1) = y_1$, $y(x_2) = y_2$, and the problem is solved.

10.3.1 Lagrange's Equation

General problems in the calculus of variations also use the approach of the last few equations. The quantity to be made stationary is called the *action*:

$$S = \int_{x_1}^{x_2} L(y, y')\, \mathrm{d}x\,. \tag{10.17}$$

The action usually depends both on the curve $y(x)$ and its derivative $y'(x)$. With the curve anchored at its end points, perturb it to write

$$S(\lambda) \equiv \int_{x_1}^{x_2} L\big(y + \lambda\eta, (y + \lambda\eta)'\big)\, \mathrm{d}x\,, \tag{10.18}$$

and require $S'(0) = 0$ for all functions $\eta(x)$, as well as $\eta(x_1) = \eta(x_2) = 0$:

$$S'(\lambda) = \int_{x_1}^{x_2} \left[\frac{\partial L}{\partial(y+\lambda\eta)} \frac{\mathrm{d}(y+\lambda\eta)}{\mathrm{d}\lambda} + \frac{\partial L}{\partial(y'+\lambda\eta')} \frac{\mathrm{d}(y'+\lambda\eta')}{\mathrm{d}\lambda} \right] \mathrm{d}x$$

$$= \int_{x_1}^{x_2} \left[\frac{\partial L}{\partial(y+\lambda\eta)} \eta + \frac{\partial L}{\partial(y'+\lambda\eta')} \eta' \right] \mathrm{d}x\,. \tag{10.19}$$

Those partial derivatives might look troubling. Does it really make sense to differentiate with respect to η while holding η' constant? Surely, if η changes, then mustn't η' also? That's true, but it is not what partial differentiation is all about. After all, when we write an expression such as

$$\frac{\mathrm{d}f(x,y)}{\mathrm{d}\lambda} = \frac{\partial f}{\partial x}\frac{\mathrm{d}x}{\mathrm{d}\lambda} + \frac{\partial f}{\partial y}\frac{\mathrm{d}y}{\mathrm{d}\lambda}\,, \tag{10.20}$$

it might well be that there is some functional dependence between x and y. But that doesn't matter; we can still calculate $\mathrm{d}f/\mathrm{d}\lambda$ by applying this expression to the function $f(x,y)$. That's because if $z = f(x,y)$ is plotted in three dimensions to give a 2-surface, then this surface also contains any points that have been singled out by a dependence between x and y. So rates of increase of f over this subset are calculated in just the same way as rates of increase over the general surface. The two partial derivatives of (10.20) are really telling us to differentiate f with respect to its first argument and then with respect to its second argument.

In that respect, partial derivative notation such as $\partial f(x,y)/\partial x$ is used in physics to mean "differentiate mechanically with respect to the first argument of f, disregarding the fact that y might depend on x". It differs from that used by many mathematicians, who generally explicitly indicate the same procedure by writing $D_1 f(x,y)$ and will avoid an expression like $\partial f(x,y)/\partial x$ if there is any possibility of ambiguity.

Thus, in (10.19) the variable λ appears in both arguments of the function, and all we are doing is applying the chain rule (10.20) by differentiating with respect to both arguments.

Now set $\lambda = 0$ and integrate (10.19) by parts to eliminate η', as we did for the straight-line calculation in (10.14):

$$S'(0) = \int_{x_1}^{x_2} \left[\frac{\partial L}{\partial y}\eta + \frac{\partial L}{\partial y'}\eta' \right] \mathrm{d}x$$

$$= \int_{x_1}^{x_2} \frac{\partial L}{\partial y}\eta\,\mathrm{d}x + \underbrace{\frac{\partial L}{\partial y'}\eta\Big|_{x_1}^{x_2}}_{\substack{\text{Boundary}\\ \text{condition:}\\ \text{this vanishes!}}} - \int_{x_1}^{x_2} \frac{\mathrm{d}}{\mathrm{d}x}\left(\frac{\partial L}{\partial y'}\right)\eta\,\mathrm{d}x$$

$$= \int_{x_1}^{x_2} \left[\frac{\partial L}{\partial y} - \frac{\mathrm{d}}{\mathrm{d}x}\frac{\partial L}{\partial y'} \right] \eta\,\mathrm{d}x\,. \tag{10.21}$$

Requiring $S'(0)$ to be zero for all $\eta(x)$ implies that

$$\boxed{\frac{\partial L}{\partial y} - \frac{\mathrm{d}}{\mathrm{d}x}\frac{\partial L}{\partial y'} = 0\,.} \tag{10.22}$$

This is *Lagrange's equation*, the fundamental equation of variational calculus. The quantity $L(y, y')$ in the action is called the *lagrangian* for the scenario being considered.

10.3.2 Other Variational Approaches

For the sake of pedagogy, we have pursued a slightly different route to the lagrangian than what is usually followed. Other notations are frequently used, however, so let's pause for a moment to look at them. The most common writes our $S(\lambda)$ as $S + \delta S$, where S is identical to our $S(0)$. The variation is labelled δy, equalling our $\lambda\eta(x)$. A Taylor expansion shows the variation to be

$$\delta S = S'(0)\lambda + O(\lambda^2)\,. \tag{10.23}$$

Working only to first order in the variation (since it eventually goes to zero anyway), and following the same line of argument that produced (10.21), gives an equation analogous to (10.21):

$$\delta S = \int_{x_1}^{x_2} \left[\frac{\partial L}{\partial y} - \frac{\mathrm{d}}{\mathrm{d}x}\frac{\partial L}{\partial y'}\right] \delta y \, \mathrm{d}x\,. \tag{10.24}$$

Notice that since δL is defined such that

$$\delta S = \int_{x_1}^{x_2} \delta L(y, y') \, \mathrm{d}x\,, \tag{10.25}$$

the integrands of these last two equations are usually equated to write

$$\frac{\delta L}{\delta y} = \frac{\partial L}{\partial y} - \frac{\mathrm{d}}{\mathrm{d}x}\frac{\partial L}{\partial y'}\,, \tag{10.26}$$

although this only holds when L is a function of y and y' only. Although expressions of the form $\delta(\cdot)/\delta(\cdot)$, known as *functional derivatives*, are routinely written in variational calculations (see, for example, the general relativistic analogue on p. 515), we should not forget that they presuppose that an integration will eventually be performed using the boundary condition, such as shown in (10.21). Unfortunately, some authors write $\delta S/\delta y$ in place of $\delta L/\delta y$. This gains nothing, producing only an equation with incorrect units.

An interesting point in this second approach is a possible ambiguity. When the variation is written

$$\delta S = \int \delta L(y, y') \, \mathrm{d}x = \int \left[\frac{\partial L}{\partial y}\,\delta y + \frac{\partial L}{\partial y'}\,\delta y'\right] \mathrm{d}x\,, \tag{10.27}$$

we need to realise that $\delta y'$ could be either $(\delta y)'$ or $\delta(y')$. But these two are identical since

$$\delta(y') = (y + \delta y)' - y' = y' + (\delta y)' - y' = (\delta y)' . \tag{10.28}$$

A third type of variational calculus notation employs the Dirac delta function. In this case, the basic variation $\eta(x)$ is much more constrained, being a delta function with its spike at some arbitrary value of x. We demand a stationary value of S over all of these spike positions:

$$\eta(x) = \delta(x - a) \quad \text{with } S'(0) \overset{\text{req.}}{=\!=} 0 \quad \forall a . \tag{10.29}$$

Here, the delta function derivative is needed, defined by how it acts on a sufficiently well-behaved test function $T(x)$:

$$\begin{aligned} \int_{-\infty}^{\infty} T(x)\, \delta'(x - a)\; \mathrm{d}x &= T(x)\, \delta(x - a)\Big|_{-\infty}^{\infty} - \int_{-\infty}^{\infty} T'(x)\, \delta(x - a)\; \mathrm{d}x \\ &= -T'(a) . \end{aligned} \tag{10.30}$$

Using this approach, the Lagrange equation is very economically recovered—although the groundwork has really been hidden inside the mechanism of the delta function and its derivative. Begin with

$$S_a(\lambda) \equiv \int_{x_1}^{x_2} L\Big(y + \lambda\, \delta(x - a),\ y' + \lambda\, \delta'(x - a)\Big)\; \mathrm{d}x , \tag{10.31}$$

so that

$$\begin{aligned} S_a'(0) &= \int_{x_1}^{x_2} \left[\frac{\partial L}{\partial y}\, \delta(x - a) + \frac{\partial L}{\partial y'}\, \delta'(x - a)\right] \mathrm{d}x \\ &= \left.\frac{\partial L}{\partial y} - \frac{\mathrm{d}}{\mathrm{d}x}\frac{\partial L}{\partial y'}\right|_{x=a} \overset{\text{req.}}{=\!=} 0 \quad \forall a , \end{aligned} \tag{10.32}$$

which again gives Lagrange's equation.

10.3.3 Application to Mechanics: Hamilton's Principle

When a system is constrained in some way, Newton's laws, with their vector forces and accelerations, might not be economical for producing its *equation of motion*. For example, a bead sliding down a curved wire can never leave that wire, so the situation is one-dimensional, leading to a hope that a knowledge of the force applied by the wire on the bead might not be necessary to solve for the resulting motion.

Such a *lagrangian approach* to mechanics is catered for by variational calculus, and it can be shown that such an approach is equivalent to applying Newton's laws. The variable x that we have used up until now in equations such as (10.17) is replaced by time, while the range y is replaced by any quantity useful for describing the system. This is often a space coordinate but can just as well be anything else that quantifies the system dynamics. Corresponding to (10.17), the action for some set of coordinates $\{x^i\}$ is

$$S = \int_{t_1}^{t_2} L(x^i, \dot{x}^i)\, \mathrm{d}t\,, \tag{10.33}$$

where if there are two particles, each with three degrees of freedom, then the index i counts 1 to 6, and so on. For a conservative field (i.e. one for which we can define a potential), the nonrelativistic lagrangian $L(x^i, \dot{x}^i)$ turns out to be

$$L = \text{total kinetic energy} - \text{total potential energy.} \tag{10.34}$$

A justification for L having this form is that it leads to Newton's equation of motion, which can be seen in the following way. Consider the simplest case: a particle moving subject to some potential energy $V(x, y, z)$ that is a function of position only. The lagrangian is

$$L(x, y, z, \dot{x}, \dot{y}, \dot{z}) = \frac{1}{2}mv^2 - V(x, y, z)\,. \tag{10.35}$$

Lagrange's equation for the x coordinate is

$$\frac{\partial L}{\partial x} - \frac{\mathrm{d}}{\mathrm{d}t}\frac{\partial L}{\partial \dot{x}} = 0\,, \tag{10.36}$$

which leads to

$$-\partial V/\partial x = m\ddot{x}\,. \tag{10.37}$$

Similar equations are produced for the y and z coordinates, and the three can be combined to give

$$-\nabla V = m(\ddot{x}, \ddot{y}, \ddot{z})\,. \tag{10.38}$$

But this is just Newton's force law if we identify the force with the rate of loss of potential energy with position as in (10.6). Conversely, it can also be shown that Newton's force law gives rise to Lagrange's equation, although we won't do that here.

Whereas Newton's theory of mechanics is couched in terms of forces acting on an object to produce an acceleration, the lagrangian approach takes an entirely different view, speaking only of energy, while considering all at once the entire path of the object in space and time. Nature chooses this path to minimise the action. So while Newton considers that a particle responds to whatever push it receives from one moment to the next, the Principle of Least Action considers the particle's path holistically. This way of describing the physical world works very well—and not just for mechanical problems but for all of physics. In a way, modern physics is a search for lagrangians to describe ever more fundamental systems, along with finding ways to extract information from those lagrangians.

10.3.4 Nöther's Theorem and Lagrangian Invariances

A prime importance of the lagrangian lies in its ability to embody invariances of the system, since these can be shown to give rise to conserved quantities. This famous theorem was expressed in general form by the mathematician Emmy Nöther.

In the case of space and time invariances, Nöther's theorem produces two conserved quantities of fundamental importance in classical and quantum mechanics: the generalised momentum $\boldsymbol{b}$ and the hamiltonian H. These actually date from early in the history of lagrangian mechanics, long before Nöther wrote down her theorem.

The Generalised or Canonical Momentum $\boldsymbol{b}$

If we write the Lagrange equation for the coordinate x^i as

$$\frac{\partial L}{\partial x^i} = \frac{\mathrm{d}}{\mathrm{d}t}\frac{\partial L}{\partial \dot{x}^i}\,, \tag{10.39}$$

then it suggests the existence of a conserved quantity. To see this, define the *canonical* or *generalised momentum* for the coordinate x^i as

$$b_i \equiv \frac{\partial L}{\partial \dot{x}^i}\,, \tag{10.40}$$

in which case the Lagrange equation implies that

$$\boxed{\dot{b}_i \equiv \frac{\partial L}{\partial x^i}\,.} \tag{10.41}$$

So if the lagrangian is independent of some coordinate x^i, then the corresponding canonical momentum b_i will be a *constant of the motion*: that is, conserved over time. A case in point is a single particle with some potential energy $V(t)$ that depends at most on time. Here the canonical momentum equals the usual more elementary momentum $m\boldsymbol{v}$. The lagrangian in this case has no spatial dependence, so it must be that $m\boldsymbol{v}$ is conserved in time, as we know is true from Newton's laws. As we shall see later, the canonical momentum is particularly important in the transition from classical mechanics to quantum mechanics.

The single-particle canonical momentum $\boldsymbol{b}$ is really just a generalisation of the elementary momentum $\boldsymbol{p} \equiv m\boldsymbol{v}$ to more sophisticated lagrangians, and both are usually denoted by $\boldsymbol{p}$. This is a reasonable thing to do, but can sometimes cause confusion when we need to deal with both quantities in one equation, which we'll do in (10.130). It is the canonical momentum that finds especial use in quantum mechanics, and in this text $\boldsymbol{b}$ has been used for canonical momentum purely to highlight the care needed in such a situation.

The Hamiltonian H

The time derivative of the lagrangian is

$$\begin{aligned}\frac{\mathrm{d}L}{\mathrm{d}t} &= \frac{\partial L}{\partial x^i}\dot{x}^i + \frac{\partial L}{\partial \dot{x}^i}\ddot{x}^i + \frac{\partial L}{\partial t} \quad \text{(sum over } i \text{ and over all particles)} \\ &= \frac{\mathrm{d}}{\mathrm{d}t}\left(\frac{\partial L}{\partial \dot{x}^i}\dot{x}^i\right) + \frac{\partial L}{\partial t} \\ &= \frac{\mathrm{d}}{\mathrm{d}t}(b_i v^i) + \frac{\partial L}{\partial t} \qquad \left(v^i \equiv \dot{x}^i\right),\end{aligned} \tag{10.42}$$

so that

$$\frac{\mathrm{d}}{\mathrm{d}t}\left(L - b_i v^i\right) = \frac{\partial L}{\partial t}\,. \tag{10.43}$$

This relation motivates the definition of a *hamiltonian*,

$$H \equiv b_i v^i - L \quad \text{(sum over } i \text{ and all particles)}, \tag{10.44}$$

which implies that

$$\boxed{\dot{H} = \frac{-\partial L}{\partial t}\,.} \tag{10.45}$$

Compare this with (10.41). If the lagrangian is independent of time, the hamiltonian becomes a constant of the motion and is then called the *total energy* of the system. This, then, is a principal use of the lagrangian: through (10.40) and (10.44) it produces candidates for constants of the motion, and similarly for more general "internal" coordinates via Nöther's theorem. In Sect. 10.5, we will see how $\boldsymbol{b}$ and H find a central use in quantum mechanics.

10.3.5 Continuous Systems: First Steps to a Field Theory

The fact that the total lagrangian is just the sum of the lagrangians for each particle enables us to construct the lagrangian for a continuous system. For example, what is the lagrangian of a vibrating string, and how does it produce the expected equation of motion?

Parametrise the string by x along its length, where its displacement from equilibrium is ϕ. Consider it as made up of many particles, each of infinitesimal mass $\mu\,\mathrm{d}x$, where μ is the string's linear mass density. Finally, calculate the total lagrangian by summing the contributions of all the masses. A realistic simplification is that the vibrating particles only move transversely to the string's length, so that the kinetic energy of each will be

$$\text{kinetic energy} = \frac{1}{2}\mu\,\mathrm{d}x\,\dot{\phi}^2, \tag{10.46}$$

where $\dot{\phi} \equiv \partial\phi/\partial t$. The potential energy of each small mass is the work done against the tension T to stretch the string by an amount $\mathrm{d}s - \mathrm{d}x$, where $\mathrm{d}s$ is

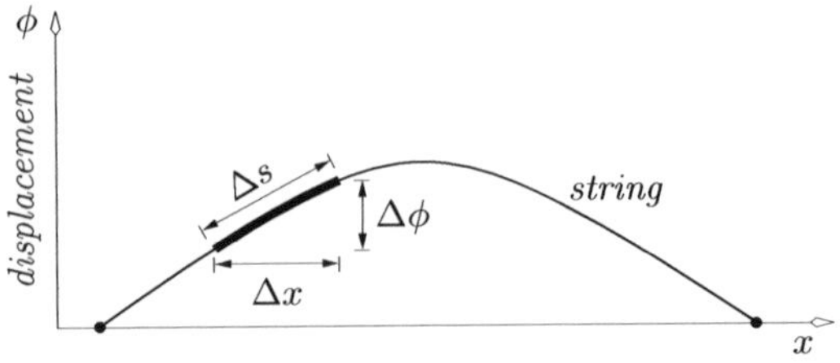

Fig. 10.4. Calculating the potential energy of a guitar string, shown here with a heavily exaggerated displacement. Any element of the string is stretched by an amount $\Delta s - \Delta x$, where $\Delta s^2 \simeq \Delta x^2 + \Delta\phi^2$. In the infinitesimal limit, this stretch becomes $\mathrm{d}s - \mathrm{d}x = \left(\sqrt{1+(\phi'(x))^2} - 1\right)\mathrm{d}x$. Small displacements mean the string's slope ϕ' is also small, so that the stretch becomes $\mathrm{d}s - \mathrm{d}x = 1/2\,\phi'^2\,\mathrm{d}x$. The resulting potential energy is the work done by the tension T in displacing the element by this amount, or $1/2\,T\phi'^2\,\mathrm{d}x$.

its stretched length, given by (10.8) and shown in Fig. 10.4. For $\phi' \equiv \partial\phi/\partial x$ small, we then have

$$\text{potential energy} = \frac{1}{2}T\phi'^2\,\mathrm{d}x\,. \tag{10.47}$$

The lagrangian for each mass is therefore

$$\frac{1}{2}\mu\,\mathrm{d}x\,\dot{\phi}^2 - \frac{1}{2}T\phi'^2\,\mathrm{d}x\,. \tag{10.48}$$

Hence, the lagrangian for the whole string of length L is, by summing the individual elements,

$$L = \int_0^L \frac{1}{2}\left[\mu\,\dot{\phi}^2 - T\phi'^2\right]\mathrm{d}x\,. \tag{10.49}$$

The lagrangian is now itself an integral, unlike that for the single-particle case (10.35). So, we can interpret the integrand in (10.49) as a lagrangian per unit distance, the *lagrangian density* $\mathcal{L}$:

$$\mathcal{L}(\phi,\dot{\phi},\phi') = \frac{1}{2}\left[\mu\,\dot{\phi}^2 - T\phi'^2\right]. \tag{10.50}$$

Thus, even in this nonrelativistic case, the action now puts space and time on an equal footing:

$$S = \int L\,\mathrm{d}t = \iint \mathcal{L}(\phi,\dot{\phi},\phi')\,\mathrm{d}t\,\mathrm{d}x\,. \tag{10.51}$$

How do we vary this action? Follow the same ideas as before, except now the perturbation function is $\eta(t,x)$ with a more stringent anchoring:

$$\eta(t_1, x) = \eta(t_2, x) = \eta(t, x_1) = \eta(t, x_2) \overset{\text{req.}}{=\!=} 0\,. \tag{10.52}$$

The perturbed action is

$$S(\lambda) \equiv \iint \mathcal{L}\big(\phi + \lambda\eta,\ \partial_t(\phi + \lambda\eta),\ \partial_x(\phi + \lambda\eta)\big)\ \mathrm{d}t\ \mathrm{d}x\,. \tag{10.53}$$

In that case, the calculation analogous to (10.21) is

$$\begin{aligned}
S'(0) &= \iint \left[\frac{\partial\mathcal{L}}{\partial\phi}\eta + \frac{\partial\mathcal{L}}{\partial\dot\phi}\dot\eta + \frac{\partial\mathcal{L}}{\partial\phi'}\eta'\right] \mathrm{d}t\ \mathrm{d}x \\
&= \iint \frac{\partial\mathcal{L}}{\partial\phi}\eta\ \mathrm{d}t\,\mathrm{d}x + \int \Bigg[\underbrace{\frac{\partial\mathcal{L}}{\partial\dot\phi}\eta\bigg|_{t_1}^{t_2}}_{\substack{\text{Boundary conditions:}\\ \text{this vanishes}}} - \int \frac{\mathrm{d}}{\mathrm{d}t}\left(\frac{\partial\mathcal{L}}{\partial\dot\phi}\right)\eta\ \mathrm{d}t\Bigg]\mathrm{d}x \\
&\quad + \int \Bigg[\underbrace{\frac{\partial\mathcal{L}}{\partial\phi'}\eta\bigg|_{x_1}^{x_2}}_{\text{This also vanishes}} - \int \frac{\mathrm{d}}{\mathrm{d}x}\left(\frac{\partial\mathcal{L}}{\partial\phi'}\right)\eta\ \mathrm{d}x\Bigg]\ \mathrm{d}t \\
&= \iint \left[\frac{\partial\mathcal{L}}{\partial\phi} - \frac{\mathrm{d}}{\mathrm{d}t}\frac{\partial\mathcal{L}}{\partial\dot\phi} - \frac{\mathrm{d}}{\mathrm{d}x}\frac{\partial\mathcal{L}}{\partial\phi'}\right]\eta\ \mathrm{d}t\,\mathrm{d}x \overset{\text{req.}}{=\!=} 0 \quad \forall\,\eta\,.
\end{aligned} \tag{10.54}$$

It follows that the equation of motion for the string is

$$\frac{\partial\mathcal{L}}{\partial\phi} - \frac{\mathrm{d}}{\mathrm{d}t}\frac{\partial\mathcal{L}}{\partial\dot\phi} - \frac{\mathrm{d}}{\mathrm{d}x}\frac{\partial\mathcal{L}}{\partial\phi'} = 0\,. \tag{10.55}$$

Applying this to the lagrangian density of (10.50) easily produces the usual wave equation for a string that has been given a small sideways displacement:

$$\frac{\partial^2\phi}{\partial x^2} = \frac{\mu}{T}\frac{\partial^2\phi}{\partial t^2}\,, \tag{10.56}$$

which leads to the usual wave speed of $c = \sqrt{T/\mu}$.

Finally: The Variational Principle for a Field Theory

The previous example dealt with a function of one space and one time dimension, $\phi(t, x)$. It generalises quite easily to a field $\phi(t, x, y, z)$ with cartesian coordinates x^α, as well as no gravity, because it turns out that gravity alters the volume element in the integration in a way that we'll leave for Chap. 12. We will also only use a scalar field here since it has no indices to keep track of, but the same formalism also applies to tensor fields. Beginning with the action

$$S = \int \mathcal{L}(\phi, \phi_{,\alpha})\, \mathrm{d}^4x\,, \quad \alpha = 0 \to 3\,, \tag{10.57}$$

use the summation convention to write

$$S'(0) = \int \left[\frac{\partial \mathcal{L}}{\partial \phi} \eta + \frac{\partial \mathcal{L}}{\partial \phi_{,\alpha}} \eta_{,\alpha} \right] \mathrm{d}^4x\,, \tag{10.58}$$

remembering that $\partial\phi_{,\alpha} \equiv \partial\,(\phi_{,\alpha})$. The second term can be integrated by parts as we have done previously, and requiring η to vanish on all boundaries leads to

$$\boxed{\frac{\partial \mathcal{L}}{\partial \phi} - \partial_\alpha \frac{\partial \mathcal{L}}{\partial \phi_{,\alpha}} = 0\,.} \tag{10.59}$$

Last, we wish to generalise (10.59) to arbitrary coordinates. This is not really any more difficult or different from what we have already done, but it's a task more suited to Chap. 12, where it will be needed to investigate how gravity affects other fields. The relevant details are explained in Sects 12.8 and 12.8.1. There we will find the correct expression to be any one of (12.117)–(12.119).

10.3.6 Nöther's Theorem for a Scalar Field

The hamiltonian and generalised momentum that we saw previously become fused into a single tensor in the case of a scalar field. This comes about by our almost mimicking the previous derivation of the hamiltonian for point particles, although to shorten the calculation we'll assume here that the lagrangian density has no explicit coordinate dependence. Hence,

$$\begin{aligned}\partial_\beta \mathcal{L} &= \frac{\partial \mathcal{L}}{\partial \phi} \phi_{,\beta} + \frac{\partial \mathcal{L}}{\partial \phi_{,\alpha}} \phi_{,\alpha\beta} \overset{(10.59)}{=\!=} \partial_\alpha \left(\frac{\partial \mathcal{L}}{\partial \phi_{,\alpha}} \right) \phi_{,\beta} + \frac{\partial \mathcal{L}}{\partial \phi_{,\alpha}} \phi_{,\alpha\beta} \\ &= \partial_\alpha \left(\frac{\partial \mathcal{L}}{\partial \phi_{,\alpha}} \phi_{,\beta} \right).\end{aligned} \tag{10.60}$$

We can get the same differentiation subscript on each side by rewriting the last line above using a Kronecker delta,

$$\delta^\alpha{}_\beta\, \partial_\alpha \mathcal{L} = \partial_\alpha \left(\frac{\partial \mathcal{L}}{\partial \phi_{,\alpha}} \phi_{,\beta} \right), \tag{10.61}$$

in which case

$$\partial_\alpha \underbrace{\left(\frac{\partial \mathcal{L}}{\partial \phi_{,\alpha}} \phi_{,\beta} - \delta^\alpha{}_\beta \mathcal{L} \right)}_{\equiv T^\alpha{}_\beta} = 0\,. \tag{10.62}$$

In a generalisation of the energy and momentum discussions for the point particle lagrangian, the quantity $T^\alpha{}_\beta$ inside the parentheses is termed the *energy–momentum tensor*.

The discussion around (8.63) is very pertinent here. If we wish to lower the α in $T^{\alpha}{}_{\beta}$ to produce an expression for $T_{\alpha\beta}$, then we need to realise that the $\delta^{\alpha}{}_{\beta}$ in (10.62) becomes the metric, which in Minkowski spacetime is $\eta_{\alpha\beta}$. Plus, we have deliberately written the Kronecker delta not as δ^{α}_{β} but instead with indices spaced apart, which amounts to a choice of index order in the definition of the energy–momentum tensor. Also, it's desirable for the energy–momentum tensor to be symmetric; the definition here is not explicitly so. Alternative approaches to constructing a symmetric energy–momentum tensor are found in the literature.

What is the field generalisation of the conservation of energy and momentum for point particles? Because the *four-divergence* $T^{\alpha}{}_{\beta,\alpha}$ is zero, the energy–momentum tensor is called *divergence-free*, and in a cartesian basis (no covariant derivatives needed) we can write

$$-T^{0}{}_{\beta,0} = T^{i}{}_{\beta,i} = \nabla\cdot\left(T^{1}{}_{\beta}, T^{2}{}_{\beta}, T^{3}{}_{\beta}\right) \equiv \nabla\cdot\boldsymbol{T}_{\beta}\,. \tag{10.63}$$

Alternatively, note that (10.63) could have been written

$$-T^{0}{}_{\beta,0} = T^{i}{}_{\beta,i} \equiv \partial_i\left(\boldsymbol{T}_{\beta}\cdot\boldsymbol{e}^i\right) \overset{\boldsymbol{e}^i\ \text{const.}}{=\!=\!=} \left(\partial_i\boldsymbol{T}_{\beta}\right)\cdot\boldsymbol{e}^i = \nabla\cdot\boldsymbol{T}_{\beta}\,. \tag{10.64}$$

Constant cobasis vectors have been assumed, which is why we stipulated that the calculation must use cartesian coordinates.

Integrating (10.63) over a closed volume gives

$$-\int_{\text{volume}} T^{0}{}_{\beta,0}\,\mathrm{d}V = \int_{\text{volume}} \nabla\cdot\boldsymbol{T}_{\beta}\,\mathrm{d}V\,, \tag{10.65}$$

and Gauss's theorem from vector calculus then converts the integral of the divergence on the right-hand side of (10.65) to a surface integral:

$$\underbrace{-\partial_0\int_{\text{volume}} T^{0}{}_{\beta}\,\mathrm{d}V}_{\substack{\text{Rate of loss of}\\ \text{“something” in the}\\ \text{enclosed volume}}} \quad=\quad \underbrace{\int_{\text{surface}} \boldsymbol{T}_{\beta}\cdot\boldsymbol{n}\,\mathrm{d}S\,.}_{\substack{\text{Rate of flow out of that}\\ \text{volume, where } \boldsymbol{T}_{\beta} \text{ is a flux}\\ \text{density (i.e. flow of something}\\ \text{per unit area per unit time)}}} \tag{10.66}$$

Figure 10.5 shows the meaning of the terms. What has resulted is a conserved quantity, which in this case defines the *field energy*, related to the *field momentum* flux entering the volume. So the field has an *energy density* of T_{00} and three *momentum densities* of T_{0i} (where the first index has been lowered for convenience). In general, the quantity $\int T_{0\beta}\,\mathrm{d}V$ is called a *Nöther charge*, while the flux term is the associated *Nöther current*. So the lagrangian formalism for fields allows us to predict the forms of these conserved quantities. We will meet a different formulation of the energy–momentum tensor in the context of general relativity in Chap. 12.

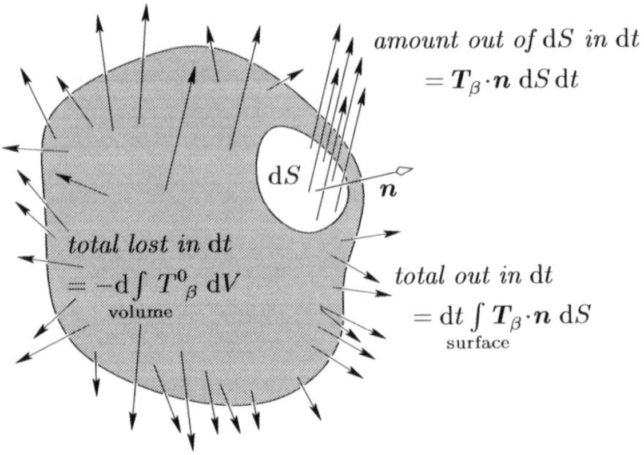

Fig. 10.5. Interpreting (10.66). As an aid, multiply both sides of that equation by $\mathrm{d}t$ to give $-\mathrm{d}\int_V T^0{}_\beta \, \mathrm{d}V = \mathrm{d}t \int_S \boldsymbol{T}_\beta \cdot \boldsymbol{n} \, \mathrm{d}S$. These can now be interpreted as the loss of "something" in a time $\mathrm{d}t$ from a closed volume, along with the amount of that something escaping through the surface. Thus $T^0{}_\beta$ is a density, while $\boldsymbol{T}_\beta$ is a flux density, being the flow of that something per unit area per unit time. These two quantities form a natural pair, as we saw at the start of Chap. 6, in particular (6.11).

10.4 Building a Lagrangian

Why should the lagrangian have the form that it does? A clue is provided by the motion of a free particle in special relativity (and also, as it turns out, in general relativity). Such a particle moves along a path that maximises a particular scalar, the proper time between the starting and ending events:

$$\int_{\text{path taken}} \mathrm{d}\tau \quad \text{is a maximum.} \tag{10.67}$$

We can see that the path really must maximise the proper time since, for example, in just one space dimension with no gravity, the metric is

$$\mathrm{d}\tau^2 = \mathrm{d}t^2 - \mathrm{d}x^2. \tag{10.68}$$

In the frame in which the particle is at rest, its worldline runs along the time axis and $\mathrm{d}\tau = \mathrm{d}t$. Any departure from this axis will introduce a $\mathrm{d}x^2$ into the proper time interval, being subtracted from $\mathrm{d}t^2$ and so acting to reduce the total proper time experienced by the particle. So the worldline that runs along the time axis must maximise the particle's proper time. Now notice what happens if we multiply the proper time by the particle's negative rest mass $-m$:

$$-m \int \mathrm{d}\tau = -\int \frac{m\,\mathrm{d}t}{\gamma} \simeq \int \left(\tfrac{1}{2}mv^2 - m\right) \mathrm{d}t \tag{10.69}$$

The m in the parentheses is a constant and so does not affect any extremisation. So relativity explains why the nonrelativistic lagrangian for a free particle is just its kinetic energy. Notice that the actual lagrangian to be integrated over the time coordinate t is not a scalar, because $\mathrm{d}t$ is not a scalar either. But the action in (10.69) certainly is a scalar (composed of rest mass and proper time, both scalars), at least insofar as the binomial approximation doesn't change it too much.

More generally, being a scalar is a demand we place on the action because that way its stationary value will be guaranteed independent of the coordinate system. Plus, having all observers agree on its value is an important property that surfaces in other areas, such as in Sect. 10.4.1, where we use it to build the electromagnetic potential four-vector, and in Sect. 10.8, where the action assumes the central role in the path-integral formulation of quantum mechanics.

Most particles are not free but subject to some force, which suggests including a potential energy term in (10.69). To look more closely at such a term, remain for the moment nonrelativistic and write down the lagrangian for a mass m in a newtonian gravitational field with potential Φ. The discussion of Sect. 10.3.3 prescribes a lagrangian

$$L = \frac{1}{2}mv^2 - m\Phi\,. \tag{10.70}$$

But potential energy is not a property of the mass alone; rather, it is a property of the two gravitating bodies. It's an *interaction*, and as is clear from (10.70), it is linear in both mass and potential. This suggests that we could make the lagrangian more complete and symmetrical by adding a new term that involves the gravitational potential Φ alone. Because $mv^2/2$ is quadratic in the time derivative of the coordinate describing the particle, we'll try adding a term quadratic in space derivatives of Φ since, in newtonian gravity, Φ depends only on space. We suggest the following in cartesian coordinates:

$$\sum_i \left(\Phi_{,i}\right)^2\,, \quad \text{or alternatively} \quad \left|\nabla\Phi\right|^2. \tag{10.71}$$

The field is spread over space, so we integrate this new term over all space, including a relative weighting factor of $-1/(8\pi G)$ in hindsight. The lagrangian becomes

$$L = \underbrace{\frac{1}{2}mv^2}_{\text{free particle}} \underbrace{- \frac{1}{8\pi G}\int \left|\nabla\Phi\right|^2 \mathrm{d}^3x}_{\text{free field}} \underbrace{- m\Phi}_{\text{interaction}}\,. \tag{10.72}$$

The new term involving the field alone now has no space dependence, so varying the action with respect to the space coordinates of the particle's trajectory will give Newton's force law $-\nabla\left(m\Phi\right) = m\boldsymbol{a}$ as before. This law describes how the mass is acted upon by the field. But by symmetry we expect a new result. By varying the lagrangian with respect to Φ, we might obtain the "equation of motion" of the field: how the *field* is influenced by the *mass*. This is indeed what results. To show this, write the lagrangian as a density, as was done in Sect. 10.3.5. The $mv^2/2$ term has no relevance here, being independent of Φ; so extract from the rest of the lagrangian (10.72) the density

$$\mathcal{L}_{\text{grav}} = \frac{-1}{8\pi G} \left|\nabla\Phi(\boldsymbol{x})\right|^2 - \varrho(t,\boldsymbol{x})\,\Phi(\boldsymbol{x})\,, \tag{10.73}$$

where the density of the mass is now specified by $\varrho(t,\boldsymbol{x})$, which for a particle will be a time-dependent delta function. Now apply the Lagrange equation (10.59), writing

$$\frac{\partial\mathcal{L}_{\text{grav}}}{\partial\Phi} - \partial_\alpha \frac{\partial\mathcal{L}_{\text{grav}}}{\partial\Phi_{,\alpha}} = 0\,, \tag{10.74}$$

ignoring the case $\alpha = 0\,(\equiv t)$ since $\mathcal{L}_{\text{grav}}$ has no time dependence. Equations (10.73) and (10.74) combine to produce *Poisson's equation*,

$$\nabla^2\Phi(t,\boldsymbol{x}) = 4\pi G\,\varrho(t,\boldsymbol{x})\,. \tag{10.75}$$

We will show that this has a spherically symmetric solution $\Phi(r)$ for a point mass at rest at the origin: $\varrho = m\,\delta(\boldsymbol{x})$. Use (8.227) to write Poisson's equation as

$$\frac{1}{r^2}\left(r^2\Phi_{,r}\right)_{,r} = 4\pi G\, m\,\delta(\boldsymbol{x})\,. \tag{10.76}$$

For a potential required to vanish at $r \to \infty$, a solution for $r > 0$ is

$$\Phi = a/r \quad \text{for some constant } a, \text{ with } r \neq 0\,. \tag{10.77}$$

The constant a can be fixed by first integrating (10.75) over space:

$$\int \nabla^2\Phi \,\mathrm{d}^3x = 4\pi G\, m\,. \tag{10.78}$$

Use Gauss's theorem to convert (10.78) to a flux integral over a sphere of radius R with surface element $\mathrm{d}S$:

$$\begin{aligned} 4\pi G\, m &= \int \nabla\!\cdot\!\nabla\Phi\,\mathrm{d}^3x \overset{\text{Gauss's theorem}}{=\!=\!=} \int \nabla\Phi\cdot\boldsymbol{e}_r\,\mathrm{d}S \\ &\overset{(10.77)}{=\!=\!=} \int \frac{-a}{r^2}\,\mathrm{d}S = \frac{-a}{R^2}\,4\pi R^2 = -4\pi a\,. \end{aligned} \tag{10.79}$$

So $a = -G\,m$ and the solution is

$$\Phi = \frac{-G\,m}{r}\,, \quad \text{with the field} \quad \boldsymbol{g} = -\nabla\Phi = \frac{-G\,m}{r^2}\,\boldsymbol{e}_r\,, \tag{10.80}$$

as expected. The programme is a success: a lagrangian can be built as the sum of terms for the free mass, the free field, and an interaction.

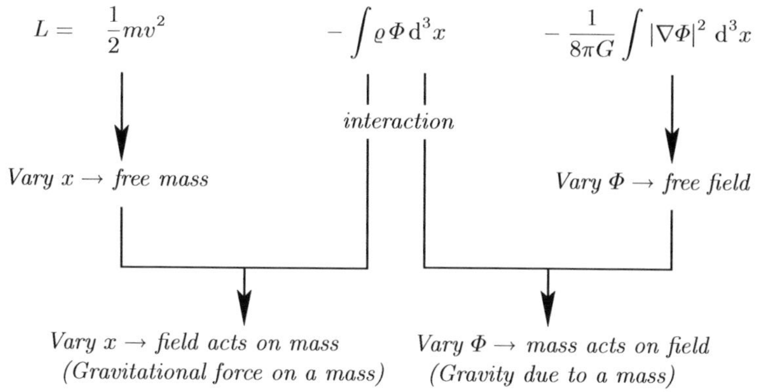

Fig. 10.6. The variation of terms that gives the two "equations of motion" when a mass interacts with a newtonian gravitational field. The lagrangian is (10.72).

Something really new has emerged here. We only added a free field to the lagrangian describing how the field affects the particle, and yet the formalism has told us correctly how the particle affects the field. The single interaction term $-m\Phi$ describes correctly how the mass and the gravitational field affect each other. The overall scenario is shown in Fig. 10.6.

And therein lies some of the beauty of the lagrangian approach to field theory. Where previously Newton's force law was seen as something completely separate from his law describing the gravitational field produced by a point mass, the lagrangian (10.72) has united these two things into one coherent whole. Finally, the current philosophy of modern physics emerges: to search for a lagrangian describing as much of the physical world as possible and, in choosing terms that lead to experimentally verifiable results, to predict new laws in just the same way as we have here been led to the law of gravity for a point mass.

10.4.1 A Relativistic Lagrangian for a Charge in an EM Field

The next step in our study of lagrangians is to make everything relativistic. We'll leave gravity until Chap. 12. Instead, here we will focus on the lagrangian for a charge q in an electromagnetic field. Again, the free particle part of the scalar lagrangian is $-m$. Statically, the interaction term is the potential energy $q\Phi$. But the presence of Φ reminds us that in electromagnetism, Φ tends to appear along with the vector potential $\boldsymbol{A}$ in a highly symmetrical way, and these two together give the electric and magnetic fields by way of (10.5):

$$\boldsymbol{E} = -\nabla\Phi - \partial\boldsymbol{A}/\partial t\,, \quad \boldsymbol{B} = \nabla\times\boldsymbol{A}\,. \tag{10.81}$$

Because the physical fields depend on the derivatives of Φ and $\boldsymbol{A}$, these two potentials have a certain amount of freedom in their definitions. We'll consider the possibility that there might well be a valid set of Φ and $\boldsymbol{A}$ that belong together in a lagrangian, and indeed that they might even form a vector $A^\alpha \boldsymbol{e}_\alpha$, the *electromagnetic potential*, whose time component is Φ and whose space projection is $\boldsymbol{A}$. So, for example, in cartesian coordinates

$$\left(A^t, A^x, A^y, A^z\right) \equiv (\Phi, \boldsymbol{A}) . \tag{10.82}$$

Let's see where such an assumption leads. If the interaction is required to be a scalar in the action integration over τ, then we wish to include *all* of the A^α components in the lagrangian and not just the potential Φ. In analogy to the $m\Phi$ interaction for gravity, but with an eye toward introducing the charge's velocity (which we know interacts with the magnetic field), try a scalar interaction of $qu^\alpha A_\alpha$, including a factor from the metric of $\operatorname{sgn}\eta_{00}$ for extra clarity, as we'll explain in a moment:

$$\begin{aligned} S &= \int (-m - qu^\alpha A_\alpha \operatorname{sgn}\eta_{00})\, d\tau \\ &= \int \left(\frac{-m}{\gamma} - \frac{q}{\gamma} u^\alpha A_\alpha \operatorname{sgn}\eta_{00}\right) dt . \end{aligned} \tag{10.83}$$

Actually the term u^α/γ in (10.83) just equals $(1, \boldsymbol{v})$, so we will write it as v^α for conciseness, where $v^0 \equiv 1$ (although v^α is not a tensor!). The (fully relativistic) lagrangian is thus

$$L = -m/\gamma - \operatorname{sgn}\eta_{00}\, qv^\alpha A_\alpha . \tag{10.84}$$

The factor of $\operatorname{sgn}\eta_{00}$ has been included because raising or lowering indices involves the metric, but unfortunately there is no common convention for the choice of signs in the metric. It's useful to leave that sign choice unspecified for now, partly because the different choices make for confusing reading across different books. However, since we'll want to raise and lower indices in the following discussion, we'll work with cartesians for simplicity, where $\operatorname{sgn}\eta_{00} = \eta_{tt}$ and the metric is simply $\eta_{tt}\operatorname{diag}(1, -1, -1, -1)$, where η_{tt} can be chosen as either ± 1. The end result will be covariant for cartesian coordinates, and so will easily be modified to apply to all other coordinates by converting all ordinary partial derivatives into covariant derivatives.

Thus, for all vector components V^α,

$$\begin{aligned} V^k &= -\eta_{tt} V_k , \quad V_k = -\eta_{tt} V^k, \\ V^t &= \eta_{tt} V_t , \quad V_t = \eta_{tt} V^t. \end{aligned} \tag{10.85}$$

Regardless of the choice of η_{tt}, the nonrelativistic limit of L in (10.84) is easily found to be

$$L \to \frac{1}{2} mv^2 - q\Phi + q\boldsymbol{v}\cdot\boldsymbol{A} , \tag{10.86}$$

which we can see is no longer a simple kinetic energy minus potential energy expression. Showing how the equation of motion for the nonrelativistic charge results from this is only slightly simpler than doing the same for the relativistic version (10.84); so we will only consider the relativistic case. Lagrange's equations require

$$\frac{\partial L}{\partial x^k} = -\eta_{tt}\, q v^\beta A_{\beta,k}\,, \quad (k = x \to z)\,, \tag{10.87}$$

$$\begin{aligned}\frac{\partial L}{\partial v^k} &= \gamma m v^k - \eta_{tt}\, qA_k \\ &= p^k - \eta_{tt}\, qA_k = -\eta_{tt}\,(p_k + qA_k)\,,\end{aligned} \tag{10.88}$$

where $p^k \equiv \gamma m v^k$ is the relativistic momentum defined in (5.49), and where the superscript of p^k was lowered in (10.88) to match the qA_k there, ensuring that the sum $p_k + qA_k$ is covariant (i.e., the k appears either as a superscript or a subscript, but not a mixture of both). This is a good rule of thumb to use when aiming for simplification. The required time derivative is

$$\frac{\mathrm{d}}{\mathrm{d}t}\frac{\partial L}{\partial v^k} = -\eta_{tt}\left(p_{k,t} + qA_{k,t} + qA_{k,\ell}\, v^\ell\right), \quad (\ell = x \to z)\,. \tag{10.89}$$

The Lagrange equations combine (10.87) with (10.89) to give

$$\frac{1}{q}\frac{\mathrm{d}p_k}{\mathrm{d}t} = v^\beta A_{\beta,k} - A_{k,t} - A_{k,\ell}\, v^\ell. \tag{10.90}$$

This expression will be more covariant if we write it using the proper time τ, where

$$\mathrm{d}t = \gamma\,\mathrm{d}\tau = u^t \mathrm{d}\tau\,, \tag{10.91}$$

so that

$$\frac{\mathrm{d}p_k}{\mathrm{d}\tau} = qu^\beta\left(A_{\beta,k} - A_{k,\beta}\right). \tag{10.92}$$

But we have seen the expression in the last parentheses before: it's part of the generalised curl, $F_{\alpha\beta}$, that was foreseen in (8.239), whose cartesian components are listed in (10.94) and (10.95) in the box on the next page.

Note that in (10.94) we have included an arbitrary factor of $s \in \{\pm 1\}$ since again sign conventions differ, although the most common choice (including this book) seems to be $s = +1$. We'll set $s = 1$ in what follows. The factor s can always be included in the following calculations with the replacements $F^{\alpha\beta} \to sF^{\alpha\beta}$ and $F_{\alpha\beta} \to sF_{\alpha\beta}$. Equivalently, a sign change to $s = -1$ simply swaps the indices of both $F^{\alpha\beta}$ and $F_{\alpha\beta}$.

Equation (10.92) becomes

$$\frac{\mathrm{d}p_k}{\mathrm{d}\tau} = qu^\beta F_{k\beta}\,. \tag{10.93}$$

The Components of the Faraday Tensor are $\boldsymbol{E}$ and $\boldsymbol{B}$

The Faraday tensor was defined in (8.239), where A^α was the electromagnetic potential. Sometimes it's defined as the negative of our definition. In general, if we define it as

$$F_{\alpha\beta} \equiv s\,(A_{\beta,\alpha} - A_{\alpha,\beta})\,, \tag{10.94}$$

where the sign s is usually $+1$ but sometimes -1, then inspection of (10.81) shows that the Faraday tensor has the following cartesian components written in matrix form, where $\alpha =$ row index and $\beta =$ column index:

$$(F_{\alpha\beta}) = s\,\eta_{tt}\begin{bmatrix} 0 & E^x & E^y & E^z \\ -E^x & 0 & -B^z & B^y \\ -E^y & B^z & 0 & -B^x \\ -E^z & -B^y & B^x & 0 \end{bmatrix}, \quad (F^\alpha{}_\beta) = s\begin{bmatrix} 0 & E^x & E^y & E^z \\ E^x & 0 & B^z & -B^y \\ E^y & -B^z & 0 & B^x \\ E^z & B^y & -B^x & 0 \end{bmatrix},$$

$$(F_\alpha{}^\beta) = s\begin{bmatrix} 0 & -E^x & -E^y & -E^z \\ -E^x & 0 & B^z & -B^y \\ -E^y & -B^z & 0 & B^x \\ -E^z & B^y & -B^x & 0 \end{bmatrix}, \quad (F^{\alpha\beta}) = s\,\eta_{tt}\begin{bmatrix} 0 & -E^x & -E^y & -E^z \\ E^x & 0 & -B^z & B^y \\ E^y & B^z & 0 & -B^x \\ E^z & -B^y & B^x & 0 \end{bmatrix}. \tag{10.95}$$

Note that while $\boldsymbol{E}$ appears as three simple components, $\boldsymbol{B}$ is more complicated: it appears with alternating signs in a form like the "cross matrix" of (4.23), alluding to the fact that $\boldsymbol{B}$ has a rotational nature. ($\boldsymbol{E}$ is traditionally called a *polar* vector, while $\boldsymbol{B}$ is an *axial* vector.)

Transforming electric and magnetic fields under frame and coordinate changes is easily done using the Faraday tensor. Less useful is to transform A^α itself, since there is some ambiguity in the definition of A^α; given $\boldsymbol{E}$ and $\boldsymbol{B}$, we would first need to find A^α by solving (10.81), then transform the A^α, and then apply (10.81) to the resulting new components to get the transformed $\boldsymbol{E}$ and $\boldsymbol{B}$. Contrast this with transforming $F_{\alpha\beta}$ in the usual tensor way, which is much more direct.

This is the *Lorentz force* in cartesian coordinates. We can show it reduces to the more familiar expression by setting $k = x$; the expressions for $k = y, z$ follow similarly:

$$\frac{\mathrm{d}p^x}{\mathrm{d}t} = \frac{-\eta_{tt}}{\gamma}\frac{\mathrm{d}p_x}{\mathrm{d}\tau} = \frac{-\eta_{tt}}{\gamma} q u^\beta F_{x\beta} \overset{(10.95)}{=\!=\!=} q\big[E^x + (\boldsymbol{v}\times\boldsymbol{B})^x\big]\,. \tag{10.96}$$

This combines with the matching expressions for $k = y, z$ to give the familiar vector form of the Lorentz force,

$$\mathrm{d}\boldsymbol{p}/\mathrm{d}t = q(\boldsymbol{E} + \boldsymbol{v}\times\boldsymbol{B})\,. \tag{10.97}$$

Experiments verify that this equation of motion also holds relativistically (even though it contains t instead of τ!). The nonrelativistic lagrangian (10.86) actually gives almost the same equation of motion, except that the force term $\mathrm{d}\boldsymbol{p}/\mathrm{d}t$, which has the relativistic form $m\,\mathrm{d}(\gamma\boldsymbol{v})/\mathrm{d}t$, is replaced by its non-relativistic version $m\,\mathrm{d}\boldsymbol{v}/\mathrm{d}t$. So the lagrangian that we have chosen, (10.84), does indeed give the correct dynamics for a relativistic charged particle.

Equation (10.93) looks as if the index k ($=x$ to z) might be replaceable by α ($=t$ to z). Could this be true? Again using (10.95), we have

$$qu^\beta F_{t\beta} = q\gamma\,\boldsymbol{v}\cdot(F_{tx}, F_{ty}, F_{tz}) = \eta_{tt}\gamma\,\boldsymbol{v}\cdot q\boldsymbol{E}\,, \tag{10.98}$$

where $\boldsymbol{v}\cdot q\boldsymbol{E}$ is the rate of work done on a charge. (The Lorentz force is everywhere at right angles to the magnetic field, so that $\boldsymbol{B}$ contributes nothing to this work.) The rate of work done must equal the increase in the charge's energy, so

$$qu^\beta F_{t\beta} = \eta_{tt}\gamma\,\frac{\mathrm{d}p^t}{\mathrm{d}t} = \frac{\mathrm{d}p_t}{\mathrm{d}\tau}\,, \tag{10.99}$$

which makes it evident that (10.93) holds generally in an almost covariant form (that is, but still only for cartesians):

$$\frac{\mathrm{d}p_\alpha}{\mathrm{d}\tau} = qu^\beta F_{\alpha\beta}\,. \tag{10.100}$$

This is the cartesian expression for the rate of work done ($\alpha = t$) and the Lorentz force ($\alpha = x, y, z$) on a charge. For full covariance—that is, to write (10.100) in general coordinates—we require that all commas become semicolons, as discussed on p. 328. This doesn't change $F_{\alpha\beta}$ due to (8.239), but it certainly does change the derivative in (10.100). Comparing (9.100) with (9.102) means that, in arbitrary coordinates, the Lorentz force and rate of work done are given by

$$\boxed{\frac{\mathrm{D}p_\alpha}{\mathrm{d}\tau} = qu^\beta F_{\alpha\beta}\,.} \tag{10.101}$$

Adding a Term for the Free Electromagnetic Field

The lagrangian (10.84) is still one-sided in that it only describes the effect of the field on the charge. As in the gravity case, a term must be added for the free field, which will then conspire with the interaction term to quantify how the charge affects the field (i.e. Maxwell's equations). Just as we were able to write down a gravitational field lagrangian density using a quadratic term in first derivatives of the potential, we can do the same for the electromagnetic potentials $(\Phi, \boldsymbol{A})$. This should be a term like $A^{\alpha,\beta}A_{\alpha,\beta}$. But retaining covariance demands a semicolon, so an expression like $A^{\alpha;\beta}A_{\alpha;\beta}$ would be more suitable (see Sect. 8.9.5). However, this turns out not to give the correct physical theory. We know that Maxwell's equations involve the Faraday tensor, so it might come as no surprise that the correct quadratic expression turns out to be the scalar

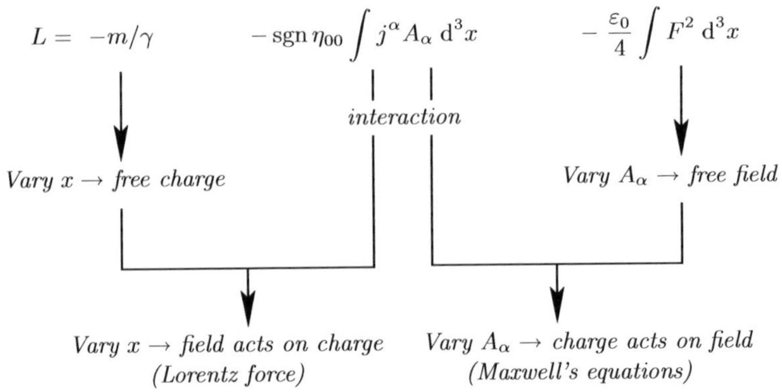

Fig. 10.7. The variation of terms that gives the two equations of motion when a charge interacts with an electromagnetic field. The lagrangian is (10.103), incorporating the j^α notation of (10.104).

$$F^2 \equiv F^{\alpha\beta} F_{\alpha\beta} \,. \tag{10.102}$$

Integrating (10.102) over space and adding to the lagrangian (10.84) gives the overall lagrangian for a charge in an electromagnetic field. Ever-useful hindsight provides a weighting factor of the permittivity constant:

$$L = \underbrace{-m/\gamma}_{\text{free charge}} \quad \underbrace{-\frac{\varepsilon_0}{4}\int F^2 \,\mathrm{d}^3x}_{\text{free field}} \quad \underbrace{-\operatorname{sgn}\eta_{00}\, q v^\alpha A_\alpha}_{\text{interaction}} \,. \tag{10.103}$$

The term for the free charge here is not the quadratic expression that we might have expected. In fact, in Sect. 10.6.2, we will find that in the more correct version (10.175) that treats the charged particle as representable by its own quantum field ψ, this field does occur quadratically.

In (10.87)–(10.100) we varied the charge and interaction terms of the lagrangian (10.103) with respect to the space coordinates of the particle's trajectory and obtained the Lorentz force. Now we will vary the field and interaction terms of that lagrangian with respect to the field coordinates, hoping that the field equations of motion—Maxwell's equations—will result. The programme is shown in Fig. 10.7. Again, work in cartesian coordinates for simplicity, in which case $\operatorname{sgn}\eta_{00}$ is replaced by η_{tt}.

Just as we did for gravity, express the terms in (10.103) as densities. Replace the charge q by a charge density ϱ, noting that $\varrho v^\alpha = \varrho_0 u^\alpha = j^\alpha$, the current density defined back in Sect. 6.4. The free-charge lagrangian $-m/\gamma$ has no relevance here, so corresponding to (10.73), write the electromagnetic part of the lagrangian density alone as

$$\mathcal{L}_{\text{em}} = \frac{-\varepsilon_0}{4} F^2 - \eta_{tt} j^\alpha A_\alpha \,. \tag{10.104}$$

There will be four Lagrange equations for the field. Equation (10.59) produces

$$\frac{\partial \mathcal{L}_{\rm em}}{\partial A_\beta} - \partial_\alpha \frac{\partial \mathcal{L}_{\rm em}}{\partial A_{\beta,\alpha}} = 0\,. \tag{10.105}$$

The first required derivative is easy:

$$\frac{\partial \mathcal{L}_{\rm em}}{\partial A_\beta} = -\eta_{tt} j^\beta. \tag{10.106}$$

Differentiating $\mathcal{L}_{\rm em}$ with respect to $A_{\beta,\alpha}$ requires more work. We must express F^2 in terms of $A_{\beta,\alpha}$ terms, treating raised indices as shorthand notation for terms involving lowered indices:

$$F^2 = \eta^{\mu\varrho}\eta^{\nu\sigma}\left(A_{\sigma,\varrho} - A_{\varrho,\sigma}\right)\left(A_{\nu,\mu} - A_{\mu,\nu}\right), \tag{10.107}$$

so that with some care we can write

$$\begin{aligned}\frac{\partial\left(F^2\right)}{\partial A_{\beta,\alpha}} &= \eta^{\mu\varrho}\eta^{\nu\sigma}\left[\left(\delta^\beta_\sigma \delta^\alpha_\varrho - \delta^\beta_\varrho \delta^\alpha_\sigma\right) F_{\mu\nu} + F_{\varrho\sigma}\left(\delta^\beta_\nu\, \delta^\alpha_\mu - \delta^\beta_\mu\, \delta^\alpha_\nu\right)\right] \\ &= 4F^{\alpha\beta}\,. \end{aligned} \tag{10.108}$$

It follows that

$$\partial_\alpha \frac{\partial \mathcal{L}_{\rm em}}{\partial A_{\beta,\alpha}} = -\varepsilon_0 F^{\alpha\beta}{}_{,\alpha}\,, \tag{10.109}$$

and so the four Lagrange equations (10.105) become

$$F^{\alpha\beta}{}_{,\alpha} = \eta_{tt}\frac{j^\beta}{\varepsilon_0}\,. \tag{10.110}$$

That these really are equivalent to Maxwell's equations can be shown by writing time and space components explicitly, remembering that we are using cartesian coordinates to allow for a simpler treatment of the vector notation. The time component ($\beta = t$) expands to

$$\Phi^{,\alpha}{}_{,\alpha} - \eta_{tt} A^\alpha{}_{,t\alpha} = \eta_{tt}\frac{\varrho}{\varepsilon_0}\,. \tag{10.111}$$

The identities

$$\begin{aligned}\Phi^{,\alpha}{}_{,\alpha} &= \eta^{\beta\alpha}\Phi_{,\beta\alpha} = \eta_{tt}\left(\Phi_{,tt} - \nabla^2\Phi\right), \\ \eta_{tt}A^\alpha{}_{,t\alpha} &= \eta_{tt}\left(\Phi_{,tt} + \partial/\partial t\;\nabla\cdot\boldsymbol{A}\right)\end{aligned} \tag{10.112}$$

convert (10.111) to vector form,

$$\nabla\cdot\underbrace{\left(-\nabla\Phi - \frac{\partial\boldsymbol{A}}{\partial t}\right)}_{\equiv\boldsymbol{E}} = \frac{\varrho}{\varepsilon_0}\,, \tag{10.113}$$

which is the first of Maxwell's equations. The remaining three equations of motion are the spatial parts of (10.110):

$$A^{i,\alpha}{}_{,\alpha} - A^{\alpha,i}{}_{,\alpha} = \eta_{tt} \frac{j^i}{\varepsilon_0} \,. \tag{10.114}$$

Lowering comma indices eliminates η_{tt} so that the resulting equations can be converted to a single vector expression,

$$\frac{\partial^2 \boldsymbol{A}}{\partial t^2} - \nabla^2 \boldsymbol{A} + \nabla \frac{\partial \Phi}{\partial t} + \nabla \left(\nabla \cdot \boldsymbol{A}\right) = \frac{\boldsymbol{j}}{\varepsilon_0} \,, \tag{10.115}$$

where $\nabla^2 \boldsymbol{A}$ conventionally means $(\nabla^2 A^x, \nabla^2 A^y, \nabla^2 A^z)$, and $\boldsymbol{j}$ is the spatial current density $\varrho \boldsymbol{v}$. Defining $\boldsymbol{B} \equiv \nabla \times \boldsymbol{A}$ and using the identity

$$\nabla \times \boldsymbol{B} = \nabla \times (\nabla \times \boldsymbol{A}) = \nabla(\nabla \cdot \boldsymbol{A}) - \nabla^2 \boldsymbol{A} \tag{10.116}$$

converts (10.115) to the simpler

$$\nabla \times \boldsymbol{B} = \frac{\partial \boldsymbol{E}}{\partial t} + \frac{\boldsymbol{j}}{\varepsilon_0} \,. \tag{10.117}$$

This is the second of the four Maxwell equations. The remaining two equations follow easily from the definitions of $\boldsymbol{E}$ and $\boldsymbol{B}$:

$$\nabla \cdot \boldsymbol{B} = \nabla \cdot (\nabla \times \boldsymbol{A}) = 0$$
$$\text{(since the divergence of a curl is always zero)} \,,$$

$$\nabla \times \boldsymbol{E} = \nabla \times \left(-\nabla \Phi - \frac{\partial \boldsymbol{A}}{\partial t}\right) = \frac{-\partial \boldsymbol{B}}{\partial t}$$
$$\text{(since the curl of a gradient is always zero)} \,. \tag{10.118}$$

It seems that our initial guess has paid off: if a Φ and $\boldsymbol{A}$ can be specified that together form a vector $A^\alpha \boldsymbol{e}_\alpha$, then they will be part of an action that leads to the Lorentz force and Maxwell's equations. We'll see more of the electromagnetic potential in later sections and will solve for it explicitly in the next chapter.

It's a curious fact that experimentally Maxwell's equations hold relativistically (although the form in (10.110) is changed when gravity is present), even though the free field lagrangian density F^2 was simply written in a quadratic analogy to the nonrelativistic $mv^2/2$ for a free particle. The reason is beyond our scope here, but its roots may be sought in a quantum field theoretic treatment using an arbitrary function of F^2 in the lagrangian density.

A convenient mnemonic for remembering div-curl-grad identities runs like this. Write "*d c g*" (for div, curl, grad), and consider neighbouring pairs only, placing a zero before and after each right-hand letter:

$$d\,0\,c\,, \quad d\,c\,0\,, \quad c\,0\,g\,, \quad c\,g\,0\,. \tag{10.119}$$

Now read each from left to right. The first expression, $d\ 0\ c$, reminds us that if a div is zero, then the relevant function must be a curl. The second one says that the div of a curl is zero. The third says that if a curl is zero, then the relevant function must be a grad. The fourth says that the curl of a grad is zero.

10.5 Producing the Schrödinger Equation

The lagrangian approach to classical mechanics provides a good path to the Schrödinger equation of quantum mechanics. At the end of Chap. 2, we showed how the basics of Fourier theory could accommodate the ideas of x- and p-space so central to quantum mechanics, using (2.230) to show how the x-representation of the momentum operator $\widehat{p}_x$ might be written as

$$\widehat{p}_x \overset{x\text{-rep.}}{=\!=\!=} -i\hbar\,\partial_x\,, \tag{10.120}$$

at least in the realm of plane waves that forms de Broglie's original idea. This spatial differentiation reminds us of the canonical momentum in Sect. 10.3.4; specifically (10.40) and (10.41):

$$b_i \equiv \frac{\partial L}{\partial \dot{x}^i}\,, \quad \dot{b}_i \equiv \frac{\partial L}{\partial x^i}\,. \tag{10.121}$$

Spatial differentiation of the lagrangian indicates whether the corresponding component of the *canonical* momentum is conserved. So perhaps (10.120) only holds for plane waves, and in general it might better be replaced with an expression involving the *canonical* momentum,

$$\boxed{\widehat{b}_x \overset{x\text{-rep.}}{=\!=\!=} -i\hbar\,\partial_x\,,} \tag{10.122}$$

or, for one particle,

$$\widehat{\boldsymbol{b}} \overset{x\text{-rep.}}{=\!=\!=} -i\hbar\,\nabla\,. \tag{10.123}$$

Paired with the canonical momentum was the hamiltonian H, which we saw in (10.45) varies temporally as

$$\dot{H} = \frac{-\partial L}{\partial t}\,. \tag{10.124}$$

Now just as (10.121) and (10.122) imply that $\widehat{b}_i$ is an operator that acts on the lagrangian to give the time derivative of b_i, or $\widehat{b}_i L = -i\hbar\,\dot{b}_i$, we can define an operator $\widehat{H}$ that acts on the lagrangian to give the time derivative of H, or $\widehat{H} L = -i\hbar\,\dot{H}$, giving rise to

$$\boxed{\widehat{H} \equiv i\hbar\,\partial_t\,.} \tag{10.125}$$

Canonical Versus Elementary Momentum in Quantum Mechanics

When reading textbooks in classical mechanics, quantum mechanics, and relativity, it is possible to confuse canonical momentum with the more elementary momentum $m\boldsymbol{v}$, considering that the same symbol $\boldsymbol{p}$ is usually reserved for both. In quantum mechanics it's the *canonical* momentum, denoted here by $\boldsymbol{b}$, that is represented by the spatial operator $-\mathrm{i}\hbar\nabla$, as opposed to the more elementary momentum $\boldsymbol{p} = m\boldsymbol{v}$ of classical mechanics and its relativistic version $p^\alpha = mu^\alpha$. The famous commutation relation is then $[x, b_x] = \mathrm{i}\hbar$. The canonical momentum is almost always called $\boldsymbol{p}$ in quantum mechanics texts. We have chosen to use $\boldsymbol{b}$ in this text to highlight the care needed when canonical and elementary momenta are related in one equation, such as the expression of energy conservation in (10.130). We hope that the use of $\boldsymbol{b}$ makes an equation like (10.130) more transparent.

Equations (10.123) and (10.125) were used in quantum mechanics from early on in its history to give a prescription for constructing a quantum system's equation of motion: the Schrödinger equation. The equation of energy conservation for the system to be quantised is written down and the replacements (10.123) and (10.125) are made, converting it to an operator equation that acts on the wave function $\Psi(\boldsymbol{x}, t)$. The simplest case in point is that of a single particle of mass m with some potential energy V dependent only on space coordinates. Such a particle has identical canonical and elementary momenta, as well as a time-independent lagrangian, so H is its energy and we can write

$$H = \frac{\boldsymbol{b}\cdot\boldsymbol{b}}{2m} + V\,. \tag{10.126}$$

The Schrödinger equation prescription makes the relevant operator replacements (10.123) and (10.125):

$$\mathrm{i}\hbar\,\frac{\partial\Psi}{\partial t} = \left(\frac{-\hbar^2}{2m}\nabla^2 + V\right)\Psi\,. \tag{10.127}$$

In Sect. 10.8.1, we'll see how the Schrödinger equation comes about through a totally different approach: that of Feynman's path-integral formalism.

As a side point, in analogy to (2.230) and (2.232), we could first write the Schrödinger equation as

$$H\Psi(\boldsymbol{x}, t) = \mathrm{i}\hbar\frac{\partial\Psi}{\partial t}\,, \quad \text{or } H\langle\boldsymbol{x}|\Psi, t\rangle = \mathrm{i}\hbar\frac{\partial}{\partial t}\langle\boldsymbol{x}|\Psi, t\rangle\,, \tag{10.128}$$

and then omit $\langle\boldsymbol{x}|$ to give

$$H|\Psi, t\rangle = \mathrm{i}\hbar\frac{\partial}{\partial t}|\Psi, t\rangle\,. \tag{10.129}$$

But as with the discussion around (2.222) and (2.232), we should always remember that the state $|\Psi, t\rangle$ is not really being differentiated here. Rather, $\langle \boldsymbol{x}|$ is understood to be invisibly present, and the differentiation applies to the wave function, the amplitude $\langle \boldsymbol{x}|\Psi, t\rangle$, as opposed to the state ket.

The Schrödinger Equation for a Charged Particle in an EM Field

While the canonical and elementary momenta happen to be equal in the previous example, they will not be so in general. As an example, construct the Schrödinger equation for a charged particle in an electromagnetic field. The nonrelativistic lagrangian (10.86) and other relevant quantities are

$$\begin{aligned} L &= \frac{1}{2}mv^2 - q\Phi + q\boldsymbol{v}\cdot\boldsymbol{A}\,, \\ \boldsymbol{p} &= m\boldsymbol{v}\,, \\ \boldsymbol{b} &= \boldsymbol{p} + q\boldsymbol{A}\,, \\ H &= \frac{\boldsymbol{p}\cdot\boldsymbol{p}}{2m} + q\Phi = \frac{(\boldsymbol{b}-q\boldsymbol{A})\cdot(\boldsymbol{b}-q\boldsymbol{A})}{2m} + q\Phi\,. \end{aligned} \tag{10.130}$$

Replacements (10.123) and (10.125) give the Schrödinger equation for this system:

$$i\hbar\frac{\partial\Psi}{\partial t} = \left[\frac{(-i\hbar\nabla - q\boldsymbol{A})\cdot(-i\hbar\nabla - q\boldsymbol{A})}{2m} + q\Phi\right]\Psi(\boldsymbol{x}, t)\,. \tag{10.131}$$

So the canonical momentum $\boldsymbol{b}$ has introduced the field $\boldsymbol{A}$ into the dynamics. This forms the starting point for a quantum mechanical treatment of the interaction of atoms with radiation.

10.6 Quantising Field Theory: Fields Describe Particles, Too!

To illustrate the ideas of field theory using a lagrangian approach, we have until now concentrated on constructing the relativistic lagrangian (10.103) for a charged particle interacting with an electromagnetic field. In the context of quantum mechanics, it was realised historically that since the electromagnetic field has a particle nature (as demonstrated by photon experiments such as the photoelectric effect), particles such as electrons might well have a field nature. Ascribing a particle character to fields is the domain of *quantum field theory.*

The quantisation of field theory has its roots in classical mechanics, with the idea of the energy ascribable to degrees of freedom. When we write the kinetic energy of a particle as

$$E = \frac{1}{2}mv^2 = \frac{1}{2}mv_x^2 + \frac{1}{2}mv_y^2 + \frac{1}{2}mv_z^2\,, \tag{10.132}$$

thanks to the magic of Pythagoras, the energy can be decomposed into three parts, corresponding to the particle's motion along each of the cartesian axes. The same sort of idea is true of the *Equipartition of Energy Principle* of classical statistical mechanics, which states that for a certain reasonable set of assumptions (such as Maxwell–Boltzmann statistics), each degree of freedom in a system at temperature T will contribute, on average, $\frac{1}{2}kT$ to the system's total energy, where k is Boltzmann's constant.

This idea of being able to add the energies for different motions of a system as if they were really separate is even more impressive in the case of a vibrating string. The motion of a plucked guitar string can be Fourier-decomposed into a sum of eigenfunctions of the basic wave equation, called *modes*, as was done in (2.211). On paper, the string's motion is a sum of these modes, and yet that same motion looks nothing like a set of sinusoids. Plucking a guitar string by stretching and releasing it near one end forms a peak in the string's displacement that bounces back and forth along the string, reversing its sign at each end. This sign reversal is easily seen by pulling the string upward over the guitar's sound hole and then releasing; the initial peak evolves to a trough over the frets and slaps against them. The same sort of procedure when done over the frets themselves produces no slap at all, since the trough then forms harmlessly over the sound hole. So instead of really looking like an oscillation of sinusoids, the motion of the string is much more readily seen as composed of two peaked waves travelling in opposite directions on the string, reversing their signs as they bounce off the fixed ends, and always superposing to give the actual string shape. This shape might be quite skewed if the string was plucked far from its midpoint.

But, despite appearances, the energy of the string can also be written simply as the sum of the energies of its sinusoidal modes. To see why, return to the guitar string's motion (2.211), where $\phi(x,t)$ describes the string's small sideways displacement from equilibrium:

$$\phi(x,t) = \sum_{n=1}^{\infty} A_n \sin\frac{n\pi x}{L}\cos\frac{n\pi ct}{L}\,. \tag{10.133}$$

Let's calculate the total energy of vibration of this string. As we saw in (10.46) and (10.47), its kinetic and potential energies can be summed to give

$$E = \int_0^L \mathrm{d}x \left[\frac{\mu}{2}\left(\phi_{,t}\right)^2 + \frac{T}{2}\left(\phi_{,x}\right)^2\right], \tag{10.134}$$

which with $c = \sqrt{T/\mu}$ becomes

$$E = \frac{\mu}{2}\int_0^L \mathrm{d}x \left[\left(\phi_{,t}\right)^2 + c^2\left(\phi_{,x}\right)^2\right]. \tag{10.135}$$

Substituting ϕ as a sum of modes from (10.133) into (10.135) produces various products of sinusoids. They are easily integrated using the orthogonality relations

$$\int_0^L \sin\frac{m\pi x}{L}\sin\frac{n\pi x}{L}\,\mathrm{d}x = \int_0^L \cos\frac{m\pi x}{L}\cos\frac{n\pi x}{L}\,\mathrm{d}x = \frac{L}{2}\,\delta_{mn} \tag{10.136}$$

to give the simple expression

$$E = \frac{\mu\pi^2 c^2 L}{4}\sum_n n^2\left(\frac{A_n}{L}\right)^2, \tag{10.137}$$

where A_n/L is unitless. The important point here is that because the summation and integration of (10.133) and (10.134) commute, each term of the energy sum (10.137) is just the energy of each mode of (10.133). So in this sense each mode has a separate existence, and the total energy of the string's motion is the sum of the energies of each of those modes. Classical guitarists are used to treating the modes as a collection of entities: plucking a string near its centre excites the lower-frequency modes (smaller n values of the Fourier series), producing a very mellow sound, while plucking it nearer the string's end excites higher-frequency modes (larger n), giving a harsher sound.

We can perhaps think of the nth mode as having a basic unit of energy $\mu\pi^2c^2Ln^2/4$, and the amount of this energy present—the loudness of the mode—is determined by the square of the relative amplitude A_n/L. Typically this square will be proportional to $1/n^4$.

Second Quantisation: Quantising the String Modes

So far, our discussion of the string's motion has been purely classical. But the sinusoidal modes remind us of a basic exercise in quantum mechanics: the quantisation of a harmonic oscillator. Is it meaningful to apply a quantum mechanical treatment to the string modes? Let's reiterate the quantisation of a single mass m, oscillating over a position variable $q(t)$ with spring constant k, so that more generally it can be considered to have a potential energy of $V = kq^2/2$. Classically, the force on the mass is $-kq$, so its equation of motion will be

$$\ddot{q} = \frac{-k}{m}q \equiv -\omega^2 q\,. \tag{10.138}$$

When the system is quantised, the Schrödinger equation for the wave function $\Psi(q,t) = \psi(q)\,\mathrm{e}^{-\mathrm{i}Et/\hbar}$ becomes

$$\left(\frac{kq^2}{2} - \frac{\hbar^2}{2m}\frac{\mathrm{d}^2}{\mathrm{d}q^2}\right)\psi(q) = E\psi\,. \tag{10.139}$$

A change of variables

$$z \equiv \sqrt{\frac{m\omega}{\hbar}}\,q \equiv \alpha\,q \tag{10.140}$$

produces

$$\psi''(z) + \left(\frac{2E}{\hbar\omega} - z^2\right)\psi = 0\,. \tag{10.141}$$

The solution of this differential equation is a standard exercise in the mathematics of quantum theory; there turn out to be multiple wave functions $\psi_N(z)$, each written as products of gaussians and hermite polynomials.

Although the precise forms of these functions are not needed here, it's useful to write them down. In terms of q, the wave functions are

$$\psi_N(q) = \sqrt{\frac{\alpha}{2^N N! \sqrt{\pi}}}\, H_N(\alpha\, q)\, \mathrm{e}^{-\alpha^2 q^2/2}, \tag{10.142}$$

where H_N are *hermite polynomials*:

$$H_N(z) = (-1)^N\, \mathrm{e}^{z^2} \frac{\mathrm{d}^N}{\mathrm{d}z^N}\, \mathrm{e}^{-z^2}. \tag{10.143}$$

The important point is that the total energy E of the oscillator turns out to be quantised with values $E_N \equiv (N + 1/2)\hbar\omega$.

Return now to the string modes (10.133), treating each of them heuristically like an oscillator of mass m_n. The nth mode is

$$\phi_n(x,t) = A_n \sin\frac{n\pi x}{L} \cos\frac{n\pi ct}{L}\,, \tag{10.144}$$

whose time-dependent part is

$$q_n(t) \equiv A_n \cos\frac{n\pi ct}{L}\,. \tag{10.145}$$

Writing $\omega_n \equiv n\pi c/L$, equation (10.145) gives

$$\ddot{q}_n = -\omega_n^2\, q_n\,, \tag{10.146}$$

which is reminiscent of (10.138), so that this $q_n(t)$ is the string mode equivalent of the variable $q(t)$ for the oscillator of mass m of (10.138). Without looking too closely at the meaning of the mass m_n, we now see that the nth string mode will have quantised energy levels of

$$E_{n,N} \equiv (N + 1/2)\,\hbar\omega_n\,. \tag{10.147}$$

It appears that each mode has a *zero-point* energy of $\hbar\omega_n/2$, and if the nth mode is excited to the energy level $E_{n,N}$, then we can consider that mode to be composed of N *phonons*, each of energy $\hbar\omega_n$ and frequency $\omega_n/(2\pi)$, that in some sense reside in the string. So the excitation of the string as a whole is tied to the presence of phonons of different frequencies, and the mix of these frequencies in various proportions gives the string its "colour". (Colour is an apt musical term, being entirely analogous to the visual colour of a hot body that is caused by photons of different frequencies.) Of course, in a vibrating string, the modes don't actually have a mass m_n, which renders our treatment of the oscillating string somewhat heuristic; but then neither does the phonon energy depend on m_n. The idea of phonons being present in crystal excitations is important in solid state physics.

This idea of decomposing a string motion, or more generally a field, into modes that are then quantised via the Schrödinger equation is the key idea of quantum field theory. In general, we are solving a wave equation (be it Schrödinger's or a classical one) and then *second quantising* by treating each degree of freedom as a new variable whose equation of motion will be the Schrödinger equation. These degrees of freedom might be the field value at each spacetime point, in the case of a field not tied to a boundary, or they might be field modes if the field is constrained at a boundary, as was the case with our guitar string. In *quantum electrodynamics* (QED), the phonons of the vibrating string become the photons of the quantised electromagnetic field, and indeed quantum electrodynamics is one of the most successfully tested theories in all of physics. We profit by regarding *all* particles, not just phonons and photons, as quanta of their associated fields. The fields' properties must be chosen to give those quanta their appropriate properties (mass, spin, etc.) since, except for electromagnetism and gravity, we don't actually observe the field itself.

But there is one problem. Already, with the guitar string, the $\hbar\omega_n/2$ zero-point energy of each mode implies that if an infinite number of modes are considered to be present, then the string must have infinite energy even without vibrating. In practice, of course, a guitar string cannot support modes of arbitrarily high frequency; these would necessitate its being able to bend arbitrarily tightly, which a real nylon string cannot do. The situation for a quantum field is not so clear-cut, but the idea of infinite zero-point energies is better understood nowadays, and researchers' confidence in its technicalities continues to grow.

Zero-point energy aside, when particles can be created or absorbed (such as in nuclear reactions), the simple single-particle wave function description breaks down, and the idea of associating particles with fields becomes more necessary. Historically, this need to describe interactions led to attempts to write down a relativistic version of the Schrödinger equation, and the first such attempt produced the *Klein–Gordon equation.*

10.6.1 First Steps: The Klein–Gordon Equation

As we have seen in the previous pages, the nonrelativistic Schrödinger equation for a particle in a velocity-independent potential with a time-independent lagrangian is obtained from the nonrelativistic conservation of energy by the substitutions of (10.123) and (10.125).

Is it possible to produce a *relativistic* version of the Schrödinger equation by converting a relativistic energy expression into an operator expression? Let's try to do that. Begin with the relativistic conservation of energy, and for simplicity choose a lagrangian for which the canonical and elementary momenta are equal, so that $\boldsymbol{b} = \boldsymbol{p}$. Also set $\hbar = c = 1$, although for clarity we'll include those two constants in the following two equations. First is the relativistic conservation of energy, from (5.59):

$$E^2 = p^2c^2 + m^2c^4\,, \quad \text{or just } \; E^2 = p^2 + m^2 = \boldsymbol{p}\cdot\boldsymbol{p} + m^2\,, \tag{10.148}$$

where m is the particle's (rest) mass. Now make the substitutions $\boldsymbol{p} \to -i\hbar\nabla$ and $E \to i\hbar\,\partial_t$ that each act on a wave function $\Psi(\boldsymbol{x}, t)$, giving

$$-\partial_t^2\,\Psi = \left(-c^2\nabla^2 + \frac{m^2c^4}{\hbar^2}\right)\Psi\,, \quad \text{or just } \; -\partial_t^2\,\Psi = \left(-\nabla^2 + m^2\right)\Psi\,. \tag{10.149}$$

This is the Klein–Gordon equation. It can be written covariantly for cartesian coordinates by noting that

$$\partial_t = \eta_{tt}\partial^t\,, \quad \partial_k = -\eta_{tt}\partial^k\,, \tag{10.150}$$

which together produce

$$\left(\eta_{tt}\partial^\mu\partial_\mu + m^2\right)\Psi = 0\,. \tag{10.151}$$

This equation predicts negative energies, but more importantly, the quantum mechanical probability derived from it is not relativistically invariant, a fact that is related to its being of second order in time. Historical attempts at a cure postulated different forms for the probability, but these all failed; for example, forms that were relativistically invariant were not always positive, and so on.

10.6.2 A Route to the Dirac Equation

The Klein–Gordon equation dates from the early days of quantum mechanics, but almost from its inception, the second time derivative was seen as not in keeping with fundamental quantum mechanical ideas, which suggested a first derivative in time might be necessary. Acknowledging the energy–momentum four-vector, we can group the energy and momentum together in (10.148), write the momentum as a covector to make eventual use of (10.122) (since, e.g., $p^{x\,2} = p_x^2$), and finally take the square root to give

$$\pm\sqrt{E^2 - p_x^2 - p_y^2 - p_z^2} = m\,. \tag{10.152}$$

(The $\pm$ means that in principle the mass, or energy, might be negative; but in practice the minus sign is not written, and instead the energy eigenvalues that the theory produces are given both signs, leading to the concept of antiparticles.) The plan was to convert (10.152) to a wave equation by first writing

$$\sqrt{E^2 - p_x^2 - p_y^2 - p_z^2}\;\Psi(\boldsymbol{x}, t) = m\,\Psi(\boldsymbol{x}, t) \tag{10.153}$$

and then making the replacements $E \to i\partial_t$, $p_k \to -i\partial_k$. But how do we take the square root to give an expression involving the energy and momenta linearly? Dirac did this by taking a cue from the behaviour of the Pauli matrices, which in effect allow the square root of a sum of squares to be taken. That is, for any numbers a^1, a^2, a^3 (see Table 4.1),

$$\left(a^1\sigma_1 + a^2\sigma_2 + a^3\sigma_3\right)^2 = \left(a^1\right)^2 + \left(a^2\right)^2 + \left(a^3\right)^2. \tag{10.154}$$

The three Pauli matrices were too few in number to take a square root of the *four* terms in (10.152), so Dirac just introduced a new set of quantities that were essentially equivalent to the modern notation of $\gamma^0 \to \gamma^3$, where the indices are raised in order to keep covariance with the lowered momenta indices. The properties of the γ^α needed only to be such as to copy the idea of (10.154). Write

$$\sqrt{E^2 - p_x^2 - p_y^2 - p_z^2} = \gamma^0 E - \gamma^1 p_x - \gamma^2 p_y - \gamma^3 p_z\,, \tag{10.155}$$

where we have used minus signs with the p_k to take eventual account of the sign difference in energy and momentum operators [compare (10.122) with (10.125)]. The only demands made on the set of γ^α are that

$$\left(\gamma^0\right)^2 \equiv 1\,, \quad \left(\gamma^k\right)^2 \equiv -1\,, \quad \gamma^\alpha\gamma^\beta \equiv -\gamma^\beta\gamma^\alpha, \tag{10.156}$$

since these are sufficient to allow both sides of (10.155) to square to the same quantity. Then

$$\left(\gamma^0 E - \gamma^1 p_x - \gamma^2 p_y - \gamma^3 p_z\right)\Psi(\boldsymbol{x},t) = m\Psi(\boldsymbol{x},t)\,, \tag{10.157}$$

in which case with $E \to \mathrm{i}\partial_t = \mathrm{i}\partial_0$, $p_k \to -\mathrm{i}\partial_k$ we obtain

$$\left(\mathrm{i}\gamma^0\partial_0 + \mathrm{i}\gamma^k\partial_k\right)\Psi = m\Psi\,, \tag{10.158}$$

or more covariantly, but still using cartesian coordinates,

$$\boxed{\left(\mathrm{i}\gamma^\alpha\partial_\alpha - m\right)\Psi = 0\,.} \tag{10.159}$$

This is the celebrated *Dirac equation.* It can be convenient to use the *anticommutator* $\{\gamma^\alpha, \gamma^\beta\} \equiv \gamma^\alpha\gamma^\beta + \gamma^\beta\gamma^\alpha$, along with a $+---$ metric, so that (10.156) could be written as $\{\gamma^\alpha, \gamma^\beta\} = 2\eta^{\alpha\beta}$. This is not quite a good idea, though, because it suggests that the definition of the γ^α depends on the metric signature, which is not true. The algebra of the γ^α, (10.156), is a definition rooted in history and was not made to follow metric sign choices.

What matrix representation of the γ^α might be found to mimic the Pauli choices in Table 4.1? Dirac decided to go for matrices that used the Pauli matrices in a block form and so settled on 4×4, where in the following, "0" and "I" are 2×2 zero and unit matrices, respectively:

$$\gamma^0 = \begin{bmatrix} I & 0 \\ 0 & -I \end{bmatrix}, \quad \gamma^k = \begin{bmatrix} 0 & \sigma_k \\ -\sigma_k & 0 \end{bmatrix}. \tag{10.160}$$

In fact, there are an infinite number of matrix forms for the γ^α, but the really important identity is the algebra that they must all satisfy, (10.156).

Note that if $\alpha \neq \beta$, then clearly

$$\gamma^\alpha \gamma^\beta = -\gamma^\beta \gamma^\alpha . \tag{10.161}$$

Forgetting for the moment the size that Dirac postulated for the matrices, if we consider them as $n \times n$, then taking the determinant of each side of (10.161) gives

$$|\gamma^\alpha||\gamma^\beta| = (-1)^n |\gamma^\alpha||\gamma^\beta| , \tag{10.162}$$

in which case n must be even if the matrices are not all to have zero determinant, which might be seen as an undesirable property. Since there is no set of four anticommuting 2×2 matrices, the next size up to try would have to be 4×4.

The multiplication of a four-component vector by the set of γ^α occurs so frequently in Dirac's theory that a new symbol was made up by Feynman from the typewriters of the day, as well as lowering the index on γ being done in the usual way using the metric:

$$\not{A} \equiv \gamma^\alpha A_\alpha = \gamma_\alpha A^\alpha . \tag{10.163}$$

The wave function is conventionally written bold to emphasise its four components, and in lowercase, so the Dirac equation is thus written

$$(\mathrm{i}\not{\partial} - m)\,\boldsymbol{\psi} = \mathbf{0} , \quad \text{or} \quad (\mathrm{i}\hbar\not{\partial} - mc)\,\boldsymbol{\psi} = \mathbf{0} \text{ with the correct units.} \tag{10.164}$$

This "first quantised" Dirac equation (i.e. a Schrödinger-type equation) turned out to have four-component solutions $\boldsymbol{\psi}$ that were interpreted as fields describing spin-½ particles and their antiparticles. Just as the Klein–Gordon equation exhibited negative energies, so, too, did these antiparticles, giving rise to an idea that antiparticles might be the physical manifestation of "holes" in a sea of energy states that could be equated with the vacuum. So even a vacuum began to suggest a multiparticle theory, and that suggested the idea of second quantising the Dirac equation.

Although the second quantisation is beyond our scope here, its effect was that the Dirac field $\boldsymbol{\psi}$ no longer represented just one particle; it described many such particles that were the equivalent of the phonons and photons we described earlier with the guitar string. Either way, the Dirac particles correspond to electrons and positrons, but these are free particles; they have no interaction with any electromagnetic field at all. In order to fully describe them, the idea of a current of such particles still needed to be carefully defined.

When the wave function in basic quantum mechanics is a scalar Ψ, the probability density is postulated to be $|\Psi|^2 = \Psi^*\Psi$. It stands to reason that when the wave function becomes a vector $\boldsymbol{\psi}$ (still a function of space and time, but conventionally written lowercase), we must use the hermitian adjoint and work with expressions such as $\boldsymbol{\psi}^\dagger\boldsymbol{\psi}$, since a probability defined this way will certainly be positive. But any expression for probability forces us to consider the *continuity equation*, which is a statement of conservation relating the amount of material crossing any closed surface to the loss of that material from the volume enclosed, using Gauss's theorem. We first met this discussion in the context of Nöther's theorem in Sect. 10.3.6. If the flux density of material out of the volume is $\boldsymbol{J}$ (i.e., $\boldsymbol{J}$ = material flow rate per unit area per unit time), and the material itself has density $J^0 \equiv \varrho$, then J^α will be a proper vector and

$$\text{rate of loss of material in volume} = \text{rate of material emerging}$$

$$\text{i.e., } -\partial_t \int \varrho \, \mathrm{d}V = \int \boldsymbol{J} \cdot \boldsymbol{n} \, \mathrm{d}S$$

$$= \int \nabla \cdot \boldsymbol{J} \, \mathrm{d}V \quad \text{(Gauss's theorem)},$$

$$\text{or} \quad \int (\partial_t \varrho + \nabla \cdot \boldsymbol{J}) \, \mathrm{d}V = 0 \,, \tag{10.165}$$

implying that $\partial_t \varrho + \nabla \cdot \boldsymbol{J} = 0$. Tensor notation shortens this to the statement that the four-divergence equals zero:

$$J^\alpha{}_{,\alpha} = 0 \,. \tag{10.166}$$

Suppose that we define J^α for the Dirac field ψ in the following way:

$$J^\alpha \equiv \eta_{tt} \, \psi^\dagger \gamma_0 \gamma^\alpha \psi \,. \tag{10.167}$$

This J^α immediately gives a density of $\varrho \equiv J^0 = \psi^\dagger \psi$, as well as satisfying the continuity equation, and so is of key importance in describing the movement of particles governed by the Dirac equation. (J^α can also be predicted from Nöther's theorem.) The term $\psi^\dagger \gamma_0$ is so common in this theory that a new symbol was introduced:

$$\bar{\psi} \equiv \psi^\dagger \gamma_0 \,, \quad \text{so that } J^\alpha = \eta_{tt} \, \bar{\psi} \gamma^\alpha \psi \,. \tag{10.168}$$

With this modified treatment of probability, the Dirac equation was able to accommodate spin-½ particles into a working field formalism. But it still remained to write a lagrangian corresponding to (10.103) for the interaction of an electron field ψ with an electromagnetic field A^μ. This marked the beginning of quantum electrodynamics, one of the most successfully tested theories in physics.

In the end, although the problems of the Klein–Gordon equation pushed research onward to the Dirac equation, the particle-hole picture, and finally second quantisation, the wheel has turned full circle: second quantisation can also be applied to the Klein–Gordon equation, giving it a physical meaning after all—although it's the Dirac equation that applies to electrons.

Local Versus Global Conservation

Equation (10.165) embodies the idea of *local* conservation, which is stronger than global conservation. Globally, something like energy could well be conserved in that it might disappear in one place only to reappear in another a long way away. But this seems never to be observed in Nature; if energy does disappear in one place and reappear in another, we always observe a current of energy in between those places. That is, energy is conserved *locally*, which is a much stronger idea than mere global conservation. Even so, it might well be that something can appear from nowhere in an apparent example of nonconservation. "Flatlanders"—beings who are confined to a 2-surface—might observe the arrival of a 2-sphere (i.e. a common garden-variety sphere that needs to be embedded in three dimensions) that passes through their world. What will they see? First, a dot appears, which rapidly grows into a circle before growing smaller again to eventually vanish. The Flatlanders have witnessed a higher-dimensional object passing through their world; they might well be perplexed, since the circle seemed to come out of the void before vanishing back into it.

A Lagrangian for the Dirac Equation. The Dirac equation is derivable from the lagrangian density

$$\mathcal{L} = \bar{\psi}(\mathrm{i}\partial\!\!\!/ - m)\psi\,, \tag{10.169}$$

which can be shown by applying Lagrange's equation for a field, (10.59):

$$\frac{\partial \mathcal{L}}{\partial \psi} - \partial_\alpha \frac{\partial \mathcal{L}}{\partial \psi_{,\alpha}} = 0\,. \tag{10.170}$$

For this, the expressions

$$\gamma^{0\dagger} = \gamma^0\,, \quad \gamma^{k\dagger} = -\gamma^k, \tag{10.171}$$

are needed, and also ψ and $\bar{\psi}$ must be treated as independent quantities in order to apply the partial differentiation. The relevant results are

$$\frac{\partial \mathcal{L}}{\partial \psi} = -m\bar{\psi}\,, \quad \frac{\partial \mathcal{L}}{\partial \psi_{,\alpha}} = \mathrm{i}\bar{\psi}\gamma^\alpha, \tag{10.172}$$

and the Dirac equation results. (Alternatively, the same result comes out much more directly by varying $\mathcal{L}$ with respect to $\bar{\psi}$, but it's wise to show that both variations agree.)

It seems, then, that from our previous work, the lagrangian density for a charge together with a noninteracting electromagnetic field must be

$$\mathcal{L}_{\text{no int.}}(x) = \bar{\psi}(\mathrm{i}\partial\!\!\!/ - m)\psi - \frac{\varepsilon_0}{4}F^2, \tag{10.173}$$

where the old current density interaction $-\eta_{tt} j^\alpha A_\alpha$ from (10.104) does not quite fit now, since it refers to a classical particle. But we can re-express the interaction in the language of the electron field ψ. To do so, remember that j^α is a current density for the electron, so write

$$\begin{aligned} &\text{current density} = \text{charge} \times \text{probability current (10.168)} \\ &\text{or} \quad j^\alpha = qJ^\alpha = \eta_{tt}\, q\, \bar{\psi}\gamma^\alpha\psi\,. \end{aligned} \tag{10.174}$$

In that case, the old interaction becomes $-\eta_{tt} j^\alpha A_\alpha = -q\,\bar{\psi}\not{A}\psi$, and the total lagrangian is

$$\mathcal{L} = \underbrace{\bar{\psi}(\mathrm{i}\not{\partial} - m)\psi}_{\text{free particle}} - \underbrace{\frac{\varepsilon_0}{4}F^2}_{\text{free field}} - \underbrace{q\,\bar{\psi}\not{A}\psi}_{\text{interaction}}\,. \tag{10.175}$$

In fact, the charge-field interaction can be calculated by following a completely different route that shows an elegant aspect of how Nature behaves. We follow that path next.

10.7 Gauge Theory and Quantum Electrodynamics

In previous chapters, we have seen that physical laws can be written in a frame-independent way, and we have the freedom to choose a frame and coordinates that simplify the equations we are trying to solve.

In Sect. 7.4, we discussed the fact that this sort of invariance with respect to a choice of frame appears in another way in Nature. This time, though, it is an invariance not with respect to a choice of spatial frame but rather with respect to a quantum mechanical phase. Historically, the term "gauge" was originally used to describe this choice, as when speaking of the width of a railway track in the sense of a choice of scale. But it's just a historical label; we are not really choosing a scale for our system.

This idea of choosing a gauge, or really a phase, turns out to have stellar significance in quantum electrodynamics. By starting out with a known theory and then postulating that Nature behaves in a certain way, we arrive at something genuinely new and, importantly, verifiable experimentally to spectacular accuracy.

10.7.1 The Starting Point: Classical Gauge Theory

Gauge theory has its roots in electromagnetism, where it was noticed that if we change the scalar and vector potentials in a particular way, then the $\boldsymbol{E}$ and $\boldsymbol{B}$ fields calculated from these will be unaltered. Suppose the potentials $\Phi, \boldsymbol{A}$ describe a given electromagnetic field. Given an arbitrary function $\chi(t, \boldsymbol{x})$, define new potentials $\Phi', \boldsymbol{A}'$ with the following *gauge transform*:

$$\begin{aligned} \Phi' &= \Phi + \partial\chi/\partial t\,, \\ \boldsymbol{A}' &= \boldsymbol{A} - \nabla\chi\,. \end{aligned} \tag{10.176}$$

In calculating the derived fields $\boldsymbol{E}', \boldsymbol{B}'$ from these new potentials, the contributions from χ cancel and we arrive at

$$\begin{aligned}\boldsymbol{E}' &\equiv -\nabla\Phi' - \partial\boldsymbol{A}'/\partial t = -\nabla\Phi - \partial\boldsymbol{A}/\partial t = \boldsymbol{E}\,,\\ \boldsymbol{B}' &\equiv \nabla\times\boldsymbol{A}' = \nabla\times\boldsymbol{A} = \boldsymbol{B}\,,\end{aligned} \tag{10.177}$$

so that the physical fields are certainly unaltered by a gauge transform.

The derivatives with respect to space and time here are a good indication that we are dealing with something that could better be written covariantly. So write $A^\alpha \equiv (\Phi, \boldsymbol{A})$, and raise and lower indices with the Minkowski metric:

$$A_\alpha = \eta_{\alpha\beta}A^\beta = \eta_{\alpha\alpha}A^\alpha \quad \text{(no sum)}. \tag{10.178}$$

With cartesian coordinates and $\eta_{\alpha\beta} = \eta_{tt}(1, -1, -1, -1)$, equation (10.176) becomes the much more elegant[1]

$$A'_\alpha = A_\alpha + \eta_{tt}\chi_{,\alpha}\,. \tag{10.179}$$

The sign η_{tt} can always be absorbed into χ, so we will do this from now on, essentially setting $\eta_{tt} = 1$ in (10.179). The new field A'_α gives us precisely the same physics as the old A_α. Choosing an appropriate χ to simplify the form of the potentials—"choosing a gauge"—is the direct analogy to choosing an appropriate frame to solve a given problem in mechanics.

Gauge choices can lead to very strange-looking functions for the potentials. The most obvious example occurs if we consider a charge q moving in the field produced by a stationary point charge Q. Here the potentials have the well-known forms, with $k \equiv 1/(4\pi\varepsilon_0)$,

$$\Phi = \frac{kQ}{r}\,, \quad \boldsymbol{A} = \boldsymbol{0}\,. \tag{10.180}$$

A gauge choice $\chi = -kQ\,t/r$ leads to new potentials

$$\Phi' = 0\,, \quad \boldsymbol{A}' = \frac{-kQ\,t}{r^2}\,\boldsymbol{e}_{\hat{r}}\,. \tag{10.181}$$

It's simple and instructive to verify that these produce the usual forms of the electric and magnetic fields. Nevertheless, they don't seem to match our intuitive idea of what form a potential should take! In particular, the total energy of the charge q is certainly the usual expression $mv^2/2 + q\,\Phi$, but this does not equal $mv^2/2 + q\,\Phi'$. Does this suggest a conflict with the new choice of gauge as regards an expression for the energy?

[1] It should be stressed that A'_α is not the same as $A_{\alpha'}$. The tensor components that interest us, A'_α, are the values of a *new* field $(\Phi', \boldsymbol{A}')$. Contrast these with the components $A_{\alpha'}$ of the *old* field $(\Phi, \boldsymbol{A})$ in a different frame. The latter are not relevant to this discussion.

Referring to Sect. 10.3.4, the hamiltonian H is called the total energy *provided* H is a constant of the motion. This will be the case as long as the lagrangian is time-independent. This is indeed the case for the potentials (10.180), so the hamiltonian is then $mv^2/2 + kQq/r$, conserved, and thus called the total energy. For the potentials (10.181), the lagrangian is a function of time, so the hamiltonian is not guaranteed to be conserved and so cannot be called an energy.

Just as choosing a suitable frame within which to solve a set of equations can simplify them enormously, choosing an appropriate gauge can simplify electromagnetic calculations. If we raise the indices in (10.179), then differentiate and contract, we obtain

$$A'^{\alpha}{}_{,\alpha} = A^{\alpha}{}_{,\alpha} + \chi^{,\alpha}{}_{,\alpha} \,. \tag{10.182}$$

The term $\chi^{,\alpha}{}_{,\alpha}$ is just the four-dimensional laplacian of χ in cartesian coordinates (in general coordinates it's $\chi^{;\alpha}{}_{;\alpha}$). Equation (10.182) is a wave equation for χ, and it turns out that this equation will always have a solution if $\boldsymbol{A}$ and $\boldsymbol{A}'$ are well-behaved. In particular, we can choose $A'^{\alpha}{}_{,\alpha}$ to be anything that helps to simplify the problem at hand. The prime is always dropped when such a choice is made, since the point of choosing a gauge is to specify one useful expression that gives the electromagnetic field. For magnetostatics, perhaps the best choice is the so-called *Coulomb gauge* (remember the prime has been dropped):

$$A^{\alpha}{}_{,\alpha} \equiv \partial \Phi/\partial t \,, \quad \text{or} \quad \nabla \cdot \boldsymbol{A} = 0 \,. \tag{10.183}$$

In general nonstatic situations, a more useful gauge is the *Lorenz gauge* (also called the *Lorentz gauge*, since although the Danish physicist Ludwig Lorenz specified it originally in 1867, the Dutch physicist Hendrik Lorentz also worked with it):

$$A^{\alpha}{}_{,\alpha} \equiv 0 \,, \quad \text{or} \quad \nabla \cdot \boldsymbol{A} = -\partial \Phi/\partial t \,. \tag{10.184}$$

We'll use this gauge to solve Maxwell's equations in Sect. 11.3.

10.7.2 A Gauge Transformation for the Dirac Lagrangian

From the basic postulates of quantum mechanics, we know that if a wave function $\Psi(t, \boldsymbol{x})$ is multiplied by any phase factor, the actual probability density that it produces is unchanged because the only physically relevant quantity is $|\Psi|^2$. So, the Dirac equation and its lagrangian density are unchanged by the transform

$$\psi \longrightarrow \psi \, \mathrm{e}^{\mathrm{i}\theta} \tag{10.185}$$

for any real number θ independent of x. This is a "global" gauge transform. Historically, a real scale factor of e^{θ} was initially considered, which was a change of scale, and hence of gauge, in the same sense that the gauge of a railway track is a measure of its size. With the factor now changed to $e^{i\theta}$, a more appropriate term would be a phase transform, but the name gauge has stuck. It is a global transform because the number θ is independent of spacetime position. We first came across this idea of a gauge transform back in Sect. 7.4.

Now, in the context of the solution ψ of the Dirac equation, we decide to do something new and investigate what results: we make θ depend upon x. This is a "local" gauge transform and is the heart of gauge theory. To make the notation a little more useful later on, rewrite θ as some arbitrary constant $-q$ times a function of spacetime $\chi(t, x, y, z)$ [shortened to $\chi(x)$]. So make the transform

$$\psi \longrightarrow \psi' \equiv \psi \, e^{-iq\chi(x)} . \tag{10.186}$$

Unfortunately, the Dirac field lagrangian density (10.169) is not invariant under this transform. The spacetime dependence of χ complicates the derivatives, destroying the invariance. Omitting the straightforward details, the transformed lagrangian density is

$$\begin{aligned} \mathcal{L}' &\equiv \overline{\psi}'(i\not{\partial} - m)\psi' \\ &= \overline{\psi}(i\not{\partial} - m)\psi + q\,\overline{\psi}(\not{\partial}\chi)\psi , \end{aligned} \tag{10.187}$$

and the extra term can only vanish in the trivial case of $q = 0$ or a constant χ. However, we have such a strong feeling that the gauge transform (10.186) should make no difference to the physics, that we accept it and change our equations—our model of the physics—so that it really does not make a difference! The invariance under (10.186) can be restored *if* the partial derivative ∂_α of (10.169) is replaced by a new type of covariant derivative D_α, reminiscent of (8.203) for coordinate transforms but here defined for gauge transforms as

$$D_\alpha \equiv \partial_\alpha + iqA_\alpha(x) , \tag{10.188}$$

where $A_\alpha(x)$ is a real four-vector with dimension of the electromagnetic potential (in which case the dimension of q turns out to be that of electric charge), *provided that*, when we transform the ψ according to (10.186), we also transform the A_α as follows:

$$A_\alpha(x) \longrightarrow A'_\alpha(x) \equiv A_\alpha(x) + \chi_{,\alpha} . \tag{10.189}$$

So in our stubborn requirement that Nature be invariant under the local gauge transform (10.186), we must modify our equations by replacing ∂_α with D_α (which is okay, as the equations always only constitute a working model). We have now been forced to include an extra quantity A_α suspiciously resembling an electromagnetic field undergoing the gauge change of (10.179), together with a constant q with the role of an electric charge. Thus the total gauge transform is given by (10.186) and (10.189), together with a modified Dirac lagrangian

$$\mathcal{L}_{\text{modified}} = \overline{\psi}(\mathrm{i}\not{D} - m)\psi\,. \tag{10.190}$$

This new lagrangian will be invariant if we change ψ, A_α to ψ', A'_α. But notice how this looks on expanding $\not{D}$:

$$\mathcal{L}_{\text{modified}} = \overline{\psi}(\mathrm{i}\not{\partial} - m)\psi - q\,\overline{\psi}\not{A}\psi\,. \tag{10.191}$$

This is just the lagrangian (10.175) but without the free-field term $-\varepsilon_0 F^2/4$. Yet $-\varepsilon_0 F^2/4$ is certainly required for a complete description of the electromagnetic field and should be added in. But notice from (10.189) that F^2 is *unchanged* by the substitution of A_α with A'_α (since $\chi_{,\alpha\beta} = \chi_{,\beta\alpha}$), which means that the lagrangian for the free electromagnetic field is already invariant under the gauge transform (10.176). We conclude then that the lagrangian (10.175)

$$\mathcal{L} = \overline{\psi}(\mathrm{i}\not{\partial} - m)\psi - \frac{\varepsilon_0}{4}F^2 - q\,\overline{\psi}\not{A}\psi \tag{10.192}$$

is gauge invariant. But, we can put this in another way—and this is where it all comes together: we can say that *if* the sum of the lagrangian densities of a free electron and an EM field

$$\mathcal{L}_{\text{free}} = \overline{\psi}(\mathrm{i}\not{\partial} - m)\psi - \frac{\varepsilon_0}{4}F^2 \tag{10.193}$$

is *required* to be gauge invariant, then we must alter it, and the new version will be obtained by adding the term $-q\,\overline{\psi}\not{A}\psi$, *which will then describe the* interaction *between a charge q and the electromagnetic field.* Equation (10.193) is the classical lagrangian that forms the starting point for the study of quantum electrodynamics. While it specifies a quantum field, it is still in a sense classical in that the modes of the field must yet be (second) quantised, just as we did for the guitar string in Sect. 10.6.

This way of treating the lagrangian density takes gauge invariance as fundamental, and we derive the extra interaction term as a consequence of replacing a partial derivative by a covariant derivative, just as is done to embody frame and coordinate invariances. So by postulating that Nature has this requirement of local gauge invariance, we have derived the electron–photon interaction! This new approach enables us to write down the lagrangian for a system without having to deal with a classical limit first, and so it becomes a new way of building theories for new particles and their interactions.

The quantities $\mathrm{e}^{\mathrm{i}\alpha(x)}$ can be considered as unitary 1×1 matrices, and in the conventional language, at each point x in spacetime they form a one-dimensional *abelian Lie group* $U(1)$, so we describe quantum electrodynamics as a $U(1)$ *gauge theory.*[2] Other gauge theories use more complicated transforms than (10.186) and (10.189).

[2] The group operation is ordinary multiplication. "Abelian" means the group elements commute, while a "Lie group" has a continuum of elements.

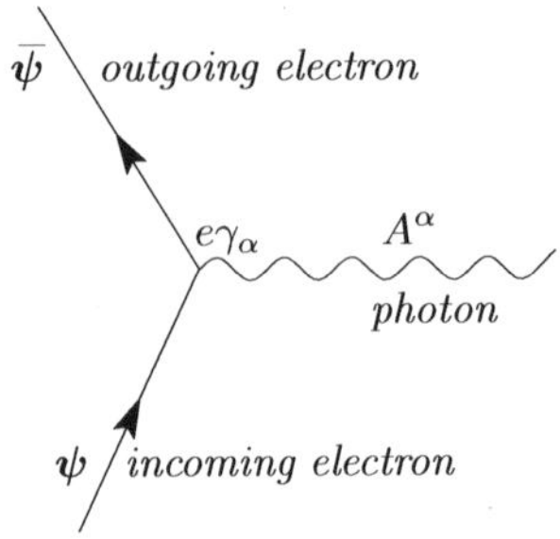

Fig. 10.8. A basic Feynman diagram showing an electron interacting with a photon. The incoming and outgoing electrons are represented by their Dirac fields ψ and $\bar{\psi}$, respectively. The photon field is A^α, while at the vertex the interaction is represented by $e\gamma_\alpha$. Diagrams like these are computational aids that help us to calculate cross sections for such interactions to occur.

It remains to place an electron into the electromagnetic field, in which case $q = -e$ and the interaction is $e\bar{\psi}\not{A}\psi$, or $e\bar{\psi}\gamma_\alpha A^\alpha\psi$. As shown in the *Feynman diagram* in Fig. 10.8, the various terms can be attached to a picture of the process in spacetime, and this sort of diagram is very useful for calculating cross sections for this and other interactions.

10.8 The Path-Integral Approach to Quantum Mechanics

The lagrangian viewpoint was extended within the realm of quantum mechanics by Richard Feynman in the 1940s. He applied a kind of democratisation principle to mechanics through the following reasoning.

We are aware that, within classical mechanics, a particle will follow a path for which the action is stationary. If we knew nothing of physics and drew all the paths that the particle might conceivably follow—one of which will be the actual one it takes, with the remainder unphysical—we could work out the action for each path and choose the one about which the action is stationary. That path would turn out to be the actual classical path that the particle follows. Because of the continuous nature of the quantities involved in the action, the multitude of spacetime paths that differ by only a small amount from the actual path taken by the particle all have similar action values, with the fractional difference in action from the classical value tending to zero as the paths deviate less and less from the classical path.

What, then, will result if we allocate a vector to each path, setting the vector's angle (from some arbitrary reference direction) to be proportional to the action of that path, and then add all the vectors corresponding to an uncountably infinite number of paths? We'll find that for paths resembling the classical one, the vectors have very similar angles, so that they add constructively. On the other hand, since the action changes more and more quickly as the paths depart ever more from the classical one, the directions in which the vectors of those paths point essentially become randomised, leading to a destructive interference over those paths. This procedure treats all paths equally, while singling out the classical one from the infinite set of paths that are drawn.

Actually, and equivalently, what Feynman did was ascribe to each path a complex number whose argument was proportional to the action and then add these complex numbers. This tallies with the idea mentioned in Chap. 2 that quantum mechanics allocates an amplitude for each way that a process can occur, so long as there is no way of distinguishing that way from another way. The absolute square of the summed amplitudes is the probability (or probability density for a continuous system) that the process occurs. On the other hand, if a process can happen in several distinguishable ways, then we must add the *probabilities* for each to obtain the final probability for it to occur.

From a quantum mechanical perspective, a particle can be viewed as travelling from one spacetime point to another over all possible paths in some kind of ghostly simultaneous sense. Since we cannot distinguish one path from another, we must add the amplitudes together to get the final amplitude for the particle to make the specified trip, and form its absolute square to obtain the associated probability density. So Feynman's complex number is a candidate for the amplitude of each path.

Here is what we do to formalise this procedure. Refer to Fig. 10.9, which shows several paths of the set that a particle might take to go from some initial event (x_0, t_0) to some final event (x_n, t_n), where we have broken the path up into n intervals. The reason for this division into intervals is twofold. First, breaking the path up into infinitesimal intervals hearkens back to the discussion of the Clock Postulate on p. 239. Despite the fact that the paths are in general curved, we are postulating that an amplitude can be ascribed to each infinitesimal segment of the path, in effect by treating that segment as straight, since we are defining it only by its start and end points. Of course, the segment is not straight, not even in the limit, and the fact that we can ascribe an amplitude in this way is purely a postulate.

The second reason for the division of paths into intervals is that by considering the intermediate events, we can expand the transition amplitude $\langle x_n, t_n \,|\, x_0, t_0\rangle$. Do this by repeatedly applying the position basis completeness relation:

$$\langle x_n, t_n \,|\, x_0, t_0\rangle = \int_{-\infty}^{\infty} \dots \int_{-\infty}^{\infty} \langle x_n, t_n \,|\, x_{n-1}, t_{n-1}\rangle \times \\ \langle x_{n-1}, t_{n-1} \,|\, x_{n-2}, t_{n-2}\rangle \dots \langle x_1, t_1 \,|\, x_0, t_0\rangle \, \mathrm{d}x_1 \dots \mathrm{d}x_{n-1} \,. \tag{10.194}$$

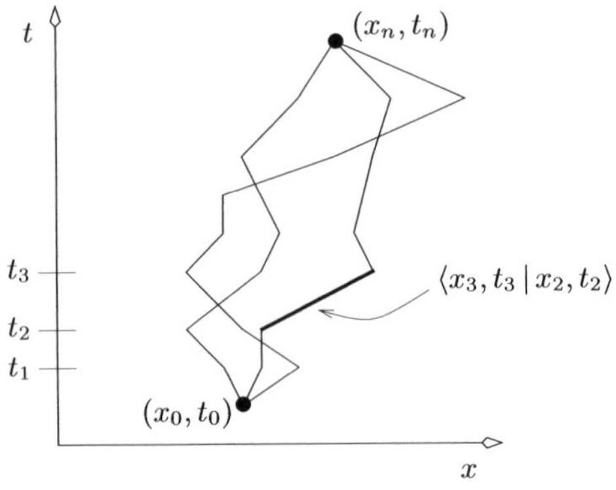

Fig. 10.9. Some of the many spacetime paths that a particle can take to go from (x_0, t_0) to (x_n, t_n). Each path can be broken up into infinitesimal segments, which enables a completeness relation to be applied as in (10.194).

The amplitude has been expressed as an uncountable number of integrals—no small task to compute!

Next we look closely at the segment joining two infinitesimally separated events (x_{k-1}, t_{k-1}) and (x_k, t_k), calculate the action $S(k, k-1)$ for that segment, and associate this with the amplitude for the particle to traverse the segment. Do this by making S proportional to the argument of a complex number as discussed above:

$$\langle x_k, t_k \,|\, x_{k-1}, t_{k-1} \rangle \propto \mathrm{e}^{\frac{\mathrm{i}}{\hbar} S(k,k-1)} . \tag{10.195}$$

Including $\hbar$ in the exponent leads to the usual results of quantum mechanics, but it's also a very reasonable thing to do since it allows the quantum theory to transit smoothly to the classical realm. That's because when included in this way, $\hbar$ takes on its usual role of determining the degree to which the system behaves quantum mechanically. If $\hbar$ were large in (10.195), there would be only little variation in the argument of the right-hand side over all paths, so that *all* paths would contribute significantly to the total amplitude and the world would be very quantum mechanical. Conversely, if $\hbar$ were vanishingly small, then the argument of the complex amplitude would effectively be randomised as soon as the paths began to depart from the classical one, so that only paths immediately adjacent to the classical one would contribute to the final amplitude—which is just the classical limit. This is one meaning of the oft-quoted phrase "$\hbar \to 0$ gives the classical limit."

Now combine (10.194) with (10.195) to write the total amplitude as

$$\begin{aligned} \langle x_n, t_n \,|\, x_0, t_0 \rangle &\propto \int \ldots \int \mathrm{e}^{\frac{\mathrm{i}}{\hbar} S(n,n-1)} \mathrm{e}^{\frac{\mathrm{i}}{\hbar} S(n-1,n-2)} \ldots \mathrm{e}^{\frac{\mathrm{i}}{\hbar} S(1,0)} \, \mathrm{d}x_1 \ldots \mathrm{d}x_{n-1} \\ &= \int \ldots \int \mathrm{e}^{\frac{\mathrm{i}}{\hbar} S(n,0)} \, \mathrm{d}x_1 \ldots \mathrm{d}x_{n-1} \, . \end{aligned} \tag{10.196}$$

The nett result is that the amplitude has been written as a multiple integral over all paths. Effectively, the action is being calculated for every path $x(t)$ that can be drawn joining (x_0, t_0) to (x_n, t_n), and these actions are being added with an appropriate infinitesimal weighting. Here we are not concerned with the details of computing this weighting, but we can include it in a generic way with the $\mathrm{d}x_1 \ldots \mathrm{d}x_{n-1}$ measure in (10.196), and rewrite the new weighting as $\mathcal{D}x$ or $\mathcal{D}x(t)$ (or even $\mathrm{D}x$, but don't confuse this with the covariant derivative in (9.102); the two are not related). Finally, a general path-integral expression for the amplitude for a system to evolve from some start point to some end point can be written as

$$\langle x_n, t_n \,|\, x_0, t_0 \rangle = \int_{(x_0,t_0)}^{(x_n,t_n)} \mathrm{e}^{\frac{\mathrm{i}}{\hbar} S[x(t)]}\, \mathcal{D}x\,. \tag{10.197}$$

Path integrals are examples of a more general type of integral over spaces of infinite dimension, known as a *functional integral.* In fact, despite the usefulness of path integrals, functional integration is not an entirely well-defined subject, in the sense that any attempts to define it completely rigorously fall short of being generally applicable.

10.8.1 Path Integrals Give the Schrödinger Equation

Section 10.5 showed how the Schrödinger equation arises when we substitute certain derivatives for the canonical momentum and the hamiltonian in an energy conservation equation. One of the great successes of the path-integral formalism was its ability to reproduce the Schrödinger equation via an apparently simpler, yet very different, approach to the prescription of Sect. 10.5.

Let's see how this is done for a single particle with potential energy V. Since the Schrödinger equation is a differential equation, consider the amplitude for a particle to arrive at (x, t) given a state $|\Psi\rangle$ by focussing on the infinitesimal time interval $t - \varepsilon$ just before it arrives. Refer to Fig. 10.10, where we make the approximation of only considering each *straight* path shown, since for $\varepsilon \to 0$ the potential V won't change the fact that the amplitude for a straight path will approximately dominate all other paths in its vicinity.

Start by expanding the amplitude $\Psi(x, t)$ via the position basis completeness relation at the earlier time $t - \varepsilon$:

$$\Psi(x,t) = \langle x, t \,|\, \Psi \rangle = \int_{-\infty}^{\infty} \langle x, t \,|\, x+\xi, t-\varepsilon \rangle \langle x+\xi, t-\varepsilon \,|\, \Psi \rangle \; \mathrm{d}(x+\xi)\,. \tag{10.198}$$

In this integral, x is being treated as a constant, in which case $\mathrm{d}(x + \xi) = \mathrm{d}\xi$ and (10.198) becomes

$$\Psi(x,t) = \int_{-\infty}^{\infty} \langle x, t \,|\, x + \xi, t - \varepsilon \rangle \; \Psi(x + \xi, t - \varepsilon)\; \mathrm{d}\xi\,. \tag{10.199}$$

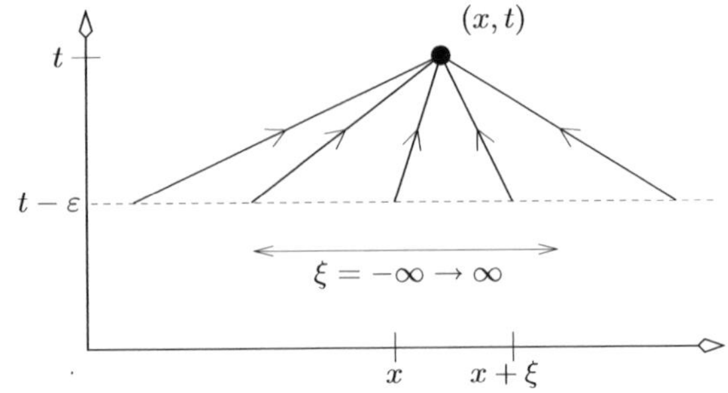

Fig. 10.10. Deriving the Schrödinger equation for a particle by focussing on the small time interval just before it arrives at (x, t). All arrival paths must be considered, although we make the approximation that each *straight* one is dominant over all other paths in its immediate vicinity.

The amplitude $\langle x, t \,|\, x+\xi, t-\varepsilon\rangle$ to travel the last short leg can be approximated by the action due to the particle's travelling at constant velocity between these two points. The action for such a short interval is just the lagrangian times the elapsed time, so

$$\begin{aligned}\langle x, t \,|\, x+\xi, t-\varepsilon\rangle &\simeq N \exp\left[\frac{\mathrm{i}}{\hbar}\left(\frac{m}{2}\frac{\xi^2}{\varepsilon^2} - V\right)\varepsilon\right], \quad (N \equiv \text{normalisation}) \\ &= N \exp\left[\frac{\mathrm{i}m\xi^2}{2\hbar\varepsilon}\right] \exp\left[\frac{-\mathrm{i}V\varepsilon}{\hbar}\right]. \end{aligned} \tag{10.200}$$

This amplitude oscillates increasingly rapidly as $\xi \to \infty$, so when integrating (10.199) over ξ, the main contribution will come from the ξ-domain over which these oscillations are not too rapid. (This is known as the method of *stationary phase* and is a powerful tool in calculating some difficult integrals of mathematical physics.) So we are only concerned with values of ξ and ε for which ξ^2/ε is not large. In that case, Taylor-expand (10.199) and (10.200) to orders ξ^2 and ε around $\Psi(x, t)$:

$$\begin{aligned}\Psi(x,t) \simeq N \int_{-\infty}^{\infty} &\exp\left[\frac{\mathrm{i}m\xi^2}{2\hbar\varepsilon}\right] \times \\ &\left[\Psi(x,t) + \varepsilon\left(\frac{-\mathrm{i}V\Psi}{\hbar} - \Psi_{,t}\right) + \xi\Psi_{,x} + \frac{1}{2}\xi^2\Psi_{,xx}\right] \mathrm{d}\xi\,. \end{aligned} \tag{10.201}$$

The $\xi\Psi_{,x}$ term can be dropped since it contributes an integral of an odd function, which is zero. For the other integrals, we need the following:

$$\int_{-\infty}^{\infty} \mathrm{e}^{\mathrm{i}\lambda\xi^2}\,\mathrm{d}\xi = \sqrt{\frac{\pi\mathrm{i}}{\lambda}}, \quad \int_{-\infty}^{\infty} \xi^2 \mathrm{e}^{\mathrm{i}\lambda\xi^2}\,\mathrm{d}\xi = \frac{\mathrm{i}}{2}\sqrt{\pi\mathrm{i}}\,\lambda^{-3/2}\,. \tag{10.202}$$

The *Fresnel integrals* of (10.202) are also used frequently in the theory of Fresnel diffraction. How are they calculated? The second can be written by differentiating the first with respect to λ. The first can be done in various ways. We can use (3.20), taking some care with the convergence, which is borderline. Alternatively, we can use Cauchy–Riemann integration to integrate from 0 to ∞, by following a pie-shaped path in the complex plane that first runs straight from the origin out to some distance R, then follows a circular arc for 45°, and finally returns along a straight path to the origin. Calculate or bound the integral on each leg, and let $R \to \infty$.

The integrals in (10.202) allow (10.201) to be calculated, giving

$$\Psi \simeq \underbrace{N\sqrt{\frac{2\pi i\hbar\varepsilon}{m}}}_{\text{Infer this term} = 1} \left[\Psi + \varepsilon \underbrace{\left(\frac{-iV\Psi}{\hbar} - \Psi_{,t} + \frac{i\hbar}{2m}\Psi_{,xx}\right)}_{\text{Infer this term} = 0}\right]. \tag{10.203}$$

This equation produces the normalisation N. Also, setting the coefficient of ε to zero produces

$$\left(V - \frac{\hbar^2}{2m}\frac{\partial^2}{\partial x^2}\right)\Psi(x,t) = i\hbar\frac{\partial\Psi}{\partial t}, \tag{10.204}$$

which is, of course, Schrödinger's famous equation in one dimension. Were the two inferences in (10.203) really justified, considering that the equation itself is just an approximation? In hindsight the answer is apparently yes, although the approximations made in the whole analysis underline the complexity of the path-integral approach.

The measure of classicality shown by the path-integral method is actually determined not so much by $\hbar \to 0$ as by $\hbar/m \to 0$. To see this, observe that because the action S is linear in the particle mass, the factor $\hbar$ only appears as $\hbar/m$ in the path-integral formalism, including the normalisation N in (10.203). This same ratio is also what appears in the Schrödinger equation, since that equation can always be written in terms of the potential energy per unit mass (i.e. the potential):

$$\left(\frac{V}{m} - \frac{\hbar^2}{2m^2}\nabla^2\right)\Psi(\boldsymbol{x},t) = \frac{i\hbar}{m}\frac{\partial\Psi}{\partial t}. \tag{10.205}$$

It seems, then, that the classicality emerging when $\hbar \to 0$ might also appear for large enough masses, even if $\hbar$ were not small. Certainly a heavy ball thrown through the air looks only classical; $\hbar/m$ is so small here as to ensure that only spacetime paths *extremely* close to the classical one will contribute to its dynamics. (The same sort of conclusion can be drawn from de Broglie's $\lambda = h/p = h/(mv)$.) Even so, it cannot quite be argued that a large mass will always ensure classicality. After all, the mass of Schrödinger's famous paradoxical cat—whose life depends on the outcome of a fully quantum process and so would seem to have to exist in a superposition of "alive" and "dead" states—does not enter into that paradox at all. Perhaps it is the mass of the constituents of the quantum process upon which the cat's life hangs that enters into the factor of $\hbar/m$, but the transition to classicality is really not understood at all. In the next section, we outline the ideas for a tool that can help to further the study of the quantum-to-classical transition.

10.9 Density Matrices: The Language of Decoherence

Quantum mechanics uses the ideas of probability theory to predict outcomes when we make a measurement of a system. The system might be an electron with spin "z up", by which we mean that the z-component of its spin—which can only take on the values $\pm\hbar/2$—has been measured to be $+\hbar/2$. Such an electron is in an eigenstate of the z-spin measurement operator, so label its state $|z+\rangle$. If we wish to know the probability that a measurement of the x-component of its spin will be, say, $+\hbar/2$, then we write the associated amplitude as $\langle x+|z+\rangle$, giving the probability as $|\langle x+|z+\rangle|^2$. (Further theory of quantum mechanics is needed to set a value for $\langle x+|z+\rangle$, but that's not relevant here.)

If the system only consists of one electron in a state $|z+\rangle$, then in order to use the language of probability theory, we need to appeal to the idea of an *ensemble*, a collection of such electrons all in the state $|z+\rangle$. To each measurement, quantum mechanics ascribes an operator that has for its eigenvalues all of the possible results. The expected value of that operator is by definition the expected value of those measurements. When we make a measurement of the x-spin of these z-up electrons, then we expect to find that 50% of the set are projected onto the state $|x+\rangle$, with the remaining 50% becoming $|x-\rangle$. And so it is that we can then consider just one electron and say that there is a 50% chance of its being measured to have x-spin up, and 50% x-spin down.

Consider an ensemble, and suppose first that it's *pure*, meaning that each of its electrons is described by the same state $|\psi\rangle$ (which needn't be an eigenstate of any operator). We make a measurement on this ensemble with some apparatus described by an operator A that has a set of eigenvalues a_i, meaning $A|a_i\rangle \equiv a_i|a_i\rangle$. We ask: what answer can be expected for the measurement? By a quantum mechanical postulate, only the eigenvalues of A can result, so the expected value $\langle A\rangle$ is a sum over the eigenvalues of A weighted by their associated probabilities; it can be concisely expressed using a completeness relation:

$$\begin{aligned}\langle A\rangle &\equiv \sum_i a_i|\langle a_i|\psi\rangle|^2 = \sum a_i\langle\psi|a_i\rangle\langle a_i|\psi\rangle \\ &= \sum\langle\psi|\,A\,|a_i\rangle\langle a_i|\psi\rangle = \langle\psi|\,A\,|\psi\rangle\,. \end{aligned} \tag{10.206}$$

This expectation is a standard expression in quantum mechanics (and we came across it previously in Sect. 2.10 and specifically (2.178), where we needed the idea of a mean, which is also just an expected value). But it can only be used for a pure ensemble, since only then will all the electrons be described by the same state ket $|\psi\rangle$. What if the electrons exist in a multitude of states? Some have been prepared with z-spin up while others have x up, and others have had no preparation at all. Such a *mixed ensemble* cannot be described by a single ket. (Note that pure and mixed ensembles are sometimes erroneously called pure and mixed states.) Right from the beginnings of quantum mechanics, the need to describe such a mixture using state vectors was recognised. Luckily, (10.206) can be rewritten in a way suited to describing mixed ensembles. First, consider a pure ensemble and reverse the factors in the second line of (10.206):

$$\begin{aligned}\langle A\rangle &= \sum \langle a_i|\psi\rangle\langle\psi|\, A\,|a_i\rangle \\ &= \operatorname{tr}(\varrho A)\,, \quad \text{where } \varrho \equiv |\psi\rangle\langle\psi| \\ &= \operatorname{tr}(A\varrho)\,, \quad \text{since } \operatorname{tr}(AB) = \operatorname{tr}(BA) \text{ for any } A, B\,. \end{aligned} \tag{10.207}$$

This alternative form of the expected value $\langle A\rangle$ is written in terms of the *density matrix* for the state $|\psi\rangle$:

$$\text{density matrix } \varrho \equiv |\psi\rangle\langle\psi|\,. \tag{10.208}$$

Suppose now that the system is a mixed ensemble comprising electrons in states $|\psi_i\rangle$ with proportions w_i, where these states can, in general, be linear combinations of eigenstates of many different operators. In that case, the electrons in state $|\psi_i\rangle$ have density matrix $\varrho_i \equiv |\psi_i\rangle\langle\psi_i|$ and expected value $\langle A\rangle_i$, so

$$\begin{aligned}\langle A\rangle &\equiv \sum_i w_i\,\langle A\rangle_i = \sum w_i \operatorname{tr}\left(A\varrho_i\right) \quad \text{from (10.207)} \\ &= \sum w_i \operatorname{tr}\left(A\,|\psi_i\rangle\langle\psi_i|\right) \\ &= \operatorname{tr}\left(A\sum w_i\,|\psi_i\rangle\langle\psi_i|\right) \\ &\quad\text{(since the trace is a sum, and sums commute)} \\ &\equiv \operatorname{tr}(A\varrho)\,, \quad \text{where } \varrho \equiv \sum_i w_i\,|\psi_i\rangle\langle\psi_i|\,. \end{aligned} \tag{10.209}$$

Through defining a mixed ensemble density matrix ϱ that encapsulates the whole ensemble as a weighted sum of the density matrices for each state in the mixture, the general expression $\langle A\rangle = \operatorname{tr} A\varrho$ becomes valid for both pure and mixed ensembles. This makes it very useful for multiple-particle quantum mechanics.

The density matrix of an ensemble of electrons all with z-spin up is given by $\varrho = |z+\rangle\langle z+|$. In the z-basis $\{|1\rangle, |2\rangle\} \equiv \{|z+\rangle, |z-\rangle\}$, this matrix has components

$$\varrho_{ij} \overset{z\text{-basis}}{=\!=} \langle i|\,\varrho\,|j\rangle = \begin{bmatrix}\langle 1|\,\varrho\,|1\rangle & \langle 1|\,\varrho\,|2\rangle \\ \langle 2|\,\varrho\,|1\rangle & \langle 2|\,\varrho\,|2\rangle\end{bmatrix} = \begin{bmatrix}\langle z+|\,\varrho\,|z+\rangle & \langle z+|\,\varrho\,|z-\rangle \\ \langle z-|\,\varrho\,|z+\rangle & \langle z-|\,\varrho\,|z-\rangle\end{bmatrix} = \begin{bmatrix}1 & 0\\ 0 & 0\end{bmatrix}. \tag{10.210}$$

Obviously, the matrix for such a pure ensemble, in the basis of its measured states, will have a one somewhere on its main diagonal with zeroes everywhere else. If the electrons have been prepared as a mixed ensemble, with, say, 10% z-spin up and the rest with x up, then the density matrix for the mixture must be

$$\varrho = {}^1\!/_{10}\,|z+\rangle\langle z+| \;+\; {}^9\!/_{10}\,|x+\rangle\langle x+|\,. \tag{10.211}$$

The appearance of explicit probabilities as weightings in density matrices, as well as the ability of density matrices to describe whole sets of systems, gives an indication that perhaps they might shed light on the most fundamental conundrum of quantum mechanics: how a quantum system becomes classical when it interacts with a detector. This is the *Measurement Problem*, and with a conspicuous lack of insight provided by the accepted rules of quantum mechanics, the Measurement Problem has in recent years been worked on in various ways.

For example, if a z-spin measurement is made on a mixed ensemble of electrons, then they will all be projected onto one or the other of the two eigenstates of the z-spin operator, with the result that their density matrix in the z-basis will have become diagonal. So the idea that a zeroing of the off-diagonal entries of the density matrix might signal the emergence of classicality has prompted the use of density matrix formalism in studies of the Measurement Problem.

Probably leading the field of competing ideas is the notion that the density matrix is able to encode the action of an environment on a system in a way that enables us to use the matrix to answer questions about the system without having to work explicitly with the environment.

To see how all of this might come about, consider a system labelled by x, interacting with an environment labelled by q (really, for example, a bath of oscillators labelled by $q_1, \dots, q_n$). Suppose we require the expected value of a system observable $A(x)$. The system–environment combination is described by a density matrix ϱ. A straightforward extension of the ideas of this section is that tracing over both the system x and environment q gives the expected value of $A(x)$:

$$\langle A\rangle = \mathrm{tr}_{q,x}\,(\varrho A) = \int \mathrm{d}x\,\mathrm{d}q\;\langle x\,q|\,\varrho A\,|x\,q\rangle\,. \tag{10.212}$$

Expand using completeness:

$$\langle A\rangle = \int \mathrm{d}x\,\mathrm{d}x'\,\mathrm{d}q\,\mathrm{d}q'\;\langle x\,q|\,\varrho\,|x'\,q'\rangle\langle x'\,q'|\,A\,|x\,q\rangle\,. \tag{10.213}$$

Because A is independent of the environment and so doesn't depend on q, the bracket containing it can be split into constituent brackets:

$$\langle A\rangle = \int \mathrm{d}x\,\mathrm{d}x'\,\mathrm{d}q\,\mathrm{d}q'\;\langle x\,q|\,\varrho\,|x'\,q'\rangle\langle x'|\,A\,|x\rangle\,\delta(q-q')$$

$$= \int dx\, dx' \underbrace{\int dq\, \langle x\, q|\, \varrho\, |x'\, q\rangle}_{\equiv \langle x|\, \varrho_r\, |x'\rangle} \langle x'|\, A\, |x\rangle$$
$$= \int dx\, dx'\, \langle x|\, \varrho_r\, |x'\rangle\langle x'|\, A\, |x\rangle$$
$$= \mathrm{tr}_x\, (\varrho_r A)\,. \tag{10.214}$$

A new quantity has been defined here: the *reduced density matrix* ϱ_r that encapsulates how the system interacts with its environment. Equipped with ϱ_r, we need only trace over the system variable x and need never refer to the environment at all. Of course, the difficulty lies in constructing the reduced density matrix, but once we have it, we can begin to investigate the system's evolution.

An example of such a system is one with two eigenstates, $|1\rangle$, $|2\rangle$, currently in some state $|\psi_s\rangle$:

$$|\psi_s\rangle = \alpha|1\rangle\ +\ \beta|2\rangle\,, \quad \text{with} \quad |\alpha|^2 + |\beta|^2 = 1\,. \tag{10.215}$$

Suppose this system interacts with a detector that itself has two eigenstates: $|d_1\rangle$, indicating the system has been measured to be in state 1; and $|d_2\rangle$, which shows the system to be in state 2. Just before the interaction, the detector is in some state $|d\rangle$. Measurement theory suggests that just before the measurement occurs, the composite system–detector is in a new state,

$$|\psi_{sd}\rangle \equiv |\psi_s\rangle\, |d\rangle = \alpha|1\rangle\, |d\rangle\ +\ \beta|2\rangle\, |d\rangle\,. \tag{10.216}$$

The act of measurement itself is a mysterious process on which the standard interpretation of quantum mechanics, the*Copenhagen Interpretation*, is silent. But suppose that after it has occurred, the system–detector is in a new *correlated state* in which the detector recognises the system eigenstates:

$$|\psi'_{sd}\rangle = \alpha|1\rangle\, |d_1\rangle\ +\ \beta|2\rangle\, |d_2\rangle\,. \tag{10.217}$$

(The system and detector are also called *entangled* since their kets in (10.217) are not able to be factorised into a product of system and detector states.) This process is shown in Fig. 10.11. The density matrix for this correlated state is

$$\varrho'_{sd} \equiv |\psi'_{sd}\rangle\langle\psi'_{sd}| = |\alpha|^2|1\rangle\langle 1|\ |d_1\rangle\langle d_1|\ +\ \beta\alpha^*|2\rangle\langle 1|\ |d_2\rangle\langle d_1|$$
$$+\ \alpha\beta^*|1\rangle\langle 2|\ |d_1\rangle\langle d_2|\ +\ |\beta|^2|2\rangle\langle 2|\ |d_2\rangle\langle d_2|\,. \tag{10.218}$$

Early in the history of quantum mechanics, it was postulated that there now occurs another mysterious process called *decoherence*, which destroys the off-diagonal elements of ϱ'_{sd} to leave the new density matrix

$$\varrho''_{sd} \equiv |\alpha|^2|1\rangle\langle 1|\ |d_1\rangle\langle d_1|\ +\ |\beta|^2|2\rangle\langle 2|\ |d_2\rangle\langle d_2|\,. \tag{10.219}$$

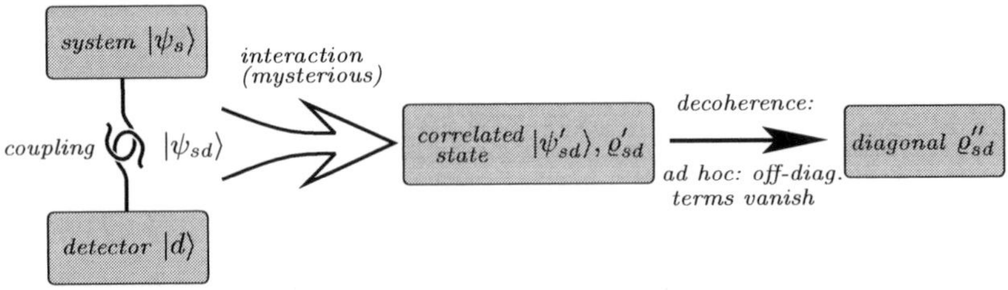

Fig. 10.11. Ad hoc decoherence applied to the correlated state of a system plus a detector. Refer to equations (10.216) to (10.219).

All that remains are the two probabilities $|\alpha|^2, |\beta|^2$: the hoped-for signature of a transition to classicality, as well as a correlation between the detector's measuring of state 1 and the system really being in state 1 (and similarly for state 2).

But this postulated decoherence is completely ad hoc. It can be improved upon by introducing an environment that brings a further interaction with the system–detector correlated state $|\psi'_{sd}\rangle$, as shown in Fig. 10.12. With the environment in a state $|E\rangle$, the new composite system–detector–environment state will be

$$|\psi'_{sdE}\rangle \equiv |\psi'_{sd}\rangle\,|E\rangle = \alpha|1\rangle\,|d_1\rangle\,|E\rangle \;+\; \beta|2\rangle\,|d_2\rangle\,|E\rangle\,. \tag{10.220}$$

Another mysterious interaction produces a correlated state of system–detector–environment:

$$|\psi''_{sdE}\rangle \equiv \alpha|1\rangle\,|d_1\rangle\,|E_1\rangle \;+\; \beta|2\rangle\,|d_2\rangle\,|E_2\rangle\,. \tag{10.221}$$

Now, the *reduced* density matrix for this last state is, from (10.214), just a trace over the environment:

$$\begin{aligned}\operatorname{tr}_E |\psi''_{sdE}\rangle\langle\psi''_{sdE}| &= \sum_i \langle E_i|\psi''_{sdE}\rangle\,\langle\psi''_{sdE}|E_i\rangle \\ &= \alpha|1\rangle\,|d_1\rangle\,\alpha^*\,\langle d_1|\,\langle 1| \;+\; \beta|2\rangle\,|d_2\rangle\,\beta^*\,\langle d_2|\,\langle 2| \\ &= \varrho''_{sd} \text{ again!}\end{aligned} \tag{10.222}$$

But ϱ''_{sd} is exactly what was obtained through the earlier ad hoc procedure. It seems then that we can eliminate a very arbitrary setting-to-zero of off-diagonal density matrix elements by keeping track of the system–detector's interaction with its environment, and such an interaction must inevitably happen in any real system.

Of course, it might be argued that the off-diagonal terms are being "swept under the rug" as it were, since what exactly defines the environment is a somewhat arbitrary choice. We might attempt to get around this objection by introducing a succession of ever-larger environments that describe the

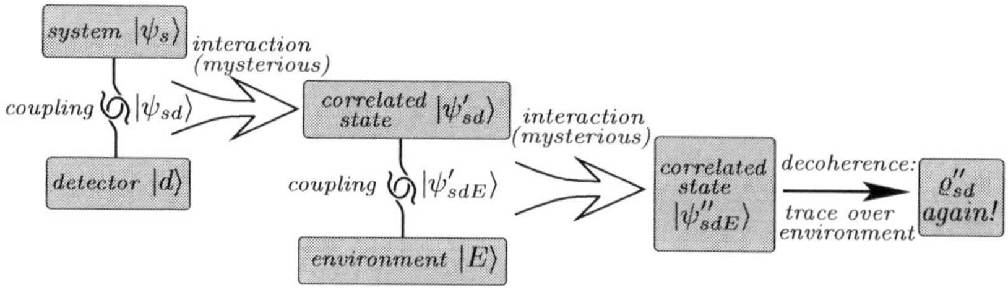

Fig. 10.12. Introducing an environment allows for a mathematical description of decoherence, laying the foundation for further work in quantum measurement theory.

interaction more and more comprehensively, but of course this cannot be done indefinitely. In the modern study of *quantum cosmology*, the quantum mechanics of the universe as a whole comes to the fore, and we run out of rugs under which to sweep the off-diagonal terms. Various refinements on the model of decoherence we have described here have been put forward, but the process is still very much a mystery.

Nevertheless, we'll describe here one result of some of the work being done in this new field. The concept of entropy was defined in Chap. 3 as the logarithm of the number of accessible states of a system. This is fine for a system with a finite number of degrees of freedom, but for systems such as fields that have an infinite number of degrees of freedom, we must resort to another definition. Luckily, statistical mechanics already has another construction that we can bend to our needs. Suppose a classical system can exist in one of any number of states $|s\rangle$, each with probability of occupation p_s. In that case, for a large number of accessible states, a careful count of their multiplicity determines the entropy of the system to be

$$\sigma = -\sum_s p_s \ln p_s \,. \tag{10.223}$$

But the density matrix ϱ for such a system will be a matrix with $p_1, p_2, \ldots$ along its diagonal, so that the entropy can also be written as

$$\sigma = -\operatorname{tr} \varrho \ln \varrho \,, \tag{10.224}$$

where the logarithm of a matrix can be found by diagonalising it. This expression now defines the entropy of a completely quantum system. It is zero for a pure ensemble, since this has a density matrix composed of only a one somewhere on its diagonal. But in general, this entropy σ will be greater than zero.

What is the entropy of the universe as a whole? If the universe evolves as a closed system, then its entropy is always zero. This is not entirely useful, and more appropriate is an attempt to relate some kind of local entropy growth to an Arrow of Time, as we discussed in Sect. 3.5. If we decompose the universe into a system of modes (modelled as a set of oscillators in its lagrangian) interacting with an environment (modelled by more oscillator modes), then the system's entropy alone can be calculated by using its *reduced* density matrix in (10.224). Path integrals can be used to do this, and the result is that the entropy of the system is indeed found to grow over time as expected, while it interacts with its environment. So the language of density matrices seems to be useful in shedding light on some of the mystery that surrounds the Arrow of Time.

11 The Green Function Approach to Solving Field Equations

11.1 The Idea of a Green Function

The differential equations that govern most physical laws tend to be second order in space and time. A good example is the pair of equations that govern how a body falls in a nonrelativistic field. The motion of the body is given by Newton's law $\boldsymbol{F}(\boldsymbol{x}) = m\ddot{\boldsymbol{x}}$, which is comparatively easy to solve. Much harder to solve is the equation that describes the force $\boldsymbol{F}(\boldsymbol{x})$, or equivalently the gravitational field. In Sect. 10.4, we produced Poisson's equation (10.75) as the *equation of motion* for the newtonian gravitational field:

$$\nabla^2 \Phi(t, \boldsymbol{x}) = 4\pi G \varrho(t, \boldsymbol{x}) . \tag{11.1}$$

We assumed a radial solution for a point mass m at the origin, and calculated the form that Φ should take. The result was (10.80):

$$\Phi = -Gm/r + \text{constant.} \tag{11.2}$$

However, because we demanded early on that Φ vanish as $r \to \infty$, the constant never explicitly appeared in our solution. But in general it is needed to satisfy any boundary conditions, such as the one we imposed.

The solution to (11.2) is composed of two terms: $-Gm/r$ is known in differential equation literature as the *particular integral* (we'll call this the PI), while the constant is the *complementary function* (here called the CF). Elementary differential equation theory shows that the general solution to a linear differential equation is the sum of these two things. The complementary function is the general solution for the homogeneous equation (that is, for no source), and always involves arbitrary constants. Often, it's not difficult to find, and it can *always* be found for linear differential equations. On the other hand, the particular integral is one solution for the actual source, involves no arbitrary constants at all, and can be very difficult to find. But once we have it, the most general solution of a linear differential equation will be the sum of the particular integral and the complementary function. The proof of this is very easy. Suppose we have a linear differential equation $L(y) = f$ for a linear differential operator L and source term f. Certainly it's true that $y = \text{PI}+\text{CF}$ is one solution, because

$$L(y) = L(\mathrm{PI} + \mathrm{CF}) = L(\mathrm{PI}) + L(\mathrm{CF}) \equiv f + 0 = f\,, \tag{11.3}$$

as required. Now suppose there is another solution y_1. Then

$$L(y - y_1) = L(y) - L(y_1) = f - f = 0\,, \tag{11.4}$$

in which case $y - y_1$ equals some constant times the CF. Thus,

$$y_1 = y + \text{constant} \times \mathrm{CF}\,, \tag{11.5}$$

which means that y_1 itself is a sum of the PI and the CF—since the CF involves an arbitrary constant that allows any amount of it to be added to the PI. So the most general solution to the equation has the form PI + CF, just as we set out to prove.

It's a curious fact that for all of the usual difficulty in finding a particular integral, it can actually creep up on us. For example, suppose we set out to solve Poisson's equation for no source at all, expecting to find the constant that got set to zero in the gravity case. This constant is the complementary function by definition. Let's see what really happens. Assuming a radial solution, use (8.227) to write Poisson's equation as

$$\nabla^2 \Phi(r) = \frac{1}{r^2}\left(r^2 \Phi_{,r}\right)_{,r} = 0\,. \tag{11.6}$$

This is easily integrated by inspection to give

$$\Phi = \frac{a}{r} + b\,, \quad \text{where } a, b \text{ are constants.} \tag{11.7}$$

We have found the complementary function b, but another solution has also slipped in—the very solution we found on p. 406 for a point source at the origin. Of course, this can be removed by setting $a = 0$, but nevertheless a particular integral for a point source has somehow surfaced in the solution. As we'll see in the next chapter on p. 497, the same thing happens when solving Einstein's equations for empty space: a constant of integration appears whose physical effect is that of a point mass at the origin. It's as if our use of polar coordinates has enabled a point mass to appear almost unnoticed at $r = 0$, a point that in a sense is right on the edge of our coordinate system. So point sources have a way of appearing unannounced, and it might be to our advantage to harness their power. Including them in the form of delta functions might render more complicated differential equations solvable. This is indeed what happens, and in this chapter we make it our aim to use the delta function in solving Maxwell's equations for a general source. But first we'll tackle a much simpler problem: we will again solve Poisson's equation for gravity but this time for a general mass source, and we'll follow a more general route than the one used for the point source in equations (10.75)–(10.80).

So return to (11.1), and for simplicity, absorb the $4\pi G$ term by dividing it into the field Φ; we are free to multiply the solution by $4\pi G$ at the end. Hence, essentially we wish to solve

$$\nabla^2 \Phi(\boldsymbol{x}) = \varrho(\boldsymbol{x}) \tag{11.8}$$

for an arbitrary mass distribution $\varrho(\boldsymbol{x})$. Remember that in Sect. 10.4 we solved this equation for Φ due to $\delta(\boldsymbol{x})$, and then multiplied the solution by m (because of linearity) to give the required solution for a point mass at rest at $\boldsymbol{x} = \boldsymbol{0}$.

To apply this approach to a general mass distribution, again make use of the additivity of solutions to linear differential equations. The required solution is simply the sum of the solutions due to infinitesimal masses placed at every point in space. The solution for the infinitesimal mass $\varrho(\boldsymbol{x}')\,\mathrm{d}^3x'$ at $\boldsymbol{x} = \boldsymbol{x}'$ is given by solving for Φ due to a unit mass $\delta(\boldsymbol{x} - \boldsymbol{x}')$, and then weighting this solution by the actual mass $\varrho(\boldsymbol{x}')\,\mathrm{d}^3x'$. The bottom-line solution is then given by adding all of these solutions over the whole space of $\boldsymbol{x}'$.

The solution for Φ due to a unit source $\delta(\boldsymbol{x} - \boldsymbol{x}')$ is called the *Green function* $G(\boldsymbol{x}, \boldsymbol{x}')$ for the differential operator ∇^2. The final value of the field is thus

$$\Phi(\boldsymbol{x}) = \underbrace{\int \underbrace{\underbrace{G(\boldsymbol{x}, \boldsymbol{x}')}_{\text{Solution for unit source at } \boldsymbol{x}'} \quad \varrho(\boldsymbol{x}')\,\mathrm{d}^3x'}_{\text{Solution for infinitesimal source at } \boldsymbol{x}'}}_{\text{Superposition of solutions due to all sources}}. \tag{11.9}$$

We can verify that this solution does indeed yield Φ by substituting it into the original equation (11.8). Remember that ∇^2 acts on $\boldsymbol{x}$, not $\boldsymbol{x}'$ (often indicated by writing $\nabla^2_{\boldsymbol{x}}$):

$$\begin{aligned} \nabla^2 \Phi(\boldsymbol{x}) &= \int \nabla^2 G(\boldsymbol{x}, \boldsymbol{x}')\varrho(\boldsymbol{x}')\,\mathrm{d}^3x' \\ &= \int \delta(\boldsymbol{x} - \boldsymbol{x}')\varrho(\boldsymbol{x}')\,\mathrm{d}^3x' \\ &= \varrho(\boldsymbol{x}), \quad \text{as required.} \end{aligned} \tag{11.10}$$

The solution to (11.8) is often written as $\Phi = \nabla^{-2}\varrho$, so that the Green function is $\nabla^{-2}\delta(\boldsymbol{x} - \boldsymbol{x}')$. We can use this notation to recreate the Green function definition in case we forget it. Begin with Poisson's equation, writing

$$\begin{aligned} \Phi(\boldsymbol{x}) &= \nabla_{\boldsymbol{x}}^{-2}\varrho(\boldsymbol{x}) = \nabla_{\boldsymbol{x}}^{-2} \int \varrho(\boldsymbol{x}')\,\delta(\boldsymbol{x} - \boldsymbol{x}')\,\mathrm{d}^3x' \\ &= \int \varrho(\boldsymbol{x}')\,\underbrace{\nabla_{\boldsymbol{x}}^{-2}\,\delta(\boldsymbol{x} - \boldsymbol{x}')}_{\equiv G(\boldsymbol{x},\boldsymbol{x}')}\,\mathrm{d}^3x', \end{aligned} \tag{11.11}$$

which is just (11.9) again.

The Green function for the radial part of ∇^2 was easy to calculate, which is really what we did in Sect. 10.4; on dividing by $4\pi G$, equations (10.75) and (10.80) imply that

$$\nabla^2_{\boldsymbol{x}} \left(\frac{-1}{4\pi|\boldsymbol{x}|} \right) = \delta(\boldsymbol{x}) \,, \quad \text{so that } \nabla^2_{\boldsymbol{x}} \left(\frac{-1}{4\pi|\boldsymbol{x}-\boldsymbol{x}'|} \right) = \delta(\boldsymbol{x}-\boldsymbol{x}') \,, \tag{11.12}$$

where the second expression in (11.12) follows from the general result for a linear change of variables: $\nabla^2_{\boldsymbol{x}} f(a\boldsymbol{x}+\boldsymbol{b}) = a^2 \nabla^2_{\boldsymbol{x}} f(\boldsymbol{x})\big|_{\boldsymbol{x}\to a\boldsymbol{x}+\boldsymbol{b}}$. The Green function for ∇^2 is then

$$\nabla^{-2}_{\boldsymbol{x}} \delta(\boldsymbol{x}-\boldsymbol{x}') = \frac{-1}{4\pi|\boldsymbol{x}-\boldsymbol{x}'|} \,. \tag{11.13}$$

Next, substituting $G(\boldsymbol{x},\boldsymbol{x}')$ into (11.9) yields the full solution for the gravitational potential Φ—remembering to multiply by the gravitational constant factor $4\pi G$ absorbed earlier. Additionally, because Newton's theory assumes that changes in the gravity field propagate at infinite speed, we have the freedom to insert a time dependence into the density and hence the field in the following expression, without changing our analysis in any way:

$$\Phi(t,\boldsymbol{x}) = \int \frac{-G\varrho(t,\boldsymbol{x}')\, \mathrm{d}^3x'}{|\boldsymbol{x}-\boldsymbol{x}'|} \,. \tag{11.14}$$

(The G in this equation is the gravitational constant, not the Green function.) As a check, we can see that for a point mass at the origin, $\varrho(t,\boldsymbol{x}') = m\,\delta(\boldsymbol{x}')$, and the potential becomes $\Phi = -Gm/|\boldsymbol{x}| = -Gm/r$ as expected.

This is all well and good, but a general linear operator can be complicated. There are more general ways to calculate $G(\boldsymbol{x},\boldsymbol{x}')$, and we'll concentrate on one such method that employs Fourier theory. For this, we deal not with $G(\boldsymbol{x},\boldsymbol{x}')$ but with its Fourier transform, in effect writing the Green function as a superposition of plane waves and focusing attention on their amplitudes. This method is well illustrated by calculating the Green function for ∇^2. The procedure treads a very thin line mathematically because we know already that the Green function for ∇^2 diverges as $|\boldsymbol{x}-\boldsymbol{x}'| \to 0$, and Fourier theory is not applicable to functions that are not square integrable, let alone functions that diverge! Nevertheless, we hope that the ever-useful ideas of generalised functions will lend a hand in resolving any difficulties. In the next section, we rederive the Green function for ∇^2 using this Fourier approach and show that it agrees with (11.13).

11.2 Deriving the Green Function for ∇^2 via Fourier Theory

The Green function for ∇^2 is defined by

$$\nabla^2 G(\boldsymbol{x}, \boldsymbol{x}') \equiv \delta(\boldsymbol{x} - \boldsymbol{x}'). \tag{11.15}$$

Begin by Fourier-decomposing G into exponential functions of $\boldsymbol{x}$. This gives the laplacian something concrete on which to operate.

$$G(\boldsymbol{x}, \boldsymbol{x}') \equiv \int \mathrm{e}^{\mathrm{i}\boldsymbol{k}\cdot\boldsymbol{x}} g(\boldsymbol{k}, \boldsymbol{x}')\, \mathrm{d}^3 k. \tag{11.16}$$

Substituting the Fourier decomposition of $G(\boldsymbol{x}, \boldsymbol{x}')$ into (11.15) allows the laplacian ∇^2 to act only on the plane wave components $\mathrm{e}^{\mathrm{i}\boldsymbol{k}\cdot\boldsymbol{x}}$, producing

$$\int -k^2\, \mathrm{e}^{\mathrm{i}\boldsymbol{k}\cdot\boldsymbol{x}} g(\boldsymbol{k}, \boldsymbol{x}')\, \mathrm{d}^3 k = \delta(\boldsymbol{x} - \boldsymbol{x}'), \quad \text{where } k \equiv |\boldsymbol{k}|. \tag{11.17}$$

This Fourier-inverts to give

$$-k^2\, g(\boldsymbol{k}, \boldsymbol{x}') = \frac{1}{(2\pi)^3} \int \mathrm{e}^{-\mathrm{i}\boldsymbol{k}\cdot\boldsymbol{x}} \delta(\boldsymbol{x} - \boldsymbol{x}')\, \mathrm{d}^3 x = \frac{\mathrm{e}^{-\mathrm{i}\boldsymbol{k}\cdot\boldsymbol{x}'}}{(2\pi)^3}. \tag{11.18}$$

Finally, isolating g and substituting it into (11.16) produces G:

$$G(\boldsymbol{x}, \boldsymbol{x}') = \frac{-1}{(2\pi)^3} \int \frac{\mathrm{e}^{\mathrm{i}\boldsymbol{k}\cdot(\boldsymbol{x}-\boldsymbol{x}')}}{k^2}\, \mathrm{d}^3 k. \tag{11.19}$$

To perform this integration, switch to $\boldsymbol{k}$-space polar coordinates (k, θ, ϕ), which come with the freedom to set the $\boldsymbol{k}$-space "z-axis" parallel to $\boldsymbol{x} - \boldsymbol{x}'$. The squared radial distance k^2 in the denominator presents no problem near the $\boldsymbol{k}$-space origin, since it is weighted by the infinitesimal volume $\mathrm{d}^3 k = k^2 \sin\theta\; \mathrm{d}k\, \mathrm{d}\theta\, \mathrm{d}\phi$. For simplicity, set $X \equiv |\boldsymbol{x} - \boldsymbol{x}'|$, and write

$$\begin{aligned} G(\boldsymbol{x}, \boldsymbol{x}') &= \frac{-1}{(2\pi)^3} \int_0^{2\pi} \mathrm{d}\phi \int_0^{\pi} \mathrm{d}\theta\; \sin\theta \int_0^{\infty} \mathrm{d}k\; \mathrm{e}^{\mathrm{i}kX\cos\theta} \\ &= \frac{-1}{4\pi^2} \int_0^{\pi} \mathrm{d}\theta\; \sin\theta \int_0^{\infty} \mathrm{d}k\; \mathrm{e}^{\mathrm{i}kX\cos\theta}. \end{aligned} \tag{11.20}$$

In which order might it be best to evaluate these integrals? The simplest approach is to integrate first over θ, since the integration is then quite well defined. Harder is to integrate first over k, an exercise that we'll leave until Sect. 11.2.1. The integration over θ gives

$$G(\boldsymbol{x}, \boldsymbol{x}') = \frac{-1}{4\pi^2} \int_0^{\infty} \mathrm{d}k \int_0^{\pi} \mathrm{d}\theta\; \sin\theta\, \mathrm{e}^{\mathrm{i}kX\cos\theta}$$

$$= \frac{-1}{4\pi^2} \int_0^\infty \mathrm{d}k \left[\frac{\mathrm{e}^{ikX\cos\theta}}{-ikX} \right]_{\theta=0}^{\pi} = \frac{-1}{2\pi^2 X} \int_0^\infty \mathrm{d}k \, \frac{\sin kX}{k} \,. \tag{11.21}$$

We'll calculate the last integral shortly, but for now we just use the result that it equals $\pi/2$. Using this, the Green function for ∇^2 becomes

$$G(\boldsymbol{x}, \boldsymbol{x}') = \frac{-1}{4\pi|\boldsymbol{x} - \boldsymbol{x}'|} \,, \tag{11.22}$$

which agrees with (11.13), as expected. So we have some confidence in this method of calculating Green functions.

The integral in (11.20) can also be evaluated by integrating over k first. But before we do that, it will prove useful to examine in some detail how the value of $\pi/2$ was derived for the last integral in (11.21):

$$\int_0^\infty \frac{\sin ax}{x} \, \mathrm{d}x = \frac{\pi}{2} \operatorname{sgn} a \,. \tag{11.23}$$

This is a standard example of an integral calculated using complex variable theory in undergraduate mathematics courses, but it forms a very good test bed to begin a discussion of the complex integration usually used in Green function theory. Without loss of generality, set $a = 1$, as it's a factor that can always be absorbed via a change of variables. We wish then to calculate the real integral

$$I \equiv \int_{-\infty}^{\infty} \frac{\sin x}{x} \, \mathrm{d}x \,. \tag{11.24}$$

This integral certainly exists (and we expect to show that it equals π since the integrand is even), because while the integrand has a discontinuity at $x = 0$, this discontinuity is *removable*. That is, the left and right limits of the integrand as $x \to 0$ are equal, so that the single point absent in a plot of the integrand has no effect on the integral. This easy-to-miss singularity can always be removed by redefining the integrand with the limit value; hence the term removable, and although the singularity is quite benign in that the integration is insensitive to its presence, it *is* still a singularity of the integrand as it stands.

In the next few pages we'll examine various approaches to calculating I that all require converting the integrand into a complex exponential. By so doing, we can appeal to the following lemma from complex analysis. The paths used in a complex integration can be drawn fairly arbitrarily, but of course some paths are more useful than others. The following lemma concerns two of these useful paths.

Jordan's lemma: *Suppose we are Fourier-integrating a rational function composed of polynomials $P(z), Q(z)$, where the degree of Q is at least one more than the degree of P:*

$$\int e^{iaz} \frac{P(z)}{Q(z)}\, dz = ? \tag{11.25}$$

If $a > 0$, draw as a path of integration a semicircle in the upper half complex plane with radius R. Call this path C_R^+. Then

$$\lim_{R\to\infty} \int_{C_R^+} e^{iaz} \frac{P(z)}{Q(z)}\, dz = 0\,. \tag{11.26}$$

If $a < 0$ and the path of integration is a semicircle C_R^- in the lower half plane, then the same thing happens: the integral over C_R^- tends toward zero as $R \to \infty$.

Jordan's lemma can be proved by using the triangle inequality (often used in complex analysis) to put an upper bound on the integral (11.25). This inequality states that

$$\left| \int e^{iaz} \frac{P(z)}{Q(z)}\, dz \right| \le \int \left| e^{iaz} \right| \left| \frac{P(z)}{Q(z)} \right| |dz|\,. \tag{11.27}$$

For the $a > 0$ case, parametrise the path by $z = Re^{i\theta}$ for $\theta = 0 \to \pi$, which traces the path counterclockwise. When R is large enough, the highest-order term in each of the polynomials P, Q dominates them, rendering P/Q bounded above by M/R for some constant M. In that case, the right-hand side of (11.27) is then also bounded above:

$$\int \left| e^{iaz} \right| \left| \frac{P(z)}{Q(z)} \right| |dz| \le \int_0^\pi e^{-aR\sin\theta} \frac{M}{R} R\, d\theta\,. \tag{11.28}$$

Since $a > 0$, the exponential factor $e^{-aR\sin\theta}$ mostly reduces the size of the last integrand as $R \to \infty$, except near $\theta = 0$ and π; but these parts of the domain contribute less and less to the integral as $R \to \infty$ anyway. So the upper bound that is the right-hand side of (11.28) shrinks to zero, proving the lemma. The $a < 0$ case is proved similarly. It's apparent that the lemma could be stated more generally, since P/Q could be replaced by any function of z whose large-R behaviour is sufficient to offset $R\, e^{-aR\sin\theta}$; but the lemma as it stands will suffice for our needs.

Using this lemma and the path choices it provides, let's look at each of several approaches to evaluating I in (11.24), some of which are successful and others of which fail. Always and everywhere, we'll be guided by the following Golden Rule:

> *When evaluating a complex integral, the path of integration must never be drawn through a pole, since this makes no sense. Also, creating such a path but swerving around the pole does not validate such a procedure.*

We will evaluate the integrals using the *Cauchy residue theorem*. This theorem is perhaps the most famous and useful in complex analysis. It says that for any function $f(z)$ that is analytic on and inside some non-self-intersecting curve Γ except for finitely many singularities $\sigma_1, \ldots, \sigma_n$ inside Γ,

$$\oint_{\Gamma} f(z)\,\mathrm{d}z = 2\pi\mathrm{i} \sum_{i=1}^{n} \operatorname*{Res}_{z=\sigma_i} f(z) \,. \tag{11.29}$$

Residues are straightforward to calculate—and for everything that follows here they are very easy, because all of the poles σ that we encounter will be simple, meaning that $(z-\sigma)f(z)$ is analytic and nonzero at σ. In such cases, the residue is

$$\operatorname*{Res}_{z=\sigma} f(z) = (z-\sigma)f(z)\Big|_{z=\sigma} . \tag{11.30}$$

First Successful Approach to Evaluating I in (11.24)

We will invoke the Cauchy residue theorem to evaluate the integral (11.24). The theorem requires a closed contour. We will limit the contour running along the real axis in (11.24) to have a finite extent ($-R \to R$), and then close it by adding another contour. The required integral will result in the limit $R \to \infty$. It's expedient for the added contour to give a zero contribution, so we refer to Jordan's lemma to arrange for this to happen. Jordan's lemma requires the integrand to have a complex exponential, which we can arrange by splitting (11.24) into its two exponential parts:

$$\begin{aligned} I &= \lim_{\varepsilon\to 0} \int_{-\infty}^{\infty} \frac{\sin x}{x+\mathrm{i}\varepsilon}\,\mathrm{d}x = \frac{1}{2\mathrm{i}} \lim_{\varepsilon\to 0} \int \frac{\mathrm{e}^{\mathrm{i}x}-\mathrm{e}^{-\mathrm{i}x}}{x+\mathrm{i}\varepsilon}\,\mathrm{d}x \\ &= \frac{1}{2\mathrm{i}} \lim_{\varepsilon\to 0} \Bigg[\underbrace{\int \frac{\mathrm{e}^{\mathrm{i}x}}{x+\mathrm{i}\varepsilon}\,\mathrm{d}x}_{\equiv I_1} - \underbrace{\int \frac{\mathrm{e}^{-\mathrm{i}x}}{x+\mathrm{i}\varepsilon}\,\mathrm{d}x}_{\equiv I_2} \Bigg] . \end{aligned} \tag{11.31}$$

The integrals I_1 and I_2 are effectively being treated as complex, so that the dummy variable x might better be replaced by z for clarity. Each integrand thus has a pole at $z = -\mathrm{i}\varepsilon$. The paths of integration are shown in Fig. 11.1. The semicircular parts added are guaranteed to give zero contribution as $R \to \infty$, by Jordan's lemma.

Neither path runs into the poles, so the Golden Rule is satisfied and we can apply the Cauchy residue theorem with confidence. Hence $I_1 = 0$ trivially (since no poles are enclosed), while

$$I_2 = -2\pi\mathrm{i} \operatorname*{Res}_{z=-\mathrm{i}\varepsilon} \frac{\mathrm{e}^{-\mathrm{i}z}}{z+\mathrm{i}\varepsilon} = -2\pi\mathrm{i}\,\mathrm{e}^{-\varepsilon} . \tag{11.32}$$

Equation (11.31) then yields the correct result $I = \pi$.

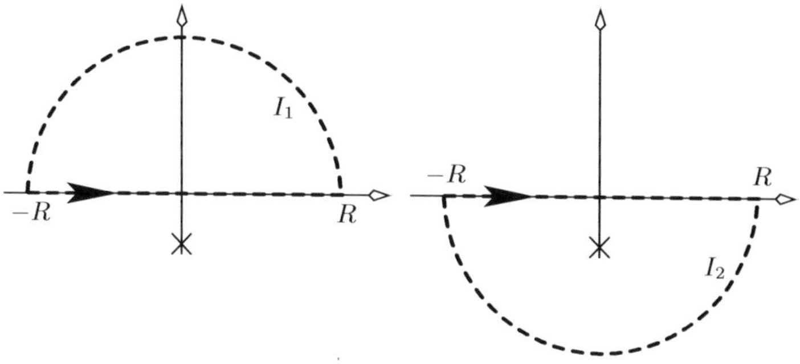

Fig. 11.1. Contours for I_1 and I_2 used in the first approach to calculating I, equation (11.31).

Second Successful Approach to Evaluating I in (11.24)

Another way of arranging to use Jordan's lemma is to convert the integrand $\sin x/x$ into a complex exponential directly by adding a cosine contribution. Unfortunately, $\int_{-\infty}^{\infty} \frac{\cos x}{x}\,\mathrm{d}x$ diverges, but because its integrand is odd, we know that the *principal value* of this latter integral is zero. The principal value defines a symmetrical integration, with particular attention to the cancellation happening around a singularity:

$$\mathrm{PV}\int_{-\infty}^{\infty} f(x)\,\mathrm{d}x \equiv \lim_{\substack{\varepsilon\to 0\\ R\to\infty}} \left[\int_{-R}^{-\varepsilon} + \int_{\varepsilon}^{R}\right] f(x)\,\mathrm{d}x\,. \tag{11.33}$$

(Linguistically it may seem a little odd that the principal value of an undefined integral can itself exist, but the term is quite standard.) We also know that because I is well defined, the precise manner of integration shouldn't matter, so in particular it must equal its principal value. Hence

$$\begin{aligned} I &= \int_{-\infty}^{\infty} \frac{\sin x}{x}\,\mathrm{d}x = \mathrm{PV}\int_{-\infty}^{\infty} \frac{\sin x}{x}\,\mathrm{d}x \\ &= \frac{1}{\mathrm{i}}\Bigg[\underbrace{\mathrm{PV}\int_{-\infty}^{\infty} \frac{\cos x}{x}\,\mathrm{d}x}_{=0} + \mathrm{i}\,\mathrm{PV}\int_{-\infty}^{\infty} \frac{\sin x}{x}\,\mathrm{d}x\Bigg] \\ &= \frac{1}{\mathrm{i}}\,\mathrm{PV}\int_{-\infty}^{\infty} \frac{\mathrm{e}^{\mathrm{i}x}}{x}\,\mathrm{d}x\,. \end{aligned} \tag{11.34}$$

We will calculate the last principal value integral of (11.34) by again adding an extra path to close it up and then applying the Cauchy residue theorem. A contour is shown in Fig. 11.2. As usual, the real integration is initially

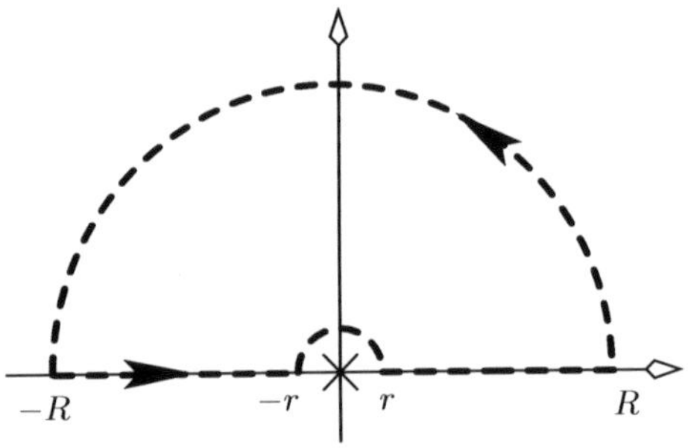

Fig. 11.2. Contour used in the second approach to calculating I in (11.34). Note that we are abiding by the Golden Rule on p. 451: the contour is *not* set to run through the pole, and the small semicircle is *not* a last-minute swerve to avoid a collision. Although the pole was deliberately inserted from the start to convert the integral to something that could make use of the Cauchy residue theorem, by construction it was *never* on the path of integration. See the further discussion after (11.34).

limited to $-R \to R$. Jordan's lemma ensures that the large semicircle gives a vanishing contribution as $R \to \infty$.

The presence of the *small* semicircle in Fig. 11.2 might make it seem that we have gone against the Golden Rule on p. 451, because that semicircle seems to be swerving around the origin to avoid the pole there. But we have *not* gone against the Golden Rule, because the small semicircle's role was never to sidestep a pole on our path. The pole was simply *never* on our path; the semicircle came into being as soon as we decided to use a principal value approach. This is a very important point, but also a very subtle one that is missed in most treatments of contour integration. Let's examine it in detail by reviewing what has been done.

1. We wished to invoke the Cauchy residue theorem to calculate the integral (11.24). This theorem uses a closed contour. So we needed to add a contour to the one running along the real axis in (11.24). We decided to add a contour that gave a zero contribution, so planned to make use of Jordan's lemma. That lemma required a complex exponential in the integrand.
2. One way to convert the integrand of (11.24) to a complex exponential is to add a cosine term. But the integral of this added term diverges. However, the *principal value* of the cosine integral equals zero. Since the principal value of the original integral I in (11.24) *equals* I, we can add the principal value of a cosine integral to it, which converts (11.24) to the principal value of an integral of a complex exponential.

3. By definition, the principal value of an integral excludes, symmetrically, any point of discontinuity. So the pole at the origin was *never* on our path.

Thus there is nothing ad hoc about the path in Fig. 11.2 because it is *not* set to run through the pole, and we are *not* making a last-minute swerve to avoid a collision. We have adhered to the Golden Rule.

The integral I is sometimes calculated by converting it to a complex exponential for no apparent reason; suddenly a pole has appeared at the origin toward which the path is headed, and that must be avoided in an ad hoc manner by going around it—which begs the question of why such symmetrical integration limits should be better, necessary, or any more correct than asymmetrical ones. We now see why this approach gives the right answer. But our careful analysis will come in very handy later in this chapter when we use a complex integration to solve Maxwell's equations.

Return now to the last principal value integral of (11.34). In the large-R limit, the principal value is the integral along the x-axis, going around the pole at the origin (intentionally!). Ignoring the vanishing contribution from the large semicircle (by Jordan's lemma), we have

$$\mathrm{PV}\int_{-\infty}^{\infty} \frac{\mathrm{e}^{\mathrm{i}x}}{x}\,\mathrm{d}x + \lim_{r\to 0} \int_{\substack{\text{semicircle } r\\ \text{clockwise}}} \frac{\mathrm{e}^{\mathrm{i}z}}{z}\,\mathrm{d}z = 0\,. \tag{11.35}$$

Parametrise the small semicircle by $z = r\mathrm{e}^{\mathrm{i}\theta}$ with $\theta = \pi \to 0$. Then

$$\int_{\substack{\text{semicircle } r\\ \text{clockwise}}} \frac{\mathrm{e}^{\mathrm{i}z}}{z}\,\mathrm{d}z = \int_{\pi}^{0} \exp\left(\mathrm{i}r\,\mathrm{e}^{\mathrm{i}\theta}\right)\mathrm{i}\,\mathrm{d}\theta \overset{r\to 0}{\longrightarrow} -\mathrm{i}\pi\,. \tag{11.36}$$

Equations (11.34)–(11.36) then combine to give the correct result $I = \pi$.

It's worth pointing out that if the integration path used in Fig. 11.2 were to be changed to include the pole by running the small semicircle *under* the x-axis, the answer would be unchanged. Whether or not we include a pole in a complex integration cannot affect the answer, since the nett difference between the two paths amounts to completely encircling the pole—which is exactly balanced by the fact that the residue is included in one calculation but not the other. The moral of this story is that we should include no poles at all if possible, since that way no residues will need to be calculated. The proverbial free lunch can indeed be eaten by evaluating a contour integral without enclosing any poles at all.

Unsuccessful Approaches to Evaluating I in (11.24)

Care is required to avoid using naïve contours when evaluating complex integrals. The following analyses fail, but are worth discussing to show how *not* to draw contours. As in (11.34), write

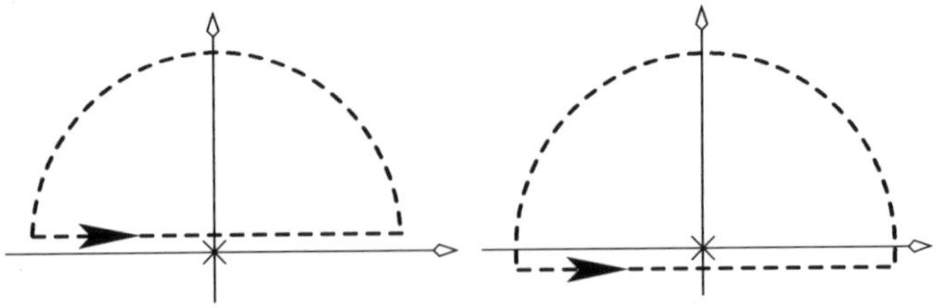

Fig. 11.3. Invalid contours for evaluating the principal value integral in (11.34). Neither of these recognises the fact that we are taking a principal value. Contours like these are conventionally used to "evaluate" the divergent integral that we'll encounter later in (11.66).

$$I = \frac{1}{\mathrm{i}} \, \mathrm{PV} \int_{-\infty}^{\infty} \frac{\mathrm{e}^{\mathrm{i}x}}{x} \, \mathrm{d}x \neq \underbrace{\frac{1}{\mathrm{i}} \int_{-\infty}^{\infty} \frac{\mathrm{e}^{\mathrm{i}x}}{x} \, \mathrm{d}x}_{\text{this diverges!}} . \tag{11.37}$$

Because the last integral is not defined, we must not be tempted to try to evaluate it as the limit

$$\lim_{\varepsilon \to 0} \int_{-\infty}^{\infty} \frac{\mathrm{e}^{\mathrm{i}x}}{x + \mathrm{i}\varepsilon} \, \mathrm{d}x . \tag{11.38}$$

This integral is just I_1 from the first approach, and is well defined (equalling zero). But we have run into a problem by discarding the use of the principal value.

Another unsuccessful approach on the same theme is to obtain the correct equation (11.34), but then attempt to evaluate it using contours that don't run along the x-axis. Two such contours are shown in Fig. 11.3. Neither of these gives the correct answer since neither calculates the principal value. The first contains no poles and so produces $I = 0$, while the second contains the pole at the origin and produces $I = 2\pi$. And neither of these values changes as we make the separation of the path from the real axis tend toward zero. In fact, we'll see later in this chapter (p. 469) that these two contours are conventionally used in solving Maxwell's equations in a mathematically ill-defined way. The fact that they give the right answers is not so much due to mathematical rigor, as to the fact that they mimic a physically more meaningful approach where the integrals are always ensured to be well defined.

11.2.1 The Other Way of Calculating the Integral (11.20)

When we derived the Green function for ∇^2 at the start of Sect. 11.2, we needed to evaluate the integral (11.20). We did this by the very standard approach of first integrating over θ. It is possible to first integrate over k instead, but a convergence subtlety arises. The integral to evaluate is

$$\int_0^\infty e^{ikX\cos\theta}\, dk\,. \tag{11.39}$$

This diverges, and so presents something of an obstacle since we know that the double integral (11.20) is well defined. But at this point we remember that the integral (11.39) was first mentioned on p. 58, where we asked:

$$\int_0^\infty e^{ikx}\, dk = ? \tag{11.40}$$

One way of giving this integral physical meaning is to introduce a damping factor inspired by other such factors that exist in physical theories; this damping can then be set to zero in the limit. Such a procedure ultimately gives rise to the fundamental Fourier identity (2.174). Define the damping by

$$\int_0^\infty e^{ikx}\, dk \equiv \lim_{\varepsilon\to 0+} \int_0^\infty e^{ikx-\varepsilon k}\, dk\,. \tag{11.41}$$

This new integral converges:

$$\begin{aligned}\int_0^\infty e^{ikx} dk &\equiv \lim_{\varepsilon\to 0+} \int_0^\infty e^{ik(x+i\varepsilon)} dk = \lim_\varepsilon \frac{i}{x+i\varepsilon}\\ &= \underbrace{\lim_\varepsilon \frac{\varepsilon}{x^2+\varepsilon^2}}_{\equiv L_1(x)} + i\underbrace{\lim_\varepsilon \frac{x}{x^2+\varepsilon^2}}_{\equiv L_2(x)}\,.\end{aligned} \tag{11.42}$$

The two real limits L_1, L_2 can be evaluated by introducing a square-integrable test function $T(x)$. We'll integrate $T(x)$ with L_1 and L_2 in turn, freely swapping limits and integrations where necessary. First, L_1:

$$\begin{aligned}\int_{-\infty}^\infty T(x) L_1(x)\, dx &= \lim_{\varepsilon\to 0+} \int \frac{T(x)\,\varepsilon}{x^2+\varepsilon^2}\, dx\\ &= \lim_\varepsilon \left[T(x)\tan^{-1}\frac{x}{\varepsilon}\right]_{x=-\infty}^{\infty} - \lim_\varepsilon \int T'(x)\tan^{-1}\frac{x}{\varepsilon}\, dx\,.\end{aligned} \tag{11.43}$$

The first term equals zero by virtue of the test function's tending to zero at $x = \pm\infty$. The second tends toward $\pm\pi/2$:

$$\begin{aligned}\int_{-\infty}^\infty T(x) L_1(x)\, dx &= -\int_{-\infty}^\infty T'(x)\,\frac{\pi}{2}\,\operatorname{sgn} x\, dx = \pi\, T(0)\\ &= \int_{-\infty}^\infty T(x)\,\pi\,\delta(x)\, dx\,.\end{aligned} \tag{11.44}$$

$L_1(x)$ has the same effect on a test function as does $\pi\,\delta(x)$, so that in the language of generalised functions they are defined to be equal.

A similar approach gives L_2. Here we break the limit up symmetrically, paying particular attention to $x = 0$:

$$\begin{aligned}
\int_{-\infty}^{\infty} T(x) L_2(x)\, dx &= \lim_{\varepsilon\to 0+} \int_{-\infty}^{\infty} \frac{T(x)\, x}{x^2+\varepsilon^2}\, dx \\
&= \lim_{\varepsilon} \lim_{\eta\to 0+} \left[\int_{-\infty}^{-\eta} + \int_{-\eta}^{\eta} + \int_{\eta}^{\infty} \right] \frac{T(x)\, x}{x^2+\varepsilon^2}\, dx \\
&= \lim_{\varepsilon} \lim_{\eta} \left[\int_{-\infty}^{-\eta} + \int_{\eta}^{\infty} \right] \frac{T(x)\, x}{x^2+\varepsilon^2}\, dx + \lim_{\varepsilon} \lim_{\eta} \int_{-\eta}^{\eta} \frac{T(x)\, x}{x^2+\varepsilon^2}\, dx \\
&= \mathrm{PV} \int_{-\infty}^{\infty} \frac{T(x)}{x}\, dx + \lim_{\varepsilon} T(0) \underbrace{\lim_{\eta} \int_{-\eta}^{\eta} \frac{x}{x^2+\varepsilon^2}\, dx}_{=0 \text{ as integrand is odd}} \\
&= \mathrm{PV} \int_{-\infty}^{\infty} \frac{T(x)}{x}\, dx \equiv \int_{-\infty}^{\infty} T(x)\, \mathrm{P}\,(1/x)\, dx\,.
\end{aligned} \tag{11.45}$$

The *principal-part* generalised function, conventionally written $\mathrm{P}\,(1/x)$, is defined so that it produces a principal value when integrated with a test function. A more useful notation is $\sigma(x) \equiv \mathrm{P}\,(1/x)$, in which case the final expression for the original one-sided integral is, from (11.42),

$$\boxed{\int_0^{\infty} \mathrm{e}^{\mathrm{i}kx}\, dk = \pi\, \delta(x) + \mathrm{i}\, \sigma(x)\,.} \tag{11.46}$$

The principal part has effectively been defined as a limit of a sequence of functions,

$$\sigma(x) = \mathrm{P}\,(1/x) \equiv \lim_{\varepsilon\to 0+} \frac{x}{x^2+\varepsilon^2}\,, \tag{11.47}$$

so that it forms a natural partner to the *lorentzian form* of the delta function calculated in (11.42)–(11.44):

$$\pi\, \delta(x) = \lim_{\varepsilon\to 0+} \frac{\varepsilon}{x^2+\varepsilon^2}\,. \tag{11.48}$$

Just as the delta function can be visualised as a spike, being the limit of a sequence of bell-shaped functions that become ever more peaked as $\varepsilon \to 0$, the principal part can be visualised as an odd function identical to $1/x$ for $x \neq 0$, while continuous at $x = 0$ and equal to zero there. Figure 11.4 shows it as the limit of a sequence of functions determined by (11.47) for various values of $\varepsilon > 0$. It's common practice in engineering and signal processing texts to identify $\sigma(x)$ completely with $1/x$. But this is only true for $x \neq 0$, and such an error will cause convergence difficulties that tend to be ignored in the integrations that result. This does nothing for the pedagogy of the subject.

Our introduction of the exponential damping in (11.41) has evidently been useful. In particular, we can recover the usual Fourier identity (2.174) by

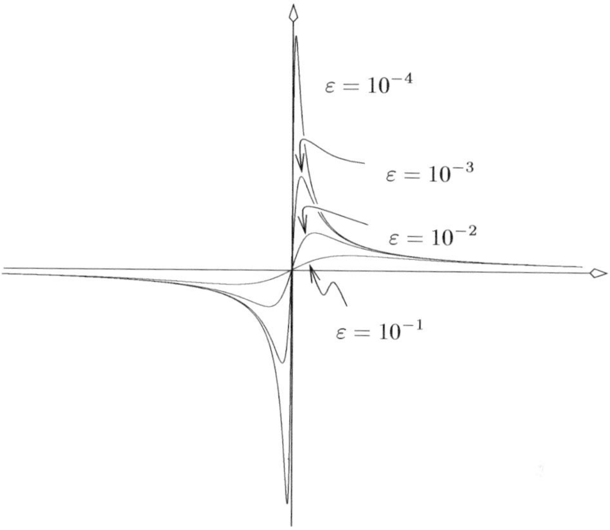

Fig. 11.4. Sequence of functions $y = \frac{x}{x^2+\varepsilon^2}$ that tend toward the principal-part generalised function $y = \sigma(x)$. The functions are all odd, intersecting at the origin. The generalised function $y = \sigma(x)$ is identical to $1/x$ for $x \neq 0$, but continuous at $x = 0$ with $\sigma(0) = 0$.

splitting it into two integrals. Being odd, $\sigma(x)$ vanishes through cancellation by a sign change, and only the even $\delta(x)$ survives:

$$\begin{aligned}\int_{-\infty}^{\infty} \mathrm{e}^{\mathrm{i}kx}\,\mathrm{d}k &= \int_{-\infty}^{0} \mathrm{e}^{\mathrm{i}kx}\,\mathrm{d}k + \int_{0}^{\infty} \mathrm{e}^{\mathrm{i}kx}\,\mathrm{d}k \\ &= \int_{0}^{\infty} \mathrm{e}^{-\mathrm{i}kx}\,\mathrm{d}k + \int_{0}^{\infty} \mathrm{e}^{\mathrm{i}kx}\,\mathrm{d}k \\ &= \pi\,\delta(-x) + \mathrm{i}\,\sigma(-x) + \pi\,\delta(x) + \mathrm{i}\,\sigma(x) \\ &= 2\pi\,\delta(x)\,, \quad \text{as expected.}\end{aligned} \tag{11.49}$$

To use the principal part in evaluating the Poisson Green function (11.20), where the integral over k is done first, begin by writing

$$\begin{aligned}\int_{0}^{\infty} \mathrm{e}^{\mathrm{i}kX\cos\theta}\,\mathrm{d}k &= \pi\,\delta(X\cos\theta) + \mathrm{i}\,\sigma(X\cos\theta) \\ &= \frac{1}{X}\left[\pi\,\delta(\cos\theta) + \mathrm{i}\,\sigma(\cos\theta)\right],\end{aligned} \tag{11.50}$$

where the identities used in the last line are well known for the delta function and obvious for the principal part. The original integral (11.20) then becomes

$$G(\boldsymbol{x}, \boldsymbol{x}') = \frac{-1}{4\pi^2 X}\int_{0}^{\pi} \mathrm{d}\theta\ \sin\theta\left[\pi\,\delta(\cos\theta) + \mathrm{i}\,\sigma(\cos\theta)\right]$$

$$= \frac{-1}{4\pi^2 X} \left[\pi \int_0^{\pi} \sin\theta\, \delta(\cos\theta)\, d\theta + i\, \mathrm{PV} \int_0^{\pi} \tan\theta\, d\theta \right]. \tag{11.51}$$

The delta function integral equals 1 (use a change of variable $u = \cos\theta$), while the principal value equals zero, since the tangent function is odd around $\theta = \pi/2$. Thus the Green function is

$$G(\boldsymbol{x}, \boldsymbol{x}') = \frac{-1}{4\pi X}, \tag{11.52}$$

just as we found earlier in (11.13) and (11.22). So the principal-part generalised function is very useful in that it can handle convergence subtleties, and enable double integrations like (11.20) to be performed without regard to integration order.

11.3 Solving Maxwell's Equations via the Green Function Approach

Now that we have some practice in the complex integrals and convergence subtleties of Green functions, along with calculating the gravitational field for an arbitrary mass distribution, it's time to tackle the harder problem of solving Maxwell's equations for an arbitrary charge distribution.

Back in Chap. 6 when discussing the Lorentz transform, we saw that electric and magnetic fields could be unified once the proper-vector nature of the charge–current density $j^{\alpha}(t, \boldsymbol{x})$ was recognised. This led to the charge–current density becoming part of the lagrangian density for the interacting electromagnetic field, equation (10.104). The resulting field equations were those of Maxwell (10.110), as required or expected. We wish now to solve these equations for a general source $j^{\alpha}(t, \boldsymbol{x})$.

Begin with (10.113) and (10.115), and remember that these two equations really contain the whole set of four Maxwell equations because of the identities $\boldsymbol{E} = -\nabla\Phi - \partial\boldsymbol{A}/\partial t$ and $\boldsymbol{B} = \nabla\times\boldsymbol{A}$.

$$\begin{aligned} -\nabla^2\Phi - \frac{\partial}{\partial t}\nabla\cdot\boldsymbol{A} &= \frac{\varrho}{\varepsilon_0}, \\ \frac{\partial^2\boldsymbol{A}}{\partial t^2} - \nabla^2\boldsymbol{A} + \nabla\frac{\partial\Phi}{\partial t} + \nabla\left(\nabla\cdot\boldsymbol{A}\right) &= \frac{\boldsymbol{j}}{\varepsilon_0}. \end{aligned} \tag{11.53}$$

Note that in the following discussion, ∇^2 will be the 3-space laplacian. Although expressions such as (8.103) and (8.225) show that ∇^2 is a sufficient notation for the laplacian in *any* number of dimensions, we will reserve it here for three dimensions to allow the following analysis to be compared with electromagnetism texts, which tend to use ∇^2 to mean only the 3-space laplacian.

Working in the Lorenz gauge ($\nabla\cdot\boldsymbol{A} = -\partial\Phi/\partial t$, Sect. 10.7.1) simplifies these and allows them to be combined into one equation in cartesians:

$$\left(\partial_t^2 - \nabla^2\right) A^\alpha = j^\alpha/\varepsilon_0\,, \quad \text{i.e.}\ \ \eta^{\mu\nu} A^\alpha{}_{,\mu\nu} = \eta_{tt}\, j^\alpha/\varepsilon_0\,. \tag{11.54}$$

The elegance of the tensor symbolism becomes apparent here: the component equations in (11.54) for the scalar and vector potentials both have the same form. In arbitrary coordinates, (11.54) becomes

$$g^{\mu\nu} A^\alpha{}_{;\mu\nu} = j^\alpha/\varepsilon_0 \operatorname{sgn} g_{00}\,, \tag{11.55}$$

but we'll restrict ourselves to cartesian coordinates only, and will use a Green function approach to solve (11.54). First, define *position* vectors

$$\vec{x} \equiv (t, x, y, z) \equiv (t, \boldsymbol{x})\,, \quad \vec{x}' \equiv (t', x', y', z') \equiv (t, \boldsymbol{x}')\,. \tag{11.56}$$

We are not Lorentz-transforming here; $\vec{x}$ and $\vec{x}'$ are sets of coordinates in the same frame. Despite the arrow, $\vec{x}$ and $\vec{x}'$ are merely position vectors and not proper vectors; we denote them with an arrow only to show that they use all four spacetime indices.

Next, electromagnetism texts will often write the four-dimensional laplacian $\partial_t^2 - \nabla^2$ as $\Box$ (or even $\Box^2$), calling it the *d'Alembertian.* For comparison with those texts, we will do likewise, noting that $\Box$ will always differentiate with respect to unprimed coordinates, and we'll write it as $\Box_{\vec{x}}$ as a reminder.

Last, the Green function approach gives the particular integral only. For the full solution, we need to add the complementary function, which will describe any field that has its source beyond the region of interest—such as waves that are just passing through, as it were.

We can see this in another way as follows. Equation (11.54) can be written as

$$\Box_{\vec{x}} A^\alpha = j^\alpha/\varepsilon_0\,. \tag{11.57}$$

Suppose we make a Lorentz transform to new, barred, coordinates. We learnt back in Sect. 6.4 that the charge–current density j^α is a vector. Also, the d'Alembertian is just the contraction over the μ, ν in (11.54), and so is coordinate-independent: $\Box_{\vec{x}} = \Box_{\vec{\bar{x}}} \equiv \Box$. (Alternatively, it's trivial to show that $\partial_{\bar{t}}^2 - \partial_{\bar{x}}^2 - \dots = \partial_t^2 - \partial_x^2 - \dots$ by using the Lorentz transform.) Thus,

$$\begin{aligned} \Box A^\alpha &= j^\alpha/\varepsilon_0 = \Lambda^\alpha_{\bar{\beta}}\, j^{\bar{\beta}}/\varepsilon_0 = \Lambda^\alpha_{\bar{\beta}}\, \Box A^{\bar{\beta}} \\ &= \Box\left(\Lambda^\alpha_{\bar{\beta}}\, A^{\bar{\beta}}\right), \end{aligned}$$

$$\text{so that} \quad \Box\left(A^\alpha - \Lambda^\alpha_{\bar{\beta}}\, A^{\bar{\beta}}\right) = 0\,. \tag{11.58}$$

Calling the term in parentheses ξ^α, we can only conclude that $A^\alpha = \Lambda^\alpha_{\bar{\beta}}\, A^{\bar{\beta}} + \xi^\alpha$, where $\Box\xi^\alpha = 0$. That is, ξ^α is a field with no source in the region of interest; i.e., the complementary function that was already inherent in (11.57). We can see here that the electromagnetic potential A^α only transforms as a vector if these extra source-free terms are excluded.

In what follows, we only deal with the particular integral as if it were the whole solution of Maxwell's equations, all the while remembering the omitted complementary function. Of course, since the field of the complementary function is just due to charges that are a great distance away, we can always include it in the particular integral by simply widening the region of integration.

In analogy with the discussion of solving Poisson's equation at the start of this chapter, we can introduce the Green function for the d'Alembertian by inverting (11.54). (And, in the equations that follow, all integrals lacking explicit limits are understood to run from $-\infty$ to ∞.)

$$\begin{aligned} A^\alpha(t,\boldsymbol{x}) &\equiv \Box_{\vec{x}}^{-1} \frac{j^\alpha(\vec{x})}{\varepsilon_0} = \Box_{\vec{x}}^{-1} \int \mathrm{d}^4x' \, \frac{j^\alpha(\vec{x}\,')}{\varepsilon_0} \, \delta(\vec{x}-\vec{x}\,') \\ &= \int \mathrm{d}^4x' \, \frac{j^\alpha(\vec{x}\,')}{\varepsilon_0} \times \underbrace{\Box_{\vec{x}}^{-1} \delta(\vec{x}-\vec{x}\,')}_{\equiv G(\vec{x},\vec{x}\,'),\ \text{the Green function for } \Box_{\vec{x}}} . \end{aligned} \tag{11.59}$$

Now the task has been reduced to finding G. Again, transform to Fourier space by defining a new function g with the four-dimensional version of (11.16). We will make use of new variables $\vec{k} \equiv (\omega, \boldsymbol{k})$ (a frequency–wavenumber vector, but this is not really important). For conciseness, write

$$\vec{k}\cdot\vec{x} \equiv \omega t - \boldsymbol{k}\cdot\boldsymbol{x} . \tag{11.60}$$

This is just notation since $\vec{x}$ is not a four-vector, although it does resemble the Minkowski dot product referred to on p. 304. In analogy with (11.16), g is defined by

$$G(\vec{x},\vec{x}\,') = \int \mathrm{e}^{\mathrm{i}\vec{k}\cdot\vec{x}} g(\vec{k},\vec{x}\,') \, \mathrm{d}^4k . \tag{11.61}$$

Similar to (11.17), use (11.61) to write $\Box_{\vec{x}} \, G(\vec{x},\vec{x}\,') = \delta(\vec{x}-\vec{x}\,')$ as (remember that $k \equiv |\boldsymbol{k}|$)

$$\int -\left(\omega^2 - k^2\right) \mathrm{e}^{\mathrm{i}\vec{k}\cdot\vec{x}} g(\vec{k},\vec{x}\,') \, \mathrm{d}^4k = \delta(\vec{x}-\vec{x}\,') . \tag{11.62}$$

Inverse Fourier transforming (11.62) produces the analogue to (11.18):

$$-\left(\omega^2 - k^2\right) g(\vec{k},\vec{x}\,') = \frac{\mathrm{e}^{-\mathrm{i}\vec{k}\cdot\vec{x}\,'}}{(2\pi)^4} . \tag{11.63}$$

Substituting g back into (11.61) gives the analogue to (11.19):

$$G(\vec{x},\vec{x}\,') = \frac{-1}{(2\pi)^4} \int \frac{\mathrm{e}^{\mathrm{i}\vec{k}\cdot(\vec{x}-\vec{x}\,')}}{\omega^2 - k^2} \, \mathrm{d}^4k . \tag{11.64}$$

Finally, it's customary to separate this integral into integrations over frequency and wavenumber:

$$G(\vec{x}, \vec{x}') = \frac{-1}{(2\pi)^4} \int \mathrm{d}^3 k \, \mathrm{e}^{-\mathrm{i}\boldsymbol{k}\cdot(\boldsymbol{x}-\boldsymbol{x}')} \int \mathrm{d}\omega \, \frac{\mathrm{e}^{\mathrm{i}\omega(t-t')}}{\omega^2 - k^2} \,. \tag{11.65}$$

Once this integral is evaluated, G can be reinserted into (11.59) to determine the field A^α, and Maxwell's equations will then be solved.

The Integral over ω in (11.65)

The first integral to tackle is that over ω:

$$I \equiv \int_{-\infty}^{\infty} \mathrm{d}\omega \, \frac{\mathrm{e}^{\mathrm{i}\omega(t-t')}}{\omega^2 - k^2} \,. \tag{11.66}$$

This has two singularities at $\omega = \pm k$. Thus I diverges, and a real problem has arisen. Our functions have become just too singular, so that a Fourier analysis approach is losing its applicability. Although the generalised functions of Fourier analysis can handle some non-square-integrable functions, for more general situations they appear to be insufficient.

Conventionally, the poles in (11.66) are avoided by using the contours of Fig. 11.3, which then gives two solutions to Maxwell's equations. But this cannot make any sense because I diverges, and no amount of careful contouring can rescue a divergent integral. It also goes completely against our Golden Rule of how not to do a complex integral on p. 451.

Luckily, the problem can be overcome through a physical approach. Suppose that we temporarily perturb Maxwell's equations in some physically reasonable way that can be reduced to zero when required. The equations as written in (11.54) are time-symmetric, implying that the solutions should not only include waves radiating out from a point source, but also waves radiating in from infinity to that source. In the real world, waves don't seem to do this, and this is a clue that we might experiment by adding a dissipative term to Maxwell's equations; perhaps then (11.66) will be well defined. Dissipation is not a fundamental part of Maxwell's theory, so we must arrange for it to vanish in the limit once the equations are solved.

Add, then, a damping term to (11.54). We take our cue from the case of damped harmonic motion in classical mechanics, where the effect of damping, for example, a mass suspended from a spring and moving in an oil bath, is incorporated by adding a velocity term to its equation of motion. For Maxwell's equations, the corresponding "velocity" term is a first derivative of A^α with respect to time, with some weight λ:

$$(\partial_t^2 + \lambda\partial_t - \nabla^2) A^\alpha = j^\alpha/\varepsilon_0 \,. \tag{11.67}$$

Consequently, (11.59) becomes, in the zero-damping limit,

$$A^\alpha(t, \boldsymbol{x}) = \lim_{\lambda\to 0} \int \mathrm{d}^4 x' \, \frac{j^\alpha(\vec{x}')}{\varepsilon_0} \times \underbrace{(\partial_t^2 + \lambda\partial_t - \nabla^2)^{-1} \delta(\vec{x} - \vec{x}')}_{\overset{\lambda\to 0}{=\!=\!=} G(\vec{x},\vec{x}'),\ \text{our new Green function}} \,. \tag{11.68}$$

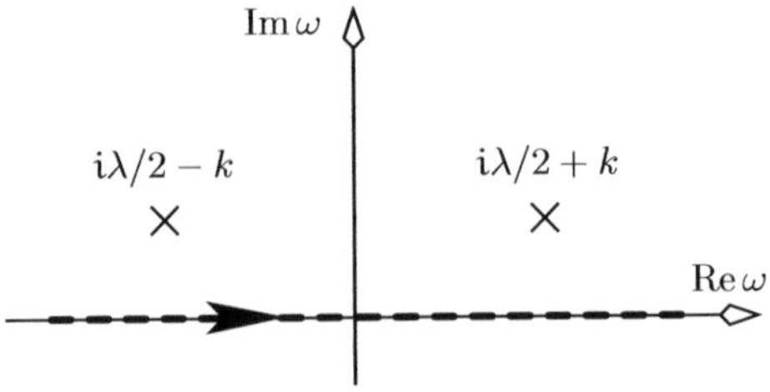

Fig. 11.5. Location of the poles of (11.70) for $\lambda > 0$.

Proceeding as in the discussion following (11.59), the new Green function becomes, in analogy with (11.65),

$$G(\vec{x}, \vec{x}') = \frac{-1}{(2\pi)^4} \int \mathrm{d}^3k \, \mathrm{e}^{-\mathrm{i}\boldsymbol{k}\cdot(\boldsymbol{x}-\boldsymbol{x}')} \lim_{\lambda\to 0} \int \mathrm{d}\omega \, \frac{\mathrm{e}^{\mathrm{i}\omega(t-t')}}{\omega^2 - k^2 - \mathrm{i}\lambda\omega} . \tag{11.69}$$

Compare this with (11.65). The dissipation has pushed the singularities off the real axis in the complex ω-plane, rendering the integrals well defined. The sign of λ should be positive to model dissipation—the expected "real-world" effect. A negative λ is unphysical in the sense that the opposite of dissipation would cause runaway effects not seen in Nature. We will evaluate the integral over ω in (11.69) for both signs of λ. Corresponding to these two signs are the integrals $I_\pm$ in the new version of (11.66):

$$I_\pm \equiv \lim_{\lambda\to 0^\pm} \int \mathrm{d}\omega \, \frac{\mathrm{e}^{\mathrm{i}\omega(t-t')}}{\omega^2 - k^2 - \mathrm{i}\lambda\omega} . \tag{11.70}$$

The integrand's singularities occur at

$$\omega = \mathrm{i}\lambda/2 \pm \sqrt{k^2 - \lambda^2/4} . \tag{11.71}$$

To be strictly rigorous, we should consider separate cases where k^2 is either larger or smaller than $\lambda^2/4$, especially because these choices turn out to affect the large-time behaviour of the equations. But the discussion is greatly simplified if we assume that the limit $\lambda \to 0$ is ultimately dominant; this simplifies the singularities to

$$\omega = \mathrm{i}\lambda/2 \pm k . \tag{11.72}$$

When $\lambda > 0$ (real physical dissipation), the poles are as drawn in Fig. 11.5. Evaluate the integral in the usual way by changing the integration path to a semicircle in the complex ω-plane. For $t - t' > 0$, Jordan's lemma (p. 450) dictates closing the real axis contour with a semicircle above the two poles, whose contribution vanishes as usual, as its radius tends to infinity. The residues must then be calculated, but they are straightforward because the poles are simple; see (11.29) and (11.30). For the $t - t' < 0$ case, the vanishing-contribution semicircle is below the real axis; this leads to the trivial result of zero since no poles are then included. Both of these time regimes can be incorporated into one expression with a step function:[1]

$$I_+ = -\theta(t - t')\,\frac{2\pi}{k}\,\sin k(t - t')\,, \tag{11.74}$$

where, as noted previously, we have assumed that the limit $\lambda \to 0$ dominates the large-time behaviour.

A similar calculation can be done for I_- ($\lambda < 0$)—which must be remembered as being unphysical in that it implies amplification, the opposite of damping. The final result is

$$I_- = \theta(t' - t)\,\frac{2\pi}{k}\,\sin k(t - t')\,. \tag{11.75}$$

Returning to (11.69) and (11.70), we see that I_+ and I_- lead to two values $G_\pm$ of the Green function:

$$G_\pm \equiv \frac{-1}{(2\pi)^4}\int \mathrm{d}^3k\;\mathrm{e}^{-\mathrm{i}\boldsymbol{k}\cdot(\boldsymbol{x}-\boldsymbol{x}')}\,I_\pm\,. \tag{11.76}$$

This integral can be evaluated by following the same procedure that was used for (11.19). Switch to $\boldsymbol{k}$-space polar coordinates (k, θ, ϕ) with the $\boldsymbol{k}$-space "z-axis" parallel to $\boldsymbol{x} - \boldsymbol{x}'$. Also set $X \equiv |\boldsymbol{x} - \boldsymbol{x}'|$, to give

$$\begin{aligned} G_\pm &= \frac{-1}{8\pi^3}\int_0^\infty \mathrm{d}k\;k^2\,I_\pm \int_0^\pi \mathrm{d}\theta\;\sin\theta\,\mathrm{e}^{-\mathrm{i}kX\cos\theta} \\ &= \frac{\pm\theta\big(\pm(t-t')\big)}{2\pi^2 X}\int_0^\infty \mathrm{d}k\;\sin k(t-t')\sin kX\,. \end{aligned} \tag{11.77}$$

The integral over k is a generalised function, and can be evaluated using the real parts of two complex exponentials. A general identity is

$$\int_0^\infty \mathrm{d}x\;\sin ax\;\sin bx = \frac{-1}{2}\int_0^\infty \mathrm{d}x\;\left[\cos(a+b)x - \cos(a-b)x\right]$$

[1] The step function $\theta(x)$ is defined by

$$\theta(x) \equiv \begin{cases} 0 & \text{for } x < 0 \\ 1 & \text{for } x > 0 \end{cases}\,. \tag{11.73}$$

Ascribing a value to $\theta(0)$ is a little problematical, and it may or may not be useful to any specific instance. Setting $\theta(0) \equiv 1/2$ is a good choice if required in a physical problem, since this is the value produced if $\theta(x)$ is expanded as a Fourier series. This allows the function to more meaningfully represent physical quantities.

$$= \frac{-1}{2} \operatorname{Re} \int_0^\infty \mathrm{d}x \, \left[\mathrm{e}^{\mathrm{i}(a+b)x} - \mathrm{e}^{\mathrm{i}(a-b)x} \right]$$

$$= \frac{-\pi}{2} \left[\delta(a+b) - \delta(a-b) \right] , \tag{11.78}$$

with the last result following from either (2.174) or (11.46). Finally, the Green functions are

$$G_\pm = \frac{\delta(t - t' \mp X)}{4\pi X} , \tag{11.79}$$

where the step functions have eliminated each of the deltas of (11.78) in the appropriate time regimes. Substituting $G_\pm$ into (11.59) gives two potentials $A^\alpha_\pm$:

$$A^\alpha_\pm(t, \boldsymbol{x}) = \int \mathrm{d}^3x' \int \mathrm{d}t' \; \frac{j^\alpha(t', \boldsymbol{x}')}{\varepsilon_0} \, \frac{\delta(t - t' \mp X)}{4\pi X} . \tag{11.80}$$

The delta functions render the time integrals easy, and (11.80) reduces to

$$A^\alpha_\pm(t, \boldsymbol{x}) = \frac{1}{4\pi\varepsilon_0} \int \mathrm{d}^3x' \; \frac{j^\alpha(t \mp |\boldsymbol{x} - \boldsymbol{x}'|, \boldsymbol{x}')}{|\boldsymbol{x} - \boldsymbol{x}'|} . \tag{11.81}$$

The two solutions A^α_+ and A^α_- are the celebrated retarded and advanced solutions, respectively, of Maxwell's equations, corresponding to the physical and unphysical regimes of λ. They have these names because of the way in which they allow the field at one point in spacetime to be influenced by that at another, as shown in Fig. 11.6. In the retarded case, the field at $(t, \boldsymbol{x})$ is determined by all events $(t', \boldsymbol{x}')$ that are separated in space from $\boldsymbol{x}$ at just such a distance that they can be connected by a light ray that travels forward in time from $(t', \boldsymbol{x}')$ to $(t, \boldsymbol{x})$. This shows that the field is not determined instantaneously at any event, but is composed of all of the effects that reach it from all points on the event's "past light cone", where these effects travel at the speed of light. The electromagnetic field adjusts itself as charges move, by propagating their influence outward in space and forward in time at the speed of light. This behaviour is entirely reasonable and accords with the Special Relativistic notion of causality.

The advanced solution has been problematic historically. It implies that the field at any event $(t, \boldsymbol{x})$ is determined by all *future* events $(t', \boldsymbol{x}')$ that lie on its "future light cone", whose influence propagates backward in time to arrive at $(t, \boldsymbol{x})$. This seems to be highly unphysical, and the advanced solution is usually discarded when solving Maxwell's equations—although its value has been discussed extensively ever since it was first discovered. But we can see here a reason why it might well be considered unphysical. It is, after all, the limiting solution of a very unphysical set of equations: Maxwell's equations with a term that gives the opposite effect of damping. Nevertheless, Feynman and Wheeler researched solutions comprising half-advanced,

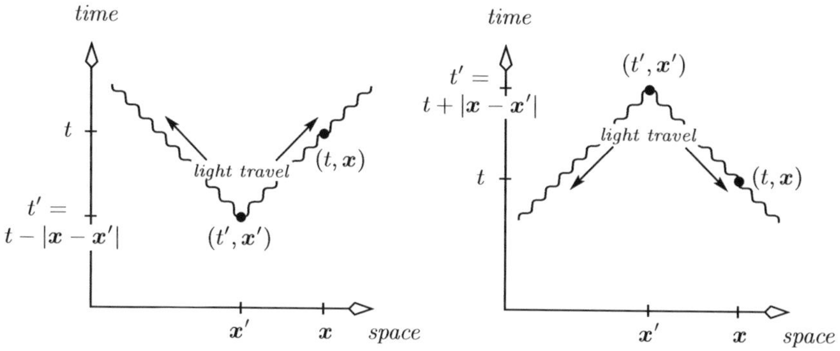

Fig. 11.6. Left: The retarded field $A_+^\alpha(t, \boldsymbol{x})$ is determined by any accelerations of the charges at all $(t', \boldsymbol{x}')$ on the past light cone of $(t, \boldsymbol{x})$. Thus we must have knowledge of the past in order to apply (11.81). The shaken charge at $(t', \boldsymbol{x}')$ sends out light that influences all events on its future light cone as shown. This light is perceived by us as "ripples" expanding outward from the charge that created them, like rings on a pond when a stone hits the water. **Right**: Likewise, the advanced field $A_-^\alpha(t, \boldsymbol{x})$ is determined by any accelerating charges at all events $(t', \boldsymbol{x}')$ on the future light cone of $(t, \boldsymbol{x})$. This requires knowledge of the *future* to apply (11.81). In some sense, the shaken charge at $(t', \boldsymbol{x}')$ sends light backward in time, which influences all events on its *past* light cone, again as shown. This light would presumably be seen by us as ripples that contract and converge onto the charge that created them, although nothing of the kind has ever been observed.

half-retarded expressions, from which they derived some novel results in electrodynamics. They concluded that the advanced solution needn't present the difficulties that one might expect.

Maxwell's theory of electromagnetism has a long history of being applied to calculations of the radiation emitted by an accelerating charge, but with mixed success. Some scenarios that are otherwise quite valid and well behaved give rise to nonphysical solutions; causality can also be violated on timescales of the order of 10^{-23} seconds. In fact, including a suitably damped advanced solution can restore order to some of the calculations, rendering them mathematically convergent. But in Chap. 7, and specifically Fig. 7.4, we saw the paradoxes that can arise by taking too simplistic a model of an accelerating charge. A charge that accelerates uniformly forever may be easy to deal with, but its worldline becomes like that of the $\bar{S}$ observer in Fig. 7.4, and problems of causality are bound to occur. Even a subject as apparently straightforward as using Maxwell's equations to calculate the radiation produced by a single accelerating charge is fraught with difficulty!

Normally, in the absence of a derivation using a vanishing damping factor, the two solutions of Maxwell's equations are tied to boundary conditions in spacetime. In the retarded case, the motion of a charge here and now affects the behaviour of all the charges in the universe at spatial and temporal infinity, which is expected and reasonable. But in the advanced case where a wave coming in from spatial infinity and temporal minus-infinity is required to converge on a charge just as that charge accelerates, there is an implication of a careful act of setting up being necessary for the motions of the multitude of charges at spatial infinity and temporal minus-infinity, in order to achieve the observed effect. This setting up of far-away charges would seem to have to work in such a way as to produce just the right incoming wave, which converges on the charge at just the right moment. This intuitive notion of a possible problem in getting boundary conditions just right does, however, use the retarded solution as part of its argument, which is not necessarily a reasonable thing to do.

11.4 Variations on the Green Function Solution of Maxwell's Equations

The approach we have discussed here differs somewhat from the more usual method of solving Maxwell's equations. There, the Fourier approach is used without adding any damping term, and the divergent integral (11.64) is tackled directly via (11.65). But we know that (11.64) is divergent, because it has singularities in its domain. No amount of complex integration theory can mend an integral that was broken from the start: a singularity is a singularity. Part of the problem is that the basic potentials of electromagnetism are not square-integrable, and so are not always amenable to Fourier analysis. The same could be said of the newtonian gravitational potential; certainly we encountered difficulties in evaluating (11.20) for Poisson's equation describing the gravity due to a static charge, but the generalised functions that came to our aid turned out to be sufficient for the job. But those same functions seem not to be sufficient for the more complicated advanced and retarded solutions of Maxwell's equations.

In spite of this, various approaches are conventionally taken to solve Maxwell's equations using a Fourier approach to arrive at (11.65), and we'll outline three of them here. The first (the principal-value approach) is an attempt to rescue something from the divergent integrals, while the two other approaches shift the contour slightly to move it away from the poles.

Principal-Value Approach

There is no a priori reason to use a principal-value approach, in contrast to the $\sin x/x$ case that we considered earlier, where in (11.34) we deliberately introduced a sort of "helper" integral, $\int \cos x/x \, \mathrm{d}x$, whose principal value was zero. Introducing this principal value was purely a means to an end, in that it enabled us to use the Cauchy residue theorem.

However, if we do attempt to calculate the principal value of I in (11.66), then we'll close the contours in the complex ω-plane again with a large semicircle, while avoiding the poles with small semicircles whose radii will tend toward zero. Omitting the details, the principal value of I turns out to be

$$\text{PV}\, I = \frac{-\pi}{k} \sin k|t - t'| \,, \tag{11.82}$$

while the Green function becomes

$$G(\vec{x}, \vec{x}') = \frac{-1}{8\pi X}\left[\delta\big(|t - t'| + X\big) - \delta\big(|t - t'| - X\big)\right]. \tag{11.83}$$

The final result for the field A^α turns out to be the average of the advanced and retarded solutions. This "half advanced, half retarded" expression found a use historically in the study of electrodynamics, as a way of investigating difficulties in Maxwell's theory having to do with why accelerating charges radiate. But we should realise that because the advanced and retarded solutions of Maxwell's equations are both particular integrals (i.e. solutions that are completely determined by the physical charge distribution j^α), we have no a priori freedom to simply add them. And in terms of actually solving Maxwell's equations, it should be borne in mind that this solution has resulted from the use of a principal value, for which there is really no mathematical justification.

Shifting the Contour

As we discussed after (11.66), the divergent integral I in that equation is usually given some kind of meaning using the contours of Fig. 11.3, where the separation of the straight paths from the real axis tends toward zero.

In fact, the effect of running the integration path above the real-ω axis is similar to introducing a negative damping term as we have done, because the negative damping term pushes the poles down below that axis. Shifting the poles down *via physical reasoning* is legitimate because it keeps the integral along the real-ω axis well defined. But, in contrast, running the integration path above the axis doesn't change the fact that any integral that includes a nonremovable singularity is undefined from the outset. So the paths of Fig. 11.3 cannot fix the fact that the integral I in (11.66) is undefined.

Because running the integration path above the real-ω axis ultimately has the same effect as introducing the negative damping term—all arguments about validity of the integrals aside—what results is the advanced solution that we found in (11.81). Similarly, running the path below the axis is akin to introducing a positive damping, and the retarded solution certainly results. So this approach gives the right answers with an incorrect method; and because of its no-fuss approach, it has become the accepted way of solving Maxwell's equations using Green functions. But a quick arrival at an answer that looks good in hindsight should never be taken to justify the method used, and the continued use of these contours only makes it difficult for a new generation of physicists to understand the mathematics of Green function theory. Perhaps the biggest criticism of the contours in Fig. 11.3 is that they give no insight into the two solutions that result, and do not explain why one solution should be physical and the other apparently not.

Shifting the Poles

Instead of shifting the contour, the poles are sometimes shifted one up and one down by a small amount. The calculation here is more complicated than previously because it involves principal parts, and we will not include it. Nevertheless, a half-advanced half-retarded solution can be produced by this approach, which is not surprising since, in essence, we have included a vanishing damping that is half positive and half negative—whatever this might mean physically. However, the solution is only one of many that can be produced that involve principal parts.

11.5 Fluctuation–Dissipation and Time's Arrow

Adding a vanishing damping that alters Maxwell's equations just enough to enable their solution via Fourier analysis suggests a consequence for the Arrow of Time. In Sects 3.5 and 10.9, we discussed the use of entropy growth in defining such an Arrow of Time. Entropy growth itself is a product of the statistical processes that play such an integral part in the physical world. These processes go hand in hand with the phenomenon of fluctuation–dissipation such as we saw in the last part of Sect. 3.2.1.

Fluctuation gives rise to entropy growth that provides an Arrow of Time, while its partner, dissipation (or damping), is what appears to single out the retarded solution of Maxwell's equations over the advanced solution. The direction of time's arrow is certainly shown by the fact that we only observe waves coming out from an oscillating charge, and never waves impinging on it from all directions that cause it to accelerate. Perhaps underlying this apparent asymmetry of the world are the forever-entangled phenomena of fluctuation and dissipation, together with the entropy growth that they produce.

12 Airliners, Black Holes, and Cosmology: The ABC of General Relativity

A defining moment in the early age of the jet airliner occurred in 1954, when the four-engined prototype of the Boeing 707 first took to the skies, flown by the expert test pilot Tex Johnston. But the inaugural flight of the "Dash-80" is probably less remembered than its most famous moment the following year. Tex was to fly the big jet over a crowd of many thousands of spectators at a fair, to which officials from major airlines had been invited in a bid by Boeing to sell the new jet transport commercially. The astonished spectators couldn't believe their eyes when Tex appeared, barrel-rolling the aircraft through a full 360°, and not once but twice. Company officials reached for their heart pills; prospective buyers reached for their cheque books.

Later, hauled over the coals, Tex couldn't understand what all the fuss was about. He was an expert pilot who could initiate and hold a $1\,g$ roll, during which the plane was under no more stress than when flying straight and level. As Tex remarked, the plane never even knew it was upside down. The Dash-80, given a voice, might have been the first to acknowledge the significance of Einstein's Equivalence Principle, which we first discussed in Chap. 7. The difficult manoeuvre combined just enough free fall (partly annulling gravity onboard) with just enough rotation about a point far beyond the body of the plane (as though it were following a spiral on the surface of a barrel, creating a centrifugal force onboard), to give an overall $1\,g$ force from the plane's point of view that was always directed perpendicular to its floor, just as if it had been at rest on the tarmac within Earth's $1\,g$ gravity.

12.1 The Equivalence Principle

The correspondence between uniform acceleration in deep space and apparent gravity forms one side of the Equivalence Principle coin. The other side, shown in Fig. 12.1, is the correspondence between free fall in a gravity field, and inertial motion. If we jump into the air, we are weightless for the whole time we're off the ground, and relative to us, other objects thrown into the air in our immediate vicinity move in straight lines at constant speed. The idea that the frame of a freely falling observer is approximately inertial gives an easy answer to an old question. A monkey sits in a tree, and a hunter who knows nothing of parabolic motion fires a dart *directly* at the monkey. The

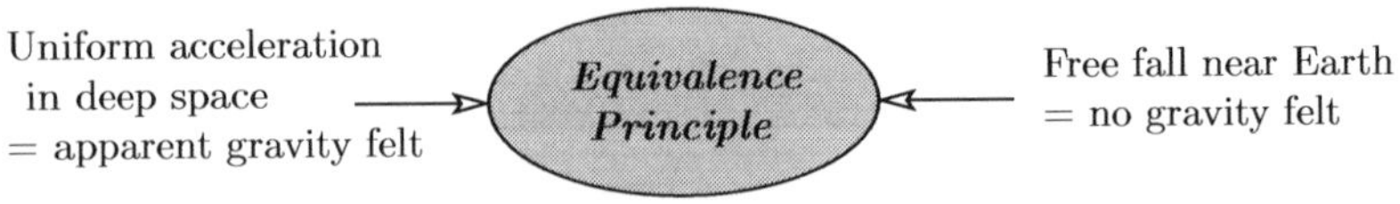

Fig. 12.1. The Equivalence Principle has two sides. One is that uniform acceleration far from any source of real gravity will be felt as an apparent gravity, "apparent" because the pseudo-potential is constant. The other side of the principle says that during free fall in a real gravity field, no gravity at all will be felt; in a small region, our frame will be inertial.

monkey sees the dart leave the gun and immediately drops to the ground to avoid it. Will the dart hit the monkey? While we can analyse the kinematics of both dart and monkey as they accelerate to the ground (the dart in a parabola and the monkey in a straight line), it's far easier to switch to the monkey's inertial frame after it drops. In this frame, the dart has a constant velocity, and since it was initially fired directly at the monkey, it will fly in a straight line in the monkey's inertial frame and indeed hit the monkey.

The inertial nature of the free fall associated with jumping in the air is routinely used to train astronauts to cope with weightlessness. An aircraft with astronauts aboard flies in an arching parabolic path, which is nothing more than free fall with a constant sideways component of the velocity that has no effect. (The plane, of course, must be powered to overcome air resistance.) During the half minute or so that it follows the parabola, including the climb, the aircraft's occupants float freely in the cabin, which is small enough to approximate an inertial frame very well. Far from the aircraft, things are not that simple. Freely falling objects on either side of Earth certainly don't measure each other as moving with constant velocity. The inertial frame attached to a freely falling body has only a limited extent.

But free fall is not an everyday activity; the ground soon rushes up to meet us catastrophically, reminding us that life seemingly is lived within an accelerated frame, not an inertial one. In Chap. 7 we looked closely at accelerated frames using the Clock Postulate, and found that in our rocket-laboratory reference frame in deep space, far from any gravity and accelerating "upward" at, say, 1 g due to a rocket engine below us, time runs at different rates at different heights. Clocks above us (i.e. higher up in the pseudo-gravitational potential that we feel) run more quickly, while those below us run more slowly, eventually slowing to a stop about one light-year below us, where there is an event horizon. So the physics of accelerated reference frames provides a glimpse into the workings of gravity, paving the way for the ideas of general relativity.

But, on first glance, the Clock Postulate might seem to contradict the Equivalence Principle. After all, if a clock's rate does not depend on its acceleration, then how can it be that it *does* depend on the strength of gravity, as verified by experiment? No, there is no conflict at all with the Equivalence

Principle. The difficulty here arises because of the confusion between acceleration and the effect of acceleration: changing velocity. It's precisely what was referred to on p. 237 when we spoke about the wind chill factor. Let's look a little more closely to see just what is happening.

Sitting on a launch pad is a rocket with no fuel and carrying two occupants, astronauts who cannot see outside and who believe they're accelerating at $1\,g$ in deep space, far from any gravity. One of the astronauts sits at the base of the rocket and the other sits at its top, and each sends a light beam to the other.

Energy conservation demands that light loses energy as it climbs up a gravitational field, so we know that the top astronaut will see a redshifted signal. Likewise, the bottom astronaut will see a blueshifted signal, because the light coming down has fallen down the gravitational well and gained some energy en route.

The astronauts believe they are accelerating in deep space, so how do they describe what is happening? The top astronaut reasons "By the time the light from the bottom astronaut reaches me, I will have gained some speed relative to my original inertial frame, so that I'll be receding from the light at a higher speed than previously as I receive it. So it should be redshifted, as indeed it is." The bottom astronaut reasons very similarly: "By the time the light from the top astronaut reaches me, I will have gained speed relative to my original inertial frame, and I'll be approaching the light at a higher speed than previously as I receive it. So I predict that it should be blueshifted; and so it is."

Despite the fact that they've started from an incorrect assumption—that they're accelerating in deep space when in fact they are really at rest in a gravitational field—the Equivalence Principle ensures that they both calculate just the right amount of red- or blueshift in the light they receive. But their analysis only used their *speed*, not their acceleration as such. So just like the wind chill factor that we spoke of earlier, applying the Equivalence Principle to the case of the rocket doesn't depend on acceleration per se, but it *does* depend on the result of acceleration: changing speeds. We discuss this further at the end of the next section, after describing the Pound–Rebka–Snider experiments that measured this predicted frequency shift.

Mach's Principle

Ideas of what can be considered absolute have never really been straightened out within the context of relativity. In his creation of the theory, Einstein was much intrigued by *Mach's Principle*: the idea that inertia—the tendency of mass to keep moving at constant velocity in an inertial frame—is due to the mass somehow "knowing" of the existence of the rest of the universe. Yet whether or how this principle finds a place within general relativity is still a matter of debate.

The absoluteness of acceleration also results from applying Ockham's Razor to experimental observations. The astronomer Jean Foucault's famous demonstration in 1851 that Earth was turning used a 67-metre-long pendulum suspended inside the vast dome of the stately Panthéon in Paris. As the pendulum slowly swung to and fro, its swing plane gradually rotated, as predicted by the idea that Earth turns within some larger inertial frame within which that plane does *not* rotate; this seems to be the frame of the distant stars. On a turning globe, the pendulum changes its plane of swing at a rate *that depends on its latitude*—the crucial point of which modern visitors to the Panthéon are not made aware. After all, why should a pendulum's swing plane *not* rotate? But if we really were to insist that Earth does not turn, then we would need to build a theory of the universe that included a new force making the pendulum change its swing plane at a rate that depends on the latitude at which it's placed. This is not something anyone bothers to do because it's far simpler to assume that there is no such force, that the pendulum maintains a fixed plane of swing relative to the distant stars, and that the world really does turn within that larger, apparently fixed and inertial frame.

12.2 The Pound–Rebka–Snider Experiments

Ideas of the Equivalence Principle and accelerated frames were put to the test in the 1960s by Pound, Rebka, and Snider in a set of experiments that used the Mössbauer Effect, a sophisticated technique that measures the energy of γ rays emitted by an ^{57}Fe source. In this case, the rays were sent from a stationary source at the bottom of a 22.5 metre tower to a stationary detector at its top. In the Mössbauer Effect, the ^{57}Fe source is actually made to vibrate, and the tiny periodic changes in its velocity alter the frequency of its emitted γ rays via a Doppler shift (by very small amounts!). Searching for a detection resonance allows the frequency of those rays to be measured after they have travelled to the detector. In the Pound–Rebka–Snider experiments, the correctly predicted frequency drop of the emitted γ rays was indeed observed after the rays had climbed up the gravitational potential.

We can predict the value of this redshift by using the idea that this experiment on Earth is equivalent to the same one performed in an accelerated frame far from any gravity. The basic idea is shown in Fig. 12.2. A clock sits on the floor of an accelerating rocket (i.e. closest to the rocket's engine), and sends light signals up to us who sit on the ceiling with an identical clock. Both clocks have been designed to "tick" (send out a light pulse) at intervals of their proper time of T. As we saw in Chap. 7, this accelerated frame can be given a global time coordinate, which we choose to be our own, identical to the time τ_{ceil} shown on the ceiling clock next to which we sit. Finally, suppose for the sake of labelling the figure that the floor clock is ageing at half the rate of the ceiling clock. We know this rate is certainly reasonable

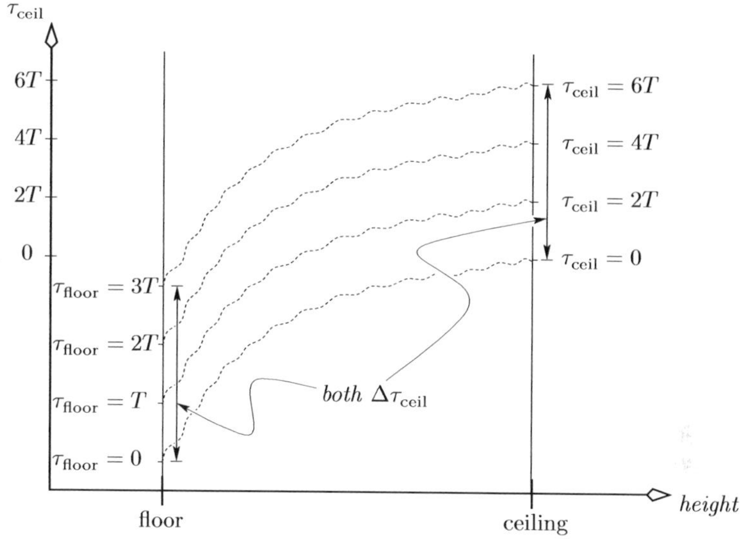

Fig. 12.2. Using the redshift of received signals to infer that clocks on the floor of a building are ageing slower than clocks on the ceiling, by appealing to an accelerated frame far from any gravity. The floor and ceiling clocks are now the floor and ceiling clocks, respectively, of an accelerating rocket. For the sake of illustration, suppose that the floor clock ages half as fast as the ceiling clock, and that the light signals are slowing as in Fig. 7.12, although their precise shape (logarithmic) isn't relevant here. We sit next to the ceiling clock and dictate a global time coordinate τ_{ceil} for the frame—which we know from Chap. 7 can always be done in a uniformly accelerated frame.

based on the work of Chap. 7. There we saw that floor clocks do indeed age more slowly in such a frame.

Everyone must agree on how many pulses were emitted (the sort of argument used in Sect. 5.1). If f_{rec} is the received frequency and f_{em} the emitted frequency, this number will be

$$\text{total pulses emitted} = f_{rec}\,\Delta\tau_{ceil} = f_{em}\,\Delta\tau_{floor}\,. \tag{12.1}$$

(In Fig. 12.2, this amounts to $3 = \frac{1}{2T} \times 6T = \frac{1}{T} \times 3T$.) The ratio of frequencies is then

$$\frac{f_{rec}}{f_{em}} = \frac{\Delta\tau_{floor}}{\Delta\tau_{ceil}} \overset{(7.19)}{=\!=} \frac{\bar{x}_{floor}}{\bar{x}_{ceil}}\,, \tag{12.2}$$

where $\bar{x}$ measures the clocks' positions in the accelerated frame relative to the horizon at $\bar{x} = 0$. Our position (the ceiling clock) is $\bar{x} = c^2/g \simeq 0.97$ light-years, while the floor clock is 22.5 metres below us. The predicted redshift is thus

$$\frac{f_{rec}}{f_{em}} = \frac{0.97\,\text{l.y.} - 22.5\,\text{m}}{0.97\,\text{l.y.}} \simeq 1 - \frac{22.5}{9.15 \times 10^{15}} \simeq 1 - 2.5 \times 10^{-15}, \tag{12.3}$$

which agrees with the Pound–Rebka–Snider results to about 1% accuracy. The Equivalence Principle works very well. Later, we'll redo this calculation to the same level of accuracy using the more accurate theory of general relativity, and will obtain the same result, in (12.56).

This calculation of the differing clock rates in an accelerated frame shows once again why a clock can be influenced by gravity yet not by acceleration, while still obeying the Equivalence Principle. In measuring the different ageings between the floor clock and ceiling clock in a uniformly accelerated frame, two effects are occurring, both of which can be seen in Fig. 7.5 as we compare all events along the common line of simultaneity in that figure. The first is that, at any time t, the ceiling clock (farther from the t-axis in the figure) is moving more slowly than the floor clock. Their velocities are only equal along the common line of simultaneity. Thus the ceiling clock is less affected by the special relativistic time-slowing γ-factor. The second effect is that the comparison of ageings is made when the ceiling clock has been moving for longer in the S-frame, allowing it to have aged even more.

These two effects add, ensuring that all observers will measure the ceiling clock to be ageing faster than the floor clock. So acceleration was not involved per se; rather, the *result* of acceleration—changing speeds—is solely responsible for the different ageings. The Pound–Rebka–Snider experiment compares clocks in a real gravitational field, but they are still just like the floor and ceiling clocks of an accelerated frame, and they age in just the same way.

12.3 A Space or Spacetime Description of Gravity?

We certainly expect that a proper description of gravity should be relativistic, given the success of the Equivalence Principle in predicting the outcome of the Pound–Rebka–Snider experiments. There is another way of seeing this, too, based on something less esoteric: the curvature of the trajectory of an arbitrary mass in both space and spacetime.

Throw a projectile and analyse its path with two space dimensions: height and horizontal distance. Figure 12.3 shows the trajectories, in both space and spacetime, of a mass that is airborne for some time t. In this time it travels some horizontal distance d, and attains some maximum height h. (None of these numbers are prior constraints; any thrown mass will do.) The "height above ground" axis points out of the page, as does the initial part of the worldline. The dashed curve is the parabolic path in space of the projectile. This is the projection of its (solid) worldline onto the plane spanned by the two space axes. Clearly, the space trajectory is a parabola and can have any amount of "squashing", depending on how the mass is originally thrown.

We'll calculate the approximate radius of curvature of the worldline by approximating it as a small piece of a large circle. It's very easy to show (using Pythagoras's theorem) that this radius is given by $r \simeq \ell^2/(8h)$, where ℓ and h

are defined in the figure. It's also easy to see, using basic newtonian kinematics, that the maximum height of the projectile will be $h \simeq gt^2/8$, where g is the acceleration due to gravity. These expressions for r and h imply that the radius of curvature r of the worldline is

$$r \simeq \frac{\ell^2}{gt^2} = \frac{d^2 + c^2t^2}{gt^2} = \frac{d^2}{gt^2} + \frac{c^2}{g} \simeq \frac{c^2}{g} \simeq 0.97 \text{ light-years near Earth.} \quad (12.4)$$

This radius, c^2/g, is now familiar to us from our accelerated frame work: it's the distance to the event horizon of the accelerated frame, the length scale beyond which that frame's coordinates begin to break down, as mentioned in (7.29). Loosely speaking, it's as if, while airborne, the mass is orbiting a point on the event horizon.

Perhaps surprisingly, the radius of curvature r of the worldline is independent of the initial conditions. It shows that while a ball and a bullet, or even two identical balls, can have completely different trajectories in space, their worldlines in *spacetime* near Earth's surface will always have a radius of curvature of about one light-year. This suggests that gravity will almost certainly look simpler when studied within spacetime as opposed to space and time separately. It also quantifies how close our laboratory frame on Earth's surface really is to being inertial: a curve with a radius of one light-year is very close to being straight over the spacetime of a typical laboratory experiment. So gravity near our planet's surface is actually *very* weak.

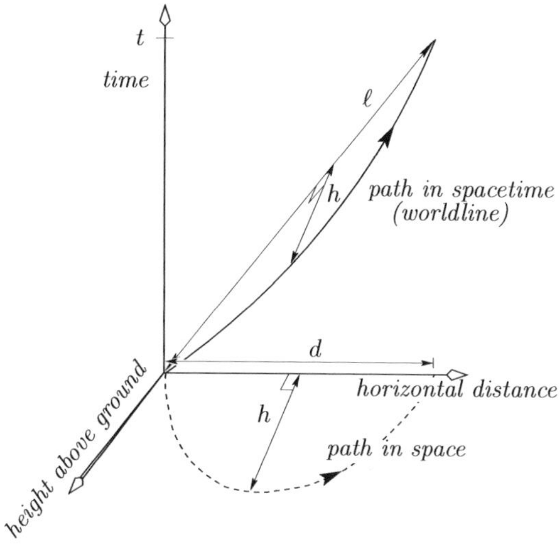

Fig. 12.3. Suppressing one horizontal space dimension, throw a mass into the air and plot its trajectory in both space and spacetime. Its space trajectory can be any shape of parabola, depending on the initial conditions. In contrast, a back-of-the-envelope calculation shows its spacetime trajectory to have an invariable curvature: the famous c^2/g, or about one light-year near Earth's surface.

12.3.1 A Route to Curved Spacetime from Lagrangian Mechanics

In previous chapters, we considered the fact that special relativity creates a single entity from space and time via the concept of a metric. But the lagrangian formalism shows that the Minkowski metric of special relativity appears to have some competition from another metric when we are near a source of gravity, in the following way.

In Sect. 10.3.3, we discussed the fact that the lagrangian that gives the correct kinematics for a simple system is its kinetic energy minus potential energy. In Sect. 10.4, we saw that a free particle in special relativity follows a path in spacetime that maximises the proper time between its start and end points. (This is just another illustration of the Twin Conundrum, since in the inertial frame of the free particle it is always at rest and so, like the stay-at-home twin, ages the most.)

Now, we know that in the absence of all forces, including gravity, a free particle follows a straight line in both space and spacetime. When gravity is present, the Equivalence Principle tells us that although the particle's path in space might not be straight as seen by an observer who feels the gravity, an inertial observer must certainly measure it to be straight over a short time interval. But the analysis of the last two pages shows that while a ball and a bullet might follow paths in space with different radii of curvature, their tracks in spacetime have the *same* radius of curvature. This suggests that it might be more useful to focus on the particle's path through spacetime. This path will also be as straight as it can be; a geodesic, but now with the spacetime having some non-Minkowski metric (i.e. $\mathrm{d}\tau^2 \neq \mathrm{d}t^2 - \mathrm{d}x^2 - \mathrm{d}y^2 - \mathrm{d}z^2$). So the particle still follows a spacetime path of both an extremal action and a maximal proper time between its start and end events. Let's explore this further. As in (10.70), the action for a single particle can be written as

$$S = \int m\left(\frac{v^2}{2} - \Phi\right)\mathrm{d}t = \int m\left(\frac{1}{2}\frac{\mathrm{d}\ell^2}{\mathrm{d}t^2} - \Phi\right)\mathrm{d}t\,. \tag{12.5}$$

We wish to take the integrand of (12.5) and relate it to the proper time $\mathrm{d}\tau$ by which the particle ages. Any other action that gives the same dynamics as (12.5) can be used; in particular we can multiply S by $-2/m$ and consider extremising

$$\int\left(2\Phi - \frac{\mathrm{d}\ell^2}{\mathrm{d}t^2}\right)\mathrm{d}t\,. \tag{12.6}$$

The potential Φ is defined up to an additive constant, so if we set $2\Phi \to 1$ in the zero-gravity limit, as well as demanding low speeds, then $2\Phi - \mathrm{d}\ell^2/\mathrm{d}t^2$ is positive. In that case it's more useful to deal with its square root, extremising

$$\int\sqrt{2\Phi - \frac{\mathrm{d}\ell^2}{\mathrm{d}t^2}}\,\mathrm{d}t\,, \quad \text{or} \quad \int\sqrt{2\Phi\,\mathrm{d}t^2 - \mathrm{d}\ell^2}\,. \tag{12.7}$$

But in the zero-gravity limit, $\mathrm{d}\tau^2 = \mathrm{d}t^2 - \mathrm{d}\ell^2$. Perhaps (12.7) suggests that in the presence of gravity, or at least the weak gravity with low particle speeds in which the laws of mechanics were developed, a more correct expression involving proper time might be

$$d\tau^2 \simeq 2\Phi\, dt^2 - d\ell^2\,, \tag{12.8}$$

which does indeed allow the elapsed proper time $\Delta\tau$ to be maximised, as required. The potential outside any spherically symmetric mass M is usually written $-GM/r$ up to an additive constant. To obtain a limit of $2\Phi \to 1$ as $r \to \infty$, we'll set $\Phi = 1/2 - GM/r$, so that the metric becomes

$$\begin{aligned} d\tau^2 &\simeq \left(1 - \frac{2GM}{r}\right) dt^2 - d\ell^2 \\ &= \left(1 - \frac{2GM}{r}\right) dt^2 - dr^2 - r^2\, d\theta^2 - r^2 \sin^2\theta\, d\phi^2\,, \end{aligned} \tag{12.9}$$

in spherical polar coordinates, defined in (9.27) and Fig. 8.3.

Equation (12.7) tells us that the right-hand side of (12.8) or (12.9) is a metric *exactly* describing newtonian gravity; it is being postulated here to approximately equal $d\tau^2$, to ensure the correct zero-gravity limit.

This result is quite remarkable: it's a description of spacetime within which the trajectories of newtonian mechanics have become geodesics. We expect the expression for proper time in (12.9) to be only approximately correct, since this equation has been derived from expressions for kinetic and potential energy that are based on our experience of the world at low speeds and weak gravity. (In fact, we'll see later that (12.9) is actually very similar to the correct metric predicted by general relativity for the same situation, known as the Schwarzschild solution.) But even so, we have been led to a new metric for spacetime that describes motion in a gravity field in a very elegant way.

Suppose that we know nothing of the lagrangian origin of (12.9), and wish to show that it predicts an inverse-square gravitational force law. We might examine the case of a freely falling particle by working through the intricacies of calculating geodesics, to show finally that the resulting motion is equivalent to the particle's being acted on by an inverse-square force. However, a shorter route is just to run the preceding analysis backward. We began with a lagrangian $L = mv^2/2 - m\Phi$ in (12.5), and decided to focus on

$$\frac{-2L}{m}\, dt^2 = 2\Phi\, dt^2 - d\ell^2. \tag{12.10}$$

We postulated this to be at least approximately $d\tau^2$, in which case

$$L \simeq \frac{-m}{2} \frac{d\tau^2}{dt^2} = \frac{-m}{2\gamma^2}\,, \tag{12.11}$$

where $\gamma \equiv dt/d\tau$, as used in special relativity. Equation (12.11) is an approximate relation that has been derived in the low-speed, low-gravity limit, to be applied to the metric (12.9) in order to generate a potential energy whose gradient will give the force on a particle. Do this by writing, from (12.9),

$$\frac{d\tau^2}{dt^2} \simeq 1 - \frac{2GM}{r} - v^2 , \tag{12.12}$$

which inserts into (12.11) to give

$$L \simeq \frac{-m}{2} + \frac{GMm}{r} + \frac{mv^2}{2} . \tag{12.13}$$

A potential energy of $m/2 - GMm/r$ can be read off from this lagrangian, whose gradient produces a "downward" force of GMm/r^2, as expected.

This force can be derived with only minor effort. The force is the negative *spatial* gradient of the (purely radial) potential energy: $-\nabla\,(m/2 - GMm/r)$, or just $-\boldsymbol{e}^r \partial_r\,(m/2 - GMm/r)$. Take care to write the spatial part of (12.9) with the correct signs; it is $d\ell^2 = dr^2 + r^2\, d\theta^2 + r^2 \sin^2\theta\, d\phi^2$, which is diagonal. For this spatial metric, $g^{rr} = +1$, so that $\boldsymbol{e}^r = \boldsymbol{e}_r = \boldsymbol{e}_{\hat{r}}$. Hence

$$\begin{aligned} \text{force} &= -\nabla\,(m/2 - GMm/r) = -\boldsymbol{e}^r \partial_r\,(m/2 - GMm/r) \\ &= -\boldsymbol{e}_{\hat{r}}\, GMm/r^2 , \end{aligned} \tag{12.14}$$

which is GMm/r^2 toward $r = 0$, or "downward".

As a side point, note that, unlike in special relativity, the particle's total energy is not simply given by $\gamma m \simeq m + GMm/r + mv^2/2$; this clearly has the wrong sign for its potential energy part.

To reiterate, (12.11) is merely a way to estimate the potential energy from a given metric in a newtonian limit. Other lagrangians could also be used. We could, for example, refer to (10.69) to experiment with the lagrangian of a relativistic free particle, $-m/\gamma$; to first order, this produces a potential energy from (12.9) of $m - GMm/r$, which still gives the expected inverse-square force. Aside from the fact that some approximation has been made in the last two pages in identifying the newtonian metric (12.9) with proper time, we see again that newtonian gravity is weak enough to allow an inverse-square force law to emerge even from the free-particle lagrangian.

We have changed our view of gravity by modelling a mass in free fall in a gravity field as following a geodesic on a spacetime endowed, at least approximately, with the metric (12.9). But is this fundamentally any different from a simple change of coordinates, such as was done when we switched to an accelerated frame in Chap. 7? To see that it is, we need only calculate the curvature associated with (12.9). The Minkowski metric has zero curvature, as does any metric calculated from it by a change of coordinates, such as the accelerated-frame metrics of Chap. 7. But for (12.9), not all of the Riemann tensor components vanish. Here is one of those components, along with the Ricci scalar (where G has been absorbed into M; i.e. GM is now written as M):

$$R_{trtr} = \frac{M}{r^3}\frac{2r-3M}{r-2M}, \qquad R = \frac{-2M^2}{r^2(r-2M)^2}. \tag{12.15}$$

Being nonzero, these show that the new metric actually describes a curved spacetime. (A zero Ricci scalar can also be attached to a curved spacetime if there are nonzero Riemann components that happen to cancel in just the right way, as we'll see occurs later for the Schwarzschild metric.)

The curvatures in (12.15) show that the metric diverges as $r \to 2M$, or $2GM/c^2$ in conventional units. The idea that this value of r is special was, in fact, first put forward by Laplace in 1795. Using basic newtonian dynamics, he found it to be the radius of a (spherically symmetric) sphere of mass M from which a particle requires the speed of light to escape to infinity. This suggested that there might exist objects that emit no light. As we'll see in Sect. 12.6, the same idea and value of r are also predicted by general relativity.

Just as a rotating Earth is not mandatory to describe why a pendulum changes its swing plane as a function of latitude (except that it simplifies the required explanation enormously), a curved spacetime is not mandatory for describing gravity—except that it points to a simple basis underlying gravitational phenomena. Curved spacetime also turns out to be very successful in making predictions that have been tested to fantastic accuracy in astronomy. The arguments of this section show that the notion of curved spacetime is quite reasonable, an idea that Einstein placed onto a very firm footing, as described in the sections to follow. As well, on p. 371 we saw that Gauss's *Theorema Egregium* releases us from any obligation to specify an appropriate embedding for a curved spacetime. This means that questions as to the "real" nature of spacetime curvature don't affect the theory as it stands.

It needs to be stressed that Rindler spacetime—the spacetime seen by an accelerated observer such as Eve in Chap. 7, and pictured in Fig. 7.12—is flat. We know that it is through calculating the Riemann tensor, which is *trivially* done. Why? Because the Riemann tensor is just that, a tensor, so that its components in Rindler coordinates [$\bar{t}, \bar{x}, \bar{y}, \bar{z}$ of (7.25)] are linear combinations of its components in any other set of coordinates. In particular, its components as calculated in Minkowski spacetime are all zero, because the Minkowski metric has no spacetime dependence (Minkowski spacetime is flat!). Thus the Riemann components in any other set of coordinates must also be zero: a spacetime that's flat for one is flat for all.

An all too common misconception of the Twin Conundrum is that its events as recorded by the accelerated observer require general relativity to be properly understood. But the spacetime of the Twin Conundrum is flat, and so general relativity is certainly not needed, being the study of *curved* spacetime. In contrast, covariant language is useful in any study of the Twin Conundrum—but covariant language on its own is not general relativity.

Einstein postulated that the metric of spacetime can always be written as a quadratic form such as in (12.9): a sum of squares of infinitesimal changes in the coordinates. (Such a metric with all signs positive is called *riemannian*, or *pseudo-riemannian* if not all of its signs are positive.) Could the universe have been created with a different metric—perhaps a sum of fourth powers? Possibly; it's a difficult question and, like discussions of Mach's Principle, involves imagining a universe quite different from our own. A quadratic spatial metric is so ingrained in our notion of geometry that it's difficult to imagine how we might evolve in, or perceive, a world in which Pythagoras's theorem involved fourth powers. But the fact that the Lorentz transform embodies an invariance involving a sum of squares that gives rise to the Minkowski metric, whether a happy fluke or otherwise, allows us to use the Equivalence Principle to geometrise space and time in a way that's close to our experience, by being able to write an arbitrary spacetime metric as a sum of squares, too. The situation has a chicken-or-egg character about it. If spacetime is inherently geometric, then it quite naturally admits a metric. Conversely, if a metric can be inferred or formed from a fortuitous invariance of the Lorentz transform, then we are able to geometrise space and time to "make" spacetime. Which came first, real geometry or fortuitous invariance, is not clear at all.

Why have we only allowed a *gravitational* potential to be present in the discussion of the last few pages, as opposed, for example, to an electric potential? The answer is because we know that although a free mass in a gravitational field will follow a spacetime geodesic, a free mass carrying *charge* in an electric field will be deflected from that geodesic, so that proper time will not be extremised on its spacetime path, even though the action *is* extremised for this path. So for a general nongravitational force, extremising the action does not correspond to extremising proper time; but when gravity alone is present, both proper time and the action are extremised together.

In fact, if a primed-coordinate frame can be defined in which a clock is at rest, leading to an effective gravitational potential Φ_{eff} that includes the effect of any pseudo-forces (which are an artifact of the coordinates chosen, as opposed to being real like an electric field), then (12.8) becomes $d\tau^2 \simeq 2\Phi_{\text{eff}}\, dt'^2$. Thus, for weak gravity, the rate of ageing of a clock is $d\tau/dt' \simeq \sqrt{2\Phi_{\text{eff}}}$, which depends on the effective potential alone. More conventionally, remember that we shifted the potential by 1/2 on p. 479 to allow the appropriate large-r limit needed in that discussion. Writing the conventional potential (vanishing at infinity) with a tilde so that $\Phi_{\text{eff}} = 1/2 + \widetilde{\Phi}_{\text{eff}}$, the rate of ageing of a clock becomes

$$\frac{d\tau}{dt'} \simeq \sqrt{2\Phi_{\text{eff}}} = \sqrt{1 + 2\widetilde{\Phi}_{\text{eff}}} \simeq 1 + \widetilde{\Phi}_{\text{eff}} < 1\,. \tag{12.16}$$

We have here the prediction that clock rates are everywhere equal on an equipotential surface (or at least approximately so), and ageing more slowly than those at spatial infinity by a factor of $1 + \widetilde{\Phi}_{\text{eff}}$. It has been verified experimentally to a very high accuracy for clocks at rest on the nonspherical rotating Earth, whose surface does approximately follow an equipotential.

Such a prediction is also reasonable when referred to the discussion of Sect. 12.2. After all, if two clocks are at rest in different places on the same equipotential surface and a photon is sent from one clock to the other, then energy conservation demands that the photon's frequency be measured as unchanged on arrival. This implies that the two clocks tick at *exactly* the same rate. We'll meet this idea again when using the Schwarzschild metric to calculate clock rates in Sect. 12.5.1.

Einstein's geometric description of spacetime uses a metric based on the presence of gravity alone; other forces, such as electromagnetism, only affect spacetime curvature insofar as their energy density creates gravity (as we'll investigate further in Sect. 12.4). The use of an extremal action in classical mechanics—as opposed to the extremal proper time of spacetime geodesics—is what differentiates the kinematics of, e.g., charged particles from the kinematics of masses that only respond to spacetime curvature. Einstein did attempt a unification of other forces with gravity, but never fully succeeded in his programme.

Finally, the Twin Conundrum, with its maximisation of proper time for inertial motion, is not just the stuff of stories about space travellers. It is happening all around us every time a leaf falls from a tree, a frog leaps from the ground, or Earth moves along her orbit to enter a new season. It lies at the very heart of motion in a gravitational field.

12.3.2 Free Particles, Geodesics, and Locally Inertial Frames

Free particles are postulated to follow the geodesics of a possibly curved spacetime. (They may still curve in a flat spacetime, which is exactly what they do in an accelerated frame, but this is only because the choice of coordinates has rendered the geodesics as curved.) How does this geodesic motion relate to the traditional newtonian view that they experience a zero force and hence a zero acceleration?

Recall that the components of the four-acceleration were defined for Minkowski space in (7.13) as $a^\alpha \equiv \mathrm{d}u^\alpha/\mathrm{d}\tau$. With hindsight, we see that the (constant) basis vectors could also have been included by writing the definition in terms of complete vectors. That won't change (7.13) since it was defined for the Minkowski metric; but for arbitrary coordinates and an arbitrary metric, we can define the four-acceleration more generally to incorporate (7.13) as a special case. (Here we write a four-dimensional vector as, e.g., $\vec{a}$, to distinguish it from its three-dimensional counterpart $\boldsymbol{a}$ used later in this section; however, we *always* write the basis vectors bold.)

$$\vec{a} \equiv \frac{\mathrm{d}\vec{u}}{\mathrm{d}\tau} = \frac{\mathrm{d}\left(u^\alpha \boldsymbol{e}_\alpha\right)}{\mathrm{d}\tau} \overset{(9.103)}{=\!=} \frac{\mathrm{D}u^\alpha}{\mathrm{d}\tau}\, \boldsymbol{e}_\alpha \equiv a^\alpha \boldsymbol{e}_\alpha\,. \tag{12.17}$$

So, the more general expression for the four-acceleration in terms of components is $a^\alpha = \mathrm{D}u^\alpha/\mathrm{d}\tau$. Now we are incorporating any more general metric describing a gravitational field. Of course, $\mathrm{D}u^\alpha$ and $\mathrm{d}u^\alpha$ are equal in the absence of gravity, because then the Christoffel symbols vanish globally and covariant differentiation reduces to partial differentiation. And by the Equivalence Principle they are also equal in the vicinity of a momentarily comoving inertial observer.

What, then, is the motion of a particle whose four-acceleration vanishes?

$$
\begin{aligned}
0 = a^\alpha &= \frac{\mathrm{D}u^\alpha}{\mathrm{d}\tau} \overset{(9.102)}{=\!=} u^\alpha{}_{;\beta}\, u^\beta = u^\alpha{}_{,\beta}\, u^\beta + \Gamma^\alpha{}_{\gamma\beta}\, u^\gamma\, u^\beta \\
&= \frac{\mathrm{d}^2 x^\alpha}{\mathrm{d}\tau^2} + \Gamma^\alpha{}_{\beta\gamma} \frac{\mathrm{d}x^\beta}{\mathrm{d}\tau} \frac{\mathrm{d}x^\gamma}{\mathrm{d}\tau}\,,
\end{aligned}
\tag{12.18}
$$

which is just the geodesic equation (9.32) again! (There is a slight clash of symbols: the u in (9.32) is our x here.) The steps of (12.18) also work in reverse, so that a vanishing four-acceleration is equivalent to geodesic motion. Thus, free particles in a gravity field are postulated via the Equivalence Principle to have no four-acceleration. Finally, the definition of three-force on p. 209, $\boldsymbol{F} \equiv \mathrm{d}\boldsymbol{p}/\mathrm{d}t$, translates to four dimensions by introducing a *four-force* $\vec{F}$ through the relation $\vec{p} = m\vec{u}$:

$$
\vec{F} \equiv \frac{\mathrm{d}\vec{p}}{\mathrm{d}\tau} = \frac{\mathrm{d}\left(p^\alpha \boldsymbol{e}_\alpha\right)}{\mathrm{d}\tau} = \frac{\mathrm{D}p^\alpha}{\mathrm{d}\tau}\,\boldsymbol{e}_\alpha = m\frac{\mathrm{D}u^\alpha}{\mathrm{d}\tau}\,\boldsymbol{e}_\alpha = m a^\alpha \boldsymbol{e}_\alpha = m\vec{a}\,. \tag{12.19}
$$

So although $\boldsymbol{F} \neq m\boldsymbol{a}$ relativistically because of complications with the changing relativistic mass γm, it's certainly true that $\vec{F} = m\vec{a}$, where m is constant.

This geometrical view of gravity turns Newton's view of force on its head. While Newton would maintain that a particle in free fall in a gravity field accelerates due to the gravity it experiences, Einstein's view is quite different: a freely falling particle follows a geodesic, and hence has *no* (four-)acceleration and feels no (four-)force, which is why freely falling observers are (and feel!) weightless. The only force on a particle comes from other things such as the electromagnetic force; all reference to a gravity force has been dropped entirely.

The Equivalence Principle states that momentarily comoving inertial observers will observe events in a small enough volume to be governed by the laws of special relativity only, regardless of what gravity is present. Any other observer moving at constant velocity relative to the momentarily comoving inertial observer will also make such observations in that small volume, and in particular will measure freely falling particles there to be momentarily following straight lines. Such observers form a superset to the MCIFs of Chap. 7 and are called *locally inertial*, and each carries with it a locally inertial frame.

Locally inertial frames bring us as close as possible to annulling gravity; they correspond to the Equivalence Principle idea that there is no gravity inside a *small*, freely falling laboratory, but that gravity cannot be made to vanish everywhere inside a *large* laboratory.

To obtain the straight-line motion followed by a freely falling particle in a small laboratory—a small neighbourhood of the particle—a locally inertial observer must be able to write the geodesic equation (12.18) as $\mathrm{d}^2x^\alpha/\mathrm{d}\tau^2 = 0$.

But this means that at least at that point of interest, the observer needs the ability to set the Christoffel symbols all to be zero. That this can really be done is quantified as follows. At any point P in a riemannian or pseudo-riemannian space, a coordinate system x^α can always be found whose origin is at P, such that the metric at P is almost Minkowski in the sense that first-order corrections in the coordinates x^α vanish:

$$g_{\alpha\beta}(P) = \eta_{\alpha\beta}(P) + \text{second-order corrections.} \tag{12.20}$$

This can be shown by expanding the metric as a Taylor series in arbitrary coordinates and counting the number of free parameters versus specified constants. It turns out that we have ample freedom to obtain $\eta_{\alpha\beta}$, only just enough freedom to ensure there are no first-order terms, and no freedom to make all of the second-order terms vanish. (Alternatively, the same theorem can be proved by specifying a transformation whose Christoffel symbols vanish and applying some linear algebra arguments.)

By differentiating (12.20) once, we see that $g_{\alpha\beta,\gamma}(P) = 0$, so that the Christoffel symbols can always be made to vanish at P with an appropriate coordinate choice. Again, this demonstrates that they don't form a tensor, as was discussed on p. 326. But differentiating (12.20) a second time shows that, in general, $g_{\alpha\beta,\gamma\delta}(P) \neq 0$. Since the Riemann tensor (i.e. curvature) is a function of $g_{\alpha\beta,\gamma\delta}$, this implies that curvature cannot be made to vanish by a suitable coordinate choice. Christoffel symbols *can* be made to vanish; Riemann components cannot. So if a spacetime is flat, it's flat for all coordinate choices; and if curved, it's curved for all coordinate choices. Curvature can no more be made to go away than an orange peel can be laid out flat, and this is precisely what puts the idea of curvature on a higher rung than mere coordinate choices. It also means that our metric (12.9) is not just about a change in coordinates; it's also about a spacetime curvature that is curvature for all.

Theorem (12.20) is sometimes said to describe "local flatness", and now we see why this is a misnomer. It can give the impression that somehow curvature has been removed at P, which is not the case. After all, the nonvanishing second-order terms are precisely what determine curvature! A simple analogy is that of taking ever-smaller arcs of a given circle. Each arc departs from a straight line less and less, but nevertheless each arc has the same curvature (equal to one divided by its radius), even in the limit as its length goes to zero. Never does this constant curvature approach the line's zero curvature. Perhaps the term "locally Minkowski" for the frame of a locally inertial observer is useful, but the term "locally flat" is quite misleading.

We have spoken about locally inertial observers and their frames. The use of the word "frame" can be a little vague in relativity. If we envisage a separate frame—a separate set of axes—at every event, then the locally inertial frame might better be called a locally inertial *moving* frame, or maybe simply referred to as a locally inertial observer, because it's something that must be defined over a time interval. But beside this is a related notion: the orthonormal frame, or rather set of orthonormal frames, one at each point, where each comprises four orthonormal vectors (axes) that can always be constructed at every event. We can make an orthonormal frame at any point by joining three space axes together (three orthogonal rulers), along with a clock. An orthonormal frame or basis need not belong to a locally inertial observer (although unfortunately some authors do equate the two). Orthonormal frames are used to make physical measurements; after all, we don't employ freely falling observers to make our day-to-day measurements. It would be a fine thing if we had to jump in the air every time we wanted to measure the length of a table.

But just how do measurements made in an orthonormal frame relate to physical (i.e. proper) measurements? Imagine that, for a moment, right next to an observer A who uses the orthonormal rulers and clock to make measurements, there is another observer B who has jumped up from a trampoline and reached his maximum height, momentarily coming to rest next to A. By the free-fall part of the Equivalence Principle (Fig. 12.1), B is inertial. Next, by the acceleration part of the Equivalence Principle, A can be likened to an accelerated observer in flat spacetime. The Clock Postulate then says that B is the MCIF of A, and so both make identical measurements for the brief time that B is this MCIF. Since the measurements made by B are by definition proper, the measurements made by A must be also; so measurements made in an orthonormal frame are proper measurements.

We'll see more of these orthonormal frames soon. In the meantime, the fact that the measurements made by observer A and trampoline expert B are identical underlines something that's inherent in the whole formalism of special and general relativity by way of the Clock Postulate (although Einstein took it as self evident). That is that when two observers are at rest relative to and right next to each other, even if only for a moment, each observes events in their immediate vicinity in the same way, and both age at the same rate.

Geodesic Deviation

Free particles that move on different straight lines on a flat surface will naturally move toward or away from each other. But the rate of change of their separation increase is constant. That is, if they are separated by ξ^α, then $\mathrm{d}^2\xi^\alpha/\mathrm{d}\tau^2 = 0$.

On a curved surface, things are different since the separations are only well defined for geodesics that are infinitesimally close. In such a case, for coordinates x^α and where each geodesic uses an affine parameter τ, it turns out through differentiating twice and using careful Christoffel bookkeeping (which we omit to avoid a digression), that

$$\frac{\mathrm{D}^2\xi^\alpha}{\mathrm{d}\tau^2} = R^\alpha{}_{\beta\gamma\delta}\,\frac{\mathrm{d}x^\beta}{\mathrm{d}\tau}\,\frac{\mathrm{d}x^\gamma}{\mathrm{d}\tau}\,\xi^\delta\,. \tag{12.21}$$

This *equation of geodesic deviation* is the relativistic version of Newton's *gravitational tidal force*, and employs the full Riemann tensor as opposed to the Ricci tensor used in Einstein's equation. Newton would attribute the bulges of water that line up with the Moon on each side of Earth to the falloff of the inverse-square gravity force across Earth's diameter due to the Moon. (Internal friction caused by the motion of the corresponding bulges in the Moon, anciently induced by Earth, across the Moon's surface as it rotates, have long since frozen its rotation with respect to Earth, so that now it presents the same face to us perpetually.) Einstein, on the other hand, would drop all reference to this gravity force. Instead, he would attribute the bulges on Earth to the natural separation of geodesic worldlines of the water molecules in a curved spacetime, combined with electromagnetic forces of their neighbours that push them off these geodesics. Again, all mention of a gravity force is absent in general relativity.

The Equivalence Principle and Covariant Derivatives

Locally inertial frames and the Equivalence Principle tell us how to formulate the laws of physics in arbitrary coordinates, whether they belong to a flat or a curved spacetime. We saw this sort of idea before with the comma-goes-to-semicolon rule on pages 328 and 411. Because the Christoffel symbols vanish in a locally inertial frame, any law that can be written in terms of tensor components and partial derivatives (commas on the tensor indices) will equally well be written using covariant derivatives (semicolons on the tensor indices). But a tensor expression that uses covariant derivatives is valid in *all* frames, and so is the generalisation of that law to arbitrary coordinates and a curved spacetime. This idea lies at the heart of writing the equations of physics for gravitational fields.

But there is one problem with converting commas to semicolons in a curved spacetime: although the order of partial derivatives makes no difference, the order of covariant derivatives certainly does, as evidenced by (9.97), where we see that swapping that order is equivalent to coupling physical quantities to spacetime curvature. This is a difficult problem with no known general solution, although it's often possible to argue that a coupling to curvature might be unreasonable.

What Goes Up Must Come Down

The idea that free particles travel on geodesics can be graphically illustrated by imagining we are on Earth and throw a ball into the air. If it follows a geodesic on a curved spacetime, can we show in a simple schematic way why it might go up and then come down again?

The idea is shown in Fig. 12.4. We are constrained to draw the figure in three dimensions. The spacetime surface must curve into one of those dimensions. Another of the three dimensions is used for time. That leaves us with only one space axis able to be plotted, which will be vertical height. We are at

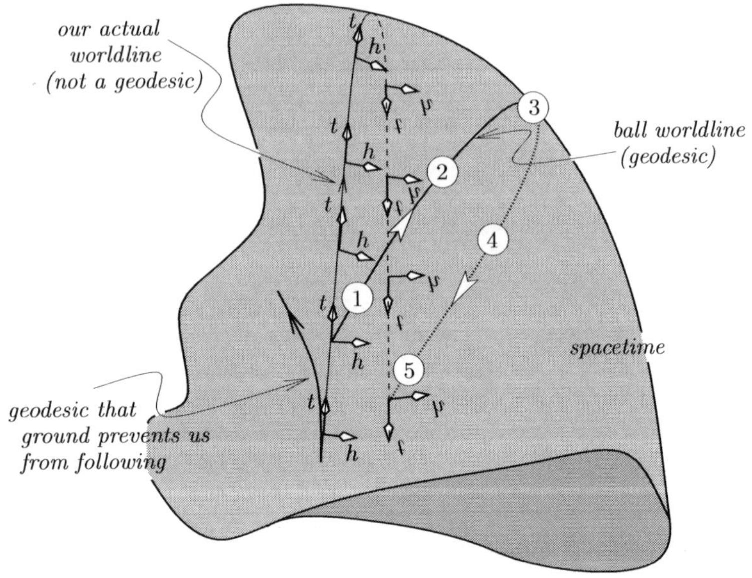

Fig. 12.4. Schematic with one space dimension of why things that go up eventually come down, in a curved spacetime. Once the ball is released, it follows a geodesic worldline. We on Earth's surface cannot follow a geodesic worldline due to the solid ground stopping us from falling. Instead we "take the long way around" to again eventually meet with the ball; this we interpret as its having fallen back down to the ground. Two axes are shown, height h and time t, which are upside down on the opposite side of the spacetime surface. The ball is thrown up at 1, ascends at 2, reaches maximum height at 3, descends at 4, and lands at 5. Its motion is simple in spacetime; *we* are the ones who have a complicated (i.e. nongeodesic) motion.

rest on Earth's surface, so that our worldline is not a geodesic. If the ground were suddenly to vanish from beneath our feet, we would drop, following the geodesic indicated in Fig. 12.4 as the one we normally are prevented from following.

Initially we hold the ball and it follows our worldline. When thrown, it's free to take the "short way around" the curved spacetime surface by following a geodesic worldline. We can draw a new set of spacetime axes at every event that allow us to track where the ball is. Its separation from us on the space axis (i.e. its height) increases initially, and then begins to decrease as our worldline heads for a rendezvous with that of the ball. Finally, our worldlines intersect: the ball lands at our feet. On Earth we thought we didn't move and that the *ball* followed a complicated motion in space and time governed by Newton's laws. In spacetime, however, the ball moved in the simplest way possible, and it was *we* who followed a complicated path, constrained by Earth's surface to forever veer away from a geodesic path in spacetime.

12.3.3 Quantities That Are Conserved on Geodesics

When we looked at the lagrangian approach to classical mechanics in Chap. 10, we found that if the lagrangian is independent of a coordinate x^a, then the corresponding *canonical* momentum b_a is conserved. It turns out that a similar law holds in arbitrary coordinates using the metric instead of the lagrangian, which points to a close relationship between the two.

To see how it comes about, focus on a free particle of rest mass m moving in a possibly curved spacetime, and ask for the rate of increase of four-momentum over proper time: $\mathrm{d}p^\alpha/\mathrm{d}\tau$. We know that a free particle has $\mathrm{D}p^\alpha/\mathrm{d}\tau = 0$ (that's just the geodesic equation). But what is $\mathrm{d}p^\alpha/\mathrm{d}\tau$? Start with the geodesic equation, writing

$$\frac{\mathrm{D}p^\alpha}{\mathrm{d}\tau} = p^\alpha{}_{;\beta}\, u^\beta = 0\,. \tag{12.22}$$

Bringing in Christoffel symbols and cancelling symmetric terms leads to

$$\frac{\mathrm{d}p^\alpha}{\mathrm{d}\tau} = \frac{-m}{2}\, g^{\alpha\mu} g_{\mu\beta,\gamma}\, u^\beta u^\gamma\,. \tag{12.23}$$

It's not clear what this is telling us, so instead we ask for $\mathrm{d}p_\alpha/\mathrm{d}\tau$, beginning with the lowered-index form of the geodesic equation:

$$\frac{\mathrm{D}p_\alpha}{\mathrm{d}\tau} = p_{\alpha;\beta}\, u^\beta = 0\,. \tag{12.24}$$

Writing this in terms of Christoffel symbols and cancelling terms produces

$$\frac{\mathrm{d}p_\alpha}{\mathrm{d}\tau} = \frac{m}{2}\, g_{\mu\nu,\alpha}\, u^\mu u^\nu\,. \tag{12.25}$$

Now, if the metric is independent of x^α, then it follows that $\mathrm{d}p_\alpha/\mathrm{d}\tau = 0$ on a geodesic. So p_α is conserved rather than p^α. In this case, the basis vector $\boldsymbol{e}_\alpha$ is called a *Killing vector*, and the conserved quantity is just the dot product of the four-momentum with the relevant Killing vector.

An example is a time-independent metric, for which p_0 is conserved on a geodesic; hence it makes sense in general relativity to define the energy of a system to be p_0 rather than p^0. (Of course, in flat spacetime with cartesian coordinates, these two will only possibly differ by a sign.) The metric of an expanding universe depends on time, in which case energy is generally *not* conserved on geodesics.

Quantities that are conserved on geodesics give an easy way to compute geodesics, as opposed to the use of the geodesic equation (although of course the geodesic equation will have to be used if the metric has no coordinate independence to exploit). As an example, the spacetime of the solar system can be approximated by a metric we'll encounter in Sect. 12.5, called the *Schwarzschild metric*. By symmetry, in polar coordinates we can confine attention to the orbital plane of a planet (or a passing light ray) and set $\theta = \pi/2$.

The metric turns out to be independent of t and ϕ, so that p_t and p_ϕ are conserved on geodesics in this plane. (Equivalently, $\boldsymbol{e}_t$ and $\boldsymbol{e}_\phi$ are Killing vectors.) We've already seen that p_t is the planet's energy. Also, p_ϕ can be interpreted as its angular momentum. This is because for the Schwarzschild metric (12.50) with $\theta = \pi/2$, we have $g_{\phi\phi} = -r^2$, giving

$$p_\phi = g_{\phi\phi}\, p^\phi = -r^2\, m u^\phi = -mr^2 \frac{\mathrm{d}\phi}{\mathrm{d}\tau}\,, \tag{12.26}$$

which is thus conserved on geodesics. So, setting $r^2 \mathrm{d}\phi/\mathrm{d}\tau = constant$ is one equation of a set that can be solved simultaneously to produce geodesics. Evident here is a close similarity with the newtonian angular momentum for the planet, $mr^2\, \mathrm{d}\phi/\mathrm{d}t$, as in (8.128). (The negative sign in (12.26) is a relic of the metric signature and has no significance.)

Finally, the conservation of energy or momentum on geodesics that is due to some coordinate independence of the metric matches the ideas of Sect. 10.3.4. There, we found that the hamiltonian or canonical momentum are conserved when the *lagrangian* is independent of a coordinate:

$$\begin{aligned} \text{metric independent of } x^\alpha &\implies \mathrm{d}p_\alpha/\mathrm{d}\tau = 0 \text{ on a geodesic, versus} \\ \text{lagrangian independent of } t &\implies \mathrm{d}H/\mathrm{d}t = 0\,, \\ \text{lagrangian independent of } x^i &\implies \mathrm{d}b_i/\mathrm{d}t = 0\,. \end{aligned} \tag{12.27}$$

This close similarity between the metric and lagrangian is not altogether surprising. In classical mechanics (Sect. 10.3), extremising the action over all paths in space and time produces an equation for the actual path followed. In the differential geometry of general relativity (Sect. 9.2), maximising the proper time over all paths in spacetime produces an equation for a geodesic. And indeed, as we saw in (10.69), there is a close relationship between the action and the proper time.

12.4 A Path to Einstein's Equation

As well as giving a prescription for how free particles move, any candidate for a theory of curved spacetime must specify the spacetime curvature, which we expect to be related to the distribution of mass. Recall Poisson's equation (10.75) that relates the newtonian gravity potential Φ to the mass density ϱ:

$$\nabla^2 \Phi(t, \boldsymbol{x}) = 4\pi G\, \varrho(t, \boldsymbol{x})\,. \tag{12.28}$$

We wish to search for some sort of relativistic version of this, but with the potential replaced by a function related to the curvature, and the mass density altered to be more appropriately relativistic:

$$\text{spacetime curvature} = \text{mass density distribution.} \tag{12.29}$$

Consider first the mass density distribution. A simple mass distribution is an ideal gas, also called a *perfect fluid* in relativity, and the simplest perfect fluid, such as the rain of Chap. 6, is one that has no internal pressure, called *dust.* Dust can have internal random motions, so that the whole of it does not have to be at rest in one frame. We only demand that it can be partitioned into possibly overlapping subunits, in each of which the drops have no random velocities. This allows us to use the flux ideas that we previously applied to rainfall. It's a conceptual simplification that will help us begin to think about mass distributions in general relativity.

To quantify mass density, or equivalently energy density, our work with raindrops in Chap. 6 suggests that a fully relativistic description of dust must combine two things:

- its **four-momentum** $\vec{p} \equiv m\vec{u}$ of (6.27), since p^0 is the relativistic mass (energy) per particle, and
- its **number–flux density** $\vec{N} \equiv n\vec{u}$ in (6.16), since N^0 is the number of particles per unit volume.

In particular, the product of these two zero-components in the laboratory frame is

$$\begin{aligned} p^0 N^0 &= \text{energy per particle} \times \text{number of particles per unit volume} \\ &= \text{total energy per unit volume} \\ &\equiv \text{energy density}. \end{aligned} \tag{12.30}$$

The Lorentz transform alters both energy and volume by a factor of γ, and the resulting γ^2 needed to transform energy density indicates that a *second-order* tensor is necessary to encode information about the mass distribution. This prompts us to define the *stress–energy tensor* (also called the energy–momentum tensor) $T^{\alpha\beta}$ for dust as

$$T^{\alpha\beta} \equiv p^\alpha N^\beta = mn\, u^\alpha u^\beta \equiv \varrho\, u^\alpha u^\beta = T^{\beta\alpha}, \tag{12.31}$$

so that, for example, T^{00} is the dust's energy density $\gamma^2 mn$. In general, the components of the stress–energy tensor for dust can be calculated by noting that in the laboratory frame,

$$\begin{aligned} p^\alpha &= mu^\alpha = m\gamma(1, \boldsymbol{v}) = (\text{energy/particle}, \ \ \text{momentum/particle}), \\ N^\alpha &= nu^\alpha = n\gamma(1, \boldsymbol{v}) \\ &= \left(\begin{bmatrix} \text{number of particles} \\ \text{per unit volume} \end{bmatrix}, \begin{bmatrix} \text{(three) numbers of particles} \\ \text{passing through the planes} \\ x,\ y, \text{ and } z = \textit{constant} \text{ per} \\ \text{unit area per unit time.} \end{bmatrix} \right), \end{aligned} \tag{12.32}$$

and the individual components are listed in the box on the following page. Although these components apply from first principles just to dust, they can

Components of the Stress-Energy Tensor for Dust

The general stress–energy tensor can be defined such that its components match those calculated using (12.31) for dust, which are listed here, where a, b are space indices. The "a-momentum" is just p^a, while the "b-plane" is the plane $x^b =$ *constant*. It's useful to remember that $T^{\alpha\beta}$ is symmetric, so that two different interpretations of its components are possible when the indices are unequal.

$$
\begin{aligned}
T^{00} = p^0 N^0 &= \text{energy/particle} \times \text{number of particles/unit volume} \\
&= \text{energy density.} \\
T^{0b} = p^0 N^b &= \text{energy/particle} \times \text{number of particles passing through } b\text{-plane} \\
&\quad \text{per unit area per unit time} \\
&= \text{energy flux density through } b\text{-plane.} \\
T^{a0} = p^a N^0 &= a\text{-momentum/particle} \times \text{number of particles/unit volume} \\
&\equiv a\text{-momentum density.} \\
T^{ab} = p^a N^b &= a\text{-momentum/particle} \times \text{number of particles passing through} \\
&\quad b\text{-plane per unit area per unit time} \\
&= \text{total } a\text{-momentum passing through } b\text{-plane} \\
&\quad \text{per unit area per unit time} \\
&\equiv a\text{-momentum flux density through } b\text{-plane.}
\end{aligned}
\tag{12.33}
$$

be used to *define* the stress–energy tensor for a general mass distribution. We encountered the field version of this tensor in Sect. 10.3.6.

A more general perfect fluid results when the particles are able to produce an internal pressure P. How does this relate to $T^{\alpha\beta}$? Consider a box of sides ℓ_x, ℓ_y, ℓ_z, containing a perfect fluid whose random motions exert a pressure on the walls of the box. What is the pressure on the wall $x =$ *constant*? Partition the fluid into subunits indexed by i, where the ith cell contains dust with rest mass per particle m_i, four-velocity per particle of u_i^α, and proper number density n_i, so that the number–flux density of this cell is $N_i^\alpha = n_i u_i^\alpha = n_i \gamma_i (1, \boldsymbol{v}_i)$. The subunits are allowed to overlap in space. We'll drop the i subscript for clarity in the following calculation.

Now, in a time $2\ell_x/v^x$, every particle has bounced off the wall, transferring a momentum of $2\gamma m v^x$ in the process. So the total momentum transferred is this momentum multiplied by the total number of particles in the cell, or

$$
\text{total momentum transferred} = 2\gamma m v^x \times \gamma n \, \ell_x \ell_y \ell_z \,. \tag{12.34}
$$

Thus, the pressure due to cell i is

$$
\text{pressure} = \frac{\text{force}}{\text{area}} = \frac{\text{total momentum transferred/time taken}}{\ell_y \ell_z}
$$

$$= \frac{2\gamma m v^x \, \gamma n \, \ell_x \ell_y \ell_z}{\ell_y \ell_z \times 2\ell_x / v^x} = \gamma m v^x \times \gamma n v^x \,. \tag{12.35}$$

The total pressure P is the sum of this over all the cells. But this sum is just T^{xx}! Why? Because

$$T^{xx} = \sum_{\text{all cells}} p^x N^x = \sum_i \gamma m v^x \times \gamma n v^x = P \,. \tag{12.36}$$

What are the other stress–energy components for the perfect fluid?

$$T^{00} = \sum_i p_i^0 N_i^0 = \sum \text{energy densities} \,, \tag{12.37}$$

which is just the total energy density as before. Finally,

$$T^{0b} = \sum_i p_i^0 N_i^b = \sum_i \gamma_i m_i \, \gamma_i n_i v_i^b \,. \tag{12.38}$$

Statistically, we can say that in the laboratory frame (in which we take the fluid's centre of mass to be at rest), for every cell with velocity $\boldsymbol{v}$, there is an overlapping cell with velocity $-\boldsymbol{v}$. This gives a pairwise cancellation in (12.38) that results in $T^{0b} = T^{b0} = 0$. The off-diagonal elements such as T^{xy} describe, for example, the total x-momentum passing through the plane $y = constant$, which is related to shear forces that are assumed absent in a perfect fluid. Also, in a perfect fluid, we expect the pressure to be the same in all directions, so that the stress–energy tensor is diagonal and contains only one pressure, P. In the locally inertial frame of the fluid, we can write

$$T^{\alpha\beta} = \begin{bmatrix} \varrho & & & 0 \\ & P & & \\ & & P & \\ 0 & & & P \end{bmatrix} = (\varrho + P)\, u^\alpha u^\beta - \eta^{\alpha\beta} P \operatorname{sgn} \eta_{00} \,, \tag{12.39}$$

so that in a general frame it must be that

$$\boxed{T^{\alpha\beta} = (\varrho + P)\, u^\alpha u^\beta - g^{\alpha\beta} P \operatorname{sgn} g_{00} \,.} \tag{12.40}$$

Referring to Sect. 8.5.1, we can attach a basis $\boldsymbol{e}_{\alpha\beta}$ to the tensor components in (12.40) to write it as $\boldsymbol{T} = (\varrho + P)\, \boldsymbol{u} \otimes \boldsymbol{u} - \boldsymbol{g} P \operatorname{sgn} g_{00}$. Some texts will write the $\boldsymbol{g}$ as $\boldsymbol{g}^{-1}$ here. But recall the discussion of Sect. 8.5.2, which stresses that there is no such thing as an inverse of the metric tensor, but that there *is* such a thing as the inverse of the matrix of metric components $g_{\alpha\beta}$.

For more complex energy distributions that can be described by a lagrangian, such as fields, an alternative form of the stress–energy tensor will be given in (12.115).

On p. 426, and specifically (10.166), we saw the idea of local conservation as the expression of a vanishing divergence. The stress–energy tensor also has a vanishing divergence $\nabla \cdot \boldsymbol{T}$. To see why, first calculate the components of the divergence:

$$\nabla \cdot \boldsymbol{T} = \boldsymbol{e}^\alpha \partial_\alpha \cdot \left(T^{\mu\nu} \boldsymbol{e}_\mu \boldsymbol{e}_\nu\right) = T^{\mu\nu}{}_{;\alpha}\, \boldsymbol{e}^\alpha \cdot \boldsymbol{e}_\mu \boldsymbol{e}_\nu = T^{\mu\nu}{}_{;\alpha}\, \delta^\alpha_\mu \boldsymbol{e}_\nu = T^{\mu\nu}{}_{;\mu} \boldsymbol{e}_\nu \,. \tag{12.41}$$

Now note that in the locally inertial frame, the integral of the divergence of the stress–energy over an arbitrary volume, using the divergence theorem (8.244) to convert the volume integral to a surface integral with measure $\boldsymbol{n}\,\mathrm{d}S$, is

$$\begin{aligned}
\int T^{\mu\nu}{}_{,\mu}\, \mathrm{d}V &= \int \left(T^{0\nu}{}_{,0} + T^{a\nu}{}_{,a}\right) \mathrm{d}V \\
&= \partial_t \int T^{0\nu}\, \mathrm{d}V + \int \nabla\cdot\left(T^{1\nu}, T^{2\nu}, T^{3\nu}\right) \mathrm{d}V \\
&= \partial_t \int T^{0\nu}\, \mathrm{d}V + \int \left(T^{1\nu}, T^{2\nu}, T^{3\nu}\right)\cdot\boldsymbol{n}\,\mathrm{d}S \\
&= \partial_t \left(\text{total} \begin{Bmatrix}\text{energy or} \\ \text{momentum}\end{Bmatrix}\right) + \int \left(\begin{Bmatrix}\text{energy or} \\ \text{momentum}\end{Bmatrix} \text{flux density}\right)\cdot\boldsymbol{n}\,\mathrm{d}S \\
&= 0\,,
\end{aligned} \tag{12.42}$$

since the total flux of energy or momentum out of the surface equals the rate of drop of energy or momentum contained in the volume, a local conservation argument we've used before in (10.66) and (10.165) (and see the box on p. 48). Since (12.42) holds for an arbitrary volume, it must follow that in a locally inertial frame $T^{\mu\nu}{}_{,\mu} = 0$ everywhere. In that case, the Equivalence Principle says that in an arbitrary frame $T^{\mu\nu}{}_{;\mu} = 0$, so that the divergence vanishes. If we postulate that the right-hand side of (12.29) will be the stress–energy tensor, then we can infer that the left-hand side of that equation should also be divergence-free, and this is an important prompt for what that left-hand side might be.

Different Forms of Einstein's Equation, and the Einstein Tensor

The left-hand side of (12.29) is expected to be something related to the spacetime curvature. After some work, Einstein suspected that it might just be the Ricci tensor $R_{\alpha\beta}$, and using this he was able to explain the precession of Mercury's orbit. This was a great achievement since this precession had been attributed to a hitherto unknown planet, christened Vulcan, that was presumed to orbit close enough to the Sun to both perturb Mercury's orbit and forever be lost in the Sun's glare.

But with $R_{\alpha\beta}$ on the left-hand side (12.29), Einstein found that some hypothesising about $T_{\alpha\beta}$ was necessary. He eventually found that this hypothesising could be abandoned if he changed the right-hand side of his equation by adding a term involving the trace of the stress–energy $T \equiv T^\alpha{}_\alpha$, and this became the final form of his equation:

$$R_{\alpha\beta} = 8\pi \left(T_{\alpha\beta} - \frac{1}{2} T\, g_{\alpha\beta} \right), \tag{12.43}$$

where the 8π allows newtonian theory to be correctly recovered in the weak field limit. So this, Einstein's famous and foremost equation of general relativity, governs how spacetime is affected by matter. That is, given some arbitrary distribution of matter encoded by $T_{\alpha\beta}$, equation (12.43) allows the metric to be calculated. Einstein was never happy that the geometrical simplicity of curvature on the left should be equated with an apparently messy stress–energy tensor for matter on the right, but this is the way the equation remains.

Einstein's equation is usually written in a slightly different way. Raising the first index of (12.43) and contracting gives the Ricci scalar $R = -8\pi\, T$. Inserting this expression for T back into (12.43) gives

$$R_{\alpha\beta} - \frac{1}{2} R\, g_{\alpha\beta} = 8\pi\, T_{\alpha\beta}\,. \tag{12.44}$$

The left-hand side of this is known as the *Einstein tensor* $G_{\alpha\beta}$, producing the most well-known form of Einstein's equation:

$$\boxed{G_{\alpha\beta} = 8\pi\, T_{\alpha\beta}\,, \quad \text{or simply } \boldsymbol{G} = 8\pi \boldsymbol{T}\,.} \tag{12.45}$$

The Einstein tensor turns out to be unique in a sense: $G_{\alpha\beta}$ plus a constant times $g_{\alpha\beta}$ is the only symmetric second-order tensor with zero divergence, vanishing in flat spacetime, that can be built from $g_{\alpha\beta}$, $g_{\alpha\beta,\mu}$, and $g_{\alpha\beta,\mu\nu}$. Alternatively, $G_{\alpha\beta}$ is the only tensor able to be built from the Riemann tensor and the metric while being linear in the Riemann tensor. These requirements have a certain simplicity, to which we'll add weight by showing in Sect. 12.8 that the Einstein tensor can be derived using a lagrangian approach that varies an action based on the Ricci scalar with respect to the metric. But when all is said and done, the justification for the Einstein equation lies in its considerable experimental success.

The metric and its derivatives are contained inside $G_{\alpha\beta}$ in a highly nonlinear way, and as a result, Einstein's equation has not been solved for anything beyond simple cases. We'll see some of those cases in the following sections.

12.5 Solving Einstein's Equation for an Empty Spacetime: The Schwarzschild Metric

Let's outline how Einstein's equation is solved for the simplest of all scenarios: a *vacuum* spacetime, being one that's empty of all stress–energy. We'll also look for a solution with spherical symmetry, so begin with a generic metric that embodies this symmetry (writing the radial coordinate as r' since we'll redefine it in a moment) using three undetermined functions:

$$d\tau^2 = f(r')\,dt^2 - g(r')\,dr'^2 - h(r')\,r'^2\left(d\theta^2 + \sin^2\theta\,d\phi^2\right). \tag{12.46}$$

This can be simplified by redefining the radial coordinate as $r^2 \equiv h(r')\,r'^2$, giving

$$d\tau^2 = a(r)\,dt^2 - b(r)\,dr^2 - r^2\left(d\theta^2 + \sin^2\theta\,d\phi^2\right), \tag{12.47}$$

which now has just two functions to be found. Is this redefinition of the radial coordinate meaningful, and, if so, what new meaning does it give that coordinate? More generally, what meaning can be given to the spatial part of any metric?

Write (12.47) as $d\tau^2 = a(r)dt^2 - d\ell^2$, so that the interval between two events measured as simultaneous by an observer A (i.e. that share the same t) is $d\tau^2 = -d\ell^2$. Now, as we discussed on p. 486, observer A measures identically to his momentarily comoving inertial observer B, and so in particular both measure the same interval. But B uses a Minkowski interval $d\tau^2 = dT^2 - dX^2 - dY^2 - dZ^2$, and also measures the two events as simultaneous. Thus B records the same T for them to write the interval as $d\tau^2 = -dX^2 - dY^2 - dZ^2$. In that case,

$$d\ell^2 = dX^2 + dY^2 + dZ^2. \tag{12.48}$$

But $dX^2 + dY^2 + dZ^2$ is the proper (i.e., physical) distance between the events, which is therefore the spatial part $d\ell^2$ of the original metric. So using (12.47), an observer at constant r and $\theta = 90°$ will measure the proper circumference of a circle of radius r to be

$$\oint d\ell = \int_0^{2\pi} r\,d\phi = 2\pi r\,. \tag{12.49}$$

So, the new radial coordinate r is just $1/(2\pi)$ of the circumference of a circle of radius r. This is not a trivial geometrical result, because r does *not* have its usual meaning of the proper radial distance. For two points separated by dr at constant θ and ϕ, (12.47) gives the proper radial distance as $d\ell = \sqrt{b(r)}\,dr$, which is not necessarily equal to dr.

Now that we have a suitably simplified metric, we can set about solving Einstein's equation (12.43). The zero stress–energy means $T_{\alpha\beta} = 0$, in which case (12.43) gives a zero Ricci tensor, $R_{\alpha\beta} = 0$. This, in turn, determines the two unknown functions $a(r), b(r)$ of (12.47). The method consists of calculating the Ricci tensor components, setting them all to zero, and then solving for $a(r)$ and $b(r)$. The straightforward details are omitted here but can be found in most books on general relativity, and the resulting metric is the famous *Schwarzschild solution*, where M is a constant of integration introduced in the process of solving the equations:

$$\boxed{d\tau^2 = \left(1 - \frac{2M}{r}\right)dt^2 - \frac{dr^2}{1 - \frac{2M}{r}} - r^2\left(d\theta^2 + \sin^2\theta\,d\phi^2\right).} \tag{12.50}$$

This is the most studied solution to Einstein's equation, which, considering it's the spherically symmetric solution for empty space, shows how difficult the equation is to solve for more realistic mass distributions.

Compare the Schwarzschild metric with the one that we calculated from lagrangian mechanics, (12.9). The only difference is in the coefficient of $\mathrm{d}r^2$. It appears that the mechanics of Newton and Lagrange is really only sensitive to the temporal part of the metric. Presumably this is because the worldlines of everyday "slow" particles are more closely aligned with the time axis, so that, in a manner of speaking, they "sample" the temporal part of the metric much more than the spatial part.

Given that the Schwarzschild metric describes an empty spacetime, what meaning can be given to M? An analysis of the geodesics in this spacetime shows that, at large r, test particles orbit as if there were a point mass M at $r = 0$ that attracts them with a newtonian gravitational force that drops as r^{-2}. Although we arranged for an empty space, it seems that a point mass has crept in unnoticed, exactly as occurred in the previous chapter when solving Poisson's equation for nonrelativistic gravity in (11.7). What has really happened is that Einstein's equation is *local*, and if we assume that the gravitational field due to a spherically symmetric mass is everywhere radial, then the Schwarzschild solution must also give the metric at any *empty* point in a space with such a spherically symmetric mass distribution. This is known as *Birkhoff's theorem*, and in particular it means that the Schwarzschild metric also describes a spacetime containing one nonrotating star. We'll assume in the next section that it does also at least approximately describe the spacetime around the very slowly rotating Earth.

There are of course problems with (12.50) at $r = 0$ and $r = 2M$ (which is $r = 2GM/c^2$ in conventional units), where the metric becomes singular. The difficulty at $r = 0$ is not altogether different from that of standard newtonian gravity near a point mass, where the inverse-square force begins to diverge. (The Schwarzschild singularity can be argued as being a stronger type in that, as we'll see, normal matter lying within $r = 2M$ *must* fall toward $r = 0$, which need not be the case in newtonian theory given a sufficient counteracting pressure.) Rather, the new problem occurs at $r = 2M$, and it shows that the difficulty in choosing an appropriate radial coordinate when solving Einstein's equation in (12.47) should not be underestimated. In fact, the radial coordinate originally introduced by Schwarzschild differs somewhat from the r in the solution that now bears his name, and the complexities surrounding what happens when $r = 2M$ took many years to iron out. For typical astronomical bodies such as stars and planets, $r \gg 2M$ and no problem arises, although we'll look more closely at this in Sect. 12.6.

The inverse-square law for newtonian gravity is recoverable from the Schwarzschild metric just as we found in Sect. 12.3.1 for the metric (12.9), provided $r \gg 2M$ so that we can approximately relate a velocity to the spatial polar coordinates. Write, with $\dot{} \equiv \mathrm{d}/\mathrm{d}t$,

The Magical Inverse-Square Force

Newtonian mechanics is mostly sufficient to describe planetary orbits, and in particular the inverse-square force of newtonian gravity governs orbital motion to a very high accuracy. But, in fact, besides being special from the point of view of allowing field lines to be drawn (as discussed in the box on p. 337), an inverse-square central force is really extremely special for a universe that has stars with orbiting planets. It turns out that stable planetary orbits are not something we get for free for just *any* central force.

There are two general requirements we might make for a well-behaved orbit. The first is that it's a stable equilibrium: any perturbations of the planet will not throw it off its orbit. The second requirement is that the orbit maps onto itself: it doesn't precess.

Analysis using newtonian mechanics shows that a central force law such as r^n (with n not necessarily an integer) will only produce a stable equilibrium when $n > -3$. But the requirement for no precession is stronger. For this, n must equal the square of a natural number $(1, 2, \ldots)$ minus 3. So we expect stable, nonprecessing orbits for $n = 1^2 - 3 = -2$, $n = 2^2 - 3 = 1$, $n = 3^2 - 3 = 6$, etc. That is, the forces look like $1/r^2$, r, r^6, etc. The only one of these that decreases with r is $1/r^2$. So the inverse-square force really is quite special. Real gravity is not quite $1/r^2$, and sure enough, orbits do precess; but quite slowly since the departure from inverse square is extremely small.

$$\begin{aligned} d\tau^2/dt^2 &= 1 - 2M/r - (1 - 2M/r)^{-1}\,\dot{r}^2 - r^2\,\dot{\theta}^2 - r^2 \sin^2\theta\,\dot{\phi}^2 \\ &\simeq 1 - 2M/r - v^2, \end{aligned} \tag{12.51}$$

which brings us back to (12.12) and its attendant discussion.

12.5.1 Deriving Gravitational Redshift Again

Earlier in this chapter, when describing the Pound–Rebka–Snider experiments, we calculated the redshift of light that climbs up a 22.5 metre well, using the Equivalence Principle to bring in the accelerated-frame ideas of Chap. 7.

By now it should come as no surprise to find that the same result is produced by the Schwarzschild metric. The same analysis of emitted and received frequencies is again used to arrive at (12.3), but now using the Schwarzschild metric instead of the accelerated-frame metric. To show this, it's permissible to use the small-time limit since the frequency ratios are quite static:

$$\frac{f_{rec}}{f_{em}} = \frac{\Delta\tau_{floor}}{\Delta\tau_{ceil}} = \frac{d\tau_{floor}}{d\tau_{ceil}}\,. \tag{12.52}$$

To a first approximation, we'll suppose that Earth does not rotate, so that the emitters and receivers in the Pound–Rebka–Snider experiments are at rest in the Schwarzschild spacetime. It follows that $\mathrm{d}r = \mathrm{d}\theta = \mathrm{d}\phi = 0$ for each clock, so use the Schwarzschild metric (12.50) to write the interval between successive pulse emissions or receptions as

$$\mathrm{d}\tau^2 = \left(1 - \frac{2M}{r}\right) \mathrm{d}t^2, \tag{12.53}$$

so that

$$\frac{f_{\text{rec}}}{f_{\text{em}}} = \frac{\mathrm{d}\tau_{\text{floor}}}{\mathrm{d}\tau_{\text{ceil}}} \simeq \frac{\left(1 - \frac{M}{r_{\text{floor}}}\right) \mathrm{d}t_{\text{floor}}}{\left(1 - \frac{M}{r_{\text{ceil}}}\right) \mathrm{d}t_{\text{ceil}}} \quad \text{These are equal! (See below.)} \tag{12.54}$$

Why are the two time intervals $\mathrm{d}t_{\text{floor}}$ and $\mathrm{d}t_{\text{ceil}}$ equal? Since the Schwarzschild metric has no time dependence, the scenario of pulses being sent and received is just like that of Fig. 12.2. But the Schwarzschild t defines a global time coordinate, being analogous to the global time coordinate $\bar{t}$ in the accelerated frame metric (7.27), which in turn was defined as the time τ_{ceil} in Fig. 12.2. So just as Fig. 12.2 shows the pulses being sent and received at equal intervals of the *global* time coordinate $\bar{t} \equiv \tau_{\text{ceil}}$, so, too, here in the Schwarzschild case, these equal time intervals are $\mathrm{d}t_{\text{floor}}$ and $\mathrm{d}t_{\text{ceil}}$, respectively, and so $\mathrm{d}t_{\text{floor}} = \mathrm{d}t_{\text{ceil}}$.

The correspondence between the accelerated frame's $\bar{t}$ and Schwarzschild's t can also be seen by comparing the lines of constant $\bar{t}$ in the accelerated frame, in Fig. 7.6, with those of constant t in Schwarzschild spacetime, in Fig. 12.5.

With a floor-to-ceiling distance of $h = 22.5\,\text{m}$ for the Pound–Rebka–Snider experiments, (12.54) becomes

$$\frac{f_{\text{rec}}}{f_{\text{em}}} \simeq \frac{1 - \dfrac{M}{r_{\text{floor}}}}{1 - \dfrac{M}{r_{\text{floor}} + h}} \simeq 1 - \frac{Mh}{r_{\text{floor}}^2}. \tag{12.55}$$

In conventional units, this is

$$\begin{aligned} \frac{f_{\text{rec}}}{f_{\text{em}}} &\simeq 1 - \frac{GMh}{c^2 r_{\text{floor}}^2} = 1 - \frac{gh}{c^2}, \quad (g \text{ is the usual } 10\,\text{ms}^{-2}) \\ &\simeq 1 - \frac{10\,\text{ms}^{-2} \times 22.5\,\text{m}}{9 \times 10^{16}\,\text{m}^2\text{s}^{-2}} \simeq 1 - 2.5 \times 10^{-15}, \end{aligned} \tag{12.56}$$

agreeing with the result of (12.3).

For more precision, we should realise that Earth's gravity has been approximated by the Schwarzschild metric in a "bigger", almost inertial frame in which Earth rotates but its centre is at rest at Schwarzschild $r = 0$. So the receiver and emitter are actually moving within the Schwarzschild spacetime, and with different velocities, owing to their different distances from Earth's centre. This more precise calculation was confirmed in 1971 by Hafele and Keating, who compared the time elapsed on clocks flown aboard two aircraft, one of which travelled east and the other west. Even with equal speeds relative to the ground, we expect different clock rates in these aircraft because they have different speeds in the almost inertial frame we just discussed, within which the Schwarzschild solution has been used as an approximation to the spacetime metric. For flight at a constant height and latitude, the Schwarzschild metric gives

$$d\tau^2 = \left(1 - \frac{2M}{r}\right) dt^2 - r^2 \sin^2\theta \, d\phi^2 . \tag{12.57}$$

The square of the ageing rate of a flying clock compared with a laboratory clock is then

$$\frac{d\tau^2_{\text{plane}}}{d\tau^2_{\text{lab}}} = \frac{1 - 2M/r_{\text{plane}} - \sin^2\theta_{\text{plane}}\, v^2_{\text{plane}}}{1 - 2M/r_{\text{lab}} - \sin^2\theta_{\text{lab}}\, v^2_{\text{lab}}} , \tag{12.58}$$

where the speeds of the clocks *in the almost inertial frame above* are given by $v \simeq r\, d\phi/dt$. Since the speeds of the east- and westbound clocks will in general be different in this frame even if the two aircraft have the same ground speed, their clocks will age differently, and these different rates of ageing were indeed observed in the Hafele–Keating experiment. A moment's thought shows that the eastbound clock should age less than the westbound clock, since while both are at the same height, the eastbound clock is travelling faster in the almost inertial frame. And that is precisely what happened:

- the **eastbound** clock was predicted to lose 40±23 ns during the trip, and was measured to have actually lost 59±10 ns, and
- the **westbound** clock was predicted to gain 275±21 ns, and actually gained 273±7 ns.

So the agreement with theory is very good. The different ageings in the Twin Conundrum have been verified experimentally.

The gain/lose nature of the results above is interesting. A plane's height serves to *increase* its ageing rate compared with the laboratory, while its flight speed serves to *decrease* that ageing rate. The fact that one of the flying clocks gained time while the other lost time shows that at the heights and speeds of commercial aircraft, these effects compete. The effect of height on a clock is similar in magnitude, but opposite in sign, to the effect of the flight speed.

12.6 The Schwarzschild Black Hole

The breakdown of the Schwarzschild metric at $r = 2M$ suggests that either there is a problem with spacetime there, or perhaps the coordinates t, r, θ, ϕ

are "bad" in some sense and the spacetime is perfectly well behaved. We must decide which is the case. In this section, we'll meet an alternative set of coordinates for Schwarzschild spacetime that settles the question, while in Sect. 12.6.1 we'll follow an alternative path of calculating the curvature at $r = 2M$.

First, given the Equivalence Principle, we might expect to draw a picture of Schwarzschild spacetime related to that of an accelerated frame. As we saw in Fig. 7.12, in the accelerated frame there is a horizon close to which time slows and light cones close up. This is exactly what happens with t, r, θ, ϕ coordinates in Schwarzschild spacetime. A clock hovering just above $r = 2M$ ages at a rate of $\mathrm{d}\tau/\mathrm{d}t = \sqrt{1 - 2M/r}$, which is close to zero. Furthermore, light cones are determined by setting $\mathrm{d}\tau = 0$, so that light moving radially obeys $\mathrm{d}r/\mathrm{d}t = \pm(1 - 2M/r)$, which means that as far as the coordinates are concerned, it slows to zero as it approaches $r = 2M$, which thus defines the *Schwarzschild horizon.*

What can be done to ascertain what is happening here? Just as in flat space the accelerated frame is not a preferred frame (in the sense that everything is simpler when viewed from the frame of an inertial observer, such as Adam instead of Eve in Chap. 7), in 1960 there were discovered more suitable coordinates for describing Schwarzschild spacetime, called *Kruskal–Szekeres coordinates*, T, R, θ, ϕ (where θ and ϕ are the same as in Schwarzschild coordinates). The T and R are defined differently above and below the $r = 2M$ horizon, so combine the two cases by writing, for $r \gtrless 2M$,

$$T \equiv \sqrt{\left|1 - \frac{r}{2M}\right|}\, \exp\left(\frac{r}{4M}\right) \begin{Bmatrix}\mathrm{sh}\\ \mathrm{ch}\end{Bmatrix} \frac{t}{4M}\,,$$

$$R \equiv \sqrt{\left|1 - \frac{r}{2M}\right|}\, \exp\left(\frac{r}{4M}\right) \begin{Bmatrix}\mathrm{ch}\\ \mathrm{sh}\end{Bmatrix} \frac{t}{4M}\,. \tag{12.59}$$

The metric in these T, R, θ, ϕ coordinates has no problems at $r = 2M$, where r is now considered to be a function of R and T:

$$\mathrm{d}\tau^2 = \frac{32M^3}{r} \mathrm{e}^{-r/(2M)} \left(\mathrm{d}T^2 - \mathrm{d}R^2\right) - r^2 \left(\mathrm{d}\theta^2 + \sin^2\theta\, \mathrm{d}\phi^2\right). \tag{12.60}$$

Although the metric does break down at $r = 0$, the more serious problem at $r = 2M$ has vanished. In this set of coordinates there is no horizon, as can be seen in Fig. 12.5, which shows a picture of Schwarzschild spacetime using Kruskal–Szekeres coordinates. Also, the light cones don't close up anymore. In fact, all light cones are once again 45° lines since when $\mathrm{d}\tau = 0$ we have $\mathrm{d}R/\mathrm{d}T = \pm 1$. So Kruskal–Szekeres coordinates are analogous to the usual polar coordinates of an inertial frame in flat spacetime. The analogies between the various coordinates are shown in Fig. 12.6.

Notice that when $r < 2M$ *all* light rays will eventually hit $r = 0$; there is no escaping the singularity within this *Schwarzschild radius.* And because

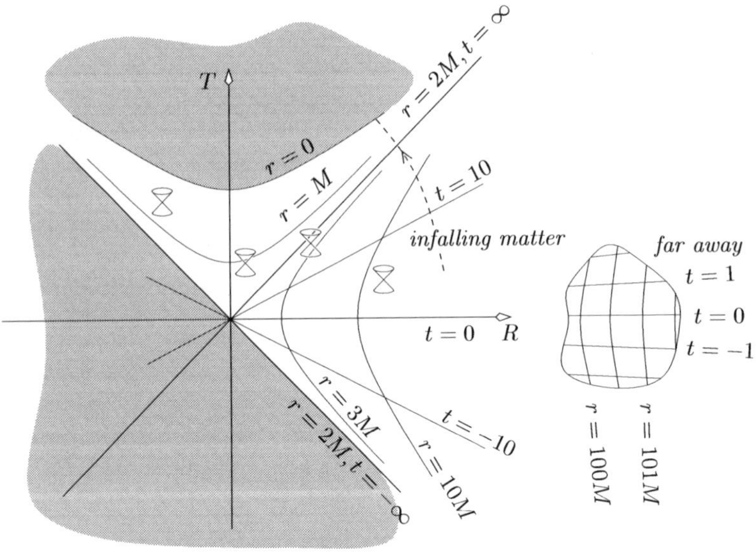

Fig. 12.5. Schwarzschild spacetime drawn with the Kruskal–Szekeres T, R coordinates, for constant θ, ϕ. Only the white regions correspond to $r > 0$ and $-\infty < T < \infty$. What was the horizon in t, r coordinates is now two 45° lines. Light cones are everywhere identical and open at 45°, so we can see immediately that light whose worldline coincides with either of the $r = 2M$ lines will never leave it, again indicating that $r = 2M$ forms a horizon. Matter falling in toward $r = 0$ crosses $r = 2M$ in finite proper time. Far away the space and time axes are almost perpendicular and almost straight, tending toward those of Minkowski space as $r \to \infty$. Compare this picture with Fig. 7.6 to see a similarity with the uniformly accelerated observer as drawn in an inertial frame.

all particle worldlines must stay within the light cone at each event (corresponding to local physics being that of special relativity for a freely falling particle), it must be that all particles that fall inside the Schwarzschild radius can never escape. This staying within the light cone corresponds to the fact that a particle must always move forward in the time coordinate of any locally inertial observer; that is, its worldline must always be *timelike* (as opposed to *spacelike*, where the worldline strays outside the light cone). However, notice that while freely falling particles relentlessly move forward in time defined by t outside the Schwarzschild radius, within the radius they must relentlessly move toward $r = 0$. And at the horizon the signs of both g_{tt} and g_{rr} swap, forcing us to conclude that inside the Schwarzschild radius the t coordinate in fact describes *space*, while the r coordinate describes *time*. We'll see more of this in a moment.

While Schwarzschild spacetime does exist around a lone spherically symmetric mass, as long as the radius of the mass is greater than $2M$ there will

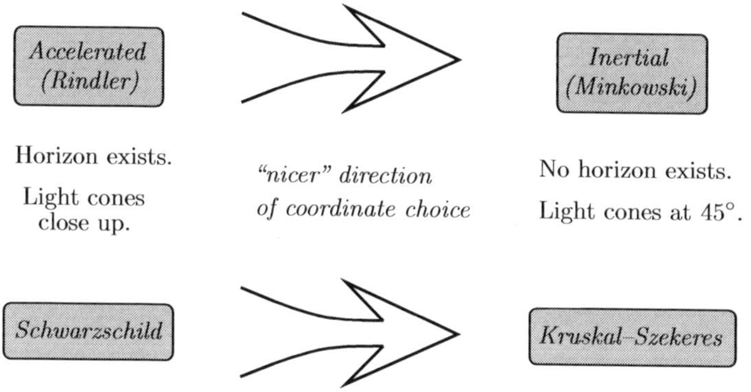

Fig. 12.6. The usual coordinates of accelerated frames (Rindler spacetime) and Schwarzschild spacetime both exhibit anomalies that are removed by a better choice of coordinates. In the accelerated case, this is trivial: we just use the inertial coordinates with which we began Chap. 7. Finding better coordinates for Schwarzschild spacetime is not so easy, but they certainly exist, as discovered by Kruskal and Szekeres.

be no horizon. This Schwarzschild radius is so small (e.g. 3 km for a solar mass) that matter is not expected to be crammed into it in any but that most cataclysmic of astrophysical processes: a supernova explosion. Supernovae are thought to mark the death of extremely massive stars, occurring when such stars' nuclear fires can no longer counteract the pull of gravity, so that a collapse occurs to what is possibly an internal bounce that results in the explosion. The physics of such a process that might squeeze the remaining stellar matter to within its Schwarzschild radius to form a *black hole* is by no means understood, but it is conjectured to be possible in sufficiently massive stars. When the *singularity theorems* of standard general relativity are applied to matter that is in some sense well enough behaved (satisfying reasonable conditions on its energy and pressure), they state that if this matter were concentrated into a small enough volume, then a moment of no return would occur where gravity must take over and pull the matter into a point.

Like the acausal oddities that we discussed in Chap. 7 and Fig. 7.4, how the notion of simultaneity in a Schwarzschild spacetime might be defined is problematic. Part of the difficulty lies with any ideas we might have of extrapolating time to infinity, which is not necessarily physically possible or meaningful. An observer outside the Schwarzschild radius might define simultaneity by the set of all events of equal t (which is certainly the case far from the hole); but what can be said of a particle that falls into the hole? According to the outside observer, the particle never falls past the horizon, because it can only reach the horizon at $t = \infty$, as is evident in Fig. 12.5.

The experience of the infalling particle is completely different: it falls into the singularity in a finite (and very short) proper time. Difficulties of this sort exist in general relativity, and we can only wonder what the ancient Greek philosopher Zeno would have made of them.

Representing Schwarzschild Space by a Curved Surface

How might we visualise the curved *spacetime* of the Schwarzschild metric? Apart from the heuristic picture of Fig. 12.4, this is problematic because the interval is not everywhere positive. Instead, we can draw the curved *space* at one particular time since the metric is time independent. The simplest such picture takes the slice of space at $\theta = 90°$, and draws a surface that represents the proper distance between two points (r, ϕ) and $(r + \mathrm{d}r, \phi + \mathrm{d}\phi)$. This proper distance is given by the euclidean distance between those two points as projected onto the surface.

The proper distance squared between the two points is the spatial part of the Schwarzschild metric,

$$\mathrm{d}\ell^2 = \left(1 - \frac{2M}{r}\right)^{-1} \mathrm{d}r^2 + r^2 \mathrm{d}\phi^2, \tag{12.61}$$

and we wish to draw the surface $z = z(r, \phi)$ in cylindrical polar coordinates such that $\mathrm{d}\ell$ equals the euclidean distance between those two points when they are projected vertically onto the surface:

$$\mathrm{d}\ell^2 = \mathrm{d}r^2 + r^2 \mathrm{d}\phi^2 + \mathrm{d}z^2. \tag{12.62}$$

Equating (12.61) with (12.62) yields

$$\left(\frac{\mathrm{d}z}{\mathrm{d}r}\right)^2 = \frac{2M}{r - 2M}. \tag{12.63}$$

We're only seeking a surface that encodes distances on the plane, so we are free to take the positive square root of (12.63), solving for z to give

$$z(r, \phi) = 2\sqrt{2M}\sqrt{r - 2M}\,. \tag{12.64}$$

The rotational symmetry ensures that z is really only a function of r. But we have been forced into considering only $r \geqslant 2M$, as shown by the plot of the surface $z = z(r, \phi)$ in Fig. 12.7. The proper distance between any two points increases as they approach $r = 2M$, although it certainly remains finite. But the breakdown of the usual Schwarzschild coordinates in this region is apparent: no surface at all has come out of the mathematics for $r < 2M$, and the surface we have drawn simply stops there.

If both signs were used for the square root in (12.64), then, rather than stopping abruptly, the surface would fold symmetrically beneath the xy-plane. This sort of topology is called a *wormhole*. Although this folding

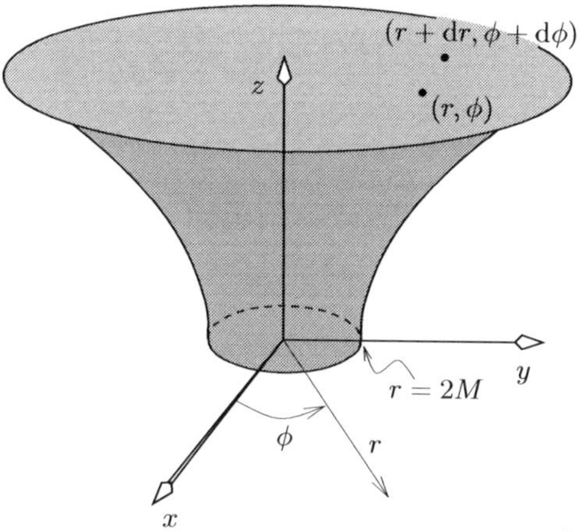

Fig. 12.7. A surface in three-dimensional euclidean space that gives the proper distance between any two points (r, ϕ) and $(r + \mathrm{d}r, \phi + \mathrm{d}\phi)$ in Schwarzschild space for $\theta = 90°$. It terminates at $r = 2M$; the t, r, θ, ϕ Schwarzschild metric doesn't allow anything further to be plotted for $r < 2M$.

might suggest a more complicated global structure than at first appears from the metric (which simply tells us the line element at any point), it does not get around the fact that our calculation has produced no surface for drawing inside the horizon, and yet presumably space still does exist there. Attempting to solve the problem by swapping the roles of t and r inside the horizon is possible but not very clear cut. The nature of Schwarzschild spacetime is still not well understood, and there is no reason why it should have anything to do with wormholes; certainly the Kruskal–Szekeres diagram in Fig. 12.5 does not suggest them. Whether wormholes exist in spacetime because of other mass distributions, or are even able to exist at all, is another question that currently remains unanswered.

The curved space surface (12.64) (i.e. for the positive square root only) is sometimes erroneously drawn with its funnel stretching down to $z = -\infty$ at $r = 0$. As we can see, that's not correct, and its walls are already vertical at the horizon. The surface we have drawn is often related to similar-looking ones made of plaster with rolling balls, which have been used many times to portray the orbit of a planet around a star according to general relativity. While suggestive, they don't really reflect the true situation, because the actual geodesic motion of the planet through a curved space*time* is being modelled somewhat trivially by an angled surface that simply uses Earth's gravity to give a centripetal force, directing the ball to swing about the central throat. What analogy to relativity there is here arises from the dependence of the ball's motion on the shape of the surface around it, as opposed to any invisible spring pulling it toward the centre of its motion—although gravity and the inward-curving plaster surface still conspire to provide that spring!

12.6.1 Tensor Components and Physical Measurements

Let's return to the problem of the Schwarzschild horizon and ask the question: if we did not have a good set of coordinates (such as Kruskal–Szekeres), how could we tell if the horizon was real or not? One way might be to calculate the value of the curvature on it. But if we do this using the t, r, θ, ϕ coordinates—which we know from hindsight are bad—then it comes as no surprise to find that the Riemann components diverge at $r = 2M$. For example,

$$R^t{}_{rtr} = \frac{2M}{r^3}\left(1 - \frac{2M}{r}\right)^{-1}, \quad R^\theta{}_{\phi\theta\phi} = \frac{2M}{r}\sin^2\theta \,. \tag{12.65}$$

Does this mean the curvature really diverges at both $r = 0$ and $r = 2M$?

In fact, while the complete Riemann tensor (sum of the products of components and basis vectors) does turn out to diverge at $r = 0$, it doesn't really diverge at $r = 2M$; there, the component divergence of (12.65) is just an artifact of the bad coordinates, and goes hand in hand with shrinking basis vector lengths that cancel component divergences. But we cannot know this just by looking at the Riemann components. The way to deal with apparent component divergences is to keep track of basis vector lengths. This is related to the question of how tensor components reflect physical measurements. When we measure anything at all, ultimately we are using rulers and clocks. The height of a chair is measured using a ruler physically laid against it; we certainly don't go about laying down a set of possibly bad coordinates centred on the Sun with arbitrarily defined behaviour. Rather, we use a frame with normalised basis vectors that are usually orthogonal: the *orthonormal frame* that was discussed back in Sects 8.7 and 12.3.2.

While normalised basis vectors are described more generally by (8.121), the most convenient set is the normalised version of the already orthogonal Schwarzschild coordinate basis, so that the simpler (8.117) can be used, allowing a vector $\boldsymbol{v} = v^\alpha \boldsymbol{e}_\alpha = v_\beta \boldsymbol{e}^\beta$ to be written over the normalised basis and cobasis as

$$\boldsymbol{v} = v^{\widehat{\alpha}} \boldsymbol{e}_{\widehat{\alpha}} = v^{\widehat{\alpha}} \frac{\boldsymbol{e}_\alpha}{|\boldsymbol{e}_\alpha|}, \quad \text{and } \boldsymbol{v} = v_{\widehat{\beta}}\, \boldsymbol{e}^{\widehat{\beta}} = v_{\widehat{\beta}} \frac{\boldsymbol{e}^\beta}{|\boldsymbol{e}^\beta|}, \tag{12.66}$$

which implies

$$v^{\widehat{\alpha}} = v^\alpha \left|\boldsymbol{e}_\alpha\right|, \quad v_{\widehat{\beta}} = v_\beta \left|\boldsymbol{e}^\beta\right| \quad \text{(no sums)}. \tag{12.67}$$

The orthonormal set of four basis vectors—our set of measuring rods and a clock—is often called a *tetrad* or *vierbein* in relativity. The corresponding normalised components of vectors or tensors are the proper quantities that we actually measure. If *these* diverge at $r = 2M$ in Schwarzschild spacetime, then there must be some sort of real divergence at $r = 2M$, because the normalised basis vectors are now well behaved and not shrinking to zero length. For example, if the normalised Riemann components diverge at $r = 2M$, then we can conclude that there is a real curvature divergence there.

To illustrate the procedure, focus on just one of the Riemann components, $R^t{}_{rtr}$ in (12.65). It diverges at both $r = 0$ and $r = 2M$, so we must check the corresponding behaviour of $R^{\hat{t}}{}_{\hat{r}\hat{t}\hat{r}}$. A point of notation is important here: putting carets on each of the indices is tedious and guarantees tired muscles; we'll write this component as $\widehat{R}^t{}_{rtr}$ instead, even though that necessitates writing $\boldsymbol{e}_{\hat{\alpha}}$ as $\widehat{\boldsymbol{e}}_\alpha$ to allow the summation convention still to apply. (First and foremost, notation should always be useful as opposed to taxing, and writing dozens of carets is definitely taxing.)

First, we need to know the corresponding versions of (12.67) for higher-order tensors. In general, everything is calculated from (8.121), but for the orthogonal coordinate basis it's easier to do it explicitly in the following way. Write a general second-order tensor as

$$\boldsymbol{T} = T^{\alpha\beta}\boldsymbol{e}_\alpha \boldsymbol{e}_\beta \overset{\text{req.}}{=\!=} \widehat{T}^{\alpha\beta}\,\widehat{\boldsymbol{e}}_\alpha \widehat{\boldsymbol{e}}_\beta = \widehat{T}^{\alpha\beta}\,\frac{\boldsymbol{e}_\alpha}{|\boldsymbol{e}_\alpha|}\,\frac{\boldsymbol{e}_\beta}{|\boldsymbol{e}_\beta|}\,, \tag{12.68}$$

so that

$$\widehat{T}^{\alpha\beta} = T^{\alpha\beta}\,|\boldsymbol{e}_\alpha|\,|\boldsymbol{e}_\beta| \quad \text{(no sums)}, \tag{12.69}$$

and similarly $\widehat{T}^\alpha{}_\beta = T^\alpha{}_\beta\,|\boldsymbol{e}_\alpha|\,|\boldsymbol{e}^\beta|$ (no sums), and so on. Further, what are the values of $|\boldsymbol{e}_\alpha|$ and $|\boldsymbol{e}^\alpha|$? For a positive definite metric (one having all plus signs), we're familiar with the basis vector lengths calculated from $|\boldsymbol{e}_\alpha|^2 = \boldsymbol{e}_\alpha\!\cdot\!\boldsymbol{e}_\alpha = g_{\alpha\alpha}$. But this is insufficient for the mixed signs of the metrics encountered in relativity. We require that any normalisation preserve the metric signature (i.e., the signs $+---$ or $-+++$ of the metric should be preserved). In that case, demand the normalised metric to have signs given by $\widehat{g}_{\alpha\alpha} \equiv \operatorname{sgn} g_{\alpha\alpha}$. But that implies

$$\operatorname{sgn} g_{\alpha\alpha} = \widehat{g}_{\alpha\alpha} = \widehat{\boldsymbol{e}}_\alpha \cdot \widehat{\boldsymbol{e}}_\alpha = \frac{\boldsymbol{e}_\alpha}{|\boldsymbol{e}_\alpha|}\cdot\frac{\boldsymbol{e}_\alpha}{|\boldsymbol{e}_\alpha|} = \frac{g_{\alpha\alpha}}{|\boldsymbol{e}_\alpha|^2}\,, \tag{12.70}$$

in which case

$$|\boldsymbol{e}_\alpha|^2 = \frac{g_{\alpha\alpha}}{\operatorname{sgn} g_{\alpha\alpha}} = |g_{\alpha\alpha}| = |\boldsymbol{e}_\alpha \cdot \boldsymbol{e}_\alpha|\,. \tag{12.71}$$

Of course, this is no different from $|\boldsymbol{e}_\alpha|^2 = \boldsymbol{e}_\alpha \cdot \boldsymbol{e}_\alpha = g_{\alpha\alpha}$ in a positive definite metric, but additionally it ensures that the signature of the normalised metric is unchanged. Thus, the required basis vector lengths are

$$\boxed{|\boldsymbol{e}_\alpha| = \sqrt{|g_{\alpha\alpha}|}\,, \quad \text{and similarly } |\boldsymbol{e}^\alpha| = \sqrt{|g^{\alpha\alpha}|}\,.} \tag{12.72}$$

Now that we know how to normalise tensor components, focus once more on $R^t{}_{rtr}$ and $\widehat{R}^t{}_{rtr}$.

$$\widehat{R}^t{}_{rtr} = R^t{}_{rtr}\, |\mathbf{e}_t|\, |\mathbf{e}^r|\, |\mathbf{e}^t|\, |\mathbf{e}^r| \quad \text{(no sum)}$$
$$= R^t{}_{rtr}\sqrt{|g_{tt}\, g^{rr} g^{tt} g^{rr}|} = R^t{}_{rtr}\, |g^{rr}| = \frac{2M}{r^3}\,\text{sgn}\left(1 - \frac{2M}{r}\right). \qquad (12.73)$$

The final expression in (12.73) doesn't diverge at $r = 2M$, but unfortunately it switches sign there! However, we remember that the roles of t and r also swap inside the Schwarzschild radius. This suggests we should also calculate $\widehat{R}^r{}_{trt}$. The calculation is straightforward: $R^r{}_{trt}$ is related to $R^t{}_{rtr}$ via the metric and Riemann symmetries (9.91), and the result is

$$\widehat{R}^r{}_{trt} = \frac{-2M}{r^3}\,\text{sgn}\left(1 - \frac{2M}{r}\right). \qquad (12.74)$$

Equations (12.73) and (12.74) can be combined if we define a coordinate τ (not to be confused with proper time) and a coordinate ϱ, that swap time and space inside the Schwarzschild radius:

$$\tau, \varrho \equiv \begin{cases} t, r & r > 2M \\ r, t & r < 2M \end{cases}. \qquad (12.75)$$

Since $g_{\tau\tau}$ is always positive both outside and inside the Schwarzschild radius, τ is timelike: it describes the spacetime direction in which a free particle moves. Similarly, ϱ is spacelike because $g_{\varrho\varrho}$ is always negative both outside and inside the Schwarzschild radius. Using these, the Riemann components become

$$\widehat{R}^\tau{}_{\varrho\tau\varrho} = \frac{2M}{r^3}, \quad \widehat{R}^\varrho{}_{\tau\varrho\tau} = \frac{-2M}{r^3} \qquad (12.76)$$

over the whole of Schwarzschild spacetime. So with this time–space swap incorporated, the normalised Riemann components don't diverge at $r = 2M$, and we know for certain that spacetime is perfectly well behaved there. The normalised Riemann components still diverge at $r = 0$, meaning that this point is a spacetime *singularity*. But what happens when a real star collapses is as yet unknown, and this singularity that signals a breakdown of general relativity might presumably fail to form for some other reason that is part of new physics yet to be discovered.

12.7 Calculating Curvature More Efficiently: Cartan's Structural Equations

Solving Einstein's equation requires us to begin with some general metric, perhaps simplified due to an imposition such as spherical symmetry, and then calculate the Ricci tensor from it. The procedure that was developed in Chap. 9 first calculates the Christoffel symbols and then uses them to build the Riemann components. It is of course quite labour-intensive; but luckily another approach, due to Cartan, exploits symmetries in the Riemann tensor to make some of the work easier. We'll omit the proof of Cartan's method, and instead will focus here on how to use it to calculate the Riemann tensor given the Schwarzschild metric.

Not surprisingly, because the Riemann tensor is antisymmetric in various pairs of its indices (9.91), an approach using the wedge product (also antisymmetric) shortens the work required. The notation of this section uses the exterior derivative $\nabla\wedge$ that we introduced in Sect. 8.10. In texts, the exterior derivative is more usually written as "d", but for the reasons given in Sect. 8.10 we'll avoid that notation. As mentioned on p. 347, a suitable alternative might be $\eth$. Here we'll continue to write it as the bulkier $\nabla\wedge$ only to emphasise the inner workings of the exterior derivative.

The relevant equations are *Cartan's first and second structural equations.* Essentially, the Cartan approach relies on calculations being done in an orthonormal frame, so that index raising and lowering can be done using the Minkowski metric. The central quantities of Cartan's equations are a set of vectors $\boldsymbol{\Gamma}^{\widehat{\alpha}}{}_{\widehat{\beta}}$ that are indexed by the orthonormal frame and related to Christoffel symbols. (For the sake of careful clarity, we'll include all carets on the indices here as opposed to the more relaxed approach of Sect. 12.6.1.) Cartan's first equation allows us to calculate the $\boldsymbol{\Gamma}^{\widehat{\alpha}}{}_{\widehat{\beta}}$ indirectly by way of

$$\boxed{\boldsymbol{e}^{\widehat{\beta}}\wedge\boldsymbol{\Gamma}^{\widehat{\alpha}}{}_{\widehat{\beta}}=\nabla\wedge\boldsymbol{e}^{\widehat{\alpha}}\,.} \tag{12.77}$$

The $\boldsymbol{\Gamma}^{\widehat{\alpha}}{}_{\widehat{\beta}}$ have the following properties, for spatial indices a, b:

$$\begin{aligned}&\boldsymbol{\Gamma}^{\widehat{t}}{}_{\widehat{a}}=\boldsymbol{\Gamma}^{\widehat{a}}{}_{\widehat{t}}\,,\qquad \boldsymbol{\Gamma}^{\widehat{a}}{}_{\widehat{b}}=-\boldsymbol{\Gamma}^{\widehat{b}}{}_{\widehat{a}}\,,\\ &\boldsymbol{\Gamma}^{\widehat{\alpha}}{}_{\widehat{\alpha}}=\boldsymbol{0}\quad\text{(no sum, for all }\alpha).\end{aligned} \tag{12.78}$$

The second of Cartan's equations uses the $\boldsymbol{\Gamma}^{\widehat{\alpha}}{}_{\widehat{\beta}}$ to give a prescription for calculating the Riemann components. We'll employ a small device for writing antisymmetric tensors $F_{\alpha\beta}$ that saves tedious index manipulation or the need to include numerical factors. This defines

$$F_{|\alpha\beta|}\,\boldsymbol{e}^{\alpha}\wedge\boldsymbol{e}^{\beta}\equiv\begin{bmatrix}\text{the sum over all } \textit{combinations} \text{ of } \alpha,\beta,\\ \text{as opposed to } \textit{permutations}.\end{bmatrix} \tag{12.79}$$

As an example of this notation using two coordinates x, y for simplicity, the usual wedge expression for an antisymmetric tensor $F_{\alpha\beta}$ is, following the argument of (8.247),

$$\begin{aligned}\boldsymbol{F}&={}^1\!/_2\,F_{\alpha\beta}\,\boldsymbol{e}^{\alpha}\wedge\boldsymbol{e}^{\beta}={}^1\!/_2\left(F_{xy}\,\boldsymbol{e}^{x}\wedge\boldsymbol{e}^{y}+F_{yx}\,\boldsymbol{e}^{y}\wedge\boldsymbol{e}^{x}\right)\\ &=F_{xy}\,\boldsymbol{e}^{x}\wedge\boldsymbol{e}^{y}=F_{|\alpha\beta|}\,\boldsymbol{e}^{\alpha}\wedge\boldsymbol{e}^{\beta}.\end{aligned} \tag{12.80}$$

In general, an n-index antisymmetric tensor $\boldsymbol{F}$ (i.e., an n-multivector) can be written in any of the following three ways, each of which has its uses:

$$
\begin{aligned}
\boldsymbol{F} &= F_{\alpha\beta\ldots\omega}\, \boldsymbol{e}^{\alpha}\boldsymbol{e}^{\beta}\ldots\boldsymbol{e}^{\omega} \\
&= \frac{1}{n!}\, F_{\alpha\beta\ldots\omega}\, \boldsymbol{e}^{\alpha}\wedge\boldsymbol{e}^{\beta}\wedge\cdots\wedge\boldsymbol{e}^{\omega} \\
&= F_{|\alpha\beta\ldots\omega|}\, \boldsymbol{e}^{\alpha}\wedge\boldsymbol{e}^{\beta}\wedge\cdots\wedge\boldsymbol{e}^{\omega}.
\end{aligned}
\tag{12.81}
$$

With this notation, the second of Cartan's equations is

$$
\boxed{R^{\widehat{\alpha}}{}_{\widehat{\beta}|\widehat{\mu}\widehat{\nu}|}\, \boldsymbol{e}^{\widehat{\mu}}\wedge\boldsymbol{e}^{\widehat{\nu}} = \nabla\wedge\boldsymbol{\Gamma}^{\widehat{\alpha}}{}_{\widehat{\beta}} + \boldsymbol{\Gamma}^{\widehat{\alpha}}{}_{\widehat{\lambda}}\wedge\boldsymbol{\Gamma}^{\widehat{\lambda}}{}_{\widehat{\beta}}\,.}
\tag{12.82}
$$

Let's see how (12.77), (12.78), and (12.82) are used to calculate the curvature generated by the Schwarzschild metric. Setting $\kappa \equiv 1 - 2M/r$, the nonzero metric elements are

$$
\begin{aligned}
&g_{tt} = \kappa\,, && g_{rr} = -1/\kappa\,, && g_{\theta\theta} = -r^2, && g_{\phi\phi} = -r^2\sin^2\theta\,, \\
&g^{tt} = 1/\kappa\,, && g^{rr} = -\kappa\,, && g^{\theta\theta} = \frac{-1}{r^2}\,, && g^{\phi\phi} = \frac{-1}{r^2\sin^2\theta}\,.
\end{aligned}
\tag{12.83}
$$

Using (12.72), these give cobasis vector lengths of

$$
\left|\boldsymbol{e}^{t}\right| = |\kappa|^{-1/2}, \quad \left|\boldsymbol{e}^{r}\right| = |\kappa|^{1/2}, \quad \left|\boldsymbol{e}^{\theta}\right| = \frac{1}{r}\,, \quad \left|\boldsymbol{e}^{\phi}\right| = \frac{1}{r\sin\theta}\,,
\tag{12.84}
$$

so that the simplest orthonormal cobasis is

$$
\boldsymbol{e}^{\widehat{t}} = |\kappa|^{1/2}\boldsymbol{e}^{t}\,, \quad \boldsymbol{e}^{\widehat{r}} = |\kappa|^{-1/2}\boldsymbol{e}^{r}\,, \quad \boldsymbol{e}^{\widehat{\theta}} = r\boldsymbol{e}^{\theta}\,, \quad \boldsymbol{e}^{\widehat{\phi}} = r\sin\theta\,\boldsymbol{e}^{\phi}.
\tag{12.85}
$$

The exterior derivatives of these orthonormalised vectors are needed for (12.77). Remembering that κ is a function of r only, the first is

$$
\nabla\wedge\boldsymbol{e}^{\widehat{t}} = \boldsymbol{e}^{\alpha}\partial_{\alpha}\wedge\left(|\kappa|^{1/2}\boldsymbol{e}^{t}\right) = \partial_{r}\left(|\kappa|^{1/2}\right)\boldsymbol{e}^{r}\wedge\boldsymbol{e}^{t} = \partial_{r}\left(|\kappa|^{1/2}\right)\boldsymbol{e}^{\widehat{r}}\wedge\boldsymbol{e}^{\widehat{t}}.
\tag{12.86}
$$

Similarly, the three other derivatives are

$$
\begin{aligned}
\nabla\wedge\boldsymbol{e}^{\widehat{r}} &= 0\,, \\
\nabla\wedge\boldsymbol{e}^{\widehat{\theta}} &= \frac{|\kappa|^{1/2}}{r}\,\boldsymbol{e}^{\widehat{r}}\wedge\boldsymbol{e}^{\widehat{\theta}}\,, \\
\nabla\wedge\boldsymbol{e}^{\widehat{\phi}} &= \frac{|\kappa|^{1/2}}{r}\,\boldsymbol{e}^{\widehat{r}}\wedge\boldsymbol{e}^{\widehat{\phi}} + \frac{\cot\theta}{r}\,\boldsymbol{e}^{\widehat{\theta}}\wedge\boldsymbol{e}^{\widehat{\phi}}.
\end{aligned}
\tag{12.87}
$$

The $\boldsymbol{\Gamma}^{\widehat{\alpha}}{}_{\widehat{\beta}}$ are found by writing (12.77) out in full, remembering from (12.78) that all $\boldsymbol{\Gamma}^{\widehat{\alpha}}{}_{\widehat{\alpha}} = 0$ (no sum):

$$
\begin{aligned}
\nabla\wedge\boldsymbol{e}^{\widehat{t}} &= \boldsymbol{e}^{\widehat{r}}\wedge\boldsymbol{\Gamma}^{\widehat{t}}{}_{\widehat{r}} + \boldsymbol{e}^{\widehat{\theta}}\wedge\boldsymbol{\Gamma}^{\widehat{t}}{}_{\widehat{\theta}} + \boldsymbol{e}^{\widehat{\phi}}\wedge\boldsymbol{\Gamma}^{\widehat{t}}{}_{\widehat{\phi}} &&= \partial_{r}\left(|\kappa|^{1/2}\right)\boldsymbol{e}^{\widehat{r}}\wedge\boldsymbol{e}^{\widehat{t}}\,, \\
\nabla\wedge\boldsymbol{e}^{\widehat{r}} &= \boldsymbol{e}^{\widehat{t}}\wedge\boldsymbol{\Gamma}^{\widehat{r}}{}_{\widehat{t}} + \boldsymbol{e}^{\widehat{\theta}}\wedge\boldsymbol{\Gamma}^{\widehat{r}}{}_{\widehat{\theta}} + \boldsymbol{e}^{\widehat{\phi}}\wedge\boldsymbol{\Gamma}^{\widehat{r}}{}_{\widehat{\phi}} &&= 0\,, \\
\nabla\wedge\boldsymbol{e}^{\widehat{\theta}} &= \boldsymbol{e}^{\widehat{t}}\wedge\boldsymbol{\Gamma}^{\widehat{\theta}}{}_{\widehat{t}} + \boldsymbol{e}^{\widehat{r}}\wedge\boldsymbol{\Gamma}^{\widehat{\theta}}{}_{\widehat{r}} + \boldsymbol{e}^{\widehat{\phi}}\wedge\boldsymbol{\Gamma}^{\widehat{\theta}}{}_{\widehat{\phi}} &&= \frac{|\kappa|^{1/2}}{r}\,\boldsymbol{e}^{\widehat{r}}\wedge\boldsymbol{e}^{\widehat{\theta}}\,, \\
\nabla\wedge\boldsymbol{e}^{\widehat{\phi}} &= \boldsymbol{e}^{\widehat{t}}\wedge\boldsymbol{\Gamma}^{\widehat{\phi}}{}_{\widehat{t}} + \boldsymbol{e}^{\widehat{r}}\wedge\boldsymbol{\Gamma}^{\widehat{\phi}}{}_{\widehat{r}} + \boldsymbol{e}^{\widehat{\theta}}\wedge\boldsymbol{\Gamma}^{\widehat{\phi}}{}_{\widehat{\theta}} &&= \frac{|\kappa|^{1/2}}{r}\,\boldsymbol{e}^{\widehat{r}}\wedge\boldsymbol{e}^{\widehat{\theta}} + \frac{\cot\theta}{r}\,\boldsymbol{e}^{\widehat{\theta}}\wedge\boldsymbol{e}^{\widehat{\phi}}.
\end{aligned}
\tag{12.88}
$$

The identities of (12.78) allow us to study (12.88) to find the values of the various $\boldsymbol{\Gamma}^{\widehat{\alpha}}{}_{\widehat{\beta}}$. Some detective work is required, since usually two equations of the set (12.88) need to be compared to find each $\boldsymbol{\Gamma}^{\widehat{\alpha}}{}_{\widehat{\beta}}$. The results are

$$\begin{aligned}
&\boldsymbol{\Gamma}^{\widehat{t}}{}_{\widehat{r}} = \partial_r\left(|\kappa|^{1/2}\right) \boldsymbol{e}^{\widehat{t}} = \boldsymbol{\Gamma}^{\widehat{r}}{}_{\widehat{t}}\,, && \boldsymbol{\Gamma}^{\widehat{t}}{}_{\widehat{\theta}} = \boldsymbol{0} = \boldsymbol{\Gamma}^{\widehat{t}}{}_{\widehat{\phi}} = \boldsymbol{\Gamma}^{\widehat{\theta}}{}_{\widehat{t}} = \boldsymbol{\Gamma}^{\widehat{\phi}}{}_{\widehat{t}}\,,\\
&\boldsymbol{\Gamma}^{\widehat{\theta}}{}_{\widehat{r}} = \frac{|\kappa|^{1/2}}{r}\,\boldsymbol{e}^{\widehat{\theta}} = -\boldsymbol{\Gamma}^{\widehat{r}}{}_{\widehat{\theta}}\,, && \boldsymbol{\Gamma}^{\widehat{\phi}}{}_{\widehat{r}} = \frac{|\kappa|^{1/2}}{r}\,\boldsymbol{e}^{\widehat{\phi}} = -\boldsymbol{\Gamma}^{\widehat{r}}{}_{\widehat{\phi}}\,,\\
&\boldsymbol{\Gamma}^{\widehat{\phi}}{}_{\widehat{\theta}} = \frac{\cot\theta}{r}\,\boldsymbol{e}^{\widehat{\phi}} = -\boldsymbol{\Gamma}^{\widehat{\theta}}{}_{\widehat{\phi}}\,.
\end{aligned} \tag{12.89}$$

The terms, though many, are straightforward to write and the work is not onerous. All of the required quantities are now in place to calculate the Riemann components. Suppose we wish to find $R^t{}_{r\alpha\beta}$ for all α, β. Apply (12.82) to first calculate the normalised components:

$$\begin{aligned}
R^{\widehat{t}}{}_{\widehat{r}|\widehat{\alpha}\widehat{\beta}|}\,\boldsymbol{e}^{\widehat{\alpha}} \wedge \boldsymbol{e}^{\widehat{\beta}} &= \nabla \wedge \boldsymbol{\Gamma}^{\widehat{t}}{}_{\widehat{r}} + \text{zero terms}\\
&= \boldsymbol{e}^{\alpha}\partial_\alpha \wedge \left[\partial_r\left(|\kappa|^{1/2}\right)|\kappa|^{1/2}\,\boldsymbol{e}^{t}\right] = \frac{2M}{r^3}\,\operatorname{sgn}\kappa\;\, \boldsymbol{e}^{\widehat{t}} \wedge \boldsymbol{e}^{\widehat{r}}\,,
\end{aligned} \tag{12.90}$$

and this implies that $R^{\widehat{t}}{}_{\widehat{r}\widehat{t}\widehat{r}} = \dfrac{2M}{r^3}\operatorname{sgn}\kappa$ as we saw earlier in (12.73), along with $R^{\widehat{t}}{}_{\widehat{r}\widehat{r}\widehat{t}} = -R^{\widehat{t}}{}_{\widehat{r}\widehat{t}\widehat{r}}$. Finally, (12.73) converts between the two bases:

$$R^t{}_{rtr} = \frac{R^{\widehat{t}}{}_{\widehat{r}\widehat{t}\widehat{r}}}{|\boldsymbol{e}_t|\;|\boldsymbol{e}^r|\;|\boldsymbol{e}^t|\;|\boldsymbol{e}^r|} = \frac{R^{\widehat{t}}{}_{\widehat{r}\widehat{t}\widehat{r}}}{\sqrt{|g_{tt}g^{rr}g^{tt}g^{rr}|}} = \frac{2M}{r^3\kappa}\,. \tag{12.91}$$

And, of course, all of the other Riemann components are now easily found because the hard work has all been done in calculating the $\boldsymbol{\Gamma}^{\widehat{\alpha}}{}_{\widehat{\beta}}$. The Cartan approach shows how orthonormal bases are very useful in simplifying the calculations of differential geometry, along with their use in interpreting measurements in general relativity.

12.8 The Variational Approach to Einstein's Equation

Einstein originally postulated his field equation (12.45), that governs how spacetime curvature is produced by matter, by considering what the most reasonable tensors describing these quantities might be. In fact, at about the same time that these ideas were publicised, and building on Einstein's earlier work, Hilbert derived the field equation quite differently by way of a variational approach. We'll describe that approach in this section. The usefulness of Hilbert's method lies in its opening the door to other theories of gravity, still based on Einstein's but governed by different field equations. These alternative theories still drive experiments today. Even so, Einstein's equation is possibly the simplest that can result from a variational approach, and as yet no experimental evidence has decided against it in favour of competing, but more complex, field equations.

For simplicity, begin by considering an empty universe; later we'll add a term to the lagrangian to incorporate any other fields present. It's easy to see the elegance of Hilbert's approach immediately, because the lagrangian density that produces Einstein's equation is just the Ricci curvature multiplied by a factor related to the metric:

$$\mathcal{L} = R\sqrt{-g}\,. \tag{12.92}$$

The use of the Ricci scalar is elegant, but before we focus on it, let's examine why the extra factor of $\sqrt{-g}$ must be present. In order to have a fully covariant lagrangian, we must ensure that the correct integration measure has been included. This is not simply $\mathrm{d}^4x \equiv \mathrm{d}x^0 \ldots \mathrm{d}x^3$ for any coordinates x^α; there will need to be a scale factor derived from the metric, since it is the metric that defines the notion of a volume. We calculated this measure when studying tensors, finding in (8.165) that the correct measure in primed coordinates as compared with unprimed coordinates is

$$\mathrm{d}x^0 \ldots \mathrm{d}x^3 \longleftrightarrow \sqrt{g'/g}\, \mathrm{d}x^{0'} \ldots \mathrm{d}x^{3'}\,, \tag{12.93}$$

where g, g' are the metric determinants. We know that the left-hand side of (12.93) is the correct spacetime volume for a locally inertial observer, whose metric is that of Minkowski space: $g_{\alpha\beta} = \eta_{\alpha\beta} = \eta_{tt}(1,-1,-1,-1)$. In that case, $g = -1$ and the right-hand side of (12.93) becomes the correct volume for an arbitrary observer: $\sqrt{-g'}\ \mathrm{d}x^{0'} \ldots \mathrm{d}x^{3'}$. We'll drop the primes on the indices from now on.

Returning to Hilbert's approach, let's see how Einstein's equation results from varying the *Hilbert action* S with respect to the metric,

$$S = \int_{\text{4-volume}} R\sqrt{-g}\, \mathrm{d}^4x\,, \tag{12.94}$$

where the constraint on the boundary 3-surface is that variations in the metric and its first derivatives go to zero there. That is, $\delta g_{\alpha\beta} = 0 = \delta \Gamma^{\alpha}{}_{\beta\gamma}$ on the boundary. (The factor of $\sqrt{-g}$ is so commonplace in variational calculations that it's often absorbed into the rest of the integrand, which is then written in a Gothic font in some texts.)

The approach we'll take is to set the action variation δS to zero, as was described in Sect. 10.3.2. But because the lagrangian is a very complicated function of the metric and its first and second derivatives, we'll follow a different line than the usual Lagrange equation approach of calculating lots of partial derivatives, and instead will break the Ricci scalar and volume element up into more manageable units. Begin with

$$
\begin{aligned}
\delta S &= \int \delta\left(R\sqrt{-g}\right) \mathrm{d}^4x = \int \delta\left(g^{\alpha\beta} R_{\alpha\beta} \sqrt{-g}\right) \mathrm{d}^4x \\
&= \int \left[\delta g^{\alpha\beta} R_{\alpha\beta} \sqrt{-g} + g^{\alpha\beta}\, \delta R_{\alpha\beta} \sqrt{-g} + R\,\delta\sqrt{-g}\right] \mathrm{d}^4x\,. \qquad (12.95)
\end{aligned}
$$

Our strategy is to bring the metric variation $\delta g^{\alpha\beta}$ outside the brackets, just as was done in (10.24). It will need to be factored out of each of the three terms in the brackets of (12.95).

The first term in the brackets of (12.95) already has $\delta g^{\alpha\beta}$ present and so needs no further attention. **The second term in the brackets of** (12.95) includes a variation in the Ricci tensor, defined in (9.122) as

$$
R_{\alpha\beta} = R^{\lambda}{}_{\alpha\lambda\beta} \overset{(9.89)}{=\!=} \quad -\Gamma^{\lambda}{}_{\alpha\lambda,\beta} - \Gamma^{\mu}{}_{\alpha\lambda}\, \Gamma^{\lambda}{}_{\beta\mu}\,. \qquad (12.96)
$$

A first-order variation in the Ricci tensor is then

$$
\delta R_{\alpha\beta} = -\delta\Gamma^{\lambda}{}_{\alpha\lambda,\beta} - \delta\Gamma^{\mu}{}_{\alpha\lambda}\, \Gamma^{\lambda}{}_{\beta\mu} - \Gamma^{\mu}{}_{\alpha\lambda}\, \delta\Gamma^{\lambda}{}_{\beta\mu}\,, \qquad (12.97)
$$

remembering that, as in (10.28), we need not include parentheses in the first term on the right-hand side of (12.97), because an expression such as $\delta\Gamma^{\lambda}{}_{\alpha\lambda,\beta}$ is unambiguous. Further tensor analysis will be easier if we convert the commas of that first term to semicolons by including Christoffel symbols in the usual way. But that can only be done if $\delta\Gamma^{\lambda}{}_{\alpha\lambda}$ is a tensor. In fact, it certainly *is* a tensor, as can be shown by starting with (8.195) and varying the metric as $g_{\mu\nu} \to g_{\mu\nu} + \delta g_{\mu\nu}$. Omitting the indices in (8.195) suffices to prove the point:

$$
\begin{aligned}
\Gamma &= \Lambda\Lambda + \Lambda\Lambda\Lambda\, \Gamma'\,, \\
\Gamma + \delta\Gamma &= \Lambda\Lambda + \Lambda\Lambda\Lambda\, (\Gamma' + \delta\Gamma')\,, \qquad (12.98)
\end{aligned}
$$

where the indices are the same in both expressions, and the term $\Lambda\Lambda\Lambda\, \Gamma'$ is the normal tensor transformation term. A subtraction then ensures that only the last term,

$$
\delta\Gamma = \Lambda\Lambda\Lambda\, \delta\Gamma'\,, \qquad (12.99)
$$

survives, which means $\delta\Gamma$ transforms in the way required of a tensor. So it certainly is a tensor. It follows that we're able to write

$$
\delta\Gamma^{\lambda}{}_{\alpha\lambda;\beta} = \delta\Gamma^{\lambda}{}_{\alpha\lambda,\beta} + \Gamma^{\lambda}{}_{\mu\beta}\, \delta\Gamma^{\mu}{}_{\alpha\lambda} - \Gamma^{\mu}{}_{\alpha\beta}\, \delta\Gamma^{\lambda}{}_{\mu\lambda} - \Gamma^{\mu}{}_{\lambda\beta}\, \delta\Gamma^{\lambda}{}_{\alpha\mu}\,. \qquad (12.100)
$$

The last term here vanishes due to the symmetry of its first Christoffel symbol's lower indices. What remains gives an expression for $\delta\Gamma^{\lambda}{}_{\alpha\lambda,\beta}$, and on substituting this into (12.97), nearly everything cancels to give

$$
\delta R_{\alpha\beta} = \delta\Gamma^{\lambda}{}_{\alpha\beta;\lambda}\,. \qquad (12.101)
$$

With hindsight, this bulk cancellation is quite reasonable. Because $\delta R_{\alpha\beta}$ is a tensor (using the same argument as was done for the Christoffel variations $\delta\Gamma$), we are free to calculate (12.97) in any coordinates. Specifically, choose ones in which the Christoffel symbols vanish at the point of interest, so that in these coordinates $\delta R_{\alpha\beta} = \delta\Gamma^{\lambda}{}_{\alpha\underset{\sqcup}{\beta},\lambda}$. This can then be generalised to any other coordinate system by changing the comma to a semicolon to give (12.101).

Finally, the second term in (12.95) is (incorporating a $\beta \leftrightarrow \lambda$ swap in the following brackets for readability)

$$g^{\alpha\beta}\,\delta R_{\alpha\beta}\sqrt{-g} = \sqrt{-g}\left[g^{\alpha\beta}\,\delta\Gamma^{\lambda}{}_{\alpha\beta} - g^{\alpha\lambda}\,\delta\Gamma^{\beta}{}_{\alpha\beta}\right]_{;\lambda} \equiv \sqrt{-g}\,A^{\lambda}{}_{;\lambda} \qquad (12.102)$$

for some A^{λ}. Although the metric variation $\delta g^{\alpha\beta}$ has not been brought outside the brackets, we'll have no need to do so, since the last expression in (12.102) is a four-dimensional divergence, and so can be integrated via the four-dimensional version of Gauss's theorem:

$$\int_{\text{4-volume}} A^{\lambda}{}_{;\lambda}\sqrt{-g}\,\mathrm{d}^4x = \int_{\text{3-surface}} A^{\lambda}n_{\lambda}\sqrt{-g}\,\mathrm{d}^3x\,, \qquad (12.103)$$

where again proper measures are used, and n_{λ} is a unit normal of the 3-surface.

This four-dimensional Gauss theorem is not the same as the generalised Stokes–Gauss theorem in Sect. 8.10, which would integrate a trivector—a 3-index antisymmetric tensor—over the 3-surface, along with its exterior derivative over the enclosed 4-volume.

But remember that the Christoffel variations (making up A^{λ}) are stipulated to vanish on the 3-surface, which means that the right-hand side of (12.103) vanishes. Hence the second term of (12.95) is zero.

Last, focus on **the third term in the brackets of** (12.95), where $\sqrt{-g}$ is varied with respect to $g^{\alpha\beta}$. Because $\sqrt{-g}$ is just a function of the metric but not the metric's derivatives, this variation is relatively easy, since we can just make use of partial derivatives by writing the analogue of (10.26) as

$$\frac{\delta\sqrt{-g}}{\delta g^{\alpha\beta}} = \frac{\partial\sqrt{-g}}{\partial g^{\alpha\beta}}\,. \qquad (12.104)$$

The differentiation is accomplished with the aid of results derived in Chap. 8. There, in (8.221), we found that

$$\frac{\partial g}{\partial g_{\alpha\beta}} = g\,g^{\alpha\beta}\,, \quad \text{and} \quad \frac{\partial g}{\partial g^{\alpha\beta}} = -g\,g_{\alpha\beta}\,. \qquad (12.105)$$

Using (12.104) and (12.105), the third term of the action variation integrand (12.95) becomes

$$R\,\delta\sqrt{-g} = \frac{-R}{2}\sqrt{-g}\,g_{\alpha\beta}\,\delta g^{\alpha\beta}. \tag{12.106}$$

Now the three bracketed terms of (12.95) can be gathered together. The first required no work on our part because the metric variation was already present. The second term integrated to zero back in (12.102) and (12.103), while the third term is in (12.106). Putting it all together,

$$\begin{aligned}\delta S &= \int \left(R_{\alpha\beta} - \frac{R}{2}g_{\alpha\beta}\right)\delta g^{\alpha\beta}\sqrt{-g}\,\mathrm{d}^4x \\ &= \int G_{\alpha\beta}\,\delta g^{\alpha\beta}\sqrt{-g}\,\mathrm{d}^4x\,. \end{aligned} \tag{12.107}$$

Requiring $\delta S = 0$ for all metric variations $\delta g^{\alpha\beta}$ that go smoothly to zero at the boundary thus produces

$$G_{\alpha\beta} = 0\,, \tag{12.108}$$

which is exactly Einstein's equation describing the curved spacetime of an empty universe! This has resulted from varying what is perhaps the most rendered-down description of curvature: the Ricci scalar.

A comment about the notation is needed here in the same vein as was discussed earlier on p. 395. Because the beginning and end of our variational calculation are written as

$$\delta S = \int \delta\left(R\sqrt{-g}\right)\mathrm{d}^4x = \cdots = \int G_{\alpha\beta}\,\delta g^{\alpha\beta}\sqrt{-g}\,\mathrm{d}^4x\,, \tag{12.109}$$

a convention has arisen that equates the integrands to give

$$\delta\left(R\sqrt{-g}\right) = G_{\alpha\beta}\,\delta g^{\alpha\beta}\sqrt{-g} \quad \text{(summation implied)}, \tag{12.110}$$

and even "divides" by $\delta g^{\alpha\beta}$ to give

$$\frac{\delta\left(R\sqrt{-g}\right)}{\delta g^{\alpha\beta}} = G_{\alpha\beta}\sqrt{-g}\,, \tag{12.111}$$

which has no summation, but is still covariant. But always remember that these expressions are rather like identities involving the Dirac delta function: they always presuppose an integration to be done wherein the boundary terms will disappear, such as occurred with (12.102) and (12.103). Only in this sense can they be considered as equalities. Notice how useful the tensor notation is here. By "dividing" each side of (12.110) by $\delta g^{\alpha\beta}$ and ignoring the implied summation, the result (12.111) holds, even though it contains no sum. We saw this previously, in the discussion just after (9.104).

12.8.1 Adding Extra Field Terms to the Lagrangian Density

What results if further fields are added to the lagrangian density in (12.94)? We saw this idea in Chap. 10 when discussing field theories, and we can do the

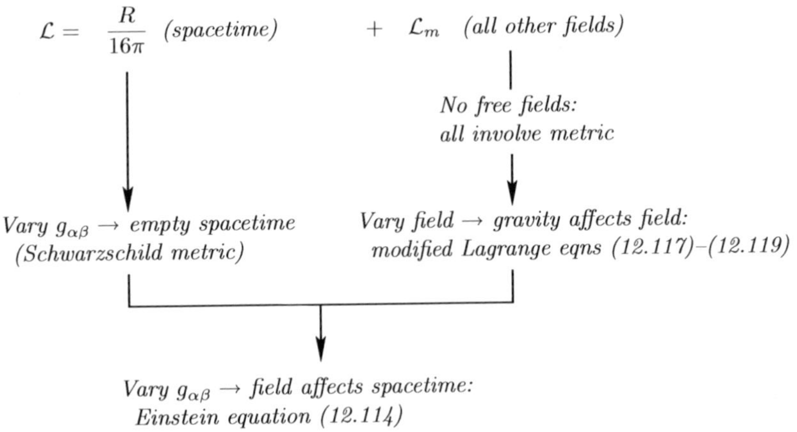

Fig. 12.8. Adding extra terms $\mathcal{L}_m$ to the Hilbert lagrangian density. Varying the metric determines how the fields affect spacetime. Conversely, varying each field determines how curved spacetime affects that field.

same for gravity to investigate how something like an electromagnetic field might behave in a curved spacetime. Because we're now dealing solely with fields, the added terms will be density terms, so that for example incorporating an electromagnetic field necessitates including $-\varepsilon_0 F^2/4$ as opposed to the integral of this that appeared in Fig. 10.7. A map of the process of adding a new term is shown in Fig. 12.8. With hindsight, to produce the correct field equations, while keeping the established weightings for lagrangian terms describing other fields, requires that the newcomer R be scaled by $1/(16\pi)$. Suppose then that all other matter/field contributions are described by a lagrangian density $\mathcal{L}_m$, so that the new action is

$$S = \int \left(\frac{R}{16\pi} + \mathcal{L}_m \right) \sqrt{-g}\, \mathrm{d}^4 x \,. \tag{12.112}$$

The variation of this is

$$\begin{aligned} \delta S &= \int \left(\frac{1}{16\pi} G_{\alpha\beta}\, \delta g^{\alpha\beta} \sqrt{-g} + \delta\left(\mathcal{L}_m \sqrt{-g} \right) \right) \mathrm{d}^4 x \\ &= \int \left(\frac{G_{\alpha\beta}}{16\pi} + \frac{1}{\sqrt{-g}} \frac{\delta\left(\mathcal{L}_m \sqrt{-g}\right)}{\delta g^{\alpha\beta}} \right) \delta g^{\alpha\beta} \sqrt{-g}\, \mathrm{d}^4 x \,, \end{aligned} \tag{12.113}$$

where the variation of the matter lagrangian might well require some work dealing with a forest of indices, as we'll soon see in Sect. 12.8.3. Demanding that δS vanish for all metric variations $\delta g^{\alpha\beta}$ produces the field equation

$$G_{\alpha\beta} = \frac{-16\pi}{\sqrt{-g}} \frac{\delta\left(\mathcal{L}_m \sqrt{-g}\right)}{\delta g^{\alpha\beta}} \,. \tag{12.114}$$

Comparing this with Einstein's original form, $G_{\alpha\beta} = 8\pi T_{\alpha\beta}$, suggests that corresponding to the extra matter/field lagrangian density, there is a new stress–energy tensor defined as

$$T_{\alpha\beta} \equiv \frac{-2}{\sqrt{-g}} \frac{\delta\left(\mathcal{L}_m \sqrt{-g}\right)}{\delta g^{\alpha\beta}} . \tag{12.115}$$

So this lagrangian approach enables a stress–energy tensor to be calculated for more than just a perfect fluid. We are free to include terms in the lagrangian density for other fields, and by coupling them to spacetime, new models can be produced of how gravity might interact with those fields.

The stress–energy tensor is not confined to general relativity, and in fact the *canonical energy–momentum tensor* is defined using the lagrangian for a field in the classical mechanics of flat spacetime, which we did in Sect. 10.3.6. (The names stress–energy and energy–momentum are interchangeable.) But the canonical stress–energy tensor is not related to the one defined in (12.115). The canonical tensor is not guaranteed to be symmetrical, and this causes problems in its implementation. Comparisons between both tensors, along with prescriptions for symmetrising the canonical one, are guided by experiments and can be found in the literature.

Determining how the new field affects spacetime means finding the resulting metric, which requires solving (12.114). Conversely, how does spacetime affect the field? Varying the field just follows the same procedure we used in Sect. 10.3.5 for a field in flat spacetime. Now, however, in general coordinates we must use the correct volume element:

$$S = \int \mathcal{L}\left(\phi, \phi_{,\alpha}\right) \sqrt{-g}\, \mathrm{d}^4 x . \tag{12.116}$$

What is being varied is not $\mathcal{L}$ but $\mathcal{L}\sqrt{-g}$, and so the Lagrange equation (10.59) becomes

$$\frac{\partial\left(\mathcal{L}\sqrt{-g}\right)}{\partial \phi} - \partial_\alpha \frac{\partial\left(\mathcal{L}\sqrt{-g}\right)}{\partial \phi_{,\alpha}} = 0 . \tag{12.117}$$

Here, ϕ can be any tensor field (i.e. it may contain indices). Since $\sqrt{-g}$ is independent of ϕ and $\phi_{,\alpha}$, we can divide (12.117) by $\sqrt{-g}$ to get

$$\frac{\partial \mathcal{L}}{\partial \phi} - \frac{1}{\sqrt{-g}}\, \partial_\alpha \left(\sqrt{-g}\, \frac{\partial \mathcal{L}}{\partial \phi_{,\alpha}}\right) = 0 . \tag{12.118}$$

This can be made to resemble (12.117) by using the notation $\mathrm{D}_\alpha f \equiv f_{;\alpha}$ for any f, which we first met in (8.229), and then referring to (8.212) to write

$$\boxed{\frac{\partial \mathcal{L}}{\partial \phi} - \mathrm{D}_\alpha \frac{\partial \mathcal{L}}{\partial \phi_{,\alpha}} = 0\,.} \tag{12.119}$$

Any of (12.117)–(12.119) is the spacetime generalisation of the flat-space cartesian lagrangian (10.59). So general coordinates require modifying (10.59) with either $\mathcal{L} \to \mathcal{L}\sqrt{-g}$ (12.117), or $\partial_\alpha \to \mathrm{D}_\alpha$ (12.119) (but not both!). And, of course, since the Ricci scalar doesn't contain the field of interest, we can replace $\mathcal{L}$ in these equations by $\mathcal{L}_m$.

Note that common usage denotes our $\mathrm{D}_\alpha f$ by $\nabla_\alpha f$. The ∇ symbol is quite overloaded with different meanings in relativity texts, and we won't use it that way. Since we have already written a vector operator $\nabla = \boldsymbol{e}^\alpha \partial_\alpha$ several chapters back, it must follow that its αth covariant component ∇_α is just ∂_α for us. In that case, it's more reasonable to refrain from giving ∇_α any additional meaning. Instead, define $\mathrm{D}_\alpha f \equiv f_{;\alpha}$ since this recalls two things: first, the D of (9.102), which *is* common notation both in this book and elsewhere; and second, the covariant derivative D of gauge theory, defined in (10.188) and probably universally used.

Generally, the field ϕ is specified as *minimally coupled* to gravity, meaning that its lagrangian has no curvature terms such as $R\,\phi^\alpha\phi_\alpha$. Including terms like this that contain curvature make the field *nonminimally coupled.* Such terms are excluded by the Equivalence Principle, but can certainly be included in the lagrangian in a search for new physics.

12.8.2 Adding a Simple Field: The Cosmological Constant

Consider the simplest example of an extra term, $\mathcal{L}_m$, which when added to the lagrangian density produces a new physical model. In newtonian dynamics, we cannot know of the possible existence of a spatially constant potential field pervading all of space because such a field can produce no force—since the force equals the rate of loss of field potential energy with distance, which is zero for such a field.

But the situation turns out to be different in a curved spacetime. Add a constant term conventionally called $-\Lambda/(8\pi)$ to the lagrangian density, where Λ is the *cosmological constant*, and ask whether it can have any effect on spacetime. That is, set $\mathcal{L}_m = -\Lambda/(8\pi)$ in (12.115). Taking the metric variation from (12.106) (i.e. just ignore R in that equation), we are easily able to calculate the stress–energy tensor (times 8π for convenience):

$$8\pi T_{\alpha\beta} = \frac{-16\pi}{\sqrt{-g}} \cdot \frac{-\Lambda}{8\pi} \cdot \frac{\delta\sqrt{-g}}{\delta g^{\alpha\beta}} \overset{(12.106)}{=\!=\!=} -\Lambda\, g_{\alpha\beta}\,, \tag{12.120}$$

so that Einstein's field equation becomes

$$G_{\alpha\beta} + \Lambda\, g_{\alpha\beta} = 0\,. \tag{12.121}$$

In general relativity, a constant potential really does change the spacetime physics by unevenly "stretching" the spacetime manifold, and this manifests as a force. We saw a similar idea back in Sect. 7.4 when discussing accelerated frames and gauge theory.

Whether there really is any such field Λ in the universe is difficult to ascertain. Current observational evidence based on supernovae and the cosmic microwave background radiation suggests that in fact there is, and that $\Lambda \simeq 10^{-34}\,\mathrm{s}^{-2}$. It turns out that this might well account for as much as 70% of the universe's energy content.

Einstein originally added the $\Lambda\, g_{\alpha\beta}$ term explicitly to his field equation so as to produce the current model of the universe at that time: a static one. This was good physics—he was constructing a theory based on current ideas that were at least partly rooted in measurements, thereby keeping his feet on the ground. But in the early part of the twentieth century, very large telescopes with good cameras were beginning to peer ever deeper into the heavens, and subsequent astronomical observations presented new evidence for a nonstatic universe. Given those observations, the cosmological constant was no longer necessary, and Einstein retracted it—again good physics, despite his wish that he'd never included it in the first place. (Einstein used sound logic all the way, but because he regretted ever including it, many books still persist in describing it as an infamous error on his part.)

But it appears that Λ may well be there after all, and there is certainly no real argument for why it should be exactly zero. In the end, there was no absolute reason to retract it, since a lagrangian approach implies that a constant potential is just one of many terms that could be present in Einstein's field equation, depending on what is included in the lagrangian.

12.8.3 Joining Electromagnetism to Gravity

We finish this discussion of lagrangians with a more concrete example: coupling the electromagnetic field to a curved spacetime, as shown in Fig. 12.9. The standard electromagnetic lagrangian density $-\varepsilon_0 F^2/4$ from (10.103) is used, although it's quite usual to absorb the ε_0 into the F^2, and we'll follow suit. Also included is a term that adds mass to the carrier of the electromagnetic field, referred to as a *massive photon*. (We'll see why later.)

There is no need to vary the lagrangian with respect to the metric to give Einstein's equation, since we have already done that in all generality to produce (12.114). Equivalently, we can calculate the stress–energy tensor from (12.115) and use this in the alternative form of Einstein's equation (12.43). As well, varying the lagrangian with respect to A_μ gives the modified version of Maxwell's equations. Without giving every detail, let's see how it all comes together.

The total lagrangian is shown in Fig. 12.9, in which case

$$\mathcal{L}_m = \frac{-F^2}{4} + \frac{1}{2} m^2 A^\alpha A_\alpha\,. \tag{12.122}$$

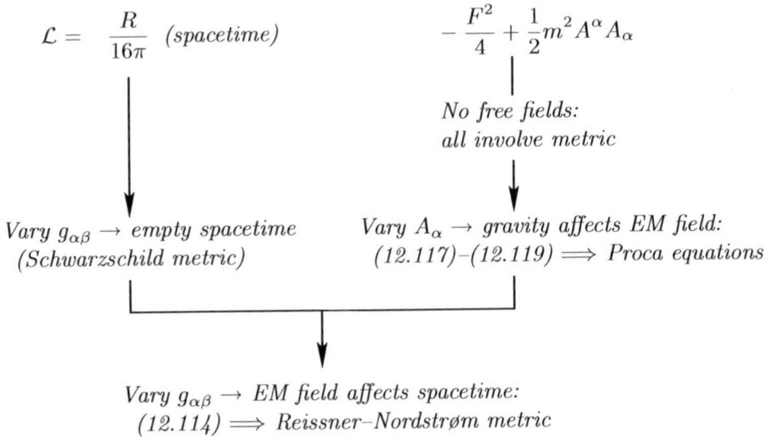

Fig. 12.9. An example of placing an electromagnetic field with a "massive photon" into a curved spacetime. Conventional electromagnetism would set $m = 0$. The resulting *Reissner–Nordstrøm* metric can be interpreted as describing a charged black hole, while the presence of the mass coupling m modifies the Maxwell equations to produce the *Proca equations*, which themselves become modified by the presence of gravity.

This will provide the stress–energy tensor, giving Einstein's equation that establishes the metric, as well as being used in Lagrange's equation establishing the electromagnetic field A_μ. Start with the stress–energy tensor (12.115), writing

$$\delta\left(\mathcal{L}_m\sqrt{-g}\right) = \delta\mathcal{L}_m\,\sqrt{-g} + \mathcal{L}_m\,\delta\sqrt{-g}\,. \tag{12.123}$$

We saw earlier in (12.106) that $\delta\sqrt{-g} = \frac{-1}{2}\sqrt{-g}\;g_{\alpha\beta}\;\delta g^{\alpha\beta}$. Now write the lagrangian with all metric terms made explicit:

$$\mathcal{L}_m = \frac{-1}{4}F_{\mu\nu}\,F_{\alpha\beta}\,g^{\mu\alpha}g^{\nu\beta} + \frac{1}{2}m^2 A_\mu A_\alpha\,g^{\mu\alpha}\,, \tag{12.124}$$

from which it follows that

$$\delta\mathcal{L}_m = \frac{-1}{4}F_{\mu\nu}\,F_{\alpha\beta}\left(\delta g^{\mu\alpha}\,g^{\nu\beta} + g^{\mu\alpha}\,\delta g^{\nu\beta}\right) + \frac{1}{2}m^2 A_\mu A_\alpha\;\delta g^{\mu\alpha}. \tag{12.125}$$

This combines with the expression for $\delta\sqrt{-g}$ a few lines up to give $\delta\left(\mathcal{L}_m\sqrt{-g}\,\right)$ in (12.123). With some index relabelling, we can single out a factor of $\delta g^{\mu\nu}$ and divide by it, while remembering the discussion on p. 515 about this division. Finally, the stress–energy tensor becomes

$$T_{\alpha\beta} = \frac{1}{2}g_{\alpha\beta}\left(\frac{-F^2}{2} + m^2 A_\mu A^\mu\right) + F_{\alpha\mu}F_\beta{}^\mu - m^2 A_\alpha A_\beta\,. \tag{12.126}$$

Contracting this gives T, and Einstein's equation (12.43) becomes

$$R_{\alpha\beta} = 8\pi \left(\frac{-F^2}{4} g_{\alpha\beta} + F_{\alpha\mu}\, F_\beta{}^\mu + m^2 A_\alpha A_\beta \right). \tag{12.127}$$

Next, we need the Lagrange equation for the electromagnetic field A_α. Use (12.119):

$$\frac{\partial \mathcal{L}}{\partial A_\alpha} - \mathrm{D}_\mu \frac{\partial \mathcal{L}_m}{\partial A_{\alpha,\mu}} = 0\,. \tag{12.128}$$

The lagrangian density $\mathcal{L}_m$ must be written in terms of A_α and $A_{\alpha,\beta}$ to allow the partial differentiations to be done. Note that not only does the simple expression

$$\frac{\partial B^\alpha}{\partial B^\mu} = \delta^\alpha_\mu = g^\alpha_\mu \tag{12.129}$$

hold for any vector B^α, but also more general expressions hold with the appropriate metric indices raised or lowered, such as

$$\frac{\partial B^\alpha}{\partial B_\mu} = \frac{\partial}{\partial B_\mu} (g^{\alpha\nu} B_\nu) = g^{\alpha\nu}\, \delta^\mu_\nu = g^{\alpha\mu}\,,$$

$$\frac{\partial B^{\alpha,\beta}}{\partial B_{\mu,\nu}} = \frac{\partial}{\partial B_{\mu,\nu}} \left(g^{\alpha\varrho}\, g^{\beta\sigma}\, B_{\varrho,\sigma}\right) = g^{\alpha\mu}\, g^{\beta\nu}, \tag{12.130}$$

and the same rules apply quite generally to tensors with any number of components, with or without commas or semicolons. This allows the differentiations of (12.128) to be done, and after some tidying the result is the curved spacetime version of the *Proca equations*, themselves being the massive-photon equivalent of the Maxwell equations:

$$F^{\alpha\beta}{}_{;\beta} - m^2 A^\alpha = 0\,. \tag{12.131}$$

Compare this with (10.110), which, with $j^\beta = 0$, gives the source-free form of the classical Maxwell equations: $F^{\alpha\beta}{}_{,\beta} = 0$. Besides the mass term in (12.131), the comma for the Minkowski metric has become a semicolon as expected, embodying the Equivalence Principle.

Equations (12.127) and (12.131) govern the entire system of gravity plus electromagnetic field. Solving them is complicated, and we'll only write the solution for two regimes. As in the Schwarzschild case, impose spherical symmetry and begin with a metric of the form (12.47) for the two unknown functions $a(r), b(r)$. The semicolon in (12.131) is converted to a comma by way of the ever-useful (8.213), using $\sqrt{-g} = \sqrt{ab}\, r^2 \sin\theta$. We'll also suppose the electromagnetic field is static, meaning that it consists only of a time-independent potential Φ. Skipping over the details, we quote solutions for the metric and Φ in two special regimes.

Solution for $m = 0$ and Spherical Symmetry. As with the Schwarzschild case, the integrations involved here not only introduce a constant M that's interpreted as a mass, but also a constant Q interpreted as a charge associated with that mass:

$$d\tau^2 = \left(1 - \frac{2M}{r} + \frac{Q^2}{r^2}\right) dt^2 - \left(1 - \frac{2M}{r} + \frac{Q^2}{r^2}\right)^{-1} dr^2 - r^2 \left(d\theta^2 + \sin^2\theta \, d\phi^2\right), \tag{12.132}$$

and

$$\Phi = \left(1 - \frac{2M}{r} + \frac{Q^2}{r^2}\right)^{-1} \frac{Q}{4\pi\varepsilon_0 r}. \tag{12.133}$$

This solution is interpreted as describing a static charged black hole. Equation (12.132) is called the *Reissner–Nordstrøm* metric.

Solution for $r \to \infty$ and Spherical Symmetry. A simplifying condition when $m \neq 0$ is to consider large r, where the metric tends toward the Minkowski form. The electric potential becomes

$$\Phi \to \frac{Q}{4\pi\varepsilon_0 r} e^{-mr}. \tag{12.134}$$

That exponential decay over distance suppresses the electrostatic force for large m, in the same way as the *Yukawa potential* that describes the strong force between nucleons. The term "massive photon" actually arises in field theory when the Proca equations are quantised, but we can already see here an indication that m will be the mass of the virtual photons that mediate the Proca field.

Motion of a Charge Placed in the Spacetime. Finally, what is the equation of motion of a charge q moving in this massive electromagnetic field in curved spacetime? Rather than add more terms to the lagrangian and re-invent the wheel, we can just use the result of (10.101):

$$\frac{Dp_\alpha}{d\tau} = q u^\beta F_{\alpha\beta}. \tag{12.135}$$

Needed for this are the metric to calculate the covariant derivative, and the electromagnetic field A^α to calculate the Faraday tensor. And, at least for the two limiting cases above, we have them both.

As a side note about the notation, an equation such as (12.135) could be written without indices by first contracting each side with $\boldsymbol{e}^\alpha$ and then remembering (9.103). If bold symbols stand for general tensors and a dot product is just a contraction over neighbouring indices, the result becomes

$$\frac{d\boldsymbol{p}}{d\tau} = -q\boldsymbol{u}\cdot\boldsymbol{F}, \tag{12.136}$$

where the minus sign arose because we needed to swap the indices of the antisymmetric Faraday tensor to get like indices to be neighbours. This is fine, although we should not mistake $\boldsymbol{p}$ for its 3-space counterpart! A safe bet is to write all tensors in bold as we have done, and carefully distinguish any 3-space counterparts if they're used. But although an equation like (12.136) might look to be more index free than (12.135), this is partly an illusion. After all, (12.135) holds in all coordinate systems, so in a sense it is already index free.

12.8.4 Path Integrals in General Relativity

In Sect. 10.8, we discussed the path-integral idea, which views the amplitude for a particle to travel from one point to another as a sum of amplitudes over all possible (and impossible!) paths in spacetime, and where each amplitude is given by a complex exponential of the action for that path. This idea has also been applied to the lagrangian approach to general relativity, although the result is certainly not in keeping with Ockham's Razor.

To illustrate the idea, imagine an infinite number of spacetimes, each with different curvature characteristics. Ascribe to each spacetime an amplitude $\exp \mathrm{i}S/\hbar$, where the action S is calculated by integrating the Ricci scalar plus some matter lagrangian $\mathcal{L}_m$ over the spacetime from (12.112). In the path-integral approach, the spacetime that is actually realised in a quantum mechanical sense is the one around which the action is stationary, and this is the one that Einstein's equation actually describes—that is, ours.

Nevertheless, *anything* to which an action principle applies can also be described in this quantum mechanical way of multiple systems, so although the idea is useful in quantum mechanics, whether it's meaningful when applied to larger systems such as the universe is unknown. Such an application is also fraught with difficulty since it's not clear that any experiments can be done, and without these the subject does not have a solid footing.

The idea of summing over an infinity of spacetimes is not the same thing as the *Many Histories* interpretation of quantum mechanics. The Many Histories approach comes in many flavours, but generally is a way of stating that when the wave function of, e.g., a two-state quantum system "collapses" with a 0.6–0.4 probability, what has really happened is that there are a multitude of universes, in 60% of which the system is forced into one of the states upon a measurement, and in 40% the system is forced into the other. This idea is not in keeping with Ockham's Razor either, and in no way does it explain why the collapse of the wave function appears to be probabilistic in the first place.

12.9 A Metric for the Universe: Proper Distances in Cosmology

We end this chapter with some remarks about cosmological metrics and how they find a use in portraying our universe. Just as the Schwarzschild metric is an especially basic solution to Einstein's equation, being determined by a zero stress–energy tensor and spherical symmetry, another basic solution is the metric approximately describing the geometry of the universe as a whole. This, the *Friedmann–Robertson–Walker*, or FRW, metric, is calculated in the same way as the Schwarzschild metric: start with a general metric having some imposed symmetry, and then consider any interesting form of the stress–energy tensor, which then determines the metric via Einstein's equation.

The most important assumption placed on how symmetrical the universe might be is the *Copernican Principle*, which postulates that we occupy no especially privileged position in the cosmos. Another assumption is that of isotropy at every point: the universe looks the same in all directions, everywhere. A consequence of this is that the universe must be homogeneous on a large scale. Of course, it's not homogeneous on galactic scales, but the FRW metric takes a far larger view that smoothens over the "small" matter peaks that comprise galaxies.

In this view, each galaxy is a coordinate marker; we are at $r = 0$, while every other galaxy lies at some constant r, θ, ϕ by definition of those coordinates. And what of the time coordinate? Could we construct one as was done for the accelerated frame in Chap. 7? There, in Fig. 7.6, we made use of the global simultaneity that could still be defined for the case of uniform acceleration, to define the time everywhere as being that which was simultaneous with the time shown on a primary observer's clock. But the universe does not have such a global standard of simultaneity based on special relativity, so we cannot define a time as being the same everywhere "now". Instead, the time coordinate t for the FRW metric is defined at each event as the proper time of the galaxy that, in principle, is present at that event. This time coordinate then *defines* a universal measure of simultaneity, and isotropy at all points is postulated to hold at each moment of this time. Using these considerations to solve Einstein's equation, the Friedmann–Robertson–Walker metric turns out to be

$$\boxed{\mathrm{d}\tau^2 = \mathrm{d}t^2 - a^2(t)\left(\frac{\mathrm{d}r^2}{1-kr^2} + r^2\mathrm{d}\theta^2 + r^2\sin^2\theta\,\mathrm{d}\phi^2\right),} \tag{12.137}$$

where the positive function $a(t)$ is to be determined, and $k \in \{0, \pm 1\}$ is the sign of the *spatial* Ricci scalar $R^a{}_a$. Thus, this metric describes three types of universes:

- those with **positive spatial curvature** ($R^a{}_a$), which, it turns out, must necessarily be closed,

- those with **zero spatial curvature**, which can be closed or open, and
- those with **negative spatial curvature**, which can also be closed or open, but are almost always assumed by cosmologists to be necessarily open.

If the matter content of the universe at large is modelled as a perfect fluid, then these three spatial geometries correspond to high, borderline, and low densities of matter, respectively. Modelling the universe in this way and relating density, pressure, curvature, and the function $a(t)$ comprise a large part of the modern subject of cosmology. We'll stop here without specifying any stress–energy tensor; that's for cosmology texts to explore. Instead we'll be content to make some remarks on proper distances as specified by the FRW metric.

The function $a(t)$ is especially interesting in that it leads to the notion of what at first appears to be a faster-than-light expansion of the universe. Let's examine how this comes about. Remember from Sect. 12.5 that the spatial part of any metric gives the proper, or physical, distance squared, $\mathrm{d}\ell^2$, between two simultaneous events. For the FRW metric, this is

$$\mathrm{d}\ell^2 = a^2(t) \underbrace{\left(\frac{\mathrm{d}r^2}{1-kr^2} + r^2 \mathrm{d}\theta^2 + r^2 \sin^2\theta \, \mathrm{d}\phi^2 \right)}_{\equiv\, \mathrm{d}L^2,\ \text{coordinate separation squared}} . \tag{12.138}$$

In that case,

$$a(t) = \frac{\mathrm{d}\ell \;\longleftarrow\; \textit{physical separation of galaxies}}{\mathrm{d}L \;\longleftarrow\; \textit{coordinate separation of galaxies.}} \tag{12.139}$$

So $a(t)$ is a ratio of physical to coordinate separations. This ratio also gives a physical meaning to the radial coordinate. We in the Milky Way Galaxy are at $r = 0$. The proper radial distance at time t to a galaxy at some point (r, θ, ϕ) is

$$\ell(t,r) = \int_0^r \mathrm{d}\ell = a(t) \int_0^r \frac{\mathrm{d}r'}{\sqrt{1-kr'^2}} \equiv a(t)\, b_k(r) . \tag{12.140}$$

The ratio of proper distances from the Milky Way out to a radial coordinate-distance r at different times t and t_0 is, from (12.140),

$$\frac{\ell(t,r)}{\ell(t_0,r)} = \frac{a(t)}{a(t_0)} . \tag{12.141}$$

Again, we see that $a(t)$ is a *scale factor*. The FRW metric allows the universe to be nonstatic. As an example, the proper distances to other galaxies can be increasing; this defines an *expanding universe*. We might wish to think of the expansion rate as manifesting in a recession velocity of those galaxies. But such an idea needs to be treated thoughtfully. For a galaxy at (t, r), this velocity is

$$v(t,r) \equiv \frac{\partial \ell(t,r)}{\partial t} = a'(t)\, b_k(r)\,. \tag{12.142}$$

The ratio of recession velocity to physical distance, the *Hubble parameter* $H(t)$, or just H, is of great importance in cosmology in that it helps quantify cosmological dynamics:

$$H \equiv \frac{v(t,r)}{\ell(t,r)} = \frac{a'(t)}{a(t)}\,. \tag{12.143}$$

Historically, H was assumed constant and is still called the Hubble constant, but this is an old term that doesn't take account of modern cosmological models. Its value is currently measured at anywhere between 15 and 30 km/s per megalight-year. Even so, it is very difficult to ascertain whether H is changing, and a constant H might well describe our universe. Equation (12.143) shows that its being constant would require $a(t) = a_0\, e^{Ht}$ for some constant a_0, and combined with a spatially flat metric ($k = 0$), this describes an important model of the universe called *de Sitter space.*

Imagine the universe is indeed de Sitter, and calculate the recession speed of a galaxy at r. Inserting $k = 0$ into (12.140) gives $b_0(r) = r$, so that (12.142) becomes

$$v = Ha_0\, e^{Ht} r\,. \tag{12.144}$$

Thus the recession speed v can become arbitrarily large. This might appear to pose a problem with travel faster than light, but in fact it does not. Although the galaxy's physical distance from us is growing ever more quickly without limit, this has nothing to do with the speed of light, which must be a constant *in an inertial frame.* The FRW metric does not specify one global inertial frame. The old picture of the universe as a pudding that expands while baking is very apt, because it says that while the galaxies (raisins embedded in the pudding) are separating as the pudding expands, they are not actually moving *through* the pudding. (Equivalently, in two spatial dimensions, galaxies are sometimes drawn on an expanding balloon whose surface represents the 2-space. They are separating, but are not moving across the surface of the balloon.) So both the Milky Way and the distant galaxy are at rest relative to the local inertial frame of each (the pudding around them, or the balloon surface). Not only is neither galaxy moving faster than light, but each is actually *at rest* in the cosmological frame! It's only the global expansion of the "pudding" that separates the galaxies at any large speed, and there is certainly no problem with that.[1]

Horizons in Cosmology

Like accelerated frames and black holes, models of the universe can also have their horizons. To see why, calculate the coordinate velocity of radially moving

[1] A very readable paper describing a classroom experiment in galactic recession using stretching rubber to represent the universe is "Cosmological expansion in the classroom" by R. Price and E. Grover in the American Journal of Physics, **69**, 125–128 (February 2001).

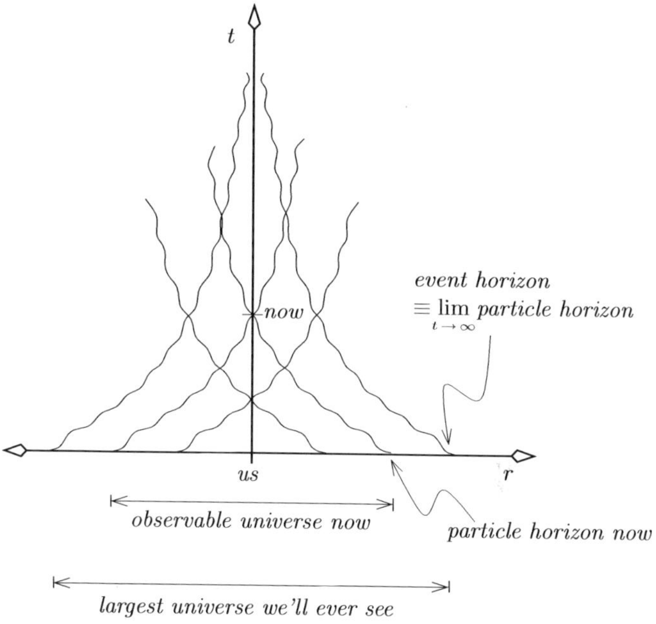

Fig. 12.10. Light arriving at Earth from distant regions of a de Sitter universe. Right "now", the farthest things we can see define our *particle horizon.* In de Sitter space, because light follows worldlines that have vertical asymptotes, there is a farthest distance beyond which we'll never see: an event horizon, just as for a black hole or for an accelerated frame in flat spacetime.

light in the de Sitter metric. Light follows a null geodesic, so setting $\mathrm{d}\tau = 0$ in (12.137) gives

$$\frac{\mathrm{d}t}{\mathrm{d}r} = \pm a_0\,\mathrm{e}^{Ht}. \tag{12.145}$$

The paths of these light rays are shown in Fig. 12.10. Corresponding to the present moment, there is a value of r whose light is only now arriving on Earth. This defines our *particle horizon*, which for the de Sitter universe is expanding outward so that we can see deeper and deeper into space as time passes. However, as shown in Fig. 12.10, the de Sitter metric produces exponential null curves having vertical asymptotes (12.145). So there is a maximum value of r at which the light emitted from a galaxy and heading toward Earth just asymptotes to our t-axis. That light will take forever to reach us, in which case that value of r denotes an event horizon beyond which we can never see, and to which the particle horizon will expand outward asymptotically. So it is that Einstein's metric description of gravity allows us to draw spacetime diagrams that aid in picturing the universe in such exotic ways as were never imagined a century ago.

With the exponential term in the de Sitter metric reminding us of the exponential first encountered many chapters back in radioactive decay, this chapter and the book draw to a close. There is always a feeling of familiarity when we meet with the same themes and functions in such vastly different arenas in mathematical physics. Sometimes that signals a deeper common level, while at other times it might just be due to simplifications that use concepts such as linearity, gaussians, and first-order differential equations, although the fact that we *can* make such simplifications in diverse areas is itself interesting. To what extent these links across different areas must grow or weaken as the language of mathematical physics evolves is something that only time will tell.

Additionally, the question of whether some of the subject's more complex notations are really necessary or useful is a difficult one. That two disparate fields share a common notation might indicate that they are related at a deep level, but it *might* not follow that further ideas from one of the fields can then be applied usefully to the other. The real test is probably the extent to which the conclusions that we draw are useful. The best conventions interrelate different fields and make abstract ideas transparent, but even the most gawdy or abominable notations have their champions who insist that the subject would be otherwise incomplete.

Still, there *are* choices in how to write mathematical physics, and the language does continue to evolve. It has been said that new ideas and fashions in science are never actually accepted by anyone; rather, the old guard retires, and the new generation does not know any better. So it is for us to ensure that the language evolves in an intelligent direction, and not be blown off course by short-term fashions similar to those that, nowadays more than ever, wreak havoc on the spoken languages of the world. (Evolution is not always positive—the current *de*volution of the English language is proof of that!)

The great thing about the language of physics is that it has been very cleverly put together over a long time, slowly but surely, and the result can sometimes be used to perform magic (and sometimes magic tricks!) that shows it at its best. Our current language often hints at deep relationships between concepts, relationships that might resist being pinned down. Can a level of language ever be attained that converts all possible theorems to a streamlined series of logical connections, rendering them all more or less obvious?

Perhaps a deeper level of understanding requires a complete shift in the way the language is constructed. Systems for calculating and expressing mathematical ideas do change slowly, of course. The ancient Egyptians had a way of dealing with fractions that only allowed unit numerators (the one exception being $^2/_3$) and no repetition, so that if asked to divide 2 by 5, they would write the answer as $^1/_3 + {}^1/_{15}$ (and not $^1/_5 + {}^1/_5$). We who write the answer as $^2/_5$ might find their way of thinking to be otherworldly, and certainly it's difficult to do arithmetic in this way. It was so difficult for the Egyptians that they needed to refer to tables of fraction identities that they had discovered one by one, either by serendipity or else through sheer hard work.

Today we still have our version of those Egyptian tables that recorded identities used time and again. For us they are tables of integrals, series, and transforms. Perhaps people of the future who have a completely different system of mathematics will wonder why *we* chose to do things in such an arduous way as indicated by these tables. But profound improvements in physics language will never happen for a new generation brought up to rely upon ever-increasing computing power. Rather, such changes require people to take the time, and *have* the time, to think about the very concepts that the language is designed to portray, elucidate, and weave together at various levels; and then maybe at some stage a quantum shift in the language and our understanding of physics will occur. Feynman once posed the question of whether advances in physics require someone to know all that has gone before—or nothing that has gone before. It's a good question, and one for which he had no answer. But we shouldn't be afraid to tinker with the language of physics if that can further our understanding, and we shouldn't be afraid to think about the basics from first principles. Such adventuring is part and parcel of creating ever finer viewpoints.

Index

Italicised page numbers denote where the entry has been defined.

4713441

Made in the USA
Lexington, KY
22 February 2010